FLORA OF TROPICAL EAST AFRICA

RUBIACEAE (Part 1)

B. VERDCOURT

Small to large trees, shrubs or less often annual or perennial herbs or woody or herbaceous climbers, sometimes spiny; tissues in many tribes containing abundant rhaphides. Leaves opposite or verticillate, decussate, almost always entire, very rarely (not in East Africa) palmatifid, toothed or finely denticulate, always obviously stipulate (save in some *Rubieae*, where the stipules may be considered foliar or almost absent according to interpretation), the stipules interpetiolar or intrapetiolar, entire or, particularly in herbaceous genera, variously divided into lobes or fimbrieae, often tipped or separated by mucilaginous hairs known as colleters* and often with colleters inside the base; the leaves in certain tribes sometimes contain small bacterial nodules. Flowers rarely solitary, mostly in various terminal or axillary inflorescences, all basically cymose but variously aggregated into panicles, etc., occasionally in globose heads to an extent that the ovaries are adnate; bracts vestigial to well developed, even conspicuous; flowers usually hermaphrodite, rarely unisexual, regular or nearly so (except in *Posoqueria* (America)) or corolla-tube rarely curved, homostylous or quite often heterostylous with 2 or rarely 3 forms (long-styled (dolichostylous), short-styled (brachystylous) or equal-styled (isostylous)). Calyx gamosepalous, the tube mostly adnate to the ovary, (3–)4–5(–8)-toothed or -lobed, sometimes only minutely so, with open, valvate, imbricate or contorted aestivation, 1 or several lobes sometimes slightly to very considerably enlarged to form a leafy often coloured lamina.** Corolla small to large and showy, gamopetalous, rotate to salver-shaped or funnel-shaped, the tube often very long, (3–)4–5(–11)-lobed, the lobes mostly contorted or valvate, sometimes valvate-induplicate, rarely imbricate or quincuncial. Stamens usually as many as the corolla-lobes and alternate with them, epipetalous; anthers basi- or dorsifixed, introrse, the thecae rarely multilocellate transversely. Pollen various, mostly simple, isopolar and 3-colporate, but sometimes porate, the number of colpi or pores varying from 2 to 25, globose, ovoid or discoid, sometimes (in some *Gardenieae*) in tetrads or rarely in polyads. Disc often present, 2-lobed or tubular. Ovary inferior, rarely half-inferior or (in *Gaertnera*) superior, syncarpous of 2–5 or more carpels, but predominantly of 2 and therefore predominantly 2-locular, but 3–5 or even 12 or more (e.g. in *Urophylleae* due to supplementary incomplete partitions); placentation axile or (in some *Gardenieae*) parietal; ovules 1–many per locule, often embedded in fleshy placentas, erect, basal or horizontal, anatropous; style simple, usually long and narrow, the " stigma " either cylindrical, clavate, or otherwise modified to form a " receptaculum pollinis " or divided into 2–many linear, spathulate or clavate lobes, the actual stigmatic surface sometimes confined to certain areas, e.g. the inner faces of the lobes. Fruit

* See Lersten, Colleter morphology in *Pavetta*, *Neorosea* and *Tricalysia* (Rubiaceae) and its relationship to the bacterial leaf nodule symbiosis, in J. L. S. 69: 125–136 (1974).
** For terminology of parts see p. 5.

small to quite large (0·2–20 cm.), a capsule, berry or drupe or indehiscent or woody, occasionally (e.g. in *Nauclea* and *Morinda*) united to form syncarps, (1–)2–many-seeded, if capsules then loculicidal or septicidal or opening by a beak. Seeds small to rather large, sometimes winged; testa cells in some tribes with very distinct pits; albumen present (save in *Guettardeae*); embryo straight or rarely curved, the radicle mostly longer than the cotyledons.*

A large family of about 500 genera and 6000 species, predominantly tropical and adapted to moist environments (except the *Rubieae* which are predominantly temperate or even arctic in distribution). Older descriptions of the family differ in some respects owing to the inclusion of elements now generally excluded, e.g. somewhat zygomorphic flowers occur in Henriquesiaceae, often considered as a tribe (*Henriquesieae*) of the Rubiaceae. In East Africa there are 100 genera and about 600 species.

The classification of this family into tribes is not really very difficult, but the basic arrangement of these tribes into subfamilies is a matter of contention. The classical division is into two groups, one with the ovary-locules containing one ovule and the other with them containing 2 or more. This is still followed by many authors, e.g. Hallé in Fl. Gabon 12 (1966), and is a partially convenient method, being subject to no more exceptions than most classifications of this type. There is no doubt, however, that this is artificial and the classification proposed by Bremekamp and myself** is more natural, though perhaps less practical in some respects, since it is partly based on the presence or absence of rhaphides. Since neither ovules nor rhaphides are easy to see, use has also been made of other characters in the key, as was first done by Hutchinson in the first edition of F.W.T.A. and later followed by Keay & Hepper in the second. A conspectus of the classification followed in this present account is given below; it differs little from that proposed by me in 1958, except for some modifications adopted from Bremekamp's latest paper.

Various pollination mechanisms exist in the family from simple protandry, where the pollen is dispersed before the stigmatic surfaces are revealed, to more sophisticated arrangements such as heterostyly, where there are 2 or even 3 types of flowers. In complete cases the style is well included and the anthers well exserted in one sort of flower and exactly the opposite in the other sort, but in some tribes the differences are far less marked. In many homostylous species (i.e. those not showing heterostyly) the stigma and the anthers are well isolated. The ixoroid pollination mechanism, mentioned quite frequently in the text, is where the usually cylindrical or clavate style-head (which could easily be taken for the stigma) acts as a pollen receptacle; it is in close contact with the anthers and pollen is deposited on it before the flower opens. After the flower opens the style elongates and the pollen is ready for insect transfer; later the 'stigma-lobes' divide to reveal the true stigmatic surfaces and are ready for pollen from other flowers. Unisexuality is rare in the Flora area, but occurs in *Anthospermum*, possibly derived from extreme cases of heterostyly. Most species of Rubiaceae are insect pollinated, many having white flowers strongly scented at night, being obviously pollinated by moths. *Anthospermum*, with its long styles and dangling anthers is clearly wind pollinated, and a few of the larger flowered *Vanguerieae* may even be bird pollinated. Doubtless, despite the special mechanisms, quite a good deal of self-pollination exists. There is a wealth of material here for studies to be made in the field by biologists on the spot.

Synopsis of subfamilies and tribes

Subfamily **Rubioideae.** Rhaphides present in the tissues (except in *Urophylleae* where these are replaced by thicker styloid crystals). Trees, shrubs or very often herbs. Corolla-lobes valvate (with some very rare exceptions none of which occur in Africa). Indumentum of stem, foliage and inflorescence, if present, usually of clearly septate hairs (see fig. 1/8, p. 6), but short undivided hairs are common in some genera. Complete heterostyly is very frequent but can be present or absent in species of the same genus. Ovules solitary to numerous in 2–many locules, erect or pendulous or attached by the middle. Fruit dry or succulent, dehiscent or indehiscent. Stipule-sheath very often divided into fimbrieae. Style usually divided into 2 linear lobes (or more in the case of multilocular ovaries), bearing the stigmatic papillae on their inner surfaces. Seeds with albumen; testa cells rarely pitted. Genera 1–42.

* Excellent more extensive accounts of the morphology and biology of the family will be found in Hallé, Fl. Gabon 12, Rubiacées: 7–22 (1966) and White, F.F.N.R.: 383–388 (1962).

** See Verdcourt in B.J.B.B. 28: 209–290 (1958) & Bremekamp in Acta Bot. Neerl. 15: 1–33 (1966).

Tribe **Psychotrieae** *Dumort.*, Anal. Fam. des Plantes: 33 (1829). Mostly shrubby, rarely herbs or trees. Ovules solitary in 2–8 locules, erect. Fruit usually succulent and containing 2–8 pyrenes, more rarely dry and separating into cocci. Pollen grains porate. Genera 1–5.

Tribe **Morindeae** *Miq.*, Fl. Ind. Bat. 2: 241 (1857) [as subtribe in DC., Prodr. 4: 342 (1830)]. Trees or shrubs. Ovary 2–12-locular; ovules solitary in the locules, affixed at or below the middle. Fruit baccate or drupaceous, 2–12-locular or containing up to 12 pyrenes, often forming a fleshy syncarp. Pollen grains porate. Genera 6, 7.

Tribe **Triainolepideae** *Bremek.* in Proc. K. Nederl. Akad. Wetensch., ser. C, 59: 3 (1956). Shrubs, with 3–7-fimbriate stipules. Ovary 2–10-locular; locules with 2(–3) collateral ovules inserted at the base of the septum. Fruits with 1 multilocular pyrene, each locule 1-seeded. Genus 8.

Tribe **Urophylleae** *Verdc.* in B.J.B.B. 28: 281 (1958). Shrubs or small trees, rarely ± herbaceous. Ovary 2–many-locular, with numerous ovules in each locule. Fruit a berry, 2–many-locular, many-seeded. Pollen grains colpate, with large apertures. Owing to the lack of true rhaphides Bremekamp has erected a subfamily *Urophylloideae* and has also placed the African members in a separate tribe *Pauridiantheae*. Genera 9, 10.

Tribe **Craterispermeae** *Verdc.* in B.J.B.B. 28: 281 (1958). Shrubs or trees, with rather marked yellow-green foliage on drying, nearly all being marked aluminium accumulators.* Ovary 2-locular, with solitary pendulous ovules. Fruit a berry, 1–2-locular. Pollen grains porate. Genus 11.

Tribe **Knoxieae** *Hook.* in G.P. 2: 9 (1873). Herbs or small subshrubs. Ovary 1–5-locular; ovules solitary, pendulous. Fruit either succulent with pyrenes, dry and indehiscent or dividing into indehiscent cocci. Flowers sometimes very bright blue. Genera 12, 13.

Tribe **Paederieae** *DC.*, Prodr. 4: 343 & 470 (1830). Usually foetid climbers. Ovary 2–5-locular; ovules solitary, erect. Fruit 2-coccous, with thin epicarp, or capsular. Genus 14.

Tribe **Hamelieae** *Kunth*, Nova Gen. Spec. 3: 413 (1820), as *Hameliaceae*, emend. DC., Prodr. 4: 342 & 438 (1830). Shrubs. Corolla-lobes imbricate. Ovary 5-locular; ovules numerous in each locule. Anthers long. Disc fleshy. Fruit a berry. Only known cultivated in East Africa, see p. 25.

Tribe **Hedyotideae** *Cham. & Schlecht.* in Linnaea 4: 150 (1829). Herbs or shrubs. Ovary 2(rarely–3–4)-locular; ovules usually numerous, rarely few or even solitary in each locule, affixed to the base of the septum. Fruit a capsule (save in a few dubiously placed genera not occurring in Africa). Genera 15–33.

Tribe **Anthospermeae** *Cham. & Schlecht.* in Linnaea 3: 309 (1828). Herbs or shrubs. Flowers often unisexual. Ovary 1–4-locular; ovules solitary in each locule, erect. Fruit dry, dividing into cocci or capsular. Genera 34, 35.

Tribe **Spermacoceae** *Kunth*, Nova Gen. Spec. 3: 341 (1820); Spreng., Anleit Kenntniss Gewachse, ed. 2, 2: 598 (1818), as "*Spermacoceen*". Herbs. Ovary 2(rarely –3–4)-locular; ovules solitary in the locules, affixed to the middle of the septum. Fruit dry, capsular or dividing into dehiscent or indehiscent cocci. Pollen grains pluricolpate. Genera 36–40.

Tribe **Rubieae.** Herbs. Stipules (with very few exceptions) foliar, there being a whorl of similar leaves and stipules at each node. Stem polygonal with collenchymatous ribs. Ovary 2-locular; ovules attached to the septum. Fruit fleshy, often didymous, or more or less dry. Genera 41, 42.

Subfamily **Cinchonoideae** *Endl.*, Gen. Pl.: 545 (1838), as *Cinchonaceae*, emend. K. Schum. in E. & P. Pf. 4 (4): 16 (1891) & re-emend. Verdc. in B.J.B.B. 28: 280 (1958). Rhaphides absent in the tissues. Trees, shrubs or rarely herbs. Corolla-lobes valvate, contorted or imbricate. Indumentum of stem, foliage and inflorescence, if present, of thick-walled non-septate or incompletely septate hairs. Complete heterostyly rare, but limited heterostyly present in a few tribes. Ovules solitary or numerous in 1–many locules, attached to the septum or pendulous or embedded in 1–several placentas. Fruit dry or succulent, dehiscent or indehiscent. Stipules usually undivided. Style with stigma fusiform, capitate or divided into linear lobes. Seeds with albumen; testa cells in some tribes with very conspicuous pits. Genera 43–99.

Tribe **Naucleeae**** (*DC.*) *Hook.* in G.P. 2: 8 (1873) [as subtribe in DC., Prodr. 4: 342 (1830)]. Trees or climbers, sometimes with hooked spines. Flowers aggregated into spherical heads. Ovary 2-locular, normally with numerous ovules in each locule (solitary

* The first two, fourth, fifth and sixth tribes of this subfamily all contain aluminium accumulators to a rather marked degree.

** This has been divided by Ridsdale into three parts (much as suggested by Bremekamp) in a paper which appeared after this part of the Flora went to press—namely *Naucleeae* sensu stricto, *Cinchoneae–Mitragyninae* and *Cephalantheae*–see Blumea 22: 541 (1975).

in *Cephalanthus*). Fruit a succulent syncarp or collection of capsules. Testa cells pitted at least in capsular-fruited species. Ixoroid pollination mechanism (see p. 2) present. Genera 43–47.

Tribe **Cinchoneae** *DC.* in Ann. Mus. Hist. Nat. 9: 217 (1807), as *Cinchonaceae*, emend. Cham. & Schlect. in Linnaea 4: 178 (1829). Trees or shrubs. Ovary 2-locular, with very numerous vertical or ascending ovules in each locule. Fruit capsular. Seeds winged; testa cell walls pitted. Ixoroid pollination mechanism (see p. 2) present in some genera. *Cinchona* has heterostylous flowers. Genera 48, 49.

Tribe **Virectarieae** *Verdc.* in K.B. 30: 366 (1975). Herbs or subshrubby herbs. Corolla-lobes valvate. Ovary 2-locular; ovules numerous, the placentas attached to the central septum. Flowers not heterostylous. Stigma capitate. Fruit capsular, ovoid or subglobose, splitting in a plane at right-angles to the central septum, very often 1 valve falling off and the other persistent (but this dehiscence is not perfect in all species). Seeds small. Genus 50 (Bremekamp includes this in the *Ophiorrhizeae*).

Tribe **Rondeletieae** (*DC.*) *Hook.* in G.P. 2: 8 (1873); as subtribe in DC., Prodr. 4: 342 (1830); including *Condamineeae*. Trees or shrubs (rarely herbs). Aestivation of the corolla-lobes various but valvate in the subtribe *Condamineineae*. Ovary 2-locular; ovules very numerous in each locule, horizontal. Fruit capsular. Seeds mostly not winged; testa walls pitted. Ixoroid pollination mechanism (see p. 2) absent. Only cultivated in East Africa, see note at beginning of subfamily.

Tribe **Mussaendeae** *Hook.* in G.P. 2: 8 (1873). Mostly trees, shrubs or woody climbers. Corolla-lobes valvate save in a very few cases. Ovary 2–many-locular, each locule with numerous ovules. Fruit fleshy or dry, indehiscent or rarely capsular (*Pseudomussaenda*). Seeds minute, not winged, angular; testa cells pitted. Ixoroid pollination mechanism absent. Genera 51–54.

Tribe **Heinsieae** (*Verdc.*) *Verdc.* in K.B. 31 (1976); as a subtribe of *Mussaendeae* in B.J.B.B. 28: 281 (1958). Shrubs or small trees. Corolla-lobes contorted or imbricate (quincuncial). Ovary 2-locular; ovules numerous on placentas attached to the septum. Flowers in terminal cymes, capitate inflorescences or spikes of lateral cymes. Style with fusiform or 2-fid stigma. Limited heterostyly present in *Heinsia*. Fruit indehiscent, ± fleshy, 1–2-locular. Stipules paired on each side of node in *Heinsia*. Genera 55, 56.

I at first maintained this as a subtribe of *Mussaendeae* differing in little but aestivation. Older workers wrongly placed it in the *Hamelieae*. Bremekamp suggests it should go in the *Urophylloideae* owing to similarities in the ornamentation of the testa cells. N. Hallé, Fl. Gabon 17: 32 (1970), treats *Bertiera* as an aberrant member of the *Gardenieae*, and F. Hallé in Adansonia 1: 266 (1961) and N. Hallé, Fl. Gabon 12: 131 (1966), put *Heinsia* in the *Mussaendeae*. I have for the present purpose maintained the tribe in an intermediate position between these two tribes.

Tribe **Gardenieae** *Cham. & Schlecht.* in Linnaea 4: 197 (1829), as *Gardeniaceae*; Dumort., Anal. Fam. des Plantes 29: 32 (1829); as subtribe in Kunth, Nova Gen. Spec. 3: 407 (1820). Trees or shrubs. Corolla-lobes contorted or imbricate. Ovary 1–many-locular, with few–numerous ovules in each locule. Fruit with a woody or leathery pericarp, the seeds embedded in a gelatinous mass. Testa cells rarely pitted. Ixoroid pollination mechanism (see p. 2) usually present. Some genera have the pollen grains in tetrads. Unisexuality occurs in a few extra-African genera. Genera 57–68.

The genus *Posoqueria*, known only in cultivation in East Africa, with horny anthers and asymmetrical corolla-limb in bud, should probably be placed in a separate tribe (see p. 19 of general key).

Tribe **Coffeeae** *DC.* in Ann. Mus. Hist. Nat. 9: 217 (1807), as *Coffeaceae*, emend. Benth., Fl. Austral. 3: 400 (1866). Syn.: *Ixoreae* (Benth.) Hook. in G.P. 2: 9 (1873). Trees or shrubs, less often climbers. Corolla-lobes contorted. Ovary 2-locular; ovules 1–few, attached to a fleshy placenta affixed near the middle of the septum, or, if solitary, then erect from the base or (in subtribe *Cremasporinae*) pendulous from the apex. Fruit a drupe with thin endocarp. Seeds with testa cells not pitted. Ixoroid pollination mechanism (see p. 2) usually present. Anthers inserted at the throat, usually just exserted, often twisted. Genera 69–85.

Tribe **Vanguerieae** *Dumort.*, Anal. Fam. des Plantes: 32 (1829), as "*Vaugnerieae*". Shrubs or trees, more rarely woody-based herbs. Corolla-lobes valvate. Ovary 2–6-locular, the ovules solitary and pendulous; stigma cylindrical or capitate. Fruit fleshy, containing pyrenes. Ixoroid pollination mechanism (see p. 2) present. Stamens inserted at the throat. Pollen grains porate, without sculpture. Genera 86–99.

Subfamily **Guettardoideae** *Verdc.* in B.J.B.B. 28: 280 (1958). Rhaphides absent in the tissues. Trees or shrubs. Corolla-lobes valvate or imbricate. Indumentum consisting of hairs which are not truly septate, Heterostyly does not occur. Ovary 2–many-locular, with a solitary pendulous ovule in each locule. Fruits drupaceous or with a woody putamen, with 2 pyrenes or rarely dicoccous. Seeds with little or no albumen; testa not sculptured or irregularly reticulate. Pollen grains porate. One tribe, *Guettardeae* DC. in Ann. Mus. Hist. Nat. 9: 217 (1807), as *Guettardaceae*, emend. Kunth, Nova Gen. Spec. 3: 419 (1820). Genus 100.

NOTE. *Gaertnera zimmermannii* K. Krause & Gilg in E.J. 48 : 430 (1912), type : Tanzania, E. Usambara Mts., Amani, by R. Kwamkuyu, *Zimmermann* in *Herb. Amani* 2926 (B, holo. †, EA, iso. !), was wrongly placed by its authors and the name is in fact a synonym of *Strychnos mellodora* S. Moore (the exact identity was established by Leeuwenberg to whom I sent the isotype some years ago, see Meded. Landbouwhogeschool Wageningen 69: 180 (1969)). No species of *Gaertnera* appears to occur in the Flora area.

NOTES ON THE KEYS

The family Rubiaceae contains so many genera and species, many of which resemble each other even when not closely related, that it is impossible to make a usable key which does not involve looking at small and difficult characters. If, after a preliminary run through the key, it is found that macroscopic characters will not suffice then it is best to examine the plant in detail and make a list of its essential characters. The aestivation of the corolla-lobes in bud is a very important character indeed and the main types are shown in fig. 1/1–5. This is fairly easy to see in fresh material under an ordinary hand-lens but when only herbarium material is available a bud should be boiled up. The arrangement of the ovules in the ovary is also a crucial character, whether there are one or several to many in each of how many locules and if solitary whether they are attached to the base, apex or middle of the locule. This is admittedly a difficult character to see and dissection under a binocular dissecting microscope is often necessary. A transverse cut across the ovary with a razor blade will often clearly show the number of locules, and in the case of solitary ovules whether they are attached at the base or apex can be ascertained by seeing which half of the cut ovule drops out easily. A good deal can be learnt by cutting gradually into the side of the ovary until the ovule or placentas are revealed. Care must be taken not to mistake a placenta covered with minute ovules for a single ovule; this mistake has been made even by professional botanists and several species have been redescribed in the wrong genus and tribe as a result. The presence or absence of heterostyly can be a very useful character and is worth noting in the field where a population is available. It is of course difficult or impossible to ascertain from a single herbarium specimen. If flowers showing long-exserted anthers and included styles are present on the single specimen then the species is almost certainly heterostylous, but the reverse is of course by no means true. Other microscopic characters are very useful and not difficult to see if adequate equipment is available. The presence of rhaphides restricts the choice of genera considerably. These are easily seen under a microscope and they break up into very numerous fine needle-shaped crystals if teased out under water; they are particularly easily seen if a polarising microscope is available,* but in many cases, e.g. those between the pyrenes of *Psychotria* fruits, they are clearly visible to the naked eye and they break down into fluffy crystals even when seen dry under a × 10 hand-lens. In other cases rhaphides can be seen in leaf and other tissues with the naked eye (fig. 1). Pollen grains vary very considerably in shape in the family and are valuable characters. Even the gross morphology easily visible at low powers of a compound microscope can restrict one's search for the identity of a difficult specimen. The following main classes may be distinguished so far as the East African tribes are concerned.

1) Colpate grains. 3–4(–5)-colpate. The usual kind of grain found in the family.
2) Pluricolpate grains. 5–25-colpate. These are divisible into two types:
   A) disc-shaped grains, usually large and with numerous short colpi on the rim—restricted to *Spermacoceae*;
   B) ellipsoidal grains with fewer longer colpi extending over the body—*Spermacoceae* and *Rubieae*.
3) Porate grains. 3–4-porate. Of common occurrence in the family particularly in *Mussaendeae, Hamelieae, Gardenieae, Guettardeae, Vanguerieae, Morindeae, Psychotrieae* and *Craterispermeae*.
4) Tetrahedral tetrads of 3-porate grains. Restricted to the *Gardenieae*.

   Examples of these types are shown in fig. 1/10–16.

Various conventions exist for describing the various parts of the calyx. For this account the part adnate to the ovary is referred to as the calyx-tube and the free joined part as the limb-tube; the limb-tube and the lobes together are referred to as the limb.

* The simplest student's microscope can be used by placing pieces of polaroid in the condenser and on top of the eyepiece.

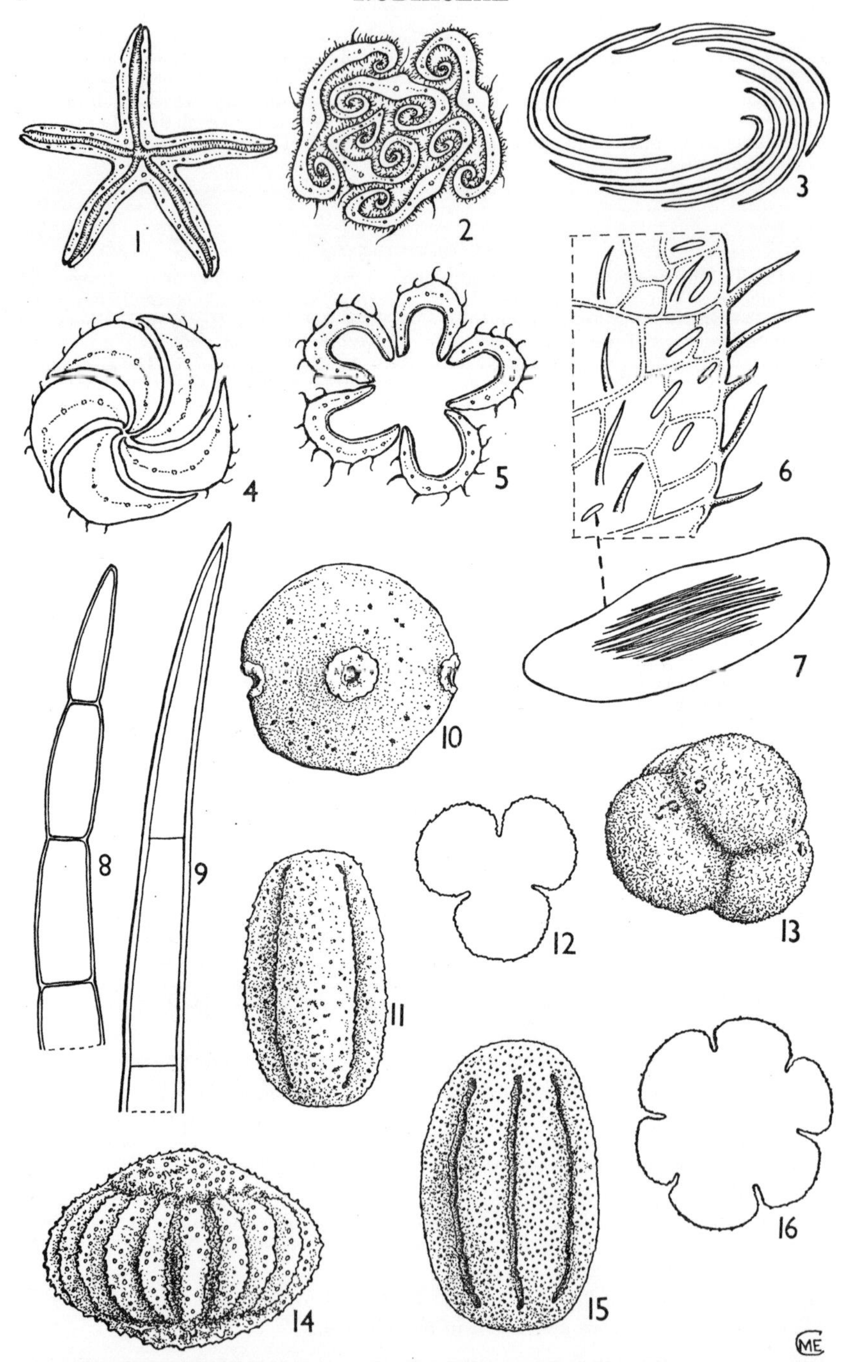

FIG. 1. Critical features in Rubiaceae. Transverse section of flower bud to show aestivation—**1,** reduplicate-valvate (*Mussaenda arcuata*, × 6); **2,** imbricate (*Heinsia crinita* subsp. *parviflora*, × 14); **3,** contorted to the left (imagine side view of bud, *Gardenia jovis-tonantis*, × 4); **4,** contorted to the right (*Bertiera racemosa*, × 14); **5,** valvate, slightly induplicate (*Pentas lanceolata*, × 12). Rhaphides and hairs—**6,** part of lower leaf surface of *Pentas lanceolata*, showing rhaphides, × 40; **7,** cell with rhaphides from same, × 360; **8,** true multicellular hair of *Pentas lanceolata*, × 80; **9,** false multicellular hair of *Virectaria multiflora*, × 80. Pollen grains—**10,** porate (*Guettarda speciosa*, × 650); **11, 12,** tricolpate (*Pentas bussei*, × 1300); **13,** tetrad of 3-porate grains (*Gardenia jovis-tonantis*, × 450); **14,** pluricolpate, disk-shaped (*Spermacoce dibrachiata*, × 650); **15, 16,** pluricolpate, ellipsoidal (*Galium stenophyllum*, × 2000).

Key to tribes*

1\. Corolla-lobes valvate 2
   Corolla-lobes imbricate or contorted 18

2\. Tree with lax inflorescences of many perfectly spherical clusters of flowers; stipules large . . . . 12. **Naucleeae** (in part)
   Inflorescences not composed of spherical elements, but some herbaceous *Spermacoceae* have flowers in spherical clusters at the nodes; many species have capitate but not spherical inflorescences 3

3\. Ovules 2 or more in each locule 4
   Ovule solitary in each locule 9

4\. Style divided into 6–8 filiform lobes at the apex; ovary with 4–10 locules, each containing 2 (rarely 3) collateral erect ovules, but only 1 developing into a seed in each locule; fruit a drupe with woody 4–10-locular putamen . . 3. **Triainolepideae,** p. 149
   Style 2-lobed or rarely with 3–4(–5) lobes but if so other characters not agreeing. 5

5\. Fruit a berry or dry and indehiscent 6
   Fruit a capsule or if indehiscent then calyx characteristic (see fig. 26/10, p. 221) 7

6\. Flowers large in terminal panicles; corolla-tube 1·3–4 cm. long . . . . . 15. **Mussaendeae**
   Flowers small in axillary or less often terminal inflorescences; corolla-tube 1·4–5 mm. long . . . . 4. **Urophylleae**

7\. Seeds very characteristically and prominently winged, 0·3–2 cm. long; trees or shrubs . . . . 13. **Cinchoneae**
   Seeds not winged or if so (e.g. *Danais*) then only 0·5–1 mm. long; herbs or small shrublets, rarely shrubs 8

8\. Style divided into 2(–3) arms which bear the stigmatic surfaces; flowers often but not always heterostylous; rhaphides present; fruit a septicidal and/or loculicidal beaked capsule . . . . . 8. **Hedyotideae,** p. 177
   Stigma capitate, the style not divided into 2 lobes; flowers not heterostylous; rhaphides absent; capsule often with 1 valve persistent and 1 valve deciduous 14. **Virectarieae**

9\. Ovules pendulous from near apex of locule 10
   Ovules erect or attached by middle to the septum 12

10\. Stigma capitate, cylindrical or various, but not divided into filiform lobes; flowers never heterostylous; rhaphides absent. 19. **Vanguerieae**
   Style with 2–5 lobes which bear the stigmatic surfaces; flowers heterostylous; rhaphides present 11

* An artificial key to genera is given on p. 9.

11.	Trees with axillary inflorescences . .	5. **Craterispermeae,** p. 161
	Herbs or small shrublets with mostly terminal but sometimes axillary inflorescences; flowers often bright blue. .	6. **Knoxieae,** p. 164
12.	Leaves and leaf-like stipules in whorls of 4–8; erect, prostrate, or climbing herbs often with ± hooked hairs rendering the plant adhesive; flowers ± rotate; fruit dry or berry-like	11. **Rubieae,** p. 380
	Leaves opposite or if in whorls of 3 or more then with other characters quite different	 13
13.	Calyx-tubes confluent; fruits ± united into a fleshy mass; 1 calyx-lobe produced into a coloured lamina in one species; trees or shrubs	2. **Morindeae** (in part), p. 134
	Calyx-tubes not confluent; fruits not forming a fleshy mass	 14
14.	Usually foetid smelling climbers; fruits flattened, the outer pericarp falling off to expose 2 compressed winged pyrenes supported by long filiform stalks .	7. **Paederieae,** p. 174
	Plants not evil-smelling; fruit not of this characteristic structure . . .	 15
15.	Fruits fleshy, indehiscent; trees or shrubs or in a few cases small forest floor herbs; flowers heterostylous	 16
	Fruits dry, usually dehiscent; herbs or small shrubs	 17
16.	Flowers terminal; ovary 2-locular or if rarely 3-locular then inflorescence with a very conspicuous involucre	1. **Psychotrieae,** p. 25
	Flowers axillary; ovary 4–12-locular, not involucrate	2. **Morindeae** (in part), p. 134
17.	Ovules erect, attached to the base of the ovary-locules; flowers unisexual or hermaphrodite; pollen 3-colpate. . . .	9. **Anthospermeae,** p. 315
	Ovules attached to the septum of the ovary; flowers hermaphrodite, mostly in globose clusters at the nodes; pollen pluricolpate	10. **Spermacoceae,** p. 333
18.	Flowers numerous in perfectly globose heads	12. **Naucleeae** (in part)
	Flowers not in globose heads (except for 2 species of *Bertiera*; many species have the flowers in capitate but not spherical inflorescences)	 19
19.	Fruit a capsule; seeds with an elegantly fimbriated marginal wing . . .	13. **Cinchoneae** (in part)
	Fruit fleshy or dry, indehiscent; seeds not winged	 20
20.	Fruit a large subglobose drupe, with ± lobed woody 4–9-locular putamen and fibrous outer layers; seeds without albumen .	20. **Guettardeae**
	Fruit not as above; seeds with albumen .	 21

21. Fruits mostly large with a woody or leathery pericarp; ovary 1–2(–4)-locular, with few to numerous ovules in each locule; pollen grains sometimes in tetrads . 17. **Gardenieae**
 Fruits mostly small drupes; ovaries 2-locular, with 1–several ovules per locule either attached to fleshy placentas affixed near the middle of the septum or often with solitary apical ovules; pollen grains never in tetrads* . . . . . 18. **Coffeeae**

ARTIFICIAL KEY TO GENERA

Apart from the keys to tribes and subsequent keys to genera a general key to what is believed to be all the genera occurring in the Flora area, both native and cultivated, is given below. Until an account of the entire family has been completed, this is of course necessarily rather tentative, and if revisional work for Part 2 necessitates extensive changes, then a revised key will be given in that volume.

1. Stems 4-angled, with some short branchlets reduced to curved or ± hooked spines; lianes with flowers in spherical heads or at least subcapitate 2
 Stems without spines or, if present, then shrubs with straight spines 3
2. Flowers in completely spherical very many-flowered heads; corolla-tube 0·8–1·5 cm. long; fruit a fusiform capsule; seeds with ribbon-like wings . . . . 47. **Uncaria**
 Flowers in few-flowered capitate clusters; corolla-tube 2·5–3·5 cm. long; fruit a globose berry; seeds not winged . . 70. **Cladoceras**
3. Nodes with whorls of 4–8 leaves and similar leaf-like stipules; herbs and herbaceous climbers with ± rotate corollas; ovules solitary in each locule; fruits globose, indehiscent; plants often adhesive due to prickles and harsh hairs (*Rubieae*) 4
 Nodes with leaves opposite or, if in whorls of 3–5, then plants either shrubs or trees and corolla not rotate 5
4. Leaf-blades ovate to lanceolate, the petiole very well developed; corolla usually 5-merous . . . . 41. **Rubia**
 Leaf-blades mostly narrow, linear or lanceolate or, if wider and ± elliptic, then petiole very short; corolla usually 4-merous . . . . 42. **Galium**
5. Climbing herbs or lianes 6
 Erect, decumbent or procumbent herbs, shrubs or trees 17
6. Some calyx-lobes in each inflorescence dilated into a large white or coloured lamina . 51. **Mussaenda**
 No calyx-lobes dilated into large laminas 7

* The tribe 16, *Heinsieae*, will key out with the *Coffeeae* (*Ixoreae*) and *Gardenieae*; a combined key to these three tribes, which are not very well delimited, will be given in part 2.

7. Corolla scarlet, salver-shaped with a very narrow tube; calyx-lobes unequal; ovules numerous in each locule . . . 17. **Otomeria volubilis**
   Corolla not scarlet or if reddish then tube not narrow 8
8. Corolla large, the tube funnel-shaped to campanulate, white and/or pink or purplish, or orange and purple, (2·5–)3–5·5 cm. long with lobes 1–2 cm. long; ovules numerous in each locule . . . . 67. **Sherbournea**
   Corolla smaller 9
9. Evil-smelling plants; fruits flattened, the outer pericarp falling off to expose 2 compressed winged pyrenes supported by long filiform stalks . . . . 14. **Paederia**
   Plants not evil-smelling; fruit not of this characteristic structure 10
10. Ovules several to numerous in each locule* 11
    Ovules solitary in each locule 15
11. Corolla salver-shaped with a narrow tube, bright yellow with an orange-red star of hairs around the throat; tube 1·3–2·7 cm. long . . . . . . 51. **Mussaenda**
    Corolla smaller 12
12. Pedicels long and graceful, 0·4–3 cm. long; inflorescences very lax . . . . 69. **Tarenna fusco-flava**
    Pedicels shorter, the inflorescences more condensed 13
13. Leaves velvety-felted beneath; flowers in sessile axillary clusters at the nodes (**U**2; **T**6) . . . . 54. **Pseudosabicea**
    Leaves not velvety-felted beneath, glabrous to densely pubescent 14
14. Seeds winged; calyx-lobes very small, 0·5–1 mm. long; flowers distinctly heterostylous; bracts absent . . . . 15. **Danais**
    Seeds not winged; calyx-lobes longer, 2–11 mm. long; flowers slightly heterostylous; inflorescences sessile or long pedunculate, bracteate . . . . . . 53. **Sabicea**
15. Flowers not heterostylous, both style and stamens exserted; stigma fusiform; rhaphides absent . . . . . 85. **Rutidea**
    Flowers heterostylous, either style or stamens exserted; stigma bifid; rhaphides present 16
16. Corolla-lobes with a median winged keel, very evident in bud; pyrenes ultimately dehiscing . . . . 5. **Chassalia cristata**
    Corolla-lobes not winged; pyrenes not dehiscent . . . . 1. **Psychotria ealaensis**
17. Calyces ± joined together; fruits fleshy, ± united into a mass; trees with in some cases 1 calyx-lobe dilated into a lamina 18
    Calyces not joined together; fruits not forming a fleshy mass 19

* See general notes on the key, p. 5.

18. Inflorescences 4–9 mm. in diameter (excluding corollas); corolla-tube 0·3–2·4 cm. long; flowers heterostylous; calyx with 1 lobe enlarged into a lamina in one species . . . . . . 7. **Morinda**
    Inflorescences 3–5 cm. in diameter (excluding corollas); corolla-tube 0·4–1·2 cm. long; flowers not heterostylous; calyx-lobes never enlarged into a lamina 43. **Nauclea***
19. Leaf-blades with distinct bacterial nodules present, sometimes restricted to the midrib (not to be confused with various fungal spots, insect galls, discolourations, etc.—if in any doubt ignore this couplet since it is merely a short cut) 20
    Leaf-blades without distinct bacterial nodules 22
20. Aestivation of the corolla-lobes valvate; flowers usually heterostylous, either the anthers or the style exserted; stigma bifid; rhaphides abundant in the leaves and especially between the pyrenes of the fruit . . . . 1. **Psychotria**
    Aestivation of corolla-lobes contorted; flowers never heterostylous, both style and anthers exserted; stigma clavate or lobed; rhaphides never present 21
21. Inflorescence axillary; ovules several in each locule . . . . 77. **Neorosea**
    Inflorescences terminal; ovules solitary in each locule . . . . . . 83. **Pavetta**
22. Flowers in spike-like inflorescences, subtended by large leafy paired stipitate venose red bracts 3·5–10 cm. long; seeds conspicuously winged . . . . 48. **Hymenodictyon floribundum**
    Flowers, if in spike-like inflorescences, then not subtended by large paired bracts; if bracts present then small or forming an involucre 23
23. One calyx-lobe often dilated into a large often stalked white or coloured lamina mostly exceeding 2 cm. long and wide 24
    Calyx-lobes equal or slightly unequal and green or if 1–2 enlarged into a coloured lamina then not attaining 2 cm. in length or width 27
24. Calyces ± joined together and fruit a compound infructescence of fused drupes; corolla-lobes valvate; ovaries with 1 ovule in each locule . . . . . 7. **Morinda**
    Calyces separate; drupes not joined to form a compound infructescence 25
25. Flowers under 5 mm. long, in very elongated

* Ridsdale, in Blumea 22: 546 (1975), has resuscitated the genus *Sarcocephalus* as distinct from *Nauclea*, both occurring in East Africa.

inflorescences bearing numerous enlarged crimson calyx-lobes (cultivated) . . **Warszewiczia**
Flowers over 1 cm. long, in much shorter inflorescences 26
26. Fruit ± succulent or at least indehiscent; buds without filiform apical appendages 51. **Mussaenda**
Fruit dry, dehiscing at the apex; buds bearing 5 apical filiform appendages . . 52. **Pseudomussaenda**
27. Anthers 5–6 cm. long, the tips just exserted; corolla white, narrowly funnel-shaped, 15–18 cm. long; flowers solitary, axillary, 1 on each side of the node; calyx-lobes leafy, 2·5 cm. long; corolla-lobes ± valvate, the edges actually very slightly imbricate (cultivated) . . **Portlandia**
Anthers shorter even if corolla as long. 28
28. Corolla-lobes valvate 29
Corolla-lobes contorted or imbricate 90
29. Style divided into 6–8 filiform lobes; flowers heterostylous; ovary with 4–10 locules, each containing 2 (rarely 3) collateral erect ovules, but only 1 develops into a seed in each locule, the undeveloped ovule often being stuck to the seed; shrub 1·8–6 m. tall; corolla white, with tube ± 1 cm. long, woolly tomentose outside; drupes red, with woody putamen; mainly littoral or in coastal bushland . . . . 8. **Triainolepis**
Style 2-fid or if with several lobes then without above characters combined 30
30. Ovules 2 or more in each locule 31
Ovules solitary in each locule 57
31. Calyx-limb spreading, eccentric, entire or ± shallowly lobed, venose and accrescent, up to 2·8 cm. wide in fruit; fruit dry and tardily or not dehiscent; corolla-tube narrowly tubular, 1·2–3(–3·5) cm. long . . . . 19. **Carphalea**
Calyx-limb not as above, usually with 4–5 distinct lobes or teeth 32
32. Fruit a capsule or splitting into 2 cocci 33
Fruit a berry or dry and indehiscent 55
33. Trees, shrubs or shrubby herbs with winged seeds; rhaphides absent except in *Bouvardia*. 34
Herbs or small shrubs; seeds not winged 37
34. Flowers in collections of dense perfectly spherical heads; stipules large and leafy 46. **Mitragyne***
Flowers in spike-like inflorescences or in lax branched inflorescences 35
35. Subshrubby herbs or herbs with leaves opposite or in whorls of 3–4; corolla 4-

* Including *Hallea* J.-F. Leroy.

merous; seeds with a circular wing (cultivated) . . . . . . **Bouvardia** (p. 25)
Trees or shrubs with 5-merous corolla. 36

36. Corolla glabrous; style always long-exserted, with a slightly clavate stigma; capsule with large lenticels, not crowned with remains of calyx-limb . . . . 48. **Hymenodictyon parvifolium**
Corolla-tube tomentose outside and lobes margined with long hairs; flowers heterostylous but style never long-exserted; capsule with ± no lenticels, crowned with the remains of the calyx-limb (cultivated) . . . . . **Cinchona**

37. Stigma capitate; flowers not heterostylous, both style and stamens exserted; capsule often with 1 persistent lobe and 1 falling lobe; corolla-lobes very narrowly lanceolate and about as long as or longer than the corolla-tube; small shrubs and herbs without rhaphides . . . . . 50. **Virectaria**
Stigma bilobed; flowers often heterostylous, with either stamens or style included (except in *Pseudonesohedyotis*); capsule mostly opening at the beak; corolla-lobes usually not so narrow and often shorter than the tube; rhaphides usually easily visible 38

38. Ovary, placentas and fruit elongated, linear; forest-floor herb with white funnel-shaped corolla 5·5–8 mm. long (**T3**) . 20. **Dolichometra**
Ovary, placentas and fruit never linear 39

39. Flowers mostly 4-merous; leaf-blades frequently narrow and uninerved or with lateral nerves obscure 40
Flowers mostly 5-merous; leaf-blades usually broad and with very obvious lateral and tertiary venation (but not always) 50

40. Corolla-lobes induplicate-valvate, covered on the inside with clavate hairs. . . . 24. **Pentanopsis**
Corolla-lobes valvate; lobes smooth or papillate inside. 41

41. Anthers and stigmas included, the latter always overtopped by the former; corolla-tube narrowly cylindrical . 22. **Kohautia**
Anthers and/or stigmas exserted or if both included then anthers overtopped by the stigma; corolla-tube cylindrical or funnel-shaped 42

42. Corolla-tube cylindrical, at least 2 cm. long; anthers included and style exserted; flowers not heterostylous . . . . 23. **Conostomium**
Corolla-tube cylindrical or funnel-shaped, always less than 2 cm. long, sometimes with included anthers and exserted style but then usually heterostylous 43

43. Capsule opening both septicidally and loculicidally 44
 Capsule opening loculicidally 48
44. Beak of the capsule as long as or longer than the rest of the capsule 45
 Beak of the capsule shorter than the rest of the capsule 46
45. Fragile annual herb with capsule of very characteristic shape, emarginate at the base, the beak much exceeding the rest of the capsule (fig. 32, p. 249); stipules triangular with 2 lobes . . . . 25. **Mitrasacmopsis**
 Robuster subshrubby herb; beak as long as the rest of the capsule; stipules 3–7-fimbriate . . . . 26. **Hedythyrsus**
46. Robust shrubby herb; stipules with a single deltoid lobe, sometimes ± lacerated but not fimbriate; style and stamens long-exserted (Uluguru Mts.) . . . . 27. **Pseudonesohedyotis**
 Herbs; stipules mostly divided into distinct fimbriae, flowers heterostylous or isostylous but both stamens and style not long-exserted 47
47. Corolla-tube bearded inside; style not shortly bifid at the apex (apart from the division into stigma-lobes); stigma-lobes subglobose or ovoid . . . . . 28. **Agathisanthemum**
 Corolla-tube glabrous inside; style shortly bifid at the apex as well as divided into ellipsoid stigma-lobes (**K4**) . . . . 29. **Dibrachionostylus**
48. Rush-like plant with linear or filiform leaves; stipule-sheath tubular, truncate or with 2 minute teeth; seeds dorsiventrally flattened . . . . 30. **Amphiasma**
 Plant not rush-like even if leaves linear; if seeds dorsiventrally flattened then leaves not linear 49
49. Capsule with thick woody wall and a solid beak, tardily dehiscent . . . . 32. **Lelya**
 Capsule with horny wall, with or without a beak but never solid, early dehiscent . 33. **Oldenlandia**
50. Leaves uninerved, mostly fairly small; decumbent herb of wet places with small white or blue flowers in very lax elongated axillary cymes . . . . 31. **Pentodon**
 Leaves larger, pinnately nerved; flowers mostly terminal. 51
51. Creeping herbs or mat herbs of the forest floor, often rooting at the nodes 52
 Erect herbs or subshrubs 54
52. Calyx-lobes conspicuously spathulate at their apices; corolla-tube ± 3 cm. long . 18. **Chamaepentas**
 Calyx-lobes not or much less spathulate; corolla-tube under 2 cm. long 53
53. Inflorescences ± several-flowered cymes, not sessile . . . . 16. **Pentas**

Inflorescences few-flowered, usually only 1 or 2 corollas open at a time, sessile in the axils of the leaves . . . . . 21. **Parapentas**

54. Flowering inflorescences capitate or lax much-branched complicated cymes, not elongating into simple spikes in fruit, although individual branches sometimes become spicate; fruit globose or obtriangular, less often ovoid-oblong . 16. **Pentas**

Flowering inflorescences capitate, later elongating into a long simple "spike", rarely with axillary spikes from the upper axils and frequently with solitary flowers at the lower nodes; fruits oblong 17. **Otomeria**

55. Flowers large, in terminal panicles, yellow or red; corolla-tube 1·3–4 cm. long . . 51. **Mussaenda**

Flowers much smaller, in axillary or less often terminal inflorescences; corolla-tube 1·4–5 mm. long 56

56. Inflorescences compact axillary clusters or terminal and axillary; ovary 2-locular beneath, 4-locular above; stigma-lobes 2 9. **Pauridiantha**

Inflorescences lax axillary cymes, ± 3 per axil; ovary 4–5-locular beneath, 8–10-locular above; stigma-lobes 4–5 . . 10. **Rhipidantha**

57. Ovule pendulous from near the apex of the locule 58

Ovule erect from near the base or attached towards the middle of the septum 75

58. Herbs with terminal inflorescences 59

Trees, shrubs or small to large subshrubs with axillary inflorescences 60

59. Slender annual, with 4-merous flowers; corolla-tube 3 mm. long; fruit breaking off to leave a small cup, which is the persistent woody flanged pedicel . . 12. **Paraknoxia**

Perennials, with mainly 5-merous flowers; corolla-tube exceeding 3 mm.; fruit not leaving a cup-like organ when falling; corolla often bright blue . . . 13. **Pentanisia**

60. Stigma with filiform lobes; flowers heterostylous; rhaphides present; trees . 11. **Craterispermum**

Stigma characteristic, capitate, globose or cylindrical, often grooved or slightly lobed; rhaphides absent; flowers never heterostylous; herbs to trees (*Vanguerieae*) 61

61. Corolla-tube slightly curved or if ± straight then ± 2·5 cm. long and broadly tubular 62

Corolla-tube straight or not so large 63

62. Cymes without bracteoles; calyx-limb truncate; corolla-tube glabrous outside . 87. **Fadogia** sect. **Temnocalyx**

Cymes bracteolate; calyx-limb with long lobes; corolla-tube adpressed ferruginous hairy outside (adult corolla not known in East African species); ovary 5-locular; young shoots adpressed ferruginous hairy . . . . 98. **Ancylanthos**

63. Flowers solitary or in simple unbranched inflorescences 64

Flowers in umbels or in conspicuously branched inflorescences 70

64. Flowers solitary and corolla-tube narrowly tubular; corolla-lobes ending in a long filiform appendage; ovary 3–4-locular; stigma 3–4-lobed . . . . . 86. **Hutchinsonia**

Flowers not solitary or if so then without the other characters combined 65

65. Leaves in whorls of 3–5; herbs or subshrubs with mostly strictly virgate stems from a woody rootstock, more rarely shrubs or even small trees; calyx-limb truncate to shortly lobed . . . . . . 87. **Fadogia**

Leaves opposite or rarely (not in Flora area) ternate 66

66. Flowering branches abbreviated; armed with opposite spines; calyx-limb lobed or slightly toothed; corolla-throat densely hairy . . . . 92. **Meyna**

Flowering branches developed; spines absent or present in a few species of *Rytigynia*. 67

67. Plant densely grey-white tomentose; calyx-limb minutely toothed, the teeth ± 1 mm. long; cymes subsessile or peduncle only 3 mm. long . . . . . 88. **Fadogiella**

Plant not densely grey-white tomentose or if so calyx distinctly lobed 68

68. Calyx-limb truncate, shortly lobed or obscurely toothed; ovary 2–5-locular . 91. **Rytigynia**

Calyx-lobes elongate, half as long as to longer than the corolla; ovary (4–)5-locular 69

69. Plants completely tomentose or felted; calyx-lobes equalling the corolla-tube or rarely longer; shrubs . . . . . . 89. **Tapiphyllum**

Plants hairy; calyx-lobes mostly longer than the corolla-tube; small subshrubs. . 90. **Pachystigma**

70. Small subshrub with subherbaceous stems; ovary 2-locular . . . . . . 94. **Pygmaeothamnus***

Trees or shrubs; ovary 2–5-locular 71

71. Calyx-lobes exceeding the corolla, 14 × 3·5 mm., ± 7 times as long as the calyx-tube 97. **Lagynias**

Calyx-lobes not as above 72

72. Bracts, bracteoles and calyx-lobes ± leafy and conspicuous, the calyx-lobes 6–15 × 4·7–6 mm. . . . . 99. **Cuviera**

Bracts, bracteoles and calyx-lobes not leafy. 73

73. Ovary 5-locular and fruits usually with 5

* It is very likely that *Pygmaeothamnus concrescens* Bullock and *Canthium crassum* Hiern will have to be merged as a section of *Canthium* characterised by the hardened corolla. The genus *Multidentia* Gilli would provide a name for this.

pyrenes, large, 0·5–3·5 cm. in diameter; inflorescences cymose with the ultimate branches elongated; calyx-lobes triangular to linear . . . . . 96. **Vangueria**

Ovary 2-locular; fruits mostly didymous or 1-locular by abortion 74

74. Inflorescence umbellate or cymose corymbose; calyx-limb truncate or minutely toothed . . . . 93. **Canthium**

Inflorescence cymose with ultimate branches always elongated; calyx-limb lobed . 95. **Vangueriopsis**

75. Ovule erect from the base of the locule 76

Ovule attached near the middle of the septum 83

76. Herbs or subshrubs with dry capsular or indehiscent fruits 77

Shrubs, trees or subshrubs with succulent fruits; flowers mostly heterostylous 78

77. Calyx-lobes unequal, 1 or more enlarged; flowers hermaphrodite, not heterostylous, both stamens and style exserted; stigmas ± smooth; corolla-tube exceedingly fine. . . . . . 34. **Otiophora**

Calyx-lobes equal, small; flowers usually unisexual or polygamous and partially heterostylous in some plants; stigmas feathery . . . . 35. **Anthospermum**

78. Flowers in axillary inflorescences; fruits mostly blue, with 4–12 pyrenes; trees or shrubs; seeds with albumen soft and oily 6. **Lasianthus**

Flowers in mostly terminal inflorescences; fruit with 2 (rarely 3) pyrenes 79

79. Pyrenes without well-marked dehiscence; seeds with a red-brown or purplish testa, often with ruminate endosperm; mostly shrubs or small trees, less often subshrubby or herbaceous . . . . 1. **Psychotria***

Pyrenes with ± well-marked dehiscence; testa of seed pale; albumen not ruminate 80

80. Herbs of the forest floor, mostly procumbent or straggling; pyrenes opening basally by 2 short dorsal slits 81

Shrubs, lianes or small trees, rarely subshrubby herbs but never procumbent or straggling 82

81. Stipules lacking a membranous sheath within; pyrenes ribbed and often rugose . 2. **Geophila**

Stipules with membranous sheath within; pyrenes smooth . . . . . 3. **Hymenocoleus**

82. Pyrenes opening by 2 marginal slits; buds never winged; shrub with leaves mostly developing after the flowers have matured; stems corky . . . . 4. **Chazaliella**

Pyrenes opening by 1 dorsal slit; buds and

* Including *Cephaelis*, with inflorescences surrounded by an involucre, and *Megalopus*, with inflorescences borne on exceedingly long peduncles and surrounded by a large boat-shaped involucre.

corolla-lobes often with longitudinal narrow wing-like keels (but not present in some species) . . . . . . 5. **Chassalia**

83. Trees with flowers in heads; the calyx-lobes ± joined; fruits fleshy, ± united into a mass . . . . 7. **Morinda**

Herbs or small subshrubs with dry capsular or ± indehiscent fruits; flowers mostly in ± globose sessile clusters at the nodes or sometimes in terminal clusters (*Spermacoceae*) 84

84. Ovary 3-locular; stigma-lobes 3; fruit with 3 cocci . . . . 40. **Richardia**

Ovary 2-locular; stigmas 2 or 1, capitate; fruits capsular with 2 valves or with 2 cocci or circumscissile 85

85. Fruit circumscissile about its middle, the top coming off like a lid; flowers minute, in globose nodal clusters; seeds with a ventral impressed x-like pattern . . 39. **Mitracarpus**

Fruit indehiscent or opening by longitudinal slits or 2-coccous 86

86. Succulent creeping plant of the seashore with imbricated leaves joined by quite broad sheathing stipules with very short processes; stems rooting at the nodes; fruits indehiscent; seeds not lobed; one record from **T3** . . . . . . 36. **Hydrophylax**

Plant not a littoral succulent or if somewhat so (*Diodia* subgen. *Pleiaulax*) then leaves not imbricated, stipules with longer processes, stems not rooting at the nodes, fruits dividing into cocci and seeds lobed; common 87

87. Capsule opening from base to apex, the calyx-limb and joined valves falling off together, leaving the oblong septum . 38. **Spermacoce** subgen. **Arbulocarpus**

Capsule opening from apex to base or fruit 2-coccous 88

88. Seeds very distinctly lobed in a characteristic manner (fig. 48/7, p. 338) . . . 37. **Diodia** subgen. **Pleiaulax**

Seeds not distinctly lobed 89

89. Fruit an obvious capsule . . . . . 38. **Spermacoce**

Fruit with 2 indehiscent or ± indehiscent cocci* . . . . 37. **Diodia**

90. Flowers in perfectly globose heads 91

Flowers in lax to dense inflorescences but even if capitate then not in perfectly globose heads 94

91. Flowers fused together by their calyx-tubes and -limbs; fruits forming a syncarp; seeds not winged . . . . . 43. **Nauclea**

* If in doubt over this couplet proceed to key to the species of *Spermacoce*, in which the species of *Diodia* are also included.

Flowers and fruits entirely free from each other even though touching 92

92. Fruits ± succulent; corolla-lobes contorted (fig. 1/4) . . . . 56. **Bertiera** (in part)

Fruits capsular; corolla-lobes imbricate 93

93. Leaves paired, broadly elliptic; calyx-lobes spathulate, persistent; seeds not winged; wood a bright orange colour . . . 44. **Burttdavya**

Leaves in whorls of 4, lanceolate; calyx-lobes oblong, at length deciduous; seeds winged at each end; wood not bright orange . . . . 45. **Breonadia** (*Adina* auctt. afr.)

94. Anthers exserted, hard and horny in texture, with the connective pubescent; buds with limb slightly asymmetric; flowers sweet-scented, in terminal inflorescences; corolla white, the tube narrowly cylindrical, 9–15 cm. long, 2–4 mm. wide with lobes 1·2 cm. long (cultivated) **Posoqueria**

Anthers if exserted then not hard and horny; bud-limbs never asymmetric 95

95. Fruit a capsule with numerous winged seeds 96

Fruit indehiscent; seeds not winged 98

96. Corolla 2–3 cm. long, pink, fragrant; calyx-lobes lanceolate, 1–1·5 cm. long; flowers slightly heterostylous, the stigma exserted or included; seeds small, with a reduced wing at each end (cultivated). **Luculia**

Corolla under 1·5 cm. long; calyx-lobes under 1 cm. long; flowers not heterostylous 97

97. Style long-exserted, the stigma clavate, slightly lobed; stamens exserted; calyx-lobes 0·5–1·5 mm. long; corolla white, under 1 cm. long; seeds with a very elegant fimbriated wing all round 49. **Crossopteryx**

Style and stamens not or scarcely exserted; calyx-lobes 0·5–6 mm. long; corolla pink or vermilion with yellow or orange throat, 0·7–1·3 cm. long; seeds with smaller much less fimbriated wing (cultivated) . . . . . . **Rondeletia**

98. Small tree or shrub growing in the littoral zone just above high tide mark, the stems covered with large leaf scars; leaves large, crowded at the ends of the branchlets; calyx truncate; corolla white, salver-shaped, velvety outside, the tube ± 2·5 cm. long with 4–9 lobes; ovary 4–9-locular; fruit globose, fibrously woody, up to 3·5 cm. in diameter 100. **Guettarda**

Not growing in the littoral zone or if so then calyx, corolla, fruit and habit all quite different 99

99. Inflorescences or solitary flowers terminal

but sometimes on lateral branchlets or appearing laterally by sympodial growth of the stem but never truly axillary—if apparently in the axils then never on both sides of the stem at the same node 100

Inflorescences or solitary flowers axillary on both sides of the node; stem growing monopodially (inflorescences when clustered at the apex can sometimes appear subterminal*) 121

100. Glabrous littoral shrub usually growing not far from high tide mark or on sea cliffs; leaves elliptic, rather shining; flowers in many-flowered terminal inflorescences; calyx ± 1 mm. long with minute lobes; corolla cream, 6–7 mm. long; style and stigma exserted; ovules 2–3 in each locule, collateral on inconspicuous placentas; fruits globose, 5–6 mm. in diameter, with a single seed containing very ruminated endosperm 71. **Enterospermum**

If a glabrous littoral shrub then other characters not agreeing 101

101. Ovules several to numerous in each locule on placentas attached to the walls or septum of the ovary 102

Ovules solitary or paired in each locule 117

102. Anthers mostly included, only the tips exserted, about ¾ the length of the corolla; ovary 5-locular; flowers in terminal branched cymes; corolla ± cylindrical, orange-red, the lobes very short; style filiform, with narrowly fusiform undivided stigma (cultivated) . . **Hamelia** (p. 25)

Anthers not ¾ the length of the corolla; ovary 2-locular 103

103. Style with 2 recurved spreading arms 104

Style-head ("stigma") usually capitate, clavate or fusiform, undivided or sometimes shortly cleft but not with spreading arms, or if slightly so (in *Heinsia*) then anther-thecae not divided into small compartments; corolla-lobes apiculate and stipules small deltoid or subulate 2 on each side of the node 105

104. Anther-thecae divided into small compartments by longitudinal and numerous transverse septa . . . . . 57. **Dictyandra**

Anther-thecae not divided into compartments; corolla-tube up to 9 cm. long; corolla-lobes not apiculate; stipules triangular to large and leafy . . . 58. **Leptactina**

* In *Lamprothamnus* the inflorescence often appears terminal but is actually subterminal, the components from both sides of the upper axils overtopping the often inconspicuous terminal shoot.

105. Inflorescences several–many-flowered, appearing laterally at alternate nodes; pollen grains not in tetrads; ovary 2-locular, with placentas on the septum 106
Inflorescences terminal or appearing laterally at successive nodes or flowers paired or solitary 107

106. Anthers and style well exserted; corolla villous inside towards the throat, glabrous or pubescent outside; tube cylindrical, the lobes often reflexed . 59. **Aidia**
Anthers and style not exceeding the corolla-tube; corolla glabrous inside or glabrous except for a villous band towards the base, silky pilose outside except at the base; tube cylindrical or funnel-shaped with mostly erect lobes . . 60. **Porterandia**

107. Flowers orange turning red, tubular, in terminal heads supported by tight involucral bracts, the inner toothed; style-lobes flattened, adhering but easily separable (cultivated) . . **Burchellia**
Inflorescence not as above 108

108. Inflorescences several–many-flowered, always terminal although sometimes on lateral branchlets 109
Inflorescences appearing laterally at successive nodes or, if terminal then only 1–2(–3)-flowered 112

109. Placentas parietal; ovary 1-locular; inflorescences terminal, several–many-flowered; corolla hairy outside; anthers completely exserted between the corolla-lobes; stigma subglobose, usually very long-exserted but only shortly so in 1 rare species; pollen grains in tetrads . . . . . 61. **Macrosphyra**
Placentas attached to the septum of the distinctly 2-locular ovary 110

110. Inflorescence either an elongated spike-like panicle of small cymes or flowers axillary, few or congested or in globose inflorescences; fruits globose; ovules numerous . . . . . . 56. **Bertiera**
Inflorescence not as above 111

111. Corolla with 1 lobe imbricate; calyx-lobes ± leafy; inflorescences few-flowered; fruit globose, ± 1·5 cm. in diameter; ovules numerous; corolla-tube 1·5–4 cm. long; corolla-lobes with small horn-like apiculum showing as 5 tails in bud; stipules 2 on each side of node, subulate or triangular . . . . . . 55. **Heinsia**
Corolla-lobes contorted; calyx-lobes mostly small and triangular; fruit not exceeding 1 cm. in diameter; ovules 1–6; in-

florescences mostly corymbose; corolla never so long . . . . . . 69. **Tarenna** (in part)

112. Branches usually armed with spines which are modified lateral branchlets; flowers 1–2(–3) together, terminating opposite pairs of very much abbreviated leafy lateral shoots; ovary 2-locular, the placentas attached to the septum which divides the seed masses in fruit; pollen grains single . . . . . . 62. **Xeromphis**

Branches not armed with spines (in some species of *Gardenia* lateral branches tend to be spine-like but these have 1-locular ovaries with 3–9 placentas); placentas 2–9, parietal; ovary 1–2-locular; seeds forming a single mass in fruit 113

113. Flowers 3 or more together in lateral cymes on only 1 side of the stem; corolla with very narrowly cylindrical tube; anthers and style both exserted; pollen grains in tetrads . . . . . . 65. **Oxyanthus**

Flowers solitary or paired or if rarely 3 or more then corolla not narrowly cylindrical, terminal or appearing lateral by sympodial growth; corolla ± funnel-shaped and often large 114

114. Stipules chaffy, persistent; leaves deciduous, crowded beneath the solitary flowers with long almost leafless internodes between the clusters; calyx-tube very short, with spreading lobes; ovary 2-locular; pollen grains in tetrads . 63. **Euclinia**

Stipules not chaffy 115

115. Ovary 1-locular, with 2–9 placentas; pollen grains in tetrads; young parts often glutinous; stipules sheathing, persistent, often truncate . . . . . 64. **Gardenia**

Ovary (1–)2-locular, with 2 placentas; pollen grains single; young parts not glutinous; stipules not sheathing, mostly soon falling 116

116. Calyx-lobes in bud open; flowers 1–10, terminal or appearing laterally on abbreviated shoots on account of the sympodial growth; corolla-lobes overlapping to left or right; ovary 2-locular or partially so; erect shrubs or trees . 66. **Rothmannia**

Calyx-lobes contorted in bud and overlapping to the right; cymes 1–several-flowered, normally axillary on one side of the stem at successive nodes, rarely terminal; corolla-lobes overlapping to the left; ovary 2-locular; lianes . . 67. **Sherbournea**

117. Flowers usually 4-merous, mostly in terminal panicles or cymes; shrubs or small

trees; ovules attached to the septum 118

Flowers usually 5-merous or 4-merous in some species of *Rutidea* but these are climbing or scrambling shrubs 119

118. Style with 2 recurved spreading arms; leaves never with bacterial nodules . 82. **Ixora**

Style clavate, subentire; leaves sometimes with bacterial nodules . . . 83. **Pavetta**

119. Corolla campanulate, constricted into a short tube at the extreme base (only in species occurring in Flora area); flowers 1–many, subtended by a single leaf, clustered together at the ends of very short terminal branchlets, often appearing laterally at the nodes owing to the sympodial growth of the stem; ovules paired in each locule, immersed in a pendulous placenta; style narrowly clavate . . . . 81. **Aulacocalyx** (including *Heinsenia*)

Corolla salver-shaped; flowers mostly in terminal panicles 120

120. Ovules 1–6 in each locule, attached to the septum; endosperm not ruminate; erect or rarely climbing shrubs . . 69. **Tarenna** (in part)

Ovules 1 in each locule, attached to the base; endosperm ruminate; nearly always climbing or scrambling shrubs. 85. **Rutidea**

121. Corolla large, 4–7·6 cm. long, the tube trumpet-shaped, longitudinally ribbed; solitary flowers supra-axillary on both sides of the monopodially growing stem; ovules numerous on 2 parietal placentas which often meet to make the ovary partially 2-locular; stems often armed with spines (shortened lateral shoots); style stoutly fusiform or clavate, entire . . . . . 68. **Didymosalpinx**

Corolla not so large nor ribbed; inflorescences several- to many-flowered; ovary 2-locular; ovules 1–many in each locule; placentas not parietal; style ± filiform or narrowly clavate or fusiform, entire or 2-armed 122

122. Flowers apparently in dense terminal inflorescences, but actually subterminal, the uppermost axillary components overtopping the inconspicuous terminal shoot; shrub of coastal regions with subsessile often ± cordate-based glabrous ± coriaceous shining leaves; calyx bracteolate at the base; corolla-tube 6 mm. long, with 6–7 lobes ± 8 mm. long; style and anthers exserted, the stigma fusiform and hairy, bifid at the apex; ovary 1–2-locular, the ovules

solitary and pendulous; fruit crowned by the persistent calyx . . . 80. **Lamprothamnus**
Flowers in clearly axillary inflorescences; without the other characters combined 123
123. Corolla almost rotate, ± 7–8 mm. long, of which only 0·5–1 mm. is tube; style and ± cohering lamellate stigma-arms hairy; ovary 2-locular, with 2 collateral pendulous ovules in each locule . . 72. **Galiniera**
Corolla-tube more developed, usually ± funnel-shaped 124
124. Leaves densely finely velvety buff tomentose beneath; flowers in sessile bracteate axillary inflorescences, the outer bracts ovate up to 1·2 cm. long; corolla-tube 0·5–1 cm. long . . 54. **Pseudosabicea**
Leaves not finely velvety beneath or if slightly so then bracts and corolla different 125
125. Stigma-lobes cohering, hairy, with longitudinal wing-like lamella, much longer than the short style; flowers in lax ± elongate bracteate cymes on either side of the node; ovules 1–2 in each locule, almost completely immersed in the placentas; corolla-tube campanulate, shorter than the lobes . . . 75. **Kraussia**
Stigma-lobes not lamellate; flowers solitary or in congested inflorescences 126
126. Style undivided or shortly bifid or arms only detachable with difficulty; ovules solitary and pendulous in each locule or 2–4 and attached to the septum 127
Style distinctly divided into 2 linear to ovoid spreading arms; ovules solitary and attached to the septum or 1–16 and inserted in the placentas; bracteoles often cupular 130
127. Anthers included or only tips showing; bracteoles cupular; flowers 4-merous; calyx truncate, subtruncate or very shortly toothed; corolla densely hairy at the throat; ovules solitary and pendulous in each locule; seeds with ruminate endosperm. . . . 79. **Polysphaeria**
Anthers exserted; bracteoles small or at least not cupular; calyx-limb with distinct teeth 128
128. Ovules solitary and pendulous in each locule; flowers 5(–6)-merous; corolla mostly hairy outside and inside; fruit oblong-ovoid, drawn out into a constricted beak at the apex; branchlets inserted above the nodes . . . 78. **Cremaspora**
Ovules 3–4 in each locule; fruit globose, not beaked 129

129. Calyx-lobes triangular, under 0·5 mm. long; ovules 2–3 together, collateral in each locule; anthers with very short filaments . . . . 73. **Zygoon**
Calyx-lobes lanceolate, 2–3 mm. long; ovules 2–4, attached to the septum; flowers fasciculate on lateral branchlets, usually opening before they elongate and often when quite leafless; stipules apiculate, imbricated at the base of the lateral shoots; anthers sessile . . . . 74. **Feretia**

130. Corolla glabrous inside and outside; ovules solitary in each locule, attached to the lower half of the septum, very rarely 2 immersed in each placenta; bracteoles usually cupular and often with a foliaceous appendage; fruit mostly ovoid or oblong, clearly stipitate . 84. **Coffea**
Corolla pubescent or pilose inside and glabrous to hairy outside; ovules 1–16, inserted on the outer face of each placenta, the placentas peltately attached to the septum; bracteoles cupular; fruit globose, less stipitate **131**

131. Anthers medifixed or ± medifixed, the connective not enlarged; anther-thecae ± contiguous and subparallel . . 76. **Tricalysia**
Anthers basifixed, the connectives enlarged so that the anther-thecae are not contiguous but diverge; flowers often appearing before the leaves develop . 77. **Neorosea**

Subfamily **RUBIOIDEAE**

Very few genera belonging to the *Rubioideae* are cultivated in East Africa. The most popular are various species of the native genus *Pentas* which are mentioned in the treatment of that genus. Specimens of only two introduced genera have been seen, namely *Hamelia* Jacq. and *Bouvardia* Salisb., both of which are included in the general key. Several very different looking specimens of *Hamelia* have been seen, but all are probably best referred to *H. patens* Jacq. (*H. erecta* Jacq.), e.g. Uganda, Kampala, Botanic Garden of Makerere University, 5 May 1971, *Lye* 6035 ! and Tanzania, Amani Nursery, 28 Oct. 1933, *Greenway* 3672 !; see also U.O.P.Z.: 290 (1949). There is a great deal of variation in the indumentum of this plant, some specimens, e.g. *Greenway* 3672, being very densely hairy and perhaps referable to *H. lanuginosa* Mart. & Gal., which is probably not specifically distinct. These very hairy specimens have been referred to *H. sphaerocarpa* Rúiz & Pavón, probably following Rehder in Bailey, Standard Cycl. Hort. (1939), but there is no authentic material available and the original illustration shows a plant with round fruits. Various *Bouvardia* hybrids have been grown in Nairobi. Jex-Blake, Gardening in East Africa, ed. 4: 105 (1957), reports *B. angustifolia* Kunth and *B. humboldtii* Hort. as grown near Nairobi. The genus *Serissa* Juss. (*Anthospermeae*) has been advertised by nurseries in Nairobi, but no specimens have been seen in cultivation.

Tribe 1. **PSYCHOTRIEAE**

Flowers in mostly terminal inflorescences; fruit with 2(–3) pyrenes; trees, shrubs, lianes or subshrubby herbs:

Pyrenes without well marked dehiscence; seeds with a red-brown or purplish testa and often with a ruminate endosperm; herbs to trees but mostly shrubs . . . . 1. **Psychotria***
Pyrenes with ± well marked dehiscence; testa of seed pale; endosperm not ruminate:
Herbs of the forest floor, mostly procumbent or straggling; pyrenes opening basally by 2 short dorsal slits:
Stipules with a membranous sheath within; pyrenes smooth . . . . . . 2. **Hymenocoleus**
Stipules lacking a membranous sheath within; pyrenes ribbed, often rugose . . . 3. **Geophila**
Shrubs, lianes or small trees, rarely subshrubby herbs, but never procumbent or straggling; dehiscence of pyrenes not as above:
Pyrenes opening by 2 marginal slits; buds never winged; shrub with leaves mostly developing after the flowers have matured; stems corky; flowers small, in small capitate inflorescences . . . . . . 4. **Chazaliella**
Pyrenes opening by 1 dorsal slit; buds and corolla-lobes often (but not always) with longitudinal narrow wing-like keels; inflorescences mostly branched and extensive 5. **Chassalia**
Flowers in axillary inflorescences; fruits mostly blue with 4–12 pyrenes; trees or shrubs; seeds with albumen soft and oily . . . . . . 6. **Lasianthus****

1. PSYCHOTRIA

L., Syst. Nat., ed. 10: 929 (1759); Hiern in F.T.A. 3: 193 (1877), pro parte; Petit in B.J.B.B. 34: 1–229 (1964) & 36: 65–190 (1966); Steyerm. in Mem. N.Y. Bot. Gard. 23: 406 (1972),*** *nom. conserv.*

Mapouria Aubl., Pl. Guian. 1: 175, t. 67 (1775)

Grumilea Gaertn., Fruct. & Sem. 1: 138, t. 1/7 (1788); Hiern in F.T.A. 3: 215 (1877)

Cephaelis Sw., Prodr. Veg. Ind. Occ.: 45 (1788); Hiern in F.T.A. 3: 222 (1877); Hepper in K.B. 16: 153 (1962) & in F.W.T.A., ed. 2, 2: 202 (1963), *nom. conserv.*

Camptopus Hook. f. in Bot. Mag. 95, t. 5755 (1869)

Uragoga [L., Gen. Pl., ed. 1: 378 (1737)] Baill. in Adansonia 12: 323 (1879); F.W.T.A. 2: 127 (1931)

Megalopus K. Schum. in E.J. 28: 490 (1900)

Apomuria Bremek. in Verh. K. Nederl. Akad. Wet., Afd. Natuurk., ser. 2, 54: 88 (1963)

Mostly shrubs but less often trees, lianes or subshrubs, some even ± herbaceous. Leaves opposite, petiolate, frequently drying reddish-brown in

* Including *Cephaelis*, with the inflorescences surrounded by an involucre, and *Megalopus*, with inflorescences borne on exceedingly long peduncles and surrounded by a large boat-shaped involucre.
** Often included in the *Psychotrieae*, but in this account I have followed Petit in transferring it to *Morindeae*.
*** See this reference for extensive extra-African generic synonymy.

colour, either with domatia or with bacterial nodules* the former mostly in the main nerve-axils but the latter either scattered throughout the lamina or variously restricted, sometimes only a very few at the base of the midrib beneath; less often neither domatia nor bacterial nodules are present; stipules either entire or variously fimbriated, usually deciduous; colleters** present. Flowers mostly small, sessile or pedicellate, 4–5-merous, hermaphrodite, heterostylous in terminal or axillary (in some American species) capitate or paniculate inflorescences; bracts and bracteoles small or large, variously placed, sometimes one bract subtending 2 or 3 flowers or only 1 flower or absent; often there is an involucre of bracts (sometimes large) and in some groups (particularly those previously considered to be worth recognising as *Cephaelis*) united and surrounding the whole inflorescence. Calyx-limb usually a short tube with mostly minute or less often shortly linear or ovate lobes. Corolla-tube mostly shortly cylindric, hairy at the throat save in a few species; lobes always valvate, often thickened at the apex, sometimes with little projections which show as small horns in bud. Stamens with long filaments in short-styled flowers and shorter filaments in long-styled flowers, mostly inserted about the middle of the tube. Ovary 2(rarely –3–4)-locular, each locule containing a single erect ovule; style filiform, long or short with 2 (rarely 3–4) linear stigmas; disc small, surrounding the base of the style. Fruit a drupe, very often red or in bracteate species often blue or black, with 1–2 (rarely 3 or 4) pyrenes. Seeds mostly ½-ellipsoid with plane ventral face and convex dorsal face; testa usually reddish brown; endosperm horny, entire, strongly ruminate or with a variously shaped median ventral fissure.

A very large genus of over 900 described species, although 500 may be a better estimate of the actual number; over 200 occur in tropical Africa. In the Flora area the genera *Cephaelis* and *Camptopus* could well be maintained, but both Petit and Steyermark (in the references cited above) after a study of many hundreds of species have demonstrated that these genera cannot really be kept separate on any logical grounds. Petit has given a detailed revision of the African species. His classification is used here with certain corrections suggested by Steyermark. Petit has placed most of the African species in subgenus *Psychotria*, but since this must of necessity include *P. asiatica* L., the type of the genus which belongs to the group formerly recognised as *Mapouria* Aubl. not occurring in Africa, Steyermark has erected a new subgenus to include all the other species except those in subgenus *Tetramerae* (Hiern) Petit. The classification used in the following account is summarised in the following conspectus.

Subgen. **Heteropsychotria** *Steyerm.* in Mem. N.Y. Bot. Gard. 23: 484 (1972)

Trees shrubs, herbs or climbers. Stipules various, rounded, acute, truncate or variously divided into teeth or lobes or erose or pectinate, the scars without ferruginous fibrillate hairs. Leaves without bacterial nodules. Inflorescences various, with small to large bracts and bracteoles or sometimes distinct involucres. Flowers 4- or mostly 5-merous. Seeds ribbed or not; endosperm markedly to scarcely or not ruminate, ventral surface sometimes fissured.

* A study of these nodules has been made by J. F. Gordon and is available in an unpublished thesis " The nature of and distribution within the plant of the bacteria associated with certain leaf-nodulated species of the families Myrsinaceae and Rubiaceae " 370 pp. (1963). N. R. Lersten and his colleagues are engaged in a more exhaustive study and some results may be found in *Journ. Bacteriol.* 94: 2027–2036 (1967), *Amer. Journ. Bot.* 55: 1089–1099 (1968) & 59: 89–96 (1972) & *Internat. Journ. Syst. Bacteriol.* 22: 117–122 (1972).

** Colleters in *Psychotria* are of three types—standard unbranched, dendroid and intermediate types. Curiously in species from Madagascar and the rest of the world the colleters are remarkably uniform and of the standard type, but in those from Africa all three types are found although only in two groups, namely sect. *Flaviflorae* and subgen. *Tetramerae*, is the presence of non-standard types marked. See also Lersten, Morphology and distribution of colleters and crystals in relation to the taxonomy and bacterial leaf nodule symbiosis of *Psychotria* (Rubiaceae), in *Amer. Journ. Bot.* 61: 973–981 (1974).

Sect. **Flaviflorae** *Petit* in B.J.B.B. 34: 41 (1964)

Stipules broadly triangular or ovate-triangular, entire at the apex or rarely bifid for less than 1 mm. Inflorescences paniculate, with minute bracts and bracteoles. Seeds often not ribbed; endosperm mostly deeply ruminate on the ventral face, less deeply or not at all ruminate on the dorsal face. Corolla often yellow. Species 1–7.

Sect. **Holostipulatae** *Petit* in B.J.B.B. 34: 70 (1964)

Stipules elliptic, ovate or obovate, often large with rounded or obtuse rarely slightly incised apex or with 2 rounded lobes. Inflorescences paniculate, often large and lax, with minute bracts and bracteoles. Seeds often ribbed, the endosperm always uniformly deeply ruminate. Corolla white. Species 8–23.

Sect. **Paniculatae** *Hiern* in F.T.A. 3: 193 (1877)

Stipules variously shaped but always deeply divided into acuminate lobes. Inflorescences paniculate; bracts and bracteoles small. Seeds ribbed or smooth; endosperm always ruminate. Corolla usually white. Species 24–26.

Sect. **Bracteatae** *Hiern* in F.T.A. 3: 193 (1877)

Stipules variously shaped but always deeply divided into acuminate lobes. Inflorescences paniculate or panicles of capitula or simply capitate; bracts conspicuous. Seeds ribbed or rarely smooth; endosperm not or scarcely ruminate. Species 27–28.

Species 29–38 are of uncertain or isolated position.

Sect. **Involucratae** *Petit & Verdc.* in K.B. 30: 255 (1975)

(*Cephaelis* sensu auctt. Afr., *non* L.; *Uragoga* auctt. Afr., *non* Baill.*)

Stipules mostly large, divided into 2 acuminate lobes. Inflorescences capitate, surrounded by large involucral bracts sometimes united to form an involucral cup, sessile or on short to very long peduncles. Fruits usually blue. The species with very long peduncles and 3–4-locular ovaries have been put in the genus *Camptopus* (*Megalopus*), but despite the striking peduncles there is no real character to separate them from *Psychotria*. Species 39–43.

Subgen. **Tetramerae** (*Hiern*) *Petit* in B.J.B.B. 34: 28 (1964)

Mostly shrubs or subshrubs, less often almost herbaceous, rarely small trees. Stipules ovate-triangular, mostly divided into 2 acuminate lobes or in one instance multifid. Leaves with bacterial nodules either dispersed throughout the lamina or sometimes restricted to the midrib (occasionally reduced to only very few at its base) and not easy to see. Inflorescence laxly to densely paniculate or subumbellate; bracts and bracteoles minute or obsolete. Flowers 4- or 5-merous. Seeds never ribbed; endosperm entire or with a fissure on the ventral face which is I-, T-, or V-shaped in section. Species 44–68.

Psychotria nigrifolia Gilli (Tanzania, Njombe District, Uwemba, 28 July 1958, *Gilli* 422 (W, holo. !)) is a variant of *Rutidea odorata* K. Krause.

NOTE. The imperfectly known species numbered 34–36, 38 and 69–81 cannot be keyed.

KEY 1

Inflorescences ± capitate, surrounded by a conspicuous involucre of separate or fused bracts (sect. *Involucratae*) or in a few cases individual ± capitate usually stalked components of compound inflorescences with involucres of rather small bracts (notably in sp. 27); fruits usually blue . . . Key 2

Inflorescences not surrounded by an involucre of bracts; bracts, if present, usually small and scattered, rarely more conspicuous but not then forming a distinct involucre (a ring of small bracts is present at inflorescence junctions in sp. 13):

* There is no doubt that the African species usually placed in *Cephaelis* are *not related* to the American species referred to the same genus including the type but are more closely related to species always maintained in *Psychotria*. See Schnell in B.J.B.B. 30: 357–73, fig. 48–56 (1960). This section badly needs a thorough revision. I have examined all the " species " of *Cephaelis* described by De Wildeman from Zaire but the elucidation of their correct relationships would be a lengthy task. When this has been done certain changes may be necessary to the treatment I have used here.

Stems climbing* . . . . . . 24. *P. ealaensis*
Stems not climbing:
Small subshrubs or woody herbs mostly under 50 cm. tall and often only 10–20 cm. . . . . Key 3 (p. 30)
Shrubs or small trees:**
Flowers in very tight globose inflorescences, the separate components (if divided) ± sessile; nodules absent:
Leaf-blades elliptic to oblanceolate, 2·5–6·5 cm. wide, acuminate; stipules obtuse; fruits ellipsoid, strongly grooved in the dry state; bracts well developed. . . 30. *P. peteri*
Leaf-blades oblong or oblong-elliptic, 3–13 cm. wide, very shortly acuminate; stipules entire, acute or very slightly bifid; fruits and pyrenes not grooved; bracts small . . 29. *P. megistantha*
Flowers in laxer inflorescences, although the separate components may be dense, or if ± capitate then leaves clearly nodulated:
Nodules absent (note—some lesions and fungus spots, etc., can closely simulate them) . . . . Key 4A (p. 32)
Nodules present, either in the lamina or situated only along the midrib or petiole and often difficult to see*** (subgen. *Tetramerae*):
Nodules linear or tuberculiform, situated on the midrib or on the petiole, sometimes only 1–2 on either side near the base and sometimes (e.g. in *P. heterosticta*) with a few in the lamina as well Key 5 (p. 36)
Nodules spot-like, linear or even branched, dispersed in the lamina Key 6 (p. 37)

KEY 2

Peduncles 12–180 cm. long; involucre forming an entirely closed keeled pouch-like cup in bud; ovary 3-locular; leaf-blades 28–43 × 9·6–24 cm. . . . . 43. *P. megalopus*
Peduncles much shorter, 0–12 cm. long; involucre of separate bracts or if joined then peduncles quite short; ovary 2-locular; leaf-blades much smaller:

* See also *Chassalia cristata* which is a climber sometimes confused with *Psychotria*.
** The distinction is sometimes difficult—if in doubt try both couplets.
*** In some species the nodules are very difficult to observe and the species will probably be keyed into Key 4; to enter several of the species two or three times in the keys is confusing. If any plant is not keyed out satisfactorily in Key 4 then try Key 5 before assuming it is omitted. A specimen mentioned at the end of sp. 57 also has linear nodules along the midrib and round ones dispersed in the lamina.

Inflorescences consisting of (1–)5–9 stalked capitate components, each supported by a small involucre of irregular bracts 3–5 mm. long, all parts ferruginous-pubescent; buds ± 5 lobed at apex; fruits white (**U2**) . 27. *P. vogeliana*

Inflorescences of single heads supported by an involucre of larger bracts; buds not so distinctly lobed at the apex; fruits mostly blue, but some whitish:

Calyx-limb quite conspicuous, spreading, 5–10 mm. wide, the lobes ovate, fringed with long ferruginous hairs 3 mm. long:

Bracts imbricate, not joined; shrub 0·3–3 m. tall . . . . 42. *P. tanganyicensis*

Bracts joined for ⅔ of their length; herb 40–50 cm. tall (**T6**, Nguru Mts.) . 41. *P. sp. K*

Calyx-limb less conspicuous, the lobes glabrous to hairy:

Bracts of inflorescence small, 7 × 3 mm., not joined to form an involucre; inflorescences sessile; leaf-blades ferruginous pubescent beneath; calyx-lobes small, glabrous (**T3**) . 39. *P. usambarensis*

Bracts of inflorescence larger, over 1 cm. long; inflorescences subsessile to pedunculate; leaf-blades glabrous to ferruginous pubescent beneath; calyx-lobes glabrous to hairy on the margins . . . . . 40. *P. peduncularis*

KEY 3

Leaf-blades more or less panduriform, cordate at the base; herb 5–15 cm. tall (**T3**, E. Usambaras) . . . . 52. *P. pandurata*

Leaf-blades not of this characteristic shape:

Leaves with bacterial nodules dispersed throughout the lamina, rather obscure to very distinct:

Flowers 4-merous; leaf-blades with short hairs on the main venation beneath . 68. *P. brevipaniculata*

Flowers 5-merous:

Lamina with spot-like and elongate nodules intermixed . . . 60. *P. kirkii* var. *diversinodula*

Lamina with spot-like nodules only:

Leaf-blades glabrous:

Seeds with T- or mushroom-shaped ventral slit . . . . 55. *P. sp. M*

Seeds with V-shaped ventral slit . 53. *P. pumila* (abnormal variant)

Leaf-blades hairy:

Leaf-blades mostly densely hairy and paler beneath, the fine venation being obvious as a dark reticulation when dry . . . 66. *P. spithamea*

Leaf-blades glabrescent to densely hairy (at least in subshrubby variants) but venation not as above . . . . . 60. *P. kirkii** (some variants)

Leaves without bacterial nodules or if these present then restricted to petiole or base of main nerve beneath and often very obscure:

Fruit didymously subglobose when dry; seeds ventrally deeply ruminate; leaves obtuse to acuminate at the apex, usually much paler beneath . . . . 5. *P. eminiana* (unusually small forms will key here)

Fruits globose; seeds with median ventral fissure or cleft but not deeply ruminate:

Flowers 4-merous; leaf-blades elliptic, oblong-elliptic or obovate-elliptic, 7–28 × 3–12 cm. (**U**2, 4). . . . 33. *P. lucens*

Flowers 5-merous:

Leaf-blades small, narrowly elliptic or lanceolate, narrowly acute at the apex, 1–3·2 × 0·3–0·9 cm., with thin often crinkly margins; calyx glabrous; corolla-tube scaly-tomentose outside . . . . . 32. *P. miombicola*

Leaf-blades mostly larger or if as small then blunter and corolla glabrous:

Seeds subglobose, with a deep median fissure which is usually branched within; leaf-blades rather distinctly margined, 3·5–10 × 0·7–4·5 cm.; petiole 0–1 cm. long, usually short (**T**4) . . . . 31. *P. butayei*

Seeds usually semi-globose, without a furrow, or if present V- or Y-shaped; leaf-blades not so distinctly margined; petiole 0·1–6 cm. long:

Leaf-blades large, ± obovate, 5–18 × 2–8 cm., only slightly paler beneath; petiole 2·5–6 cm. long (**T**3, E. Usambaras) . 51. *P. brevicaulis*

Leaf-blades smaller, ± elliptic, 2–14 × 0·5–5 cm., mostly very distinctly paler beneath; petiole 0·1–1·5 cm. long:

Peduncles 0·5–6·5 cm. long; pedicels 1–5(–7) mm. long . 53. *P. pumila*

Peduncles 1–1·5 cm. long; pedicels obsolete . . . . 54. *P. sp. L*

* What is probably a hairy form of sp. 55 will also key here; it is much smaller in habit than all but a few aberrant variants of *P. kirkii*.

KEY 4A

Fruits with a persistent disc forming a beak 2–3 mm. long projecting beyond the calyx-limb . . . 2. *P. sp. A*
Fruits without such a persistent beak:
  Flowers 4-merous; leaf-blades glabrous, drying reddish brown beneath; more or less simple stems from a rhizome; stipules narrowly bifid . . . . . . 33. *P. lucens*
  Flowers 5-merous:
    Bracts well developed, elliptic to oblong-lanceolate or linear-elliptic, the lowermost up to 2 × 0·5 cm., acuminate and entire or 3-toothed at the apex . . 28. *P. schweinfurthii*
    Bracts not well developed or if present then differently shaped and shorter or hidden by the inflorescence:
      Stipules entire or only very slightly divided:
        Mature fruits large, ± 1·6 cm. in diameter, with a row of spaces between the ridges of the pyrenes and outer wall (**K7**, Teita area) . 21. *P. petitii*
        Mature fruits much smaller:
          Inflorescence very condensed or inflorescence branched but then the individual stalked components capitate:
            Inflorescence components 1·5–3 cm. in diameter; corolla-tube 0·8–1·2 cm. long; lobes 3–4·5 mm. long (Tanzania, Uluguru Mts.) . . . . . . 29. *P. megistantha*
            Inflorescence components ± 1 cm. in diameter; corolla-tube 1·2–3 mm. long; lobes 2 mm. long:
              Stipules 2·5–3 cm. long, 2 cm. wide; corolla-lobes thickened and inflexed-appendiculate at apex but without a fine terminal appendage (Kenya, Mt. Kasigau) . . . . 13. *P. taitensis*
              Stipules smaller, 1·1 cm. long, 8 mm. wide; corolla-lobes similar but with a fine terminal appendage at the thickened inflexed apex (Kenya, Ngangao Forest) . 14. *P. sp. B*
          Inflorescence laxer, the individual components lax or ± dense but not capitate:
            Leaf-blades with very distinct raised reticulate venation in the dry state (**T6**, Uluguru Mts.) . . . . . . 9. *P. elachistantha*

Leaf-blades without such distinct raised venation:
Calyx-limb either obconic, 3 mm. wide, or almost tubular, 2–4 mm. long:
Buds very finely papillate giving a yellowish grey appearance in the dry state; calyx 3–5 mm. wide; corolla-lobes thick, 4·7–6 mm. long; fruits not markedly grooved when dry:
Leaf-blades broadly elliptic, up to 15 × 9 cm. (**K**7) . . . 3. *P. crassipetala*
Leaf-blades elliptic-oblanceolate, up to 11 × 3(–4) cm. (**T**3, W. Usambaras) . . . 4. *P. tenuipetiolata*
Buds not drying yellowish, smooth; calyx 2 mm. wide, the limb 2–3 mm. long, excluding the lobes; corolla-lobes 2 mm. long; fruits grooved when dry (**T**3) . . . . . 20. *P. cyathicalyx*
Calyx-limb cupuliform, ± 1·5 mm. long or less, never over 3 mm. wide . . . Key 4B
Stipules distinctly divided or usually so:
Peduncles scarcely developed, ± 2 mm. long, rarely longer, to 2·5 cm.; leaf-blades glabrous (**T**3, E. Usambaras) . . . . . . 7. *P. triclada**
Peduncles longer:
Young stems and undersurfaces of leaf-blades very shortly pubescent, usually densely and velvety so; fruits didymous, with distinct median groove, pubescent; leaf-blades mostly ± obtuse; seeds with flat ventral face; corolla mostly yellow . . . 5. *P. eminiana*
Without the above combination of characters:
Stipules with 2 rounded lobes; leaf-blades often drying a distinct greenish yellow colour 23. *P. succulenta*
Stipules with triangular-acuminate lobes:
Calyx and/or corolla slightly to densely scaly tomentose and sometimes densely pubescent as well; fruits

* Petit says stipules not divided but the few I have seen are divided.

not grooved; tertiary venation not very evident beneath . . . . 25. *P. mahonii*

Calyx and corolla glabrous to pubescent but not scaly-tomentose; fruits grooved when dry (where known); tertiary venation sometimes very prominent beneath:

Peduncles quite noticeably 2-winged and flattened even in living material . 26. *P. lauracea*

Peduncles not so distinctly 2-winged:

Corolla-tube ± 2 mm. long; leaf-blades glabrous beneath . . . 26. *P. lauracea*

Corolla-tube longer; leaf-blades with main venation mostly hairy beneath but sometimes quite glabrous:

Fruiting calyx ± funnel-shaped, 2–3 mm. long; corolla-tube 4·5–7 mm. long; lobes 3–6 mm. long; leaf-blades often with domatia and with lateral nerves often zig-zag; tertiary venation raised beneath . . . 15. *P. fractinervata*

Fruiting calyx not so obvious, ± 1 mm. long; corolla-tube 2·5–5 mm. long, lobes 1·8–4 mm. long; leaf-blades usually without domatia; lateral nerves less zig-zag; tertiary venation mostly not raised beneath . 12. *P. orophila*

Key 4B

Stipules elliptic-obovate to oblong-elliptic, (0·5–) 1–3·5 cm. long or if shorter then inflorescences densely ferruginous hairy or eastern species:

Inflorescences very densely ferruginous hairy:

Leaf-blades glabrous beneath, coriaceous; corolla glabrous . . . . . 17. *P. sp. D*

Leaf-blades ferruginous hairy beneath or at least some hairs on the midrib, thinner; corolla hairy:

Leaf-blades 13·5 × 8 cm., ferruginous hairy beneath, at least on the main venation; stipules 1·5–2·4 × 0·8–1·5 cm. . . . . 18. *P. brucei*
Leaf-blades 3–11 × 1–4 cm., only the midrib hairy beneath and above near the petioles; stipules 8 mm. long . 19. *P. sp. E*
Inflorescences glabrescent or with only rather sparse hairs:
Limb part of calyx (excluding the lobes) 1–1·5 mm. long (up to 2·5 mm. long including the irregular lobes; known only in very young fruit; **K**6) . . 16. *P. sp. C*
Limb part of calyx shorter:
Corolla-tube 1·5–2·5(–3·5) mm. long; leaf-blades often very large, 23 × 13·5 cm. 8. *P. goetzei*
Corolla-tube 4–6 mm. long:
Leaf-blades usually more cuneate and shortly acuminate with usually a few sparse hairs on the nerves (S. Tanzania) . . . . 10. *P. meridiano-montana*
Leaf-blades often more rounded at the base and apex, with fairly dense ferruginous hairs on the nerves beneath (N. Tanzania) . . 11. *P. pseudoplatyphylla* (see also 12. *P. orophila*)

Stipules mostly less than 1 cm. long, mostly western species or if eastern then stipules triangular or triangular-ovate; inflorescences not very densely ferruginous hairy:
Subshrubs:
Young stems and undersurfaces of leaf-blades very shortly pubescent, usually densely and velvety so; fruits didymous with distinct median groove, pubescent; leaf-blades ± obtuse; seeds with flat ventral face; corolla mostly yellow . 5. *P. eminiana*
Stems and leaves, etc. glabrous to pubescent, if pubescent then with longer hairs; fruits subglobose; seeds subglobose with a deep branched ventral cleft; corolla usually white (**T**4) . . . . 31. *P. butayei*
Shrubs:
Young and old shoots markedly different in woodiness; fruit didymous . . . 5. *P. eminiana*
Young and old shoots not markedly different in woodiness; fruits ± subglobose:
Plant ± pubescent all over . . . 1. *P. riparia*
Plant glabrescent:
Leaf-blades 3–6·5 × 0·9–2 cm.; panicles very short, totalling only 2·5 cm. long with the peduncle even in fruit 22. *P. hemsleyi*
Leaf-blades larger; inflorescences much more developed:
Pedicels ± obsolete or up to 1 mm. long:

Corolla yellow . . . . 1. *P. riparia*
Corolla white . . . . 6. *P. bagshawei*
Pedicels quite distinct, 1–1·5 mm. long; corolla white . . . 37. *P. sp. I*

KEY 5

Seeds subglobose with a deep branched cleft on the ventral surface; joints of inflorescence without tufts of ferruginous hairs (**T**4) . 31. *P. butayei*
Seeds semi-globose or semi-ellipsoid, without a cleft or with a T-, V- or U-shaped cleft (seeds not known in sp. 65 but in that species the joints of the inflorescence bear tufts of ferruginous hairs):
  Calyx-lobes linear, 2–4 mm. long, longer than the rest of the calyx . . . . 50. *P. linearisepala*
  Calyx-lobes usually shorter, mostly not as long as the rest of the calyx (except in *P. holtzii*):
    Leaf-blades smaller, discolorous, 1–8 × 0·2–3·3 cm., usually well below the upper limits:
      Stems ± glabrous save for the youngest internodes, which, with the associated petioles, midribs and stipules, are adpressed ferruginous hairy (**T**3, W. Usambaras) . . . . . 47. *P. porphyroclada*
      Stems, if hairy, then with spreading pubescence which is on older internodes as well:
        Limb of calyx cupuliform, 1·5–2 mm. long, the lobes pale, shorter, 0·25–0·8 mm. long; plant glabrous except in one variety . . . 44. *P. amboniana*
        Limb of calyx 0·5–0·75 mm. long; lobes elongate, narrow, 0·5–2·5 mm. long; plant pubescent . . 49. *P. holtzii*
    Leaf-blades mostly larger and less discolorous:
      Inflorescences distinctly umbellate; nodules scarcely if at all evident . 53. *P. pumila*
      Inflorescences ± paniculate:
        Inflorescences subsessile . . . 46. *P. schliebenii* var. *sessilipaniculata*
        Inflorescences with long peduncles:
          Stipules entire or very shortly toothed; nodules usually difficult to see:
            Calyx-limb ± 0·5 mm. long . . 46. *P. schliebenii*
            Calyx-limb 1–1·5 mm. long . . 45. *P. leucopoda*
          Stipules bilobed or bifid:
            Leaves glabrous, drying pale grey-green; nodules obvious; calyx, etc. obviously speckled with rhaphides . . . . 64. *P. heterosticta*

Leaves with ferruginous hairs beneath or at least midrib margined with ferruginous hairs in very young leaves; nodules obscure; rhaphides not so obvious:
Foliage drying reddish brown, the very young leaves with a ferruginous cobwebby indumentum beneath; inflorescence glabrous or sparsely pubescent . . 48. *P. castaneifolia*
Foliage drying pale olive-green, the very youngest leaves with a dense edging of ferruginous hairs along the midrib beneath; inflorescence with tufts of ferruginous hairs at the joints 65. *P. iringensis*

KEY 6

Corolla-tube glabrous inside:
Young stems, midrib, calyx, etc. ferruginous-pubescent; corolla-tube tomentose outside (**T3**, Pare) . . . . . . 62. *P. petiginosa*
Stems with only traces of ferruginous hairs on the stipules and at nodes of the inflorescence; corolla glabrous outside (**T3**, W. Usambaras) . . . . . 61. *P. alsophila*
Corolla-tube hairy inside:
Nodules very small black spots usually present on lower leaf-surface; peduncles distinctly winged; fruit strongly grooved when dry due to ribbed pyrenes . 26. *P. lauracea*
Nodules mostly larger; peduncles not distinctly winged; fruit not ribbed:
Nodules dark green, linear, spot-like or often branched, up to 3 mm. long . . 63. *P. cryptogrammata*
Nodules blackish, spot-like (save in one var. of *P. kirkii* already separated in Key 3):
Plant practically glabrous:
Plant of sea-shore or at least very low altitudes near the sea; leaf-blades ± obtuse . . . . . 56. *P. punctata*
Plant of higher altitudes; leaf-blades acute or acuminate:
Inflorescences much-branched pyramidal panicles with ± 8 nodes; leaf-blades (6–)10–25 × (2·7–)4–12 cm. . . . . . . 57. *P. faucicola*
Inflorescences less branched; leaf-blades 2–18 × 0·5–9 cm., usually quite small:
Inflorescences elongate, with ± 5 nodes; leaf-blades oblong-

elliptic, widest at the middle, 8·5–13·5 × 2·5–3·6 cm.; corolla-tube 2 mm. long. . . . 59. *P. sp. N*
Inflorescence paniculate or umbellate, not so elongate; leaf-blades variously shaped; corolla-tube 2·5–6 mm. long . . . 60. *P. kirkii* (variants)

Plant hairy to densely velvety:
Indumentum on stems, undersurfaces of leaves and inflorescences consisting of dense short often ferruginous hairs; leaf-blades glabrous above; inflorescence components very slender (**T1**) . . . 67. *P. cinerea*
Indumentum on stem, leaves and inflorescence much coarser, of longer paler hairs; leaf-blades hairy above; inflorescence components coarser:
Inflorescence a much-branched pyramidal panicle with ± 8 nodes; leaf-blades 10–22(–32) × 4–10 (–13) cm. . . . . . 58. *P. griseola*
Inflorescences less branched; leaf-blades 2–18 × 0·5–9 cm. . . . 60. *P. kirkii* (variants)

1. **P. riparia** (*K. Schum. & K. Krause*) *Petit* in B.J.B.B. 34: 43, photo. 1/B (1964) & in Distr. Pl. Afr. 4, map 92 (1972). Type: Tanzania, Morogoro District, Liwale R., *Busse* 557 (B, holo.†, EA, lecto.!)

Shrub 2·5–4·5 m. tall or tree up to 6(–20*) m.; stems glabrous. Leaf-blades elliptic or obovate-elliptic, 3–20 cm. long, 1–10 cm. wide, ± rounded, acute or shortly acuminate at the apex, cuneate at the base, glabrous, sometimes bright green on drying and somewhat shiny, papery to coriaceous; nerves and veins ± impressed above; nodules absent, domatia present; petiole 0·5–4 cm. long, glabrous; stipules triangular or ovate-triangular, 3–8 mm. long, acute or obtuse, glabrous, deciduous. Flowers heterostylous, 5-merous, in much-branched inflorescences 4–15 cm. long; peduncle glabrous to pubescent or puberulous, 1–10 cm. long; secondary branches 0·7–2·2 cm. long; pedicels ± obsolete or up to 1 mm. long; bracts and bracteoles minute. Calyx-tube obconic, 1–2 mm. long, glabrous or puberulous; limb cupuliform, 1–1·5 mm. long, lobes ± obsolete or up to 0·5 mm. long. Corolla yellow or white or greenish-yellow, glabrous or very rarely shortly puberulous; tube 3–4·5 mm. long; lobes triangular, 2–3 mm. long, 2 mm. wide, with a thickened horn-like appendage at the apex. Stamens with filaments 3 mm. long in short-styled flowers, 0·3 mm. long in long-styled flowers. Style 3 mm. long in short-styled flowers, 4·5–5 mm. long in long-styled flowers; stigma-lobes 0·5 mm. long. Drupes red, with 2 pyrenes, subglobose, 6·5–7 mm. in diameter, glabrous; pyrenes ½-globose, 5–6·5 mm. long and wide, 3–3·5 mm. thick, dorsal face not ribbed. Seeds red, 5·5 mm. long and wide, 2·5 mm. thick; ventral face plane, dorsal face not ribbed; albumen ventrally deeply and dorsally shallowly ruminate.

* This needs confirmation; it seems a little unlikely.

var. **riparia**; Petit in B.J.B.B. 34: 43 (1964) & in B.J.B.B. 42: 354 (1972)

Stems, leaf-blades, inflorescences, and calyces all glabrous or nearly so or with sparse to rather ferruginous pubescence on the peduncles, etc.

UGANDA. Acholi District: Chua, Feb. 1938, *Eggeling* 3514!*; Mbale District: Bugisu, Bufumbo, July 1966, *Maitland* 1258!
KENYA. Fort Hall District: Mabaloni Hill, 11 Dec. 1953, *Bally* 8530!; Kiambu/ Machakos Districts: Fourteen Falls, near where Donyo Sabuk road crosses the R. Athi, 2 Jan. 1960, *Verdcourt* 2605!; Teita District: Taveta, 13 Dec. 1961, *Polhill & Paulo* 981!; Kwale District: Vanga, Majoreni, Nov. 1929, *R. M. Graham* in *F.D.* 2185!
TANZANIA. Pare District: by R. Ruvu, 5 Nov. 1955, *Milne-Redhead & Taylor* 7235!; Mpwapwa, 9 Dec. 1930, *Hornby* 334!; Morogoro District: Mgeta R., 12 Nov. 1952, *Paulo* 21!; Zanzibar I., Marahubi, 21 Aug. 1963, *Faulkner* 3255!; Pemba I., Ngezi Forest, 29 Oct. 1929, *Vaughan* 742!
DISTR. **U**1, 3; **K**4, 7; **T**?1, 2–8; **Z**; **P**; Mozambique, Zambia and Malawi
HAB. Evergreen bushland, coastal bushland, rocky places, wooded grassland and forest-edges and especially in riparian forest (often with *Phoenix*); 0–1800 m.

SYN. *Grumilea riparia* K. Schum. & K. Krause in E.J. 39: 560 (1907); De Wild. in B.J.B.B. 9: 55 (1923); T.T.C.L.: 497 (1949); K.T.S.: 443 (1961), pro parte
G. bussei K. Schum. & K. Krause in E.J. 39: 562 (1907); De Wild. in B.J.B.B. 9: 30 (1923); T.T.C.L.: 497 (1949). Type: Tanzania, Songea District, R. Mironji [Milonyi], *Busse* 986 (B, holo. †, EA, iso.!, B, fragment of iso.!)
[*Psychotria schefferi* sensu K.T.S.: 467 (1961), pro parte, *non* K. Schum. & K. Krause]

NOTE. There is a good deal of variation in the indumentum of the inflorescence branches and corollas. *Polhill & Paulo* 802 (Kenya, Kilifi District, 40 km. NW. of Malindi, 19 Nov. 1961) has the corolla distinctly shortly pubescent outside and in some other specimens it is slightly so. *Magogo & Glover* 228 (Kenya, Kwale District, Shimba Hills, Marere Hill area, 7 Mar. 1968) is a poor fruiting specimen with pedicels up to 4 mm. long. No flowering material has been seen from the locality—whether it is a variant of *P. riparia* or an allied species must await further information. The pyrenes are 6–6·5 mm. long and wide and 2·5–3 mm. thick, convex dorsally, flat ventrally, finely ruminate dorsally.

var. **puberula** *Petit* in B.J.B.B. 34: 46 (1964). Type: Kenya, Kitui Boma, *Bally* 1529 (K, holo.!)

Young stems, undersurfaces of leaf-blades, petioles, inflorescences, calyces, corollas and drupes ± densely ferruginous pubescent.

KENYA. Kitui Boma, 18 Jan. 1942, *Bally* 1529!; Kwale District: 12·8 km. WSW. of Gazi, Buda Mafisini Forest, 22 Aug. 1953, *Drummond & Hemsley* 3951! & Shimba Hills, Marere, 7 Mar. 1968, *Magogo & Glover* 224!
TANZANIA. Handeni District: Swagilo, along Mligasi R., Dec. 1949, *Semsei* in *F.H.* 2937!; Morogoro District: Mtibwa Forest Reserve, Aug. 1952, *Semsei* 892!; Bagamoyo District: Mandera, *Sacleux* 713& 1043; Ulanga District: Ifakara, Itula, June 1960, *Haerdi* 384/OB!
DISTR. **K**4, 7; **T**3, 6, ? 8;? Ethiopia
HAB. Coastal forest and thicket; 80–1140 m.

SYN. [*Grumilea platyphylla* sensu Haerdi in Acta Trop., suppl. 8: 139 (1964), *non* K. Schum.]

NOTE. *Chakwala* in *C.A.W.M.* 5341 (Tanzania, Chunya District, 1 km. W. of Itigi–Mbeya road, 28 Jan. 1969) may be a specimen of *P. riparia* but is very poor; it could just possibly be the first record of *P. capensis* (Eckl.) Vatke from the Flora area but I would hesitate to add it on the basis of this specimen. The two species are in any case exceedingly closely related.

2. **P. sp. A**

Laxly branched tree to 12 m.; stems glabrous, purplish brown, with pale lenticels and petiole-base scars. Leaf-blades ± elliptic, 4–11 cm. long, 1·4–3·6 cm. wide, subacute at the apex, cuneate at the base, glabrous,

* So determined by Petit although not cited in his revision.

drying a pale rather yellowish green, coriaceous, without nodules but mostly with domatia; petioles 0·6–1·5 cm. long, glabrous or with a few hairs; stipules ovate, 5–6 mm. long, not bifid. Flowers not known; fruits in branched panicles; peduncle 2·3 cm. long; secondary branches 0·5–1·5 cm. long; pedicels 1–8 mm. long; all parts glabrous. Fruits ± longitudinally striped, subglobose, 4·5–6 mm. tall, 5–8 mm. wide, the ± persistent calyx-rim 1 mm. tall, undulate; beak of drupe (? persistent disc) marked, 2–3 mm. long, projecting beyond the calyx-limb; pyrene reduced to 1 in the fruits examined. Seed ± 4 mm. in diameter, strongly ruminate.

TANZANIA. Arusha District: Ngurdoto Crater National Park, near Longil Swamp, 6 June 1965, *Greenway & Kanuri* 11901 !
DISTR. T2; not known elsewhere
HAB. Dry evergreen forest of *Olea* spp., *Euclea* and *Diospyros abyssinica* on very stony grey volcanic loam; 1500 m.

3. **P. crassipetala** *Petit* in B.J.B.B. 34: 191, photo. 6/F (1964) & in Distr. Pl. Afr. 6, map 168 (1973). Type: Kenya, Teita Hills, NNE. of Ngerenyi, Ngangao, *Drummond & Hemsley* 4336 (K, holo. !, BR, EA, iso.)

Shrub or tree 3·5–10 m. tall, with glabrous branches. Leaf-blades elliptic to broadly elliptic, 6–15 cm. long, 3–9 cm. wide, ± rounded to shortly acuminate at the apex, cuneate at the base, glabrous on both surfaces, thinly coriaceous, without nodules but with domatia; petiole 1–5 cm. long, glabrous; stipules ovate-triangular, 5 mm. long, shortly acuminate at the apex, glabrous, deciduous. Flowers probably heterostylous, 5-merous, in much-branched panicles, 3–7 cm. long; peduncle 1–3 cm. long, glabrous; secondary branches 0·5–3 cm. long, with tufts of ferruginous hairs at junctions; pedicels thick, 1–2 mm. or up to 1 cm. long in fruit, glabrous; bracts and bracteoles small. Calyx-tube subglobose, 1·5 mm. long; limb campanulate, 3–4 mm. long, glabrous, the margin irregularly lobed or almost truncate, the lobes 1 mm. long, minutely ciliate. Corolla white or cream, usually densely minutely papillate outside when dry but smooth in life; tube funnel-shaped, ± 4–5 mm. long, glabrous inside; lobes thick, 6 mm. long, 2·5 mm. wide, inflexed at the apex. Stamens with filaments 3–4 mm. long in short-styled flowers, 0·5–1 mm. long in long-styled flowers. Style 2 mm. long in short-styled flowers, 7·5–9 mm. long in long-styled flowers; stigma-lobes 1 mm. long. Drupes greenish blue, with 2 pyrenes, subglobose, 8 mm. in diameter, glabrous, crowned with the ± conspicuous calyx, not grooved; pyrenes semi-globose, 5–6·5 mm. long and wide, 2·8 mm. thick, the ventral surface plane, the dorsal surface entire. Seeds similar in size and shape, ventral surface plane, the dorsal surface not ribbed; albumen ruminate on both sides.

KENYA. Teita District: Teita Hills, 8 km. NNE. of Ngerenyi, Ngangao, 10 Apr. 1971, *Faden et al.* 21/258 ! & Mbololo Hill, July 1937, *Dale* in *F.D.* 3814 in part!; Lamu District: Witu, Utwani Forest, Oct. 1937, *Dale* in *F.D.* 3831 !
DISTR. K7; not known elsewhere
HAB. Evergreen forest; 30–1850 m. (see note)

SYN. [*Grumilea riparia* sensu K.T.S.: 445 (1961), pro parte, *non* K. Schum. & K. Krause]

NOTE. Petit gives the altitudinal range as 1320–1850 m. but he cites *Dale* in *F.D.* 3831 from the Utwani Forest which can be no more than 100 ft. This is a curious distribution but I see no reason to suspect the labels.

4. **P. tenuipetiolata** *Verdc.* in K.B. 30: 254 (1975). Type: Tanzania, Lushoto District, W. Usambara Mts., 1·6 km. W. of Gologolo, *Drummond & Hemsley* 2845 (K, holo. !, EA, iso.)

Much-branched shrub 4·5–6 m. tall; stems glabrous, greyish, rather corky and longitudinally rugulose when older. Leaf-blades oblanceolate-elliptic, 3·5–11 cm. long, 1–3(–4) cm. wide, shortly acuminate at the apex, narrowly cuneate at the base, glabrous; nodules absent but small domatia present; petioles 1·2–3·5(–4) cm. long, relatively long and slender in proportion to the size of the leaf; stipules transversely oblong, 2–3 mm. long, almost immediately deciduous. Flowers heterostylous, 5-merous, in small few-flowered branched panicles; peduncles 1–1·8 cm. long; secondary branches 0·3–1 cm. long; pedicels 1–2(–3 in fruit) mm. long; articulations of inflorescence margined with sparse ferruginous hairs. Calyx glabrous, broadly obconic, 2–3 mm. long (mostly limb), subtruncate or very shallowly lobed, the lobes mostly under 0·5 mm. long. Buds depressed globose, ± truncate, appearing very finely yellowish papillate. Corolla creamy white; tube 2·2 mm. long, glabrous outside; lobes obovate-oblong, 4·7 mm. long, 2 mm. wide, thick and cucullate at the apex. Stamens with filaments 0·7 mm. long in long-styled flowers. Style 3·6 mm. long in long-styled flowers; stigma-lobes 1 mm. long. Fruits globose or ellipsoid, ± 6–7 mm. long and wide, crowned with 2 mm. long persistent calyx-limb; pyrenes ½-ovoid, 6·5 mm. long, 5·5 mm. wide, 3 mm. thick, slightly rugulose particularly on ventral surface but not grooved. Seed dark red, similar in outline, dorsally quite strongly ruminate, ventrally with a T-shaped groove.

TANZANIA. Lushoto District: W. Usambara Mts., 1·6 km. W. of Gologolo, at top of escarpment, World's View, 4 June 1953, *Drummond & Hemsley* 2845! & Shagayu [Shagai] Forest, Sept. 1955, *Mgaza* 77! & Magamba Forest, 27 Sept. 1934, *Pitt-Schenkel* 404!

DISTR. **T3**; not known elsewhere

HAB. Edge of *Podocarpus-Juniperus* and *Ocotea* forest; 1900–2000 m.

NOTE. I am practically certain that the fruiting specimen *Mgaza* 77 is conspecific with the type but confirmation would be reassuring.

5. **P. eminiana** (*Kuntze*) *Petit* in B.J.B.B. 34: 48, photo. 1/E (1964); Petit in Distr. Pl. Afr. 4, map 93 (1972). Type: Sudan, Niamniam, SW. Bendo, *Schweinfurth* 3902 (K, lecto.!, BM, P, isolecto.!)

Subshrub or small tree, with woody or herbaceous stems 0·15–2·4(–5) m. tall from a woody rhizome; stems glabrous to densely pubescent, sparsely branched. Leaf-blades variable, ovate, oblong, obovate or elliptic to almost round or oblate, 3–25 cm. long, 2–13 cm. wide, emarginate to acuminate at the apex, cuneate to rounded at the base, glabrous to scabrid-puberulous or pubescent above, glabrous to densely pubescent beneath, often strongly discolorous, pale beneath, often with a characteristic pattern of dark veinlets beneath on drying, thin to ± coriaceous; nodules absent but domatia present; petiole 0·3–3(–3·5) cm. long, glabrous to densely pubescent; stipules triangular or ovate-triangular, 0·4–1·1 cm. long, acute to acuminate at the apex, obscurely 2-toothed, glabrous or pubescent, deciduous. Flowers heterostylous, (4–)5(–6)-merous, in usually trichotomous or much-branched inflorescences, each panicle rather dense and many-flowered; peduncle (1–)4–13 cm. long, glabrous to densely pubescent, secondary branches 0·5–4 cm. long; pedicels obsolete or 0·5 mm. long; 2 inferior bracts sometimes developed, others and bracteoles minute. Calyx-tube obconic, ± 1 cm. long; limb cupuliform, 0·75–1 mm. long, glabrous or pubescent with unequal triangular lobes 0·5–1·75 mm. long, acuminate. Corolla yellow, cream or greenish yellow, glabrous or puberulous outside; tube 3·5–5 mm. long; lobes 1·7–2·5 mm. long, 1–1·2 mm. wide, thickened at the apex. Stamens with filaments 1·5–2·5 mm. long in short-styled flowers, 0·25 mm. long in long-styled flowers. Style 1–2·5 mm. long in short-styled flowers, 6 mm.

long in long-styled flowers; stigma-lobes 0·5–1 mm. long. Drupes red, with 2 pyrenes, didymously subglobose, 5–6(–7)* mm. long, 7–8(–10)* mm. wide, glabrous or puberulous or shortly pubescent; pyrenes subglobose, 5 mm. long and wide, 4 mm. thick, dorsal face entire, not folded. Seeds dark brown, subglobose, 4·5 mm. long and wide, 3·5 mm. thick, ventral face flat, dorsal face not ribbed; endosperm with ventral face deeply and dorsal face slightly ruminate.

SYN. *Uragoga eminiana* Kuntze, Rev. Gen. Pl. 2: 955 (1891)
Grumilea sulphurea Hiern in F.T.A. 3: 218 (1877); De Wild. in B.J.B.B. 9: 56 (1923). Type: as for *Psychotria eminiana, non P. sulphurea* Rúiz & Pavón (1794), *nec P. sulphurea* Seemann (1865–73)
Psychotria humilis Hutch., Botanist in S. Afr.: 514 (1946). Type: Zambia, Mwinilunga, Matonchi Farm, *Milne-Redhead* 1029 (K, holo. !)

NOTE. Other synonymy is given by Petit (*loc. cit.*)

KEY TO INFRASPECIFIC VARIANTS

Leaf-blades ± coriaceous, mostly small, only rarely attaining 15 × 8 cm.:
- Stems herbaceous or somewhat woody, the young and old growths not markedly dissimilar; all parts ± velvety pubescent . . . . var. **stolzii**
- Stems becoming distinctly woody with age, the old growths markedly dissimilar from the young ones; stems, etc. glabrous to distinctly hairy . var. **heteroclada**

Leaf-blades thin, oblate to broadly elliptic, 11–18 × 9–13 cm. . . . . var. **tenuifolia**

var. **stolzii** (*K. Krause*) *Petit* in B.J.B.B. 34: 51 (1964) & in B.J.B.B. 42: 354 (1972) & in Distr. Pl. Afr. 4, map 94 (1972). Type: Tanzania, Rungwe District, Mulinda, *Stolz* 1455 (K, lecto. !, B, S, WAG, isolecto.)

Stems ± woody, together with petioles, peduncles and undersurfaces of leaf-blades distinctly ± velvety pubescent; leaf-blades paler, greyish or pale brown beneath, often almost bullate above. Corolla puberulous outside. Drupes pubescent.

TANZANIA. Biharamulo District: Bwanga, 10 Oct. 1960, *Tanner* 5174 !; Buha District: Kasulu to Kibondo, km. 112, 15 Nov. 1962, *Verdcourt* 3326 !; Ufipa District: Rukwa Escarpment, new road to Mpui, 5 Nov. 1956, *Richards* 6883 !; Rungwe District: Mwalesi, 29 Dec. 1932, *R. M. Davies* 802 !
DISTR. T1, 4, 7, 8; Burundi, Zambia and Mozambique
HAB. *Brachystegia* and similar woodland; 850–2100 m.

SYN. ?*Grumilea ungoniensis* K. Schum. & K. Krause in E.J. 39: 563 (1907); De Wild. in B.J.B.B. 9: 56 (1923); T.T.C.L.: 498 (1949). Type: Tanzania, Songea District, Kwa Kihingi, *Busse* 1358 (B, holo. †)
G. stolzii K. Krause in E.J. 57: 48 (1920); De Wild., Pl. Bequaert. 2: 480 (1924); T.T.C.L.: 498 (1949)
[*Psychotria humilis* sensu Hutch., Botanist in S. Afr.: 514 (1946), pro parte, *non* sensu stricto]

NOTE. Petit cites *Tanner* 5329 (Biharamulo), *Richards* 6883 (mentioned above), *Stolz* 449 (Kyimbila) and *Milne-Redhead & Taylor* 7869 (Tunduru) under "var. incertae sedis" but I feel that all can be retained within the circumscription of var. *stolzii*.

var. **heteroclada** *Petit* in B.J.B.B. 34: 52 (1964) & in Distr. Pl. Afr. 4, map 95 (1972). Type: Zambia, Lake Mweru, *Fanshawe* 3917 (K, holo. !, BR, iso.)

Stems becoming very distinctly woody with age, markedly dissimilar from the young growths. Leaf-blades with few hairs beneath on the nerves to distinctly velvety tomentose but glabrous above. Stems, etc. glabrous to distinctly hairy. Inflorescences more branched. Corolla mostly densely velvety tomentose outside or rarely glabrescent.

* The measurements in parenthesis are from spirit material.

TANZANIA. Mpanda District: Lugungwisi [Lunvugwise] R., 4 Nov. 1959, *Richards* 11712 ! & just N. of Kasogi, near the Kasiha R., 29 Aug. 1959, *Harley* 9433 ! & Kapapa Camp, 28 Oct. 1959, *Richards* 11602 !
DISTR. T1, 4; Zaire (Katanga), Zambia
HAB. Forest-edges, riverine forest, scrub and grassland on lake-shores; 765–1650 m.

NOTE. The indumentum on the peduncles and midnerves is sometimes ferruginous but one specimen apparently densely ferruginous all over proved to be merely covered with red murram dust.

var. **tenuifolia** *Verdc.* in K.B. 30: 248 (1975). Type: Tanzania, Buha District, Rutanga Valley, *Pirozynski* 269 (K, holo. !, EA, iso. !)

Tree 5 m. tall, with woody branches, together with petioles, peduncles and undersurfaces of leaf-blades pubescent with short hairs, those on the stem being pale ferruginous. Leaf-blades thin, oblate to broadly elliptic, 11–18 cm. long, 9–13 cm. wide, rounded at both ends.

TANZANIA. Buha District: Rutanga valley, 21 Jan. 1964, *Pirozynski* 269 !
DISTR. T4; not known elsewhere
HAB. Valley forest; 820 m.

NOTE. This is just a variant of *P. eminiana*—the fruits and indumentum being identical but the leaves are very different. It is possibly due to some injury or physiological conditions since *Pirozynski* 138 from the nearby Kakombe valley, 31 Dec. 1963, is a shrub 2 m. tall with leaf-blades 6–15 cm. long, 1·8–8·2 cm. wide and is clearly only a large-leaved form of var. *stolzii*. The inflorescence in var. *tenuifolia* appears to be axillary and much-branched but may be a branch with leaves suppressed.

6. **P. bagshawei** *Petit* in B.J.B.B. 34: 55 (1964) & in Distr. Pl. Afr. 4, map 96 (1972). Type: Uganda, Toro District, Isunga, *Bagshawe* 1092 (BM, lecto. !)

Shrub 1–3 m. tall or small tree; young stems glabrous. Leaf-blades elliptic, 4–20 cm. long, 2–8 cm. wide, acute or shortly acuminate at the apex, narrowly cuneate at the base, glabrous, thin; nodules absent, domatia present; petiole 0·6–5 cm. long, glabrous or puberulous above on angles; stipules ovate to triangular, 5–10 mm. long, acute at the apex, shortly ferruginous-puberulous or rarely glabrous, deciduous. Flowers heterostylous, 5-merous, in much-branched panicles, the individual clusters congested; peduncles 2–6 cm. long, glabrous to ± shortly pubescent; secondary branches 0·2–3·5 cm. long, ± pubescent; pedicels ± obsolete, becoming 1–3 mm. long in fruit; bracts and bracteoles minute. Calyx-tube obconical, 1 mm. long, glabrous; limb broadly cupuliform, 1 mm. long, glabrescent or puberulous outside; lobes very short or obsolete. Corolla yellow or white, glabrous or shortly puberulous outside; tube 2·5–3·5 mm. long; lobes oblong 2·5–3 mm. long, 0·8 mm. wide, thickened at the apex. Stamens with filaments 1·75 mm. long in short-styled flowers, 0·5 mm. long in long-styled flowers. Style 2 mm. long in short-styled flowers, 3·5 mm. long in long-styled flowers; stigma-lobes 0·5–0·75 mm. long. Drupes red, with 2 pyrenes, subglobose, 7 mm. in diameter; pyrenes depressed semiglobose, 4·5 mm. in diameter, 3 mm. thick, ventral face flat, dorsal face not ribbed. Seeds similar to the pyrenes in shape and size; ventral face plane, dorsal face not ribbed; albumen ventrally deeply, dorsally shallowly ruminate.

UGANDA. Toro District: Mpanga Forest, 25 Apr. 1906, *Bagshawe* 1006 ! & Isunga, 8 July 1906, *Bagshawe* 1092 !; Ankole District: Kyamahungu, July 1939, *Purseglove* 849 !
KENYA. Nandi District: 6 km. SE. of Yala R. Bridge, 19 Apr. 1965, *Gillett* 16721 !; N. Kavirondo District: Kakamega Forest, Apr. 1934, *Dale* in *F.D.* 3282 ! & near Kakamega Forest Station, 3 Jan. 1968, *Perdue & Kibuwa* 441 !
DISTR. U2; K3, 5; E. Zaire
HAB. Evergreen forest including gallery forest; 1200–1700 m.

SYN. *Grumilea saltiensis* S. Moore in J.B. 45: 44 (1907), *non Psychotria saltiensis* (S. Moore) Guillaumin. Type: as for species

NOTE. I have excluded *Maitland* 1258 (Uganda, Bugisu, Bufumbo) cited by Petit—it seems to me to be *P. riparia*, if indeed the two are truly separable.

7. **P. triclada** *Petit* in B.J.B.B. 34: 57, photo. 1/F (1964) & in Distr. Pl. Afr. 4, map 97 (1972). Type: Tanzania, E. Usambara Mts., N. Bulwa, Misolai, *Peter* 21923 (B, holo. !)

Shrub ± 2–4·5 m. tall; stems glabrous, soon becoming woody. Leaf-blades elliptic or oblong-elliptic, 3–16 cm. long, 1·1–7·5 cm. wide, shortly acuminate at the apex, cuneate at the base, glabrous, thin; nodules absent, domatia present; petiole 0·4–3·5 cm. long, glabrous; stipules ovate, 4–7 mm. long, acute, bifid or emarginate at the apex (entire *fide* Petit), glabrous or ferruginous-puberulous and ciliate, deciduous. Panicles with (2–)3 principal branches; peduncle very short or obsolete, 2–3 mm. long, or up to 2·5 cm. long; branches 1·5–4 cm. long, glabrous or sparsely ferruginous pubescent; pedicels very short but up to 7 mm. long in fruit; bracts and bracteoles minute. Calyx-tube conical or globose, 1–1·5 mm. long; limb broadly cupuliform, 0·5–1·5 mm. high, truncate or with minute lobes, ± glabrous. Open corolla not known; buds white, glabrous. Drupes with 2 pyrenes, didymously subglobose, ± 6–9 mm. in diameter, glabrous; pyrenes semi-globose, 6·5–7 mm. in diameter, 4 mm. thick, rugulose; ventral face plane, dorsal face entire. Seeds similar in size to pyrenes, not ribbed; albumen with ventral face deeply and dorsal face shallowly ruminate.

TANZANIA. Lushoto District: E. Usambara Mts., Amani, *Zimmermann* in *Herb. Amani* 7898 & Kwamkoro to Potwe, 16 Dec. 1936, *Greenway* 4789 ! & W. Usambara Mts., near Garaya [Ngaraya], 4 Mar. 1916, *Peter* 15817 !
DISTR. T3; not known elsewhere
HAB. Lowland rain-forest; 850–1050 m.

NOTE. It is curious that not one of the 13 specimens is in open flower.

8. **P. goetzei** (*K. Schum.*) *Petit* in B.J.B.B. 34: 81 (1964) & in Distr. Pl. Afr. 4, map 111 (1972). Types: Tanzania, S. Uluguru Mts., *Goetze* 175 (B, syn.†) & upper Mgeta valley, *Stuhlmann* 9279 (B, syn.†) & Uluguru Mts., without exact locality, *E. M. Bruce* 460 (K, neo. !, BM, BR, EA, isoneo. !)

Shrub or small tree 1–7·5 m. tall; stems glabrous. Leaf-blades narrowly elliptic to elliptic, oblong- or obovate-elliptic or obovate, 5–27 cm. long, 1·5–15·5 cm. wide, rounded to acute at the apex, cuneate to rounded at the base, glabrous on both surfaces or slightly hairy on main nerve beneath, thin; nodules absent and domatia inconspicuous or absent; petiole 1–3·5 cm. long, glabrous; stipules obovate-elliptic, (0·8–)1·2–2·5 cm. long, obtuse, glabrous, deciduous. Flowers heterostylous, 5-merous, in broad lax much-branched panicles 14–25 cm. long; peduncle 2–12·5 cm. long, glabrous; branches 0·5–9 cm. long, glabrous or shortly pubescent; pedicels up to 0·5 mm. long or obsolete; lower bracts 3·5–8 mm. long, clasping the stem or sometimes forming an involucre at point of initial branching, rest and bracteoles minute. Calyx-tube obconic, 0·5 mm. long; limb cupuliform, 0·4–0·5 mm. long, glabrous; lobes triangular, 0·3–0·5 mm. long. Corolla white or tinged lilac, glabrous outside; tube 1·5–3·5 mm. long; lobes ovate to triangular, 0·75–1·75 mm. long, 0·5–0·7 mm. wide. Stamens with filaments 0·1–0·25 mm. long in short-styled flowers and ± obsolete in long-styled flowers. Style 0·25–1 mm. long in short-styled flowers and 2 mm. long in long-styled flowers; stigma 0·25–0·5 mm. long. Drupes red with 2 pyrenes, ellipsoid, ± 4·5–6 mm. long, 3·5–4 mm. wide, or subglobose, 4–5 mm. in diameter, strongly grooved when dry; pyrenes pale brown, semi-ellipsoid, 3·5–5·5 mm. long, 3·5–4 mm. wide, 2–2·2 mm. thick, with dorsal face 5-grooved. Seeds very dark blackish red, 4·2 mm. long, 3 mm. wide, 2 mm. thick, with ventral surface plane and dorsal surface ± 6-ribbed; albumen ruminate.

KEY TO INFRASPECIFIC VARIANTS

Inflorescence-branches and calyces mostly finely tomentose (but sometimes glabrous); leaf-blades large, with up to 26 pairs of lateral nerves; corolla 2·5–3·5 mm. long . . . . . . b. var. **platyphylla**

Inflorescence-branches mostly glabrous or pubescent; calyx glabrous save for ciliate margins:

Leaf-blades larger and broader, with up to 21 pairs of lateral nerves; corolla 1·6–2·5 mm. long . a. var. **goetzei**

Leaf-blades smaller and narrower, with 8–16 pairs of lateral nerves; corolla ± 1·5 mm. long. . c. var. **meridiana**

a. var. **goetzei**; Petit in B.J.B.B. 34: 82, photo. 2/E (1964) & in Distr. Pl. Afr. 4 map 111 (1972)

Shrub 2–4 m. tall or moderately sized tree 6 m. tall; lateral nerves 10–21 pairs; panicles 14–22 cm. long, with glabrous or pubescent branches; calyx glabrous, limb 0·4–0·5 mm. long; corolla 1·6–2·5 mm. long; drupes ellipsoid, 6 mm. long, 4 mm. wide.

TANZANIA. Morogoro District: Uluguru Mts., 20 Mar. 1953, *Drummond & Hemsley* 1701! & Kitundu, 6 Nov. 1934, *E. M. Bruce* 104!; Iringa District: Kalimbili, Mar. 1954, *Carmichael* 410!

DISTR. **T3**, 6, 7; not known elsewhere

HAB. Evergreen forest; 1100–1800 m.

SYN. *Grumilea goetzei* K. Schum. in E.J. 28: 497 (1900); De Wild. in B.J.B.B. 9: 36 (1923); T.T.C.L.: 498 (1949)
[*G. platyphylla* sensu K. Schum. in E.J. 28: 495 (1900), *non* K. Schum. sensu stricto (quoad *Stuhlmann* 1278)]

b. var. **platyphylla** (*K. Schum.*) *Petit* in B.J.B.B. 34: 83, photo. 2/D (1964) & in Distr. Pl. Afr. 4, map 112 (1972). Type: Tanzania, Usambara, Mashewa [Maschewa], *Holst* 8725 (B, syn. †*, K, lecto. !)

Shrub or small tree up to 7·5 m. tall; lateral nerves 15–26 pairs; panicles 10–25 cm. long, with glabrous or finely tomentose branches; calyx glabrous or finely tomentose. limb 0·5–1·75 mm. long; corolla 2·5–3·5 mm. long; drupes subglobose, ± 5 mm. in diameter.

TANZANIA. Lushoto District: E. Usambara Mts., Uberi, 25 Nov. 1935, *Greenway* 4182! & W. Usambara Mts., Lushoto–Mkuzi road, 1·6 km. E. of Magamba, 3 June 1963, *Drummond & Hemsley* 2832!; Tanga District: Ngua Estate, W. slope of E. Usambara Mts., 19 July 1953, *Drummond & Hemsley* 3353!

DISTR. **T2**, 3, ?6**; not known elsewhere

HAB. Rain-forest and drier evergreen forest; 950–1650 m.

SYN. *Grumilea platyphylla* K. Schum. in P.O.A. C: 392 (1895), pro parte; De Wild. in B.J.B.B. 9: 43 (1923), pro parte; T.T.C.L.: 499 (1949), pro parte

NOTE. A fairly well-defined variety distinguishable by the fine tomentose indumentum on the inflorescences. Unfortunately the name *platyphylla* is not available in *Psychotria* for use to cover this species as a whole because of a prior *P. platyphylla* DC.

c. var. **meridiana** *Petit* in B.J.B.B. 34: 84 (1964) & in Distr. Pl. Afr. 4, map 113 (1972). Type: Tanzania, Njombe District, upper Ruhudje, Lupembe, *Schlieben* 123 (BR, holo., B, BM!, K!, S, iso.)

* The other syntype *Volkens* 1119 is *P. pseudoplatyphylla*.
** Petit has annotated *Peter* 52088 (Tanzania, forest above Morogoro, 12 Nov. 1914) as " ? subsp. *platyphylla* " but has not cited the specimen under that variety.

Shrub 1–4 m. tall; leaves smaller, narrower, and rather more coriaceous than in other varieties; lateral nerves 8–16 pairs; panicles 6–14 cm. long, with glabrous to pubescent branches, the hairs in lines; calyx-limb 0·25–0·5 mm. long, glabrous save for ciliate margins; corolla small, ± 1·5 cm. long; drupes subglobose, ± 4 mm. long.

TANZANIA. Iringa District: Sao Hill, 20 Oct. 1947, *Brenan & Greenway* 8240A! & near Kigogo R., 4 May 1968, *Renvoize & Abdallah* 1919! & Mufindi, Kigogo Forest Reserve, 18 Dec. 1961, *Richards* 15761 in part!
DISTR. T6, 7; Malawi
HAB. Evergreen forest; 900–1890 m.

NOTE. This is a most unsatisfactory variety and scarcely worth distinguishing save that some small-leaved specimens from the Iringa area are very different from the large-leaved material from the Usambaras and Ulugurus. The colour differences of dried leaves given by Petit are totally useless and the size varies greatly. A number of specimens from Iringa District, e.g. *Renvoize & Abdallah* 1919, *St. Clair-Thompson* 648 and *Gane* 4 are either quite indistinguishable from var. *goetzei* or are intermediate.

9. **P. elachistantha** (*K. Schum.*) *Petit* in B.J.B.B. 34: 85 (1964) & in Distr. Pl. Afr. 4, map 114 (1972). Types: Tanzania, Uluguru Mts., Ng'hwenn, *Stuhlmann* 8794 & 8806 (both B, syn.†); Uluguru Mts., *E. M. Bruce* 502 (K, neo.)*

Shrub or small tree 4·8–15 m. tall; stems glabrous. Leaf-blades elliptic, oblong-elliptic or obovate-elliptic to oblanceolate, 8–16 cm. long, 2·5–6 cm. wide, acute to shortly acuminate at the apex, cuneate at the base, glabrous, ± coriaceous, usually drying a russet-brown; venation ± prominent above in dried material; nodules and domatia absent; petiole 0·7–3 cm. long, glabrous; stipules ovate, 0·7–1(–1·7) cm. long, entire and obtuse, glabrous but fringed with ferruginous hairs, deciduous, hairy within the base. Flowers ? heterostylous, 5-merous, in much-branched panicles 6–10 cm. long; peduncles 2–5 cm. long, glabrous or slightly puberulous towards apex; secondary branches 0·5–3 cm. long, glabrous to densely finely pubescent; pedicels 0–3 mm. long, finely pubescent; main bracts clasping the stem, ± 2 mm. long, rest small or minute. Calyx-tube campanulate, 1 mm. long, puberulous; limb cupuliform, glabrous or finely puberulous, 0·75 mm. long; lobes rounded-triangular, 0·5 mm. long, finely puberulous. Corolla "yellowish-white", "white and pale pink" or "reddish", glabrous outside; tube ± 2 mm. long; lobes oblong-triangular, 1·5 mm. long, 0·9 mm. wide, thickened at the apex. Stamens with filaments 0·2 mm. long in short-styled flowers. Style ± 1 mm. long in short-styled flowers; stigma-lobes ± 0·5 mm. long. Drupes orange-red, with 2 pyrenes, subglobose, ± 6 mm. in diameter, glabrous, ± grooved; pyrenes semiglobose, 5 mm. long and wide, 2·5–3 mm. thick, with ventral surface plane, dorsal surface 5-grooved. Seeds flattened ventrally, dorsally 6-ribbed; albumen ruminate.

TANZANIA. Morogoro District: Uluguru Mts., Bondwa Mt., 24 Nov. 1932, *Schlieben* 3008! & Morningside, 28 Nov. 1934, *E. M. Bruce* 235! & Uluguru Mts., without precise locality, 23 Nov. 1932, *Wallace* 479!
DISTR. T6; not known elsewhere
HAB. Upland rain-forest; 1350–1950 m.

SYN. *P. elachistacantha* K. Schum. in Engl., P.O.A. A: 92 (1895); De Wild., Pl. Bequaert. 2: 364 (1924), *nom. nud.*
Grumilea elachistantha K. Schum. in E.J. 28: 497 (1900); De Wild. in B.J.B.B. 9: 34 (1923); T.T.C.L.: 498 (1949)

* There are 4 sheets of *Bruce* 502 at Kew—2 of this taxon and 2 of *P. goetzei* var. *goetzei* (cited by Petit)—502/II & 502/IV are isoneotype and neotype of *P. elachistantha* respectively.

NOTE. One of the sheets of *Bruce* 502 bears the note *P. elachistantha* compared with type. This species is scarcely distinguishable from *P. goetzei*. The characters given by Petit "smaller inflorescence, calyx and ovary larger and leaf-blades reddish-brown with ± prominent venation above (in dry state)" are all of dubious value. I am maintaining them until somebody has compared them in the field and collected adequate spirit material of the flowers. The dried specimens are fairly easily separable.

10. **P. meridiano-montana** *Petit* in B.J.B.B. 34: 88 (1964) & in Distr. Pl. Afr. 4, map 116 (1972). Type: Tanzania, Rungwe District, Poroto Mts., Ngozi, *Richards* 6499 (K, holo.!)

Shrub or small tree 0·8–3·6(–? 6) m. tall; stems glabrous. Leaf-blades elliptic, oblong-elliptic or slightly obovate-elliptic, 3–18 cm. long, 1–8 cm. wide, acute or very sharply acuminate at the apex, narrowly cuneate at the base, glabrous above and beneath or with hairs along the edges of the midrib, rather thinly coriaceous; nodules absent; domatia absent or inconspicuous; petiole 0·7–2·5 cm. long, glabrous or with few scattered hairs; stipules elliptic or obovate, 0·4–2·2 cm. long, obtuse, glabrous or with some marginal hairs, deciduous. Flowers heterostylous, 5-merous, in branched panicles which are often reflexed, 2·5–20 cm. long; peduncles 3–12 cm. long, glabrous or sparsely pubescent; secondary branches 2–3·5 cm. long, ± pubescent; pedicels obsolete or up to 2 mm. long; main bracts 2–6 mm. long, clasping the peduncle and ± forming an involucre, ± lobed, ± ferruginous hairy as are most of the axils of the inflorescence-branches; other bracts and bracteoles small or minute. Calyx-tube obconic, ± 1 mm. long; limb ± 1 mm. long; lobes triangular, 0·25–0·5 mm. long, with some reddish marginal hairs. Corolla white, glabrous or rarely shortly puberulous outside; tube 3–6 mm. long; lobes oblong-elliptic, 1·7–2·5 mm. long, 0·8–1·1 mm. wide, appendaged at the apex. Stamens exserted ± 0·5–1 mm. in short-styled flowers, included in long-styled flowers. Style in short-styled flowers 1·5–2·5(–3·5) mm. long; stigma-lobes 0·5–1 mm. long; exserted about 1 mm. in long-styled flowers. Drupes red, with 2 pyrenes, subglobose, ± 5 mm. in diameter, strongly grooved in the dry state, glabrous; pyrenes semiglobose, 4·5 mm. long and wide, 2 mm. thick, 4–5-grooved on the dorsal surface. Seeds 3·5 mm. long and wide, 1·8 mm. thick; ventral face plane, dorsal face obtusely ribbed; albumen conspicuously ruminate.

var. **meridiano-montana**; Petit in B.J.B.B. 34: 88, photo. 2/G (1964)

Leaf-blades broadly elliptic, mostly 7–18 cm. long and 2–8 cm. wide; stipules 1–2·2 cm. long; panicles up to 20 cm. long.

TANZANIA. Mbeya Mt., 19 May 1958, *Kerfoot* 206! & 3 Dec. 1961, *Kerfoot* 3210!; Iringa District: Imagi Mt., 15 Dec. 1961, *Richards* 15676!; Rungwe District: between upper Kiwira R. & Tukuyu road, Siwago, Pangundutani, 25 Oct. 1947, *Brenan & Greenway* 8207!
DISTR. **T7**, 8; Malawi
HAB. Evergreen forest and thicket, also *Arundinaria* forest; 1800–2650 m.

NOTE. Most of the material seen has the anthers exserted but one sheet, *Milne-Redhead & Taylor* 8175 (Songea District, Matengo Hills, Lupembe Hill, 10 Jan. 1956), cited by Petit is long-styled; it differs, however, by having the corolla-tube only about 2 mm. long, well outside the range of 5–6 mm. given by Petit. In some other specimens showing short-tubed flowers they may be artefacts due to the buds opening under pressure in the press but genuinely short-tubed flowers certainly exist.

var. **angustifolia** *Petit* in B.J.B.B. 34: 89 (1964). Type: Tanzania, Njombe, *Lynes* 64 (K, holo.!, EA, iso.!)

Leaf-blades narrowly elliptic, 3–12 cm. long, (0·8–)1–2·5 cm. wide; stipules 4–8 mm. long; panicles 2·5–8 cm. long.

TANZANIA. Iringa District: Dabaga Highlands, Kibengu, 13 Feb. 1962, *Polhill & Paulo* 1460! & near Sao Hill, 20 Oct. 1947, *Brenan & Greenway* 8240!; Njombe District: Msima Stock Farm, *Emson* 289!

DISTR. **T7**; not known elsewhere
HAB. Evergreen forest (including upland mist-forest and isolated evergreen thickets in hillside-grassland); 1800–2100 m.

NOTE. Another variety, var. *glabra* Petit, occurs in NE. Zambia.

11. **P. pseudoplatyphylla** *Petit* in B.J.B.B. 34: 90 (1964) & in B.J.B.B. 42: 356 (1972) & in Distr. Pl. Afr. 4, map 117 (1972). Type: Tanzania, SE. side of Kilimanjaro, *Schlieben* 4362 (BR, holo., BM!, HBG, P, S, iso.)

Shrub or tree 3–4 m. tall; stems glabrous or with sparse ferruginous hairs at extreme apex. Leaf-blades broadly elliptic or with a tendency to be slightly obovate, 6–22 cm. long, 3–11·5 cm. wide, obtuse to shortly acuminate at the apex, cuneate at the base, glabrous above, sparsely to densely ferruginous pubescent on the main and lateral nerves beneath, rarely glabrous, thin; nodules absent, domatia sometimes present; petiole 1–6 cm. long, sparsely to densely pubescent; stipules elliptic, 1·7–3·5 cm. long, obtuse and entire at the apex, ciliate, glabrous or pubescent all over or hairy only at the base. Flowers heterostylous, 5-merous, in much-branched panicles 10–15 cm. long; peduncle 4–10·5 cm. long, ± glabrous to ferruginous pubescent; branches 0·5–2·5 cm. long, ferruginous pubescent; pedicels up to 2 mm. long or obsolete; lower bracts clasping the stem, 5 mm. long; others and bracteoles minute. Calyx-tube conic, 0·5 mm. long; limb cupuliform, 0·6 mm. long, glabrous or ± pubescent; lobes triangular, 0·5 mm. long. Corolla greenish yellow, greenish white or white, glabrous outside; tube 4·5–5 mm. long; lobes ovate, 2–2·5 mm. long, 1·5 mm. wide. Stamens with filaments 0·5 mm. long in short-styled flowers, exserted ± 1·5 mm. Style 2–2·5 mm. long in short-styled flowers; stigma-lobes ± 0·5 mm. long. Drupes red, with 2 pyrenes, ellipsoid or subglobose, ± 4·5–10 mm. in diameter, glabrous, strongly grooved; pyrenes ellipsoid, up to 9 mm. long, 6·5 mm. wide, 3 mm. thick, ventral surface longitudinally acutely 5-grooved. Seeds similar in size and shape, ventral surface plane, dorsal surface 6-ribbed; albumen ventrally deeply grooved and dorsally 5-sulcate.

KENYA. Teita District: Teita Hills, 8 km. NNE. of Ngerenyi, Ngangao, 15 Sept. 1953, *Drummond & Hemsley* 4334! & same locality, July 1937, *Dale* in *F.D.* 3816! & Mbololo Hill, Mraru Ridge, 9 Apr. 1971, *Faden et al.* 71/233!
TANZANIA. Arusha District: E. slopes of Mt. Meru, Nasolo R., 2 Apr. 1968, *Greenway & Kanuri* 13293!; Moshi District: Kilimanjaro, near Mandara [Bismarck] Hut, on the way from Marangu, 20 Jan. 1955, *Verdcourt* 1246! & Marangu, Kilimanjaro Forest Reserve, Nov. 1960, *Steele* 139!
DISTR. **K7**; **T2, 3**; not known elsewhere
HAB. Evergreen forest, especially *Hagenia*, *Agauria*, *Schefflera*, *Podocarpus* formations; 1450–2150 m.

SYN. [*Grumilea platyphylla* sensu K. Schum. in P.O.A. C: 392 (1895), pro parte; De Wild. in B.J.B.B. 9: 43 (1923), pro parte; T.T.C.L.: 499 (1949), pro parte, quoad syn. *Volkens* 1119 (Tanzania, Kilimanjaro, Marangu, B, syn., BM, K, isosyn.!), *non* K. Schum. sensu stricto]

NOTE. Petit has chosen the lectotype of *G. platyphylla* so that the name is now applied to a variety of *P. goetzei* (see p. 45). K. Schumann had written up *Volkens* 1119 as *G. platyphylla* var. *angustior*, a name apparently never published, but sufficient evidence to show which syntype should be considered the lectotype. The indumentum on the leaf-blades and inflorescences of the Uluguru specimen cited above is dense and ± velvety; the specimen is only in young bud and more confirmatory material is required. The Kenya specimens have more condensed inflorescences and hairier flowers. Some Mau specimens of *P. orophila* are almost identical and the two may need combining. Field studies of stipule variation are needed. *Peter* 8930 (Tanzania, S. Pare Mts., Tona, 14 Feb. 1915) has narrow leaves, very condensed inflorescences with capitate components (very immature) and small stipules 6–7 mm. long. Whether this represents a distinct species must await a range of mature material. *Peter* 8996 (Tanzania, S. Pare Mts., Shengena [Schengena], 16 Feb. 1915) has very strongly ribbed fruits 9 × 6 mm. and is clearly conspecific with *Peter* 9006 determined by Petit as *P. pseudoplatyphylla*.

12. **P. orophila** *Petit* in B.J.B.B. 34: 92, photo. 2/J (1964) & in B.J.B.B. 42: 356 (1972) & in Distr. Pl. Afr. 4, 118 (1972). Type: Kenya, Masailand, near Nairobi, *Elliott* 27 (K, lecto. !)

Shrub or small tree 1·5–10 m. tall, with glabrous stems. Leaf-blades elliptic or oblong-elliptic, 7–18 cm. long, 3·5–7(–10) cm. wide, acute, obtuse or slightly acuminate at the apex, cuneate at the base, glabrous above, pubescent beneath on the midnerve and sometimes on the lateral nerves as well, or entirely glabrous beneath save for the hairy domatia; nodules absent; petiole 0·5–3 cm. long, glabrous or pubescent; stipules ovate or ovate-elliptic, 1–2(–? 2·5) cm. long, shortly bifid for 1–2 mm. or ± entire, glabrous save for a few marginal hairs, deciduous. Flowers heterostylous, 5-merous, in much-branched panicles, lowest branches usually 4; peduncles 2–7(–15) cm. long, glabrous or slightly pubescent; secondary branches 0·3–3·5 cm. long, ferruginous pubescent; pedicels ± obsolete or 1–1·5 mm. long in fruit; main bracts lanceolate with sheathing lobed bases, 5 mm. long, ± ciliate; other bracts and bracteoles smaller, ± ferruginous pubescent. Calyx-tube obconic, ± 1 mm. long, glabrous; limb cupuliform, 1–1·5 mm. long, ± glabrous, lobes very short or obsolete, margined with ferruginous hairs. Corolla white or salmon-tinged above in bud, ± white when developed; tube 2·5–5 mm. long; lobes oblong-elliptic, 1·8–4 mm. long, 0·8 mm. wide, thickened at the apices. Stamens with filaments 1·5 mm. long in short-styled flowers and 0·5 mm. long in long-styled flowers. Style 2·5 mm. long in short-styled flowers and 3·5–7 mm. long in long-styled flowers; stigma-lobes 0·5–1 mm. long. Drupes red, with 2 pyrenes, ellipsoid, 6–7 mm. long, 4–5·5 mm. wide; pyrenes semi-ellipsoid, 6 mm. long, 4·5 mm. wide, 2 mm. thick; dorsal face ± 5-grooved. Seeds blackish red, semiglobose, 4 mm. long, 3·5 mm. wide, 1·5 mm. thick; ventral face plane, dorsal face 5-grooved; albumen ruminate.

UGANDA. Mbale District: Bugisu, Mt. Elgon by Sasa Stream, 23 Mar. 1951, *G. Wood* 113 ! & Mt. Elgon, bamboo zone, Jan. 1918, *Dummer* 3582 ! & Mt. Elgon, Butandiga–Bulambuli, 19 Aug. 1927, *Snowden* 1196 !

KENYA. Northern Frontier Province: Mt. Kulal, Apr. 1959, *T. Adamson* K10B !; Meru District: NW. Mt. Kenya, Marimba Forest,14 Oct. 1960, *Verdcourt & Polhill* 2984 !; Masai District: summit of the Ngong Hills, 1 Mar. 1953, *Bally* 8808 !

TANZANIA. Mbulu District: Mangati, between Mdungara and Dareda, *Peter* 43963 !

DISTR. **U**1, 3; **K**1, 3–6; **T**2; Ethiopia (*fide* Petit)

HAB. Evergreen forest; 1650–2670(–? 3000) m.

SYN. *Grumilea elliottii* K. Schum. & K. Krause in E.J. 39; 560 (1907); De Wild. in B.J.B.B. 9: 34 (1923); K.T.S.: 444 (1961), *non Psychotria elliottii* Bremek. (1963)

[*G. riparia* sensu K. Krause in N.B.G.B. 10: 608 (1929), *non* K. Schum. & K. Krause]

NOTE. I have seen this taxon in the field on several occasions and the stipules are quite definitely not always bifid at the apex, thus breaking down the only character separating this from its allies. Several specimens, e.g. *Kabuye* 49 (Kenya, S. Nyeri District, Castle Forest Station, 12 Dec. 1966) and *Verdcourt, Polhill & Lucas* 3037 (Kenya, Naivasha District, South Kinangop, Sasumua Dam pipeline road, 11 Dec. 1960) have much larger leaf-blades than usual, up to 27 × 13 cm. Three specimens from Karamoja District, Mt. Debasien, Jan. 1936, *Eggeling* 2760 !, Oct. 1944, *Dale* U.408 ! and *Wilson* 768A do not seem to differ from *P. orophila* in any significant characters. Petit has annotated the Wilson sheet *P. fractinervata* but has not cited it. The calyx is not correct for that species.

13. **P. taitensis** *Verdc.* in K.B. 30: 248 (1975). Type: Kenya, Teita District, Mt. Kasigau, *Faden* 71/153 (EA, holo.!, K, MO, iso.!)

Small understorey tree; branchlets wrinkled, glabrous. Leaf-blades elliptic, oblong-elliptic or obovate-elliptic, 7·5–14 cm. long, 3·75–7 cm. wide, very shortly ± bluntly acuminate or subacute at the apex, cuneate at the base, glabrous, said to be thick and lustrous; petiole 1–2·7 cm. long; stipules large, broadly elliptic or almost round, 2·5–3 cm. long, 2 cm. wide, ± acute, entire, soon falling, with a line of hairs at the nodes within the stipules. Flowers in branched inflorescences, each component ± capitate; peduncles 2–7 cm. long, glabrous; secondary peduncles 1·2–2·5 cm. long; pedicels obsolete; bracts forming a small lobed ciliate involucre at the junction of the secondary peduncles, 1·3 cm. wide; secondary bracts small, ciliate. Calyx-tube obconic to semiglobose, 0·7–1·2 mm. long; free part of limb 1 mm. long; lobes rounded, 0·8–1 mm. long, ciliate; disc prominent. Corolla white; tube 2·5 mm. long; lobes ovate-lanceolate, 2 mm. long, 1 mm. wide, thickened and ± appendiculate at the apex. Anthers exserted 0·8 mm. in long-styled flowers. Style 2·5 mm. long in long-styled flowers; stigma-lobes flattened and lobulate, 0·5 mm. long. Fruits red, 7–8 mm. in diameter, crowned by the persistent calyx-lobes; pyrenes almost round in outline, 4 mm. in diameter, 1·5 mm. thick, slightly concave ventrally, strongly 4-ribbed dorsally.

KENYA. Teita District: Mt. Kasigau, pipeline route from Rukanga, 6 Feb. 1971, *Faden et al.* 71/153!
DISTR. **K7**; not known elsewhere
HAB. Mist forest mainly with *Syzygium* and *Rapanea*; 1400–1600 m.

14. **P. sp. B**

Shrub 1·2 m. tall; stems glabrous save at nodes where ferruginous hairs within the stipules are apparent after the latter have fallen. Leaf-blades elliptic-oblong to oblanceolate, 3·2–12 cm. long, 1·3–4·6 cm. wide, acute or slightly acuminate at the apex, narrowly cuneate at the base, glabrous, without nodules or domatia but with some deceptive looking lesions; petiole 0·5–2·3 cm. long, glabrous; stipules ovate-oblong, 1·1 cm. long, 8 mm. wide, entire, sparsely ciliolate around the margins, soon deciduous. Flowers in rather dense inflorescences made up of several dense ± spherical components; peduncle 2·5 cm. long, with sparse ferruginous hairs towards the apex; secondary peduncles up to 5 mm. long, ± ferruginous-pubescent and with longer ferruginous hairs at the junctions; bracts and secondary bracts triangular to lanceolate, 1–2·5 mm. long, ciliolate. Calyx tubular, the total length 4 mm.; tube and annular part of limb 3 mm. long; lobes rounded, 1 mm. long, ciliolate. Corolla white; tube 1·2–3 mm. long; lobes oblong, 2 mm. long, 1 mm. wide, thickened at the apex, inflexed and with a fine apical appendage. Anthers just exserted in short-styled flowers, the style and stigmas very short, 0·8 mm. long. Drupes not known.

KENYA. Teita District: Ngangao Forest, 4 May 1972, *Faden et al.* 72/211!
DISTR. **K7**; not known elsewhere
HAB. Upland evergreen forest, mostly with *Albizia gummifera* and *Cola greenwayi* dominant; 1650–1800 m.

15. **P. fractinervata** *Petit* in B.J.B.B. 34: 127 (1964) & in B.J.B.B. 42: 357 (1972) & in Distr. Pl. Afr. 5, map 137 (1973). Type: Tanzania, Kilimanjaro, Marangu, *Volkens* 1120 (K, lecto.!, BM!, BR, isolecto.)

Shrub or small tree 2–6(–7·5) m. tall, with glabrous or pubescent stems, often with quite dense ferruginous hairs when young. Leaf-blades narrowly elliptic to elliptic or oblong-elliptic, (3–)6–20 cm. long, (1–)2–7·3 cm. wide, acute to acuminate at the apex, cuneate at the base, glabrous above but usually with ferruginous hairs on the main and lateral nerves beneath, less often ± glabrous, thin, usually drying brownish, ± bullate above in life; lateral nerves usually quite distinctly zigzag and tertiary venation usually ± conspicuous; nodules absent, domatia usually present and often on lateral nerves; petiole 1–2·5 cm. long, glabrous or pubescent; stipules obovate, 1–2 cm. long, bilobed at the apex, the lobes 2–6 mm. long, glabrous or pubescent, particularly at the base, deciduous. Flowers heterostylous, 4–5-merous, slightly scented, in much-branched panicles 5–15 cm. long; peduncles 3–10 cm. long, glabrescent or pubescent; secondary branches 0·7–2·5 cm. long, mostly ferruginous pubescent; pedicels obsolete or up to 2·5 mm. long, pubescent; main bracts ± 5 mm. long, clasping the branches, hairy, others small. Calyx-tube ± hemispherical, 1–2 mm. long, glabrous or pubescent; limb cupuliform, 2–3 mm. long, glabrous or pubescent; lobes irregularly triangular, 0·5–1 mm. long, pubescent at least on the margins. Corolla white or cream; tube 4·5–7 mm. long, glabrous or pubescent outside; lobes oblong, 3–6 mm. long, 1·8 mm. wide, thickened and horned at the apex, densely finely papillate inside, glabrous outside, rugulose. Stamens with filaments 3–3·5 mm. long in short-styled flowers, 0·5 mm. long in long-styled flowers. Style 3·5–5 mm. long in short-styled flowers, 6–7 mm. long in long-styled flowers; stigma-lobes 0·5–1 mm. long. Drupes red, slightly but distinctly grooved when dry, with 2 pyrenes, ellipsoid or globose, 5–7 mm. long, 4–5·5 mm. wide, conspicuously crowned by the funnel-shaped calyx remains, 2–2·5 mm. long; pyrenes semi-elliptic or hemispherical, ± 5·5 mm. long, ± 4·5 mm. wide, 2·2 mm. thick, with dorsal face 5-grooved. Seeds similar, 4 mm. long and wide, 2 mm. thick, ventral face plane, dorsal face 5-grooved; albumen sulcate.

UGANDA. Mbale District: Mt. Elgon, Bugisu, 23 Apr. 1951, *G. Wood* 113b! (see note)
KENYA. Naivasha District: S. Kinangop, Sasumua Dam pipe-line road, 11 Dec. 1960, *Verdcourt, Polhill & Lucas* 3038!; Kiambu District: Lari, May 1930, *Dale* in *F.D.* 2361!; Meru District: Mt. Kenya, Marimba Forest, 14 Oct. 1960, *Verdcourt & Polhill* 2985! & 2990!
TANZANIA. Arusha District: Mt. Meru, Dec. 1927, *Haarer* 940!; Moshi District: S. slopes of Kilimanjaro, N. of Moshi, 23 Feb. 1953, *Drummond & Hemsley* 1288! & Kilimanjaro, Marangu, 2 Jan. 1955, *Verdcourt* 1245!
DISTR. **U**3; **K**3, 4, 6; **T**2; not known elsewhere
HAB. Upland rain-forest, including *Podocarpus*, bamboo, etc. and reaching *Hypericum* zone; (1500–)1800–2600 m. (see note)

SYN. *Grumilea exserta* K. Schum. in P.O.A. C: 392 (1895); De Wild. in B.J.B.B. 9: 34 (1923); K. Krause in R.E. & T.C.E. Fries in N.B.G.B. 10: 607 (1929); T.T.C.L.: 497 (1949); K.T.S.: 444 (1961), *non Psychotria exserta* DC.

NOTE. This is very similar to *P. orophila*, but the leaf-blades usually have domatia often on the lateral nerves and are hairier beneath; the flowers are mostly larger and the fruits are crowned with much longer more funnel-shaped calyx remnants; apart from this the lateral nerves have a curious tendency to be zigzag rather than straight. I have seen both in the field and had noted that in the Meru District of Kenya *P. fractinervata* had the stipules much more obviously bifid, the leaf-blades more bullate above and more hairy beneath. On Mt. Meru in Tanzania collectors state that it grows at altitudes of 1500–1800 m., but elsewhere it is usually 2100–2400 m. The figures 1000–2000 m. and 3000 m. appear on several specimens collected by Grote on Kilimanjaro but both limits need confirmation. Petit's maximum figure of 10 cm. for leaf-blade width is I think derived from a specimen (*Wilson* 768A) from Mt. Debasien which he annotated as *P. fractinervata* but did not cite; I have referred this specimen to *P. orophila.*

16. **P. sp. C**

Shrub to 3 m. tall; stems glabrous, finely ridged. Leaf-blades broadly elliptic, up to 13·7 cm. long, 8·2 cm. wide, ? subacute at the apex, cuneate or rounded at the base, glabrous above, with a few scattered ferruginous hairs on the midrib beneath; nodules and domatia absent; petiole 1·2 cm. long; stipules ovate-lanceolate, 1·4 cm. long, 6 mm. wide, entire, acute. Flowers in branched panicles ± 5 cm. long and wide; peduncle 5 cm. long; secondary branches 0·5–2 cm. long; pedicels 2·5–4 mm. long; bracts up to 4 mm. long; all parts with ± sparse ferruginous hairs. Calyx-tube ovoid, 1·8 mm. long, glabrous; limb ± tubular, 2·5 mm. long including the irregular teeth 0·2–1 mm. long, with a few stiff marginal cilia.

KENYA. Masai District: Shabal Tarakwa, 4·8 km. from Ol Pusimoru sawmill, 20 May 1961, *Glover, Gwynne & Samuel* 1411!
DISTR. **K6**; not known elsewhere
HAB. *Juniperus, Podocarpus* forest; 2550 m.

NOTE. Although clearly allied to *P. orophila, P. pseudoplatyphylla* and *P. fractinervata* it differs from all in either stipules or calyx-limb.

17. **P. sp. D**

Shrub to 1·8 m., with thick purplish, longitudinally wrinkled stems, glabrous, rather prominently scarred by the petiole and stipule bases, possibly slightly succulent in life. Leaf-blades elliptic to oblanceolate-elliptic, 3–10·5 cm. long, 1·2–5 cm. wide, obtuse or obscurely acute at the apex, glabrous, rather thick, probably ± coriaceous in life, purplish beneath, margins reflexed; petiole 0·2–1·5 cm. long; stipules oblong-obovate, 1·2–2·2 cm. long, 0·8–1·4 cm. wide, entire, drying purplish. Flowers pale green, known only from buds, in branched umbel-like panicles; peduncles ± 8·5 cm. long, glabrous below, ferruginous hairy towards the apex; secondary branches 0·6–1·4 cm. long, with long ferruginous hairs; pedicels obsolete; bracts narrowly ovate, 4–5 mm. long, ferruginous pubescent, the bracts of the other rays forming a sort of stipule to the main bracts. Calyx ± 4–5·5 mm. long; tube obconic, 2 mm. long, ferruginous pubescent; limb campanulate, the lobes triangular, 1–1·5 mm. long. Corolla (very undeveloped) with lobes distinctly thickened and cucullate at the apex.

TANZANIA. Morogoro District: Uluguru Mts., Lukwangule Plateau, 30 Jan. 1935, *E.M. Bruce* 743! & same locality, 19 Sept. 1970, *Thulin & Mhoro* 1026 B!
DISTR. **T6**; not known elsewhere
HAB. Upland rain-forest; 2300–2406 m.

NOTE. *Thulin & Mhoro* 1026 consists of two fruiting portions and one flowering portion, the latter matching *Bruce* 743 but the leaf-blades have a very distinctive margin 0·3 mm. wide and drying orange-brown like the midrib.

18. **P. brucei** *Verdc.* in K.B. 30: 249, fig. 1 (1975). Type: Tanzania, Uluguru Mts., 4·8 km. S. of Bunduki, *Drummond & Hemsley* 1620 (K, holo.!, EA, iso.)

Shrub 1·8–3 m. tall; stems with short matted almost velvety dark ferruginous pubescence, at least on the young stems seen. Leaf-blades elliptic, oblong-elliptic or rarely oblanceolate-elliptic, 8–13·5 cm. long, 2·5–8 cm. wide, mostly obtuse or sometimes obscurely acute at the apex, broadly cuneate at the base, glabrous above, pubescent beneath, particularly on the venation, with bright dark ferruginous hairs; nodules and domatia absent; petiole 1·5–4 cm. long, velvety ferruginous pubescent; stipules white, elliptic or obovate-oblong, 1·5–2·4 cm. long, 0·8–1·5 cm. wide, entire, ciliate and ferruginous pubescent, soon deciduous. Flowers heterostylous, 5-merous, in branched panicles and often with lateral panicles from the same node; peduncles 1·5–7 cm. long; secondary branches 0·3–1·2 cm. long; pedicels 2–3 mm. long, all these components velvety ferruginous; corolla in bud whitish pubescent contrasting with the ferruginous calyx and pedicels; bracts ovate or irregularly cuneiform, up to 6 mm. long; secondary bracts ± 2 mm. long, almost forming an involucre; bracteoles 1 mm. long, all hairy. Calyx-tube obconic, 1 mm. long, glabrescent or densely ferruginous; limb ± 1·5 mm. long; lobes triangular to broad and subtruncate, 0·5 mm. long. Corolla white or yellowish white, greyish or whitish pubescent outside; tube 6·5 mm. long; lobes ovate-triangular, 2·2 mm. long, 1·3 mm. wide. Stamens with filaments 0·5 mm. long in short-styled flowers. Style 1·5–2 mm. long in short-styled flowers; stigma-lobes 1·5 mm. long. Drupes ellipsoid, 6 mm. long, 5 mm. wide, strongly grooved in dry state, crowned with the conspicuous calyx remnants, with a few scattered short hairs; pyrenes strongly ribbed dorsally.

TANZANIA. Morogoro District: Uluguru Mts., 4·8 km. S. of Bunduki, Salaza Forest, 15 Mar. 1953, *Drummond & Hemsley* 1620! & Lukwangule Plateau, 30 Jan. 1935, *E.M. Bruce* 748! & Tanana, Matombo road, 1 Feb. 1935, *E. M. Bruce* 801!
DISTR. T6; known only from Uluguru Mts.
HAB. Shrub layer in upland rain-forest; 1700–2400 m.

NOTE. The material cited above is not entirely uniform. *Drummond & Hemsley* 1620 has the calyx, bracts and bracteoles glabrous save for the margins, whereas *Bruce* 749 and 801 have the calyx entirely ferruginous. There are also differences in the calyx-lobes and possibly also in flower colour. Without further material it is not possible to assess the importance of these differences. *Harris et al.* 6076/AM (Uluguru Mts., W. slope of Lukwangule Plateau above Chenzema) is probably a variant of this species differing in larger leaves (to 14 × 10 cm.), apparently ± sessile panicles which may be abnormal, glabrous stipules and glabrescent buds.

19. **P. sp. E**

Shrub 3–6 m. tall, the young stems with shaggy ferruginous hairs, the older pale grey-brown, ridged longitudinally and ± corky. Leaf-blades elliptic or oblong-elliptic, 3–11 cm. long, 1–4 cm. wide, very shortly obtusely or ± acutely acuminate, cuneate at the base, drying with a dull purplish tinge, stomata appearing like very dense pale scales on lower surface, glabrous above save for extreme base of midrib and on the margins, pubescent on the midrib beneath; nodules and domatia probably absent but there are lesions and also black spots, some on the edges of the blade beneath and some along the midrib; petiole 0·3–1·3 cm. long, densely pubescent; stipules ovate-elliptic, 8 mm. long, with a strong projecting midrib, with long shaggy ferruginous hairs on the margins and hairy at base and inside, soon deciduous. Flowers in branched panicles, densely ferruginous hairy, the individual components of the inflorescence very dense; peduncles 1·7–4·5 cm. long; secondary branches 0·5–1·2 cm. long; pedicels obsolete; bracts 3–5 mm. long, those supporting the ultimate inflorescence-components almost like an involucre; bracteoles 1–2 mm. long; all parts densely ferruginous hairy. Calyx densely ferruginous pubescent; tube obconic, 1·2 mm. long; limb ± 1·5–2 mm. long including the triangular ± 0·5–1 mm. long lobes. Corolla not seen in situ but described as cream; tube 2·5 mm. long; lobes ovate-triangular, 1·5 mm. long, 0·8 mm. wide, cucullate at the apex and densely hairy outside. Style 2·5 mm. long, thickened upwards; stigma-lobes 0·5 mm. long. Fruits not known.

TANZANIA. Morogoro District: S. Nguru Mts., Ruhamba Peak, 2 Apr. 1953, *Drummond & Hemsley* 1959! & Turiani, Lusunguru Forest, Apr. 1953, *Semsei* 1154!
DISTR. T6; known only from the Nguru Mts.
HAB. Upland rain-forest; 1900 m.

NOTE. The corolla described was adhering to the specimen but not in situ; in this corolla the style and stigma combined were 1·5 mm. long. The style and stigma mentioned in the description above are in situ. It is possible this might be *Grumilea blepharostipula* K. Schum. but the description of that is very short and the inflorescences in particular do not agree (see sp. 69 on page 105).

20. **P. cyathicalyx** *Petit* in B.J.B.B. 34: 93, t. 2 & photo. 2/K (1964) & in Distr. Pl. Afr. 4, map 119 (1972). Type: Tanzania, S. Kilimanjaro, *Schlieben* 4649 (BR, holo., B, EA !, HBG, K !, iso.)

Shrub 2–5 m. tall or tree up to 10 m., with glabrous stems. Leaf-blades elliptic or elliptic-oblanceolate to elliptic-obovate, 2·8–8(–10) cm. long, 1–5 cm. wide, obtusely acuminate, acute or ± obtuse at the apex, mostly narrowly acuminate at the base, glabrous on both surfaces but lower surface looking as if covered with very closely packed minute scales (due to the stomata), moderately thick; nodules and domatia absent; petiole 0·3–3 cm. long, glabrous; stipules whitish, obovate-elliptic to rounded, 0·5–1·5(–1·7) cm. long, obtuse at the apex, glabrous or margin sometimes ciliate, very soon deciduous; nodes with zone of hairs inside at base of stipules. Flowers heterostylous, 5-merous, in much-branched panicles 2·5–6 cm. long; peduncles 0·5–3 cm. long, glabrous or shortly pubescent; branches 0·3–1·2 cm. long, shortly sparsely pubescent; pedicels short or ± 1–2 mm. long, very shortly pubescent or tomentose; main bracts ovate, 2–3 mm. long, clasping the stem, rest small, often apically ciliate. Calyx-tube rounded-conical, 0·5–1 mm. long, glabrous or tomentose; limb campanulate to tubular, 2–3 mm. long, glabrous, often split for ± 1 mm.; lobes obtusely triangular, ± 0·5–1 mm. long, often ciliate on the apical margins. Corolla white or cream or reddish in bud, glabrous outside; tube 4–5 mm. long; lobes oblong-elliptic to ovate, 2 mm. long, 1–1·8 mm. wide, ± uncinate at the apex. Stamens with filaments ± 0·3 mm. long in short-styled flowers. Style 2·5–3 mm. long in short-styled flowers, 4·5 mm. long in long-styled flowers; stigma-lobes ± 0·5 mm. long. Drupes orange-red, with 2 pyrenes, subglobose or ellipsoid, 5–7 mm. long and wide, glabrous, grooved, crowned with the calyx-limb; pyrenes pale brown, semi-globose, 6 mm. long, 5–5·5 mm. wide, 2·5 mm. thick, dorsally 6-grooved. Seeds 5 mm. long, 4·5 mm. wide, 2·2 mm. thick, ventrally flat, dorsally obtusely ribbed; albumen all strongly ruminate. Fig. 2.

TANZANIA. Moshi District: Kilimanjaro, Bismarck Hill, 28 Feb. 1934, *Greenway* 3874 !; Lushoto District: W. Usambara Mts., Magamba Peak, 7 Nov. 1947, *Brenan & Greenway* 8302 ! & 8302A ! & Kwai valley, Matondwe Hill, 28 Feb. 1953, *Drummond & Hemsley* 1346 !; Morogoro District: Uluguru Mts., S. Uluguru Forest Reserve, edge of Lukwangule Plateau, 17 Mar. 1953, *Drummond & Hemsley* 1663 !

DISTR. T2, 3, 6; not known elsewhere

HAB. Upland evergreen forest, including *Ocotea* forest and *Podocarpus-Erica* associations at upper edges, often a dominant where it occurs; 1600–3000 m.

SYN. [*Grumilea buchananii* sensu T.T.C.L.: 499 (1949), *non* K. Schum.]

NOTE. Certain specimens from the Uluguru Mts. are distinctly atypical and could probably be separated off as a variety. All have a reduced calyx-limb in fruit and some have other variant characteristics, but all have the same close pattern on the undersurfaces of the leaves due to stomata simulating silvery scales which is also evident in all typical specimens. *Drummond & Hemsley* 1639 (S. Uluguru Forest Reserve, edge of Lukwangule Plateau, 2200 m., 17 Mar. 1953) has been annotated by Petit as sp. nov. aff. *P. cyathicalyx*; it has the reduced calyx and the leaf venation raised above. *Thulin & Mhoro* 1026 (same locality, 2300 m., 19 Sept. 1970) has the young inflorescences and midribs on undersides of the very young leaves ferruginous pubescent. *Harris et al.* 5086 and *Gibbon & Pócs* 6051/A (both from Uluguru Mts., Bondwa Peak, 26 Sept. 1970 & 12 Oct. 1969) have the reduced calyx-limb. Detailed field investigations are needed to establish the relations between these and typical specimens which also occur in the same areas of the Uluguru Mts.

21. **P. petitii** *Verdc.* in K.B. 30: 249 (1975). Type: Kenya, Teita Hills, Ngangao, 8 km. NNE. of Ngerenyi, *Drummond & Hemsley* 4343 (K, holo. !, EA, iso.)

Shrub or small tree 3–9 m. tall, with glabrous branches save for tufts of ferruginous hairs within the stipules, visible after they have fallen; petiole-

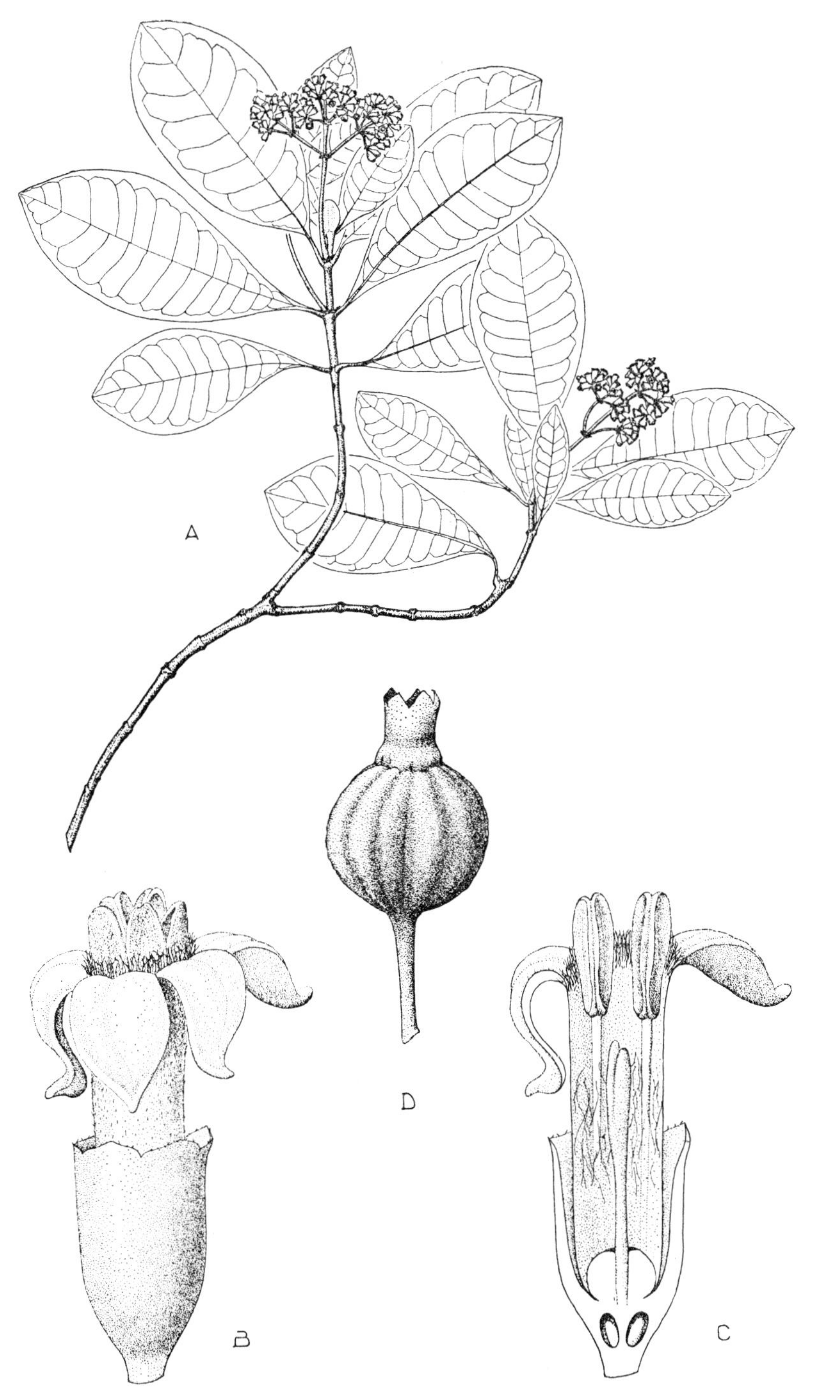

FIG. 2. *PSYCHOTRIA CYATHICALYX*—**A**, habit, × ½; **B**, short-styled flower, × 8; **C**, longitudinal section of same, × 8; **D**, drupe, × 4. A, from *Schlieben* 4649; B, C, from *Drummond & Hemsley* 1346; D, from *Peter* 52100B. Reproduced by permission of the Director of the Jardin Botanique National de Belgique.

bases leaving conspicuous raised round scars, at least in the dried material. Leaf-blades elliptic to oblong-elliptic, 4·5–13·5 cm. long, 2·2–5 cm. wide, mostly shortly acuminate at the apex, cuneate at the base, glabrous on both surfaces, without nodules or domatia; petiole 0·8–3 cm. long; stipules ovate-elliptic, 7–10 mm. long, 5–7 mm. wide, entire, distinctly or obscurely ciliate with short hairs, soon deciduous. Flowers ? heterostylous, 5-merous, in branched panicles, each component rather dense, ± 1 cm. long; peduncle 6 cm. long; secondary branches 0·5–1·2 cm. long; bracts ± triangular, distinctly ciliate. Calyx-tube subglobose, 1 mm. long, glabrous; limb 1 mm. long; lobes oblong or triangular, 0·6–1 mm. long, ciliate; disc prominent. Corolla white; tube 4 mm. long; lobes ovate-elliptic, 2 mm. long, 1·2 mm. wide, acuminate. Stamens with filaments 1 mm. long in short-styled flowers. Style 1·2 mm. long in short-styled flowers; stigma-lobes ellipsoid, 0·4 mm. long. Drupes globose, 1·6 cm. in diameter, smooth in fresh state, ± grooved in the dry state, when dry with a ring of lacunae between ridges of pyrenes and the outer wall; pyrenes 5–6 mm. thick, shiny and grooved within, with ± 15 wing-like ridges. Seeds dark purplish brown, semi-ellipsoid or almost circular, 9–13 mm. long, 7–11 mm. wide, 3·5–4 mm. thick, ventrally flattened, dorsally ± 5-ribbed, deeply ruminate on both surfaces.

KENYA. Teita District: Ngangao Hill, July 1937, *Dale* in *F.D.* 3815! & same locality, 6 Feb. 1953, *Bally* 8751! & Mbololo Hill, Mraru Ridge, 18 Oct. 1970, *Faden & Githui* 70/739!

DISTR. **K7**; not known elsewhere

HAB. Evergreen forest; 1410 m.

NOTE. *R. B. & A. J. Faden* 72/256 (Kenya, Teita Hills, Vuria Hill, 8 May 1972) has been named *P. sp. aff. cyathicalyx* but the calyx is not of the correct shape. The dried fruits are black, 6 mm. in diameter and grooved strongly; they appear mature but I think are juvenile and that the specimen actually belongs to the above species. The stems and foliage, etc. agree well.

22. **P. hemsleyi** *Verdc.* in K.B. 30: 252 (1975). Type: Tanzania, Lushoto District, W. Shagayu Forest, *Drummond & Hemsley* 2542 (K, holo.!, EA, iso.)

Shrub 3–4 m. tall, with whitish or brownish strongly ridged nodose glabrous branches, the petiole-scars quite conspicuous. Leaf-blades narrowly elliptic, 3–6·5 cm. long, 0·9–2 cm. wide, acute at the apex, cuneate at the base, glabrous; nodules and domatia absent; petiole short, 2–5 mm. long; stipules ovate-oblong, 3 mm. long, entire, fimbriate with ferruginous hairs, soon deciduous. Flowers heterostylous, 5-merous, in very short panicles only totalling 2·5 cm. even in fruit; peduncle ± 5 mm. long; secondary branches 0·5–1 cm. long; pedicels 0·5 mm. long, attaining 2–3 mm. in young fruit; bracts minute, with a few ferruginous hairs. Calyx glabrous; tube campanulate, 1 mm. long; limb 0·5 mm. long, truncate or only obscurely lobed. Corolla white, glabrous; tube 4·2 mm. long; lobes oblong-ovate, 2 mm. long, 1·2 mm. wide. Stamens with filaments 0·5 mm. in short-styled flowers. Style 0·9 mm. long in short-styled flowers; stigmas 0·9 mm. long. Young drupes 4·5–5 mm. in diameter, ± grooved.

TANZANIA. Lushoto District: W. Usambara Mts., Shagayu Forest, near Sunga, 18 May 1953, *Drummond & Hemsley* 2612! & Western Shagayu Forest, 15 May 1953, *Drummond & Hemsley* 2542!

DISTR. **T3**; not known elsewhere

HAB. Understorey of *Podocarpus-Ocotea* forest and *Podocarpus-Lachnopylis-Balthasaria* (*Adinandra*) forest; 1950–2000 m.

23. **P. succulenta** (*Hiern*) *Petit* in B.J.B.B. 33: 382 (1963) & in B.J.B.B. 34: 97, photo. 2/L (1964) & in B.J.B.B. 42: 356 (1972) & in Distr. Pl. Afr. 4, map 121 (1972). Type: Sudan, Niamniam, Ngananje, *Schweinfurth* 2900 (K, holo. !, BM !, P, iso.)

Shrub or small tree 1·5–10 m. tall, with glabrous stems; trunk warty and knobby, with reddish brown bark. Leaf-blades elliptic or oblong-elliptic, 4–25 cm. long, (1·1–)2–10 cm. wide, acute to shortly acuminate at the apex, rounded to cuneate at the base, entirely glabrous, with narrowly hyaline, sometimes revolute margins, coriaceous, often drying the characteristic yellow of an aluminium accumulating plant or young leaves drying purplish, often ± shining above; nodules absent; domatia present, not margined with hairs; petiole 0·5–2 cm. long, glabrous; stipules obovate, 1–1·5 cm. long, emarginate at the apex, the lobes rounded, glabrous, soon deciduous, a few hairs present on nodes inside the stipules at the base. Flowers sweet-scented, fleshy, heterostylous, 5-merous, in much-branched panicles 6–20 cm. long; peduncle 2·5–13 cm. long, glabrous; secondary branches 0·8–1·7 cm. long, glabrous; pedicels obsolete; bracts and bracteoles small with a few cilia. Calyx rounded-conic (subglobose in life), ± 1 mm. long, glabrous; limb cupuliform, 2–2·5 mm. long, glabrous; lobes small or obsolete. Corolla white or ? greenish yellow, glabrous outside; tube 4–6 mm. long; lobes oblong-triangular, 2–2·5(–3) mm. long, 1·2(–2) mm. wide, margined in dry state, thickened and inflexed at the apex. Stamens with filaments 2·5–3 mm. long in short-styled flowers, 0·5–1 mm. long in long-styled flowers. Style 3·5–4 mm. long in short-styled flowers, 6 mm. long in long-styled flowers; stigma-lobes 0·5–1 mm. long. Drupes red, with 2 pyrenes, subglobose or ellipsoid, 6–7 mm. in diameter, glabrous, only slightly grooved even in dry state; pyrenes depressed-semiglobose, 5·5–6 mm. long, 5–5·5 mm. wide, 2·2–3 mm. thick, the dorsal face scarcely grooved. Seeds dark, 5 mm. long and wide, 2·8 mm. thick, semiglobose, ventral face rugose, dorsal face with 3 basally joined obtuse ribs and rugose; albumen strongly ruminate.

UGANDA. W. Nile District: Adumi, May 1936, *Eggeling* 3021!; Mengo District: Old Entebbe, Oct. 1931, *Eggeling* 76! & 20·8 km. on Entebbe–Kampala road, Dec. 1937, *Chandler* 2034!

TANZANIA. Mwanza District: Rubya Forest Reserve, 17 June 1962, *Carmichael* 876!; Ufipa District: top of Kawa R. Falls, 2 Oct. 1956, *Richards* 6343! & Kalambo R., Katundo village, 15 Dec. 1958, *Richards* 10355!

DISTR. **U1**, 4; **T1**, 4; Nigeria, Cameroun, Central African Republic, Sudan, Zaire, Rwanda, Burundi, Zambia and Angola

HAB. Evergreen forest, woodland and thicket, particularly at the edges of lakes and rivers; 1100–1650 m.

SYN. *Grumilea succulenta* Hiern in F.T.A. 3: 216 (1877); K. Krause in Fries, Wiss. Ergebn. Schwed. Rhod.-Kongo-Exped. 1911–12, Bot. Erg.: 16 (1921); Z.A.E.: 68 (1922); De Wild. in B.J.B.B. 9: 55 (1923); F.F.N.R.: 408 (1962)
Uragoga succulenta (Hiern) Kuntze, Rev. Gen. Pl. 2: 962 (1891)
Psychotria giorgii De Wild., Pl. Bequaert. 2: 369 (1924). Type: Zaire, Bas-Katanga, Lubunda, *De Giorgi* 921 (BR, holo.)

24. **P. ealaensis** *De Wild.*, Miss. Laurent 1: 348 (1906) & 2, t. 94 (1906) & in Ann. Mus. Congo, Bot., sér. 5, 2: 182 (1907), pro parte; Th. & H. Dur., Syll. Fl. Congo: 281 (1909), pro parte; Petit in B.J.B.B. 34: 103, photo. 3/A (1964) & in B.J.B.B. 42: 357 (1972) & in Distr. Pl. Afr. 5, map 123 (1973). Type: Zaire, near Eala, *Laurent* 224 (BR, holo.)

Climbing shrub with stems several m. long and attaining 3 cm. in diameter near the base, at first fairly densely covered with very short pubescence but later glabrous. Leaf-blades elliptic, 3·5–11 cm. long, 1–5·5 cm. wide, distinctly acuminate at the apex, cuneate at the base, glabrous save sometimes for some pubescence on the lower part of the main nerve beneath, thin; nodules absent; domatia present; petiole 0·2–1·5 cm. long, pubescent; stipules ovate-triangular, 4–6·5 mm. long, bilobed at the apex, the lobes 1–2·5 mm. long, pubescent or glabrous, deciduous; nodes with long hairs within the stipules. Flowers heterostylous, (4–)5(–6)-merous, in much-branched panicles 2·5–10 cm. long; peduncle 1–4 cm. long, glabrous or pubescent; secondary branches 0·2–1 cm. long, pubescent; pedicels obsolete or ± 1 mm. long, very shortly pubescent; main bracts ± 3 mm. long, lobed at the base, clasping the stem, rest small, ± pubescent. Calyx-tube conic, ± 1 mm. long, glabrescent; limb cupuliform, 0·75–1 mm. long, glabrous or sparsely covered with very short almost papilla-like hairs; lobes very short, broadly triangular, ± 0·5 mm. long; corolla yellowish, greenish or lilac, glabrous to finely densely tomentose outside; tube 3·25–5·5 mm. long; lobes ovate-triangular, 2–2·5 mm. long, 1–1·5 mm. wide. Stamens purple, with filaments 2–2·5 mm. long in short-styled flowers, 0·5–0·75 mm. long in long-styled flowers. Style 1·5–3 mm. long in short-styled flowers and 4–5 mm. long in long-styled flowers; stigma-lobes 0·75–1 mm. long. Drupes red, ellipsoid, 5–7(–10 in living state) mm. long, 7(–8) mm. wide, with 2 pyrenes; pyrenes depressed semi-ellipsoid, 5·5 mm. long, 4·2 mm. wide, 2·5 mm. thick, ventral face plane, dorsal face obscurely ribbed. Seeds dark blackish-red, of similar shape, 4·5 mm. long, 4 mm. wide, 2 mm. thick, ventral face rugose, dorsal face irregularly 7-ribbed; albumen ruminate on both faces but particularly between the dorsal ribs.

UGANDA. Mengo District: Kipayo, Aug. 1914, *Dummer* 998! & Entebbe road, Kajansi Forest, May 1935, *Chandler* 1211! & Mabira Forest, Dec 1922, *Maitland* 490!; Mubende District: Lwentaama, 9 Oct. 1970, *Katende* 645!

TANZANIA. Buha District: Kibondo area, Nyaviyumbu, Nov. 1956, *Procter* 597!; Mbeya District: Bundali Hills, Tewe, 6 Nov. 1966, *Gillett* 17584! (see note)

DIST. **U4**; **T4, 7**; Cameroun, Gabon and Zaire

HAB. Edges and thinner parts of lowland evergreen forest, also riverine forest; 1110–1500(–2050) m.

SYN. *Grumilea ealaensis* (De Wild.) De Wild. in B.J.B.B. 9: 33 (1923) & Pl. Bequaert. 2: 456 (1924)

NOTE. Dummer notes on his field label "rare" which is supported by the fact that only four specimens have been seen from Uganda. Petit's doubts about the identity of the first Tanzanian collection are I believe unfounded; despite the collector's note "shrub", the small specimen has a distinctly 'climbing' look. The one from **T7** is also atypical but appears to be correctly placed here.

25. **P. mahonii** *C. H. Wright* in K. B. 1906: 106 (1906); De Wild., Pl. Bequaert. 2: 386 (1924); Petit in B.J.B.B. 34: 210 (1964). Type: specimen grown at Kew from material collected in Malawi, Likangala stream, *Mahon* 597–1898 (K, holo.!)

Tree 5–15(–24) m. tall or shrub 1·5–5 m. tall; stem soon rugose or striate or the bark sometimes corky and transverse fissured, glabrous or densely shortly velvety pubescent on the young shoots. Leaf-blades elliptic, oblong-elliptic or obovate, 3–23 cm. long, 1·5–10 cm. wide, acuminate at the apex, cuneate to rounded at the base, glabrous or sometimes with a mixture of long multicellular and papilla-like hairs on the venation beneath and sometimes with what look like scattered minute pale scales beneath (actually stomata glistening), surface often rugulose beneath, rather thin to ± coriaceous; nodules absent but sometimes there are lesions and bumps which closely simulate them; domatia usually very marked, hairy; petiole 0·2–3·5 cm. long, glabrous to pubescent; stipules obovate, 0·4–1·7 cm. long, bilobed at the apex, the lobes 1–5 mm. long, glabrous or margins ciliate and basal parts pubescent, deciduous. Flowers sweet-smelling, heterostylous, 5-

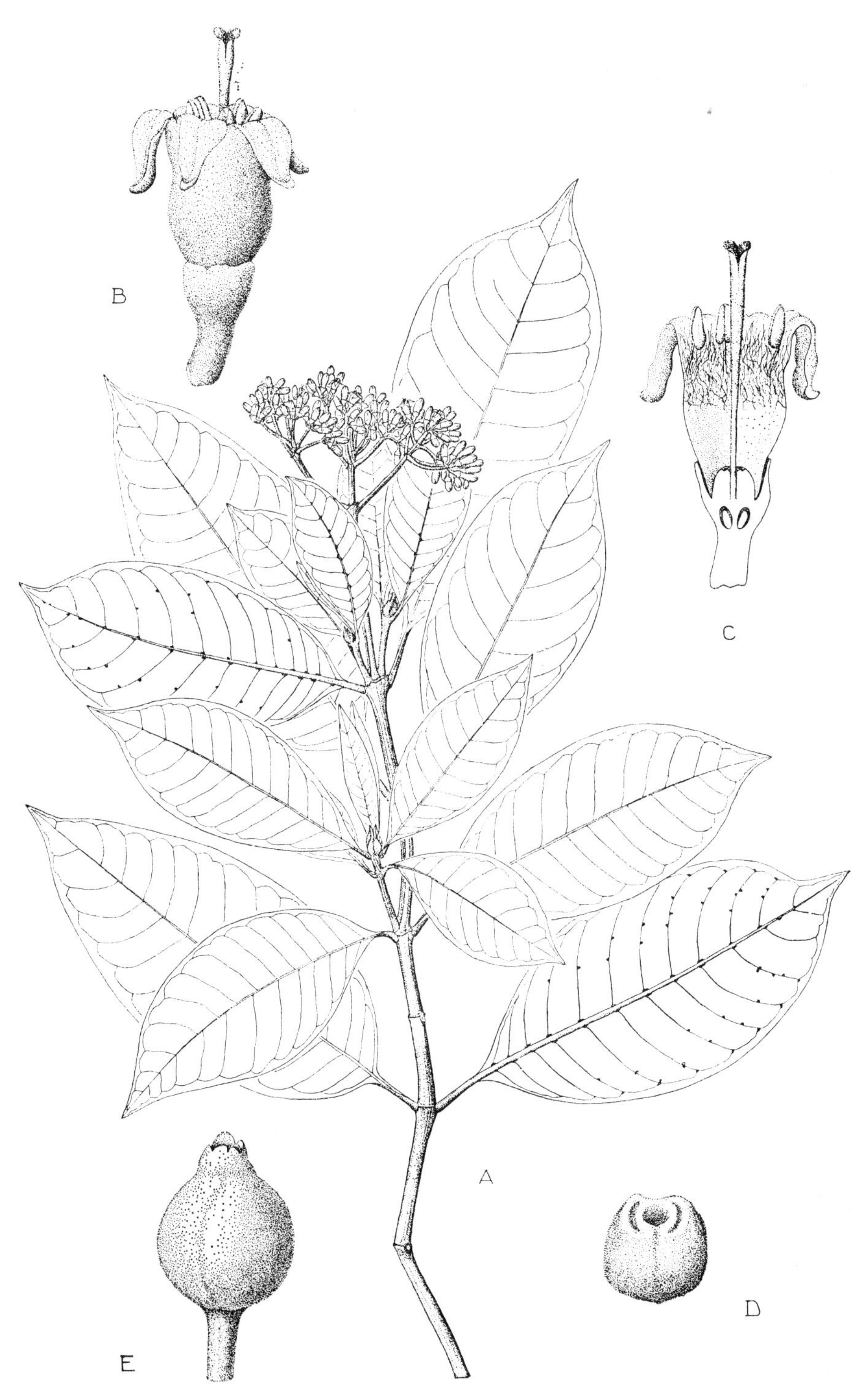

FIG. 3. *PSYCHOTRIA MAHONII* var. *PUBERULA*—**A,** habit, × ½; **B,** long-styled flower, × 4; **C,** longitudinal section of same, × 4; **D,** disc, × 10; **E,** drupe, × 3. All from *de Witte* 1318. Reproduced by permission of the Director of the Jardin Botanique National de Belgique.

merous, in much-branched panicles 4–18 cm. long; peduncle 1·5–8 cm. long, glabrous or densely covered with minute papilla-like hairs; secondary branches 0·4–1·4 cm. long, similarly hairy; pedicels 0–3 mm. long, usually glabrous or similarly velvety-papillose; bracts triangular to linear (very reduced leaves), 3–8 mm. long; other bracts and bracteoles small, usually margined with red hairs. Calyx-tube obconic, 1 mm. long, glabrous to densely papillate; limb cupuliform, 1–1·5 mm. long, glabrous to velvety-papillate; lobes bluntly triangular or rounded, 0·5–0·75 mm. long, glabrous to densely velvety-papillate. Corolla white, greenish or yellow, glabrous to densely velvety-papillose or pubescent outside; tube 4–6 mm. long; lobes elliptic-oblong, 2·75–3·5 mm. long, 1·3–1·5 mm. wide, incrassate and inflexed at the apex, densely grey papillose inside. Stamens with filaments 2–4 mm. long in short-styled flowers, 0·5 mm. long in long-styled flowers. Style 2–3 mm. long in short-styled flowers, 6–7 mm. long in long-styled flowers; stigma-lobes 1–2 mm. long. Drupes red, with 2-pyrenes, subglobose, 5–6 mm. in diameter, scarcely grooved; pyrenes subglobose, 4 mm. long, 4·5 mm. wide, 2 mm. thick, ventral surface plane, dorsal surface slightly grooved. Seeds dark blackish red, subglobose, about the size of the pyrenes, plane ventrally, with narrow deep dorsal fissures but not noticeably grooved; albumen ruminate.

var. **puberula** (*Petit*) *Verdc.* in K.B. 30: 253 (1975). Type: Tanzania, Njombe District, Ruhudji R., Lupembe, *Schlieben* 1168A (BR, holo. !, BM, K, iso. !)

Leaf-blades mostly smaller, 3–16 cm. long, 1.5–9 cm. wide, with 8–13 pairs of lateral nerves. Fig. 3, p. 59.

UGANDA. Toro District: Ruwenzori, Nyinabitaba, Aug. 1933, *Eggeling* 1359! & Bwamba Pass, 22 Sept. 1932, *A.S. Thomas* 692!; Kigezi District: Kachwekano Farm, Jan. 1950, *Purseglove* 3220!

KENYA. Elgeyo Forest, *Gardner* 1398a!; S. Mt. Kenya, *Wimbush* 1513!; Kiambu District: Limuru, near Girls' High School, 26 Mar. 1961, *Polhill* 366!

TANZANIA. Mpanda District: below summit of Kungwe Mt., 7 Sept. 1959, *Harley* 9551!; Iringa District: Ihangana Forest Reserve, near Kibengu, 14 Feb. 1962, *Polhill & Paulo* 1470!; Rungwe Forest Reserve, Feb. 1954, *Semsei* 1616!

DISTR. **U**2, 4; **K**3–5; **T**1, 4, 6, 7; Zaire, Rwanda, Malawi, Zambia and Rhodesia

HAB. Evergreen forest (including very wet, secondary and bamboo), also *Syzygium-Rapanea* swamp forest, derived thickets; 1230–2700 m.

SYN. *Grumilea macrantha* K. Schum. in E.J. 28: 102 (1899); De Wild. in B.J.B.B. 9: 42 (1923). Type: Malawi, Mt. Malosa ('Matosa'), *Whyte* (B, holo. †, ?K, iso. !), *non Psychotria macrantha* Muell. Arg.

G. megistosticta S. Moore in J.L.S. 38: 256 (1908); De Wild. in B.J.B.B. 9: 42 (1923); I.T.U., ed. 2: 346 (1952). Type: Uganda, Ruwenzori, *Wollaston* (BM, holo. !)

G. punicea S. Moore in J.L.S. 40: 101 (1911); De Wild. in B.J.B.B. 9: 45 (1923); Goodier & Phipps in Kirkia 1: 64 (1961). Type: Rhodesia, Chimanimani Mts., *Swynnerton* 563 (BM, holo. !, K, iso. !)

Psychotria ficoidea K. Krause in Z.A.E.: 336 (1911); De Wild., Pl. Bequaert. 2: 365 (1924); F.P.N.A. 2: 365 (1947). Types: Zaire, Ruwenzori, Butagu valley, *Mildbraed* 2493 (B, syn. †) & Rwanda, Sabinyo Mt. Forest, *Mildbraed* 1729 (B, syn. †, BR, fragment) & Rugege Forest, *Mildbraed* 1019 (B, syn. †, BR, fragment)

Grumilea bequaertii De Wild. in B.J.B.B. 9: 28 (1923) & Pl. Bequaert. 2: 453 (1924); F.P.N.A. 2: 366 (1947); K.T.S.: 443 (1961). Types: Zaire, Kivu, Butagu, *Bequaert* 3618 & Lamia, *Bequaert* 4308 & Ruwenzori, *Bequaert*, (BR, syn.)

G. sp. sensu I.T.U., ed. 2: 346 (1952)

Psychotria megistosticta (S. Moore) Petit in B.J.B.B. 34: 112, t. 3, photo. 3/F (1964) & in Distr. Pl. Afr. 5, map 129 (1973)

P. megistosticta (S. Moore) Petit var. *puberula* Petit in B.J.B.B. 34: 116 (1964)

P. megistosticta (S. Moore) Petit var. *punicea* (S. Moore) Petit in B.J.B.B. 34: 117 (1964)

NOTE. Petit has divided this species into three varieties, var. *megistosticta* with glabrous flowers and var. *puberula* and var. *punicea* both with tomentose or pubescent flowers. In Uganda and Kenya specimens with glabrous or ± glabrous flowers occur, whereas nearly all specimens from Tanzania, particularly in the south, have densely tomentose or pubescent flowers. Specimens cited as var. *megistosticta*, e.g. *Eggeling* 1359, have flowers just as tomentose as other Uganda specimens referred to var. *puberula*. In the Mbeya area specimens from the same place, Mbozi, which are clearly quite identical have been annotated differently, e.g. *Jessel* 68, annotated var. *punicea*, and *St. Clair-Thompson* 1071, annotated as var. *puberula*. The variation is complicated and only partly correlated geographically. I am therefore not formally recognising these varieties. Those who wish could refer any southern plants with densely pubescent flowers to var. *punicea*, glabrous flowered or with only the corolla tube slightly papillate-pubescent to a var. *megistosticta* and papillate-pubescent flowered plants to var. *puberula*. *Lucas & Polhill* 77 (Kenya, Kiambu District, Gatamayu Forest, 30 Mar. 1961) bears the note "divided and undivided stipules present".

var. **pubescens** (*Robyns*) *Verdc.* in K.B. 30: 254 (1975). Type: Zaire, Kivu, between Lulenga and Sake, *Lebrun* 5019 (BR, syn.) & Lake Mokoto, *Ghesquière* 4999 (BR, syn., S, isosyn.)

Leaf-blades mostly larger, 7–22 cm. long, 2·5–10 cm. wide, with 7–22 pairs of lateral nerves, glabrous to distinctly hairy beneath (densely ferruginous hairy or even velvety when very young); domatia mostly small or absent. Inflorescences larger, 8–18 cm. long with peduncles 2·5–8 cm. long, all mostly pubescent, sometimes with quite dense ferruginous hairs.

UGANDA. Ankole District: Kyamahungu, June 1938, *Eggeling* 3726 !; Masaka District: SW. of Lake Nabugabo, Bugabo, 1 Feb. 1969, *Lye et al.* 1903 ! & Sese Is., Bugala I., Sozi Point, Nov. 1931, *Eggeling* 114 !

KENYA. Trans-Nzoia/Elgeyo Districts: Cherangani, Kapolet [Kabolet] R., Nevile's Farm, May 1964, *Tweedie* 2801 !; S. Kavirondo District: Kisii, Sept. 1933, Napier 3004 in *C.M.* 5281 !; Kericho District: 8·5 km. due S. of Kericho, 22 Nov. 1967, *Perdue & Kibuwa* 9171 !

TANZANIA. Bukoba District: near Kitwe, Oct. 1931, *Haarer* 2213 ! & Nyakato, Apr. 1935, *Gillman* 241 ! & Minziro Forest Reserve, July 1950, *Watkins* 473 !

DISTR. **U**2, 4; **K**3, 5; **T**1, ?6, ?8; Zaire

HAB. Mostly fringing forest, woodland and thicket by lakes and rivers, also secondary scrub; (1030–)1140–2150 m.

SYN. *Grumilea bequaertii* De Wild. var. *pubescens* Robyns in B.J.B.B. 17: 96 (1943); F.P.N.A. 2: 367 (1947)

Psychotria robynsiana Petit in B.J.B.B. 34: 120 (1964) & in Distr. Pl. Afr. 5, map 132 (1973). Type: Zaire, between Lubenge and Sake, *Lebrun* 5019 (BR, holo.)

P. robynsiana Petit var. *pauciorinervata* Petit in B.J.B.B. 34: 121 (1964). Type: Kenya, Kericho, SW. Mau Forest Reserve, *Maas Geesteranus* 5749* (BR, holo., COI, K !, S, WAG, iso.)

P. robynsiana Petit var. *glabra* Petit in B.J.B.B. 34: 122 (1964). Type: Uganda, Sese Is., Bufumira, *A. S. Thomas* 845 (K, holo. !)

NOTE. I am unable to accept this as a separate species. Petit himself mentions that it was very difficult to obtain a satisfying arrangement for *P. megistosticta* and its allies, but his division into quite a number of species is not in accordance with the facts. The amount of indumentum varies greatly in one place, e.g. Sese Is. In the text *P. robynsiana* is separated from *P. megistosticta* by having larger leaf-blades with more lateral nerves and larger inflorescences, but he describes two varieties *glabra* and *pauciorinervata* which totally blur this distinction. In the main key the two are separated on the presence or absence of distinct domatia but two of his varieties of *P. megistosticta*, *puberula* and *punicea*, usually also lack them. *P. mushiticola* Petit (Type: Zaire, Parc National de l'Upemba, R. Kamitano, near Lusinga, *Van Meel* in *de Witte* 5122), separated from *P. robynsiana* by having fewer lateral nerves and corolla puberulous outside, clearly is not distinct but may perhaps be considered another minor variant; much of the cited material of *P. robynsiana* has quite distinctly tomentose flowers. If it is not separated, Zambia must be added to the distribution. *P. mahonii* he dismissed as a dubious species, but the material is quite adequate so it must be accepted as the earliest name for this taxon as a whole; typical var. *mahonii*, from Malawi, has dense ferruginous hairs on the inflorescence branches.

26. **P. lauracea** (*K. Schum.*) *Petit* in B.J.B.B. 34: 129, photo. 3/L (1964) & in B.J.B.B. 42: 357 (1972) & in Distr. Pl. Afr. 5, map 138 (1973). Type: Tanzania, Kilimanjaro, Marangu, *Volkens* 1393 (BR, lecto., BM, K, isolecto. !)

* Petit gives 5479 in error.

Shrub or rarely a small tree 1–6(–7·5) m. tall, with glabrous stems. Leaf-blades elliptic, oblong-elliptic or ovate-lanceolate, 5–19(–23·5) cm. long, 3–10·5(–14·2) cm. wide, acute or obscurely ± obtusely acuminate or rounded at the apex, cuneate to rounded at the base, completely glabrous, mostly drying grey-green or reddish brown (" russet "), ± shining in life and said to be green above and yellow beneath, rather thin to distinctly coriaceous, with a distinct thin almost hyaline ± revolute margin, the lamina often covered with minute scattered nodule-like black dots beneath but these may be very obscure or absent; domatia inconspicuous or absent; petiole 1–6 cm. long, glabrous; stipules ± ovate or oblong, 1–2 cm. long, bilobed at the apex, the lobes 2–8 mm. long, glabrous, deciduous. Flowers sweetly scented, heterostylous, 5-merous, in much-branched often somewhat pyramidal panicles 7–18 cm. long or rarely 3 inflorescences from a node with ± undeveloped leaves resembling bracts; peduncles 4–12 cm. long, glabrous, usually quite noticeably 2-winged and flattened even in life; secondary branches (0·3–)0·6–3·3(–6·5) cm. long; pedicels 2–3·5 mm. long; bracts and bracteoles minute. Calyx glabrous; tube obconic, 0·8 mm. long; limb cupuliform, 0·3–0·7 mm. long, subtruncate, or lobes very small, rounded, 0·25 mm. long. Corolla white, cream, pinkish white, greenish yellow, brownish or pale green, glabrous; tube funnel-shaped, 2–3 mm. long; lobes ± ovate, 2–2·5 mm. long, 0·8–1·5 mm. wide. Stamen-filaments ± 1 mm. long in short-styled flowers but almost obsolete in long-styled flowers. Style 1·3–2 mm. long in short-styled flowers, 3–3·5 mm. long in long-styled flowers. Drupes red, yellow or scarlet, with 2 pyrenes, subglobose, 4–5 mm. tall, 5–6 mm. wide, glabrous, densely speckled with rhaphides, very distinctly slightly ribbed in dry state; pyrenes grey, rhaphide-packed, semiglobose, 4·5 mm. long, 4 mm. wide, 2–3 mm. thick, dorsally slightly 6-ribbed (including margins), ventrally flat. Seeds dark red, semiglobose, 4 mm. long and wide, 2·2 mm. thick; ventral surface rugulose, dorsal surface obscurely ribbed or smooth; albumen slightly to deeply ruminate on ventral face but scarcely so on dorsal face, ventrally with a ± Y-shaped groove.

UGANDA. Kigezi District: Maramagambo Forest, Feb. 1950, *Purseglove* 3301!; Mbale District: 3·2 km. SW. of Solo [Nsolo] R., W. Bugwe Forest Reserve, 10 Apr. 1951, *G. H. S. Wood* 36!; Mengo District: Busuju, Kasa Forest, 17 Nov. 1949, *Dawkins* 447!

KENYA. Meru District: Nyambeni Hills, circular road where it is cut by R. Thangatha, 10 Oct. 1960, *Polhill & Verdcourt* 276!; S. Kavirondo District: Kisii area, Bukuria [Bakeria], Sept. 1933, *Napier* 2938 in *C.M.* 5280!; Kwale District: Shimba Hills, Pengo Forest, 9 Feb. 1953, *Drummond & Hemsley* 1189!

TANZANIA. Moshi District: 11 km. from Moshi on road to Arusha, Kikafu R., 15 Dec. 1961, *Polhill & Paulo* 986!; Pangani District: near Pangani, 13 Nov. 1947, *Brenan & Greenway* 8319!; Morogoro District: Matombo, Oct. 1930, *Haarer* 1869!; Zanzibar I., Marahubi, 13 Sept. 1930, *Vaughan* 1516! & Kombeni Cave-Well, 16 Feb. 1930, *Vaughan* 1218!

DISTR. **U**2–4; **K**4–7; **T**1–3, 6, 8; **Z**; Zaire, Rwanda, Zambia

HAB. Various types of evergreen forest and derived thicket including coastal, seasonal swamp, riverine and *Podocarpus*; 0–1800 m.

SYN. *Grumilea lauracea* K. Schum. in P.O.A. C: 392 (1895); De Wild. in B.J.B.B. 9: 40 (1923); T.T.C.L.: 497 (1949); K.T.S.: 444 (1961)

?*Psychotria fuscula* K. Schum. in E.J. 34: 338 (1904); De Wild., Pl. Bequaert. 2: 368 (1924); T.T.C.L.: 524 (1949). Type: Tanzania, E. Usambara Mts., near Longuza, *Engler* 368 (B, holo. †)

P. erythrocarpa K. Krause in E.J. 43: 152 (1909); De Wild., Pl. Bequaert. 2: 364 (1924); T.T.C.L.: 524 (1949). Type Tanzania, Rufiji District, Matumbi Mts., Tamburu R., *Busse* 3125 (B, holo. †, BM!, BR, HBG, P, iso.)

P. wauensis K. Krause in Z.A.E.: 335 (1911); De Wild., Pl. Bequaert. 2: 439 (1924). Type: Zaire, Lake Kivu, Wau I., *Mildbraed* 1164 (B, holo. †, BR, iso.)

P. bequaertii De Wild., Pl. Bequaert. 2: 335 (1924). Types: Zaire, Rutshuru, *Bequaert* 5617 & 6213 (BR, syn.)

NOTE. This is a well-characterised species, easily distinguished by the compressed ± winged peduncle, the complete lack of indumentum save inside the corolla-throat, the shape of the inflorescence and usual presence of numerous scattered black nodules. As Petit points out these can be present or absent even in leaves adjacent to each other. It is the only species of the genus with ruminate albumen formerly included in *Grumilea* which also has nodules. Lersten (*in litt.*) has pointed out that this species differs from all others in sect. *Paniculatae* in having dendroid colleters. For this reason it has been placed by itself.

One specimen *Procter* 885 (Tanzania, Bukoba District, Minziro Forest, Apr. 1958) is clearly *P. lauracea* as annotated by Petit but *all* the flowers are much larger or appear so, the buds being 6 mm. long. The "pathological look" was confirmed by dissection, an insect being found in the corolla-tube. This "weak" kind of galling has apparently not prevented fertilisation since young fruits are developing. It is quite well-known that in cases of flower-galling every flower in an inflorescence may be affected.

27. **P. vogeliana** *Benth.* in Hook., Niger Fl.: 420 (1849); Hiern in F.T.A. 3: 210 (1877); De Wild. in Ann. Mus. Congo, Bot., sér. 5, 3: 302 (1910) & in B.J.B.B. 7: 291 (1921) & Pl. Bequaert. 2: 436 (1924); Schnell in Mém. Inst. Fr. Afr. Noire 50: 61, photo. I (1957); Hepper in F.W.T.A., ed. 2, 2: 198 (1963); Petit in B.J.B.B. 34: 135, photo. 4/C–D (1964) & in Distr. Pl. Afr. 5, map 140 (1973). Type: Nigeria, Niger [Quorra] R., Aboh, *Vogel* 43 (K, lecto.!)

Subshrub or shrub 1–6 m. tall; young stems densely covered with bright ferruginous hairs, the older stems glabrous. Leaf-blades elliptic or ± obovate, 6–25 cm. long, 2–12 cm. wide, acuminate at the apex, cuneate to narrowly rounded at the base, glabrous and ± shining above, ferruginous pubescent on the midrib beneath and lateral nerves very finely puberulous; domatia and nodules absent; petiole 0·2–3 cm. long, ferruginous pubescent; stipules 0·7–2·5 cm. long, the base oblong, hairy at the middle, with 2 triangular lobes ± 0·2–1·2 cm. long, margins lacerate-ciliate, soon deciduous. Flowers heterostylous, 5-merous, in panicles 3–4(–16) cm. long, the individual (1–)5–9 components capitate, 0·4–1·2 cm. in diameter; peduncles 3–10 cm. long; secondary peduncles 0·6–2·2 cm. long, all ferruginous pubescent; pedicels obsolete; main bracts hastate-lacerate, 4·5–5 mm. long, 4 mm. wide; capitula supported by a whorl of irregular free secondary bracts, 3–5 mm. long, ciliate. Calyx-tube obconic, 1–1·5 mm. long, papillate; tubular part of the limb 1–1·5 mm. long; lobes irregular, low and broad to triangular, 0·3–0·7 mm. long, ciliate. Buds finely papillate-puberulous, 5-lobed at the apex. Corolla white, cream or greenish; tube cylindrical, 3·5–6 mm. long, shortly funnel-shaped above, glabrous below, papillate above; lobes oblong-elliptic, 1·7–2·5 mm. long, 1 mm. wide, thickened and appendaged at the apex, inflexed, papillate outside. Anther tips just exserted in long-styled flowers; filaments 1·5–3·5 mm. long in short-styled flowers. Style exserted 1·5 mm. in long-styled flowers, 5–6·5 mm. long, 2·5–3·5 mm. long in short-styled flowers; stigma-lobes 1–2 mm. long, slender. Drupes white, ellipsoid, 7–8(–12 in life) mm. long, 5–6(–9) mm. wide, glabrous, containing 2 pyrenes; pyrenes 4–5-sulcate. Seeds 5–6·5 mm. long, 3·5–4·5 mm. wide, 1·3–3 mm. thick, ventral face rugulose, dorsal face 5–6-ribbed; albumen not or slightly ruminate.

UGANDA. Kigezi District: S. Maramagambo Forest, halfway along Biteriko road, 11 Oct. 1969, *Synnott* 400! & Biteriko tract, 18 Sept. 1969, *Faden, Lock & Lye* 69/297! & 4·8 km. from E. boundary of forest on Biteriko tract, 8 Nov. 1969, *Synnott* 423!

DISTR. U2; W. Africa from Guinée to Gabon, Central African Republic and Zaire

HAB. Mixed evergreen forest; 990 m.

NOTE. Schnell has described several varieties from W. Africa.

28. **P. schweinfurthii** *Hiern* in F.T.A. 3: 210 (1877); K. Krause in Z.A.E.: 334 (1911); De Wild., Pl. Bequaert. 2: 420 (1924); Petit in B.J.B.B. 34: 146, t. 4 & photo. 4/G (1964) & in B.J.B.B. 42: 358 (1972) & in Distr. Pl. Afr. 5, map 145 (1973). Type: Sudan, Niamniam, R. Fluuh, *Schweinfurth* 3736 (K, holo.!)

Shrub or subshrub 0·4–2 m. tall, the young stems ± bifariously pubescent with ferruginous hairs but at length glabrous. Leaf-blades elliptic or oblong-elliptic to elliptic-lanceolate, 4–20 cm. long, 1·5–7 cm. wide, narrowly acuminate at the apex, cuneate at the base, glabrous above and beneath save for pubescence on the nerves beneath or rarely pubescent all over and sometimes papillate on venation, thin, often drying dark green or grey-green above and reddish brown beneath, in life said to be dark and glossy above and almost silvery beneath; nodules absent and domatia inconspicuous; petiole 0·1–1·2 cm. long, glabrous or pubescent; stipules narrowly to broadly obovate, 1–1·7 cm. long, bilobed, the lobes usually narrowly lanceolate, 5–8 mm. long, the margins fringed with ferruginous hairs and base of stipule usually hairy outside, deciduous. Flowers heterostylous, 5-merous, in much-branched ± pyramidal panicles 3–13 cm. long; peduncle 2–8 cm. long, pubescent, sometimes only bifariously; secondary branches 0·3–2 cm. long, similarly pubescent or glabrous; pedicels 0–2·5 cm. long, glabrous; lowermost bracts elliptic to oblong-lanceolate or linear-elliptic, up to 2 cm. long, 5 mm. wide, acuminate and entire or 3-toothed with central tooth the longest, glabrous save for marginal hairs; other bracts smaller. Calyx-tube ± campanulate, 0·7 mm. long, glabrous; limb cupuliform, 0·5–0·7 mm. long, glabrous, teeth irregularly triangular, 0·25 mm. long. Buds distinctly 5-horned at the apex, the horns 0·3–1 mm. long. Corolla usually glabrous outside, white; tube 3·5–6 mm. long; lobes ovate-oblong, 2–2·5 mm. long, 0·9 mm. wide, conspicuously horned. Stamens with filaments 1·7–3·5 mm. long in short-styled flowers, 0·2 mm. long in long-styled flowers. Style 3–4·5 mm. long in short-styled flowers, 5·5–7·5 mm. long in long-styled flowers; stigma-lobes 0·75–1·5 mm. long. Drupes red or said to be blue turning red and drying black, with 2 pyrenes, subglobose or ± ellipsoid, 4–5 mm. in diameter, glabrous, strongly grooved in the dry state; pyrenes rhaphide-packed, semiglobose or semi-ellipsoid, 3·8–4·5 mm. long, 3·5–3·8 mm. wide, 2 mm. thick, the ventral face slightly concave with 2 shallow grooves, the dorsal face sharply 6-ribbed. Seeds red, semiglobose, 3 mm. long and wide, 1·8 mm. thick, the ventral face slightly concave with 2 shallow grooves, thickened at the middle, dorsal face obtusely 4-ribbed; albumen not ruminate.

UGANDA. W. Nile District: Madi, Metuli, Apr. 1940, *Eggeling* 3868!; Bunyoro District: Budongo Forest, May 1935, *Eggeling* 2016!; Mbale District: W. Bugwe Forest Reserve, peninsula between Kami [Khami] & Malaba [Malawa] rivers, 16 Mar. 1951, *G. H. S. Wood* 39!

DISTR. U1–3; Ivory Coast, Ghana, Togo, Nigeria, Cameroun, Congo (Brazzaville) & Zaire, Central African Republic, Sudan and Angola

HAB. Secondary and primary evergreen forest, *Celtis*, *Holoptelea* forest, near seasonal swamps and riverine grassland; 885–1260 m.

SYN. *P. obscura* Benth. in Hook., Niger Fl.: 419 (1849); Hiern in F.T.A. 3: 212 (1877); De Wild., Pl. Bequaert. 2: 397 (1924); Hepper in F.W.T.A., ed. 2, 2: 199 (1963), *non* Zoll. et Mor. (1846), *nom. illegit.* Type: Ghana, Accra, *Vogel* (K, holo.!)

P. soyauxii Hiern in F.T.A. 3: 213 (1877); De Wild., Pl. Bequaert. 2: 423 (1924). Type: Congo (Brazzaville), Loango, *Soyaux* 165 (K, holo.!)

[*P. reptans* sensu Hiern in F.T.A. 3: 211 (1877), pro parte; Good in J.B. 64, Suppl. 2: 31 (1926); F.W.T.A. 2: 124 (1931), pro parte, *non* Benth.]

Uragoga accraensis Kuntze, Rev. Gen. Pl. 2: 954 (1891). Type: as for *P. obscura*

[*Myristiphyllum reptans* sensu Hiern, Cat. Afr. Pl. Welw. 1: 493 (1898), *non* (Benth.) Hiern]

Psychotria sodifera De Wild., Pl. Bequaert. 2: 421 (1924); F.W.T.A. 2: 123 (1931); Schnell in Mém. I.F.A.N. 50: 74, fig. 4 (1957). Types: Central African Republic, Chari, Ndés, between R. Mpokou and Ungourras Plateau, *Chevalier* 6084 & Ungourras Plateau, *Chevalier* 6101 bis & Krebedje, Fort Sibut, *Chevalier* 5732 & Diouma, *Chevalier* 5920 (all P, syn.)
P. farmari Hutch. & Dalz., F.W.T.A. 2: 120, 123 (1931). Type: Ghana, without locality, *Farmar* 385 (K, holo. !)

NOTE. This is closely allied to several other W. African species but differs in indumentum, stipule shape etc.; since Bentham, Hepper and Petit have all kept them separate I have refrained from making any alterations in rank.

29. **P. megistantha** *Petit* in B.J.B.B. 34: 192 (1964) & in Distr. Pl. Afr. 6, map 169 (1973). Type: Tanzania, NW. Uluguru Mts., *Schlieben* 3414 (BR, holo., B, K !, iso.)

Shrub or tree 2·4–10 m. tall, with glabrous stems. Leaf-blades elliptic or oblong-elliptic, 7–19 cm. long, 3–13 cm. wide, very shortly acuminate at the apex, cuneate to ± rounded at the base, glabrous above and beneath save for hairs around the domatia, papery to thinly coriaceous; nodules absent; petiole 1–4(–5·5) cm. long, glabrous; stipules ovate-triangular, 0·6–1·7 cm. long, the apex entire or slightly bifid, mostly ciliate at the apex but otherwise glabrous, deciduous. Flowers heterostylous, 5-merous, in usually branched inflorescences 3–16 cm. long, the individual components being ± condensed heads 1·5–3 cm. in diameter, becoming laxer in fruit; peduncles 1·5–13(–15) cm. long, glabrous or glabrescent; secondary branches 0·3–1·3 (–2) cm. long, often with bifarious ferruginous hairs; pedicels 0–1·5 mm. long, lengthening to 4 mm. in fruit; main bracts ovate, ± 4 mm. long, aristate, others small. Calyx-tube obconic, 1 mm. long, glabrous to densely ferruginous hairy; limb ± funnel-shaped, 2–2·5 mm. long, glabrous or pubescent; lobes rounded-triangular, 1·5–2·5 mm. long, ± ciliate. Corolla yellow, glabrous outside; tube 0·8–1·2 cm. long, glabrous inside; lobes ovate-elliptic, 3–4 mm. long, 2·2–2·5 mm. wide, thickened at the apex. Stamens with filaments 1·5 mm. long in short-styled flowers, 0·2–0·5 mm. in long-styled flowers. Styles in short-styled flowers 5·5 mm. long, in long-styled flowers 0·75–1·2 cm. long; stigma-lobes 1·5–2 mm. long. Drupes steel-blue, subglobose, 7–8(–12 in spirit) mm. in diameter, with 2 pyrenes, glabrous, not grooved, crowned with the conspicuous calyx; pyrenes pale brown, depressed semiglobose, in spirit material 7·5 mm. long and wide, 4 mm. thick, smooth. Seeds similar in size with ventral face plane and dorsal face not ribbed; albumen conspicuously ruminate on both faces.

TANZANIA. Morogoro District: Uluguru Mts., Lukwangule Plateau, above Chenzema Mission, 13 Mar. 1953, *Drummond & Hemsley* 1537 ! & Bunduki, Kimbinyuko, Mar. 1953, *Semsei* 1105 ! & Lukwangule Plateau, 5 Feb. 1935, *E. M. Bruce* 784 !
DISTR. **T6**; not known elsewhere
HAB. Upland rain-forest; 1500–2500 m.

30. **P. peteri** *Petit* in B.J.B.B. 34: 193, photo. 6/G (1964) & in Distr. Pl. Afr. 6, map 170 (1973). Type: Tanzania, E. Usambara Mts., near Amani, *Peter* 17764 (B, holo.)

Shrub or small tree 1–7·5 m. tall, with glabrous branches. Leaf-blades elliptic, narrowly elliptic or broadly oblanceolate, 9–20 cm. long, 2·5–6·5 cm. wide, shortly acuminate at the apex, narrowly cuneate at the base, glabrous on both surfaces, but the stomata appearing as a dense close speckling beneath, thinly coriaceous, without nodules or sometimes (see note) covered with small nodules beneath; domatia absent; petiole 0·3–2 cm. long, glabrous; stipules ovate-triangular, 0·8–1·4 cm. long, obtuse at the apex, glabrous or pubescent at the base, soon deciduous. Flowers heterostylous,

5-merous, in capituliform 3–many-flowered inflorescences; peduncle 0–4 cm. long; secondary branches very short or up to 1·5 cm. long; pedicels obsolete; bracts ovate, elliptic or linear, 2·5–6 mm. long, up to 4 mm. wide but not forming an involucre, margined with hairs. Calyx-tube ovoid, 0·7 mm. long; limb funnel-shaped, tube glabrous, 2–2·5 mm. long; lobes triangular-ovate, 1–1·5 mm. long, hairy on the margins. Corolla green or cream; tube 4–5·5 mm. long, glabrous; lobes oblong-elliptic, 2·5–3 mm. long, 1·2 mm. wide, thickened and hooded at the apex, sparsely puberulous outside at the apex. Stamens with filaments 1·2 mm. long in short-styled flowers, 0·7 mm. long in long-styled flowers. Style 1·7 mm. long in short-styled flowers, 5 mm. long in long-styled flowers; stigma-lobes 0·5–2 mm. long. Drupes red, with 2 pyrenes, ellipsoid, 7–9 mm. long, 5 mm. wide, glabrous; pyrenes semi-ellipsoid, 7–7·5 mm. long, 3·8 mm. wide, 1·8 mm. thick, the dorsal face strongly 6-ribbed (including margins), ventral face flat. Seeds purple-red, 6·2 mm. long, 3·2 mm. wide, 1·5 mm. thick, ventral face plane, dorsal face strongly 6-ribbed; albumen ruminate on ventral surface, mostly or only in 2 ± median fissures (forming a U-shaped groove) and on the dorsal grooves.

TANZANIA. Lushoto District: E. Usambara Mts., 0.8 km. W. of Amani, 26 July 1953, *Drummond & Hemsley* 3468! & Kwamkoro Forest Reserve, 17 Jan. 1961, *Semsei* 3162!; Tanga District: Mlinga Peak, 18 Feb. 1937, *Greenway* 4901!

DISTR. T3,? 6; not known elsewhere

HAB. Rain-forest; 900–1000(–? 1900) m.

SYN. [*Grumilea blepharostipula* sensu T.T.C.L.: 498 (1949), pro parte, *non* K. Schum.]

NOTE. A specimen from N. Uluguru Mts., 31 Aug. 1913, *Brundenberg* 83 (?8) seems to be this species but is too poor for certainty; the T6 and 1900 m. cited above are solely derived from this specimen.

31. **P. butayei** *De Wild.*, Ann. Mus. Congo, Bot., sér. 5, 2: 180, t. 47 (1907) & Pl. Bequaert. 2: 346 (1924); Petit in B.J.B.B. 34: 195, photo. 6/D (1964) & in Distr. Pl. Afr. 6, map 171 (1973). Type: Zaire, Bas-Congo, *Butaye* in *Gillet* (BR, holo.)

Shrub or subshrub with simple or branched stems 0·15–1·2(–2) m. tall from a woody rhizome or with many short unbranched stems from a woody stock; stems dull purplish, glabrous or pubescent, longitudinally rugose or ribbed. Leaves often missing from the lower nodes; blades ovate to obovate or narrowly elliptic, 3·5–10 cm. long, 0·7–4·5 cm. wide, obtuse to acute at the apex, cuneate at the base, completely glabrous or sometimes pubescent on both sides, thin to thinly coriaceous, sometimes drying blackish- to yellowish-green; domatia absent; nodules absent or sometimes a few long tubercular ones along the main nerve; petiole 0–1 cm. long, glabrous; stipules ovate-triangular, 4–6 mm. long, acuminate or narrowly bifid at the apex, glabrous or pubescent outside but with ferruginous hairs within, deciduous. Flowers heterostylous, 5-merous, in sparsely branched often condensed panicles 2–6 cm. long; peduncle 1–5 cm. long, glabrous or pubescent; secondary branches 0·5–2 cm. long; pedicels 1–2 mm. long, glabrous; bracts and bracteoles minute. Calyx-tube obconical, 0·75 mm. high, glabrous; limb cupuliform, 0·75–1 mm. tall, glabrous; lobes ± triangular, 0·5–1 mm. long, glabrous or margined with hairs. Corolla white, glabrous or densely minutely papillate outside; tube 4–6 mm. long; lobes triangular, 1·75–3 mm. long, 1·5 mm. wide, thickened at the apex. Stamens with filaments 3–3·5 mm. long in short-styled flowers, obsolete in long-styled flowers. Style 3 mm. long in short-styled flowers, 4·5–6 mm. long in long-styled flowers; stigma-lobes ± 1 mm. long. Drupes red, transversely ellipsoid, divided into 2 subglobose parts by a median groove, with 2 pyrenes, 5·5–7 mm. tall, 7·5–10 mm. wide, or sometimes subglobose,

± 6 mm. in diameter and containing only 1 pyrene; pyrenes subglobose, 5–7 mm. long, 5–7 mm. wide, 4–5 mm. thick, both faces entire. Seed subglobose, ventral face deeply ruminate, usually with a deep median fissure which is branched within, dorsal face scarcely or not ruminate.

var. **butayei**; Petit in B.J.B.B. 34: 195 (1964)

Branched shrub or subshrub with glabrous stems and leaves, etc.

TANZANIA. Buha District: Gombe Stream Reserve, Kasakela valley, 4 Mar. 1964, *Pirozynski* 502 !; Ufipa District: near Zambia border, 4 Dec. 1960, *Richards* 13645 !
DISTR. **T4**; Zaire and Zambia
HAB. Rock outcrops and open woodland; 1520–1700 m.

var. **glabra** (*Good*) *Petit* in B.J.B.B. 34: 197 (1964). Type: Angola, Cassuango, R. Cuiriri, *Gossweiler* 3700 (BM, holo. !, K !, LISJ, iso.)

Branched shrub or subshrub with puberulous or pubescent stems and leaves, etc.

TANZANIA. Buha District: Kasulu to Kibondo, 112 km. from Kasulu, 15 Nov. 1962, *Verdcourt* 3325 ! & Kasulu to Kivumba, 24 Feb. 1926, *Peter* 37517 !; Mpanda District: 37 km. from Mpanda, Uruwira, 22 Sept. 1970, *Richards & Arasululu* 26104 !
DISTR. **T4**; Zaire, Zambia, Angola
HAB. *Brachystegia* woodland; 1350–1500 m.

SYN. *Grumilea flaviflora* Hiern var. *glabra* Good in J.B. 64, suppl. 2: 31 (1926)

NOTE. Owing to the rules of nomenclature this variety has to be called var. *glabra* despite the fact that it is pubescent and the typical variety is glabrous. The specimen cited above is very much more pubescent than the type of the variety.

var. **simplex** *Petit* in B.J.B.B. 34: 196 (1964). Type: Zaire, between Kianza and Kahemba, *Robyns* 3677 (BR, holo., K, iso. !)

Pyrophyte with erect simple stems 15–40 cm. tall from a woody rootstock; glabrous.

TANZANIA. Buha District: Gombe Stream Reserve, Kakombe valley, 10 Jan. 1964, *Pirozynski* 233 !
DISTR. **T4**; Zaire
HAB. Woodland bordering fire-swept grassland; 1370 m.

NOTE. Although the specimen cited above has the habit of var. *simplex* it differs from the typical material in having more abruptly cuneate leaf-blades; in the type they are very gradually long-cuneate at the base.

32. **P. miombicola** *Verdc.* in K.B. 30: 255, fig. 2 (1975). Type: Tanzania, Kigoma District, Uvinza–Mpanda road, *Verdcourt* 3433 (K, holo. !, EA, iso. !)

Small shrub 30–60 cm. tall; stems slender, blackish grey and rather strongly longitudinally ridged, shortly pubescent when young with yellowish white hairs on the new shoots and ferruginous hairs on last season's wood and with some persistent hairs even at base. Leaf-blades very narrowly elliptic or lanceolate, 1–3·2 cm. long, 3–9 mm. wide, narrowly acute at the apex, cuneate at the base, glabrous save for a few white hairs on lower parts of midnerve and main lateral nerves beneath, thinly coriaceous, margined, the margin distinctly crinkling in the dry state; domatia and nodules not evident; petiole obsolete or ± 1 mm. long; stipules pale chestnut brown, ovate-triangular, 2 mm. long, pubescent, bifid or rarely entire at the apex, the lobes linear, 1 mm. long, soon deciduous. Flowers heterostylous, 5-merous, in small subumbellate heads ± 1 cm. in diameter or in 3-branched inflorescences, each component a small head; peduncles 0·7–1·5 cm. long, pubescent; secondary branches, if present, 5 mm. long, pubescent; pedicels 1–1·5 mm. long, puberulous; bracts and bracteoles minute. Calyx glabrous; tube depressed subglobose, 0·5 mm. tall, 1 mm. wide, perhaps ribbed in life; limb cupuliform, 0·5 mm. long; lobes ± triangular, 0·2–0·5 mm. long. Corolla greenish white; tube 5 mm. long, distinctly tomentose outside with

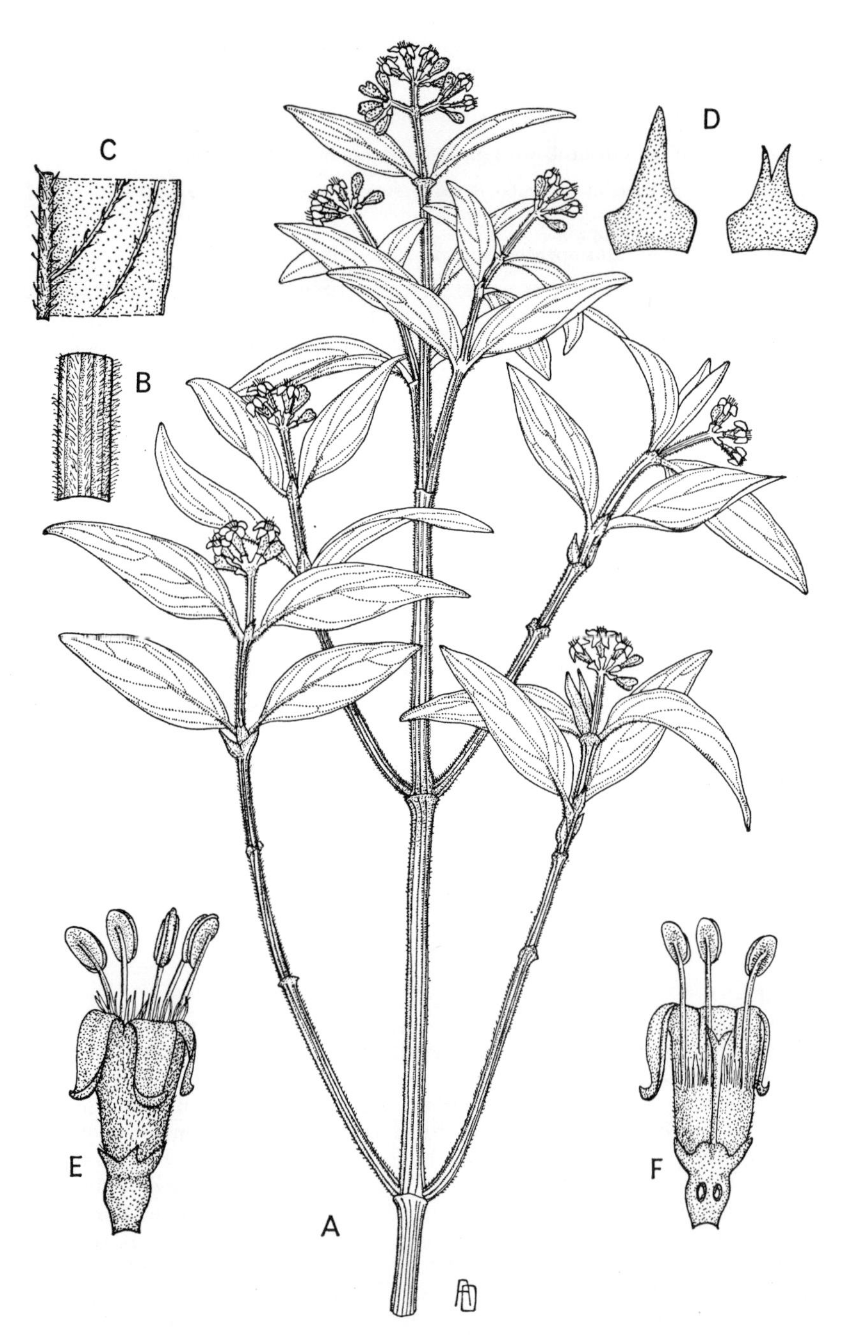

FIG. 4. *PSYCHOTRIA MIOMBICOLA*—**A,** flowering branch, × 1; **B,** detail of branchlet, × 4; **C,** detail of undersurface of leaf, × 4; **D,** stipules, × 4; **E,** short-styled flower, × 6; **F,** longitudinal section of same, × 6. A, from *Richards* 17333; B–F, from *Verdcourt* 3433. Drawn by Ann Davis.

scaly papilla-like hairs and densely hairy inside; lobes oblong, 3 mm. long, 1·5 mm. wide. Filaments 1·5–2 mm. long in short-styled flowers. Style 3 mm. long in short-styled flowers; stigma-lobes 1·5–2 mm. long. Drupes not known. Fig. 4.

TANZANIA. Kigoma District: Uvinza–Mpanda road, 56 km. from Uvinza, 23 Nov. 1962, *Verdcourt* 3433 !; Mbeya District: Sao Hill–Madibira road, 8 Dec. 1962, *Richards* 17333 !
DISTR. **T4**, 7; not known elsewhere
HAB. *Brachystegia* woodland, by sandy roadside and in rock clefts with *Xerophyta*, *Euphorbia*, etc.; 1500 m.

NOTE. Only 2 specimens, both with the stamens exserted, have been seen.

33. **P. lucens** *Hiern* in F.T.A. 3: 211 (1877); De Wild., Pl. Bequaert. 2: 385 (1924); G. Taylor in Exell, Cat. Vasc. Pl. S. Tomé: 213 (1944); Petit in B.J.B.B. 34: 199, photo. 6/H (1964) & in Distr. Pl. Afr. 6, map 173 (1973). Type: Principe, *Barter* 1961 (K, holo. !, P, iso.)

Shrub or subshrub, ? sometimes creeping, 0·3–1·5 m. tall, with glabrous stems. Leaf-blades elliptic, narrowly oblong-elliptic or obovate-elliptic, (7–)12–28 cm. long, 3–12 cm. wide, acuminate at the apex, cuneate at the base, glabrous on both surfaces, drying brownish beneath, thin, with 6–16 pairs of lateral nerves; domatia and nodules absent; petioles 0·5–2·5 cm. long, glabrous; stipules ovate, 0·8–1·8 cm. long, glabrous, often bilobed at the apex, the lobes 2·5–6 mm. long, deciduous. Flowers heterostylous, 4-merous in several- to much-branched panicles 2–20 cm. long; peduncle 0·5–10 cm. long, glabrous; secondary branches 0·5–1·6 cm. long; pedicels 0–3 mm. long, lengthening to 5 mm. in fruit; lower bracts ovate-triangular or narrowly triangular, up to 5 mm. long, glabrous, rest minute. Calyx-tube obconical, 0·75–1 mm. long, glabrous; limb cupuliform, 0·5 mm. long, glabrous, truncate or the lobes very short. Corolla white or greenish ochre, glabrous outside; tube 1·6–2 mm. long; lobes ovate, 1·5 mm. long, 1·3 mm. wide, horned at the apex. Stamens with filaments 1·5 mm. long in short-styled flowers, 0·5 mm. long or almost free in long-styled flowers. Style 1 mm. long in short-styled flowers, 2 mm. long in long-styled flowers; stigma-lobes 0·5–0·7 mm. long. Drupes red, of 2 pyrenes, ellipsoid, ± 9 mm. long, ± 6 mm. wide, glabrous, not grooved; pyrenes semi-ellipsoid, 8·5 mm. long, 5·5 mm. wide, 3 mm. thick, the ventral face plane, the dorsal face 3-ribbed or slightly undulate. Seeds slightly smaller, ventral face plane, but with a T-shaped median fissure, dorsal face 3-ribbed or slightly undulate; albumen not ruminate.

var. **minor** *Petit* in B.J.B.B. 34: 200 (1964) & in Distr. Pl. Afr. 6, map 174 (1973). Type: Zaire, Isangi, R. Lubilu, Yabwesu–Ogeto, *Germain* 8762 (BR, holo.)

Subshrub 0·3–1 m. tall, with ± underground rhizomes. Leaf-blades 7–15 cm. long, 3–8 cm. wide, with 6–9 pairs of lateral nerves. Inflorescences 2–6 cm. long; peduncle 0·5–3 cm. long.

UGANDA. Bunyoro District: Budongo Forest, *Sangster* 188 !; Mengo District; Mawokota, Mar. 1905, *E. Brown* 183 ! & Semunya Forest, Walufumbo, 16 June 1950, *Dawkins* 596 !
DISTR. **U2**, 4; Zaire
HAB. Evergreen forest including swamp forest; 1170–1200 m.

NOTE. Var. *lucens* occurs in Nigeria, S. Tomé, Principe and Cameroun.

34. **P. sp. F**

Shrub ?; stems glabrous, finely grooved. Leaf-blades elliptic to oblong, 7·5–21 cm. long, 3·6–7 cm. wide, shortly to distinctly acuminate at the apex, cuneate and sometimes asymmetric at the base, glabrous; midrib slightly impressed above, drying pale and with a very narrow pale area on either side; small domatia present; nodules absent; petioles 2·5–5 cm. long; terminal young stipules spathulate, 7 mm. long, 3·5 mm. wide, with a few scattered hairs; at lower nodes the 2 stipules are fused and since soon deciduous form a loose collar 1·3 cm. long, each tip very shortly bifid. Inflorescence known only in fruiting state, very lax; peduncle 9 cm. long; secondary branches 1·5–6 cm. long, the node with ferruginous hairs but otherwise entirely glabrous; pedicels 0·8–1 cm. long, but it is probable that some flowers are practically sessile. Young fruits ovoid, 5 mm. long and wide; free calyx-limb scarcely 0·5 mm. long; calyx-teeth triangular, ± 0·7 mm. long. Pyrene (undeveloped) discoid, ± 3–3·5 mm. in diameter, not grooved.

TANZANIA. Lushoto District: Shagayu Forest Reserve, Goka, 30 Sept. 1964, *Mgaza* 66!
DISTR. **T3**; not known elsewhere
HAB. Presumably rain-forest; ? 1900 m.

NOTE. It does not seem possible that this is in any way close to *P. tenuipetiolata* despite the locality and equally long petioles. The pedicels, stipules and leaf size are quite different. Probably related to *P. triclada* or perhaps even a variant of that species but more material is needed.

35. **P. sp. G**

Shrub with glabrous pale grey-brown finely longitudinally striate stems. Leaf-blades elliptic, 3·3–7·5 cm. long, 1·7–3·8 cm. wide, acute or acuminate at the apex, cuneate at the base and decurrent to a slight extent, appearing to be densely covered with minute scales beneath (actually stomata); petiole 0·3–1·3 cm. long; stipules joined to form a sheath 5·5 mm. long, ? entire, soon deciduous, with some hairs at the nodes beneath. Inflorescences apparently sessile; secondary branches 1·5–2 cm. long, densely shortly pubescent; pedicels 4 mm. long. Drupes 6–7 mm. in diameter; pyrenes semiglobose, 6 mm. long, 5 mm. wide, 2 mm. thick, rugulose. Seeds strongly ruminate.

TANZANIA. Lushoto District: E. end of W. Usambara Mts., Ambangulu Estate, 23 Sept. 1940, *Wallace* 941!
DISTR. **T3**
HAB. Presumably evergreen forest; ± 1300 m.

NOTE. Probably related to *P. triclada* or perhaps even a variant of that species but more material is needed.

36. **P. sp. H**

Small tree or shrub 2·7–3 m. tall; stems with spreading stiff ferruginous hairs when very young, soon glabrous, corky, rugose, with prominent petiole-bases. Leaf-blades elliptic, 2·5–9 cm. long, 0·7–3·2 cm. wide, shortly acuminate at the apex, cuneate at the base and decurrent into the petiole, glabrous save for some hairs on midrib near the base, drying purplish grey or grey-green save for the ± orange-brown midrib; nodule-like domatia present at midrib in axils of lateral nerves; petioles apparently 0·2–1 cm. long but actually leaf-blade almost decurrent to the base, hairy; stipules ovate, 6–7 mm. long, prominently bilobed. Inflorescences few-flowered, known only from young fruiting stage; peduncle 0·5–1 cm. long or almost obsolete, hairy; secondary branches 2–3 mm. long; pedicels ± 1 mm. long. Fruits crimson ± 6 mm. in diameter; pyrenes semiglobose, 5 mm. in diameter, 2·5 mm. thick, dorsally convex, ventrally flat, with 5 shallow dorsal grooves. Seeds deeply grooved on both surfaces.

TANZANIA. Mpanda District: Kasangazi, 25 July 1958, *Mahinda* 142! & Kungwe Mt., Selimweguru, 25 July 1959, *Newbould & Harley* 4552!

DISTR. **T4**; not known elsewhere
HAB. Understorey in very shady mist forest and high gallery forest; 2040 m.

NOTE. It is not quite certain that the two specimens cited above are conspecific; there are undoubtedly differences, *Newbould & Harley* 4552 having the leaves more sharply acuminate and longer petiolate and the inflorescences more distinctly pedunculate, but the general facies is so similar that I am practically sure despite the poor material.

37. **P. sp. I**

Shrub or small tree about 4–6 m. tall; stems reddish brown, finely longitudinally striate, glabrous. Leaf-blades elliptic or oblong-elliptic, 3·5–17·5 cm. long, 1·8–9·3 cm. wide, shortly acutely acuminate at the apex, broadly cuneate at the base, glabrous; nodules and domatia absent; petiole 1·5–6 cm. long; stipules forming a short sheath 3–4 mm. long, ± truncate to ± acute, the margins with small fimbriae, soon deciduous. Flowers probably heterostylous, 5-merous, in branched panicles; peduncle 2–3 cm. long, glabrous; secondary branches 1–3 cm. long, glabrous or with a few hairs; pedicels 1–1·5 mm. long; bracts and bracteoles small, margined with ferruginous hairs. Calyx-tube obconic, 1 mm. long, glabrous, grooved; limb shallow, 0·8–1 mm. long, including the very broad short teeth distinctly funnel-shaped. Corolla. glabrous, white; tube 5·5 mm. long, lobes narrowly ovate, 3·5 mm. long, 1·8–1·9 mm. wide. Stamens with filaments 0·5 mm. long in long-styled flowers Style 7 mm. long in long-styled flowers; stigma-lobes 1 mm. long. Drupes not seen.

TANZANIA. Morogoro District: Uluguru Mts., Bondwa Mt., 25 Nov. 1932, *Schlieben* 3023!
DISTR. **T6**; not known elsewhere
HAB. Riverine forest; 1500 m.

NOTE. Two sheets of this have been seen, the field labels not quite agreeing in details; one states shrub or small tree 4–6 m., gives altitude as 1500 m. and habitat as 'Urwald' and mentions green fruits. The other states shrub 4 m. tall, gives altitude as 1650 m. and habitat as 'Bachtal' and does not mention fruits.

38. **P. sp. J**

Shrub 2–4 m. tall, with dark purplish lenticellate finely longitudinally striate stems, glabrous save for some hairs within the bases of fallen young stipules, scarred by obvious petiole-bases. Leaf-blades narrowly elliptic, up to 15 cm. long, 6–6·8 cm. wide, ? obtuse at the apex, cuneate at the base, glabrous, probably thinly coriaceous, drying purplish brown; nodules absent but domatia present; petiole ± 0·7–1·6 cm. long; stipules difficult to make out, ± ovate, 6 mm. long, ± lobed or ? lacerate at apex, very soon deciduous. Flowers in a branched panicle, the individual components ± capitate; peduncle 2·4 cm. long; secondary branches at least 4–5 mm. long, ± pubescent; pedicels ± obsolete. Calyx obconic, ± 1 mm. long; limb cupular, 2–2·5 mm. long, obscurely shortly pubescent, irregularly lobed, the longest lobe ± 0·7 mm. long, margins ± lacerate-ciliate. Corolla stated to be yellowish. Open corollas and fruits not known.

TANZANIA. Lindi District: Lake Lutamba, Milola stream valley, 9 Nov. 1934, *Schlieben* 5587!
DISTR. **T8**; not known elsewhere
HAB. Not known, probably riverine woodland; 180 m.

39. **P. usambarensis** *Verdc.* in K.B. 30: 691 (1976). Type: Tanzania, Usambara Mts., Ngambo, *Peter* 55854 (K, holo.!, B, iso.!)

Shrub 50–80 cm. tall; stem pubescent with red-brown hairs at the apex, glabrescent below save for the hairy nodes. Leaf-blades lanceolate, oblanceolate or elliptic, 6·5–17 cm. long, 2·5–7·5 cm. wide, acute at the apex or shortly acuminate, narrowly cuneate at the base, glabrous above, ferruginous pubescent beneath when young but eventually only on the main nerves; petiole 0·7–3 cm. long, red pubescent. Stipules oblong, ovate or elliptic, 0·9–1·2(–2) cm. long, mostly glabrous, margins glabrous or ciliate and with 2 basal divergent ferruginous-hairy ridges, divided into 2 broadly triangular or cuspidate lobes at the apex. Inflorescences capitate, sessile or peduncle only 3 mm. long, up to 1·5 cm. broad, 20–35-flowered; bracts not joined to form an involucre, elliptic, 7 mm. long, 3 mm. wide, ± glabrous. Calyx-tube narrow, 0·5–0·8 mm. long; tubular part of limb campanulate, 1·5 mm. long, glabrous or with a few sparse hairs; lobes narrowly triangular, 1 mm. long, glabrous, the margins inrolled. Buds with 5 obtuse appendages 0·5 mm. long at the apex. Corolla white; tube cylindric, 3·5 mm. long, funnel-shaped at the apex, glabrous; lobes oblong-elliptic, 1·5 mm. long, 0·8 mm. wide, strongly appendaged at the apex, produced laterally both backwards and forwards. Tips of anthers just exserted in short-styled flowers; style 1 mm. long, mostly hidden in the large disc; stigmatic lobes flattened, ± 0·8 mm. long. Drupes ellipsoid, 5·8–6 mm. long, 3·2–3·5 mm. wide, strongly ribbed, glabrous, crowned by the persistent reflexed calyx-lobes; pyrenes semi-ellipsoid, 5·5 mm. long, 2·5 mm. wide.

TANZANIA. Lushoto District: East Usambara Mts., Bulwa to Ngambo, 31 Aug. 1916, *Peter* 55912 ! & path from Amani to Kwamkoro, 21 May 1916, *Peter* 55889 ! & Amani, Bomole Mt., 2 Jan. 1915, *Peter* 55600 !
DISTR. **T3**; not known elsewhere
HAB. Rain-forest; 850–1000 m.

40. **P. peduncularis** (*Salisb.*) *Steyerm.* in Mem. N.Y. Bot. Gard. 23: 546 (1972). Type: grown from material collected in Sierra Leone by *Smeathman* (BM, holo. !)

A very variable subshrub, shrub or subshrubby herb 0·2–3(–4·5) m. tall, the stems branched or unbranched, glabrous or sparsely to densely covered with often ferruginous hairs. Leaf-blades narrowly to broadly elliptic, oblong- to obovate-elliptic or ± lanceolate, 2·5–26·5 cm. long, 3·1–13·5 cm. wide, acute to abruptly acuminate at the apex, cuneate to ± rounded at the base, glabrous above and beneath or sparsely to densely covered with mostly ferruginous hairs, or tomentose, often thinly coriaceous; petioles 1–6·5 cm. long, pubescent or glabrous; stipules obovate-oblong to lanceolate, 1–2·2 cm. long, 0·8–1·5 cm. wide, bilobed at the apex, the lobes acuminate, 0·5–2 cm. long, with a median and 2 lateral lines of basal hairs, glabrous to ciliate on the margins, soon deciduous, the nodes hairy above the scars. Flowers usually ± numerous in involucrate capitate inflorescences; peduncles usually solitary but sometimes 2–3 together, 1–9(–15) cm. long, pubescent all round or with 2 lines of pubescence or glabrous; pedicels sometimes white, 2–11 mm. long in fruit; bracts free or joined together to form an entire to lobed involucre, white, green, pale blue-green or yellow-green, 0·6–3 cm. long, 0·4–1·5 cm. wide, rounded, glabrous, pubescent or ciliate. Calyx-tube ellipsoid or obconic, 1–3 mm. long; limb 1–1·3 mm. long; lobes almost obsolete to attenuate-triangular, 0·5–4 mm. long, pubescent or glabrous. Corolla white; tube funnel-shaped, cylindric below, 3·5–6·5 mm. long; lobes triangular or elliptic-lanceolate, 1–3 mm. long, 0·6–1·2 mm. wide, often horned outside at the apex. Anthers just included or exserted in long-styled flowers; filaments exserted (0–)1–2 mm. in short-styled flowers. Style 4·3–9 mm. long in long-styled flowers, the stigmas exserted (1–)2–3 mm.; 4·5 mm. long in short-styled flowers the stigmas 1–2·5 mm. long. Fruits blue, blue-black or waxy white, ellipsoid, 5·5–8(–10 in life) mm. long,

3·5–6(–9 in life) mm. wide, grooved in the dry state, crowned or not with the persistent calyx; pyrenes straw-coloured, ½-ellipsoid, 4·5–7 mm. long, 3·2–5 mm. wide, 1·5–2 mm. thick, grooved and convex dorsally, ventrally plane or with 2 narrow grooves. Seeds very dark red, closely conforming in size and ribbing to the pyrenes but not ruminate and with a shallow V-shaped ventral groove.

SYN. *Cephaelis peduncularis* Salisb., Parad. Lond., t. 99 (1808); Hiern, F.T.A. 3: 223 (1877); Hepper in K.B. 16: 154 (1962) & in F.W.T.A., ed. 2, 2: 204 (1963)

KEY TO INFRASPECIFIC VARIANTS

Stems, undersurfaces of leaf-blades (at least the main venation), etc. densely ferruginous pubescent:
- Peduncles ± 1–2 cm. long; calyx-lobes small, glabrous . . . . a. var. **suaveolens**
- Peduncles longer, ± 7·5 cm. long; calyx-lobes larger and more distinct, pubescent to densely hairy on the margins. . . . e. var. **A**

Stems, leaf-blades, etc. glabrous to sparsely pubescent or very finely tomentose; calyx-lobes glabrous to hairy:
- Leaf-blades with exceedingly fine short tomentum on the main venation beneath. . b. var. **semlikiensis**
- Leaf-blades glabrous beneath or with very sparse ferruginous hairs on the midrib:
  - Leaf-blades thinner, the venation not so raised above; petioles often up to 6·5 cm. long; calyx glabrous to ± hairy (**U**2, 4) . . . . c. var. **ciliato-stipulata**
  - Leaf-blades mostly somewhat coriaceous, the venation often raised above in the dry state; petioles up to ± 2·5 cm. long; calyx densely hairy on the margins (**T**4, 7, 8) . . . . d. var. **nyassana**

a. var. **suaveolens** (*Hiern*) *Verdc.* in K.B. 30: 257 (1975). Type: Sudan, Jur, Orel, *Schweinfurth* 1736 (K, holo. !, BM, iso. !)

Leaf-blades with ferruginous hairs beneath at least on the main venation. Stipules with ciliate margins or rarely glabrous. Inflorescences shortly pedunculate, the peduncle 1–2 cm. long. Calyx-lobes very small, glabrous.

UGANDA. Masaka District: NW. side of L. Nabugabo, 9 Oct. 1953, *Drummond & Hemsley* 4698 ! & Minziro Forest, path to Kagera R., Oct. 1925, *Maitland* 923 !; Mengo District: Busuju County, near Mityana, Kasa Forest, 17 Jan. 1950, *Dawkins* 490 !
KENYA. Uasin Gishu District: Sosiani R., Mar. 1963, *Tweedie* 2589 !; S. Kavirondo District: Kuja R., *Glasgow* 46/2 !
TANZANIA. Bukoba District: Kabwoba [Kabobwa] Forest Reserve, 20 Oct. 1955, *Sangiwa* 92 !; Biharamulo, Bukoba road, 13 Nov. 1948, *Ford* 855 !; Buha District: 6·4 km. N. of Kibondo, Feb. 1955, *Procter* 380 !
DISTR. **U**4; **K**3; **T**1,4; Guinée, Ghana, Cameroun, Zaire, Sudan, Malawi and Zambia
HAB. Riverine forest, swampy valley forest and edges of forest clumps; 1000–1800 m.

SYN. *Cephaelis suaveolens* Hiern in F.T.A. 3: 224 (1877)
Uragoga suaveolens (Hiern) K. Schum. in P.O.A. C: 392 (1895); T.T.C.L.: 536 (1949); K.T.S.: 477 (1961)
U. cyanocarpa K. Krause in E.J. 43: 157 (1909); T.T.C.L.: 535 (1949). Type: Tanzania, Mwanza District, W. Ukerewe, *Uhlig* 74 (B, holo.†)
Cephaelis peduncularis Salisb. var. *suaveolens* (Hiern) Hepper in K.B. 16: 156 (1962) & in F.W.T.A., ed. 2, 2: 204 (1963)

b. var. **semlikiensis** *Verdc.* in K.B. 30: 690 (1976). Type: Zaire, Lesse, *Bequaert* 4116 (BR, holo.!)

Leaf-blades with very fine short tomentum on the venation beneath. Stipules with ciliate margins or quite glabrous. Inflorescences shortly pedunculate, the peduncle 0·6–2·8 cm. long. Calyx-lobes glabrous or ± hairy.

UGANDA. Bunyoro District: Budongo Forest, track to Busingiro, 10 Dec. 1970, *Synnott* 493! & Kitoba, May 1943, *Purseglove* 1565!; Toro District: Ruimi [Wimi] Forest, 3 June 1906, *Bagshawe* 1040!
DISTR. **U2**; Zaire, Burundi
HAB. Evergreen forest, including swamp forest; 900–1200 m.

SYN. *Uragoga semlikiensis* De Wild. in Mém. Inst. Roy. Col. Belge, Sect. Sci. Nat. Med. 8°(4)4: 171 (1936) [descr. in French]; F.P.N.A. 2: 369 (1947), *nom invalid.*

c. var. **ciliato-stipulata** *Verdc.* in K.B. 30: 689 (1976). Type: Zaire, Lesse, *Bequaert* 4188 (BR, holo.!)

Leaf-blades glabrous beneath when mature or with only scattered ferruginous hairs on the nerves beneath, but sometimes quite hairy beneath when very young; petioles often long, up to 6·5 cm. long. Stipules with margins ciliate, rarely glabrous. Peduncles short, 0·9–1·3 cm. long. Calyx-lobes glabrous to shortly hairy.

UGANDA. Toro District: Semliki Forest, Bwamba, near Kirimia R., 27 Oct. 1951, *Osmaston* 1372!; Masaka District: Sese Is., Bugala, Towa Forest, July 1945, *Purseglove* 1761!; Mengo District: 3.2 km. E. of Entebbe, Kyewaga Forest, 21 Sept. 1949, *Dawkins* 387!
KENYA. N. Kavirondo District: Kakamega Forest, May 1935, *Dale* in *F.D.* 3133! & same locality, 26 Nov. 1969, *Bally* 13662!; Kericho, Sept. 1933, *Napier* in *C.M.* 6099!
DISTR. **U2, 4; K5**; Zaire, Sudan, Ethiopia
HAB. Undergrowth of evergreen forest; 885–1800 m.

SYN. *Uragoga ciliato-stipulata* De Wild. in Mém. Inst. Roy. Col. Belge, Sect. Sci. Nat. Méd. 8°(4)4: 133 (1936) [descr. in French]; F.P.N.A. 2: 369 (1947), *nom invalid.*

NOTE. *Maitland* 239 (Uganda, Entebbe, Oct. 1924) bears the information "a forest tree not of large dimensions" but this must be an error.

d. var. **nyassana** (*K. Krause*) *Verdc.* in K.B. 30: 257 (1975). Types: Tanzania, Rungwe District, Masoko to Mulinda, *Stolz* 418 (B, syn.†, BM, EA, K, isosyn.!) & *Stolz* 1405 (B, syn.†, K, isosyn.!)

Leaf-blades more coriaceous than in other varieties, glabrous; venation mostly raised on upper surface in dry state. Stipules with ciliate or glabrous margins. Inflorescence pedunculate, the peduncle 2–5·5 cm. long. Calyx-lobes mostly larger and densely margined with hairs up to ± 1 mm. long.

TANZANIA. Kigoma District: Kasoge R., Muholoholo, 17 Feb. 1963, *Carmichael* 1025!; Rungwe District: Masoko to Mulinda, 17 Nov. 1910, *Stolz* 418!; Songea District: SW. of Kitai, R. Nakawali, 7 Mar. 1956, *Milne-Redhead & Taylor* 9106!
DISTR. **T4, 7, 8**: Malawi, Zambia, Rhodesia, ? Angola
HAB. Riverine and other evergreen forest; 600–1500 m.

SYN. *Uragoga nyassana* K. Krause in E.J. 57: 50 (1920); T.T.C.L.: 536 (1949)

e. **var. A**

Similar to var. *nyassana* but peduncle mostly larger, up to 7·5 cm. long, leaf-blades ferruginous pubescent beneath at least on the venation; stems, etc. more densely ferruginous hairy

TANZANIA. Mbeya District: Mbimba Experimental Station, 28 Nov. 1966, *Robertson* 320!
DISTR. **T7**; Malawi, Zambia, Rhodesia
HAB. Swampy woodland; 1500 m.

NOTE. Merges with variants of *P. tanganyicensis* Verdc. Often wrongly identified with var. *suaveolens* in Flora Zambesiaca area.

GENERAL NOTE. The exact relationships of these variants to the many occurring in W. Africa and Zaire must await a detailed revision of the section. This would involve several years work and is clearly impracticable at this stage.

41. **P. sp. K**

Subshrubby herb about 40–50 cm. tall; stems at first ferruginous hairy but later glabrous. Leaves as in species 42 but up to 13·5 cm. long and 6 cm. wide, glabrous; petiole 0·5–1·5 cm. long, densely hairy; stipules oblong, 1·2 cm. long, 6 mm. wide, bilobed at the apex for 5 mm., the lobes asymmetrically acuminate, ferruginous pubescent at base and lobes ciliate. Inflorescences small, few-flowered; bracts oblong, 1·6 cm. long, 1·2 cm. wide, joined together for ⅔ of their length; numerous long hairs between the flowers. Calyx-tube narrowly obconic, 3·5 mm. long; free part of the limb 1 mm. long; lobes lanceolate, 3 mm. long, 1 mm. wide, with a distinct apiculum at the apex, margined with long hairs 3 mm. long; disc cylindrical, 1 mm. long. Corolla white; tube cylindrical, 1·1 cm. long, slightly widened at the apex; lobes ovate-elliptic, 3 mm. long, 1·8 mm. wide, with short papilla-like hairs at the tips outside. In long-styled flowers the tips of the anthers are just exserted from the throat; style exserted 4 mm., the stigmatic arms 1·5 mm. long, slightly papillose. Fruits not seen.

TANZANIA. Morogoro District: Nguru Mts., Liwale valley, Manyangu Forest, 27 Mar. 1953, *Drummond & Hemsley* 1848!
DISTR. **T6**; not known elsewhere
HAB. Lowland rain-forest; 600 m.

42. **P. tanganyicensis** *Verdc.* in K.B. 30: 258 (1975). Type: Tanzania, E. Usambaras Mts., Gonja Mt., *Engler* in *Herb. Amani* 3366 (B, holo.†, EA, iso.!)

Small little-branched shrub 0·3–3 m. tall; stems glabrous save for ferruginous hairs visible beneath the deciduous stipules or with some scattered ferruginous hairs; petiole scars very evident. Leaf-blades elliptic, rhomboid-elliptic or elliptic-oblanceolate, 16–32 cm. long, 7–14 cm. wide, sharply acuminate at the apex, very narrowly cuneate at the base, glabrous on both sides or pubescent beneath on the nerves or quite densely ferruginous hairy beneath; petiole 2–3·5 cm. long, glabrous to shortly pubescent; stipules elliptic-oblong to obovate, 1·5–2·5 cm. long, 1·2–1·6 cm. wide, bilobed at the apex, the lobes up to 1 cm. long, margined with ferruginous hairs and also with hairs at middle of base. Inflorescences mostly many-flowered (probably up to well over 100), terminal, sessile or pedunculate, 1·3–1·8 cm. wide when in very young stage but becoming over 4 cm. wide when older; main bracts paired, with an inner and an outer set, imbricate, 2 entirely internal, 3·5 cm. long, 2 cm. wide; peduncles 0–3(–12 in a subspecies) cm. long, angled above, glabrous or pubescent; pedicels 1–2 mm. long (up to 3–5 mm. in fruit). Calyx-tube ovoid-obconic, 1·2–2 mm. long; limb quite conspicuous, mostly spreading and over 5 mm. across (1 cm. in spirit); lobes ovate or obovate, sometimes ± 3–4-lobed, 2 mm. long, 1·2 mm. wide, fringed with long ferruginous hairs, 1·2–3 mm. long. Buds with 4 small bumps at the apex. Corolla cream or white; tube cylindric below, funnel-shaped at apex, 6–7 mm. long, the apical part 1·5 mm. long; throat hairy; lobes ovate, 2·5–3 mm. long, 1·5–2 mm. wide, thickened at the apex. Anthers up to ½-exserted; style exserted 0·5–2 mm., the stigmatic arms 1–1·8 mm. long, cylindric or flattened and oblong. Fruit purple or blue, glossy, oblong-cylindric, 7–8·5 mm. long, 3–4·2(–7) mm. wide, strongly ribbed in the dry state but smooth in life; pyrenes ellipsoid, oblong-cylindric or lanceolate in outline, compressed, 6·2–7 mm. long, 3–3·5 mm. wide, 1·9–2 mm. thick, 4-ribbed. Seeds red or pale red-brown, 4·2–5·5 mm. long, 1·3–2·5 mm. wide, 1·2–1·5 mm. thick, dorsally obtusely 6-ribbed, ventrally very flattened.

KEY TO INFRASPECIFIC VARIANTS

Peduncles 0–3·5 cm. long (subsp. *tanganyikensis*):
  Leaf-blades glabrous beneath or with only sparse hairs on the venation . . . . . . var. **tanganyicensis**
  Leaf-blades ferruginous hairy beneath . . . var. **ferruginea**
Peduncles 11–12 cm. long . . . . . . subsp. **longipes**

subsp. **tanganyicensis**

Peduncles 0–3·5 cm. long. Leaf-blades glabrous to ferruginous hairy beneath.

var. **tanganyicensis**

Leaf-blades glabrous beneath or with only sparse short hairs on the venation beneath. Peduncles 0–3 cm. long, glabrous.

KENYA. Kwale District: Shimba Hills, Pengo Forest, 9 Feb. 1953, *Drummond & Hemsley* 1194! & Lango ya Mwagandi [Longo Mwagandi] area, 24 Mar. 1968, *Magogo & Glover* 440! & Makadara, 12 Dec. 1958, *Moomaw* 1140!
TANZANIA. Lushoto District: E. Usambara Mts., Kwamkoro–Kihuhwi, 16 Dec. 1936, *Greenway* 4788! & above Magunga, 23 Sept. 1953, *Faulkner* 1249!; Morogoro District: N. Nguru Mts., above Manyangu Forest, 2 Apr. 1953, *Drummond & Hemsley* 2011!
DISTR. **K**7; **T**3, 4, 6; not known elsewhere
HAB. Rain-forest, riverine forest, etc.; 300–1530 m.

SYN. *Uragoga macrophylla* K. Krause in E.J. 39: 569 (1907); T.T.C.L.: 536 (1949); K.T.S.: 477 (1961)

var. **ferruginea** *Verdc.* in K.B. 30: 258 (1975). Type: Tanzania, Lushoto District, Makuyuni area, *Koritschoner* 753 (K, holo.!, EA, iso.!)

Leaf-blades with main venation or entire surface covered with short ferruginous hairs beneath. Peduncles 1·5–3·5 cm. long, ± ferruginous pubescent.

TANZANIA. Lushoto District: Makuyuni area, *Koritschoner* 653! & 821!; Morogoro District: Uluguru Mts., Bunduki, 11 Apr. 1936, *E. M. Bruce* 1045!; Ulanga District: near Mahenge, Issongo, 10 Feb. 1932, *Schlieben* 1743!
DISTR. **T**3, 6; not known elsewhere
HAB. Evergreen forest, including fringing forest, also savanna (*fide* Schlieben); 720–1000 m. (Koritschoner's sheets all bear the information 400–1000 m.)

subsp. **longipes** *Verdc.* in K.B. 30: 258 (1975). Type: Tanzania, Morogoro District, Uluguru Mts., Tegetero, *Drummond & Hemsley* 1707 (K, holo.!)

Peduncles 11–12 cm. long. Leaf-blades glabrous beneath.

TANZANIA. Morogoro District: Uluguru Mts., Tegetero, 20 Mar. 1953, *Drummond & Hemsley* 1707!
DISTR. **T**6; not known elsewhere
HAB. Rain-forest; 1800 m.

NOTE. This looks very distinct, but the peduncle length varies in subsp. *tanganyicensis* and until further material in spirit is available it is not possible to see if there are any correlated floral and inflorescence differences.
Tanner 4789 (Biharamulo District, Rusumo Falls, 23 Mar. 1960, 1350 m.) found in falls spray in shade of high trees has more oblong-elliptic leaves as in *P. peduncularis*, but the calyx is as in *P. tanganyicensis*. It probably represents yet another variant—the seeds are stated to be white and the petioles have lines of rather rough hairs; no other material has been seen from this area.

43. **P. megalopus** *Verdc.* in K.B. 30: 259 (1975). Type: Tanzania, Iringa District, Mufindi, *Goetze* 750 (B, holo.†, K, iso.!)

Spindly laxly branched shrub or small tree 2·4–10 m. tall; stems glabrous and longitudinally wrinkled save for long hairs at the nodes within the stipules; old petiole scars very evident. Leaf-blades oblong to obovate or obovate-oblanceolate, 28–43 cm. long, 9·6–24 cm. wide, very rounded at the apex, narrowly cuneate at the base, rather thick, glabrous, shining beneath in the living state; midnerve often reddish; lateral nerves numerous, 12–27 pairs, the venation closely but ± obscurely reticulate; a well-marked intra-marginal nerve present; petioles often very stout, 1·5–6 cm. long including a portion bordered by the narrowly decurrent base of the lamina; stipules oblong, 2·6 cm. long, 2·4 cm. wide, probably acuminate but often split at the apex, very soon deciduous. Flowers in very condensed heads, 4·5–9 cm. long, 3·5–10 cm. wide, the main enveloping bracts large, forming an entirely closed cup which bears a strong median keel in the direction of the longest dimension, which is decurrent on to the thickened peduncle but absent from the central upper part; the flowers are gathered into 3–10 groups within the cup and become more separate after the main bracts have deteriorated; secondary bracts linear-oblong to almost round, the narrowest next to the flowers, 2–2·5 cm. long, 0·6–2·5 cm. wide, the margins incurved, with long hairs outside at the apex, often in dried material the corollas appearing to emerge from holes in a surface made up of hairy bracts; peduncles attaining 1·8 m. in length, pinky red, eventually resting on the ground but often only 12–60 cm. long, up to 4 cm. wide at the apex and ± 2-angled. Calyx-tube ovoid or ellipsoid, glabrous, 1–2 mm. long, 0·7 mm. wide, ribbed; limb 3–5 mm. long, toothed at the apex, the lobes irregular or triangular, 2–5 mm. long, with long hairs outside at the apex. Corolla white; tube funnel-shaped, 1·2–1·3 cm. long, cylindrical below, 3 mm. wide at the top, 4 mm. at the bottom, glabrous outside, usually with top half hairy inside; lobes 6, green tinged, lanceolate or ovate, 4–6 mm. long, 2–2·3 mm. wide. Disc 2·5 mm. long, wrinkled. Anthers 3·8 mm. long, exserted so that bases are just included in the tube. Ovary 3-locular; style 4·5 mm. long, with 3 stigma-arms. Pyrenes straw-coloured, ellipsoid in section or oblong, 8·5–9·5 mm. long, 3·5–5 mm. wide, 2·5 mm. thick, compressed, trigonous, grooved on both surfaces. Seeds dark red, ellipsoid in section, 6·5–7 mm. long, 2·2–3·6 mm. wide, 1·5–2 mm. thick, grooved on both surfaces, rugulose.

TANZANIA. Morogoro District: Uluguru Mts., below Hululu Falls, by Mgeta R., 15 Mar. 1953, *Drummond & Hemsley* 1574!; Ulanga District: Mahenge, Kwiro Forest remnant, Oct. 1951, *Eggeling* 6310!; Iringa District: Mufindi, Kigogo R., 17 Mar. 1962, *Polhill & Paulo* 1797!

DISTR. **T**6, 7; not known elsewhere

HAB. Evergreen forest, including riverine, particularly with *Allanblackia, Chrysophyllum, Macaranga, Bridelia, Phoenix* and bamboo; 1140–1850 m.

SYN. *Megalopus goetzei* K. Schum. in E.J. 28: 491 (1900), *non Psychotria goetzei* (K. Schum.) Petit
Camptopus goetzei (K. Schum.) K. Krause in N.B.G.B. 7: 386 (1920)
Cephaelis goetzii (K. Schum.) Hepper in K.B. 16: 153 (1962)

NOTE. More information is needed on heterostyly. K. Schumann gives the length of the style as 10–14 mm., of which 3·5–4 is stigma-lobes, and anthers on filaments 1·5–2 mm. long, 11–12 mm. above the base of the tube, presumably a long-styled specimen.

44. **P. amboniana** *K. Schum.* in P.O.A. C: 390 (1895); S. Moore in J.B. 43: 353 (1905); De Wild., Pl. Bequaert. 2: 331 (1924); T.T.C.L.: 523 (1949); K.T.S.: 466 (1961). Types: Tanzania, Tanga District, Amboni, *Holst* 2716 (B, syn.†, HBG, K, isosyn.!) & same locality, *Holst* 2855a (B, syn.†)

Shrub 1–3 m. tall; young stems mostly pale, greyish white, often chestnut when older, with grooved slightly peeling epidermis, glabrous or in one variety velvety. Leaf-blades oblong-elliptic, narrowly to broadly elliptic, narrowly obovate or even ± linear-lanceolate, 1·5–8 cm. long, 0·2–3·3 cm. wide, acute to shortly acuminate at the apex, very narrowly cuneate at the base, discolorous, glabrous on both surfaces, or in one variety velvety pubescent; nodules few, situated in the basal part of the midrib and domatia often present in the axils of the lateral nerves beneath; petiole 1–3(–15) mm.

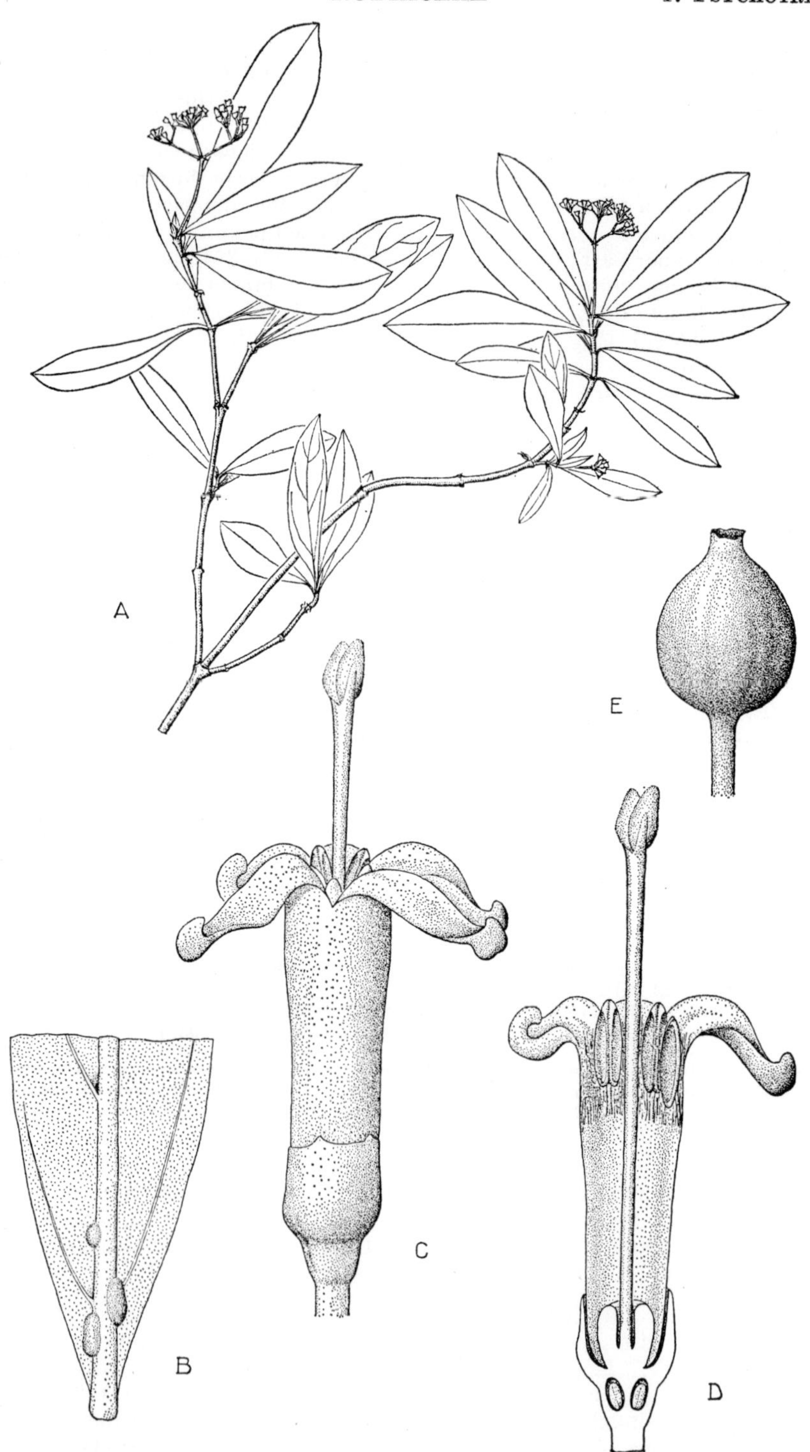

FIG. 5. *PSYCHOTRIA AMBONIANA* var. *AMBONIANA*—**A,** habit, × ½; **B,** part of underside of leaf, showing nodules, × 5; **C,** long-styled flower, × 8; **D,** longitudinal section of same, × 8; **E,** drupe, × 4. A, from *Tanner* 3443; B–D, from *Drummond & Hemsley* 4259; E, from *Peter* 24872. Reproduced by permission of the Director of the Jardin Botanique National de Belgique.

long, passing very gradually into the very narrow base of the lamina; stipules sometimes brown, glabrous, bilobed, the base ovate-triangular, 1–1·5 mm. long, the lobes 0·3–0·6(–2) mm. long. Flowers heterostylous, 5-merous, in ± many-flowered rather congested glabrous panicles or rarely subumbellate; peduncle 0·7–4 cm. long; secondary peduncles 4–10 mm. long; pedicels 1–4 mm. long; bracts minute. Calyx glabrous, tube campanulate, 0·5–1 mm. long; limb deeply cupuliform, drying pale, 1·25–2 mm. long, with ± triangular teeth 0·25–0·8 mm. long. Corolla white or cream, glabrous outside; tube 4–6 mm. long; lobes narrowly oblong, 3–4 mm. long, 0·9 mm. wide, ± appendiculate at the apex. Stamens with filaments 2·5–3·5 mm. long in short-styled flowers, 0·3–0·4 mm. long in long-styled flowers. Style 2–2·5 mm. long in short-styled flowers, 6–8 mm. long in long-styled flowers; stigma-lobes 1–2 mm. long. Drupes red, with 1–2 pyrenes, subglobose or ovoid, 4–6 mm. in diameter, glabrous, crowned with the calyx-limb; pyrenes ovoid or semiglobose, 4 mm. long, 2·7 mm. wide, 2·4 mm. thick, the dorsal surface 6–7-ribbed. Seeds ± semiglobose or ovoid, 4 mm. long, 2·7 mm. wide, 2·4 mm. thick, grooved on the ventral face; albumen not ruminate.

var. **amboniana**

Young stems and leaves glabrous. Fig. 5.

KENYA. Kwale District: Shimba Forest, 26 Apr. 1962, *Lucas, Jeffrey & Kirrika* 243!; Mombasa, Nov. 1884, *Wakefield*! & 30 Apr. 1927, *Linder* 2660!; Lamu District: Witu area, Utwani Ndogo Forest, Apr. 1957, *Rawlins* 426!
TANZANIA. Tanga District: near Kwale, 14 Nov. 1947, *Brenan & Greenway* 8324!; Pangani District: Bushiri, 11 Mar. 1950, *Faulkner* 545!; Uzaramo District: Hundogo, May 1952, *Procter* 53!
DISTR. **K7**; **T3**, 6; not known elsewhere
HAB. Coastal forest and derived thickets and scrub, grassland with scattered trees; 0–300 m.

SYN. *P. albidocalyx* K. Schum. in P.O.A. C: 390 (1895); De Wild., Pl. Bequaert. 2: 327 (1924); Petit in B.J.B.B. 36: 80, t. 5 (1966). Types: Tanzania, Tanga District, Kibafuta, *Holst* 2137 (B, holo.†, K, lecto.!, HBG, P, iso.)
P. albidocalyx K. Schum. var. *angustifolia* S. Moore in J.B. 45: 116 (1907); De Wild., Pl. Bequaert. 2: 328 (1924). Type: Kenya, Kwale District, Bome, *Kassner* 307 (BM, holo.!, K, iso.!)

NOTE. Petit uses the name *P. albidocalyx* for this species but Brenan (T.T.C.L.: 521 (1949)) clearly put this in the synonymy of *P. amboniana* (which S. Moore did not do) and the decision of the first person to unite two names of equal date must be followed.

var. **velutina** (*Petit*) *Verdc.* in K.B. 30: 260 (1975). Type: Kenya, Kwale/Kilifi District, Mazeras, *R. M. Graham* in *F.D.* 2337 (K, holo.!, EA, iso.!)

Young stems and leaves pubescent.

KENYA. Kwale/Kilifi District: Mazeras, *R. M. Graham* in *F.D.* 2337!; Tana River District: E. of Tana R., 3 km. N. of Wema, 10 km. NE. of Garsen, 15 July 1972, *Gillett & Kibuwa* 19928!; Lamu District: Utwani Ndogo Forest, Apr. 1957, *Rawlins* 427!
TANZANIA. Lushoto District: Gombelo [Gombero] to Bwiti, 19 Mar. 1918, *Peter* 22970!; Tanga District: Tanga, Apr. 1951, *Parry* 10! & Sawa, 22 Apr. 1968, *Faulkner* 4106!
DISTR. **K7**; **T3**; not known elsewhere
HAB. Coastal bushland and ? forest; 0–160 m.

SYN. *P. albidocalyx* K. Schum. var. *velutina* Petit in B.J.B.B. 36: 83 (1966)

NOTE. This variety occurs together with the typical variety.

45. **P. leucopoda** *Petit* in B.J.B.B. 36: 84 (1966). Type: Tanzania, E. Usambara Mts., Sigi R., 12·8 km. below Amani, *Verdcourt* 1748 (BR, holo., EA, K, iso.!)

Shrub 0·6–3 m. tall, with rather slender glabrous branches. Leaf-blades elliptic to narrowly oblong-elliptic, 3·5–16 cm. long, 1–6 cm. wide, acuminate at the apex, cuneate at the base, often deep brown on drying, completely glabrous or sometimes ciliate along the sides of the midrib; nodules situated in the midrib; petiole 0·3–2 cm. long, glabrous; stipules acuminate, the apex entire or very shortly bidentate, the base ovate-triangular, 3–6 mm. long, glabrous, deciduous. Flowers heterostylous, 5-merous, in branched glabrous panicles, often drying a salmon-red colour, 3–12 cm. long; peduncles 1·5–6 cm. long; secondary branches 0·5–1·5 cm. long; pedicels 3–5 mm. long, sometimes attaining 1 cm. in fruit; bracts and bracteoles minute. Calyx glabrous, broadly turbinate; tube ± 1 mm. long; limb deeply cupuliform, 1–1·7 mm. long, the teeth narrowly triangular 0·25–0·5 mm. long. Corolla white, glabrous outside; buds with 5 small " horns " at the apex; tube 3·5–5 mm. long; lobes oblong-lanceolate, 2·5–3 mm. long, 0·8–1 mm. wide, slightly appendaged at the apex. Stamens with filaments 1·5–3 mm. long in short-styled flowers, 0·3 mm. long in long-styled flowers. Style 2·5 mm. long in short-styled flowers, 5–6 mm. long in long-styled flowers; stigma-lobes 0·5–1 mm. long. Drupes bright cherry-red when ripe, with 2 pyrenes, ellipsoid, 6–8 mm. long, 4–6 mm. wide, crowned with the calyx-limb, glabrous; pyrenes semi-ellipsoid, 4·2 mm. long, 3·5 mm. wide, 2 mm. thick, the dorsal surface 4–5-ribbed. Seeds 3·5 mm. long, 3·2 mm. wide, 1·6 mm. thick, obtusely ribbed on the dorsal surface, the ventral surface ± 2-grooved; albumen not ruminate.

KENYA. Kilifi District: Rabai, Aug. 1937, *V. G. van Someren* 7180!; presumably coastal area near Mombasa,* *C. F. Elliott* 292!; Nyika Country, Jan. 1878, *Wakefield*!
TANZANIA. Tanga District: East Usambara Mts., 3·2 km. E. of site of Sigi Railway Station, 27 July 1953, *Drummond & Hemsley* 3487! & Lower Sigi R., 30 Nov. 1932, *Greenway* 3296!; Pangani District: Madanga, Mkuzi Katani, Kibubu, 24 July 1957, *Tanner* 3628!; Zanzibar I., Ufufuma, 23 Dec. 1930, *Vaughan* 1756!
DISTR. **K**7; **T**3, ?6; **Z**; not known elsewhere
HAB. Rain-forest, lowland fringing forest and also in drier coastal evergreen forest; 45–1050 m.

SYN. *P. sp.* sensu T.T.C.L.: 525 (1949)

NOTE. *Pócs et al.* 6279/F (Tanzania, E. Uluguru Mts., Kimboza Forest Reserve, 5 Nov. 1970, alt. 250 m.) has large leaves up to 17 × 9·5 cm., with obvious lateral nervation, but the calyx of *P. leucopoda*.

46. **P. schliebenii** *Petit* in B.J.B.B. 36: 86 (1966). Type: Tanzania, Ulanga District, Mahenge, *Schlieben* 1733 (BR, holo., B, BM!, HBG, K!, P, S, iso.)

Shrub 0·9–3 m. tall, with glabrous or minutely puberulous stems. Leaf-blades oblong-elliptic or narrowly elliptic, 4·5–18 cm. long, 1·5–9 cm. wide, apex with a long or short acumen, cuneate at the base, drying dark brown or blackish, glabrous save for some hairs beneath along the sides of the mid-rib; lateral nerves 9–17 on both sides; nodules often present in the midrib; petiole 0·3–8 cm. long, glabrous; stipules ovate-triangular, 3–5 mm. long, acuminate at the apex, entire or very shortly 2-toothed, glabrous. Flowers heterostylous, 5-merous, in much-branched panicles 4–13 cm. long, glabrous or minutely puberulous; peduncles 0–7 cm. long; secondary branches 0·4–2·4 cm. long; pedicels (0·5–)3–7(–11 in fruit) mm. long; bracts and bracteoles minute. Calyx glabrous; tube ovoid, ± 1 mm. long; limb cupuliform, 0·5–0·8 mm. long, lobes triangular or oblong, up to 0·2–1 mm. long. Corolla white, glabrous outside, buds acute, apiculate or rarely 5-horned at the apex; tube 3·25–4 mm. long; lobes triangular-ovate, 2–3 mm. long, 1·3 mm. wide, slightly appendaged at the apex. Stamens with filaments 1–2·5 mm. long in short-styled flowers, 0·2 mm. long in long-styled flowers.

* No. 290 was collected at Mazeras.

Style 3 mm. long in short-styled flowers, 4·5–5 mm. long in long-styled flowers; stigma-lobes 0·5 mm. long. Drupes red, with 2 pyrenes, ellipsoidal, 6·5–8 mm. long, 4–5 mm. wide, glabrous; pyrenes ellipsoidal, 6 mm. long, 4 mm. wide, the dorsal surface 6-undulate. Seeds similarly shaped, the dorsal face obscurely 6-ribbed, the ventral face with a median fissure; albumen not ruminate.

KEY TO INFRASPECIFIC VARIANTS

Peduncles developed, up to 7 cm. long:
  Leaf-blades 4·5–18 × 1·5–5 cm., with 9–17 pairs of lateral nerves; petiole 0·3–2 cm. long; inflorescences 4–13 cm. long . . a. var. **schliebenii**
  Leaf-blades 3–13 × 1–3·5 cm., with 7–13 pairs of lateral nerves; petiole 0·2–1 cm. long; inflorescences 0·7–5 (–7·5) cm. long (including the 0·5–2 (–4·5) cm. long peduncle) . . . . . . c. var. **parvipaniculata**
Peduncles obsolete or very short, 2–4(–10 in fruit) mm. long . . . . . b. var. **sessilipaniculata**

a. var. **schliebenii**; Petit in B.J.B.B. 36: 86 (1966)

Leaf-blades 4·5–18 cm. long, 1·5–5 cm. wide; lateral nerves 9–17 on either side; petiole 0·3–2 cm. long. Peduncles 2–7 cm. long. Inflorescences 4–13 cm. long.

TANZANIA. Morogoro District: Uluguru Mts., Mgeta R. gorge above the Hululu Falls, 19 Mar. 1953, *Drummond & Hemsley* 1676! & Uluguru Mts., *E. M. Bruce* 235A! & Uluguru Mts., Bunduki, 10 Jan. 1935, *E. M. Bruce* 490!; Pemba I., Ngezi Forest, *G. de C. Taylor* 80/2, in part!
DISTR. **T6**; **P**; not known elsewhere
HAB. Evergreen forest; 0–2000 m.

NOTE. Despite Petit's description as glabrous several specimens have the panicles and stems finely puberulous. *Semsei* 939 (Morogoro District, Mtibwa Forest Reserve, Sept. 1952) has a puberulous calyx and the inflorescence though in young bud does not altogether fit the other material. The species is not very well defined.
Vaughan 606 (Pemba, Ngezi Forest, 30 Aug. 1929) is probably a variant of *P. schliebenii* but the nodules are ± absent and the midribs of the leaves, young stems and inflorescence components have a very short sometimes ferruginous pubescence.

b. var. **sessilipaniculata** *Petit* in B.J.B.B. 36: 84 (1966). Type: Tanzania, Morogoro District, Turiani, *Semsei* 1499 (K, holo.!, BR, EA, iso.!)

Leaf-blades mostly wider, 7–18 cm. long, 2·5–9 cm. wide; lateral nerves 8–12 on either side; petiole 0·5–8 cm. long. Peduncles obsolete or very short, 2–4(–10 in fruit) mm. long. Drupes red, with 2 pyrenes, ellipsoid, 7–8 mm. long, glabrous, grooved in the dry state; pyrenes semi-ellipsoid, 5·5–6 mm. long, 3·5–4 mm. wide, 2 mm. thick, the ventral face slightly concave, the dorsal face 5-grooved. Seeds compressed, 4·5 mm. long, 3·5 mm. wide, 1·5 mm. thick.

KENYA. Kwale District: Shimba Hills, Makadara, Aug. 1929, *R. M. Graham* in *F.D.* 2031! & Shimba Hills, Giriama Point, 8 Feb. 1968, *Magogo & Glover* 14! & Shimba Hills, Lango ya Mwagandi [Longo Mwangandi], 11 Mar. 1968, *Magogo & Glover* 259!
TANZANIA. Tanga District: near Tanga, Siga caves, 8 Apr. 1952, *Bally* 8134! & same locality, 15 Aug. 1932, *Geilinger* 1282!; Morogoro District: Uluguru Mts., Tanana, 3 Feb. 1936, *E. M. Bruce* 767!; Pemba I., Piki, 1 Sept. 1929, *Vaughan* 629! & Ngezi Forest, *G. de C. Taylor* 80/2, in part!
DISTR. **K7**; **T3**, **6**; not known elsewhere
HAB. Evergreen forest; 0–1350 m.

c. var. **parvipaniculata** *Petit* in B.J.B.B. 36: 87, photo. 4/A (1966). Type: Kenya, Kwale District, Gazi, Gogoni Forest, *Dale* in *F.D.* 3570 (BR, holo., K, iso.!)

Leaf-blades 3–13 cm. long, 1–3·5 cm. wide; lateral nerves 7–13; petiole 0·2–1 cm. long. Inflorescences including the 0·5–2(–4·5) cm. long peduncle 0·7–5(–7·5) cm. long.

KENYA. S. Kavirondo District: Gucha [Kuja] R., Kabwoch [Kabuoch] Forest, 22 Apr. 1958, *Jarrett* SN/41! & Kuja R., 11 Jan. 1946, *Glasgow* 46/5!; Kwale District: Gazi, Gogoni Forest, Aug. 1936, *Dale* in *F.D.* 3570!
DISTR. **K5**, 7; not known elsewhere
HAB. Evergreen forest; ?60–1200 m.

NOTE. The isolated occurrence of coastal taxa in Kavirondo is not without precedence but not very usual. The species in this genus are so nondescript particularly when dry that conclusions about identity can easily be wrong. On the limited material available I see nothing to disagree with Petit's conclusions. Two poor specimens from Tanzania, Kilosa District, *Swynnerton* 744 & 759 are annotated by Petit as this variety but not cited.

47. **P. porphyroclada** *K. Schum.* in P.O.A. C: 390 (1895); De Wild., Pl. Bequaert. 2: 405 (1924); T.T.C.L.: 523 (1949); Petit in B.J.B.B. 36: 88 (1966). Type: Tanzania, W. Usambara Mts., Mtai, Shagayu [Schagai], *Holst* 2510 (B, holo. †, K, lecto.!)

Shrub 0·3–2·7 m. tall; stems slender, covered with dense almost velvety ferruginous pubescence when young, the older dark purplish brown, glabrescent. Leaf-blades narrowly elliptic, 2–11 cm. long, 0·7–3 cm. wide, acute or usually shortly acuminate at the apex, cuneate at the base, ± discolorous, usually drying dark reddish brown beneath, glabrous above and beneath save for ferruginous pubescence on the midrib beneath and sometimes also a few hairs on the lateral nerves as well; nodules situated in the midrib but linear and not readily visible; petiole 0·3–1·5 cm. long, with ferruginous pubescence similar to young stems; stipules prominently bilobed, the base ovate-triangular, 2–3·5 mm. long, the lobes acuminate, 2–3·5 mm. long, pubescent, deciduous. Flowers heterostylous, 5-merous, in sparsely branched panicles 3–6 cm. long; peduncles slender, 0·5–2 cm. long, puberulous; secondary branches ± 4 mm. long; pedicels 1–2·5 mm. long, glabrous or puberulous; bracts and bracteoles minute. Calyx puberulous or glabrescent; tube turbinate, 0·5–0·8 mm. long; limb cupuliform, ± 0·5 mm. long, the lobes triangular, 0·25 mm. long. Corolla white, glabrous outside; tube 2–2·75 mm. long; lobes oblong or ovate-triangular, 1·5–2 mm. long, 1 mm. wide, appendaged at the apex. Stamens with filaments 1·25 mm. long in short-styled flowers, included in long-styled flowers. Style 1 mm. long in short-styled flowers, 2·5 mm. long in long-styled flowers; stigma-lobes 0·5–0·75 mm. long. Drupes red, with 2 pyrenes, depressed-subglobose, ± 4·5–5·5 mm. in diameter (6·5 × 8 mm. in spirit), glabrous; pyrenes semiglobose, with thin smooth walls, 4 mm. in diameter, 3 mm. thick. Seeds semiglobose, 4 mm. long, 4·2 mm. wide, 2·8 mm. thick, with 2 parallel median grooves on the ventral surface but albumen otherwise entire.

TANZANIA. W. Usambara Mts., Mazumbai, 9 May 1953, *Drummond & Hemsley* 2448! & Shagayu Forest Reserve, 24 Mar. 1961, *Sudi* 4! & W. Shagayu Forest, 15 May 1953, *Drummond & Hemsley* 2564! & between Magamba and Gologolo, 6 Nov. 1947, *Brenan & Greenway* 8293!
DISTR. **T3**; known only from the W. Usambara Mts.
HAB. Shrub-layer of *Podocarpus-Ocotea* forest; 1500–2000 m.

48. **P. castaneifolia** *Petit* in B.J.B.B. 36: 89 (1966). Type: Tanzania, Morogoro District, Nguru ya Ndege Hill, *Schlieben* 3342 (BR, holo., B, K, iso.!)

Shrub 2–3 m. tall; stems glabrous or slightly pubescent when very young. Leaf-blades oblong-elliptic, 6–12 cm. long, 2–4 cm. wide, acute at the apex, cuneate at the base, ± coriaceous, somewhat shining above, drying reddish brown; very young leaves with ferruginous cobwebby indumentum beneath but soon entirely glabrous on both surfaces or with sparse hairs beneath;

midrib impressed above; nodules numerous, situated along the midrib; petiole 0·5–1·5 cm. long, ferruginous pubescent in young leaves but soon glabrous; stipules bilobed, the base oblong-triangular, 0·5–1·1 cm. long, the narrow acuminate lobes 3 mm. long, with indumentum similar to the petiole. Flowers heterostylous, 5-merous, in rather congested sparsely branched panicles 5–8 cm. long; peduncle 3–6 cm. long (said by Petit to be bialate), pubescent to glabrous; secondary branches up to 5 mm. long; pedicels up to 0·5 mm. long, glabrous or hairy; bracts small. Calyx hairy or glabrous; tube 0·75 mm. long; limb cupuliform, 1 mm. long, the margin scarcely or only shortly dentate. Corolla white, glabrous outside; tube 3·75–5 mm. long; lobes ± 3 mm. long. Stamens with filaments 3 mm. long in short-styled flowers, practically absent in long-styled flowers. Styles 3·5 mm. long in long-styled flowers, 5·5 mm. long in short-styled flowers; stigma-lobes 0·5–1 mm. long. Drupes not seen.

TANZANIA. Morogoro District: Nguru ya Ndege Hill, 30 Jan. 1933, *Schlieben* 3342! & Uluguru Mts., 8 Nov. 1932, *Schlieben* 2956b!
DISTR. **T6**; not known elsewhere
HAB. Rain-forest; 1200–1300 m.

49. **P. holtzii** (*K. Schum.*) *Petit* in B.J.B.B. 36: 90 (1966). Types: Tanzania, Dar es Salaam, Mogo Forest [Sachsenwald], *Engler* 2137, 2161 & 2186 (all B, syn. †); same locality, *Holtz* 301/02 (EA, neo.!)

Shrub 0·6–3 m. tall, with slender stems, densely pubescent with whitish or ferruginous hairs when young, later glabrescent, the hairs short and spreading, rather stiff on the old shoots. Leaf-blades small, narrowly elliptic to oblong, 1–4 cm. long, 0·5–1·5 cm. wide, acute or subobtuse at the apex, cuneate at the base, discolorous, glabrous above and beneath, or with some hairs bordering domatia in the nerve-axils and at the base of the midrib beneath; nodules mostly 2, situated at the base of the midrib; petiole 1–2·5 mm. long, densely pubescent with hairs similar to those on the stem; stipules prominently bilobed, becoming blackish, rather rigid and brittle but not persistent; the triangular base 1–2 mm. long, pubescent, the lobes filiform-acuminate, 1–4 mm. long. Flowers heterostylous, 5-merous, in sessile few-flowered subumbellate inflorescences ± 0·5 cm. long; pedicels 1–3 mm. long, glabrous or slightly pubescent. Calyx-tube obconic, 0·75 mm. long, ± pubescent; limb very short, 0·5–0·75 mm. long, ± glabrescent; lobes very narrowly triangular or linear, ± unequal, 0·5–2·5(–4) mm. long, glabrous or ± pubescent, occasionally bifid. Disc fleshy. Corolla white; buds hairy at apex; tube 3–3·5 mm. long; lobes oblong-elliptic, 1·7–3 mm. long, ± 1 mm. wide, thickened and with a small appendage at the apex. Stamens with filaments 2 mm. long in short-styled flowers, almost obsolete in long-styled flowers. Styles 2 mm. long in short-styled flowers, 3·5 mm. long in long-styled flowers; stigma-lobes 0·5–0·7 mm. long. Drupes crimson-red, with 2 pyrenes, subglobose, ± 5 mm. in diameter, glabrous, crowned by the persistent calyx; pyrenes with dorsal surface slightly ribbed. Seeds semiglobose, 2·5–3 mm. long and wide, 1·5–2·8 mm. thick, the ventral face with a rather broad fissure; albumen not ruminate.

var. **holtzii**

Leaf-blades with upper surface glabrous and lower surface glabrescent.

KENYA. Kwale District: 6·4 km. from junction of Mrima Hill road with Mombasa–Shimoni road, near Mwandeo, 5 Sept. 1957, *Verdcourt* 1884! & Buda Mafisini Forest, 19 Aug. 1953, *Drummond & Hemsley* 3895! & Shimba Hills, Pengo Forest, 9 Feb. 1953, *Drummond & Hemsley* 1190! & Kwale, 14 Jan. 1964, *Verdcourt* 3915!

TANZANIA. Uzaramo District: Kisiju, Kiregese [Kurekese] Forest Reserve, Sept. 1953, *Semsei* 1388!; Rufiji District: Mafia I., Utmaini–Miwani, 3 Oct. 1937, *Greenway* 5379!
DISTR. **K7**; **T6**; not known elsewhere
HAB. Evergreen forest, also *Syzygium-Pandanus* association in valley bottoms; 6–360(–450) m.

SYN. *Anthospermum holtzii* K. Schum. in E.J. 34: 340 (1904); T.T.C.L.: 482 (1949); K.T.S.: 425 (1961)

var. **pubescens** *Verdc.* in K.B. 30: 261 (1975). Type: Tanzania, 25·6 km. W. of Dar es Salaam on Morogoro road, *Welch* 397 (K, holo.!, EA, iso.)

Leaf-blades with upper surfaces finely pubescent; lower surfaces pubescent on the main nerves or all over.

KENYA. Kwale District: Shimba Hills, Marere Hill, 7 Mar. 1968, *Magogo & Glover* 227!
TANZANIA. Pangani District: Msubugwe Forest, 3 Mar. 1956, *Tanner* 2633!; Uzaramo District: 25·6 km. W. of Dar es Salaam on Morogoro road, 10 Oct. 1957, *Welch* 397! & near Kibaha, 2 Apr. 1970, *Harris & Flock* 4490!
DISTR. **K7**; **T3, 6**; not known elsewhere
HAB. Evergreen forest and dry riverine thicket; 90–435 m.

NOTE. A specimen from Tanzania, *Ngoundai* 194 (Rufiji District, Ngulakula Forest Reserve, Dimani, 19 Dec. 1968) is intermediate with *P. linearisepala* Petit. *Ludanga* 859 from the same area has the small leaves of *P. holtzii* but long calyx-lobes; the leaves are also slightly pubescent.

50. **P. linearisepala** *Petit* in B.J.B.B. 36: 91 (1966). Type: Zaire, Katanga, Lukafu, *de Witte* 61 (BR, holo.)

Shrub (or ? small tree) 0·6–2·5(–4) m. tall, with slender stems, densely pubescent with whitish or ± ferruginous hairs when young, glabrescent, blackish- or greyish-purple and longitudinally ridged when older; in some specimens the internodes are short and the shoots very nodose and roughened in appearance. Leaf-blades elliptic to oblong-elliptic or in some forms almost round, 1–9 cm. long, 0·5–4·5 cm. wide, rounded to distinctly acuminate at the apex, narrowly cuneate to rounded at the base, rather markedly discolorous, densely to rather finely puberulous or pubescent on both sides, often rather thickly so beneath in young leaves; nodules few, situated by the lower part of the midrib and visible beneath at its side; petiole 0·2–1·3 cm. long, puberulous to densely pubescent; stipules prominently bilobed, becoming rather thick and somewhat persistent, the triangular base 3–5 mm. long, the lobes acuminate, 1·5–3(–4) mm. long, drawn out into fine points when very young but soon lost in older ones, at first ± ferruginous pubescent, later glabrous. Flowers heterostylous, 5–6(–7)-merous, in few-flowered condensed subumbellate inflorescences 0·5–2·5 cm. long; peduncle 0·3–2·3 cm. long, rather densely pubescent; pedicels 1–3 mm. long, pubescent; bracts minute. Calyx pale, pubescent; tube obconic, 1 mm. long; limb very short or up to 0·75 mm. long; lobes linear or narrowly linear-triangular, 2–4 mm. long; disc fleshy. Corolla white or greenish cream, pubescent outside; tube 3–5·5 mm. long; lobes narrowly-oblong, 2·5–3 mm. long, 0·8 mm. wide. Stamens with filaments 2–2·5 mm. long in short-styled flowers, almost obsolete in long-styled flowers. Style 2·5–4 mm. long in short-styled flowers, 5–5·5 mm. long in long-styled flowers; stigma-lobes 0·5–0·75 mm. long. Drupes reddish or scarlet, with 1–2 pyrenes, subglobose, ± grooved between the pyrenes where 2, 3·5–5 mm. in diameter, glabrous or glabrescent, crowned with the persistent calyx. Seeds red-brown, subglobose, 4 mm. long, 3·5 mm. wide, 2·5 mm. thick, the ventral face with a T-shaped median fissure; albumen not ruminate.

var. **linearisepala**

Leaf-blades more oblong-ovate, acuminate, rather less densely pubescent. Young shoots whitish pubescent or often less obviously ferruginous. Shrub over 1 m. tall.

TANZANIA. Ufipa District: Kawa R., 30 Dec. 1956, *Richards* 7402! & 8 km. W. of Kate, Dec. 1954, *Procter* 323!; Songea District: Gumbiro, 24 Jan. 1956, *Milne-Redhead & Taylor* 8410!
DISTR. **T4**, 7, 8; Zaire, Mozambique, Zambia, Malawi
HAB. Riverine forest, steep rocky gorges, also *Brachystegia* woodland on sand; 630–1500 m.

SYN. *P. sp.* sensu Brenan in Mem. N.Y. Bot. Gard. 8: 453 (1954); F.F.N.R.: 418 (1962)

var. **subobtusa** *Verdc.* in K.B. 30: 261 (1975). Type: Tanzania, just S. of Songea, Matagoro Hills, *Milne-Redhead & Taylor* 10806 (K, holo. !, EA, iso.)

Leaf-blades more ovate, subacute to rounded at the apex, densely pubescent, particularly beneath. Young shoots covered with very obvious dark ferruginous pubescence. Shrub about 0·6 m. tall.

TANZANIA. Songea District: Matagoro Hills, 14 June 1956, *Milne-Redhead & Taylor* 10806!
DISTR. **T8**; not known elsewhere
HAB. Cracks in exposed rocks; 1380 m.

NOTE. This may merely be an ecotype of exposed positions but is distinctive in appearance although intermediates occur in Malawi.

51. **P. brevicaulis** *K. Schum.* in P.O.A. C: 391 (1895); De Wild., Pl. Bequaert. 2: 340 (1923); T.T.C.L.: 523 (1949); Petit in B.J.B.B. 36: 94 (1966). Type: Tanzania, E. Usambara Mts., Gonja, *Holst* 4270 (B, holo. †, HBG, K, iso. !)

Subshrub 0·1–0·5 m. tall, with simple or rarely branched glabrous erect stems, the lower parts decumbent, reddish, roughened with leaf-scars. Leaf-blades pale, obovate, oblong-obovate or narrowly to broadly elliptic, 5–18 cm. long, 2–8 cm. wide, acute to shortly acuminate at the apex, cuneate at the base, glabrous; nodules situated in petiole; petiole 2·5–6 cm. long, glabrous; stipules prominently bilobed, purplish when dry, ± scarious, deciduous, the base triangular, 2–7 mm. long, the lobes acuminate, 2–4 mm. long, glabrous. Flowers heterostylous, 5-merous, in 3–5-branched panicles, each branch with several flowers, congested, ± 7 mm. in diameter, internodes between pairs of branches 0·5–1·7 cm. long; peduncles 3–7 cm. long; pedicels 1–2 mm. long; bracts and bracteoles minute. Calyx glabrous; tube obconic, 0·8 mm. long; limb cupular, 0·75–1 mm. long; lobes obtusely triangular, 0·5 mm. long. Corolla white, glabrous outside; tube 3–4 mm. long; lobes triangular-elliptic, 2·5–3 mm. long, 1–1·2 mm. wide. Stamens with filaments ± 1 mm. long in short-styled flowers, ± obsolete in long-styled flowers. Style ± 2 mm. long in short-styled flowers, 4·5 mm. long in long-styled flowers; stigma-lobes 0·5–0·75 mm. long. Drupes green at first, later red, ovoid-subglobose, 6–7 mm. in diameter, crowned by the persistent calyx, containing 2 semi-ellipsoid pyrenes, 6 mm. long, 4–5 mm. wide, 2·5 mm. thick, rugose. Seeds similar in size and shape to the pyrenes, rugose but without any fissures; albumen not ruminate.

TANZANIA. Lushoto District: E. Usambara Mts., Amani–Monga, 30 Oct. 1935, *Greenway* 4147! & Amani, above Dodwe R., 18 Nov. 1947, *Brenan & Greenway* 8334! & Amani to Kwamkoro, 28 Dec. 1958, *Semsei* 2877!
DISTR. **T3**; not known elsewhere
HAB. Rain-forest floor; 850–1050 m.

SYN. *P. distegia* K. Schum. in E.J. 34: 337 (1904); De Wild., Pl. Bequaert. 2: 361 (1924); T.T.C.L.: 524 (1949). Types: Tanzania, E. Usambara Mts., Amani, Mt. Bomole, *Engler* 522, 536, 537 & 661 (all B, syn. †)

NOTE. No authentic material of *P. distegia* has been seen but I agree with Petit that it is almost certainly a synonym of the species described above.

52. **P. pandurata** *Verdc.* in K.B. 14: 349 (1960); Petit in B.J.B.B. 36: 95 (1966). Type: Tanzania, E. Usambara Mts., between Derema and Amani, near the Kwamkuyo Falls, *Verdcourt* 1741 (EA, holo. !, K, iso. !)

Perennial herb about (5–)10–15 cm. tall, the lower part of the mostly simple stem decumbent, the leafy apical parts erect, shortly rather densely pubescent. Leaf-blades rather glaucous blue-green, oblong-obovate, oblong-elliptic or ovate, (3–)4·5–11 cm. long, (1·5–)3–4·6 cm. wide, broadly subacute or almost rounded at the apex, cordate at the base, glabrous above, shortly pubescent on the nerves beneath, the margin irregularly and minutely undulate-crenulate; nodules very inconspicuous; petiole 0·5–3 cm. long, pubescent; stipules prominently bilobed, scarious, persistent, the bases triangular, 5–6 mm. long with 2 pubescent ribs extending into the linear acuminate lobes which are 3–8 mm. long. Flowers heterostylous, (4–)5-merous, in terminal few-flowered subcapitate panicles, 0·5–1 cm. long and wide; peduncles 1·5–5 cm. long, densely pubescent; pedicels 1–2 mm. long. Calyx-tube 2 mm. long, 0·8 mm. wide, puberulous, the free part ± glabrescent, 1 mm. long; lobes triangular, 0·5 mm. long. Corolla white, glabrous outside; tube 3–3·5 mm. long, 1·2 mm. wide at the base, 2·8 mm. wide at the apex, with a ring of hairs inside near the insertion of the filaments; lobes oblong-lanceolate, 2–2·8 mm. long, 1·5 mm. wide. Stamens included in long-styled flowers. Style (1·5–)2–2·1 mm. long; stigmas filiform, (0·5–)1 mm. long, papillate. Drupes depressed-ellipsoid, 4·5–5 mm. tall, 4–5 mm. wide, with a median rib, crowned by the persistent calyx, containing 2 pyrenes, slightly pubescent or glabrous. Seeds similar in shape to the pyrenes, 3 mm. long and wide, dorsally convex, ventrally slightly concave, rugulose, or where only one seed develops, subglobose 4–4·5 × 3·5–4 mm.; albumen not ruminate.

TANZANIA. Lushoto District: E. Usambara Mts., Kwamkuyo Falls, 17 Jan. 1917, *Zimmermann* in *E.A.H.* 7832 ! & above the Kwamkuyo R., 19 Apr. 1954, *Verdcourt* 1121 !; Morogoro District: Uluguru Mts., Kinole, *Schlieben* 3956
DISTR. T3, 6; not known elsewhere
HAB. Rain-forest; 300–1000 m.

53. **P. pumila** *Hiern* in F.T.A. 3: 207 (1877); K. Schum. in P.O.A. C: 391 (1895); De Wild., Pl. Bequaert. 2: 409 (1924); Petit in B.J.B.B. 36: 96, t. 6 (1966). Type: Mozambique, Morrumbala [Moramballa], *Kirk* (K, holo. !)

Subshrub, with erect shoots (2–)5–30 cm. tall from ± horizontal underground stems, or shrub 1–5 m. tall; stems shortly densely pubescent, glabrescent or glabrous, older ones longitudinally ridged. Leaf-blades narrowly elliptic to elliptic or sometimes ovate or obovate to oblanceolate, 2–14 cm. long, 0·5–5 cm. wide, rounded to acute or rarely acuminate at the apex, narrowly to rather broadly cuneate at the base, usually markedly discolorous, the under surface pale or quite whitish, glabrous or finely puberulous to pubescent above, glabrous or with nerves or less often whole surface pubescent beneath, the surfaces often with a minute honeycomb reticulation in the dry state; domatia absent; nodules absent or almost obsolete or rarely dispersed in the lamina; petiole 0·1–1 cm. long, glabrous or shortly pubescent; stipules prominently bilobed, the base ovate, 2·5–5 mm. long, the lobes acuminate, 1·5–4 mm. long, the base somewhat persistent. Flowers heterostylous, 5-merous, in few–many-flowered condensed subumbelliform cymes 1–1·5 cm. in diameter; peduncles 0·5–6·5 cm. long, glabrous or puberulous; pedicels 1–5(–7) mm. long, ± glabrous; bracts obsolete. Calyx turbinate, 1 mm. long, glabrous or slightly pubescent; limb cupuliform, glabrous or ± pubescent, 1–2 mm. long, the lobes unequal, usually short, subtriangular or ovate, up to 0·3 mm. long or practically obsolete. Corolla white, glabrous outside; tube (2·5–)5–6·5 mm. long; lobes oblong-lanceolate, (2·5–)3–4·5 mm. long, 1 mm. wide, usually slightly appendaged at the apex. Stamens with filaments 2·5–3 mm. long in short-styled flowers, 0·2 mm. long in long-styled flowers. Style 3–4·5 mm. long in short-styled flowers, 6–7 mm. long in long-styled flowers; stigma-lobes 0·5–1·5 mm. long. Drupes green turning yellow, orange and finally becoming bright red, with 2 pyrenes, subglobose, 5–6 mm. in diameter, glabrous, crowned with the persistent calyx-limb; pyrenes pale, semi-ellipsoid, 4·2 mm. long, 3·6 mm. wide, 2 mm. thick, ± 6-ribbed dorsally, ventrally flat, all surfaces rhaphide-packed. Seeds dark red, semi-ellipsoid or semiglobose, 3·5 mm. long, 3·5 mm. wide, 2 mm. thick, with a V-shaped median fissure on the ventral face; albumen not ruminate.

KEY TO INFRASPECIFIC VARIANTS

Small subshrub 2–15(–30) cm. tall . . . . a. var. **pumila**
Shrubs 1–5 m. tall:
  Glabrous; flowering pedicels 3–7 mm. long . . b. var. **leuconeura**
  Glabrous, slightly pubescent or densely pubescent; flowering pedicels 1–3 mm. long:
    Slightly pubescent or glabrous . . . . c. var. **buzica**
    Densely pubescent . . . . . . d. var. **puberula**

a. var. **pumila**; Petit in B.J.B.B. 36: 96, t. 6 (1966)

Stems 2–15(–30) cm. tall.

TANZANIA. Rufiji, 10 Jan. 1931, *Musk* 152!; Rungwe District: Mulinda Forest, 30 July 1912, *Stolz* 1479!; Songea District: about 7 km. W. of Songea, 18 Jan. 1956, *Milne-Redhead & Taylor* 8262!; Lindi District: 3·2 km. W. of Nachingwea, 8 May 1962, *Boaler* 572B!
DISTR. T4, ?5, 6–8; Zaire (Katanga), Mozambique, Malawi, Zambia and Rhodesia
HAB. *Brachystegia* woodland; 15–1500 m.

SYN. *P. brachythamnus* K. Schum. & K. Krause in E.J. 39: 566 (1907); De Wild., Pl. Bequaert. 2: 339 (1924); T.T.C.L.: 522 (1949). Type: Tanzania, Tunduru District, between R. Matebende and Kwa-Mtira, *Busse* 1006 (B, holo. †, EA, iso. !, B, fragment of iso. !)

NOTE. *Hornby* 485 (Mpwapwa, 27 Feb. 1933) has been annotated by Petit as ?*P. pumila* var. *pumila* but the lamina definitely contains nodules. The venation and general appearance precludes it from being placed in *P. spithamea*.

b. var. **leuconeura** (*K. Schum. & K. Krause*) *Petit* in B.J.B.B. 36: 99 (1966). Type: Tanzania, Songea District, Ruvuma R., Mt. Makorro, *Busse* 861 (B, syn. † and fragment of lecto., EA, lecto. !)

Shrub 1–5 m. tall, totally glabrous; flowering pedicels 3–7 mm. long. Leaves usually drying a dark reddish brown.

TANZANIA. Chunya District: Rungwa Game Reserve, 1 km. W. of Itigi–Mbeya road, 28 Jan. 1969, *Mpemba* in *C.A.W.M.* 4630!; Songea District: Mironji [Milonyi] R., 5 Feb. 1901, *Busse* 983!; Lindi District: Mlinguru, 21 Dec. 1934, *Schlieben* 5769!
DISTR. T7, 8; not known elsewhere (but intermediates between this and the typical variety occur in N. Zambia)
HAB. *Brachystegia* woodland, sometimes riverine; up to 1450 m.

SYN. *P. leuconeura* K. Schum. & K. Krause in E.J. 39: 554 (1907); De Wild., Pl. Bequaert. 2: 381 (1924); T.T.C.L.: 525 (1949)

NOTE. In view of his treatment of the *P. kirkii* complex it seems strange to me that Petit did not consider this to merit full specific rank. Further study may well support this idea.

c. var. **buzica** (*S. Moore*) *Petit* in B.J.B.B. 36: 99 (1966). Type: Mozambique, Manica e Sofala, Boka, R. Buzi, *Swynnerton* 561 (K, lecto.!)

Shrub about 1 m. tall, slightly pubescent; flowering pedicels 1–3 mm. long. Leaves drying brownish and not so discolorous as in var. *pumila*. Corolla-tube mostly shorter than in var. *pumila*, 2·5–3·75 mm. long.

TANZANIA. Mpanda/Chunya District: Rungwa, *Mawalla* 4962!; Lindi, 4 Dec. 1926, *Migeod* 388!; Newala, 22 Jan. 1960, *Hay* 78!
DISTR. **T**4/7, 8; Mozambique, ? Malawi
HAB. Not recorded, said to occur in deep shade

SYN. *P. buzica* S. Moore in J.L.S. 40: 100 (1911); De Wild., Pl. Bequaert. 2: 346 (1924)
P. madandensis S. Moore in J.L.S. 40: 100 (1911); De Wild., Pl. Bequaert. 2: 386 (1924). Type: Mozambique, Manica e Sofala, Madanda Forest, *Swynnerton* 561 (BM, holo.!, K, iso.!)

NOTE. Despite Petit's statement that it can be separated from var. *leuconeura* by the pedicels being short, 1–3 mm. long, the fruiting pedicels in var. *buzica* are 2–5·5 mm. long in *Swynnerton* 1900 from Mozambique, a specimen he cites.

d. var. **puberula** *Petit* in B.J.B.B. 36: 100 (1966). Type: Tanzania, Lindi District, Tendaguru, *Migeod* 41 (BM, holo.!)

Shrub. Young stems, leaves, peduncles and pedicels pubescent. Pedicels short, ± 3 mm. long. Corolla-tube 4–5·5 mm. long.

TANZANIA. Kilosa District: Tembo, Jan. 1931, *Haarer* 1942!; Lindi District: Tendaguru, 20 Dec. 1925, *Migeod* 41! & same locality, 21 May 1929, *Migeod* 507!
DISTR. **T**6, 8; not known elsewhere
HAB. Grassland, woodland and fringing bushland; 210–750 m.

NOTE. *Migeod* 507 is described as a small tree.

54. P. sp. L

A many-stemmed prostrate shrubby herb with rhizomatous stock and leaves in ± rosettes at ground-level, the whole plant ± 10–15 cm. tall; stems pubescent apically, the lower parts ridged and glabrous. Leaf-blades elliptic to obovate-elliptic, 3·5–9 cm. long, 1·5–4·5 cm. wide, obtuse or quite rounded at the apex, narrowly cuneate at the base, pubescent above with short hairs and mainly on the nerves beneath; thickenings along the basal sides of the midrib beneath are probably nodules and there are incipient nodules in the lamina beneath; petiole 1–1·5 cm. long, pubescent; stipules almost immediately deciduous, probably triangular. Inflorescence very short, known only in fruiting state, probably only 2–3-flowered in flowering state; peduncles 1–1·5 cm. long, densely pubescent; pedicels ± obsolete. Fruiting calyx-limb about 2 mm. long including the 0·8 mm. long rounded ovate lobes, ciliate at the margins. Fruits subglobose, drying dull reddish brown with obscure white stripes, 4·5–6 mm. in diameter, strongly grooved in the dry state. Seeds compressed, ± 4 × 5 × 1·5 mm. when 2 pyrenes per fruit, shallowly grooved dorsally and with 2 ventral grooves in the seed; in another fruit with 1 pyrene, subglobose 4 mm. long, 3 mm. in diameter and with a ventral Y-shaped groove extending quite deeply.

TANZANIA. Dodoma District: 35 km. S. of Itigi Station on the Chunya road, 17 Apr. 1964, *Greenway & Polhill* 11603!
DISTR. **T**5; not known elsewhere
HAB. Open *Brachystegia*, *Combretum*, *Commiphora*, *Canthium* woodland; 1380 m.

55. P. sp. M

Subshrub about 50 cm. tall; stems pale, glabrous, thinly corky. Leaf-blades elliptic, 9·5–14·5(–?) cm. long, 3·2–6·4 cm. wide, obtuse or rounded at the apex, cuneate and narrowly decurrent at the base, glabrous, with nodules scattered in the lamina, margined; domatia absent; petiole mostly short, up to 8 mm. long; stipules ± 2 mm. long, bifid, very soon deciduous. Inflorescences branched panicles; peduncle 4·5–6·5 cm. long; secondary branches 0·3–2 cm. long; pedicels 2–5 mm. long. Fruits ? red, subglobose, 4 mm. long, 6–7 mm. wide; pyrenes subglobose, 3·5 mm. in diameter, rugulose, extremely densely packed with rhaphides. Seeds with a T- or mushroom-shaped ventral slit.

TANZANIA. Tabora District: Kombe to Kaliuwa, 27 Jan. 1926, *Peter* 35678 !; Lindi District: 3·2 km. W. of Nachingwea, 8 May 1962, *Boaler* 572A !
DISTR. **T4**, ?7, 8; not known elsewhere
HAB. Secondary *Brachystegia* woodland; 450–1100 m.

NOTE. *Milne-Redhead & Taylor* 8262A (Tanzania, Songea Government Rest Camp, 3 June 1956) does not seem to be the same as 8262 although both are annotated as *P. pumila* Hiern; it is I think the same as *Boaler* 572A. *Boaler* 572B is undoubtedly *P. pumila* so the two grow together. *Meinertzhagen* 'Iringa 10' (Tanzania, Iringa, July 1957) is probably a hairy form of this species.

56. **P. punctata** *Vatke* in Oest. Bot. Zeitschr. 25: 230 (1875); Hiern in F.T.A. 3: 203 (1877); K. Schum. in P.O.A. C: 391 (1895); De Wild., Pl. Bequaert. 2: 409 (1924); Bremek. in J.B. 71: 279 (1933); T.T.C.L.: 523 (1949); K.T.S.: 466 (1961), pro parte; Petit in B.J.B.B. 36: 110, photo. 4/c (1966). Type: Zanzibar, *Hildebrandt* 1136 (W, holo., BM, K, iso. !)

Shrub or small tree (0·6–)1·2–3 m. tall, with pale glabrous stems. Leaves opposite or rarely in whorls of 4; blades elliptic to ovate-elliptic, (1–)3–13 (–15) cm. long, (0·5–)1–6·3 cm. wide, obtuse, rounded or rarely ± acute or emarginate at the apex, cuneate at the base, glabrous or shortly puberulous on both surfaces, thin to coriaceous or subsucculent; nodules numerous, often thickly scattered in the blade; petiole glabrous, 0·2–2 cm. long; stipules ovate-triangular, 2·4 mm. long, scarcely to distinctly bifid at the apex, with very obvious rhaphides, glabrous outside but with lines of hairs at the base of the petioles and inside. Flowers sweetly scented, heterostylous, 5-merous, in rather condensed rounded branched panicles, (1–)3–8 cm. long; peduncle glabrous, 1–5 cm. long; secondary branches 0·3–1 cm. long; tufts of hairs are present in the angles of the branches; pedicels 1–3 mm. long, glabrous; bracts obsolete. Calyx-tube turbinate, 0·5–1 mm. long, glabrous; limb cupuliform, 0·6–1 mm. long, glabrous, truncate or the lobes represented only as faint undulations. Corolla white, glabrous outside; tube 3·5–5·5 mm. long; lobes oblong-elliptic, 2·5–3·5 mm. long, 1·2 mm. wide. Stamens with filaments 2–3 mm. long in short-styled flowers, 0·2–0·7 mm. long in long-styled flowers. Style 1·5–3·5 mm. long in short-styled flowers and 4·5–5·5 mm. long in long-styled flowers; stigma-lobes 0·5–1 mm. long. Drupes red, with 2 pyrenes, subglobose, 5–6 mm. tall, 6–9 mm. wide, slightly 2-lobed, glabrous; pyrenes subglobose, 5 mm. tall, 5·5 mm. wide, 4·5 mm. thick, the walls packed with rhaphides. Seeds red, subglobose, 4·5 mm. long, 5 mm. wide, 4 mm. thick, with a ventral T-shaped median fissure ± 2 mm. deep; albumen not ruminate.

KEY TO INFRASPECIFIC VARIANTS

Plant glabrous; leaf-blades ± thick, up to 13(–15) cm. long, 6·3 cm. wide . . . . a. var. **punctata**
Young stems sparsely to densely pubescent but soon glabrous:
  Leaf-blades ± thick, small, 1–3·5 cm. long, 0·5–2·5 cm. wide . . . . b. var. **minor**
  Leaf-blades thinner and more oblanceolate than in the other varieties, 3–9 cm. long, 0·8–2·2 cm. wide . . . . c. var. **tenuis**

a. var. **punctata**; Petit in B.J.B.B. 36: 110, photo. 4/c (1966)

Plant glabrous; leaf-blades attaining the maximum dimensions given above.

KENYA. Mombasa, English Point, 25 May 1934, *Napier* 3252 in *C.M.* 6282 !; Kilifi District: Malindi, Oct. 1951, *Tweedie* 1045 !; Lamu District: E. side of Lamu town, 14 Feb. 1956, *Greenway & Rawlins* 8907 !
TANZANIA. Pangani District: Pangani & Bushiri, 6 Nov. 1950 & 10 Dec. 1950, *Faulkner* 764 !; Uzaramo District: Dar es Salaam, 26 Feb. 1926, *B. D. Burtt* 251 !; Rufiji, 10 Jan. 1931, *Musk* 66 !; Zanzibar I., Mangapwani, 24 Jan. 1929, *Greenway* 1149 ! & Chwaka, 26 Jan. 1960, *Faulkner* 2474 !
DISTR. **K**7; **T**3, 6, 8; **Z**; **P**; Comoro Is.
HAB. Coastal bushland, savanna and forest, mostly on sand or coral rag, sometimes only just above high-tide mark; also in open places in mangrove associations, sandy foreshores, etc.; 0–45(–280) m.

SYN. *P. melanosticta* K. Schum. in Abh. Preuss Akad. Wiss.: 16, 26 (1894), *nom. nud.*
P. pachyclada K. Schum. & K. Krause in E.J. 39: 558 (1907); De Wild., Pl. Bequaert. 2: 401 (1924); Bremek. in J.B. 71: 279 (1933); T.T.C.L.: 525 (1949). Type: Tanzania, Kilwa District, Maliwe [Mariwe], *Busse* 457 (B, holo.† & fragment of iso. !, EA, iso. !)
P. bacteriophila Valeton, Ic. Bogor, 3: 187, t. 271 (1908). Type: Java, plant cultivated at Bogor, originally from Comoro Is., collector not stated (BZF, holo., L, iso.)
Apomuria punctata (Vatke) Bremek. in Verh. K. Nederl. Akad. Wet., Natuurk., ser. 2, 54 (5): 91 (1963)

b. var. **minor** *Petit* in B.J.B.B. 36: 112 (1966). Type: Kenya, Samburu to Mackinnon Road, *Drummond & Hemsley* 4070 (K, holo. !, BR, EA, S, iso.)

Young stems densely pubescent but soon glabrous; leaf-blades small, 1–3·5 cm. long. 0.5–2.5 cm. wide.

KENYA. Kwale District: between Samburu and Mackinnon Road, 31 Aug. 1953, *Drummond & Hemsley* 4070 ! & 9 km. E. of Samburu on Mombasa–Voi road, 16 Jan. 1972, *Gillett* 19555 !; Kilifi District: Dakabuka Hill, May 1960, *Dale* in *F.D.* 1074 !
TANZANIA. Handeni District: Handeni–Korogwe road, 3 Mar. 1954, *Faulkner* 1363 !
DISTR. **K**7; **T**3; not known elsewhere
HAB. Deciduous bushland; 300–350 m.

c. var. **tenuis** *Petit* in B.J.B.B. 36: 112 (1966). Type: Kenya, Kilifi District, Arabuko Forest, *Dale* in *F.D.* 3535 (BR, holo., EA, K !, iso.)

Young stems sparsely pubescent but soon glabrous; leaf-blades thinner and more oblanceolate than in other varieties, 3–9 cm. long, 0·8–2·2 cm. wide.

KENYA. Kilifi District: Arabuko Forest, *Dale* in *F.D.* 3535! & same locality, *R. M. Graham* in *F.D.* 1932 !
DISTR. **K**7; not known elsewhere
HAB. Coastal bushland or forest; below 100m.

SYN. [*P. punctata* sensu K.T.S.: 466 (1961), quoad *Dale* in *F.D.* 3535, *non* Vatke sensu stricto]

NOTE. Several specimens from localities well inland have been written up by Petit as *P. punctata* (e.g. *Salim* in *Peter* 19330, Tanga District, Mlinga Mt., Magrotto to Magila, 1 Feb. 1917, & *Peter* 19309, Magrotto to Muheza, 1 Feb. 1917, 670 m.). These are scarcely to be distinguished from *P. kirkii* var. *nairobiensis* and are certain links between the two. *P. punctata* is normally so well characterised ecologically that I have not considered *P. kirkii* to be a variety of it although logically this is probably the correct course to follow.
Harris 1022 (Tanzania, 3·2 km. N. of Tanga, 25 Sept. 1967) is an abnormally narrow leaved variant, the leaf-blades under 1 cm. wide.

57. **P. faucicola** *K. Schum.* in E.J. 34: 336 (1904); De Wild., Pl. Bequaert. 2: 364 (1924); Bremek. in J.B. 71: 277 (1933); T.T.C.L.: 524 (1949); Petit in B.J.B.B. 36: 123, photo. 4/D (1966). Type: Tanzania, E. Usambara Mts., Amani, *Engler* 606 (B, holo.†) & same locality, *Warnecke* in *Herb. Amani* 474 (K, neo. !, BM !, EA !, P, isoneo.)

Shrub or subshrub 0·5–3·6 m. tall, with glabrous branches. Leaf-blades broadly to narrowly elliptic or ovate, (6–)10–25 cm. long, (2·7–)4–12 cm. wide, acute to shortly acuminate at the apex, narrowly cuneate or rounded and then cuneate at the base, thin, usually grey-green, glabrous on both surfaces; nodules usually numerous, up to 2 mm. long, dispersed throughout the blade; petiole 1–9 cm. long, glabrous; stipules prominently bilobed, the base ovate-triangular, 3–9 mm. long, the lobes 1–4 mm. long, glabrous. Flowers heterostylous, 5-merous, in slightly to much-branched glabrous panicles 5–14 cm. long; peduncles 2·5–5·5 cm. long; secondary branches 0·5–1·5 cm. long; pedicels up to ± 1 mm. (or even 5–8 mm. in fruit) long; bracts almost obsolete. Calyx glabrous; tube turbinate, ± 1 mm. tall; limb cupuliform, 0·75–1 mm. tall, the lobes obsolete or broadly triangular, 0·2 mm. long. Corolla white, greenish, yellowish or cream, glabrous outside; tube 2–2·75 mm. long; lobes oblong-elliptic, 1·5–2·25 mm. long, 0·8 mm. wide. Stamens in short-styled flowers with filaments 1–2 mm. long, almost obsolete in long-styled flowers. Style ± 1 mm. long in short-styled flowers, 3·5–4·5 mm. long in long-styled flowers; stigma-lobes 0·3–0·75 mm. long. Drupes red, with 2 pyrenes, subglobose, but quite didymous when young, 5–6·5 mm. tall, 7–8(–10) mm. wide, somewhat bilobed, glabrous; pyrenes semi-globose or globose when only 1 developed, 4·5–5·5 mm. in diameter, 3 mm. thick. Seeds semiglobose or globose, 4·2 mm. in diameter, the ventral face with a broad or T-shaped median groove; albumen not ruminate.

KENYA. Kwale District: Buda Forest Reserve, 3 Nov. 1959, *Napper* 1375! & Mrima Hill, 25 June 1970, *Faden* 70/242! & Shimba Hills, Kwale Forest, 2 May 1968, *Magogo & Glover* 972!

TANZANIA. Lushoto District: E. Usambara Mts., Kwamkoro Forest Reserve, 25 May 1961, *Semsei* 3200! & W. slope of E. Usambaras, between Ngua and Magunga Estates, 26 June 1953, *Drummond & Hemsley* 3018!; Morogoro District: Nguru South Forest Reserve, E. slopes above Kwamanga, near Mhonda Mission, 5 Feb. 1971, *Mabberley* 656!

DISTR. **K**7; **T**3, 6–8; not known elsewhere

HAB. Rain-forest and forest edges, both upland and lowland; (150–)750–1500 m.

SYN. *P. amaniensis* K. Krause in E.J. 43: 152 (1909); De Wild., Pl. Bequaert. 2: 331 (1924). Types: Tanzania, E. Usambara Mts., Amani, *Vosseler* in *Herb. Amani* 827 (B, syn.†, EA, isosyn.!) & *Warnecke* in *Herb. Amani* 474 (B, syn.†, BM, EA, K, P, isosyn.!) & *Busse* 2260 (B, syn.†)

NOTE. *Perdue & Kibuwa* 11487 (Tanzania, Iringa District, Mufindi, Ipafu Hill, beyond Supreme Estate, 28 Sept. 1971) is a poor fruiting specimen which may belong here but apart from very distinct thick round nodules in the lamina there are linear nodules along the midrib.

58. **P. griseola** *K. Schum.* in E.J. 34: 337 (1904); De Wild., Pl. Bequaert. 2: 371 (1924); Bremek. in J.B. 71: 277 (1933); T.T.C.L.: 522 (1949); Petit in B.J.B.B. 36: 125 (1966). Types: Tanzania, E. Usambara Mts., between Muheza and Longuza, *Engler* 395 (B, holo. †) & Morogoro District, Turiani Falls to Mahonda Sawmill, *Brenan & Greenway* 8282 (K, neo.!)

Shrub or subshrub 0·2–1(–2) m. tall; young branches ± pubescent or hairy, often becoming ± glabrous and greyish white when older. Leaf-blades broadly to narrowly elliptic or oblong-elliptic, 10–22(–32) cm. long, 4–10(–13) cm. wide, acuminate at the apex, narrowed to the base, usually thin, drying grey-green, sparsely to densely pubescent or hairy on both surfaces; nodules usually numerous, up to 2 mm. long, dispersed throughout the blade; petiole 1–7 cm. long, sparsely to densely pubescent or hairy; stipules prominently bilobed, the base ovate-triangular, 6·5–9 mm. long, deciduous, the lobes 1·5–4 mm. long, acuminate, glabrescent to densely pubescent. Flowers heterostylous, 5-merous, in slightly to much-branched panicles 4–8 cm. long; peduncles 1·5–4 cm. long, glabrescent to pubescent; secondary branches 0·7–1·3 cm. long; pedicels ± 1 mm. long, glabrous or glabrescent; bracts almost obsolete. Calyx glabrous; tube conical, 1 mm. long; limb cupuliform, ± 1 mm. tall, the lobes practically obsolete. Corolla white, glabrous outside; tube 2·5–3·5 mm. long; lobes ovate, 2·5 mm. long, 1·2 mm. wide. Stamens in short-styled flowers with filaments ± 1 mm. long. Style in short-styled flowers ± 1 mm. long; stigma-lobes ± 0·75 mm. long. Drupes red, with 2 pyrenes, broadly subglobose, 5 mm. tall, 7–8 mm. wide, somewhat 2-lobed, glabrous; pyrenes red-brown, semiglobose, 5 mm. in diameter, 3 mm. thick. Seeds semiglobose, similar in size and shape, the ventral face with a median fissure; albumen not ruminate.

TANZANIA. Lushoto District: W. Usambara Mts., Mazumbai to Baga, *Peter* 52149 & E. Usambara Mts., Monga, *Zimmermann* in *Herb. Amani* 7890; Tanga District: Mlinga Mt., 4 Feb. 1917, *Peter* 52158!
DISTR. **T3**, 6; not known elsewhere
HAB. Evergreen forests, including riverine forest; 450–1300 m.

59. **P. sp. N**

Shrub about 1 m. tall, with pale yellowish brown finely striate glabrous stems. Leaf-blades narrowly elliptic, 8·5–13·5 cm. long, 2·5–3·6 cm. wide, acute at the apex, cuneate at the base, glabrous, densely covered with black nodules, rather thin; petioles 1–2 cm. long, slender; stipules with oblong base 3 mm. long, bifid, the lobes lanceolate 1·5 mm. long. Flowers heterostylous, 5-merous, in elongated branched panicles 6 cm. long excluding the 2·5 cm. long peduncle; secondary branches 0·7–3 cm. long; pedicels 2 mm. long, all glabrous. Calyx glabrous; tube ± oblong, 0·6 mm. long; limb 0·4 mm. long, almost truncate or very shallowly lobed. Corolla yellowish; tube 2 mm. long; lobes oblong-elliptic, 2·5 mm. long, 1·1 mm. wide. Stamens with filaments 2 mm. long in short-styled form (only one seen). Style 1·2 mm. long in short-styled form; stigma-lobes 0·3 mm. long. Fruits unknown.

TANZANIA. Morogoro District: NE. Uluguru Mts., 19 Nov. 1932, *Schlieben* 3005!
DISTR. **T6**; not known elsewhere
HAB. Rain-forest; 1500 m.

NOTE. Somewhat similar to *P. lauracea* (K. Schum.) Petit but presumably not related.

60. **P. kirkii** *Hiern* in F.T.A. 3: 206 (1877); K. Schum. in P.O.A. C: 391 (1895); De Wild., Pl. Bequaert. 2: 379 (1924); Bremek. in J.B. 71: 280 (1933), pro parte; Garcia in Mem. Junta Invest. Ultramar., sér. 2, 6: 42 (1959); Petit in B.J.B.B. 36: 126 (1966); Verdc. in K.B. 30: 262 (1975). Type: Mozambique, Morrumbala [Maramballa], *Kirk* 9 (K, holo.!)

Shrub or subshrub (0·1–)0·2–6 m. tall; stems glabrous to densely velvety pubescent, sometimes with yellow-brown hairs, the older stems often becoming glabrescent or glabrous. Leaf-blades narrowly to broadly elliptic, elliptic-obovate, ± ovate or oblong-elliptic, 2–18 cm. long, 0·5–9 cm. wide, rounded to acute or ± acuminate at the apex, cuneate at the base, glabrous or densely velvety pubescent above and beneath or sometimes with only rather sparse long hairs on the midnerve beneath, mostly rather thin but sometimes ± coriaceous, the margins often thin, drying yellow and sometimes very marked, sometimes wavy; domatia absent; nodules numerous, spot-like or rarely linear, scattered in the lamina; petiole 0·1–2·5 cm. long, glabrous or pubescent; stipules usually brown and ± scarious, ovate-triangular, 2·5–11 mm. long, bilobed at the apex, the lobes acuminate or filiform, 0·5–4(–6) mm. long, glabrous to densely pubescent. Flowers heterostylous, 5(–6)-merous, in panicles or umbels or collections of umbels, 1–14 cm. long, the actual component parts often quite dense; peduncles 0·5–9 cm. long, glabrous to densely pubescent; secondary branches 0·2–2·5 cm. long, similarly hairy; pedicels 1–3(–6 in fruit) mm. long, glabrous or hairy; bracts and bracteoles small, filiform or narrowly triangular, up to 4 mm. long. Calyx glabrous or hairy; tube 0·5–1 mm. tall; limb cupuliform, 0·75–1·5 mm. long, glabrous or hairy; lobes 0·2–0·75 mm. long or ± obsolete. Corolla white, cream, greenish or rarely yellow, glabrous outside; tube 2·5–6 mm. long; lobes oblong or elliptic, 2–3(–5) mm. long, 0·8–1·2 mm. wide. Stamens with filaments 1·5–3 mm. long in short-styled flowers, 0–0·5 mm. long in long-styled flowers. Style 1·5–3·75 mm. long in short-styled flowers, 3·5–6(–10) mm. long in long-styled flowers; stigma-lobes 0·25–0·75(–2) mm. long. Drupes red, with 2 pyrenes, subglobose, 5–7 mm. diameter or didymously subglobose, 7–8 × 4–6 mm., glabrous or sparsely hairy when young; pyrenes semiglobose, 4–4·5 mm. diameter, 3 mm. thick, slightly rugulose. Seeds semiglobose, similar in size, ventral face with a T-shaped median fissure but rest of albumen not ruminate.

KEY TO INFRASPECIFIC VARIANTS*

Plant glabrous or glabrescent:
- Leaf-blades crinkly at the margins, rather thick, usually small, 2–10 × 0·5–4 but typically 6·5 × 1·7 cm. . . . . b. var. **nairobiensis**
- Leaf-blades not crinkly at the margins, mostly thinner, usually larger, 5–18 × 1·5–9 cm. . . g. var. **mucronata**

Plant pubescent to densely velvety:
- Typically suffrutescent, with mostly ± herbaceous stems from a woody rootstock:
  - Subshrub with mostly herbaceous stems (15–)20–50(–90) cm. tall; leaf-blades 6–17 cm. long, 1·5–8 cm. wide, with punctate nodules (**T**3, 6, 7) . . . . f. var. **swynnertonii**
  - Subshrub 9–25 cm. tall; leaf-blades 4–8·5 cm. long, 1·1–6·4 cm. wide, with punctate and linear nodules intermixed (**T**4) . . . . h. var. **diversinodula**
- Typically shrubby:
  - Inflorescence typically umbellate; leaf-blades usually sparsely pubescent; petiole 0·1–1·5 cm. long . . . . a. var. **kirkii**
  - Inflorescence paniculate:
    - Leaf-blades typically densely velvety or at least densely pubescent, usually discolorous:
      - Leaf-blades usually 4 × 1·5 cm. (1·5–7 × 0·5–3·5), often crinkly at the margins; petiole 1–6 mm. long c. var. **volkensii**
      - Leaf-blades usually 9·5 × 3·5 cm. (7·5–15 × 2–8), usually not crinkly at the margins; petiole 0·5–1·2(–2) cm. long . . . . d. var. **tarambassica**
    - Leaf-blades glabrescent to pubescent, not so distinctly discolorous:

* These varieties are so ill-defined that many specimens can only be named after the range of variation has been studied and appreciated—this key is a guide to typical specimens.

Leaf-blades usually densely pubescent, broadly cuneate at the base into a petiole 0·5–3 cm. long (**K7**; **T1**, 2, 7) e. var. **hirtella**
Leaf-blades glabrescent to pubescent, narrowly cuneate at the base into a petiole 0·5–2·5 cm. long (**U1**–4; **T1**) g. var. **mucronata**

a. var **kirkii**

Shrub or less often a subshrub (0·2–)0·3–4·5 m. tall, the stems at first glabrescent to densely pubescent but at length glabrous. Leaf-blades elliptic or oblong-elliptic, 3–12 cm. long, 1–6 cm. wide, usually sparsely pubescent on both surfaces, or ± glabrous above, usually with longer hairs on main nerves beneath, the margins distinct but not especially crinkled; petiole 0.1–1.5 cm. long. Flowers mostly in umbels or collections of umbels but often ± paniculate.

TANZANIA. Arusha District: Arusha National Park, Longil swamp, 25 Jan. 1969, *Richards* 23806!; Ufipa District: Ikuu, 21 Jan. 1950, *Bullock* 2303!; Mpwapwa, 6 Feb. 1932, *Hornby* 426!
DISTR. **T2**, 4, 5, 8; Malawi Mozambique, Zambia and Rhodesia
HAB. Bushland, grassland with scattered trees; 1050–1370 m.

SYN. *P. petroxenos* K. Schum. in E.J. 39: 557 (1907); De Wild., Pl. Bequaert. 2: 403 (1924); Bremek. in J.B. 71: 270 (1933); T.T.C.L. 2: 523 (1949). Type: Tanzania, Songea District, shore of Lake Nyasa, Bendera, *Busse* 905 (B, holo.†, fragment of iso.!, EA, iso.!)

NOTE. Really not capable of definition—merely what is left after all the more distinctive variants have been removed. Numerous intermediates between this and var. *volkensii*, var. *swynnertonii* and the hairy form of var. *mucronata* occur.

b. var. **nairobiensis** (Bremek.) Verdc. in K.B. 30: 262 (1975). Type: Kenya, Nairobi, *Dawson* 490 (K, holo.!)

Small shrub 1–2·4 m. tall, with usually glabrous stems. Leaf-blades mostly narrowly elliptic, usually quite small, 2–10 cm. long, 0·5–4 cm. wide, typically about 6·5 × 1·7 cm., often shining above, quite glabrous or with sparse rather long hairs on the midrib beneath or rarely in intermediates with scattered hairs elsewhere beneath, with a very distinct margin which often dries yellowish and thin and is mostly very crinkly in life and dry; petioles 1–5 mm. long.

UGANDA. W. Nile District: Madi, 14 Dec. 1862, *Grant* 694!; Teso District: Kyere, Feb. 1933, *Chandler* 1075 in part!; Mengo District: Entebbe, lake-shore, Oct. 1922, *Maitland* 164! & N. of Entebbe, Kitubulu Forest, Nov. 1935, *Chandler* 1466!
KENYA. Northern Frontier Province: Marsabit, 5 June 1960, *Oteke* 40!; Nairobi, grounds of Prince of Wales School, 9 Oct. 1953, *Verdcourt* 1017!; Teita Hills, footpath between Wusi [Wuzi]–Ngerenyi road and Bura Bluff, 17 Sept, 1953, *Drummond & Hemsley* 4398!
TANZANIA. N. Mara District: headwaters of Mara R., 10 Nov. 1953, *Tanner* 1776!; Arusha District: Ngurdoto Crater National Park, Longil, E. Lake, 24 Feb. 1966, *Greenway & Kanuri* 12389!; Lushoto District: Segoma Forest, 14 Jan. 1970, *Faulkner* 4322!
DISTR. **U1**–3; **K1**, 3, 4, 6, 7; **T1**–3, ?4, 5–8; S. Ethiopia, Malawi
HAB. Grassland, bushland, thicket, forest edges, open woodland, often in rocky places, also mist forest and upland *Tarchonanthus* thicket; (150–)450–2011 m.

SYN. *P. nairobiensis* Bremek. in J.B. 71: 278 (1933); T.T.C.L.: 525 (1949); K.T.S.: 466 (1961); E.P.A.: 1018 (1965); Petit in B.J.B.B. 36: 116 (1966)
P. marginata Bremek. in J.B. 71: 279 (1933); T.T.C.L.: 525 (1949). Type: Tanzania, Kondoa District, Kolo, *B. D. Burtt* 1258 (K, holo.!, EA, iso.!)
P. punctata Vatke var. *hirtella* Chiov. in Miss. Biol. Borana, Racc. Bot.: 231, fig. 72 (1939). Types: Ethiopia, Javello, *Cufodontis* 387 & Mega, *Cufodontis* 626 (both FI, syn.)
P. ciliatocostata Cufod. in Nuov. Giorn. Bot. Ital., n.s. 55: 90 (1948) & in Phyton 1: 149 (1949). Types as for *P. punctata* var. *hirtella* Chiov.

NOTE. The varietal epithet *hirtella* cannot be used since there is already a var. *hirtella* under *P. kirkii* (see p. 96).

c. var. **volkensii** (*K. Schum.*) *Verdc.* in K.B. 30: 262 (1975). Type: Tanzania, Kilimanjaro, Marangu, *Volkens* 604 (B, syn.†, K, lecto.!, BM, isolecto.!)

Small shrub 0·5–2 m. tall, with stems mostly densely velvety pubescent at least when young. Leaf-blades mostly elliptic or narrowly elliptic, 1·8–7 cm. long, 0·5–3·5 cm. wide, typically 4 × 1·5 cm., usually discolorous, densely pubescent with short hairs above, pale velvety pubescent beneath with longer matted hairs, the margins distinct and often crinkly as in var. *nairobiensis*; petiole 1–6 mm. long.

KENYA. S. Mt. Kenya, between Kirimiri and foot of mountain, 27 Feb. 1922, *Fries* 2096!; Machakos, 1 Jan. 1968, *Mwangangi* 520!; Teita District: Mbololo Hill, 14 Feb. 1953, *Bally* 8583!

TANZANIA. Arusha District: Engare Nanyuki R. gorge, 1 Nov. 1971, *Greenway & Kanuri* 14836!; Lushoto District: W. Usambara Mts., Soni–Baga road, 1 km. SE. of Soni, 8 May 1953, *Drummond & Hemsley* 2429!; Kondoa District: Swagaswaga, 10 Jan. 1928, *B. D. Burtt* 882!

DISTR. **K**1, 4, 6, 7; **T**1–3, 5–7; not known elsewhere

HAB. Bushland, *Acacia* thicket, montane scrub, *Brachystegia* woodland and in rocky places and on lava cliffs; (300–)660–1800(–2000) m.

SYN. *P. volkensii* K. Schum. in P.O.A. C: 390 (1895); De Wild., Pl. Bequaert. 2: 438 (1924); Bremek. in J.B. 71: 280 (1933); T.T.C.L.: 522 (1949); K.T.S.: 467 (1961); Petit in B.J.B.B. 36: 113 (1966)

P. subhirtella K. Schum., P.O.A. C: 390 (1895); De Wild., Pl. Bequaert. 2: 427 (1924). Type: Tanzania, Kilimanjaro, Marangu, *Volkens* 1455 (B, holo.†, BM, iso.!)

[*P. punctata* sensu K. Krause in N.B.G.B. 10: 607 (1929), *non* Vatke]

P. kassneri Bremek. in J.B. 71: 278 (1933); K.T.S.: 466 (1961). Types: Kenya, Machakos District, Mukaa, *Kassner* 923 in part (K, holo.!) & 923 (BM, iso.!)

[*P. kirkii* sensu Bremek. in J.B. 71: 280 (1933), quoad syn. *P. subhirtella*]

NOTE. This is more or less identical with var. *nairobiensis* except for the usually very dense indumentum but some intermediates occur. The two varieties often occur together. The Kew sheet of *Kassner* 923 bears three scrappy specimens which were all referred to different 'species' by Bremekamp namely *P. volkensii*, *P. nairobiensis* and *P. kassneri*! What is more *Dummer* 5078 from Mukaa in the same district has a few less hairs than *volkensii* and a few more than *nairobiensis* and is referred to *P. kirkii*. All four are clearly the same species. Other sheets from Tanzania namely *Bruce* 829 (**T**6) and *Tanner* 1176 have been divided by Petit between *P. volkensii* and *P. swynnertonii* on apparently nothing but leaf size. He draws attention to this himself and hints that there is some doubt as to the specific distinctness of *P. volkensii*. *Thomas* 3917 (Uganda, Kampala, 16 May 1941) cited by Petit as *P. volkensii* I prefer to refer to var. *mucronata*. *Peter* 32607 (Tanzania, Kilosa, open end of Mukondokwa Valley, 28 Nov. 1925), consisting of 5 sterile shoots bearing extremely densely velvety hairy leaves, may be an extreme form of var. *volkensii*.

d. var. **tarambassica** (*Bremek.*) *Verdc.* in K.B. 30: 263 (1975). Type: Kenya, Baringo District, Kamasia, Tarambas Forest, *Dale* in *F.D.* 2436 (K, holo.!, EA, iso.!)

Shrub or small tree 1·2–6 m. tall, the stems at least when young very densely covered with pale yellowish brown or almost white hairs, later glabrescent or shortly pubescent. Leaf-blades narrowly elliptic, elliptic or ± ovate, 7·5–15 cm. long, 2–8 cm. wide (typically 9·5 × 3·5 cm.), bluntly pointed to acute at the apex or rarely ± obtuse, cuneate at the base, mostly discolorous, densely sometimes almost velvety pubescent above with short hairs, densely velvety beneath with longer more tangled hairs, but sometimes ± glabrous above and only pubescent beneath; the margins fairly distinct, scarcely crinkly; petioles 0.5–1.2(–2) cm. long.

UGANDA. Karamoja District: about 10 km. NE. of Kokumongole, June 1955, *M.S. Philip* 712!

KENYA. Northern Frontier Province: Uaraguess, 1 Dec. 1958, *Newbould* 3002!; Turkana District: Karasuk, Karikau [Karakau], 1 July 1956, *M.S. Philip* 793!; Baringo District: Kamasia, Aug. 1959, *Dale* 1038!

DISTR. **U**1; **K**1–4; not known elsewhere

HAB. Upland dry evergreen forest, e.g. *Podocarpus*, *Juniperus*, *Teclea*, *Croton*, etc., and grassland with *Podocarpus* thickets; (1000–)1500–2550 m.

SYN. *P. tarambassica* Bremek. in J.B. 71: 280 (1933); K.T.S.: 367 (1961), pro parte

NOTE. Typically this taxon is easy to recognise but outside the areas mentioned above specimens are found which are not easy to place and are intermediate with other varieties, particularly var. *volkensii*. Petit refers the Brussels duplicate of *Rammell* in *F.D.* 3322 (Kenya, Leroghi Plateau, Naiboi) to *P. volkensii* (loc. cit. p. 113, 114). The Kew material does not seem to me to be separable from var. *tarambassica* and the specimen is so cited in K.T.S.: 467 (1961). *Tanner* 1480 (Tanzania, Mwanza District,

Mbarika, Nghoyokoyo, 12 May 1953, alt. 1140 m.) is an example of a specimen which could be referred to var. *tarambassica* technically but is ecologically incongruous with the material cited above. *Eggeling* 1737 (Uganda, Acholi District, Chua area, Mt. Madi, Mar. 1935) is probably best referred here.

e. var. **hirtella** (*Oliv.*) *Verdc.* in K.B. 28: 321 (1973). Type: Tanzania, Kilimanjaro, *Johnston* (K, holo. !, BM, iso. !)

Shrub up to 4·5 m., with hairy stems. Leaf-blades broadly elliptic to almost round, 4–13 cm. long, 2–8 cm. wide, subacute to rounded at the apex, broadly cuneate at the base, densely finely pubescent on both surfaces, with distinct but scarcely crinkly margins; petiole 0·5–2·5 cm. long.

TANZANIA. Mwanza District: Bunegezi [Bunegeji], 12 Dec. 1951, *Tanner* 511 !; Shinyanga, *Koritschoner* 2004 ! & 2206 !; Kilimanjaro, lower slopes, *Johnston* !
DISTR. **K**?7; **T**1, 2, 7 (see note); not known elsewhere
HAB. Bushland, rocky hills; 600–1350 m.

SYN. *P. hirtella* Oliv. in Trans. Linn. Soc., ser. 2, 2: 336 (1887)
P. kirkii sensu Bremek. in J.B. 71: 280 (1933), quoad syn. *P. hirtella*

NOTE. Although Bremekamp had actually reduced this to *P. kirkii*, Petit considered it best kept distinct in the absence of better material because of its large branched inflorescences and wider leaf-blades. A specimen from Kenya (**K**7, Tsavo Park East, 10 Feb. 1969, *Hucks* 1104 !), having rather coarser indumentum on the lower surfaces of the leaf-blades which are very narrowly decurrent into petioles 2–3 cm. long, is probably best referred to this variety, likewise *Fuller* 11 (Tanzania, Mbeya District, 48 km. SW. of Mbeya on the Tunduma road, Ruanda Farm, 17 Jan. 1970, 1590 m.); it clearly shows that this cannot be kept a distinct species from *P. kirkii*.

f. var. **swynnertonii** (*Bremek.*) *Verdc.* in K.B. 30: 263 (1975). Type: Tanzania, Kilosa, *Swynnerton* 770 (BM, lecto. !, K, isolecto. !)

Suffrutex with mostly ± herbaceous stems, (15–)20–50(–90) cm. tall, from a woody rootstock (see note); stems densely pubescent or less often pubescent or almost glabrous. Leaf-blades elliptic or oblong-elliptic, 6–17 cm. long, 1·5–8 cm. wide, typically 11·5 × 3·5 cm., acute or subacute at the apex, cuneate at the base, densely sometimes ± subscabridly pubescent above, sparsely to densely pubescent or thickly woolly beneath with tangled hairs, the margins distinct but only slightly crinkly; petiole mostly very short, rarely up to 8 mm. long. Fig. 6.

TANZANIA. Handeni District: Mziha–Handeni road, about 32 km. from Handeni, 5 Apr. 1953, *Drummond & Hemsley* 2031 !; Morogoro, 18 Feb. 1935, *E. M. Bruce* 829 !; Iringa District: N. Gologolo Mts., 13 Sept. 1790, *Thulin & Mhoro* 982 !
DISTR. **T**3, 6–8; Mozambique and Malawi
HAB. Grassland, bushland, *Brachystegia* woodland, hills with wooded grassland; 280–1200 m.

SYN. ?*P. collicola* K. Schum. in E.J. 33: 364 (1903); De Wild., Pl. Bequaert. 2: 353 (1924); Bremek. in J.B. 17: 279 (1933); T.T.C.L.: 522 (1948); Petit in B.J.B.B. 36: 186 (1966). Type: Tanzania, Kilosa/Uzaramo District, E. of Mtondwe, *Stuhlmann* 8281 (B, holo. †)
P. swynnertonii Bremek. in J.B. 71: 277 (1933); T.T.C.L.: 522 (1949); Petit in B.J.B.B. 36: 115, t. 7 (1966)

NOTE. This is a very poorly defined variant; Petit himself, although stating that it is usually easily distinguishable from *P. volkensii*, admitted that some gatherings exist which contain specimens partly referable to one and partly to the other, e.g. *Bruce* 829 (Tanzania, Uluguru Mts., Morogoro) and *Tanner* 1176 (Tanzania, Geita Juma Is.). Moreover he refers in his paper to *Swynnerton* 771 (BM, K) one of the syntypes of *P. swynnertonii*, stating that it is *P. volkensii*, but the Kew sheet bears two specimens, one of which he has annotated as *P. volkensii* and the other as *P. swynnertonii*. The variety is interesting as showing the path by which such small subshrubs as *P. spithamea* evolved. K. Schumann does not mention nodules in his original description of *P. collicola* but Bremekamp who must have seen the type includes it in his so-called revision of the species. I think Petit's suggestion as to its identity is probably correct.

g. var. **mucronata** (*Hiern*) *Verdc.* in K.B. 30: 263 (1975). Type: Sudan, Djur, Seriba Ghattas, *Schweinfurth* 1843 (K, holo. !, P, iso.)

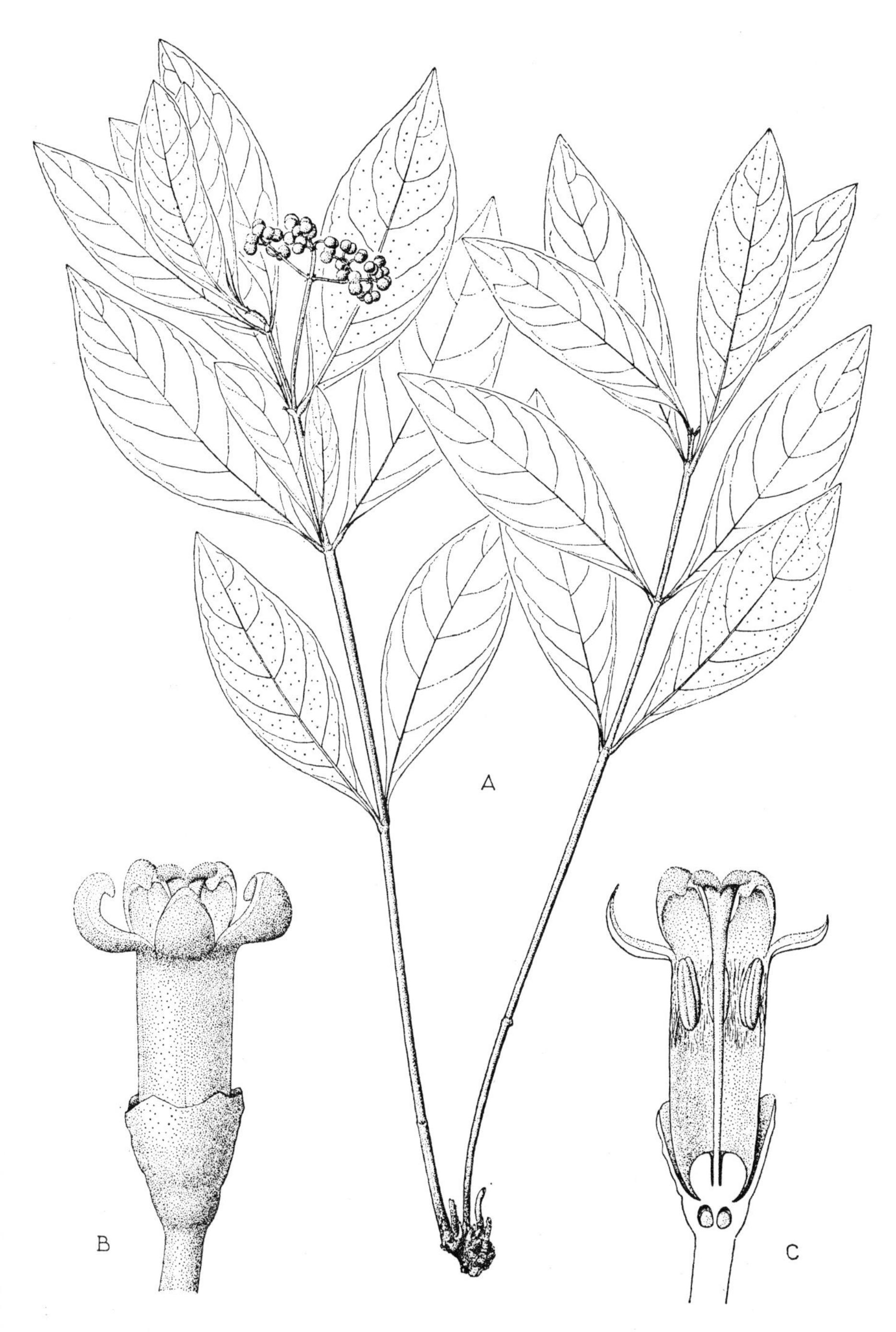

Fig. 6. *PSYCHOTRIA KIRKII* var. *SWYNNERTONII*—**A,** habit, × ½; **B,** long-styled flower, × 8; **C,** longitudinal section of same, × 8. A, from *Peter* 24272; B, C, from *Schlieben* 5909. Reproduced by permission of the Director of the Jardin Botanique National de Belgique.

Shrub (0·3–)0·6–3 m. tall, with glabrous or pubescent stems which are usually longitudinally rugose-striate. Leaf-blades elliptic to elliptic-ovate, 5–18 cm. long, 1·5–9 cm. wide, ± rounded to subacute or distinctly acute at the apex, cuneate and often very narrowly decurrent into the petiole at the base, either completely glabrous save for a very few hairs on the midrib beneath or pubescent on both surfaces; petiole 0·5–2·5 cm. long.

UGANDA. W. Nile District: Paidha [Payida], Dec. 1947, *Dale* U471!; Kigezi District: Ishasha Gorge, May 1950, *Purseglove* 3414!; Mengo District: Entebbe–Kampala road, Kajansi Forest, May 1935, *Chandler* 1233!
TANZANIA. Mwanza District: Saa Nane I., 17 Jan. 1965, *Carmichael* 1171! & Ito I., Nov. 1954, *Carmichael* 477!
DISTR. **U**1–4; **T**1; Zaire, Sudan
HAB. Riverine, lakeside and other forests, also thickets; 1125–1300 m.

SYN. *P. mucronata* Hiern in F.T.A. 3: 211 (1877); De Wild., Pl. Bequaert. 2: 390 (1924); Bremek. in J.B. 71: 279 (1933), pro parte; Petit in B.J.B.B. 36: 121 (1966)
P. maculata S. Moore in J.B. 44: 84 (1906); De Wild., Pl. Bequaert. 2: 385 (1924); Bremek. in J.B. 71: 278 (1933); F.P.N.A. 2: 364 (1947). Type: Uganda, Entebbe, *Bagshawe* 694 (BM, holo.!)
P. beniensis De Wild., Pl. Bequaert. 2: 333 (1924); Bremek. in J.B. 71: 278 (1933); F.P.N.A. 2: 364 (1947). Type: Zaire, Kivu, Beni, *Bequaert* 3422 (BR, holo.)
P. pubifolia De Wild., Pl. Bequaert. 2: 408 (1924); Bremek. in J.B. 71: 280 (1933); F.P.N.A. 2: 365 (1947). Types: Zaire, Kivu, Mayolo, *Bequaert* 3973 & Lesse, *Bequaert* 4184 & Orientale, Irumu, *Bequaert* 2860 (all BR, syn.)
P. rutshuruensis De Wild., Pl. Bequaert. 2: 417 (1924); Bremek. in J.B. 71: 278 (1933). Type: Zaire, Kivu, R. Rutshuru, *Bequaert* 6207 (BR, holo.)

NOTE. Petit admits that he has grouped together under this name plants showing considerable variability in their indumentum. I agree that the sort of geographical grouping he has made is the only compromise possible. One specimen from Kampala (*Thomas* 3917) which he has cited under *P. volkensii* is clearly better placed here. The practically glabrous obtuse-leaved specimens from S. Lake Victoria are clearly close to the type of *P. mucronata* (actually a subshrub about 0·5 m. tall), but on the other hand are so similar to *P. punctata* that they differ only in the components of the inflorescence being more dense and compact. *Dawkins* 521, from Kyewaga Forest near Entebbe, is scarcely to be distinguished from typical *P. kirkii*, having much smaller leaves than most of the Mengo material. *Purseglove* 3414 cited above is exceedingly close to *P. faucicola* and *P. griseola*, also forest species clearly derived from the same stock as *P. kirkii*. The slight difference in corolla length affords a technical separation of dubious value. Petit places them next to each other in his account.

h. var. **diversinodula** *Verdc.* in K.B. 30: 264 (1975). Type: Tanzania, Ufipa District, Chapota, *Bullock* 2043 (K, holo.!)

Subshrub 9–25 cm. tall, with dense short pubescence. Leaf-blades elliptic to triangular-ovate, 4–8·5 cm. long, 1·1–6·4 cm. wide, acute or obtuse at the apex, cuneate or truncate at the base, pubescent on both surfaces, with punctate and linear nodules intermixed in the lamina. Inflorescence condensed, oblong or ovoid, up to 2 cm. long, 1·5 cm. wide; peduncle 2–2·7 cm. long, densely pubescent.

TANZANIA. Ufipa District: Chapota, 4 Dec. 1949, *Bullock* 2043!
DISTR. **T**4; Zambia
HAB. Crevices in rocks, presumably in scarp woodland; 1950 m.

NOTE. May prove to be a distinct species but many species produce suffrutescent variants in this area.

GENERAL NOTE. I have in this genus as far as possible tried to follow the circumscriptions of species marked out by Petit in his monograph, but in the case of this group of bushland, savanna and open woodland "species" I have been quite unable to follow him. He and Bremekamp have used characters such as leaf size, indumentum and petiole length to delimit "species", some of which grow together. A number of single gatherings have been divided by these authors into two or even three species (see page 96) in cases where what is involved are probably only minor genetic variants differing only in indumentum or leaf-size, if indeed they are not merely variations due to habitat. I have kept up most of the names at varietal level, but even this may be disallowed by other workers. The variability of many species of this savanna zone is now so well known, particularly to field workers in Africa, as scarcely to need mention. *P. kirkii* is almost as variable as *Astripomoea malvacea* (Klotzsch) Meeuse or *Combretum molle* R. Br. The character of the umbelliform inflorescence used by Petit to distinguish *P. kirkii* really is of little significance; it is much more variable in this respect than he admits and he did not see much of the available material. Other workers may be tempted to sink a good deal further but I have maintained *P. kirkii* distinct from *P. punctata* very largely for ecological reasons, although intermediates can be found. Particularly puzzling is a population occurring in S. Lake Victoria which has been maintained as a variety of *P. kirkii* (see page 96).

Harris & Pócs 3219 (Tanzania, 40 km. from Morogoro on road to Kisaki, 6 Sept. 1969) is a very striking variant with leaf-blades 28 cm. long and 8 cm. wide, glabrous, and peduncle to 12 cm. long; if there is a population characterised by these outsize measurements then a varietal name would be well-merited.

61. **P. alsophila** *K. Schum.* in P.O.A. C: 390 (1895); De Wild., Pl. Bequaert. 2: 331 (1924); Bremek. in J.B. 71: 280 (1933); T.T.C.L.: 524 (1949); Petit in B.J.B.B. 36: 128 (1966). Type: Tanzania, W. Usambara Mts., Kwa Mshuza, *Holst* 9065 (B, syn. †, HBG, lecto., BR, K, P, isolecto.!)

Shrub (0·9–)1·2–3 m. tall or rarely a small tree up to 6 m. tall; stems glabrous, narrowly 2-ribbed in the dry state. Leaf-blades drying blackish, narrowly elliptic or elliptic-lanceolate, 2–12 cm. long, 0·6–4·5 cm. wide, subacute or narrowly rounded at the apex, cuneate at the base, glabrous on both surfaces, rather coriaceous; lateral nerves usually inconspicuous but midrib very prominent beneath; nodules conspicuous and raised, 0·5–1(–3) mm. long, scattered in the lamina but particularly along the midrib; petiole 0·1–0·5(–1·5) cm. long, glabrous. Stipules red-brown with 2 acuminate lobes; base ovate-triangular, 3–4 mm. long, the lobes linear 1–4 mm. long, the base glabrous or hairy with chestnut-coloured hairs on the margins or ± erose. Flowers heterostylous, 5-merous, in 2-3-branched panicles, each branch congested and subcapitate, 8 mm. long; peduncles 1–4 cm. long and secondary branches 3–8 mm. long, both glabrous, sometimes ± ribbed; bracts small; pedicels 0–3(–5 in fruit) mm. long, glabrous. Calyx glabrous; tube ovoid, ± 1 mm. long; limb cupuliform, ± 0·75 mm. long, the margin subtruncate or ± dentate, the teeth minute. Corolla white, glabrous outside and inside; tube 4–7 mm. long; lobes ovate-elliptic, 2·75–3 mm. long, 1·5 mm. wide. Stamens with filaments 1·75–4·5 mm. long in short-styled flowers, ± 0·5 mm. long in long-styled flowers. Style 5–7·5 mm. long in long-styled flowers, 2–3·5 mm. long in short-styled flowers; stigma-lobes 0·3–0·7 mm. long. Drupes red, with 2 pyrenes, subglobose, 6·5–7·5 mm. diameter, glabrous; pyrenes 5 mm. long, 4·5 mm. wide, 3·2 mm. thick. Seeds blackish, subglobose, 4·5 mm. long, 4·5 mm. wide, 2·7 mm. thick, with a median fissure on the ventral face; endosperm slightly or scarcely ruminate.

KENYA. Teita Hills, Ngangao Forest, 4 May 1972, *Faden et al.* 72/213!
TANZANIA. Moshi District: Narumuru [Naremura], Jan. 1956, *Carmichael* 540!; Lushoto District: W. Usambara Mts., Mtai–Malindi road, near Kidologwai, 19 May 1953, *Drummond & Hemsley* 2660! & W. Usambara Mts., Mkusi, Goose Green, 28 Dec. 1950, *Greenway* 8487! & Mazumbai, 9 May 1953, *Drummond & Hemsley* 2450!
DISTR. **K**7; **T**2, 3; not known elsewhere
HAB. Rather dry evergreen forest and derived pastures, cultivations, etc; 1050–1950 m.

SYN. *P. eickii* K. Schum. & K. Krause in E.J. 39: 555 (1907); De Wild., Pl. Bequaert. 2: 364 (1924); Bremek. in J.B. 71: 280 (1933); T.T.C.L.: 524 (1949). Types: Tanzania, W. Usambara Mts., near Kwai, *Eick* 127 & 330 (both B, syn.†)

NOTE. The quite extensive W. Usambara material seen is very uniform but the material from **T**2 is not identical, all having broader more elliptic leaf-blades with the lateral nerves much more obvious or even drying whitish. *F.S. Wilson* 35 (Tanzania, N. Pares, Same, Kiverenge, 25 Feb. 1957) not only has broader leaves but is a tree to 6 m. whereas W. Usambara plants never exceed 3 m. These **T**2 plants would really seem almost as distinct from *P. alsophila* as that is from *P. petiginosa*.

62. **P. petiginosa** *Brenan* in K.B. 4: 86 (1949); T.T.C.L.: 522 (1949); Petit in B.J.B.B. 36: 129 (1966). Type: Tanzania, Pare District, Kindoroko Forest Reserve, *Herring* 371 (K, holo. !, BM !, BR, FHO, iso.)

Shrub; young stems with ferruginous pubescence but becoming glabrescent or glabrous. Leaf-blades narrowly oblong-elliptic or oblanceolate, 5–15 cm. long, 1·5–4 cm. wide, acuminate at the apex, cuneate at the base, glabrous above, puberulous or glabrescent on the midrib and main nerves beneath; nodules numerous, dispersed throughout the lamina; petiole 0·5–2·5 cm. long, puberulous; stipules puberulous, with 2 acuminate lobes; base ovate-triangular, 2–4 mm. long; lobes linear, 3–4 mm. long. Flowers heterostylous, 5-merous, in panicles, the branches with flowers densely and sometimes subumbellately congested, 2–4 cm. long; peduncles 1–2 cm. long, ferruginous puberulous; secondary branches 0·3–1 cm. long; pedicels 1–1·5 mm. long, glabrous to puberulous; bracts small, ferruginous pubescent. Calyx-tube ovoid, 1 mm. tall, ferruginous puberulous; limb cupuliform, 0·6 mm. tall, glabrous or puberulous, the lobes very short or up to 0·3 mm. long. Corolla white; tube 6·5 mm. long, minutely puberulous outside, glabrous inside; lobes ovate, 3·5 mm. long, 1·5–1·8 mm. wide. Stamens with filaments 2·75 mm. long in short-styled flowers, 0·7 mm. long in long-styled flowers. Style 3 mm. long in short-styled flowers, 7·5 mm. long in long-styled flowers; stigma-lobes ± 0·75 mm. long. Drupes ?red, with 1–2 pyrenes, subglobose, 6 mm. in diameter, glabrous or sparsely pubescent but crowned with pubescent calyx-limb; pyrenes ellipsoid, semiglobose, 4·2 mm. long, 3·2 mm. wide. Seeds ellipsoid, 4 mm. long, 3 mm. wide, with a ventral groove; albumen not ruminate.

TANZANIA. Moshi District: N. Kilimanjaro, Kitenden, 1 Sept. 1950, *Carmichael* 36 !; Pare District: Kindoroko Forest Reserve, 11 Oct. 1928, *Herring* 371 ! & N. Pare Mts., Kilomeni to Kissangara, 27 June 1915, *Peter* 52123 !
DISTR. T2, 3; not known elsewhere
HAB. Floor of evergreen forest; 1500–2000 m.

63. **P. cryptogrammata** *Petit* in B.J.B.B. 36: 130, photo. 4/E (1966). Type: Tanzania, W. Usambara Mts., Balangai–Sakare, *Peter* 52148 (B, holo. !)

Shrub or subshrub to 2 m.; stems slender, at first ± bifariously pubescent but soon glabrescent or entirely glabrous and covered with thin pale grey cork. Leaf-blades narrowly elliptic to elliptic or rhomboid-elliptic, 3·5–14 cm. long, 1·5–4·5 cm. wide, acuminate or very acute at the apex, cuneate at the base, glabrous or sometimes pubescent beneath when young, thin; domatia absent; nodules dark green, numerous, linear or spot-like or often irregularly branched or radiate, up to 3 mm. long, densely dispersed in the lamina; petiole 0·5–2·5 cm. long, glabrous; stipules lanceolate to ovate-triangular, 4–7 mm. long, with 2 fine acuminate lobes at the apex 1–2 mm. long, glabrous or pubescent, soon deciduous. Flowers yellow, heterostylous, 5-merous, in sparsely branched small panicles, 1·5–4·5 cm. long; peduncles ± angular, 1–3·5 cm. long, glabrous or rarely sparsely pubescent; pedicels up to 2 mm. long; bracts almost obsolete. Calyx glabrous; tube obconic, 0·75–1 mm. long; limb cupuliform, 0·75 mm. long; lobes ± triangular, up to 1 mm. long. Corolla ?white, glabrous; tube 2–4·5 mm. long; lobes elliptic, 1·2–2·5 mm. long. Stamens with filaments 2 mm. long in short-styled flowers, 0·3 mm. long in long-styled flowers. Style 2 mm. long in short-styled flowers, not known for long-styled flowers; stigma-lobes ± 1 mm. long. Fruiting inflorescence 3-fruited, small, 1·5 cm. wide; peduncle 2-ribbed, 1·7–2 cm. long; pedicels 1–2 mm. long. Drupe pale, globose, 3–6 mm. in diameter, glabrous, not markedly ribbed, crowned with the calyx-limb consisting of 2 mm. long cylindrical part and ± 1–3 mm. long triangular lobes; pyrenes subglobose but flattened ventrally, 3·5 mm. long and wide, 2·3 mm. thick, rugulose. Seeds closely conforming to the pyrenes with a U-shaped ventral groove.

TANZANIA. Lushoto District: W. Usambara Mts., Garaya [Ngaraya]–Balangai, 9 Mar. 1916, *Peter* 15981! & Balangai–Sakare, 16 Mar. 1916, *Peter* 16127! & 16127a!; Morogoro District: Nguru Mts., between Kombola and Muskati, 20 Aug. 1971, *Schlieben* 12275!; Rungwe District: Kyimbila, Dec. 1911, *Stolz* 1043
DISTR. **T3, 6, 7**; not known elsewhere
HAB. Evergreen forest; 1200–1800 m.

64. **P. heterosticta** *Petit* in B.J.B.B. 36: 131, photo. 4/F (1966). Type: Zaire, Katanga, Parc National de l'Upemba, R. Lupiala, *de Witte* 2925 (BR, holo.)

Shrub or subshrub 0·2–2·7 m. tall, or small tree 6–7·5 m. tall, with glabrous, bifariously or less often extensively pubescent stems. Leaf-blades elliptic, narrowly elliptic, or oblanceolate to narrowly oblong-lanceolate, 3·5–19 cm. long, 0·5–5·5 cm. wide, acute or shortly acuminate at the apex, cuneate at the base, completely glabrous or shortly pubescent on lower part of main nerve beneath, thin to slightly coriaceous; domatia absent; nodules present, usually linear or spot-like along the main nerve beneath and usually but not always some spot-like ones in the lamina; petiole obsolete or 0·5–2 cm. long, glabrous or pubescent; stipules ovate or ovate-triangular, 2·5–6 mm. long, glabrous or pubescent, divided into 2 lobes at the apex, 0·5–3 mm. long, eventually deciduous. Flowers heterostylous, 5-merous, usually congested into umbel-like inflorescences, each branch with a ± head of flowers or in panicles 2–8 cm. long; peduncles often 2-ribbed, 1–9 cm. long, glabrous or pubescent; secondary branches 0·4–1·5 cm. long, glabrous or pubescent; pedicels 0·5–3(–5 in fruit) mm. long, glabrous or pubescent; bracts and bracteoles minute. Calyx-tube ovoid, 0·75–1 mm. long, glabrous; limb cupuliform, 0·5–0·75 mm. long; lobes irregular, very small or rarely 0·5 mm. long. Corolla white or yellowish green, glabrous outside; tube 2·5–4 mm. long; lobes elliptic-triangular, 2–2·5 mm. long, 0·8 mm. wide. Stamens with filaments 1·5–2·5 mm. long in short-styled flowers, 0·3 mm. long in long-styled flowers. Style 2–2·5 mm. long in short-styled flowers, 2·5 mm. long in long-styled flowers; stigma-lobes 0·5 mm. long. Drupes red, subglobose, 5–7 mm. long, 6–10 mm. wide, glabrous, with 2 pyrenes; pyrenes semiglobose, similar to the seeds. Seeds semiglobose, 4·2 mm. long, 5 mm. wide, 3 mm. thick; ventral face with a V- or U-shaped median fissure but rest of the albumen not ruminate.

KEY TO INFRASPECIFIC VARIANTS

Plant mostly glabrous:
  Leaf-blades less distinctly acuminate; lateral nerves 5–11 pairs; petioles 0–5 mm. long . . . a. var. **heterosticta**
  Leaf-blades more distinctly and more narrowly acuminate; lateral nerves 8–14 pairs; petioles 0·2–2 cm. long . . . . c. var. **plurinervata**
Plant mostly densely pubescent . . . . . b. var. **pubescens**

a. var. **heterosticta**; Petit in B.J.B.B. 36: 131 (1966)

Mostly entirely glabrous; pedicels obsolete or up to 5 mm. long, lateral nerves 5–11 pairs; leaf-blades mostly narrowly elliptic or narrowly obovate.

TANZANIA. Ngara District: Bushubi, Mumwendo, 2 July 1960, *Tanner* 5009!; Mwanza District: Geita, W. Uzinza, 8 June 1937, *B. D. Burtt* 6591!; Buha District: Kibondo, Kifura, Feb. 1955, *Procter* 402!

DISTR. **T1**, 4; Zaire, Burundi, Zambia
HAB. Riverine forest and *Brachystegia* woodland; 1200–1800 m.

b. var. **pubescens** *Verdc.* in K.B. 31 (1976). Type: Tanzania, Buha District, Kasulu–Kibondo, 112 km. from Kasulu, *Verdcourt* 3323 (K, holo.!, EA, iso.)

Stems, petioles, peduncles, etc. quite densely shortly pubescent; petioles obsolete or up to 5 mm. long, lateral nerves 6–9 pairs; leaf-blades elliptic or oblong-elliptic.

TANZANIA. Buha District, Kasulu–Kibondo, 112 km. from Kasulu, 15 Nov. 1962, *Verdcourt* 3323!
DISTR. **T4**; not known elsewhere
HAB. *Brachystegia*, *Pterocarpus* woodland; 1500 m.

c. var. **plurinervata** *Petit* in B.J.B.B. 36: 133 (1966). Type: Tanzania, Mpanda District, Kungwe–Mahali Peninsula, S. Pasagulu, *Harley* 9202 (K, holo.!)

Mostly entirely glabrous; petioles 0·2–2 cm. long; lateral nerves 8–14 pairs; leaf-blades narrowly elliptic to elliptic, more narrowly and distinctly acuminate than in the other varieties.

TANZANIA. Mpanda District: Mahali Mts., Sisaga, 4·8 km. W. of summit, 30 Aug. 1958, *Newbould & Jefford* 1941! & just below the summit of Musenabantu, 13 Aug. 1959, *Harley* 9328!; Songea District: about 60 km. E. of Songea, eastern Matagoro, 27 Mar. 1956, *Milne-Redhead & Taylor* 9410!
DISTR. **T1**, 4, 8; Malawi
HAB. Evergreen forest including riverine forest, bamboo forest and forest supported by mist; 1440–2100 m.

NOTE. I am not altogether convinced about the relationships between this variety and the others in the species.

65. **P. iringensis** *Verdc.* in K.B. 30: 264, fig. 3 (1975). Type: Tanzania, Iringa District, N. Gologolo Mts., *Thulin & Mhoro* 936 (K, holo.!, EA, iso.)

Presumably a shrub; stem glabrous, longitudinally wrinkled and with fairly marked petiole-scars. Leaf-blades lanceolate, 7–11 cm. long, 1·5–2·6 cm. wide, acute at the apex, cuneate at the base, glabrous above, drying minutely rugulose, glabrous beneath save for dense ferruginous hairs along the midrib when very young, probably ± coriaceous, said to be shining; a few nodules present along the side of the midrib which appear to break down into lesions; petioles thick, 0–8 mm. long; stipules with an ovate base 4 mm. long, bifid into 2 narrow acuminate lobes 4 mm. long. Flowers heterostylous, 5-merous, in branched purplish brown panicles; peduncle 3·5–4·5 cm. long, ± glabrous; secondary branches 1·5–10 mm. long; bracts and bracteoles small, ferruginous hairy as are the articulations of the inflorescence. Calyx purplish, with a few ferruginous hairs; tube campanulate, 1 mm. long; limb 0·8 mm. long; lobes triangular with rounded sinuses, 0·25 mm. long. Corolla white; tube 5 mm. long, glabrous outside, hairy inside; lobes oblong-lanceolate, 4 mm. long, 1·6 mm. wide. Stamens with filaments 4 mm. long in short-styled flowers, well exserted. Style 3·2 mm. long in short-styled flowers, thickened upwards; stigma-lobes 1·5 mm. long. Fruits not seen.

TANZANIA. Iringa District: N. part of Gologolo Mts., 13 Sept. 1970, *Thulin & Mhoro* 936!
DISTR. **T7**; not known elsewhere
HAB. Upland evergreen forest; 1800 m.

NOTE. This had been doubtfully named *P. heterosticta*, but the inflorescence does not agree; similarities to *P. alsophila* and its ally *P. petiginosa* are also not indicative of true relationships because in these the corolla is glabrous inside.

66. **P. spithamea** *S. Moore* in J.B. 48: 222 (1910); De Wild., Pl. Bequaert. 2: 423 (1924); Petit in B.J.B.B. 36: 134, t. 8 (1966). Type: Zambia, Katenina Hills, *Kassner* 2187 (BM, holo.!)

Subshrub with several stems 10–50 cm. tall from a mostly creeping woody rhizome; stems usually only leafy at the upper nodes, glabrous to densely pubescent, sometimes becoming covered with soft corky bark. Leaf-blades narrowly elliptic to elliptic but variable and sometimes obovate, elliptic-ovate or linear-oblong, 3–22 cm. long, 0·6–5·4 cm. wide, rounded to acute at the apex, narrowly cuneate at the base, glabrous to pubescent above and beneath, sometimes densely so on the venation beneath rendering it almost velvety to the touch, mostly distinctly paler beneath, thin to ± coriaceous, sometimes drying yellowish green; domatia absent; nodules mostly numerous, scattered in the lamina and often, at least when dry, not easy to see; petiole 0·3–2 cm. long, glabrous or pubescent; stipules ovate-triangular, 3–8 mm. long, entire at the apex or bifid for almost 1 mm., either thin and soon deciduous or thick and woody and ± persistent. Flowers heterostylous, 5-merous, in usually ± dense slightly branched inflorescences 1·5–5 cm. long; peduncle 0·5–3 cm. long, pubescent or glabrescent; secondary branches 0·3–1·3 cm. long, similarly hairy; pedicels 1–2 mm. long, attaining 5 mm. in fruit, glabrous or pubescent; bracts small, filiform, ± 3 mm. long. Calyx-tube subglobose, ± 1 mm. long, glabrous or pubescent; limb cupuliform, 0·75–1·5 mm. tall, usually glabrous, more rarely densely hairy; lobes mostly triangular-lanceolate, unequal, 1–2 mm. long, usually ± ciliate on the margins, or rarely lobes obsolete. Corolla white or cream, glabrous outside; tube 3–5 mm. long; lobes narrowly oblong, 2·5–3·5 mm. long, 0·8 mm. wide. Stamens with filaments 1·5 mm. long in short-styled flowers, 0·1–0·2 mm. long in long-styled flowers. Style 3 mm. long in short-styled flowers, 3·5–6·5 mm. long in long-styled flowers; stigma-lobes 0·5–1 mm. long. Drupes red, didymously biglobose, 3·5–6 mm. tall, 7–10 mm. wide, glabrous or pubescent, sometimes with only 1 pyrene due to abortion; pyrenes subglobose, 5 mm. long, 4 mm. wide and thick. Seeds subglobose, ventral surface with a deep Y- or V-shaped fissure and, save for the area of the junction of the 2 pyrenes, having a spherical air space between the hard outer layers and the albumen, which is not otherwise ruminate.

TANZANIA. Dodoma District: 43 km. on Itigi–Chunya road, 18 Apr. 1964, *Greenway & Polhill* 11632!; Njombe District: Mlangali–Njombe road, 3 Feb. 1961, *Richards* 14221!; Songea District: 9 km. W. of Songea, 18 Jan. 1956, *Milne-Redhead & Taylor* 8267!

DISTR. **T4**, 5, 7, 8; Zaire, Burundi, Mozambique, Malawi, Zambia and Angola

HAB. Grassland and chiefly *Brachystegia* (including mixed *Brachystegia*, *Pterocarpus* and *Combretum*) woodland, sometimes in rocky places; 960–1800 m.

SYN. [*Grumilea moninensis* sensu De Wild. in Ann. Mus. Congo, Bot., sér. 4, 1: 229 (1903) & in B.J.B.B. 9: 43 (1923) & Contr. Fl. Katanga: 215 (1921), *non* Hiern]
[*Psychotria swynnertonii* sensu Bremek. in J.B. 71: 277 (1933), quoad *Kassner* 2201, *non* Bremek. sensu stricto]

67. **P. cinerea** *De Wild.* in Ann. Mus. Congo, Bot., sér. 5, 2: 181, t. 43 (1907) & in Pl. Bequaert. 2: 353 (1924); Bremek. in J.B. 71: 278 (1933); Petit in B.J.B.B. 36: 154 (1966). Type: Zaire, near Elungu, *Dewèvre* 1063 (BR, holo.)

Subshrub or shrub 0·3–1·5 m. tall; young stems ± densely covered with short ± ferruginous hairs, the older stems glabrous. Leaf-blades narrowly to broadly elliptic, 3–10 cm. long, 0·9–3·5 cm. wide, acuminate at the apex, cuneate at the base, glabrous above, very shortly pubescent beneath on the main venation or ± all over the lower surface; domatia absent; nodules round, large, ± 0·5 mm. in diameter, dispersed throughout the lamina; petiole 0·2–1·5 cm. long, shortly densely hairy; stipules 5–7 mm. long, the base narrowly ovate-triangular, markedly bilobed, the lobes narrow, acuminate, 2·5–4 mm. long, glabrous or puberulous. Flowers heterostylous,

5-merous, in graceful panicles 2–5 cm. long; peduncles 1–3 cm. long, slender, shortly pubescent; secondary peduncles up to 1·2 cm. long; pedicels slender, 2–3 mm. long; bracts linear-triangular, 2 mm. long. Calyx-tube campanulate, 0·5 mm. long, glabrous or pubescent; calyx-limb ± 0·5 mm. long, the lobes broadly triangular (Petit's reference to calyx-tube 3 mm. long must be an error). Buds glabrous. Corolla white; tube 2·75–3 mm. long; lobes ovate-elliptic, 1·5–2 mm. long. Stamens with filaments 1·5 mm. long in short-styled flowers, almost obsolete in long-styled flowers. Style 2 mm. long in short-styled flowers, 3–4·75 mm. long in long-styled flowers. Drupes red, subglobose, 4–5 mm. in diameter, with 2 pyrenes, not ribbed; pyrenes dark purple-brown, semiglobose. Seeds white, similar in size, 3·5 mm. long, 4 mm. wide, 2·5 mm. thick, ventrally with 2 very shallow grooves, otherwise smooth; albumen not ruminate.

TANZANIA. Kigoma District: Kasakati, May 1965, *Suzuki* 242!
DISTR. **T4**; Zaire, Burundi
HAB. ? Evergreen forest; 1200 m.

68. **P. brevipaniculata** *De Wild.*, Pl. Bequaert. 2: 340 (1924); Bremek. in J.B. 71: 273 (1933); Petit in B.J.B.B. 36: 177, t. 11, photo. 5/G (1966). Type: Zaire, Mobwesa, *Lemaire* 241 (BR, lecto.)

Shrub or subshrub 20–75 cm. tall, the sometimes unbranched stems covered when young with dense short ± adpressed often ± brownish hairs. Leaf-blades oblong or elliptic-oblong, 4–14 cm. long, 2–6 cm. wide, acute to shortly acuminate at the apex, cuneate at the base, glabrous above, adpressed pubescent on the main nerves beneath and sometimes also between them, thin; domatia absent; nodules few to many, scattered throughout the lamina; petiole slender, 0·5–3 cm. long, shortly pubescent; stipules ovate-triangular, 6–8 mm. long, divided into 2 lobes at the apex 3–5 mm. long, pubescent and with hairy margins, deciduous. Flowers ± heterostylous, 4-merous, in small sparsely branched inflorescences 1·5–2·5 cm. long; peduncle obsolete, up to ± 1 cm. long, pubescent; secondary branches 1–7 mm. long, pubescent; pedicels 0–1 mm. long; bracts minute. Calyx-tube ovoid, ± 0·5–0·8 mm. long, glabrous or puberulous; limb cupuliform, 0·3–0·5 mm. long, glabrous, almost truncate. Corolla white, glabrous or sparsely puberulous outside; tube 3–3·5 mm. long; lobes triangular, 1–1·5 mm. long, 0·8 mm. wide. Stamens in long-styled flowers with filaments 0·2 mm. long, 0·5 mm. long in short-styled flowers. Style in long-styled flowers ± 3·3 mm. long, 1·2 mm. long in short-styled flowers; stigma-lobes 0·5–1 mm. long. Drupes red, depressed subglobose, ± 5 mm. diameter, glabrous, containing 2 pyrenes; pyrenes semiglobose, 4 mm. long and wide, 2·8 mm. thick, dorsally with 2–3 raised lines, ventrally concave. Seeds semiglobose, the ventral face bearing 2 faint submedian fissures; albumen not ruminate.

UGANDA. Bunyoro District: Budongo Forest, Apr. 1933, *Eggeling* 1205 in *F.D.* 1301!
DISTR. **U2**; Cameroun, Central African Republic and Zaire
HAB. Evergreen forest; 1100 m.

NOTE. Petit does not mention if this species is heterostylous or not and the evidence available is not adequate; I think it is weakly so.

DESCRIBED SPECIES OF UNKNOWN POSITION

There remain over a dozen species of which no authentic material has been seen by either Petit or myself. In such a difficult genus of rather featureless plants it has been impossible to equate them with known species. Where possible, suggestions have been made but there can be no certainty about these. No revisional work of any kind had been carried out on the genus prior to the destruction of the Berlin herbarium so that no other material had been equated with these unicate types, many collected by Stuhlmann. They have been listed in alphabetical order and left in the genera in which they were described, i.e. G = *Grumilea* and P = *Psychotria*.

69. **G. blepharostipula** *K. Schum.* in E.J. 28: 495 (1900); De Wild. in B.J.B.B. 9: 30 (1923); T.T.C.L.: 498 (1949), pro parte; Petit in B.J.B.B. 34: 214 (1964). Type: Tanzania, SE. Uluguru Mts., Ng'hweme, *Stuhlmann* 8807 (B, holo. †)

Shrub with stout subfleshy 4-angled glabrous stems. Leaf-blades oblong, 7–16 cm. long, 4–7 cm. wide, acute at the apex, cuneate at the base, glabrous on both surfaces, coriaceous; petiole 1–1·2 cm. long, glabrous; stipules ample, ovate, 1·7–2·2 cm. long and wide, obtuse, glabrous save for the ciliate margin. Flowers in short subcorymbose panicles; peduncles short. Only in bud.

TANZANIA. Morogoro District: SE. Uluguru Mts., Ng'hweme, Oct. 1894, *Stuhlmann* 8807
DISTR. **T6**
HAB. Evergreen forest; 1600 m.

NOTE. From the brief description this could be *P. sp. D*, *P. goetzei* or the very closely related *P. elachistantha*, very probably the last-named, one of the syntypes of which actually came from the same locality Ng'hweme. *P. goetzei* was described on p. 497 of the same paper and it seems unlikely K. Schumann would have described the same species twice in so short a space but with only single specimens available of perhaps different facies it is possible.

70. **P. bukobensis** *K. Schum.* in P.O.A. C: 391 (1895); De Wild., Pl. Bequaert.: 346 (1924); T.T.C.L.: 521 (1949); Petit in B.J.B.B. 34: 218 (1964). Type: Tanzania, Bukoba, *Stuhlmann* 3739 (B, holo. †)

Shrub with stout branches, pubescent with ferruginous hairs when young. Leaf-blades oblong, up to 17 cm. long, 5·5 cm. wide, acuminate at the apex, acute at the base, subcoriaceous, glabrous above, pubescent with red hairs beneath; petiole over 1·5 cm. long. Flowers subsessile in ample many-flowered sessile panicles 7 cm. tall, 10 cm. wide. Calyx 1 mm. long, repand-dentate, scarcely pilose. Corolla puberulous, tube 3·4 mm. long, lobes 1–1·5 mm. long.

TANZANIA. Bukoba, *Stuhlmann* 3739
DISTR. **T1**
HAB. Not known

NOTE. This could be *P. mahonii* var. *pubescens* (Robyns) Verdc. except for corolla size but it is impossible to be certain.

71. **P. cephalidantha** *K. Schum.* in E.J. 33: 361 (1903); De Wild., Pl. Bequaert.: 350 (1924); T.T.C.L.: 522 (1949); Petit in B.J.B.B. 34: 218 (1964). Type: Tanzania, Uluguru Mts., *Stuhlmann* (B, holo. †)

Habit not known but presumably shrubby. Stems slender, papillose rather than pilosulose. Leaf-blades lanceolate or oblong, 2·5–10 cm. long, 0·6–3 cm. wide, acuminate at the apex, narrowed to the base, glabrous above, subtomentose beneath particularly in the axils of the nerves, thin, brownish green above, more yellowish beneath; lateral nerves impressed above, prominent beneath; domatia present; petiole 5–8 mm. long, channelled above; stipules short, bicuspidate to the base. Inflorescences capitate; peduncle 1·5–3 cm. long. Drupes pale, ellipsoid, 7 mm. long, 5 mm. wide, ribbed, crowned with a short denticulate calyx.

TANZANIA. Uluguru Mts., without further locality, *Stuhlmann*
DISTR. **T6**
HAB. Presumably evergreen forest

72. **G. chaunothyrsus** *K. Schum.* in E.J. 28: 496 (1900); De Wild. in B.J.B.B. 9: 31 (1923); T.T.C.L.: 498 (1949); Petit in B.J.B.B. 34: 214 (1964). Type: Tanzania, SE. Uluguru, Ng'hweme, *Stuhlmann* 8774 (B, holo. †)

Tree with slender glabrous branches. Leaf-blades lanceolate, 4–10 cm. long, 1–2·7 cm. wide, acute at the apex, gradually narrowed to the base, glabrous on both surfaces, coriaceous; petiole 0·5–1 cm. long, glabrous, slightly bisulcate above; stipules ovate, 0·7–1 cm. long, coriaceous, deciduous. Panicles few-flowered, lax, glabrous; peduncle 3·5 cm. long; fruiting pedicels 5 mm. long. Drupes elliptic, 8 mm. long, 5 mm. wide, subcostate, crowned with the small 5-lobed calyx.

TANZANIA. Morogoro District: SE. Uluguru Mts., Ng'hweme, Oct. 1894, *Stuhlmann* 8774
DISTR. **T6**
HAB. Primary evergreen forest; 1500 m.

73. **G. diploneura** *K. Schum.* in E.J. 28: 496 (1900); De Wild. in B.J.B.B. 9: 32 (1923); T.T.C.L.: 498 (1949); Petit in B.J.B.B. 34: 215 (1964). Type: Tanzania, Uluguru Mts., Lukwangule Plateau, *Stuhlmann* 9109 (B, holo. †)

Shrub 1–2 m. tall with robust glabrous branches; bark black. Leaf-blades obovate or obovate-oblong, 6–10 cm. long, 3–6·5 cm. wide, apiculate at the apex, cuneate at the base, glabrous on both sides, very glossy; lateral nerves prominent on both sides, a double marginal nerve visible in upper half of lamina; petiole 1·2–2·5 cm. long, narrowly channelled above; stipules ovate-oblong, 1–1·2 cm. long, coriaceous, deciduous. Flowers not seen. Fruiting inflorescences 2·5–3·5 cm. wide; peduncle 4–5 cm. long. Drupes glabrous, didymously subglobose, 5–6 mm. wide, crowned by the 5-toothed campanulate calyx 2·5 mm. long.

TANZANIA. Morogoro District: Uluguru Mts., Lukwangule Plateau, Nov. 1894, *Stuhlmann* 9109
DISTR. **T6**
HAB. Upland evergreen forest; 2100 m.

74. **G. euchrysantha** *K. Schum.* in E.J. 28: 496 (1900); De Wild. in B.J.B.B. 9: 34 (1923); T.T.C.L.: 499 (1949); Petit in B.J.B.B. 34: 215 (1964). Type: Tanzania, SE. Uluguru Mts., Mt. Kikurungu, *Stuhlmann* 9245 (B, holo. †)

Shrub with fairly stout glabrous branches. Leaf-blades lanceolate, 5–9 cm. long, 1–2·5 cm. wide, acute at the apex, cuneate at the base, glabrous on both surfaces, coriaceous; petiole 7–9 mm. long, glabrous, subsulcate above; stipules very soon deciduous, not seen. Flowers sessile in terminal sparsely branched panicles 2–4·5 cm. long, the ultimate components capitulate; peduncle 1–3 cm. long. Calyx-tube glabrous, together with the truncate limb 2 mm. long. Corolla yellow-green, 4–4·5 mm. long, glabrous outside, pubescent inside at the throat.

TANZANIA. Morogoro District: SE. Uluguru Mts., Kikurungu Mt., Nov. 1894, *Stuhlmann* 9245
DISTR. **T6**
HAB. Evergreen forest; 1000 m.

75. **P. kilimandscharica** *Engl.*, Abh. Preuss. Akad. Wiss: 400 (1892) & in P.O.A. C: 391 (1895); De Wild., Pl. Bequaert.: 378 (1924); T.T.C.L.: 524 (1949); Petit in B.J.B.B. 34: 221 (1964). Type: Tanzania, Kilimanjaro, *Kersten* (B, holo. †)

Shrub with slender branches. Leaf-blades lanceolate, oblong-lanceolate or subovate-oblong, 5–7 cm. long, 1·2–2·2 cm. wide, mucronulate at the apex, acute at the base, glabrous on both surfaces, shining above, purplish-black, thin; petiole 4–7 mm. long; stipules purplish black, 3 mm. long, probably entire. Flowers in short panicles 3 cm. long and wide; peduncle 2·5–3·5 cm. long; panicle components ± capitate, 7–10-flowered; pedicels very short or obsolete. Calyx-tube 1 mm. long, glabrous; limb 0·5–0·7 mm. long, lobulate. Corolla glabrous, 5–6 mm. long, lobed for $\frac{1}{3}$–$\frac{1}{4}$ its length.

TANZANIA. Kilimanjaro, *Kersten*
DISTR. **T2**
HAB. Evergreen forest; 2100–2700 m.

NOTE. K. Schumann compares this with *P. brassii* Hiern. It is difficult to assimilate it with anything recorded from Kilimanjaro although it might possibly be *P. riparia*.

76. **G. kwaiensis** *K. Schum. & K. Krause* in E.J. 39: 561 (1907); De Wild. in B.J.B.B. 9: 40 (1923); T.T.C.L.: 499 (1949); Petit in B.J.B.B. 34: 216 (1964). Types: Tanzania, W. Usambara Mts., near Kwai, *Eick* 127 & 330 (both B, syn. †)

Shrub with slender glabrous stems, with rugose grey or grey-brown bark which peels here and there. Leaves ± conferted at apices of stems; blades oblong or obovate-oblong, 4–7 cm. long, 2–3·5 cm. wide, acute at the apex, subcuneate at the base, completely glabrous, rigidly coriaceous; petiole 0·6–1 cm. long; stipules ovate-oblong, 0·8–1 cm. long, obtuse, slightly joined at the base, glabrous on both sides but with upper margin fimbriated, very soon deciduous. Flowers sessile in terminal 6–12-flowered corymbose cymes, 2–3 cm. long. Ovary shortly turbinate, 1 mm. long, sparsely minutely papillate; calyx-limb cupuliform, just over 1 mm. long, obsoletely 5-dentate. Corolla drying black, tube 2–2·5 cm. long, glabrous outside, the throat hairy; lobes ovate, 1·8–2·2 mm. long, obtuse, cucullate-incrassate at the apex, spreading or at length reflexed. Filaments under 1 mm. long. Style thick, 3 mm. long, ± attenuate at the base, bifid at the apex. Drupes globose, 3–4·5 mm. in diameter, longitudinally ribbed.

TANZANIA. Lushoto District: W. Usambaras, near Kwai, *Eick* 127 & 330
DISTR. **T3**
HAB. Evergreen forest

NOTE. The authors say it has the habit of *G. purtschelleri* K. Schum. but narrower, harder, thicker, denser leaves and laxer inflorescence.

77. **P. lamprophylla** *K. Schum.* in P.O.A. C: 391 (1895); De Wild., Pl. Bequaert. 2: 380 (1924); T.T.C.L.: 525 (1949); Petit in B.J.B.B. 34: 221 (1964). Type: Tanzania, Usambara Mts., *Holst* 3723 (B, holo. †)

Shrub with graceful glabrous branches. Leaf-blades oblong, up to 10 cm. long, 6 cm. wide, acuminate at the apex, acute at the base, glabrous, sometimes with spots beneath and with domatia, shining, coriaceous; petiole up to 3 cm. long. Flowers in small axillary panicles, papillose; peduncle up to 1·5 cm. long. Calyx pilosulose, scarcely 0·5 mm. long, truncate. Corolla-tube glabrous, with lobes 2 mm. long.

TANZANIA. Usambara Mts., *Holst* 3723
DISTR. **T3**
HAB. Not known, presumably evergreen forest

78. **G. orientalis** *K. Schum.* in E.J. 34 : 336 (1904) ; De Wild. in B.J.B.B. 9 : 43 (1923) ; T.T.C.L. : 499 (1949) ; Petit in B.J.B.B. 34 : 216 (1964). Type : Tanzania, E. Usambara Mts., Sangerawe, *Engler* 879 (B, holo. †)

Tree 7–8 m. tall. Leaf-blades linear-oblong or subspathulate-oblong, 15–20 cm. long, 8–9 cm. wide in the upper quarter, obtuse at the apex, acute at the base, glabrous on both sides, thin ; petiole 1–2·5 cm. long ; stipules suborbicular, 1·5 cm. in diameter, thin, glabrous, deciduous. Panicles ample and many-flowered, 15 cm. long ; peduncles 10 cm. long ; pedicels very short or obsolete. Drupes spherical, 4 mm. in diameter, strongly ribbed, glabrous.

TANZANIA. Lushoto District : Sangerawe, 23 Sept. 1902, *Engler* 879
DISTR. **T3**
HAB. Rain-forest ; 1000 m.

79. **G. pallidiflora** *K. Schum.* in E.J. 28 : 497 (1900) ; De Wild. in B.J.B.B. 9 : 43 (1923) ; T.T.C.L. : 499 (1949) ; Petit in B.J.B.B. 34 : 216 (1964). Type : Tanzania, E. Uluguru Mts., Tununguo, R. Ruvu, *Stuhlmann* 8962 (B, holo. †)

Shrub with slender glabrous branches. Leaf-blades oblong-lanceolate, 10–15 cm. long, 3–6 cm. wide, acute at the apex, cuneate at the base, glabrous on both surfaces, coriaceous ; petiole 1–2·5 cm. long, channelled above, glabrous ; stipules ovate, 7 mm. long, brown, coriaceous, glabrous. Flowers in terminal glabrous many-flowered panicles, the component parts capitate ; peduncle 0·8–1·8 cm. long ; pedicels ± obsolete, together with ovary and calyx 3–3·5 mm. long. Calyx cupuliform, repand-denticulate, minutely ciliolate. Corolla coriaceous, minutely papillate outside ; tube 4 mm. long, pubescent below the throat inside ; lobes 2 mm. long.

TANZANIA. Morogoro District : E. Uluguru Mts., Tununguo, Ruvu R., Oct. 1894, *Stuhlmann* 8962
DISTR. **T6**
HAB. Foothills ; 300 m.

NOTE. Larger leaves with more lateral nerves and more numerous flowers than *G. purtschelleri*.

80. **G. rufescens** *K. Krause* in E.J. 43 : 154 (1909) ; De Wild. in B.J.B.B. 9 : 55 (1923) ; T.T.C.L. : 497 (1949) ; Petit in B.J.B.B. 34 : 216 (1964). Type : Tanzania, Uzaramo District, Pugu Hills, *Holtz* 1067 (B, holo. †)

Shrub with stout stems, densely ferruginous pilose when young but later glabrous. Leaf-blades oblong or obovate-oblong, 10–14 cm. long, 6·5–7·5 cm. wide, shortly acuminate at the apex, subacute or ± obtuse at the base, glabrous above, beneath mostly on the prominent main nerve, fuscous-pilose, coriaceous ; petiole 2–3 cm. long, stout, channelled near the base, densely pilose on young leaves ; stipules broadly ovate, 6–7 mm. long, 5 mm. wide, subacute at the apex, densely ferruginous pilose outside, mostly soon deciduous. Flowers subsessile in short terminal densely ferruginous pilose subpaniculate cymes 12–14 cm. long. Ovary globose, 1·2 mm. in diameter, densely pilose ; limb cupuliform, hardly 1 mm. long, obsoletely lobed. Corolla yellow in life, drying brown to black ; tube 3 mm. long, 1·5 mm. wide, puberulous outside, villous at the throat ; lobes ovate-oblong, ± 2 mm. long. Filaments ± 0·5 mm. long. Style 1–1·5 mm. long ; stigma-lobes 0·5 mm. long.

TANZANIA. Uzaramo District : by Pugu stream, 1 Dec. 1903, *Holtz* 1067
DISTR. **T6**
HAB. Not known

NOTE. Compared with *G. riparia* and *G. goetzei* by its author.

81. **P. scheffleri** *K. Schum. & K. Krause* in E.J. 39: 554 (1907); De Wild., Pl. Bequaert.: 420 (1924); T.T.C.L.: 523 (1949); Petit in B.J.B.B. 34: 223 (1964). Type: Tanzania, E. Usambara Mts., between Derema and Monga, *Schleffler* 173 (B, holo. †)

Shrub 3–4 m. tall, with fairly stout stems, glabrous or sparsely pilose when young. Leaf-blades oblong to oblong-elliptic, 8–15 cm. long, 2·5–6 cm. wide, shortly acuminate at the apex, subacute at the base, shining, glabrous on both surfaces, thin; petiole 2–5 cm. long; stipules triangular, 4–6 mm. long, acute, densely ferruginous-pilose outside and ciliate on the margins, soon deciduous. Flowers subsessile in terminal 6–10-flowered subcapitate cymes; peduncles 3–5 cm. long; pedicels slender, subtetragonous, glabrous. Ovary 1·5 mm. long and calyx (i.e. limb) about as long, cupuliform, thick, deeply 5-dentate. Corolla yellow, drying black; tube 2 mm. long, glabrous outside, hairy at the throat; lobes oblong, 2–2.5 mm. long, thickened at the apex. Filaments under 1 mm. long. Style 1·5–2 mm. long, included, obsoletely bifid.

TANZANIA. Lushoto District: E. Usambara Mts., between Derema and Monga, 8 Dec. 1899, *Schleffler* 173
DISTR. T3
HAB. Rain-forest, shady places near small streams; 900 m.

NOTE. The authors state that by its capitate inflorescence it stands near *P. kirkii* and *P. abrupta* (i.e. *Chazaliella abrupta*) two species not remotely related to each other.

2. GEOPHILA

D. Don, Prodr. Fl. Nepal.: 136 (1825); Hepper in Taxon 9: 88 (1960), *nom. conserv.*

Carinta W. F. Wight in Contr. U.S. Nat. Herb. 9: 216 (1905); G. Taylor in Exell, Cat. Vasc. Pl. S. Tomé, Suppl.: 25 (1956); L. B. Smith & Downs in Sellowia 7: 65 (1956)

Geocardia Standley in Contr. U.S. Nat. Herb. 17: 444 (1914)

Perennial forest-floor herbs, mostly with slender creeping stems which root at the nodes and have fibrous roots. Leaves opposite, with mostly long petioles; blades ovate-cordate to rounded-reniform; stipules interpetiolar, ovate, entire or bilobed at the apex. Flowers hermaphrodite, sometimes heterostylous, mostly in terminal umbels or sometimes solitary, often on long peduncles held erect from the main stems, occasionally with an involucre of quite conspicuous bracts. Calyx-tube obovoid, the limb short, 5–7-lobed, the lobes subulate or linear, spreading or reflexed, persistent. Corolla cylindrical or funnel-shaped; lobes 4–7, spreading or recurved; throat pilose inside. Stamens 4–7, inserted in the corolla-tube; filaments filiform; anthers dorsifixed, included or exserted. Disc swollen. Ovary 2-locular; ovules solitary in each locule, erect from the base, anatropous; style slender, included or exserted; stigma-lobes 2, linear, densely papillate or stigma subcapitate, bifid. Drupe fleshy, containing 2 pyrenes; pyrenes plano-convex, dorsally compressed, obtusely ribbed, rugulose and often with an annular area at junction of ventral and dorsal surfaces, 1-seeded. Seeds the same shape as the pyrenes, the ventral surface plane; testa membranous; endosperm corneous.

A genus of about 10–15 species in the tropics of both the Old and New Worlds; 2 species occur in the Flora area. Other species previously referred to *Geophila* have been accepted as belonging in the genus *Hymenocoleus* very recently described by Robbrecht.

Inflorescence 1(rarely –2–3)-flowered, the bracts small and never forming an involucre; leaf-blades rounded-reniform . . . . 1. *G. repens*
Inflorescences always several-flowered with a distinct involucre made up of separate bracts; leaf-blades ovate-cordate or ovate-reniform . . . 2. *G. obvallata*

1. **G. repens** (*L.*) *I. M. Johnston* in Sargentia 8: 281, 282 (1949); Brenan in Mem. N.Y. Bot. Gard. 8: 453 (1954); Hepper in F.W.T.A., ed. 2, 2: 205 (1963); Steyerm. in Mem. N.Y. Bot. Gard. 23: 395 (1972); F.P.U., ed. 2: 164, fig. 106 (1972); Verdc. in K.B. 28: 321, fig. 1/2 A–C (1973); U.K.W.F.: 407 (1974). Types: Rheede, Hort. Malabar. 10, t. 21 (1690) & Sloane, A Voyage . . . Jamaica 1: 243 (1707)

Creeping herb with stems 20–30 cm. long, densely adpressed pubescent, rooting at the nodes; leafy and flowering shoots reaching a height of 3–7·5 cm. above ground-level. Leaf-blades rounded-reniform, 1·2–3·6(–5) cm. long, 1·3–4·3(–5·5) cm. wide, very rounded at the apex, emarginate-cordate at the base, glabrous above, glabrous or pubescent beneath; petiole 0·4–11·5 cm. long, adpressed or spreading pubescent; stipules transversely elliptic, almost truncate, 1·5 mm. long, 2–3·5 mm. wide. Flowers not truly heterostylous, solitary, or more rarely up to 2 or even 3–5 in American material; peduncle 0·5–4·3 cm. long, pubescent; bracts 1–2, lanceolate, 2·5–4 mm. long. Calyx-tube obconic, 1·2–2 mm. long, pubescent; limb-tube 0·8–1·2 mm. long; lobes lanceolate, 1·5–3 mm. long, 0·5–0·7 mm. wide at base. Corolla white; tube cylindrical, 0·5–1·3 cm. long, finely pubescent; lobes oblong-elliptic, 4–9 mm. long, 2–5 mm. wide, basal part of limb funnel-shaped. Anthers and style included, the latter 3·5–7 mm. long; stigma 0·25–0·5 mm. long, either level with or reaching beyond the anthers. Berries globose, bright red or orange, glossy, glabrous, 0·5–1·2 cm. in diameter, crowned by the persistent lobes. Pyrenes greyish or straw-coloured, half-ovoid, 3·5–4·1 mm. long, 3–3·5 mm. wide, dorsal side convex, rugose, ventral side flat, rugose, with a narrow impressed smooth annular area where the 2 areas join. Seeds with chestnut coloured testa easily removable to show the white endosperm, lenticular, 3 mm. long, 2·8 mm. wide, 1 mm. thick, smooth. Fig. 7/1, 2.

UGANDA. Kigezi District: Ishasha Gorge, May 1950, *Purseglove* 3412!; Mengo District: Entebbe road, Kajansi Forest, July 1935, *Chandler* 1274! & Kyagwe [Kiagwe], Mau Forest, May 1932, *Eggeling* 433!
KENYA. Kwale District: Mrima Hill, 16 Jan. 1962, *Verdcourt* 3939A! & Buda Mafisini Forest, 19 Aug. 1953, *Drummond & Hemsley* 3891! & same locality, 3 Nov. 1959, *Napper* 1379!
TANZANIA. Lushoto District: E. Usambara Mts., Amani, 15 May 1950, *Verdcourt* 203A! & same locality, 20 Apr. 1968, *Renvoize & Abdallah* 1645!; Morogoro District: Nguru Mts., *Greenway*! (cultivated at Kew from seeds and rootstocks, Dec. 1954)
DISTR. **U**2, 4; **K**4, 5, 7; **T**1, 3, ?4, 6; pantropical, in Africa from Guinea Bissau to Angola, Zaire, Sudan, Malawi, Rhodesia and Madagascar
HAB. Evergreen forest floors; 80–1600 m.

SYN. *Rondeletia repens* L., Syst. Nat., ed. 10, 2: 928 (1759)
Psychotria herbacea Jacq., Enum. Pl. Carib.: 16 (1760). Type: W. Indies, *Jacquin* (? W, holo.)
P. herbacea L., Sp. Pl., ed. 2: 245 (1762), *non* Jacq., *nom. illegit.* Type: Jamaica, *P. Browne* (probably LINN, but specimen not labelled *Browne*)
Geophila reniformis D. Don, Prodr. Fl. Nepal.: 136 (1825); Hiern in F.T.A. 3: 220 (1877). Type: Bangladesh, Sylhet, *Wallich* (BM, holo.!)
G. uniflora Hiern in F.T.A. 3: 221 (1877); F. Hallé, Ic. Pl. Afr. 7, No. 156 (1965). Types: Nigeria, Nupe, *Barter* (K, syn.!) & Sudan, Niamniamland, Nabambiso, *Schweinfurth* 3856 (K, syn.!, BM, isosyn.!)

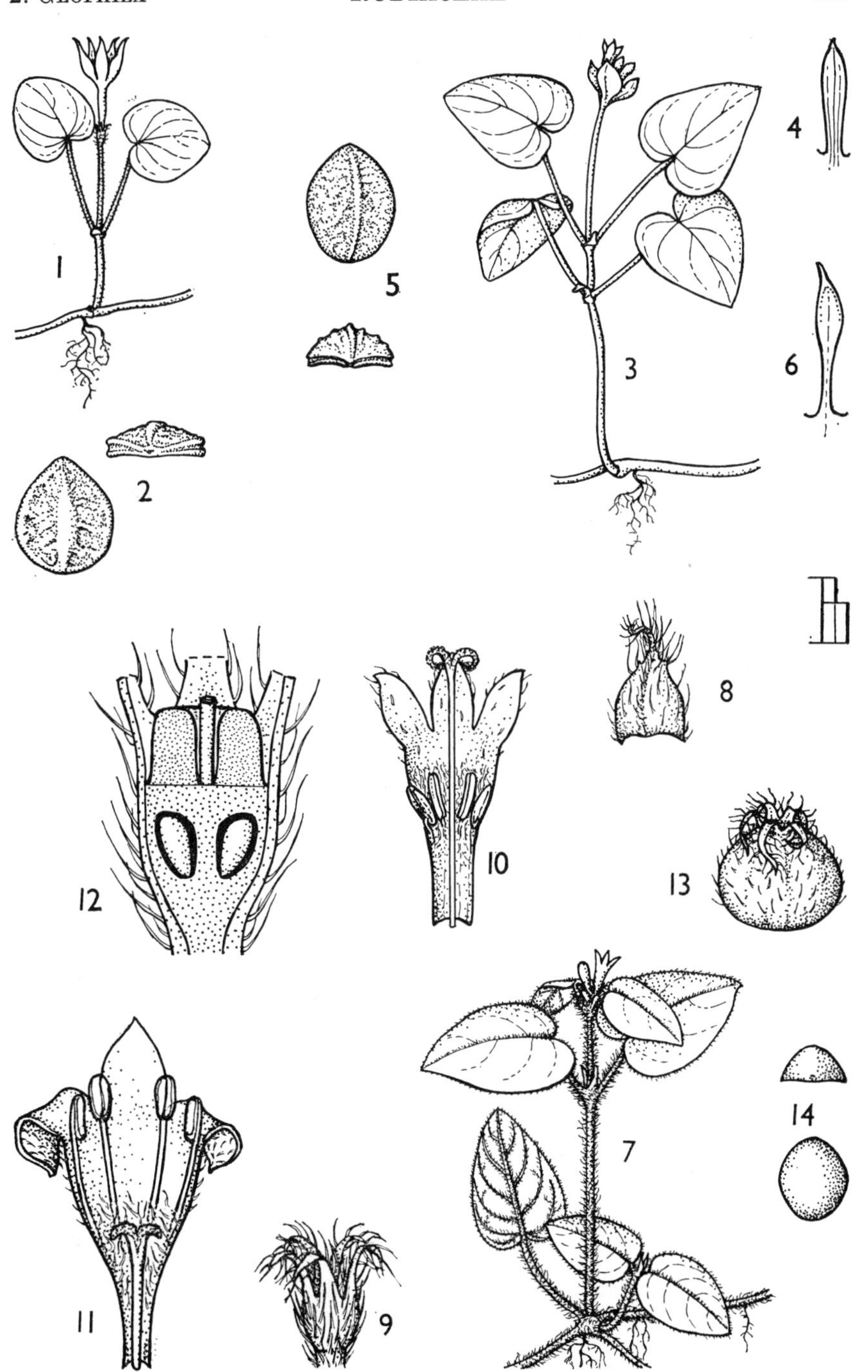

FIG. 7. *GEOPHILA REPENS*—**1,** habit, × $\frac{2}{3}$; **2,** pyrene, two views, × 4. *G. OBVALLATA* subsp. *IOIDES*—**3,** habit, × $\frac{2}{3}$; **4,** calyx-lobe, × 5; **5,** pyrene, two views, × 4. *G. OBVALLATA* subsp. *INVOLUCRATA*—**6,** calyx-lobe, × 5. *HYMENOCOLEUS HIRSUTUS*—**7,** habit, × $\frac{2}{3}$; **8,** bract, × 4; **9,** calyx, × 6; **10,** longitudinal section of upper part of a long-styled flower, × 4; **11,** same of short-styled flower, × 4; **12,** longitudinal section of ovary, × 20; **13,** berry, × 2; **14,** pyrene, two views, × 4. 1, 2, from *Drummond & Hemsley* 3891; 3, from *Drummond & Hemsley* 3414; 4, 5, from *Drummond & Hemsley* 3892; 6, from *Dawkins* 873; 7, 10, from *Symes* 442; 8, 9, 11–14, from *Dawkins* 875. Drawn by Diane Bridson.

G. herbacea (L.) K. Schum. in E. & P. Pf. 4(4): 119 (1891); S. Moore in Fl. Jam. 7: 111, fig. 32 (1936), as "(Jacq.) K. Schum."; Bremek. in Pulle, Fl. Suriname 4: 234 (1934)
Carinta herbacea (Jacq.) W. F. Wight in Contr. U.S. Nat. Herb. 9: 216 (1905)
Geocardia herbacea (L.) Standley in Contr. U.S. Nat. Herb. 17: 444 (1914)
Carinta uniflora (Hiern) G. Taylor in Exell, Cat. Vasc. Pl. S. Tomé, Suppl.: 25 (1956)
C. repens (L.) L. B. Smith & Downs in Sellowia 7: 88 (1956)
Geocardia repens (L.) Bakh. f. in Backer, Beknopte Fl. Java 15, fam. 173: 144 (1956)

NOTE. This species is fairly uniform in the Old World but New World specimens often have up to 5-flowered (and only rarely 1-flowered) inflorescences and hence a rather different appearance. Possibly the Old World material could form a subspecies with a name based on *G. reniformis*. I am unable to agree with Hepper that *G. lancistipula* Hiern (in F.T.A. 3: 221 (1877); F.W.T.A. 2: 128 (1931); type: Gabon?, Sierra del Crystal 1°N., *Mann* 1087 (K, holo. !)) is a synonym of *G. repens*; it differs in having dark blue hairy fruits, pyrenes of a very different shape and somewhat differently shaped leaf-blades.

The **T4** records are based on three sterile *Peter* specimens, 35837 (Uvinsa, SW. of Malagarasi, 30 Jan. 1926), 36316 (Uvinsa, side of Malagarasi Valley, 5 Feb. 1926) and 37145 (Ujiji, Machaso to Mkuti R., 19 Feb. 1926). Fruits are needed to confirm to which species these belong.

2. **G. obvallata** (*Schumach.*) *F. Didr.* in Vidensk. Meddel. Dansk Naturhist. Foren. Kjøbenh. 1854: 186 (1855); Hiern in F.T.A. 3: 222 (1877); Hepper in F.W.T.A., ed. 2, 2: 206, fig. 243/A (1963); F. Hallé, lc. Pl. Afr. 7, No. 155 (1965); Verdc. in K.B. 30: 265 (1975). Type: Ghana, Akwapim, *Isert* (C, holo., B, iso., K, photo. !)

Creeping herb with prostrate often underground stems (10–)30–60 cm. long, glabrous, rooting at the nodes and often forming carpets. Leaf-blades triangular-ovate, ovate or ovate-reniform, 0·8–4(–9) cm. long, 0·6–4(–5·5) cm. wide, acute to rounded at the apex, cordate at the base, glabrous above, glabrous or with some pubescence at the sides of the midnerve beneath; petiole 0·5–9·5 cm. long, often with lines of short hairs above or ± densely pubescent above at apex; stipules transversely elliptic, 1–2(–3) mm. long, not bifid. Inflorescences 0·5–1·1 cm. across, several-flowered, enclosed in a whorl of bracts; peduncles 1·3–5(–9) cm. long, glabrous, finely papillate-pubescent or densely hairy (see note); bracts leafy, obovate, rounded-elliptic or rhomboid, 0·5–1·5 cm. long, 0·35–1·1 cm. wide; flowers not heterostylous. Calyx glabrous or rarely hairy; tube and pedicel 1·5–2·5 mm. long; limb-tube 0·2–1·5 mm. long; lobes subulate, linear-lanceolate, narrowly triangular or distinctly spathulate, 0·7–6·6 mm. long and up to 1·3 mm. wide; disc 0·3–1·1 mm. tall. Corolla white; tube funnel-shaped, 3·1–6·5 mm. long, glabrous or puberulous outside; lobes ovate-oblong, 1·2–3 mm. long, 0·6–2 mm. wide, sometimes shortly joined at the base. Anthers situated near centre of the tube. Style 0·3–1 mm. long, widened above; stigma ± capitate, bifid, 0·3–0·5 mm. long. Berries black, purple or blue, (4–)7–8 mm. long, (4–)7–8 mm. wide, crowned with the persistent calyx-lobes; pyrenes dull yellowish brown, half-ovoid, 4–4·7 mm. long, 3·5–3·7 mm. wide, 1·7–2·3 mm. thick, the ventral surface fairly flat but with 2 depressed areas, bounded by the raised margin and raised median area, dorsal surface rugose, with a longitudinal median keel and a smooth depressed annular area where the dorsal and ventral surfaces meet. Seeds brown, ± lenticular, ± 3 mm. long, ± 2·6 mm. wide, ± 1 mm. thick, dorsally rounded, ventrally flattened, smooth.

SYN. *Psychotria obvallata* Schumach., Beskr. Guin. Pl.: 111 (1827)
Carinta obvallata (Schumach.) G. Taylor in Exell, Cat. Vasc. Pl. S. Tomé, Suppl.: 25 (1956)

subsp. **ioides** (*K. Schum.*) *Verdc.* in K.B. 30: 267 (1975). Type: Mozambique, Quelimane, *Stuhlmann* 711 (B, holo.†)

Leaf-blades rounded-reniform to ovate. Calyx-lobes linear or linear-subulate or very slightly spathulate at the apex, 1·4–3·4 mm. long. Corolla-tube 3–4 mm. long. Fig. 7/3–5, p. 111.

KENYA. Kwale District: Buda Mafisini Forest, 19 Aug. 1953, *Drummond & Hemsley* 3892 ! & Shimba Hills, Kwale, 14 Jan. 1964, *Verdcourt* 3920 ! & Shimba Hills, Makadara Forest, 6 May 1968, *Magogo & Glover* 1018 !

TANZANIA. Lushoto District: Amani–Monga road, 0·8 km. NNE. of Amani, 23 July 1953, *Drummond & Hemsley* 3414 !; Tanga District: lower Sigi valley, 15 May 1950, *Verdcourt* 203 !; Morogoro District: Nguru Mts., Liwale valley, Manyangu Forest, 27 Mar. 1953, *Drummond & Hemsley* 1858 !

DISTR. **K7**; **T3**, 4, 6, 8; Burundi, Mozambique (see note)

HAB. On leaf-litter of coastal evergreen forest and lowland rain-forest (e.g. of *Cephalosphaera, Chlorophylla, Strombosia*, etc., or *Conopharyngia, Lannea, Apodytes, Julbernardia, Newtonia*), also light woodland; 80–950 m.

SYN. *G. ioides* K. Schum. in P.O.A. C: 392 (1895)
G. cecilae N.E. Br. in K.B. 1906: 107 (1906). Type: Mozambique, Beira, swamps at Dondo, *Cecil* 254 (K, holo. !)

subsp. **involucrata** (*Hiern*) *Verdc.* in K.B. 30: 267 (1975). Types: Zaire, Monbuttu, Munsa, *Schweinfurth* 3412 (K, syn. !, BM, isosyn. !) & Sudan, Niamniam, Mbruole, *Schweinfurth* 3670 (K, syn. !)

Leaf-blades rounded-reniform to ovate. Calyx-lobes linear with distinctly spathulate tips, 3–6·6 mm. long, 0·8–1·3 mm. wide or rarely linear-subulate with scarcely any widening. Corolla-tube 4–6 mm. long. Fig. 7/6, p. 111.

UGANDA. Bunyoro District: Budongo Forest, June 1935, *Eggeling* 2069 !; Masaka District: Sese Is., Bugala I., Towa Forest, July 1945, *Purseglove* 1748 !; Mengo District: Sezibwa Forest, May 1915, *Dummer* 2618 !

DISTR. **U2**, 4; Zaire, Sudan

HAB. Leaf-litter of rain-forest; 1100–1200 m.

SYN. *G. involucrata* Hiern in F.T.A. 3: 222 (1877)

NOTE. A very similar plant occurs in Angola.

DISTR. (of species as a whole). **U2**, 4; **K7**; **T3**, 4, 6, 8; Guinea Bissau to Angola, Zaire, Sudan and Central African Republic, south-central Africa

NOTE (on species as a whole). Subsp. *obvallata* occurs in W. Africa and has calyx-lobes similar to those in subsp. *ioides* but differs in its triangular-ovate to ovate leaf-blades, larger bracts, etc.; some specimens from Zaire, the Central African Republic and Cameroun are practically identical with subsp. *ioides* and have been treated as forms of subsp. *obvallata* showing intermediate characters. The three subspecies have always been considered worthy of specific rank but the calyx-lobe characters are not entirely constant; ± spathulate lobes turn up in W. Africa (e.g. *Thomas* 10170 & *Marmo* 290 from Sierra Leone) and to some extent in coastal Kenya (e.g. *Verdcourt* 3920 cited above). Some specimens from the distribution area of subsp. *involucrata* have linear-deltoid calyx-lobes with no trace of spathulate apices (e.g. *Dawkins* 873 from Masaka District, Nkose I.). In Zaire, Zambia, Malawi, Rhodesia and Angola there is a plant for long named as *G. ioides*, but with less spreading stems, shorter calyx-lobes and mostly very hairy peduncles (e.g. Zambia, Mbala, Chilongowelo, 28 Jan. 1955, *Richards* 4264 !). This I believe might be considered a fourth subspecies although some material of subsp. *ioides* from Mozambique has quite hairy peduncles.

3. HYMENOCOLEUS

Robbrecht in B.J.B.B. 45: 274 (1975)

Stoloniferous creeping herbs rooting at the nodes, with erect flowering branches, or sometimes erect shrubs. Stipules bifid at the apex or divided into 2 completely separate lobes; a membranous cylindric or cupular sheath present at the nodes beneath the stipules. Inflorescences terminal, ± capitate, often with a cup-like involucre. Flowers sessile or subsessile, 5-merous, distinctly or obscurely heterostylous. Style bifid. Fruits usually orange, sometimes red, inflated, full of watery juice, crowned with the persistent calyx, with 2 pyrenes; pyrenes semi-ellipsoid, brittle, the dorsal face convex and not ribbed, the ventral face with 1 or 2 grooves, basally dehiscent by 2 short lateral fissures. Seeds semi-ellipsoid, with entire albumen.

A small genus closely allied to *Geophila* and *Psychotria*, containing 11 species restricted to tropical Africa, of which 4 occur in the Flora area.

Inflorescences with a cupuliform involucre with lobed margin; leaf-blades broadly elliptic, narrowly cordate at the base, distinctly broadest near the middle; pyrenes with 1 ventral groove . . . . . . 1. *H. rotundifolius*
Inflorescences with inconspicuous bracts, never an involucre:
  Leaf-blades mostly ovate, distinctly cordate at the base, at least in mature non-apical foliage; pyrenes with 2 ventral grooves . 2. *H. hirsutus*
  Leaf-blades elliptic, narrowed to a cuneate, rounded or almost imperceptibly cordate base:
    Petioles up to 3·5 cm. long; stipules almost round to ovate-triangular, not distinctly bifid; pyrenes with 2 ventral grooves . 3. *H. libericus*
    Petioles up to 1·1 cm. long; stipules distinctly bifid; pyrenes with 1 ventral groove . 4. *H. neurodictyon*

1. **H. rotundifolius** (*Hepper*) *Robbrecht* in B.J.B.B. 45: 287 (1975). Type: Ghana, Kwahu, *W. H. Johnson* 659 (K, holo. !)

Creeping herb with stems 10–40 cm. long, at first adpressed hairy, later glabrescent, rooting at the nodes. Leaves congested towards the apices of the stems; blades rounded, elliptic-ovate or elliptic-oblong, 3·5–6 cm. long, 1·9–4 cm. wide, acute or subacute to ± obtuse or rounded at the apex, shallowly narrowly cordate or rarely cuneate at the base, glabrous above except for hairy midnerve or pubescent to hairy all over upper surface, pubescent or hairy beneath; petiole 0·2–2(–3·5) cm. long, hairy; stipules membranous, green, ovate or triangular, 3–7·5(–10) mm. long, 2–3·5 mm. wide, shortly bifid at the apex, with ciliate margins. Flowers probably showing reduced heterostyly, ± 12 in capitate inflorescences ± 1–1·5 cm. in diameter; peduncle 0–5 mm. long; bract cupuliform, forming an involucre 0·9–1·1 cm. long, undulate, erose or lobed, the lobes 5 mm. long, 1·5 mm. wide, pubescent, the margins ciliate. Calyx ± 7 mm. long, the tubular part 5–6 mm. long; lobes ± 1 mm. long, recurved, margins ciliate. Corolla white; tube funnel-shaped, 7–8 mm. long; lobes ovate, 1·5 mm. long, 0·6 mm. wide. Anthers sessile at the top of the tube. Stigma 2-lobed. Berries orange; pyrenes yellow-brown, subglobose, 3 mm. long, 2·6 mm. wide, 2·2 mm. thick, with a narrow deep slit on the ventral side. Seeds yellow-brown, subglobose, 2·5 mm. long, 2·2 mm. wide, with 1 deep U-shaped hilar groove.

UGANDA. Bunyoro District: Budongo Forest, *Sangster* 187 !
DISTR. U2; W. Africa from Sierra Leone to Cameroun
HAB. Rain-forest; 1100 m.

SYN. *Geophila rotundifolia* Hepper in K.B. 16: 331 (1962) & F.W.T.A., ed. 2, 2: 206 (1963) & in Hook, Ic. Pl., ser. 5, 7, t. 3627 (1967), pro parte

NOTE. Hepper includes in his description several specimens with much larger leaves and involucres; these have been separated by Robbrecht as *H. petitianus* Robbrecht.

2. **H. hirsutus** (*Benth.*) *Robbrecht* in B.J.B.B. 45: 288 (1975). Type: Nigeria, Nun R., *Vogel* (K, holo.!)

Creeping herb with stems 20–75 cm. long, densely hairy, rooting at the nodes; leafy and flowering shoots reaching a height of 5–10 cm. above ground-level. Leaf-blades ovate to ovate-oblong, 1·2–7 cm. long, 0·5–5·1 cm. wide, acute or subacute at the apex, less often ± rounded, cordate at the base, glabrous above except for lower part of midnerve or latter entirely hairy and with some scattered hairs on the surface as well, pubescent beneath, particularly on the nerves; petiole 0·5–10 cm. long, hairy; stipules oblong or triangular, 3–8 mm. long, 1·5–2·5 mm. wide, bifid, hairy and ciliate. Flowers conspicuously heterostylous, in inflorescences ± 0·4–1·1 cm. in diameter; peduncles 0–1 cm. long; bracts not conspicuous. Calyx-tube 0·6–1·1 mm. long; limb-tube 0·8–1·9 mm. long; lobes lanceolate, 1·5–4·5 mm. long, 0·3–0·8 mm. wide, ciliate. Corolla white or greenish white; tube funnel-shaped, 5·5–9 mm. long, glabrous or slightly pubescent outside; lobes ovate-triangular or oblong, 2–3·5 mm. long, 1–2 mm. wide. Anthers included in long-styled flowers, just exserted in short-styled flowers. Style 6·5 mm. long (exserted 2 mm.) in long-styled flowers, 2·9–3·6 mm. long in short-styled flowers; stigma 0·7–1·1 mm. long. Berries bright orange or reddish, ± 0·7 mm. long, 0·9 mm. wide, sparsely hairy, crowned with similarly coloured persistent calyx-lobes; pyrenes greenish brown, subglobose, 2·2–2·8 mm. long, 2–2·5 mm. wide, dorsal surface strongly convex, ventral surface plane, with 2 grooves. Seeds yellow-brown, lenticular, up to 2·4 mm. long, 2·2 mm. wide, dorsal face convex, ventral face with 2 shallow grooves surrounding a slightly raised median elliptic area. Fig. 7/7–14, p. 111.

Uganda. Kigezi District: Ishasha Gorge, Mar. 1946, *Purseglove* 2001!; Masaka District: Sese Is., Bugala I., 5 Oct. 1958, *Symes* 442!; Mengo District: Busiro, Kyewaga Forest, 3 Mar. 1950, *Dawkins* 534!
Tanzania. Bukoba District: Kantale [Kantare], *Gillman* 377! & Minziro Forest, Sept. 1952, *Procter* 91A!
Distr. **U**2, 4; **T**1; W. Africa, Cameroun, Central African Republic, Zaire and Angola
Hab. Rain-forest and other forests, e.g. *Raphia-Macaranga* swamp-forest, always within or just above the leaf-litter; 1120–1200 m.

Syn. *Geophila hirsuta* Benth. in Niger Fl.: 422 (1849); Hiern in F.T.A. 3: 221 (1877); Hepper in F.W.T.A., ed. 2, 2: 205 (1963); F.P.U., ed. 2: 164 (1972)

Note. *Bagshawe* 565 is said to have brick-red flowers but I think this must be a slip.

3. **H. libericus** (*Hutch. & Dalz.*) *Robbrecht* in B.J.B.B. 45: 291 (1975). Type: Ivory Coast, Middle Cavally, Tébo and neighbourhood, *Chevalier* 19382 (P, holo., K, photo.!)

Herb with erect shoots 30 cm. tall presumably from a creeping rhizome; stems drying blackish, sparsely pubescent or at length glabrescent. Leaf-blades ovate or less often elliptic, 7–12 cm. long, 2·5–12 cm. wide, acuminate at the apex, cordate or less often rounded to cuneate (often slightly unequally so) at the base, glabrous above, almost so beneath or with pubescence bordering the venation, particularly the base of the midnerve; petiole 0·5–3·2 cm. long, pubescent; stipules almost round to ovate-triangular, 4–8 mm. long, 4–4·5 mm. wide, the margins ciliate and probably slightly lacerate, deeply bifid or almost entire, at length deciduous. Flowers subsessile in small terminal clusters ± 1 cm. wide; involucre absent; bracts linear, 3–6·5 mm. long. Heterostyly less marked, the anthers and style and stigmas included in both long- and short-styled forms. Calyx-tube veined,

1–2 mm. long. Calyx-lobes 5, ± unequal, triangular, 0·5–1·5 mm. long, glabrous or ciliate. Corolla white; tube ± cylindrical, 6–8 mm. long; lobes ± 2·5 mm. long. Fruiting inflorescences with peduncle ± 5 mm. long. Fruits ovoid-globose, 5 mm. long. Pyrenes pale yellow-brown, semi-ovoid, 4 mm. long, 3 mm. wide, 2 mm. thick, dorsal surface very convex, smooth, ventral surface with 2 deep grooves.

UGANDA. Bunyoro District: Budongo Forest, 1 Dec. 1938, *M.V. Loveridge* 163! & same locality, near Sonso R., Apr. 1933, *Eggeling* 1204!
DISTR. **U**2; Liberia, Ivory Coast, Ghana, Cameroun and Zaire
HAB. Floor of evergreen forest; 1200 m.

SYN. *Geophila liberica* Hutch. & Dalz., F.W.T.A. 2: 128 (1934) in clavi; Hepper, F.W.T.A., ed. 2, 2: 205 (1963)

NOTE. Typically *H. libericus* has more ovate leaves with a cordate base but there is a good deal of variation and M. Robbrecht has examined the 2 specimens cited above.

4. **H. neurodictyon** (*K. Schum.*) *Robbrecht* in B.J.B.B. 45: 291 (1975). Type: Cameroun, E. of Victoria, *Preuss* 1196 (B, syn. †, K, lecto.!)

Decumbent or creeping herb 15–45 cm. long, with ± erect flowering shoots; rhizome creeping; stems often rooting along the internodes, glabrous to ± adpressed hairy. Leaf-blades oblong, oblong-obovate, oblanceolate or ± elliptic to rhombic, 1·7–10 cm. long, 1·1–4·4 cm. wide, acute to usually rounded at the apex, narrowed to a rounded or minutely cordate base, glabrous or obscurely pustulate above, mostly densely shortly hairy beneath, particularly on the venation, or even velvety when young (glabrous except for sparsely pubescent nerves in var. *rhombicifolius*); venation mostly very distinctly reticulate, often impressed above; petiole 0·3–1·1 cm. long, glabrous to hairy; stipules triangular, 5 mm. long, distinctly bifid. Inflorescences subsessile or shortly pedunculate, 1 cm. in diameter, several-flowered; peduncles 0–5 cm. long; bracts not evident; flowers conspicuously heterostylous. Calyx-tube obconic, 2 mm. long; lobes lanceolate to linear, 2–2·5(–4) mm. long, 0·5–0·8 mm. wide, hairy. Corolla white; tube funnel-shaped, cylindrical at the base, 4·5–5 mm. long; lobes triangular-ovate, 2 mm. long, ± 1 mm. wide. Style exserted 1·5–3 mm. in long-styled flowers; stigma-lobes 0·5–1·5 mm. long; style 0·8 mm. long in short-styled flowers with stigma-lobes 1 mm. long; anthers exserted 1–1·5 mm. Berries fleshy, orange, 6 mm. long, 5 mm. wide, ± hairy, crowned with the persistent calyx-lobes; pyrenes yellow-brown, almost hemispherical, 2·5–3 mm. long, 2·2–3 mm. wide, dorsal surface strongly convex, smooth, ventral surface plane with a groove.

SYN. *Psychotria neurodictyon* K. Schum. in E.J. 33: 368 (1903)
Geophila neurodictyon (K. Schum.) Hepper in K.B. 16: 331 (1962) & in F.W.T.A., ed. 2, 2: 206 (1963)

var. **orientalis** (*Verdc.*) *Robbrecht* in B.J.B.B. 45: 300 (1975). Type: Sudan, Imatong Mts., Thallanga Forest, *A. S. Thomas* 1583 (K, holo.!)

Leaf-blades narrowed to an acute apex; venation not reticulate beneath.

TANZANIA. Bukoba District: Minziro Forest, Sept. 1952, *Procter* 91!
DISTR. **T**1; Ivory Coast, Ghana, Zaire and Sudan
HAB. Undergrowth of *Podocarpus-Baikiaea* forest; 1140 m.

SYN. *Geophila neurodictyon* (K. Schum.) Hepper subsp. *orientalis* Verdc. in K.B. 30: 267 (1975)

NOTE. Var. *neurodictyon* occurs from Guinée to Cameroun and in Principe, Gabon and Zaire.

4. CHAZALIELLA

Petit & Verdc. in K.B. 30: 268 (1975)

Shrubs, often flowering before the leaves are fully developed; stems usually 2-ribbed, all but the youngest shoots usually covered with pale brown soft cork. Leaves opposite or in whorls of 3–4, mostly drying pale, petiolate or subsessile, usually deciduous; nodules absent; domatia small, white-pubescent or absent; stipules mostly short, ovate or triangular, entire, bifid or sometimes with a few teeth. Flowers small, heterostylous, (4–)5(–6)-merous, hermaphrodite, in mostly small, sessile or pedunculate inflorescences, often in small heads, or sometimes paniculate, ± sessile or pedicellate; bracts and bracteoles very small or absent. Calyx-limb usually short, truncate or toothed. Corolla mostly yellow or white; tube shortly cylindrical, hairy at the throat; lobes triangular to elliptic-lanceolate, always valvate. Stamens with long filament in short-styled flowers and shorter filaments in long-styled flowers, the anthers included or only the tips exserted. Ovary 2-locular, each locule containing a single erect ovule; style filiform, usually papillate-pubescent, divided into 2 thick stigma-lobes. Fruit a drupe with 2 pyrenes; pyrenes ± flat ventrally, mostly ± 3-ribbed or 3-lobed dorsally, opening by 2 slits extending along the margins of the ventral face for about ½ its length. Seeds pale; endosperm not ruminate.

A small genus of 24 species restricted to tropical Africa and previously always included in *Psychotria*, but equally related to *Chassalia*. I had intended to treat the taxon as a subgenus of *Psychotria*, but Petit's detailed revision of the African species of *Psychotria* convinced him that the group of species involved merited generic rank. Only one species occurs in the Flora area,

C. abrupta (*Hiern*) *Petit & Verdc.* in K.B. 30: 268 (1975). Types: Mozambique, Shiramba Dembe and Chigogo [Shigogo], *Kirk* (K, syn. !)

Small shrub 0·6–4·5 m. tall, with pallid (mostly greyish or white) thinly cork-covered stems; youngest parts green, the internodes with 2 longitudinal keels, glabrous, glabrescent or pubescent. Leaf-blades elliptic to ovate-lanceolate, 0·8–17·5(–22) cm. long, 0·4–7·5(–10) cm. wide, acute to acuminate at the apex, cuneate at the base, usually rather undeveloped or even not present at the flowering stage and fully expanding in the fruiting stage, glabrous or finely pubescent, pale green but often with a bronze tinge beneath; nodules absent; domatia reduced to small white tufts or quite absent; petiole 1–4(–30) mm. long; stipules ovate or triangular, 2 mm. long, obtuse, acute, shortly bifid or even with several teeth. Flowers (4–)5(–6)-merous, in small 6–20-flowered heads, or sometimes with a few flowers beneath the heads, up to 8 mm. in diameter; peduncle 0·7–3·5 cm. long, ± papillate, finely pubescent or glabrous; pedicels 0·5–1·2 mm. long; bracts minute. Calyx glabrous to pubescent; tube oblong-conic, 1 mm. long; limb very shallow, 0·5 mm. long, truncate or lobes obsolete to distinctly triangular, 1 mm. long. Corolla bright yellow, glabrous or very slightly papillate-pubescent outside; tube 2·8 mm. long; lobes 1·5 mm. long, 1–1·3 mm. wide, slightly cucullate at the apex. Stamens with filaments 1 mm. long in short-styled flowers, 2–3 mm. long in long-styled flowers. Style 1·8 mm. long in short-styled flowers, 3·5 mm. long in long-styled flowers; stigma-lobes thick, 0·8–1·2 mm. long. Drupes ellipsoid, 6–9·5 mm. long, 4–6·5 mm. wide, glabrous or slightly pubescent, slightly ribbed in the dry state; pyrenes pale, semi-ellipsoidal, 6–7 mm. long, 4–4·5 mm. wide, 0·8–2·5

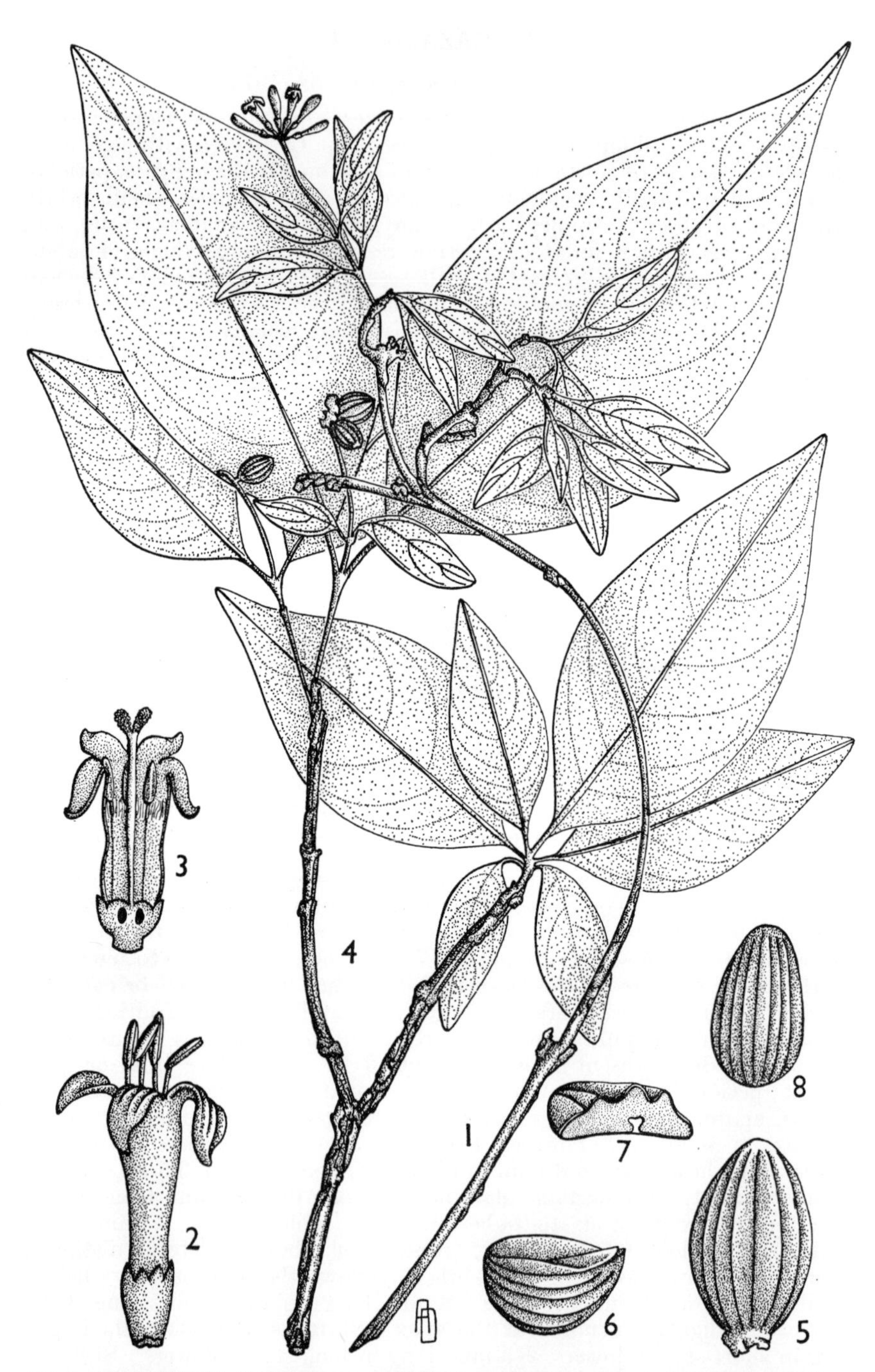

FIG. 8. *CHAZALIELLA ABRUPTA* var. *ABRUPTA*—**1,** flowering branch, × $\frac{2}{3}$; **2,** short-styled flower, × 6; **3,** longitudinal section of long-styled flower, × 6; **4,** fruiting branch, × $\frac{2}{3}$; **5,** drupe, × $3\frac{1}{3}$; **6,** pyrene, × $2\frac{2}{3}$; **7,** section of pyrene, × 4; **8,** seed, × 4. 1, 2, from *Renvoize & Abdallah* 1486; 3, from *Tanner* 2743; 4, from *Bally* 8850; 5–8, from *Drummond & Hemsley* 1892. Drawn by Ann Davis.

mm. thick. Seeds pale brown, compressed semi-ellipsoid, 5 mm. long, 4–4·5 mm. wide, 1–1·2 mm. thick.

var. **abrupta**

Mature leaf-blades attaining maximum dimensions indicated, very thin, usually narrowly acuminate. Fig. 8.

KENYA. Kwale District: Shimba Hills, Lango ya Mwagandi [Longo Mwagondi] area, 21 Mar. 1968, *Magogo & Glover* 375 ! & Cha Shimba [Pemba] Flats, 20 Mar. 1902, *Kassner* 393 !; Kilifi District: 33·6 km. N. of Mombasa, Gongoni Forest, 5 Apr. 1953, *Bally* 8850 !
TANZANIA. Lushoto District: Amani, Mt. Bomole, 21 Feb. 1950, *Verdcourt* 84 !; Morogoro District: Turiani, Nov. 1953, *Semsei* 1470 !; Lindi District: Lake Lutamba, 10 Dec. 1934, *Schlieben* 5717 !; Zanzibar I., Kombeni Cave Wells, 16 Feb. 1930, *Vaughan* 1223 !
DISTR. **K**7; **T**1, 3–8; **Z**; Mozambique, Malawi, Rhodesia
HAB. Evergreen forest including rain-forest, also *Isoberlinia, Combretum, Acacia* and *Brachystegia* woodland; 30–1200 m.

SYN. *Psychotria abrupta* Hiern in F.T.A. 3: 205 (1877); T.T.C.L.: 522 (1949); F.F.N.R.: 417 (1962)*
P. coaetanea K. Schum. in P.O.A. C: 391 (1895); T.T.C.L.: 523 (1949). Type: Tanzania, E. Usambara Mts., Derema [Nderema], *Holst* 2242a (B, holo.†, HBG, K, iso. !**)

var. **parvifolia** *Verdc.* in K.B. 30: 268 (1975). Type: Kenya, Kilifi District, Arabuko, *R.M. Graham* in *F.D.* 2339 (K, holo. !, EA, iso.)

Mature leaf-blades small, up to 4 × 1·7 cm., rather thick, mostly blunt or subacute.

KENYA. Mombasa, Nov. 1884, *Wakefield* !; Kilifi District: Fumbini, 11 Sept. 1936, *Swynnerton* 257 (K 29) ! & Arabuko, Mar. 1930, *R.M. Graham* in *F.D.* 2339 !
DISTR. **K**7; not known elsewhere
HAB. Presumably evergreen forest; ? ± 50 m.

NOTE. This species is readily recognised by its pallid corky bark covering all but the current year's growth. There is considerable variation in leaf size, pubescence and the toothing of the calyx-limb which varies from truncate to distinctly triangular-toothed. Some specimens have the leaf-blades distinctly pubescent beneath and sparsely so above, e.g. *Welch* 401 (Tanzania, Morogoro District, Kingolwira Station, 17 Oct. 1957). Varietal rank might be appropriate for these, but there is variation in the degree of hairiness and no geographical factor is apparent since similar variants occur in Malawi. Nevertheless there is a preponderance of pubescent specimens in the Morogoro area.

The syntypes and many specimens from the south-eastern part of the species' range show some nodes with the leaves in whorls of 3 or even 4. Further study is needed in the field of the distribution of this character. If it were found to merit subspecific separation then the name *coaetanea* (which cannot be upheld on the grounds of calyx-lobe development) could be used for the race occurring in most of E. Africa and W. Zambia.

Braun in *Peter* 52184 (Tanzania, Gonja Mt., near Mnyussi, 9 Apr. 1908), probably a fragment removed from an Amani Herbarium sheet, is in very young bud but has been annotated *Chazaliella sp.* by Petit. Only *C. abrupta* is known from East Africa and the fragment is certainly not that. It will have to remain unidentified until further material turns up.

*Hepper (F.W.T.A., ed. 2, 2: 201 (1963)) cites *Psychotria abrupta* as occurring in West Africa but the material he cites is mostly *Chazaliella domatiicola* (De Wild.) Petit & Verdc.

** The Kew sheet is numbered 2242, as is also the HBG sheet, of which a photograph is preserved at Brussels.

5. CHASSALIA

Poir., Encycl. Méth. Bot., Suppl. 2 : 450, in obs. (1812); Verdc. in K.B. 30: 270 (1975)

Chasalia ou *Chassalia* Poir. in Dict. Sci. Nat. 8: 198 (1817)

Chasallia Juss. in Mém. Mus. Paris 6: 379 (1820)

Chasalia DC., Prodr. 4: 431 (1830)

Chazalia DC., Prodr. 9: 32 (1845); Petit in B.J.B.B. 29: 378 (1959) & 34: 20 (1964), attributed to A. L. Juss.*

Shrubs or less often small trees or subshrubby herbs, with mostly glabrous or only finely pubescent stems. Leaves opposite or rarely ternate, mostly acuminate, usually quite thin, shortly to distinctly petiolate, usually glabrous; stipules interpetiolar, ovate to triangular or quite short and broad, sometimes united into a small sheath, entire or with 2 short fimbriae, often with colleters and hairs within the base, mostly persistent. Flowers hermaphrodite, 4–5-merous, heterostylous, mostly small, in branched panicles, the ultimate elements usually being small heads but in some few species the flowers are pedicellate; bracts small. Calyx-tube mostly ovoid or oblong, ± ribbed, the free limb mostly very short, lobes triangular or linear but mostly very short. Buds often winged. Corolla white, pink or purple, sometimes yellow inside; tube cylindrical, hairy or glabrous inside; lobes often winged; venation of corolla often curiously prominent in dry material. Stamens included or exserted. Disc cylindrical, distinct. Ovary 2-locular; ovules solitary in each locule, erect from the base; style included or exserted; stigma-lobes linear. Fruits succulent, with 2 pyrenes; pyrenes pale, semi-globose or semi-ellipsoid, the ventral surface often grooved, often with a median dorsal keel along which dehiscence takes place. Seeds concavo-convex, with a pale testa; endosperm not ruminate.

A genus of about 40–50 species, mostly in tropical Africa and Madagascar, but with a few species in China, India, Burma, Ceylon, Malay Peninsula and Malay Is., extending to the Philippine Is. Closely allied to *Psychotria* but with a distinctly different facies, the buds often characteristically winged and the pyrenes with distinct median dorsal dehiscence.

1. Flowers very distinctly pedicellate . . . 7. *C. violacea*
   Flowers usually but sometimes not all ± sessile, the ultimate elements of the inflorescence subcapitate 2
2. Plant distinctly a liane up to 6 m.; corolla-tube 0·5–1 cm. long . . . . 1. *C. cristata*
   Plant subshrubby, shrubby or a small tree 3
3. Calyx-lobes linear-oblong, (1·5–)3–5 mm. long; stipules with single central apiculum . . . . 6. *C. zimmermannii*
   Calyx-lobes shortly triangular or ovate, mostly 0·25–1 mm. long but if ± lanceolate and up to 2·5–3·5 mm. long (in some forms of *C. discolor*) then stipules with 2 apical triangular to filiform lobes** 4

* Petit has accepted this spelling considering that De Candolle had made a valid correction of spelling the genus being named after Chazal de Chamarel.

** Note also that a hybrid of *C. albiflora* and *C. zimmermannii* has been recorded which has calyx-lobes up to 5 mm. long.

4. Plant a subshrubby herb ± 12 cm. tall . 3. *C. umbraticola* subsp. *geophila*
   Plant a shrub or small tree 5
5. Leaf-blades very shortly papillose-puberulous on the main venation beneath, large, up to 24 × 8·5 cm.; corolla-tube 1 cm. long (**U**2) . . . . 2. *C. ugandensis*
   Leaf-blades glabrous 6
6. Corolla-tube 0·4–1·2 cm. long; wings of buds, corolla-tube and lobes absent or not very evident; not growing below 900 m. 7
   Corolla-tube 1·2–2·1 cm. long; wings of buds, corolla-tube and lobes very marked; petiole 0·5–4 cm. long; either coastal or highland species. 10
7. Peduncles and inflorescence-branches shortly pubescent; stipules 0·4–1 cm. long (**T**6, 220 m.) . . . . 9. *C.* sp. A
   Peduncles and inflorescence-branches glabrous 8
8. Corolla-tube 0·5–1·2 cm. long; fruits ± didymous, 4–5 mm. tall, 5 mm. wide (**K**4). . . . . 11. *C. kenyensis*
   Corolla-tube 0·5–0·6 cm. long; fruits distinctly ellipsoid 9
9. Stipules joined to form a short sheath; peduncles 2–6 cm. long (**U**2; **K**5; **T**4) . 10. *C. subochreata*
   Stipules not joined to form a short sheath; peduncles 0·6–1·7 cm. long (**K**7; **T**2, 3, 6) . . . . 8. *C. parvifolia*
10. Calyx-tube and inflorescence-components mostly papillate-puberulous but sometimes glabrous; leaf-blades elliptic, acute rather than distinctly acuminate at the apex; petioles (even the lower ones) mostly short, 0·5–2(–2·5) cm. long; stipules undivided at the apex; coastal species growing at 0–450(–800) m. . 3. *C. umbraticola* subsp. *umbraticola*
    Calyx-tube and inflorescence-components glabrous; leaf-blades mostly elliptic-oblanceolate, distinctly acuminate; petioles (particularly lower ones) usually longer; stipules with 2 lobes at the apex; upland species 11
11. Corolla-lobes with a distinct lateral triangular or rounded appendage at the apex of the lobes showing as distinct projections in the bud; rhaphides very evident on the upper surface of the leaf-blades . 4. *C. albiflora*
    Corolla-lobes distinctly winged but without appendages; buds winged but ± truncate at the apex; rhaphides not very evident on the upper surface of the leaf-blades . . . . . . 5. *C. discolor*

1. **C. cristata** (*Hiern*) *Bremek.* in B.J.B.B. 22: 104 (1952); K.T.S.: 434 (1961); Hepper in F.W.T.A., ed. 2, 2: 192 (1963); Verdc. in K.B. 30: 272 (1975). Types: Sudan, Equatoria Province, Chief Rikkete's village, Khor Atirizi (Atazilly), *Schweinfurth* 3159 (K, syn.!) & Zaire, Chief Wando's village, Khor Dyagbe, *Schweinfurth* ser. II, 7 (K, syn.!) & Munsa [Munza's village], *Schweinfurth* 3463 (BM, syn.!)

Scandent shrub or woody twiner (0·9–)1·2–6 m. tall; very young parts of stems with very short but sometimes ± dense and ferruginous scurfy pubescence, but soon quite glabrous and slightly ridged or ± smooth. Leaf-blades oblong, elliptic, elliptic-oblong or somewhat ovate, 4·5–15(–18) cm. long, 1·2–6·5(–8·8) cm. wide, very distinctly narrowly acuminate at the apex, cuneate at the base, glabrous, the lateral nerves narrowly prominent beneath, the margins narrowly revolute; petiole 0·5–1·8(–4) cm. long; stipules hairy and with colleters inside, base ± semicircular, 3·5 mm. long with 2 linear lobes ± 0·5 mm. long, ± persistent on the flowering shoots. Flowers in dense many-flowered terminal sweet-scented cymes, and often with accessory lateral cymes from the axils of the first pair of leaves, altogether 3–7·5 cm. wide; peduncles purple or pinkish white, 1–2·2 cm. long, pubescent; pedicels obsolete; bracts and bracteoles small, with erose or ciliate margins. Calyx greenish white or tinged or dotted with purple; tube ovoid, ± 1 mm. long, glabrous or rarely pubescent; limb-tube ± 0·2–0·8 mm. long; teeth pinkish, small, broadly triangular, ± 0·2–0·5 mm. long. Buds glabrous outside, of characteristic shape due to the keeling of the lobes, the tips held at ± right-angles to the tube and giving a star-like appearance in plan view. Corolla ± fleshy, pink, white, purplish or white with purple streaks at the base, often described as white inside with a yellow ring at the throat; tube 0·5–1 cm. long, sometimes curved, the veins rather prominent at the apex; lobes white or yellowish, often tipped with pink, triangular-oblong, 2 mm. long, 1·9 mm. wide, the upper half thickened inside, longitudinally keeled outside and with a terminal lateral appendage. Stamens with anthers half exserted in short-styled forms, just included in the tube in long-styled forms. Style 4·5 mm. long in short-styled forms, 8·5 mm. long in long-styled forms; stigma-lobes 2–3 mm. long. Disc conspicuous, cylindrical, ± 1 mm. long. Drupes shining black or purple, subglobose, 5 mm. long, 6 mm. wide, grooved longitudinally round the middle, each lobe ribbed longitudinally at the middle of the base; pyrenes pale, semi-globose, 4–5 mm. long, 4·5–6·5 mm. wide, 2·6 mm. thick. Seeds reddish brown, semiglobose, 3·5–5 mm. tall, 4–6 mm. wide, 2–3·5 mm. thick, slightly rugulose, flattened ventrally and with a deep excavation so that this side of the seed appears horseshoe-shaped or almost annular.

UGANDA. Kigezi District: Ruzhumbura, Bugangari, Oct. 1947, *Purseglove* 2558!; Mengo District: Entebbe, Dec. 1930, *Hansford* in *Snowden* 1886! & Entebbe, Kyiwaga Forest, 17 Feb. 1950, *Dawkins* 514!

KENYA. N. Kavirondo District: NW. Kakamega Forest, 6 May 1971, *Mabberley & Tweedie* 1092! & Kakamega, May 1935, *Dale* in *F.D.* 3387!; S. Kavirondo District: Kisii, Bukuria, Sept. 1933, *Napier* 2939 in *C.M.* 5282!

TANZANIA. Bukoba District: Rubare Forest Reserve, Oct. 1957, *Procter* 709! & Kabwoba [Kabobwa] Forest Reserve, 19 Oct. 1955, *Sangiwa* 83!; Mpanda District: Mahali Mts., Kasiha, 27 Sept. 1958, *Newbould & Jefford* 2733!

DISTR. **U**2, 4; **K**5; **T**1, 4; S. Nigeria, Cameroun, ? Fernando Po (see note), Zaire, S. Sudan and Angola

HAB. Forest, particularly fringing forest and derived thickets, sometimes in grassland with scattered trees; (990–)1050–1650 m.

SYN. *Psychotria cristata* Hiern in F.T.A. 3: 205 (1877); F.P.S. 2: 400 (1952)
Myristiphyllum cristatum (Hiern) Hiern, Cat. Afr. Pl. Welw. 1: 493 (1898)
? *Chassalia lacuum* K. Krause in E.J. 43: 156 (1909). Types: Tanzania, Bukoba District, near Kazinga, *Conrads* 97 (B, syn. †) & Burundi, near Usumbura, Ndulnura R., *Keil* 206 (B, syn. †) & Buangai, *Mildbraed* 58 (B, syn. †)

NOTE. W. Africa material is not at all typical, the buds being less characteristic in shape owing to a reduction in keeling of the lobes; the calyx-tube and buds are often shortly puberulous and the fruit more ovoid. I am particularly dubious about the material from Fernando Po, which has longer calyx-lobes and pedicels and particularly much longer bracts; it may be a distinct subspecies. I have seen no authentic material of *C. lacuum*, but it seems very probable that it belongs here despite the description of the stipules as acuminate and the ovary as sparsely hairy.

2. **C. ugandensis** *Verdc.* in K.B. 30: 272, fig. 4 (1975). Type: Uganda, Kigezi District, Kayonza Forest, *Eggeling* 4200 (K, holo.!)

Shrub to 3 m., the youngest parts of the stem slightly pubescent but soon glabrous. Leaf-blades elliptic-oblanceolate, 12–24 cm. long, 6–8·5 cm. wide, acuminate at the apex, cuneate at the base, thin, drying rather yellowish beneath, glabrous above, finely papillose-puberulous on the main venation beneath; petiole 1·5–4·5 cm. long; stipules triangular, up to 1 cm. long, ? bilobed. Inflorescence fleshy, the flowers in compact dense panicles ± 2–3 cm. wide; peduncles ± 1 cm. long; secondary peduncles 3–5 mm. long; pedicels obsolete; all parts finely papillate-puberulous. Buds winged. Calyx ± microscopically puberulous; tube obconic, 2 mm. long; limb-tube 0·5 mm. long; teeth rounded-triangular, 0·5 mm. long; disc grooved, 0·8 mm. long. Corolla pink, microscopically puberulous; tube 1 cm. long; lobes narrowly triangular, 2·5 mm. long, 1·2 mm. wide, keeled outside. Anthers just exserted about half above and half below the throat in short-styled flowers. Style 4·8 mm. long in short-styled flowers; stigma-lobes linear, 1·7 mm. long. Fruits not seen.

UGANDA. Kigezi District: Kayonza Forest, Oct. 1940, *Eggeling* 4200!
DISTR. **U**2; not known elsewhere
HAB. Evergreen forest; ? 1000 m.

3. **C. umbraticola** *Vatke* in Oest. Bot. Zeitschr. 25: 230 (1875); K. Schum. in P.O.A. C: 392 (1895); T.T.C.L.: 490 (1949); K.T.S.: 435 (1961); Verdc. in K.B. 30: 274 (1975). Type: Zanzibar I, *Hildebrandt* 1158 (B, holo. †, BM, K, iso.!)

Shrub or more rarely a small subshrubby herb, 0·12–4·5 m. tall, rarely stated to be slightly scandent; older stems pale greyish and glabrous, younger glabrous or papillate-pubescent. Leaf-blades elliptic, 3–16 cm. long, 1·35–6·3 cm. wide, acute or slightly acuminate at the apex but distinctly less so than in other species, cuneate at the base, the blade ± decurrent so that the petiole length is not clear cut, ± discolorous when dry, the margins revolute, glabrous; petioles 0·5–2(–2·5) cm. long, usually short and the lower ones distinctly shorter than in related species; stipules broad, 1–1·5 mm. long, undivided, with fine marginal hairs and longer hairs within, sometimes becoming corky. Flowers sweet-scented, in terminal branched inflorescences, the ultimate components 3–several-flowered clusters; inflorescence-components white, tinged purple; peduncles 0·5–2·5(–4) cm. long; secondary peduncles 0·3–1·3 cm. long; pedicels actually absent or very short but in reduced inflorescences may appear 1·5–4 mm. long; all parts usually finely papillate-puberulous or rarely glabrous; bracts small, but sometimes part of the inflorescence is subtended by a pair of reduced leaves. Calyx cream with purple upper margin; tube oblong-ovoid, 1·2–2 mm. long, finely papillate-puberulous or rarely glabrous, slightly ribbed; limb-tube 0·2–0·5 mm. long; lobes triangular, 0·1–0·5 mm. long. Buds distinctly winged in limb portion. Corolla cream or white, often tinged purple sometimes at the base of the tube and tips of the lobes, glabrous or finely puberulous; tube 1·5–2 cm. long; lobes linear-oblong, 5–7 mm. long, 1–2·5 mm. wide, con-

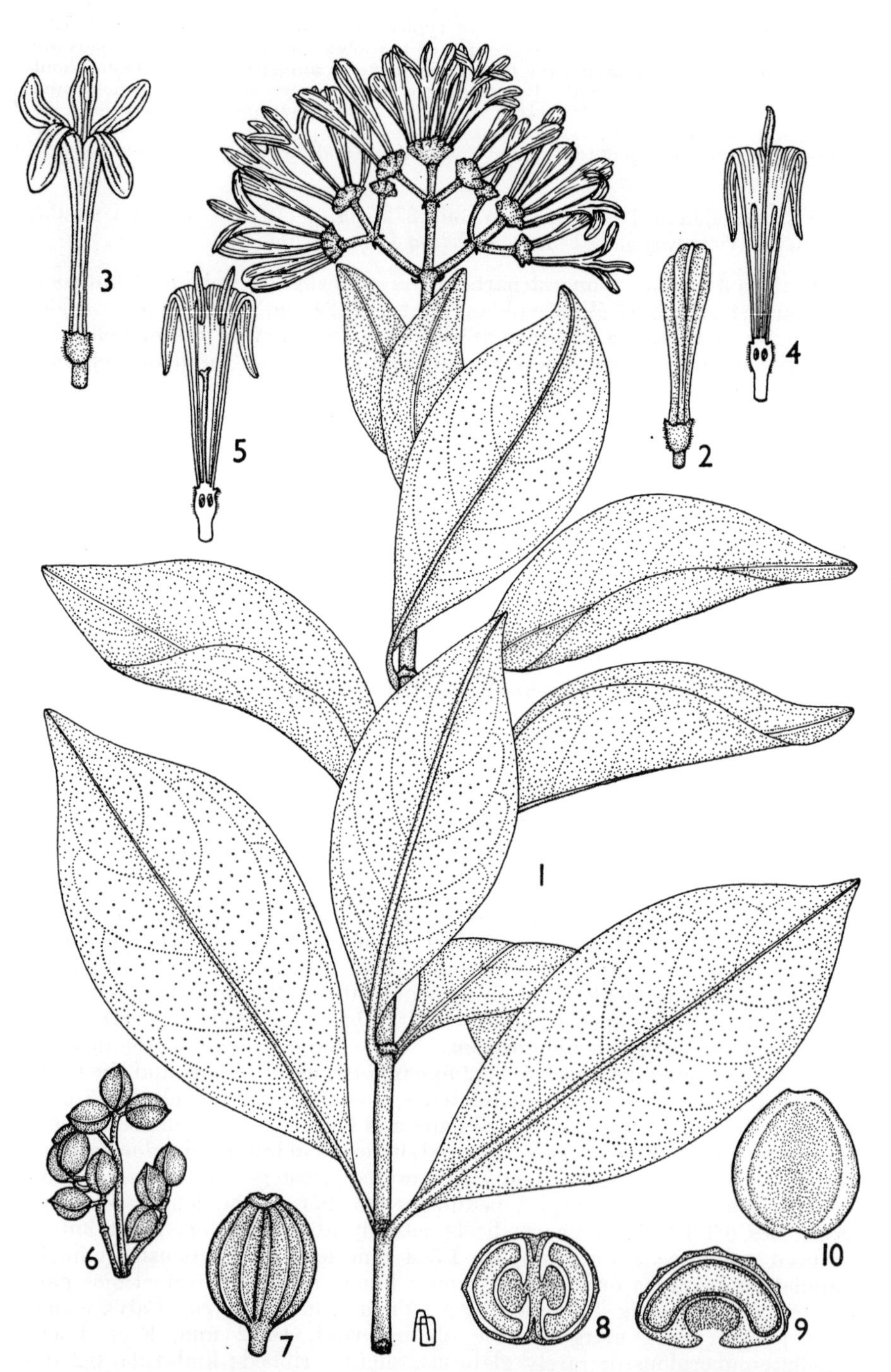

FIG. 9. *CHASSALIA UMBRATICOLA* subsp. *UMBRATICOLA*—**1,** flowering branch, × 1; **2,** flower bud, × 2; **3,** long-styled flower, × 2; **4,** longitudinal section of same, × 2; **5,** longitudinal section of short-styled flower, × 2; **6,** part of infructescence, × 1; **7,** fruit, × 2⅔; **8,** transverse section of fruit, × 4; **9,** transverse section of pyrene, × 6; **10,** seed, × 4. 1, 5, from *Tanner* 2823; 2–4, from *Tanner* 2797; 6, from *Archbold* 1003; 7–10, from *Tanner* 3014. Drawn by Ann Davis.

spicuously winged, the wings decurrent on the tube, the venation often curiously raised and prominent in dry material, particularly if flowers were picked in a fading state. Filaments with anthers just completely exserted in short-styled flowers; anthers with tips 3 mm. below the throat in long-styled flowers. Style 10 mm. long in short-styled flowers, with stigma-lobes linear, 3 mm. long, flattened, just included; ± 1·6–1·7 cm. in long-styled flowers, with stigma-lobes linear, ± 3 mm. long, flattened. Fruits black, subglobose or rounded-ovoid, ± compressed, 4–5(–7) mm. long, 4–5(–7) mm. wide, ribbed, mostly distinctly densely rugulose in the dry state; disc ± persistent; pyrenes pale, half-ovoid, 4–5 mm. long, 3·8–5 mm. wide, 1·8–2·5 mm. thick, vaguely tuberculate to densely covered with pointed rugae. Seeds concavo-convex, basin-shaped.

subsp. **umbraticola**

Shrub 0·9–4·5 m. tall. Fig. 9.

KENYA. Kwale District: Buda Forest Reserve, 3 Nov. 1959, *Napper* 1377 ! & Shimba Hills, Kwale, 14 Jan. 1964, *Verdcourt* 3921 !; Kilifi District: Marafa, 22 Nov. 1961, *Polhill & Paulo* 836 !

TANZANIA. Tanga District: Ngomeni, 30 July 1953, *Drummond & Hemsley* 3543 !; Morogoro District: Mtombozi, Oct. 1930, *Haarer* 1853 !; Uzaramo District: 9·5 km. E. of R. Ruvu, 26 Nov. 1955, *Milne-Redhead & Taylor* 7394 !; Zanzibar I., Kizimbani, 20 May 1959, *Faulkner* 2258 !; Pemba I., Ngezi Forest, 18 Feb. 1929, *Greenway* 1476 !

DISTR. **K**7; **T**3, 6, 8; **Z**; **P**; Mozambique

HAB. Coastal bushland, coastal *Brachystegia* woodland, evergreen forest and forest edges, also in old cultivations; 0–450(–800) m.

SYN. *Psychotria zanguebarica* Hiern in F.T.A. 3: 214 (1877). Types: Tanzania, Bagamoyo, Mafia I. & Rovuma Bay, *Kirk* (K, syn. !)
Uragoga zanguebarica (Hiern) O. Kuntze, Rev. Gen. Pl.: 963 (1891), as "*zangebarica*"
Psychotria umbraticola sensu U.O.P.Z.: 425 (1949)
[*Chassalia albiflora* sensu K.T.S.: 434 (1961), *non* K. Krause]

NOTE. This species is usually well characterised by its elliptic acute (not long acuminate) leaf-blades, short petioles, mostly papillate-pubescent calyx-tubes, etc. and coastal habitat. There is, however, considerable variation in indumentum and petiole length and all the distinguishing characters may not be present in all specimens. *Hay* 17 (Tanzania, Newala, 20 Dec. 1958), a much-branched shrub to 1·2 m. tall, with pink and purple flowers, may be a form of this species with glabrous ovaries. The unicuspid stipules and short petioles are in agreement but the general habit is not; the unusually high altitude of 660 m. is not without precedent. Further material is needed to confirm this suggestion.

subsp. **geophila** *Verdc.* in K.B. 30: 274 (1975). Type: Tanzania, Songea–Tunduru road, *Richards* 17718 (K, holo. !)

Small subshrub or herb, apparently not exceeding 15 cm. tall, but otherwise very similar to subsp. *umbraticola* except for a more reduced inflorescence.

TANZANIA. Songea District: Songea–Tunduru road, 3 Mar. 1963, *Richards* 17718 !; Tunduru, *Allnut* 22 !

DISTR. **T**8; not known elsewhere

HAB. Bushland and woodland; 780–1080 m.

4. **C. albiflora** *K. Krause* in E.J. 39: 566 (1907); T.T.C.L.: 489 (1949); Verdc. in K.B. 30: 274 (1975). Types: Tanzania, E. Usambara Mts., Gonja Mt., *Engler* 3378 (B, syn. †, EA, isosyn. !) & Derema, *Scheffler* 119 (B, syn. †) & Amani, *Warnecke* 490 (B, syn. †, BM, K, isosyn. !)

Much-branched evergreen shrub or small tree 1·2–4·5 m. tall; stems glabrous. Leaf-blades obovate-oblanceolate to oblong, 7–18(–20) cm. long, 3–7·5(–9) cm. wide, acuminate at the apex, cuneate at the base, thin, glabrous, the rhaphides showing very characteristically on the upper surface;

petiole 2·5–5(–6) cm. long; stipules triangular, 5 mm. long, bilobed at the apex, the lobes ± 1 mm. long, the lower ones becoming corky. Flowers in ± many-flowered panicles, very often galled; peduncles 1·3–2 cm. long; secondary peduncles 0·5–1·2 cm. long; pedicels obsolete. Buds with the limb-appendages projecting. Calyx-tube oblong, 1·2 mm. long; limb-tube 0·5 mm. long; lobes triangular, 0·2–1 mm. long. Corolla white, pink or reddish pink, glabrous; tube 1·1–1·4 cm. long; lobes triangular, 3 mm. long, 1·6 mm. wide, with a distinct apical lateral rounded or triangular appendage outside 1–2 mm. long. Anthers exserted, the bases just below the throat in short-styled flowers; tips just included in long-styled flowers. Style 8·5 mm. long, with stigma-lobes 1·8 mm. long, in short-styled flowers, 1·2–1·3 cm. long in long-styled flowers. Fruits pink, fleshy, ovoid-subglobose, 6–7·5 mm. tall, 6–8·5 mm. wide; pyrenes orange-brown, 7 mm. long, 5·5 mm. wide, 4 mm. thick, cleft apically and with a very distinct median keel at the base. Seeds yellow-brown, helmet-shaped, deeply excavated, 5 mm. long, 4·2 mm. wide, 3 mm. thick, keeled like the pyrenes.

TANZANIA. Lushoto District: E. Usambara Mts., Amani–Maramba, 30 Oct. 1935, *Greenway* 4144! & Amani, 18 Sept. 1928, *Greenway* 860! & same locality, 16 Nov. 1956, *Tanner* 2744!

DISTR. **T3**; known only from the E. Usambara Mts.

HAB. Rain-forest; 800–1050 m.

SYN. *Psychotria albiflora* (K. Krause) De Wild., Pl. Bequaert. 2: 328 (1924)

NOTE. See also 12, *C. buchwaldii* K. Schum., which seems likely to be synonymous and if so then the correct name.

Tanner 2745, said to be from Tanzania, Pangani, 30 m., is I am certain wrongly localised. It undoubtedly belongs to the above species and is probably from Amani. Despite numerous collections made by A. Peter, better material is needed to fill in the gaps of our knowledge about this surprisingly poorly known species. Part of a sheet of *C. zimmermannii* (Lushoto District, Korogwe, Magunga Estate, 19 Dec. 1952, *Faulkner* 1089!) seems to be a fruiting specimen of *C. albiflora*, but whether this is due to a confusion in the herbarium or whether the latter species actually occurs at Magunga is not clear.

This species is usually very well characterised; the only atypical specimen seen is *Zimmermann* in *Herb. Amani* 7795 (Lushoto District, Amani, Mt. Bomole, 8 Oct. 1916), the leaves of which show clearly the numerous surface rhaphides, but the corolla-lobes practically lack appendages.

5. **C. discolor** *K. Schum.* in E.J. 34: 339 (1904); T.T.C.L.: 490 (1949); K.T.S.: 435 (1961), pro parte; Verdc. in K.B. 30: 274 (1975). Types: Tanzania, W. Usambara Mts., between Kwai and Gare, *Engler* 1224 (B, syn. †) & Magamba, above Kwai, *Engler* 1285 (B, syn. †) & presumably W. Usambara Mts., *Buchwald* 102 (B, syn. †, K, isosyn.!)

Shrub (or rarely ± herbaceous) (0·5–)1–4(–6) m. tall, with finely ridged glabrous stems. Leaf-blades oblanceolate to elliptic- or ovate-oblanceolate, 4–15·5 cm. long, 1·4–5(–7) cm. wide, distinctly narrowly acuminate at the apex, cuneate at the base, thin, slightly discolorous, glabrous, the margins often slightly wavy; petiole 0·8–4 cm. long; stipules triangular, 3–4 mm. long with 2 apical colleter-tipped filiform lobes 1 mm. long; older stipules often corky, persistent. Flowers scented, in mostly trichotomously branched panicles, each ultimate branchlet 3–several-flowered; peduncles 0·7–2 cm. long, mostly purplish; secondary peduncles 0·3–2 cm. long; pedicels obsolete or up to ± 1 mm. long; bracts small, ciliate. Calyx-tube mostly magenta pink, oblong, 1–1·8 mm. long, ribbed; limb-tube 0·7–1·2 mm. long; lobes triangular to lanceolate, 0·5–3·5 mm. long, if longer then often recurved at the tips. Corolla waxy, glabrous, white, white and pink, white suffused pink outside or magenta-pink, probably usually yellow inside; tube often curved, ± cylindrical, 1·2–2·1 cm. long; lobes elliptic-lanceolate, 2–4 mm. long,

1·5–2·5 mm. wide, venose, winged (particularly in bud). Anthers just exserted for $\frac{2}{3}$ their length or tips at throat-level or entirely exserted in short-styled flowers, tips 4–6 mm. below the throat in long-styled flowers. Style 7–10 mm. long in short-styled flowers, with stigma-lobes 4–5 mm. long; 14·5 mm. long in long-styled flowers, with stigma-lobes 5 mm. long. Fruits white, reddish mauve or blackish purple, often translucent, ± didymous or depressed subglobose, 5–6 mm. long, 6–7·5 mm. wide, 5 mm. thick, mostly ribbed, with a median groove, usually crowned with the persistent calyx-lobes; pyrenes pale, ± semi-ovoid or subglobose, 5–6 mm. long, 4·5 mm. wide, 3–3·5 mm. thick, with a basal median keel, ± rugulose. Seed ± 5 mm. wide, strongly concavo-convex.

KEY TO INFRASPECIFIC VARIANTS

Calyx-lobes mostly triangular, short, under 2·5 mm. long:
  Leaf-blades up to 16 cm. long, 7 cm. wide, but usually much smaller (**T3**, 6, 7) subsp. **discolor**
  Leaf-blades up to 24·5 cm. long, 9·5 cm. wide (**T6**) . . . . subsp. **grandifolia**
Calyx-lobes linear to linear-triangular, 2·5–4 mm. long (Kenya, **K7**) . . . . . subsp. **taitensis**

subsp. **discolor**

Calyx-lobes mostly more triangular, under 2·5 mm. long. Leaf-blades up to ± 16 cm. long and 7 cm. wide but usually much smaller.

TANZANIA. Lushoto District: Mkuzi, 13 Dec. 1961, *Greenway* 10403!; Morogoro District: Nguru Mts., Nguru South Forest Reserve, E. slopes above Kwamanga Village, near Mhonda Mission, 5 Feb. 1971, *Mabberley & Pócs* 684! & Uluguru Mts., Kitundu, 25 Mar. 1935, *E. M. Bruce* 938!
DISTR. **T3**, 6, ?7; not known elsewhere
HAB. Evergreen forest; (1050–)1300–1950 m.

SYN. [*C. violacea* sensu Brenan, T.T.C.L.: 490 (1949) quoad *E. M. Bruce* 938; K.T.S.: 435 (1961), *non* K. Schum.]

subsp. **taitensis** *Verdc.* in K.B. 30: 275 (1975). Type: Kenya, Teita Hills, Ngangao Forest, *Faden & Evans* 69/877 (K, holo.!, EA, iso.!)

Calyx-lobes distinctly longer, linear to linear-triangular, 2·5–4 mm. long. Leaves as in subsp. *discolor*.

KENYA. Teita Hills, Ngangao Forest, 6 Feb. 1953, *Bally* 8765! & same locality, 7 July 1969, *Faden & Evans* 69/877! & Mwangoji Forest, July 1937, *Dale* in *F.D.* 3837!
DISTR. **K7**; not known elsewhere
HAB. Evergreen forest including mist forest; 1450–1950 m.

subsp. **grandifolia** *Verdc.* in K.B. 30: 275 (1975). Type: Tanzania, Uluguru Mts., E. foothills, Kimboza Forest Reserve, near R. Ruvu, *T. & S. Pócs* 6274/D (EA, holo.!)

Calyx-lobes short, triangular as in subsp. *discolor*. Leaves large up to 24·5 cm. long, 9·5 cm. wide.

TANZANIA. Morogoro District: E. Uluguru Mts., Milawilila Forest Reserve, 4 km. N. of Tawa Village in the Mvuha R. Valley, 13 Mar. 1971, *Pócs* 6416/A! & Uluguru Mts., Kimboza Forest Reserve, between Mkuyuni and Matombo, 19 Nov. 1969, *Pócs & Gibbon* 6060/D! & without exact locality, *Rounce* 636!
DISTR. **T6**; not known elsewhere
HAB. Lowland rain-forest; 300–400 m.

NOTE. There is considerable variation in the length and colour of the corolla-tube. There has been a tendency to use the name *C. violacea* for this species but K. Schumann distinctly says flowers pedicellate in his description of that species and flowers 1·3 cm. long. W. Usambara collections seem to be the only ones where the

FIG. 10. *CHASSALIA ZIMMERMANNII*—**1,** flowering branch, × ⅔; **2,** node showing stipule, × 4; **3,** part of inflorescence, × 3; **4,** short-styled flower, × 3; **5,** corolla, opened out, × 3; **6,** flower, with corolla removed, × 6; **7,** longitudinal section of ovary, × 6; **8,** fruit, × 5; **9,** transverse section of fruit, × 5. **All** from *Faulkner* 1050. Drawn by Mrs. M. E. Church.

flowers are said to have the corolla-lobes yellow inside. There are specimens at Kew and Brussels, *Buchwald* 161, which are labelled *C. buchwaldii* K. Schum. by a Berlin hand but do not agree with *C. albiflora* K. Krause (q.v.). There is nothing to prove this is *C. buchwaldii* since under the description of that K. Schumann merely says "Früher auch von Dr. Buchwald gesammelt" with no mention of a number. These sterile specimens could be *C. discolor*.

6. **C. zimmermannii** *Verdc.* in K.B. 30: 275, fig. 5 (1975). Type: Tanzania, Korogwe, Magunga Estate, *Faulkner* 1050 (K, holo.!)

Shrub of unknown habit, probably only a few m. tall but possibly slightly scandent; older stems pale greyish, younger blackish, all glabrous. Leaf-blades elliptic to oblanceolate, 3–17 cm. long, 1·2–5 cm. wide, shortly distinctly acuminate at the apex, cuneate at the base, glabrous, discolorous; petiole short, 0·2–2·5(–4) cm. long, scarcely separable from the decurrent leaf-base; stipules transverse with single central apiculum, 1·5–2 mm. long. Flowers in rather few-flowered subsessile or shortly pedunculate subcapitate inflorescences; peduncles and secondary peduncles up to 5 mm. long; pedicels obsolete; bracts linear, 1·2–1·3 cm. long. Calyx glabrous; tube ovoid-oblong, 1 mm. long; lobes linear, (1·5–)3–5 mm. long, 0·5–0·8 mm. wide, narrowed at the base. Buds conspicuously winged. Corolla white shaded with purple and yellow; tube 1–1·2 cm. long; lobes triangular, 2·5 mm. long, 1·5 mm. wide, winged outside. Anthers just completely exserted in short-styled flowers. Style 5·5 mm. long and stigma-lobes 2·5 mm. long in short-styled flowers. Fruits white, globose, 5 mm. in diameter, ribbed, mostly crowned with the persistent calyx-lobes; pyrenes pale, ½-ovoid, 5 mm. long, 3·6 mm. wide, 3 mm. thick, with very strong dorsal keel. Seeds strongly concavo-convex. Fig. 10.

TANZANIA. Tanga District: Korogwe, Magunga Estate, 19 Dec. 1952, *Faulkner* 1089! & E. Usambara foothills, Sigi, *Zimmermann*! & Longuza, 7 Apr. 1922, *Soleman* in *Herb. Amani* 5940! & by R. Sigi, 1·5 km. below Longuza, 6 May 1926, *Peter* 46533!
DISTR. T3; not known elsewhere
HAB. Evergreen forest; 250–450(–800) m.

NOTE. *Peter* 21840 (Lushoto District, NE. of Bulwa, 25 Oct. 1917) has the leaves and buds of *C. albiflora* but calyx-lobes up to 5 mm. long. Out of the large number of sheets of the genus collected by Peter this is the only one showing such a combination of characters and I strongly suspect it to be a hybrid between *C. albiflora* and *C. zimmermannii*.

7. **C. violacea** *K. Schum.* in E.J. 28: 498 (1900); T.T.C.L.: 490 (1949), pro parte; Verdc. in K.B. 30: 276 (1975). Type: Tanzania, Uluguru Mts., 1200–1800 m., *Goetze* 186 (B, holo. †)

Sprawling shrub or undershrub 1–3 m. tall, with slender branches, glabrous, soon covered with ridged pale grey-brown corky bark. Leaf-blades narrowly elliptic to narrowly elliptic-obovate, 2–10(–18) cm. long, 0·6–3·3(–6) cm. wide, acuminate at the apex, cuneate at the base, glabrous, ± discolorous when dry; petiole 0·1–1 cm. long; stipule-base 1·5 mm. long, glabrous with 2 close obscure median cusps 1 mm. long. Inflorescences lax, glabrous, ± 20-flowered; peduncle 2·7–5 cm. long; secondary branches 1·5–3·5 cm. long; pedicels 1–3·5(–7) mm. long; bracts acuminate, up to 3 mm. long. Calyx-tube glabrous or finely puberulous, oblong, 0·8–1 mm. long; limb-tube 0·2–0·5 mm. long; lobes triangular, 0·3–0·5 mm. long. Corolla white, pinkish white or mauve or white marked with purplish, glabrous; tube narrowly funnel-shaped, 0·5–1·8 cm. long; lobes narrowly triangular, 1·5–4 mm. long, 1·5–2 mm. wide. Disc 1 mm. tall, rather conspicuous. Filament-tips just exserted or just included in short-styled flowers, 1–2 mm. below the throat in long-styled flowers. Style up to 7 mm. long in short-styled flowers, with stigma-lobes filiform, 2·5 mm. long; exserted ± 1–1·5 mm. in long-styled flowers, with stigma-lobes 1·5–2 mm. long. Fruit dark mauve, ellipsoid or fusiform-ellipsoid, 7–8(–11 in spirit) mm. long, 3·5–5·5(–8 in spirit) mm. wide; pyrenes oblong, 6 mm. long, 3·5 mm. wide, 2 mm. thick, splitting down the median dorsal ridge and around the margin, the rugose surface often showing through the epicarp of the dry fruit.

var. **violacea**

Corolla-tube 1·1–1·8 cm. long.

TANZANIA. Morogoro District: Uluguru Mts., Bunduki, 10 Jan. 1935, *E. M. Bruce* 496! & Bondwa Peak, Jan. 1953, *Eggeling* 6451! & 4·8 km. S. of Bunduki, Salaza Forest, 19 Mar. 1953, *Drummond & Hemsley* 1681!
DISTR. **T6**; not known elsewhere
HAB. Upland evergreen forest; 1200–2000 m.

NOTE. I at first considered this might be a new species having accepted Brenan's interpretation which was probably based on the Kew-named sheet *Bruce* 938 (possibly even named by comparison with a type from Berlin since it is known types were borrowed to name this collection). K. Schumann definitely gives the flower length as 1·3 cm. (not 2·3 cm. as given by Brenan) and clearly states *flowers pedicellate*. There are, however, still some doubts about my interpretation—K. Schumann states the flowers are tetramerous (not noted), the petioles are 1·5–3 cm. and the stipules 5–6 mm. long with 1 mm. long lobes. A sheet *Herring* in *F.H.* 2008, said to be from **T4** Tabora, must I think be wrongly localized.

var. **parviflora** *Verdc.* in K.B. 30: 278 (1975). Type: Tanzania, Uluguru Mts., near Bunduki, *Drummond & Hemsley* 1616 (K, holo.!, EA, iso.)

Corolla-tube 5–6 mm. long.

TANZANIA. Morogoro District: Uluguru Mts., Tanana, 14 Feb. 1933, *Schlieben* 3445! & 4·8 km. S. of Bunduki, Salaza Forest, 15 Mar. 1953, *Drummond & Hemsley* 1616! & Tegetero, 20 Mar. 1953, *Drummond & Hemsley* 1693!
DISTR. **T6**; not known elsewhere
HAB. Evergreen forest; 1100–2000 m.

NOTE. Some specimens of this are at first sight scarcely distinguishable from *C. parvifolia* and I had considered the possibility it might be a pedicellate variety of that species. The variation in the corolla length in typical *C. violacea* makes it more reasonable to follow the above course. It might possibly be of hybrid origin involving the two species mentioned.

8. **C. parvifolia** *K. Schum.* in E.J. 28: 103 (1899); T.T.C.L.: 490 (1949); Verdc. in K.B. 30: 278 (1975). Type: Tanzania, SE. Uluguru Mts., Kikurungu, *Stuhlmann* 9253 (B, holo. †)

Much-branched bushy shrub to small tree 2–4·5(–7·5) m. tall, with 2-ribbed or ridged stems, glabrous, soon becoming grey and corky. Leaf-blades elliptic to obovate-oblanceolate or elliptic-oblong, 1–13 cm. long, 0·5–5·2 cm. wide, abruptly acuminate at the apex, cuneate at the base, glabrous, thin, the margins often ± crinkly in the dry state; petiole 0·2–1·1(–1·5) cm. long; stipules triangular or short and very rounded, 1–2 mm. long, either without cusps or with 2 rather obscure separated cusps. Flowers scented, in trichotomous or much branched ± small inflorescences, the ultimate elements being subcapitate; primary peduncles 0·6–1·7 cm. long; secondary peduncles 0·3–2 cm. long; pedicels obsolete. Calyx-tube squarish, 0·8 mm. long; limb-tube ± 0·25 mm. long; lobes ovate-triangular, 0·25 mm. long. Buds not or scarcely winged. Corolla white or greenish-white, often tinged or tipped with pink; tube 4–5(–6) mm. long, widened above; lobes ovate, 2 mm. long, 1·2 mm. wide. Filaments distinctly exserted in short-styled flowers, the bases of the anthers 0·5–1 mm. above the throat; style 1·4 mm. long in short-styled flowers, stigma-lobes 0·6 mm. long. Anthers included in long-styled flowers, the tips 1 mm. below the throat; style exserted 2 mm. in long-styled flowers, the stigma-lobes 1·5–2 mm. long. Fruit ovoid or ellipsoid, translucent greenish yellow, pink or shiny black according to field notes, 4·5–5 mm. tall, 3–3·5 mm. wide, grooved between the pyrenes; pyrenes pale, ½-ellipsoid, 4–5 mm. long, 3·8 mm. wide, 2 mm. thick, with short median dorsal keel at base and shallow ventral grooves.

KENYA. Teita District: Ngangao Forest, 7 July 1969, *Faden & Evans* 69/878! & Ngangao, 8 km. NNE. of Ngerenyi, 15 Sept. 1953, *Drummond & Hemsley* 4332! & Mbololo Hill, *Gardner* in *F.D.* 2913 in part!
TANZANIA. Moshi District: Kilimanjaro Forest, Machame, 27 Jan. 1968, *Carmichael* 1453!; Lushoto District: E. Usambara Mts., Kwamkoro, 12 Nov. 1959, *Semsei* 2941!; Morogoro, 17 Nov. 1935, *E. M. Bruce* 143!
DISTR. **K7**; **T2**, 3, 6–8; N. Malawi, NE. Zambia
HAB. Evergreen forest including rain- and mist-forest and also drier types; 600–2300 m.

SYN. *Psychotria engleri* K. Krause in E.J. 43: 153 (1909); T.T.C.L.: 524 (1949). Types: Tanzania, E. Usambara Mts., Gonja Mt., *Engler* 3361 (B, syn. †, EA, isosyn.!) & Mt. Bomole, *Braun* in *Herb. Amani* 870 (B, syn. †, EA, isosyn.!, B, fragment of isosyn.!)
P. parvifolia (K. Schum.) De Wild., Pl. Bequaert. 2: 401 (1924), *non* Oerst. (1852)
[*Chassalia discolor* sensu K.T.S.: 435 (1961), quoad specim. *Dale* in *F.D.* 3858 (? sphalm. 3838), *non* K. Schum.]

NOTE. There is great variation in the size of the leaf-blades, some specimens having uniformly small blades about 4·5 × 1·3 cm., but this is probably only an ecological state.

9. **C. sp. A**; Verdc. in K.B. 30: 279 (1975)

Shrub to 2 m. tall, with glabrous stems. Leaf-blades narrowly elliptic or somewhat oblanceolate, 2–11 cm. long, 1–4 cm. wide, apiculate at the apex, cuneate at the base, glabrous; petiole 0·3–2 cm. long; stipules elongate-triangular, 8–9 mm. long, bilobed at the apex, the lobes ± 1 mm. long, ± persistent and eventually corky. Flowers in 3–5-branched panicles; peduncles dark red-brown, 8–9·5 cm. long, finely puberulous; secondary peduncles 0·6–2·3 cm. long, puberulous to densely pubescent; pedicels obsolete; bracts linear, 5–9 mm. long; bracteoles 1–1·5 mm. long. Calyx green with brown-purple tips, glabrous; tube oblong, 1 mm. long; limb-tube 0·5 mm.; lobes triangular, 0·5–1 mm. long. Open corolla not seen, immature greenish white with a faint purple flush; buds 7·5 mm. long, not winged. Fruits not seen.

TANZANIA. Morogoro District: S. Uluguru Forest Reserve, edge of Lukwangule Plateau, 17 Mar. 1953, *Drummond & Hemsley* 1654!
DISTR. **T6**; not known elsewhere
HAB. Upland rain-forest; 2200 m.

NOTE. Possibly only a variant of *C. parvifolia* but without further material I prefer to keep it separate.

10. **C. subochreata** (*De Wild.*) *Robyns*, F.P.N.A. 2: 367 (1947); K.T.S.: 435 (1961), pro minore parte; Verdc. in K.B. 30: 279 (1975). Type: Zaire, between Masisi and Walikale, *Bequaert* 6433 (BR, holo.!)

Shrub or small tree 1·8–9 m. tall, with slender branched glabrous stems. Leaf-blades oblanceolate to narrowly elliptic, 3·5–18 cm. long, 1·2–5·5 cm. wide, narrowly acuminate at the apex, narrowly cuneate at the base, thin, glabrous, the lateral nerves prominent in the dried state beneath; petiole 0·2–3·5 cm. long; stipules joined to form a sheath, at least when young, hairy within, only very bluntly triangular, 3–4 mm. long, 3–5 mm. wide, persistent and eventually becoming yellowish and corky. Flowers in branched glabrous panicles; peduncles 2–6 cm. long, mostly ± white; secondary peduncles 0·5–2·5 cm. long, similar; ultimate components of the inflorescence compact, 20–30-flowered; pedicels obsolete or very short, white. Calyx-tube ovoid, 0·8–1 mm. long, glabrous; limb-tube 0·2–0·3 mm. long, truncate or with mostly minute triangular teeth 0·25–0·5 mm. long. Buds very slightly horned but not winged. Corolla white; tube narrowly infundibuliform, 5–6 mm. long; lobes oblong-elliptic, 4·5 mm. long, 1·2 mm. wide. Filaments 4 mm. long, exserted 2·8 mm., the anther-bases 1·2 mm. above the throat and tips almost exceeding the corolla-lobes in short-styled flowers or anthers only ½-exserted, the anther tips just exserted in long-styled flowers. Style 2·5–4·5 mm. long, with stigma-lobes 0·8–2 mm. long, in short-styled flowers, 6 mm. long in long-styled flowers. Fruits black, ellipsoid, 4·5–6·5 mm. long, 3·5–4·5 mm. wide, the fruiting pedicels up to 0·5 mm. long, with a median groove between the pyrenes and keels on the faces; pyrenes pale, ½-ellipsoid, 5 mm. long, 3·2 mm. wide, 1·8 mm. thick, with 2 close ventral grooves. Seeds thin, concavo-convex, 5 mm. long, 4·1 mm. wide, 1·2 mm. thick, keeled dorsally near the base.

UGANDA. Kigezi District: Rubuguli, Apr. 1948, *Purseglove* 2697! & Impenetrable Forest, Apr. 1969, *Hamilton* 1219! & same locality, June 1938, *Eggeling* 3681!
KENYA. Kericho District: SW. Mau Forest, Sambret Catchment, 20 Apr. 1958, *Kerfoot* in *E.A.H.* 11405! & same locality, Timbilil Stream, 16 June 1958, *Kerfoot* 159! & Camp 8, 0° 36′ 30″ S., 35° 18′ 20″ E., 14 Aug. 1949, *Maas Geesteranus* 5772!
TANZANIA. Kigoma District: 57·6 km. S. of Uvinsa, 31 Aug. 1950, *Bullock* 3257!; Mpanda District: Mahali Mts., Sisaga, 27 Aug. 1958, *Newbould & Jefford* 1820!
DISTR. **U2**; **K5**; **T4**; E. Zaire, Burundi
HAB. Evergreen forest, including riverine forest; 1650–2550 m.

SYN. *Psychotria subochreata* De Wild., Pl. Bequaert. 2: 428 (1924)

NOTE. The material from **T4** is not typical; the leaves are larger and the fruits have pyrenes with a more evident dorsal keel.

11. **C. kenyensis** *Verdc.* in K.B. 30: 279 (1975). Type: Kenya, Limuru, Limuru Girls' High School, *Faden* 70/72 (K, holo. !, EA, iso. !)

Small erect or somewhat scandent shrub 0·9–1·8 m. tall, with much-branched glabrous ridged stems. Leaf-blades elliptic, 5·5–15 cm. long, 1·7–4·1 cm. wide, thin, discolorous, distinctly abruptly acuminate at the apex, cuneate at the base, glabrous, the margins somewhat wavy; petiole 0·4–1·5 cm. long; stipules semicircular, ± 1·5–2 mm. long, entire or with obscure fimbriae. Flowers in branched panicles, the ultimate elements small 3–5-flowered capitula; peduncles 1·5–3 cm. long, glabrous save for slightly puberulous bract areas; secondary peduncles 0·2–2 cm. long; pedicels absent. Calyx-tube oblong-ovoid to transversely oblong, 0·5–1 mm. tall, 1·2 mm. wide; limb-tube very short, 0·3–0·4 mm. long; lobes broadly triangular, 0·4–0·5 mm. long. Corolla white, tinged purple or white and lobes margined and tipped with purple, the venation also sometimes reddish; tube (0·5–)1·2 cm. long; lobes oblong-lanceolate, 3·2–4 mm. long, 1·3–1·5 mm. wide, not winged or only faintly so along the margins, with incurved thickened tips. Anthers exserted in short-styled flowers, the bases 0·5–1·2 mm. above the throat. Style 3·5–7 mm. long in short-styled flowers, the stigma-lobes 1·5–2·2 mm. long. No long-styled flowers have been seen but presumably they occur. Fruits purple or deep wine-red, ± didymous, 4–5 mm. tall, 5 mm. wide, 4 mm. thick; pyrenes pale, semiglobose, 4 mm. long, 3·5 mm. wide, 2 mm. thick, not strongly keeled.

KENYA. Kiambu District: Limuru, forest next to Limuru Girls' High School, 26 Mar. 1961, *Polhill* 360!; Meru District: NE. Mt. Kenya, 12 May 1923, *Rammell* in *F.D.* 1083! & Meru, Feb. 1932, *Honoré* in *F.D.* 2761!
TANZANIA. Moshi District: Kilimanjaro, Useri, Jan. 1929, *Haarer* 1732!
DISTR. **K4**; **T2**, ?3 (see note); not known elsewhere
HAB. Evergreen forest; 1650–2250 m.

SYN. [*C. subochreata* sensu K.T.S.: 435 (1961), pro majore parte, *non* (De Wild.) Robyns]

NOTE. *Peter* 55837 (Tanzania, Pare District, N. Pare Mts., Kilomeni to Kissangara) may belong here but is in bud.

Imperfectly known species

12. **C. buchwaldii** *K. Schum.* in E.J. 34: 338 (1904); T.T.C.L.: 490 (1949); Verdc. in K.B. 30: 280 (1975). Types: Tanzania, Amani, Mt. Bomole, *Engler* 514, 528, 529 & 538 (B, syn. †)

Shrub to about 1·5 m. tall, with flattened graceful glabrous flowering branchlets. Leaf-blades oblong-lanceolate, oblanceolate or oblong, 9–17 cm. long, 4–7(–8) cm. wide, shortly acuminate at the apex, cuneate at the base, glabrous above, with sparse very minute hairs beneath; petiole 1–4 cm. long; stipules triangular, 5–6 mm. long, bicuspid at the apex. Inflorescences congested; peduncles 1–1·5 mm. long; pedicels obsolete. Calyx-tube 2 mm. long, the limb 1 mm. long, toothed; disc equalling the calyx-limb. Corolla red; tube 2·1–2·2 cm. long; lobes 4 mm. long, recurved, shortly crested outside, glabrous. Stamens inserted 1·6 cm. above the base of the tube. Style 1·3 cm. long.

TANZANIA. Lushoto District: above Amani, Mt. Bomole, 14 Sept. 1902, *Engler* 514, 528, 529 & 538 & without locality, *Buchwald*
DISTR. **T3**
HAB. Rain-forest; 915–1100 m.

NOTE. I have seen no authentic material of this—on geographical grounds it appears that it is probably the same as *C. albiflora* K. Krause and would be the correct name for the taxon if so, but there are some differences in the description, notably the presence of minute hairs on the under surface of the leaves and the longer corolla-tubes. It seems unwise to use the name in this sense without further evidence.

13. **C. sp. B**

Woody herb or shrub, ? ± 2·4 m. tall, with glabrous stems. Leaf-blades apple-green in life, oblong-elliptic, 8–18 cm. long, 4·5–7·5 cm. wide, narrowed at the apex into a very narrow acumen ± 1·5 cm. long which is rounded at its extreme tip, cuneate at the base, glabrous on both surfaces; petiole 0·7–3·2 cm. long; stipules triangular, 6·5 mm. long, divided at the apex into 2 fine apicula 1·5 mm. long. Young inflorescences only known, distinctly red in general appearance (in life), subglobose, 1·5 cm. in diameter; peduncle 7 mm. long; pedicels very short; bracteoles broadly triangular, ± 1 mm. long, papillate-pubescent. Calyx-tube obconic, 1 mm. long, papillate-pubescent; limb papillate-pubescent, the tube ± 0·5 mm. long; lobes ovate-triangular, 0·3 mm. long; disc prominent. Bud angular but not winged.

UGANDA. Toro District: Kampala–Fort Portal road, Kibale Forest, Apr. 1964, *Philip* 919!
DISTR. **U2**; not known elsewhere
HAB. Understorey of evergreen forest; 1440 m.

NOTE. This specimen has been unmatched; it is not identifiable with any of De Wildemann's species of *Chassalia* or *Psychotria* described from Zaire. It is not included in the key.

14. **C. sp. C**

Shrub with slender branches, very slightly papillose when young. Leaf-blades oblong, up to 10 cm. long, 4 cm. wide, obtuse or acute at the apex, narrowed to the base, glabrous on both surfaces, the young ones blackish above, white beneath; petioles short. Flowers in pedunculate terminal or spuriously axillary globose umbels. Calyx glabrous, irregularly denticulate. Corolla glabrous; tube 4 mm. long; lobes 1 mm. long, crested.

TANZANIA. Uzaramo [Usaramo], *Stuhlmann* 6550 & 7122
DISTR. **T6**
HAB. Not known

SYN. *Psychotria hypoleuca* K. Schum. in P.O.A. C: 391 (1895); De Wild., Pl. Bequaert. 2: 374 (1924); T.T.C.L.: 523 (1949); Petit in B.J.B.B. 34: 220 (1964). Types: Tanzania, Uzaramo, *Stuhlmann* 6550 & 7122 (B, syn. †)

NOTE. K. Schumann compares *Psychotria hypoleuca* with *P. cristata* Hiern, i.e. *Chassalia cristata* (Hiern) Bremek., and the crested corolla-lobes make it almost certain that it is a *Chassalia*, but the description hardly agrees with any known species. *C. umbraticola* Vatke is the only species occurring in the area and although the description might fit a badly dried specimen in bud K. Schumann knew this species well and is unlikely to have been misled by it. *Busse* 2996 (Tanzania, Lindi District, Yangwani, 20 June 1903) distributed from Berlin as *Psychotria hypoleuca* does not agree with the description of that species but may be a variant of *P. pumila* similar to var. *leuconeura*.

Tribe 2. MORINDEAE

Flowers not in compact heads, the calyx-tubes not confluent; enlarged lamina-like bracts never present; fruits usually blue, not forming a compound fleshy mass . . . 6. **Lasianthus**

Flowers in compact heads, the calyx-tubes mostly confluent; outer flowers with coloured stipitate lamina-like bracts in one of the species; fruits fused together to form a fleshy mass . . . . . 7. **Morinda**

6. LASIANTHUS

Jack in Trans. Linn. Soc. 14: 125 (1823); Verdc. in K.B. 11: 450 (1957), *nom. conserv.*

Shrubs or rarely small trees, sometimes foetid (usually not in the Flora area), glabrous to hairy or strigose. Leaves opposite, mostly acuminate, thin to coriaceous, petiolate, usually with numerous arching lateral nerves and close venation; stipules interpetiolar, usually broadly triangular or lanceolate, not divided, persistent or deciduous. Flowers hermaphrodite or sometimes unisexual, sometimes heterostylous, mostly small, mostly in sessile axillary fascicles or glomerules or less often in pedunculate, simple or branched inflorescences; pedicels mostly absent; bracts present, usually small. Calyx-tube subglobose, ovoid, oblong or urceolate; limb 3–6-toothed or lobed, persistent. Corolla often white or pink, salver-shaped or somewhat funnel-shaped; tube densely hairy at the throat; lobes 4–6, spreading or ± erect. Stamens 4–6, inserted in the throat of the corolla; filaments very short; anthers ± dorsifixed near their base, included or shortly exserted. Disc swollen and fleshy. Ovary 4–12-locular; ovules solitary in each locule, erect from the base, bent, anatropous; style short or elongate, glabrous or hairy, shortly 4–10-lobed at the apex, the lobes linear or obtuse. Fruits ± succulent, very often blue but sometimes pink, purple, white or black, with 4–12 pyrenes; pyrenes cartilaginous or bony, segment-shaped or pyriform, ± 3-angled with flat sides, the dorsal curved face often grooved, keeled or winged, 1-seeded. Seeds narrowly oblong, curved, with membranous testa and fleshy albumen.

A large genus, estimated at 150 species, predominantly in eastern tropical Asia (where the number has been probably much exaggerated) but about 20 species in tropical Africa and 1 in the W. Indies; 11 species occur in the Flora area predominantly in the Uluguru Mts. Much further study is needed to fully elucidate the species, well-collected material in spirit being essential.

Inflorescences pedunculate:
  Inflorescences lax; peduncles 2–5·5 cm. long; calyx-limb 2-lobed . . . . 1. *L. pedunculatus*
  Inflorescences shortly pedunculate fascicles; peduncles 0·3–1·6(–2·5) cm. long:
    Calyx-lobes 2·5 mm. long; corolla-tube slender, under 2 mm. wide in dry state; leaf-blades oblong-oblanceolate . . 8. *L. sp. A*
    Calyx-lobes 0·8–1 cm. long; corolla-tube broad, ± 5 mm. wide in dry state; leaf-blades mostly elliptic or oblong . . 3. *L. macrocalyx*
Inflorescences sessile:
  Corolla small, the tube under 4 mm. long:
    Ovary 9–10-locular; corolla-tube 2·5–3·2 mm. long (lowland Uganda lakeside forests) . 9. *L. seseensis*
    Ovary 4–5-locular (montane forest):
      Leaf-blades small, 4·5–8 × 1·8–3 cm., coriaceous; lateral nerves 5–6 pairs (Uluguru Mts.) . . . . . 11. *L. xanthospermus*
      Leaf-blades larger, 4–25 × 1·2–9·5 cm., coriaceous or papery, if under 8 × 3 cm. then papery; lateral nerves (5–)9–11 pairs:
        Leaf-blades mostly coriaceous, usually more oblong, acute at the apex; shoots with strongly ridged almost peeling brittle ± shining epidermis; pyrenes 2·5 mm. long (Uluguru Mts.). . . . . . . 6. *L. glomeruliflorus*
        Leaf-blades papery, more lanceolate, narrowly acuminate at the apex; shoots with duller corky bark, striate, but not strongly ridged; pyrenes 3·2–4 mm. long (widespread) . . . . . 10. *L. kilimandscharicus*
  Corolla larger, the tube over 4 mm. long:
    Calyx-lobes small, 0·5–3 mm. long; stems usually with pale yellowish or yellowish grey bark:
      Lobes of calyx 2–3 mm. long; corolla-tube 8 mm. long; stipules 6–7 mm. long; fruit 8–10 × 10 mm., 4–5-locular, glabrous; pyrenes lunate, 5–6 mm. long . . . . 5. *L. microcalyx*
      Lobes of calyx 0·5–2 mm. long; stipules 3–4 mm. long; fruit 3·5–5 × 4–6 mm., glabrous or pubescent; pyrenes shaped like a bird's head, 2·5–3·8 mm. long:

Calyx-lobes small, rounded, 0–1 mm. long; leaf-venation distinctly raised on upper surface in dry material; corolla-tube under 1 cm. long . 6. *L. glomeruliflorus*

Calyx-lobes more squarish, 1–2 mm. long; leaf-venation scarcely if at all raised on upper surface in dry material; corolla-tube (0·8–)1·2–1·4 cm. long . . . . . . 7. *L. cereiflorus*

Calyx-lobes larger, 5–10 mm. long; stems usually blackish:

Calyx-tube densely hairy; corolla-tube and lobes densely hairy . . . 2. *L. wallacei*

Calyx-tube glabrous or sparsely tomentose; corolla-tube ± glabrous outside:

Leaf-blades broadly oblong, truncate or emarginate at the apex save when very young, 13–16 × 9·5–10·5 cm.; calyx-lobes 5–6 mm. long; corolla-tube 1·5 cm. long, lobes 7–9 mm. long; fruits 2·3 cm. long; inflorescences sessile . . . 4. *L. grandifolius*

Leaf-blades oblong or elliptic, shortly but distinctly acuminate at the apex, 6–20 × 3·5–9 cm.; calyx-lobes 8–10 mm. long; corolla-tube 1·2–1·3 cm. long, lobes 6–8 mm. long; fruits probably 0·8–1 cm. long; inflorescences mostly shortly pedunculate . . . . . 3. *L. macrocalyx*

1. **L. pedunculatus** *E. A. Bruce* in K.B. 1936: 480 (1936); T.T.C.L.: 503 (1949). Type: Tanzania, Uluguru Mts., Bondwa, *E. M. Bruce* 1096 (K, holo.!, BM, EA, iso.!)

Shrub or small tree 1·5–6 m. tall, more or less foetid; stems glabrous, striate and lenticellate. Leaf-blades elliptic to obovate-oblanceolate, 4–17 cm. long, 2·4–9 cm. wide, shortly acuminate at the apex, cuneate at the base, thickly papery but not coriaceous, glabrous; lateral nerves 6–10 pairs; petiole 0·9–1·5 cm. long; stipules triangular or rounded, 1–3 mm. long, with tomentose margins. Flowers showing limited heterostyly, 2(–6) in pedunculate lax axillary 1–2-branched cymes, 1(–3) per axil; peduncles (0·8–)2–5·5 cm. long; secondary branches (0–)0·5–2·5 cm. long; pedicels not developed; bracts small, together with the junction of primary and secondary peduncles hairy. Calyx-tube turbinate, 1·5–3 mm. long, constricted, glabrous; limb 2-lobed; joined part 2·5–3 mm. long; lobes convex, rounded-ovate, 3–5·5 mm. long, 4–6(–8) mm. wide, glabrous save for a few hairs on the main nerve at the apex. Corolla waxy, bluish-white to mauve, lilac or deep blue; tube broadly cylindrical, 0·9–1·5(–2) cm. long, 3–8 mm. wide, densely or sparsely hairy inside, glabrescent to sparsely covered with minute hairs outside; lobes 5–7, ovate-triangular, 2–4 mm. long, 2·8–3·2(–5) mm. wide, hairy inside, glabrous to pubescent outside, particularly near the edges at the base or ± glabrous. Stamens 5, inserted in the throat; anthers included in long-styled flowers, the tips exserted 2–3 mm. in short-styled flowers. Ovary 5-locular; style 5 mm. long in short-styled flowers, 1·3–1·4 cm. long in long-styled flowers; stigma-lobes 5–6, oblong-subglobose, 2 mm. long; style-tips ± 2–3 mm. below throat in one form, just exserted in the other. Fruit subglobose, 5-lobed in dry state, 4–5 mm. tall, 6–7 mm. wide; pyrenes ± 3–5 mm. long.

TANZANIA. Morogoro District: N. Uluguru Forest Reserve, above Morningside, June 1953, *Semsei* 1223! & Uluguru Mts., Mkambaku [Mbambaku] Mts., 26 Feb. 1933, *Schlieben* 3583! & Nguru Mts., near Turiani, saddle to NW. of Mkobwe, 29 Mar. 1953, *Drummond & Hemsley* 1902!
DISTR. **T6**; not known elsewhere
HAB. Upland rain-forest; 1600–2100 m.

NOTE. The material is not uniform. The largest dimensions in the above description are taken from *Drummond & Hemsley* 1652 (S. Uluguru Forest Reserve, edge of Lukwangule Plateau, 17 Mar. 1953). This has considerably larger leaves, much longer branched inflorescences with elongated secondary peduncles and wider corolla-tubes, but it seems unlikely to be more than a minor variant despite the rather striking differences. In typical plants the inflorescence is about 4 cm. long, the branches attaining 1·5 cm., the leaf-blades 5–6 cm. wide and the corolla-tube about 4 mm. wide, whereas the corresponding measurements for *Drummond & Hemsley* 1652 are 10 cm., 6 cm., 9 cm. and 8 mm.

2. **L. wallacei** *E. A. Bruce* in K.B. 1936: 481 (1936); T.T.C.L.: 504 (1949). Type: Tanzania, Morogoro District, *Wallace* 491 (K, holo.!)

Shrub or tree 3–18 m. tall;* stems glabrous, finely ridged. Leaf-blades oblanceolate-oblong or elliptic, 9–17 cm. long, 2·5–7 cm. wide, obtuse or very shortly obtusely acuminate or subacute at the apex, cuneate at the base, ± coriaceous, glabrous above, pubescent beneath on the sides of the main nerve and on some of the lateral nerves towards the apex of the leaf; lateral nerves 10–14 pairs; petiole 0·5–1 cm. long; stipules triangular, 3–4 mm. long, hairy along the margins. Flowers 1–3, subsessile in the axils of the leaves; bracts small. Calyx white tinged purple; tube turbinate, 2–3 mm. long, quite densely covered with ferruginous hairs; limb 3–4-lobed; joined part 3 mm. long; lobes elliptic, 0·7–1 cm. long, 3·5–8 mm. wide, obtuse, venose, with rather sparse ferruginous hairs particularly about the midnerve, accrescent in fruit. Corolla white to deep violet; tube cylindrical, 1·2–1·6 cm. long, covered above outside with dense pale scaly hairs, throat hairy but lower part of the tube glabrous inside; lobes oblong-lanceolate, 5 mm. long, 2 mm. wide, with dense hairs inside and out. Stamens with tips just exserted. Ovary 3–6-locular; style 1·4 mm. long; stigma-lobes ? ± 4, linear-oblong, ± 1 mm. long. Fruit intense blue, ellipsoid, 1·7 cm. long, 1·3 cm. wide, hairy; pyrenes ± 4, woody, grey-brown or chestnut-brown, pyriform ("pip-like"), 5–5·5 mm. long, (2·2–)3–3·2 mm. wide and thick, often pointed at the base.

TANZANIA. Morogoro District: Uluguru Mts., Bondwa Hill, 23 Mar. 1953, *Drummond & Hemsley* 1765! & same locality, Jan. 1953, *Eggeling* 6449! & Lukwangule Plateau, Kinolo road, 5 Apr. 1935, *E. M. Bruce* 978!
DISTR. **T6**; not known elsewhere
HAB. Evergreen forest, sometimes in disturbed places; 1680–1900 m.

NOTE. In the specimens dissected the stigma-lobes and anther-tips have been at the same height, just exserted from the throat; the flowers may be genuinely isostylous.

3. **L. macrocalyx** *K. Schum.* in E.J. 28: 499 (1900); T.T.C.L.: 503 (1949). Type: Tanzania, Uluguru Mts., Lukwangule, *Stuhlmann* 9123 (B, holo. †)

Spreading shrub (1–)2–3 m. tall, occasionally epiphytic; stems glabrous and drying blackish when young, usually ± 2-ribbed. Leaf-blades oblong or elliptic, 6–23 cm. long, 3·5–13 cm. wide, [acute]** shortly but distinctly

* Wallace reports it to be a timber tree to 18 m., but I think this needs confirmation.
** See note.

acuminate at the apex, cuneate at the base, glabrous above, hairy on the nerves beneath; lateral nerves 8–14 pairs, sometimes appearing impressed on some young leaves but probably not genuinely so and mostly slightly raised; petiole 1–2·8 cm. long; stipules oblong-triangular, 5–8(–10) mm. long, [glabrous] shortly hairy inside and at margins. Flowers in subsessile or shortly pedunculate 2–several-flowered fascicles on older parts of stems; peduncles 0·3–1[1·5–2] cm. long; bracts and bracteoles ovate, hairy inside. Calyx violet; tube turbinate, 4 mm. long, glabrous; limb-tube 2·5 mm. long; lobes 3, ovate, 0·8–1 cm. long, 0·7–1 cm. wide, venose, glabrous save for a few scattered hairs on the midvein outside near apex [calyx elongate-tubulose, 7–8 mm. long, denticulate]. Corolla white, flushed violet, fleshy; tube 1·2–1·3 cm. long, 4·5–6 mm. wide, hairy inside; lobes 5–6, lanceolate, 6–8 mm. long, thickened at the apex, hairy inside. Stamens included in long-styled form. Ovary 6-locular; style 1·4 cm. long, finely ribbed, just exserted; stigma-lobes 6, linear-oblong, 2·2 mm. long. Fruit [pale steel blue] or cobalt blue, [0·8–1 cm. diameter], 6-locular, grooved between the pyrenes in dry state, crowned with the persistent calyx; pyrenes dark red-brown, segment-shaped, ± lunate, 7·5 mm. long, 4 mm. wide, 3 mm. thick, strongly pointed at the basal end, apically truncate, so that the general effect is that of a bird's head, excavated around the hilar area, faintly rugose.

TANZANIA. Morogoro District: Uluguru Mts., Lukwangule Plateau, above Chenzema Mission, 13 Mar. 1953, *Drummond & Hemsley* 1527! & same locality, Jan. 1975, *Polhill & Wingfield* 4646! & Bondwa Peak, 12 Oct. 1969, *Gibson & Pócs* 6051/V!
DISTR. **T6**; not known elsewhere
HAB. Shrub layer of upland rain-forest; 1800–2430 m.

NOTE. There are some discrepancies between the material I have ascribed to this species and the original description, the most serious being the description of the calyx as denticulate. I feel there is an error here, but only extensive surveys will prove this. Major discrepancies or dimensions given for the type which do not agree with the recent material are enclosed in square brackets. Unfortunately the mature fruits mentioned on the field-note of the *Drummond & Hemsley* material are not present on the specimen nor in the spirit material.

4. **L. grandifolius** *Verdc.* in K.B. 11: 450 (1957). Type: Tanzania, Uluguru Mts., Bondwa Peak, *Eggeling* 6472 (EA, holo.!, K, iso.!)

Shrub or small tree about 4·5 m. tall, with blackish rugulose glabrous branches. Leaf-blades broadly oblong, 13–16 cm. long, 9·5–10·5 cm. wide, acute at the apex in very young leaves but very soon truncate or even emarginate, cuneate at the base, distinctly coriaceous, at first sparsely puberulous on the nerves beneath, soon completely glabrous, glossy above, the margins revolute; lateral nerves 14–15 on each side, prominent on both surfaces; petiole thick, 1·5–2 cm. long; stipules rigid, triangular, 1·1 cm. long, 4·5 mm. wide. Flowers 3–4, subsessile in the leaf-axils. Calyx white tinged violet, urceolate, sparsely tomentose; tube 8 mm. long, 1 cm. wide at the apex; lobes 3, rounded-oblong, 5–6 mm. long, 1 cm. wide, rounded at the apex, recurved. Corolla fleshy, waxy white with blue-violet lobes, glabrous; tube cylindrical, 1·5 cm. long, 8 mm. wide, hairy above inside; lobes 6–7, lanceolate, 7–9 mm. long, 5 mm. wide, acute at the apex. Stamens 6–7, inserted in the throat; filaments 2 mm. long. Ovary 4–5-locular; style 5-lobed at the apex. Fruits blue, globose or ellipsoid, 2·3 cm. high, 1·9 cm. in diameter, fleshy, crowned by the persistent calyx-lobes; pyrenes brownish orange, pear-shaped, 7 mm. long, 4 mm. wide, acute at the base.

TANZANIA. Morogoro District: Uluguru Mts., above Morningside, near top of Bondwa Peak, Jan. 1953, *Eggeling* 6472!
DISTR. **T6**; not known elsewhere
HAB. Upland rain-forest; 2040 m.

NOTE. The floral measurements are all from spirit material. With so little material of each available it is not possible to comment on the possibility that this and *L. macrocalyx* are forms of one species.

5. **L. microcalyx** *K. Schum.* in E.J. 28: 107 (1899) & 499 (1900); T.T.C.L.: 504 (1949). Type: Tanzania, Uluguru Mts., Lukwangule Plateau, *Stuhlmann* 9168 (B, holo. †)*

Shrub 1·8–3 m. tall; stems angular, glabrous, blackish with distinct lenticels when young but soon covered with a ± somewhat shining yellow ± peeling bark. Leaf-blades oblong or obovate, 5·5–12·5(–16) cm. long, 2·5–6(–8·2) cm. wide, acute or rounded but apiculate at the apex, cuneate or rounded at the base, distinctly coriaceous, glabrous or slightly hairy beneath at apex of young leaves; lateral nerves 10–12, the venation closely and prominently reticulate on both surfaces; petiole 0·5–1·3(–1·8) cm. long; stipules triangular or oblong to lanceolate, 5–7 mm. long, acute, coriaceous, glabrous or with hairy margins, persistent. Flowers heterostylous, fleshy, few in sessile axillary fascicles. Calyx white, greenish-white or mauve; tube turbinate, (2–)4–6 mm. long; limb-tube 1–2 mm. long; lobes 3–4, oblong-triangular, semicircular or ovate-quadrate, 2–3·5 mm. long, 3–5 mm. wide, obtuse, truncate or emarginate, ± venose, glabrous or hairy at apex. Corolla whitish to mauve; tube 7–8 mm. long, glabrous outside, hairy inside; lobes elliptic-lanceolate, 4–6 mm. long, 1·5–2·8 mm. wide, hairy inside. Anthers exserted in short-styled flowers. Ovary 4–5-locular; style 4·5 mm. long in short-styled flowers; stigma-lobes 3–4, oblong-elliptic, 1 mm. long, included. Fruit blue, 0·8–1 cm. long and wide, 4–5-locular, grooved between the pyrenes in the dried state, crowned with the persistent 3–5 mm. long calyx-limb. Pyrenes straw-coloured, segment-shaped, ± lunate, 5–6 mm. long, 3 mm. wide, ± 2 mm. thick, ± pointed at both ends, dorsally with irregularly rugulose sides and either a median rugulose keel or a depression.

TANZANIA. Morogoro District: Uluguru Mts., Lukwangule Plateau, 30 Jan. 1935, *E. M. Bruce* 706!, 745!, 746! & 747! & same locality, 19 Sept. 1970, *Thulin & Mhoro* 996! & Uluguru Mts., above Morningside, Bondwa Peak, Jan. 1953, *Eggeling* 6476!**

DISTR. **T6**; not known elsewhere

HAB. Upland rain-forest; 2100–2430 m.

NOTE. Certain specimens, e.g. *Eggeling* 6476, cited above, and *Mgaza* 294 (said to attain 9 m.) from the same place, have very large leaves.

6. **L. glomeruliflorus** *K. Schum.* in E.J. 28: 107 (1899) & 499 (1900); T.T.C.L.: 503 (1949). Type: Tanzania, Uluguru Mts., Nghweme, *Stuhlmann* 8808 (B, holo. †)

Shrub 3–4·5 m. tall; stems ± 4-angled, 2-grooved, swollen at the nodes and often 2-ribbed just below, glabrous, covered when older with dirty yellowish green ridged and almost peeling bark. Leaf-blades elongate-oblong or subovate, 9–25 cm. long, (2·5–)4·5–9·5 cm. wide, acute at the apex, cuneate at the base, pubescent beneath with brownish hairs when very young, later glabrous, coriaceous; lateral nerves 10–11 on each side; petiole 0·6–2·5 cm. long; stipules ovate, 3 mm. long, apiculate, hairy around the margins. Flowers many, in dense sessile axillary glomerules or on the older wood, ? isostylous. Calyx-tube cupular, 2 mm. long; limb irregularly lobulate, glabrous or pubescent; lobes 0·6–1 mm. long, 1–2·5 mm. wide or limb almost truncate. Corolla white and pale violet; tube 2–4(?–8) mm. long, glabrous outside, throat densely hairy; lobes oblong, 2–2·8 mm. long, 1 mm. wide,

* Seen by Miss E. A. Bruce.

** Measurements from spirit material of this gathering have been included.

acute, densely hairy inside. Anthers ± exserted. Ovary 4–5-locular; style 4·8(–9) mm. long, exserted; stigma-lobes oblong, 0·6 mm. long. Fruit 3·5 mm. long, 4 mm. wide, strongly grooved between the pyrenes in dry state, crowned with 2 mm. long calyx-limb; pyrenes chestnut-brown, shaped like a bird's head, 2·5 mm. long, 1·5 mm. wide, 1·2 mm. thick, dorsally strongly curved, ventrally excavated, basally pointed, somewhat furrowed.

var. **glomeruliflorus**

Corolla-tube 8 mm. long; style 8–9 mm. long.

TANZANIA. Morogoro District: Uluguru Mts., Nghweme, 18 Oct. 1894, *Stuhlmann* 8808
DISTR. **T6**; not known elsewhere
HAB. Evergreen forest; 1600 m.

var ? (see note)

Corolla-tube 2–4 mm. long; style 4·8 mm. long.

TANZANIA. Morogoro District: Uluguru Mts., Tegetero, 20 Mar. 1953, *Drummond & Hemsley* 1718! & Bunduki, 27 Jan. 1935, *E. M. Bruce* 674! & Morningside, 28 Nov. 1934, *E. M. Bruce* 232! & Kitundu, 22 Nov. 1934, *E. M. Bruce* 198!
DISTR. **T6**; not known elsewhere
HAB. Evergreen forest; 1260–1830 m.

NOTE. The material I have associated as a variety of this species differs from the original description in so many small details that I have reservations about the identification, but Miss E. A. Bruce states on the label of *E. M. Bruce* 674 that she had compared it with the type and I have relied upon her judgement. K. Schumann describes the leaves as " glaberrima ", the stipules and calyx as glabrous and more particularly the corolla-tube as 8 mm. and the style as 8–9 mm. long; the material cited above has the young leaves pubescent beneath, the stipules with hairy margins, the calyx-tube pubescent, the corolla-tube 2–4 mm. long and the style under 5 mm. long. I have tentatively suggested that it might be a variety. The absolute distinctions between *L. glomeruliflorus* and *L. cereiflorus* E. A. Bruce will remain unclear until adequate material has been collected in spirit. Some apparent juvenile fruits preserved in spirit belonging to the gathering *Drummond & Hemsley* 1718 are obovate or subglobose, 0·9–1·1 cm. long, 6·5–7 mm. wide, but contain no pyrenes and are I believe galled ovaries. There is also difficulty in telling some specimens apart from *L. kilimandscharicus* K. Schum.

7. **L. cereiflorus** *E. A. Bruce* in K.B. 1936: 480 (1936); T.T.C.L.: 503 (1949). Type: Tanzania, Uluguru Mts., *E. M. Bruce* 140 (K, holo.!, BM, iso.!)

Shrub 1–6 m. tall; stems pale, glabrous. Leaf-blades oblong or narrowly obovate, 10–25 cm. long, 3–9·5 cm. wide, subacute or shortly obtusely acuminate at the apex, narrowly and sometimes unequally cuneate at the base, ± coriaceous, glabrous, paler and minutely pitted beneath; lateral nerves 9–11 on each side; petioles 1–2·5 cm. long; stipules ovate-triangular, 3–4 mm. long, obtuse. Flowers numerous, in dense sessile axillary glomerules in leaf-axils and also on older leafless parts of the stem. Calyx whitish, ± pubescent; tube depressed subglobose, 1·4 mm. long; limb-tube 1·6 mm. long; lobes 3, squarish or triangular, 0·5–1 mm. long, 1–2 mm. wide, emarginate or rounded, glabrous or slightly puberulous. Corolla waxy, whitish, pink, violet or lilac; tube cylindrical, slightly curved, 1·2–1·4 cm. long, 3 mm. wide at the throat, glabrescent outside, hairy inside; lobes 6, linear-lanceolate, 4 mm. long, acute at the apex, glabrescent outside, slightly hairy inside. Stamens 6, slightly exserted. Ovary 5–6-locular; style 1·2 cm. long, exserted ± 3 mm.; stigma-lobes 6. Immature fruit (spirit, *Drummond & Hemsley* 1719) ellipsoid, 6 mm. tall, 4·2 mm. wide. Fruit (from *Paulo* 223) blue, depressed subglobose, 5 mm. tall, 6 mm. wide, strongly grooved between the pyrenes in the dried state. Pyrenes orange-brown, shaped like a bird's head, dorsally rounded, ventrally angularly excavated, basally beaked,

3·8 mm. long, 3 mm. wide, 1·8 mm. thick, with a median dorsal groove and some lateral grooves, rugose.

TANZANIA. Morogoro District: Uluguru Mts., Tegetero, 20 Mar. 1953, *Drummond & Hemsley* 1719! & NW. Uluguru Mts., 29 Sept. 1932, *Schlieben* 2757! & without exact locality, 16 Nov. 1932, *Wallace* 438!
DISTR. T6; not known elsewhere
HAB. Shrub layer of rain-forest; 1000–1700 m.

NOTE. Nothing is known about heterostyly in this species.

8. **L. sp. A**

Small tree 4–8 m. tall; stems glabrous, the young parts blackish, the older yellowish grey-brown and 2-ribbed when dry. Leaf-blades oblong-oblanceolate, 18–22·5 cm. long, 6–9 cm. wide, probably bluntly acuminate at the apex, cuneate at the base, glabrous or with very fine hairs on the nerves when young, later entirely glabrous on both surfaces, paler beneath; lateral nerves 9–10; petiole 1·5–2·5 cm. long; stipules triangular, 4 mm. long, subacute, thick, hairy inside or near base. Inflorescences pedunculate, axillary, 5–6-flowered, glabrous; peduncle 0·4–1·6 cm. long. Calyx-tube conic, ± 2 mm. long; limb-tube ± 1 mm. long; lobes 2 (? always), rounded-quadrate, 2·5 mm. long, 3 mm. wide, emarginate or rounded, glabrous or with a few hairs in the sinus. Corolla pale lilac, seen only as buds 6–7 mm. long. Fruits not seen.

TANZANIA. Morogoro District: Uluguru Mts., N. side, 18 Dec. 1932, *Schlieben* 3112!
DISTR. T6; not known elsewhere
HAB. Rain-forest; 1550 m.

NOTE. Although agreeing with *L. macrocalyx* K. Schum. in peduncle length and leaf size there are numerous discrepancies, e.g. leaf-shape, stipule size, the nerves not impressed above and the elongate denticulate calyx. The interpretation followed (see p. 137) is much more certainly correct.

9. **L. seseensis** *M. R. F. Taylor* in K.B. 1937: 421 (1937). Type: Uganda, Sese Is., Towa Forest, *A. S. Thomas* 1340 (K, holo.! & iso.!)

Shrub 0·9–1·8 m. tall, with 1–several stems from one root; young shoots with rather sparse brownish ± adpressed hairs, later glabrous. Leaf-blades narrowly elliptic or elliptic-lanceolate, 8·5–23 cm. long, 1·8–6·1 cm. wide, shortly acuminate at the apex, narrowly cuneate at the base, not coriaceous, glabrous above save for yellowish or brownish hairs on the midrib, pubescent on the venation beneath particularly when young but soon glabrescent; lateral nerves 14–32 on each side; petiole 0·5–3·5 cm. long; stipules ovate-triangular, 5–8(–10) mm. long, 4–5·5 mm. wide, acute, hairy inside and out, the marginal hairs often giving a fimbriate appearance. Flowers bibracteate, 3–5 in sessile axillary glomerules; bracts small, triangular, acute, hairy. Calyx-tube campanulate, 1·5–2·5 mm. long; lobes 3–5, narrowly ovate-triangular, ovate-oblong or elliptic, 2–3 mm. long, 1·5–1·6 mm. wide, acute, pubescent, spreading. Corolla whitish or usually pink; tube cylindrical, 2·5–3·2 mm. long, glabrous outside or hairy at apex, the throat densely hairy inside; lobes 4–5(?–6), triangular, 2 mm. long,* 1·4 mm. wide, hairy inside particularly at the base. Stamens 5–6, inserted in the throat, the filaments very short; anthers included. Ovary ± 10-locular; disc depressed, fleshy; style 3·5–4·5 mm. long, pilose, scarcely or just exserted; stigma-lobes 8–10*, linear, ± 1 mm. long. Fruit waxy-white at first, becoming a bright " porcelain " blue when ripe, subglobose, 5 mm. in diameter (0·9–1 cm. when fresh), fleshy, with about 10 grooves between the slight lobes when dry but not evident in fresh state, ± glabrous, crowned with very fleshy calyx-lobes, 4 × 3 mm., slightly ciliate near inflexed tips; pyrenes pale waxy yellow,

* The original description is erroneous.

9–10, shaped rather like orange segments but ± pyriform, 2·2 mm. long, 1·4 mm. wide, 1·1 mm. thick, pointed at the base, ± 4-ribbed on the outer curved face.

UGANDA. Masaka District: Sese Is., Central Bugala I., July 1945, *Purseglove* 1723! & Bugala I., Kalangala, 25 Feb. 1945, *Greenway & A. S. Thomas* 7184!; Mengo District: Damba I., 22 Nov. 1949, *Dawkins* 456! & Nakiza Forest, near Nansagazi, 24 Jan. 1951, *Dawkins* 710!

DISTR. U4; not known elsewhere

HAB. Undergrowth of rain-forest, sometimes the dominant shrub; 1110–1200 m.

NOTE. I have seen only long-styled flowers but I believe the species shows limited heterostyly although I have hesitated to dissect the few flowers available on some sheets. This point could immediately be settled in the field.

10. **L. kilimandscharicus** *K. Schum.* in P.O.A. C: 396 (1895); F.P.N.A. 2: 370 (1947); T.T.C.L.: 504 (1949); K.T.S.: 450 (1961); F.F.N.R.: 410 (1962). Types: Tanzania, Kilimanjaro, above Shira, *Volkens* 1949 (B, syn. †, BM, K, isosyn.!) & presumably Kilimanjaro, *Volkens* 1555 (B, syn. †)

Shrub or small tree 1·2–7·5 m. tall, with smooth grey bark; fresh wood reported in Uganda to smell unpleasant; shoots glabrous or finely pubescent, drying black above but with pale yellowish bark, the nodes often with some persistent indumentum below the stipules but older stems quite glabrous. Leaf-blades oblong, narrowly oblong-elliptic or oblong-lanceolate, (4–)9–17(–22) cm. long, (1·2–)2–6(–7) cm. wide, narrowly acuminate at the apex, cuneate at the base, firmly papery but not coriaceous, glabrous above and usually beneath save for fine sparse adpressed pubescence on the nerves or rarely hairy; lateral nerves (5–)8–10 on each side; petiole 0·5–1(–1·8) cm. long; stipules narrowly to broadly triangular, 1·5–6 mm. long, hairy, particularly along the margins. Flowers few, sessile in the axils of the leaves, heterostylous; bracts ovate to lanceolate, 2–6 mm. long, 1–2 mm. wide (? sometimes stipules of reduced shoots), with distinctly hairy margins. Calyx pinkish white or tinged purple, particularly on the lobes, glabrous or puberulous; tube turbinate, 2 mm. long; lobes very convex in living state 0·5–3 mm. long, 2 mm. wide, sometimes with pubescent traces of intermediate accessory lobes. Corolla glistening white or violet outside, white inside; tube cylindrical, 2·5–4(–5) mm. long, glabrous to slightly pubescent outside, with densely hairy throat but tube glabrous inside or hairy only in upper half; lobes 4–5, ovate-oblong, 2·2–3·7 mm. long, 1·5–2 mm. wide, densely hairy inside with white hairs, finely hairy or glabrous outside. Stamens with anther-tips just exserted in long-styled flowers but with anthers and ± 1 mm. of filaments exserted in short-styled flowers. Ovary 4–6-locular; style 4·5–5·5 mm. long in long-styled flowers, 2·5 mm. long in short-styled flowers; stigma-lobes 4–5, oblong or subcapitate, 0·5–0·6 mm. long. Fruit intense cobalt blue, subglobose, prominently 4–6-lobed in dry state but not grooved when fresh, ± 4·5 (dry) to ± 10 (fresh) mm. in diameter, finely puberulous, the persistent calyx-lobes oblate, very obtuse, constricted at the base, 2 mm. long, 2 mm. wide; pyrenes chestnut-brown or straw-coloured, basically pyriform but with a marked ventral notch extending from the pointed end for about half the length of the pyrene, 3·2–4 mm. long, 2·3–2·5 mm. wide, 2 mm. thick.

subsp. **kilimandscharicus**

Tertiary venation of leaves very close and with a conspicuous transversely parallel element; nerves glabrous or pubescent beneath; leaves attaining maximum dimensions noted in description. Fig. 11.

UGANDA. Ankole District: Kalinzu Forest, Aug. 1936, *Eggeling* 3200!; Kigezi District: Ishasha Gorge, Kayonza Forest Reserve, 5 Aug. 1960, *Paulo* 660! & Impenetrable Forest, Sept. 1930, *Eggeling* 3294!

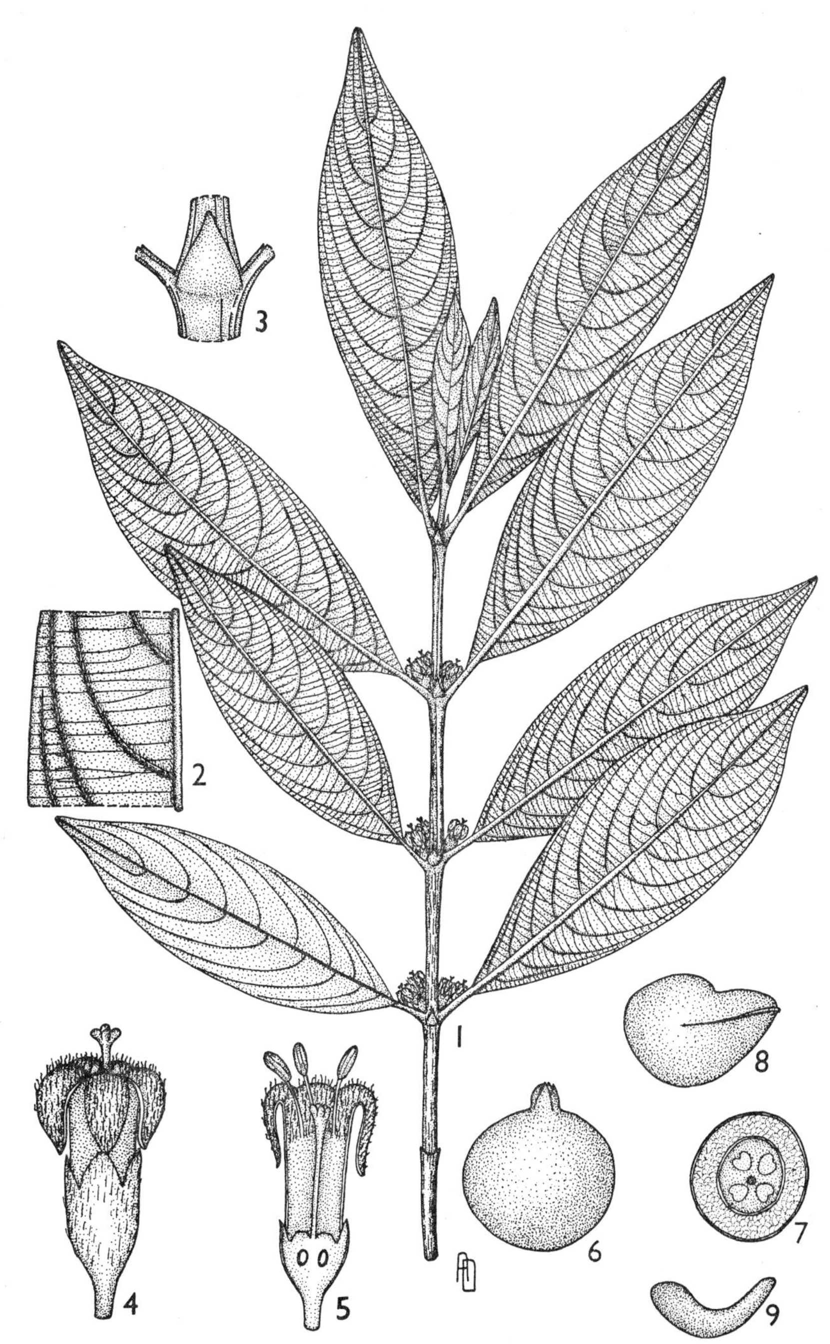

FIG. 11. *LASIANTHUS KILIMANDSCHARICUS* subsp. *KILIMANDSCHARICUS*—**1,** flowering branch, × ½; **2,** detail of undersurface of leaf, × 1⅓; **3,** node showing stipule, × 4; **4,** long-styled flower, × 3⅓; **5,** longitudinal section of short-styled flower, × 3⅓; **6,** fruit, × 2; **7,** transverse section of fruit, × 5⅔; **8,** pyrene, × 6; **9,** seed, × 9. 1, 2, from *Osmaston* 2537; 3, from *Drummond & Hemsley* 2256; 4, 7, from *Drummond & Hemsley* 885; 5, from *Brenan & Greenway* 8301; 6, 8, from *Verdcourt* 3987A; 9, from *Dowson* 76. Drawn by Ann Davis.

KENYA. Nakuru District: Mau Forest, Endabarra, 16 Jan. 1946, *Bally* 4845 !; Kiambu District: Gatamayu Forest, 16 Nov. 1958, *Verdcourt* 2310 ! & 8 Mar. 1964, *Verdcourt* 3987A !; Teita Hills, Ngangao Forest, 6 Feb. 1953, *Bally* 8748 !
TANZANIA. Moshi District: Kilimanjaro, Bismarck Hill to Marangu, 1 Mar. 1934, *Greenway* 3888 !; Lushoto District: W. Usambara Mts., Magamba Peak, 7 Nov. 1947, *Brenan & Greenway* 8301 ! & 5 Nov. 1947, *Brenan & Greenway* 8289 !; Morogoro District: S. Nguru Mts., Ruhamba Peak, 2 Apr. 1953, *Drummond & Hemsley* 1962 !
DISTR. **U2**; **K1**, 3–5, 7; **T2**, 3, 5–7; Zaire, Burundi, Mozambique, Malawi, Zambia and Rhodesia
HAB. Understorey of upland rain-forest, often dominant particularly in *Ocotea* forests; 1710–2400 m.

SYN. *L. sp.* sensu I.T.U., ed. 2: 348 (1951)

NOTE. It is interesting to note that this was collected on Kilimanjaro as long ago as Aug. 1871 by the Rev. C. New.
Newbould 3480 (Kenya, Northern Frontier Province, Mt. Nyiru, 5 Jan. 1959) has leaf-blades up to 7 cm. wide, the tertiary venation extremely close and the petioles up to 1·8 cm. long and is a variant perhaps worthy of a name but it is very close to Mt. Kenya material. Uganda, Zaire and Burundi material has much coarser hairs on the leaf venation beneath and on the inflorescences; the calyx-lobes are rather larger and more triangular than in typical material; better material may show that another subspecies should be recognised.

subsp. **laxinervis** *Verdc.* in K.B. 30: 281 (1975). Type: Tanzania, W. Usambara Mts., Dole Mt., *Greenway* 7545 (K, holo. !, EA, iso. !)

Tertiary venation of leaves much more lax and without a conspicuously transversely parallel element; nerves often glabrous beneath; leaf-blades usually much smaller than in subsp. *kilimandscharicus.*

TANZANIA. Lushoto District: W. Usambara Mts., Shume, 28 July 1958, *Carmichael* 671 ! & Sungwi Forest Reserve, 3 Mar. 1967, *Semsei* 4216 ! & Mkungaungo Forest, 20 Aug. 1952, *G. R. Williams* 505 !
DISTR. **T3**; not known elsewhere
HAB. As subsp. *kilimandscharicus*; 1400–2250 m.

SYN. *L. sp.* sensu T.T.C.L.: 504 (1949)

NOTE. This has long been annotated as *L. sp. nov.* near *L. kilimandscharicus* but I have been reluctant to give it specific rank since the venation varies considerably in certain areas. The two subspecies grow together or near, e.g. at Sungwi. It seems difficult to justify recognition at specific level but there is an immense difference between this subspecies and the variant from **K1** previously mentioned. I do not know what *L. holstii* K. Schum. var. *parvifolius* K. Schum. is (see E.J. 34: 340 (1904) and T.T.C.L.: 504 (1949)). It was based on two syntypes collected at 2400–2700 m. above Kwai, Magamba, W. Usambaras, *Engler* 1287, 1295 (B, syntypes †). As Brenan points out the only other mention of *L. holstii* is in V.E. 1(1): 336 (1910), where it is mentioned as occurring near Magamba. It does not appear to have been validly described. Gillett (*in litt.*) has suggested that it might be a misprint for *Urophyllum holstii* but I think it may well be what I have called *L. kilimandscharicus* subsp. *laxinervis.*

GENERAL NOTE. This is the most widespread of the African species, but it can scarcely be ancestral to any of the endemic Uluguru species.

11 **L. xanthospermus** *K. Schum.* in E.J. 28: 499 (1900); T.T.C.L.: 504 (1949). Type: Tanzania, Uluguru Mts., Lukwangule Plateau, *Goetze* 276 (B, holo. †)

Shrub 2–3 m. tall, with rather weak terete glabrous branches. Leaf-blades oblong, 4·5–8 cm. long, 1·8–3 cm. wide, very shortly acuminate or subapiculate at the apex, cuneate at the base, coriaceous, glabrous and shining above, subtomentose beneath when young but with scattered rusty hairs or almost glabrous later; lateral nerves 5–6 on each side; petiole 0·6–1·2 cm. long, grooved above; stipules broadly semi-elliptic, 1–2 mm. long, glabrous outside, glandular inside, deciduous. Flowers in 3-4-flowered sessile axillary fascicles; pedicels very short. Calyx-tube turbinate, 1 mm. long, glabrous;

lobes 3–4, 1·5 mm. long, very obtuse, coriaceous, with a minute fascicle of hairs at the apex in the emargination, and also with some minute accessory lobes between the main ones. Corolla white, ± 4 mm. long in bud, about equally divided into tube and lobes, the lobes with an internal apiculum at the apex, glabrous on both sides. Anthers ± 2 mm. above the base of the tube, 1·5 mm. long. Ovary 4-locular; style 1·5 mm. long; stigma-lobes 4. Fruit ultramarine blue, fleshy, 5 mm. wide, with 4 pyrenes; pyrenes yellow, cap-shaped (subovate), basally obliquely acuminate, smooth.

TANZANIA. Morogoro District: Uluguru Mts., Lukwangule Plateau, Nov. 1898, *Goetze* 276

DISTR. **T6**; not known elsewhere

HAB. Upland evergreen forest; 2400 m.

NOTE. No authentic material of this species has been seen, neither has it been possible to identify it with any described species. It should be easy to look for it in the type locality. The flowers may attain a larger size; it may prove to be a form of *L. kilimandscharicus*, which is known to occur on Lukwangule (*Drummond & Hemsley* 1533; *E. M. Bruce* 749) or of *L. glomeruliflorus*. The former is the more likely, particularly as *Harris et al.* 3727 (Uluguru Mts., W. Lukwangule Plateau) has some leaf-blades the same dimensions as given for *L. xanthospermus*, and it seems possible that it is only a small-leaved form of *L. kilimandscharicus*. Other specimens from the same place and altitude (2400 m.) have much larger leaves, but all Uluguru material has more elliptic and less oblong leaf-blades than specimens from other areas. Possibly a variety based on the name would be the correct solution.

7. MORINDA

L., Sp. Pl.: 176 (1753) & Gen. Pl., ed. 5: 81 (1754)

Trees, shrubs or less often lianes, with mostly glabrous, less often hairy or tomentose stems. Stipules leafy, undivided, free or forming a sheath with the petioles. Leaves opposite or rarely in whorls of 3, sometimes only 1 at flowering nodes. Flowers heterostylous (? always), hermaphrodite or rarely unisexual, in tight capitula, the flowers usually joined, at least by the bases of the calyces, the capitula sometimes bearing single large coloured bracts or occasionally many smaller bracts; capitula 1–several at the nodes, frequently arranged in umbels, pedunculate or rarely sessile. Calyx-tube urceolate or hemispherical, the limb short, truncate or obscurely to distinctly toothed, persistent. Corolla ± coriaceous, funnel-shaped or salver-shaped; lobes (4–)5(–7), valvate; throat glabrous or pilose. Stamens (4–)5(–7), inserted in the throat; filaments short; anthers and style included or exserted. Disc swollen or annular. Ovary 2–4-locular, sometimes imperfectly so; style with 2 short to long linear branches; ovules solitary in the locules, attached to the septum below the middle or near the base, ascending, anatropous or amphitropous. Fruit syncarpous (very rarely scarcely so), succulent, containing several pyrenes; pyrenes cartilaginous or bony, 1-seeded or joined into a 2–4-locular woody structure. Seeds obovoid or reniform, with a membranous testa and fleshy endosperm.

A genus of about 80 species throughout the tropics; 3 species occur in the Flora area, all very distinct.

Capitula not bearing large whitish bracts:
- Leaf-blades elliptic, 5·8–18 cm. long, 2·2–8·9 cm. wide; corolla-tube 1·2–1·6 cm. long; capitula (1–)3 per node . . . . 1. *M. lucida*
- Leaf-blades oblanceolate-elliptic, 18–43 cm. long, 6·5–19 cm. wide; corolla-tube 3–4·3 mm. long; capitula in umbels . . 3. *M. titanophylla*

At least some capitula bearing large whitish or greenish yellow leafy petiolate bracts; corolla green and yellow . . . . . 2. *M. asteroscepa*

1. **M. lucida** *Benth.* in Niger Fl.: 406 (1849); Hutch. in K.B. 1916: 9, fig. (1916); T.T.C.L.: 506 (1949); I.T.U., ed. 2: 351 (1952); Aubrév., Fl. For. Côte d'Ivoire, ed. 2, 3: 270, t. 348 (1959); Hepper in F.W.T.A., ed. 2, 2: 189, fig. 241 (1963); Keay, Onochie & Stanfield, Nigerian Trees 2: 396, fig. 169 (1964). Types: Nigeria, on the Quorra (specimen not traced) & Fernando Po, *Vogel* (K, syn.!)

Tree or rarely a shrub 2·4–18 m. tall, with smooth or rough scaly grey or brown bark and crooked or gnarled bole and branches; stems pubescent when young, later glabrous; cork grey but often with some distinct purple layers; slash greenish or yellow. Leaf-blades elliptic, 5·8–18 cm. long, 2·2–8·9 cm. wide, acute to acuminate at the apex, rounded to broadly cuneate at the base, shining above, glabrous, save for tufts of hairs in the axils beneath and some hairs on the midrib, or slightly finely pubescent all over when young; petiole 0·5–1·6 cm. long; stipules short, 1–2·5 mm. long, mucronate at the apex, sometimes splitting into 2 parts, hairy inside, or sometimes ovate or triangular and up to 7 mm. long. Flowers heterostylous; peduncles (1–)3 per node opposite a single leaf, alternating at successive nodes, 2·5–7·5 cm. long, often densely pubescent when young; at the base of these 3 peduncles there is a usually stalked cup-shaped gland 2–6 mm. long; capitula (8–)10–13(–14)-flowered, 4–7 mm. in diameter (excluding corollas). Calyx-tube pubescent when young, cupular, ± 2 mm. long, the truncate limb ± 0·5 mm. Corolla heavily scented, white or greenish yellow outside, white inside; tube 1·2–1·6 cm. long, widened at the apex; lobes ovate-lanceolate, 3·5–5 mm. long, 1·5–2·5 mm. wide; throat glabrous. Ovary 2-locular; style 8 or 11 mm. long; stigma-lobes ± 4 mm. or 7 mm. long. Syncarps green, hard for a long time but ultimately becoming soft and black, 0·8–2·2(–2·5) cm. in diameter; pyrenes dark reddish brown, compressed ovoid, 5·5–6·5 mm. long, 4 mm. wide, very hard. Seeds yellowish, soft, elliptic, about 3·5 mm. long, 2 mm. wide, 0·4 mm. thick.

UGANDA. Kigezi District: Kambuga, 22 Apr. 1941, *A. S. Thomas* 3800!; Busoga District: Lolui I., 16 May 1964, *G. Jackson* 88!; Masaka District: Sese Is., Bugala, Sozi Point, Nov. 1931, *Eggeling* 72b!

TANZANIA. Mwanza District: Kome I., Chikuku Forest Reserve, no collector's name 234 in *F.H.* 2680!; Buha District: Kibondo, E. of Kifura, Feb. 1955, *Procter* 408! & Gombe Stream Chimpanzee Reserve, Kasakela, 18 Nov. 1962, *Verdcourt* 3365A (fallen inflorescence only)!

DISTR. **U**2–4; **T**1, 4; W. Africa from Senegal to Angola, Zaire and Sudan

HAB. Grassland, exposed hillsides, thicket and bush, often on termite-mounds, also in *Anthocleista*, *Mussaenda-Grewia*, etc. forest; 756–1290 m.

SYN. [*M. citrifolia* sensu Hiern in F.T.A. 3: 192 (1877), pro parte, *non* L.]

2. **M. asteroscepa** *K. Schum.* in E.J. 34: 340 (1904); T.T.C.L.: 506 (1949). Type: Tanzania, E. Usambara Mts., Amani, *Engler* 775 (B, holo. †)

Small to large evergreen tree 6–25 m. tall, much branched, with a dense rounded crown and often branched low down; stems glabrous. Leaf-blades elliptic to oblong-ovate, 7·8–22·6(–25) cm. long, 3·5–15·2 cm. wide, rounded to subacute at the apex, rounded, truncate or broadly cuneate and often unequal-sided at the base, glabrous save for small tufts of hair in the nerve-axils beneath; petiole (1–)1·7–3 cm. long; stipules lanceolate, 3·6–6 cm. long, 0·5–2 cm. wide, connate at the base and enclosing the terminal bud,

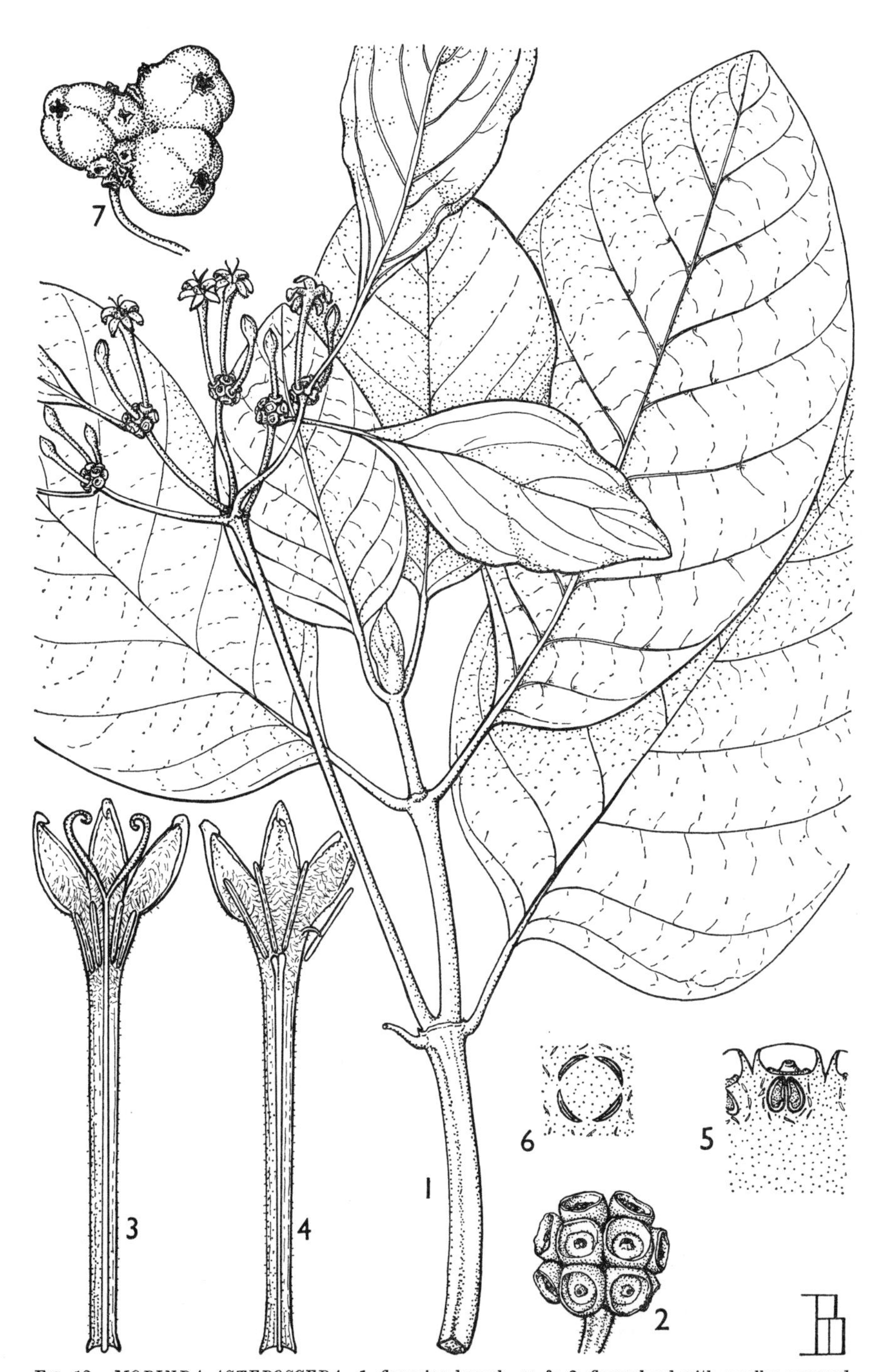

FIG. 12. *MORINDA ASTEROSCEPA*—**1,** flowering branch, × $\frac{2}{3}$; **2,** flower-head with corollas removed, × 3; **3,** longitudinal section of long-styled flower, × 3; **4,** same of short-styled flower, × 3; **5,** longitudinal section of fused ovaries, × 4; **6,** transverse section of one ovary, × 8; **7,** syncarp, × 2. 1, from *Hughes* 28; 2, 4–6, from *Verdcourt* 1730; 3, from *Peter* K. 705; 7, from *Peter* K. 684. Drawn by Diane Bridson.

soon circumscissile and deciduous. Flowers heterostylous, in umbels of (1–)2–7(–9) capitula, the capitula 4–9 mm. in diameter, 5–20-flowered; common peduncles 7–17·5 cm. long; secondary peduncles 0·8–4·5 cm. long, finely puberulous; many of the capitula bear a large coloured petiolate bract, greenish below, cream above with greenish yellow veins, produced on one of the lateral flowers, the lamina (3–)6–11·2 cm. long, 1·5–5·8 cm. wide, the petiole 0·6–2 cm. long. Calyx cupular, ± 2 mm. tall, of which 1 mm. is a free truncate rim. Corolla cream to yellow, the tube mostly greenish and the lobes bright yellow inside; tube 1·6–2·4 cm. long, widened at the apex, puberulous; lobes ovate-lanceolate, 3–7 mm. long, 1·2–4 mm. wide, margined and with thickened apices, hairy inside particularly near throat with yellow hairs; throat densely hairy. Ovary 3–4-locular; style 1·2 or 2·4 cm. long according to form; stigma-lobes 3·5–4·5 mm. long, at least exserted ones green. Syncarps ± 1·5 cm. in diameter; pyrenes broadly elliptic, much compressed, 5 mm. long, 4·5 mm. wide. Immature seeds 3·5 mm. long, 2 mm. wide, probably winged. Fig. 12 p. 147.

TANZANIA. Lushoto District: E. Usambara Mts., Amani, Mt. Bomole, 24 Jan. 1933, *Greenway* 3338! & Amani, "Zimmermann's Garden" by the club, 26 Dec. 1956, *Verdcourt* 1730!; Morogoro District: Uluguru Mts., Bunduki, 3 Jan. 1948, *Wigg* in *F.H.* 2277!

DISTR. **T**3, 6; Malawi (Nchisi Forest)

HAB. Rain-forest and marginal bushland; 870–1350 m.

NOTE. The bracts do not appear to come from the calyx in this species but studies on properly preserved material are needed; in some Asiatic species, e.g. *M. bracteata* Roxb. there are numerous smaller bracts clearly originating on the calyx rim.

3. **M. titanophylla** *Petit* in B.J.B.B. 32: 188 (1962). Type: Zaire, Walikale to Lubutu, *Bequaert* 6597 (BR, holo.)

Shrub or small tree 1–8 m. tall, not or sparsely branched; branches glabrous, mostly hollow, usually becoming corky. Leaf-blades elliptic, elliptic-obovate or elliptic-oblanceolate, 17–43 cm. long, 6·5–20 cm. wide, acuminate at the apex, very gradually attenuated into the petiole at the base, glabrous or venation puberulous beneath; petiole 1·6–5 cm. long*; stipules deciduous, ovate-triangular, 0·6–2·3 cm. long, up to 1·5(–2) cm. wide, often leafy and bent back. Flowers 4–6-merous, heterostylous, in umbels of 3–7 capitula, the capitula 4–9 mm. in diameter, 16–35- or more flowered; common peduncles 1·8–15 cm. long, glabrous or puberulous; secondary peduncles 0·4–1·5 cm. long, glabrous or puberulous; bracts mostly small or 0·5–1 cm. long. Calyx cupular, ± 1 mm. long, not lobed. Corolla white or pale yellow; tube (3–)3·75–4·25 mm. long; lobes elliptic, 1·5–2 mm. long, 1–1·1 mm. wide, thickened at the apex; throat densely hairy. Ovary 2-locular; style 1·5 or 4 mm. long; stigma 1 mm. long. Syncarps succulent, subglobose, 1–1·5 cm. in diameter; pyrenes flattened obovoid, 2·2–3 mm. long, 1·5–1·8 mm. wide. Seeds olive, oblong or elliptic, flattened, 2 mm. long, 0·75 mm. wide, ± winged at the base, the wing ± ovate, ± 0·5 mm. long.

UGANDA. Ankole District: Kashoya Forest, Aug. 1936, *Eggeling* 3215!; Kigezi District: Kayonza Forest Reserve, Ishasha Gorge, 5 Aug. 1960, *Paulo* 661! & same locality, Mar. 1946, *Purseglove* 2009!

DISTR. **U**2; Zaire

HAB. Rain-forest and semi-deciduous forest; 1200–1800 m.

* Petit's 20–30 cm. is presumably an error for 20–30 mm.

Tribe 3. **TRIAINOLEPIDEAE**

One genus only . . . . 8. **Triainolepis**

8. **TRIAINOLEPIS**

Hook. f. in G.P. 2: 126 (1873); Bremek. in Proc. K. Nederl. Akad. Wetensch., ser. C, 59: 4 (1956)

Princea Dubard & Dop in Journ. de Bot., sér. 2, 3: 2 (1925)

Shrubs or small trees. Leaves opposite, petiolate; blades lanceolate to ovate or oblong-elliptic, often curved; stipules ovate or ovate-triangular, divided into 3–5 lobes. Flowers hermaphrodite, heterostylous, (4–)5-merous in terminal corymbs; bracts small or minute. Calyx-tube campanulate, the limb cupular, unequally 5–7-toothed. Corolla white, yellowish or tinged red, salver-shaped, tomentose to woolly outside; tube with throat and sometimes upper half inside densely barbate with white hairs; lobes lanceolate, hairy outside, glabrous inside. Stamens inserted at the top of the corolla-tube, exserted or included. Disc mostly glabrous. Ovary 4–10-locular, each locule with 2(–3)* collateral erect anatropous ovules; style filiform, glabrous, exserted or included; stigma-lobes 4–10, filiform, straight or twisted, sometimes cohering. Fruits usually red, globose or depressed globose, drupaceous, containing a woody or bony 4–10-celled putamen, which is entire or slightly to deeply incised between the locules; locules 1-seeded. Seeds ellipsoid, compressed, with membranous testa and fleshy albumen.

A small genus occurring in Madagascar, Comoro Islands, Aldabra Group and also on the East African mainland (almost entirely on the coast); Bremekamp has recognised 12 species but this is I feel too high an estimate; many of the species seem very close to *T. africana* but since that is the oldest name the identity of his species has not concerned this account. Two species have always been accepted in East Africa but it is not possible to continue to keep them separate and unfortunately the older name refers to the least well-known taxon.

T. africana *Hook. f.* in G.P. 2: 126 (1873); Hiern in F.T.A. 3: 219 (1877); T.T.C.L.: 533 (1949); Bremek. in Proc. K. Nederl. Akad. Wetensch., ser. C, 59: 13 (1956). Type: Mozambique/Tanzania boundary, Rovuma Bay, *Kirk* (K, holo.!)

Weak shrub or rarely a small tree, 1·8–6 m. tall; branches pubescent or covered with dense white hairs, later glabrescent. Leaf-blades elliptic, narrowly ovate or elliptic-lanceolate, 3–11·5(–16) cm. long, 1·2–5·7 cm. wide, narrowly acuminate at the apex, cuneate at the base, almost or completely glabrous to densely adpressed pubescent above, more or less velvety beneath with ± white hairs, particularly along the main nerves so that they stand out against the rest of the surface; petioles 0·5–1·7 cm. long; stipules with ovate or triangular bases 1–4 mm. long, with 0–7 (usually 3) unequal subulate fimbriae 0·5–4(–6) mm. long, mostly with distinct colleters. Inflorescences hairy, up to 4 cm. across; peduncle 0·35–2·2 cm. long; pedicels 0–2·5 mm. long. Calyx-tube 1–1·5 mm. long; limb-tube 1–2·4 mm. long; lobes 0·3–2·9 mm. long, 0·2–1·3 mm. wide. Corolla with pale green to white tube and white limb; tube 0·7–1·05 cm. long; lobes 3·5–5·5 mm. long, 1–2 mm. wide, the inner and outer layers usually so demarcated that the outer appears as a distinct horn up to 0·6 mm. long. Anthers half-exserted in long-styled flowers, well exserted in short-styled flowers. Style 1–1·25 cm. long in long-styled flowers, 4·3–5·5 mm. long in short-styled flowers; stigma-lobes 6–8, 0·9–2 mm. long. Drupe white at first, turning dark red, subglobose, 4–6 mm. in

* Not 1 as stated in original description.

diameter when dry, ± 8·5 mm. when fresh, ribbed (at least in dry state); putamen depressed-subglobose, 4–5 mm. in diameter, grooved, ± hairy. Seeds pale brown, narrowly oblong-ellipsoid, 2–2·4 mm. long, 0·9–1 mm. wide, 0·5–0·6 mm. thick, often with the undeveloped collateral ovule stuck to the testa.

subsp. **africana**

Leaf-blades distinctly pubescent above, sparsely to densely pubescent between the hairy nerves beneath.

TANZANIA. Rufiji/Ulanga/Kilwa Districts: Selous Game Reserve, 13 Jan. 1969, *Rodgers* 571 !; Lindi District: Rondo Plateau, Mar. 1952, *Semsei* 681 ! & same locality, Nyenea, 11 Dec. 1955, *Milne-Redhead & Taylor* 7633 !

DISTR. **T6**, 8; Mozambique

HAB. Evergreen thicket in *Chlorophora*, *Albizia* woodland; 280–810 m.

subsp. **hildebrandtii** (*Vatke*) *Verdc.* in K.B. 30: 282 (1975). Type: Zanzibar I., *Hildebrandt* 1126 (B, holo. †, K, iso. !)

Leaf-blades entirely glabrous or with a few obscure hairs above and nerves hairy beneath. Fig. 13.

KENYA. Kwale District: 19·2 km. S. of Mombasa, Twiga, 22 Jan. 1964, *Verdcourt* 3961 !; Mombasa District: Likoni Beach, Dec. 1956, *Ossent* 242 !; Kilifi District: Malindi, June 1962, *Tweedie* 2377 !

TANZANIA. Pangani District: Msubugwe Forest, 4 Mar. 1956, *Tanner* 2531 !; Uzaramo District: Dar es Salaam, 27 Aug. 1926, *Peter* 46071 !; Rufiji District: Mafia I., Jibondo I., 25 Sept. 1937, *Greenway* 5318 !; Zanzibar I., Mangapwani, 24 Jan. 1929, *Greenway* 1144 ! & Chwaka, 28 July 1959, *Faulkner* 2316 ! & Pemba I., Madunga, 25 Oct. 1929, *Vaughan* 863 !

DISTR. **K7**; **T3**, 6; **Z**; **P**; Mozambique, Malawi, Zambia, Madagascar, Comoro Is., Aldabra

HAB. Mostly in coastal bushland near high tide mark, particularly in *Grewia glandulosa*, *Colubrina*, *Cordia subcordata*, *Sideroxylon*, *Thespesia*, *Hibiscus tiliaceus*, *Scaevola* and similar associations, also in derived cultivations, e.g. coconut plantations; rarely inland; 0(–150) m.

SYN. *T. hildebrandtii* Vatke in Oest. Bot. Zeitschr. 25: 230 (1875); Hiern in F.T.A. 3: 219 (1877); T.T.C.L.: 533 (1949); Bremek. in Proc. K. Nederl. Akad. Wetensch., ser. C, 59: 11 (1956); K.T.S.: 475 (1961)

Dirichletia leucophlebia Bak. in J.L.S. 25: 321 (1890). Type: NW. Madagascar, *Baron*[5777] (K, holo. !)*

Psathura fryeri Hemsl. in J.B. 54, Suppl. 2: 20 (1916). Type: Aldabra, *Fryer* 44 (K, holo. !)

Triainolepis fryeri (Hemsl.) Bremek. in Proc. K. Nederl. Akad. Wetensch., ser. C, 59: 12 (1956)

T. fryeri (Hemsl.) Bremek. var. *latifolia* Bremek. in Proc. K. Nederl. Akad. Wetensch., ser. C, 59: 13 (1956). Type: Comoro Is., Grande Comore, *Humblot* 41 (P, holo.)

NOTE. This subspecies, curiously, has been found far inland at Mpika in Zambia and by Lake Malawi [Nyasa] at Nkata and Monkey Bays, a rather strange distribution for such a very characteristic littoral shrub. The Mpika plant is stated to be a tree to 6 m. and the Lake Malawi specimens are shrubs or trees 2·4–6 m. tall; there does not, however, seem to be any reason for separating them as a distinct taxon. Neither does there seem to be any valid reason for separating *T. fryeri*; those given by Bremekamp are highly technical and minute differences in the putamen structure. Hemsley himself later annotated the specimens he had originally called *Psathura fryeri* as *Triainolepis hildebrandtii*. The " cordons " which Bremekamp claims to be characteristic of *T. fryeri* and *T. africana* can also be found in some specimens of undoubted *T. hildebrandtii*.

* Bremekamp should have taken up this name for *T. fryeri* but did not do so on account of a double error. This much is made clear from a label he has attached to the type of *Dirichletia leucophlebia* which states " this is not *Baron* 5777 which is the type of *Dirichletia leucophlebia* Baker (according to A. M. Homolle in Bull. Soc. Bot. France 83: 620 (1936) = *D. pervilleana* H. Bn.). It is *Triainolepis fryeri* (Hemsl.) Brem. var. *latifolia* Brem." Firstly Mme. Homolle did not state = *D. pervilleana* but " non *Dirichletia* ni *Carphalea* "; moreover the sheet which bears in pencil " next 5778 " is undoubtedly the type of *Dirichetia leucophlebia* being written up in Baker's own hand.

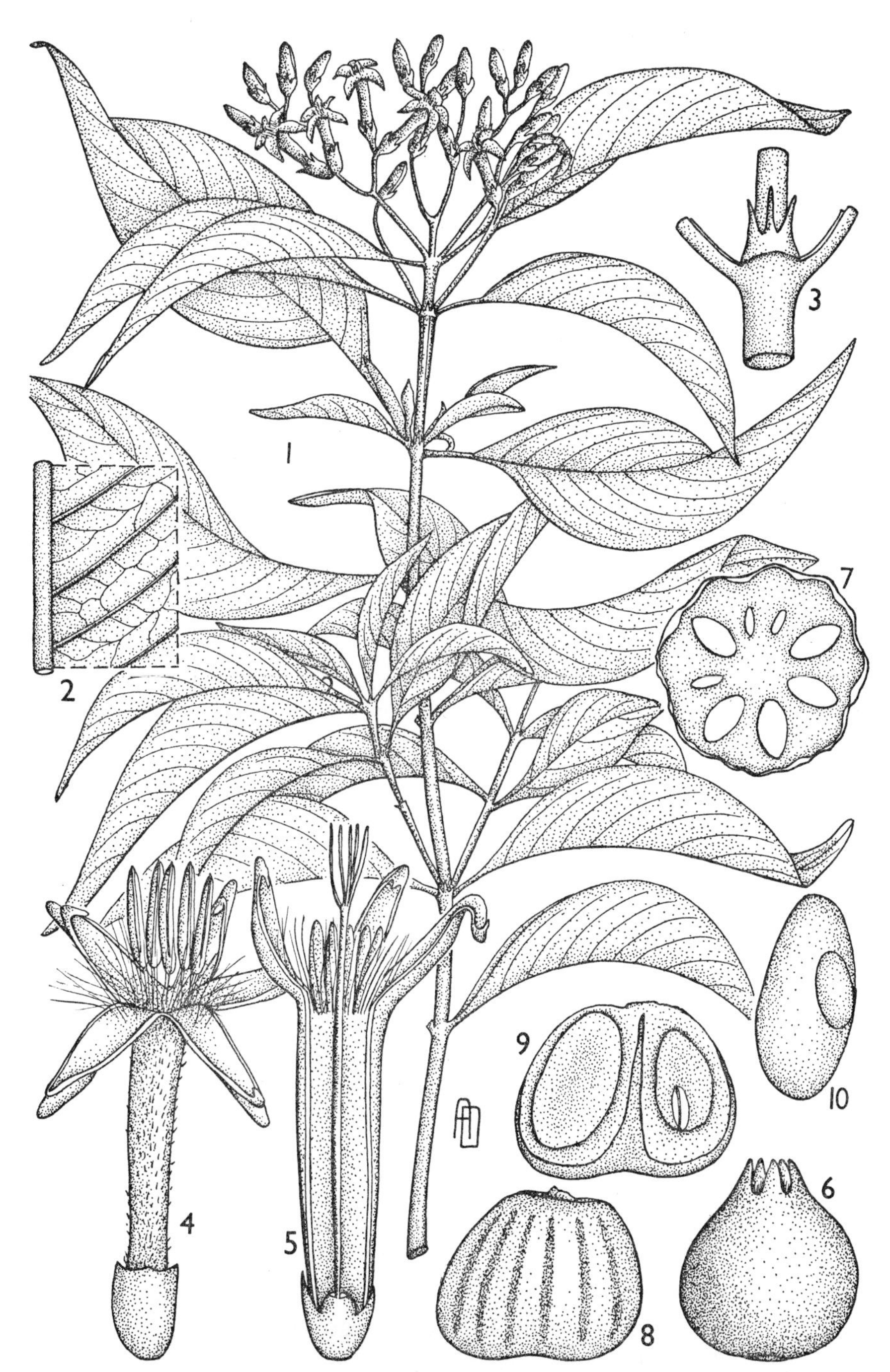

FIG. 13. *TRIAINOLEPIS AFRICANA* subsp. *HILDEBRANDTII*—**1,** flowering branch, × ⅔; **2,** detail of undersurface of leaf, × 2; **3,** node showing stipule, × 2; **4,** short-styled flower, × 4; **5,** longitudinal section of long-styled flower, × 4; **6,** fruit, × 6; **7,** transverse section of fruit, × 8; **8,** putamen, × 6; **9,** vertical section of putamen, × 8; **10,** seed, with an aborted ovule attached, × 10. 1–3, 5, from *Verdcourt* 3961; 4, from *Tweedie* 2377; 6–10, from *Faulkner* 2316. Drawn by Ann Davis.

Tribe 4. **UROPHYLLEAE**

Inflorescences compact axillary clusters or terminal and axillary; ovary 2-locular beneath, 4-locular above; stigma-lobes 2 . 9. **Pauridiantha**
Inflorescences lax axillary cymes, ± 3 per axil; ovary 4–5-locular beneath, 8–10-locular above; stigma-lobes 4–5 . . . . 10. **Rhipidantha**

9. **PAURIDIANTHA**

Hook. f. in G.P. 2: 69 (1873); Hiern in F.T.A. 3: 71 (1877); Bremek. in E.J. 71: 217 (1940)

Pamplethantha Bremek. in E.J. 71: 217 (1940)

Shrubs, small trees or subscandent woody plants. Leaves opposite or ternate, shortly petiolate, the petioles compressed, channelled; blades acuminate or caudate, usually with acarodomatia; midnerve channelled; stipules interpetiolar, triangular or ovate, entire, acute. Flowers hermaphrodite, mostly heterostylous, usually 5-merous, in axillary or terminal, sessile or pedunculate, trichotomously corymbose or subumbellate inflorescences, sometimes reduced to a few or even single flowers; peduncle with 1–2 4-parted involucels situated at the apex or middle; other bracts small or absent. Calyx-tube short, denticulate, dentate or lobed. Corolla salver-shaped, white, greenish, violet or lavender; tube short, funnel-shaped or cylindrical, the upper half densely hairy inside; lobes glabrous inside. Stamens with glabrous filaments and dorsifixed anthers, exserted in short-styled flowers. Disc cushion-shaped, papillate or shortly hairy. Ovary 2–3-locular at the base, 4–6-locular at the apex, with 2–3, sometimes lobed, placentas affixed at the middle of the true septum, the false septa incised and broadly cordate; ovules numerous; style glabrous, puberulous or hairy, included or exserted; stigmas 2, globose, mitriform or subcapitate, the apex shortly 2-lobed, the lobes cohering or rarely free and linear or lanceolate. Fruit a globose yellow or red berry, 2-locular at the base, 4-locular at the top. Seeds numerous, yellow or yellow-brown, rarely red, ovoid, testa alveolate and sometimes irregularly ribbed; endosperm oily.

A small genus of about 20–25 species confined to tropical Africa and Madagascar; formerly included in *Urophyllum* Wall. but differing from that genus in not being dioecious, in the structure of the ovary and its placentation, and the differently shaped stigma. In his most recent classification Bremekamp even maintains two separate tribes, *Urophylleae* and *Pauridiantheae*.

Inflorescences axillary, usually quite small:
 Leaf-blades smaller, 1·2–20 × 0·9–6·8 cm.; stipules narrower, 0·5–4·5 mm. wide:
  Leaf-blades glabrescent beneath or with hairs only on the main nerves:
   Leaves thin, the blade very distinctly acuminate:
    Leaf-blades oblong-elliptic to oblong-lanceolate or oblanceolate; inflorescences usually many-flowered; stigma capitate, lobes not distinctly free . . . 1. *P. paucinervis*

Leaf-blades narrowly lanceolate to lanceolate; flowers 1–3 in each axil with no or only a very short common peduncle; stigma-lobes linear, quite free from each other . 3. *P. symplocoides**
Leaves somewhat coriaceous, ± obtuse, sub-acute or very shortly acuminate; corolla slightly pubescent . . . . . . 2. *P. bridelioides*
Leaf-blades pubescent to densely velvety on the lower surface; stems densely velvety; calyx-lobes 0–1 mm. long, usually short . . 4. *P. dewevrei*
Leaf-blades large, 15–40 cm. × 3·7–12 cm.; stipules 0·9–1·9 cm. wide . . . . 5. *P. callicarpoides*
Inflorescences extensive, both axillary and terminal combined . . . . 6. *P. viridiflora*

1. **P. paucinervis** (*Hiern*) *Bremek.* in E.J. 71: 212 (1940); Hepper in F.W.T.A., ed. 2, 2: 168 (1963). Type: Fernando Po, *Mann* 577 (K, holo.!)

Forest shrub or small tree, apparently sometimes somewhat scandent, 1·5–9(–12) m. tall, with finely adpressed pubescent to distinctly hairy twigs, older stems glabrescent. Leaf-blades oblong-elliptic to oblong-lanceolate or -oblanceolate, 3·5–15·5 cm. long, 0·9–5·2 cm. wide, acuminate at the apex, cuneate at the base, glabrous above save for short hairs on the narrowly impressed midnerve, entirely glabrous or pilose on the main nerves beneath; lateral nerves ± 12 on each side, together with other venation prominent to very prominent beneath; petioles short, 2·5–10 mm. long, usually pubescent; stipules lanceolate, 0·3–1·35 cm. long, 0·8–2·5 mm. wide, adpressed pubescent. Inflorescences axillary, short, the cymes ± 1 cm. long; peduncles 1–5 mm. long; pedicels 0–3·5(–4) mm.; bracts lanceolate, 1·5–2·5 mm. long; all parts pubescent. Calyx glabrous, pubescent or hairy; tube 0·6–1·1 mm. long; limb-tube 0·3–0·9 mm. long; lobes yellow, red or purple, lanceolate-subulate, 0·4–2·8 mm. long. Corolla glabrous (or rarely puberulous) outside, greenish, yellow, cream or white but sometimes reported to be orange or yellow turning red and certainly often drying red; tube 2–5 mm. long, longest in short-styled flowers; lobes oblong-lanceolate, 1·7–3·1 mm. long, 0·8–1 mm. wide. Stamens as long as the corolla-lobes in short-styled flowers, just included in long-styled flowers. Disc depressed, glabrous. Style glabrous or sparsely papillate, 3–4·6(–5·5) mm. long in long-styled flowers, 1·4–2·8 mm. long in short-styled flowers; stigma 0·45–0·8 mm. long. Berries orange, red or tinged purple, subglobose, 3–5 mm. long, 2·5–4 mm. across. Seeds orange-brown, ellipsoid, 0·8 mm. long, 0·5 mm. wide, very strongly pitted.

SYN. *Urophyllum paucinerve* Hiern in F.T.A. 3: 74 (1877); F.W.T.A. 2: 104 (1931)

subsp. **holstii** (*K. Schum.*) *Verdc.* in K.B. 30: 283 (1975). Types: Tanzania, E. Usambara Mts., Lutindi, *Holst* 3277 (B, syn. †, K, isosyn.!) & Kilimanjaro, Shira Plateau, *Volkens* 1940 (B, syn. †, K, isosyn.!)

Leaf-blades with venation not quite so prominent beneath as in subsp. *lyallii*; inflorescences typically lax (but not in some populations); calyx-lobes 1–2·8 mm. long; seeds orange-brown. Fig. 14, p. 154.

KENYA. Kiambu District: Gatamayu Forest, near Kerita Forest Station, 16 Nov. 1958, *Verdcourt* 2303!; Kericho District: SW. Mau Forest, Saoset [Saosa] Catchment, 15 Jan. 1959, *Kerfoot* 712!; Teita District: Mbololo Hill, Sept. 1938, *Joanna* in *C.M.* 9030!

TANZANIA. Moshi District: Kilimanjaro, mountain track N. of Mweka, 11 Apr. 1968, *Renvoize* 1408!; Lushoto District: Lushoto–Shume road, Magamba Forest, 1 Mar.

* It is not certain that the E. African material (very poor) belongs to this species.

FIG. 14. *PAURIDIANTHA PAUCINERVIS* subsp. *HOLSTII*—**1**, flowering branch, × ½; **2**, node showing stipule, × 2; **3**, long-styled flower, × 4½; **4**, longitudinal section of ovary, × 8; **5**, longitudinal section of short-styled flower, × 4½; **6**, berry, × 4; **7**, seed, × 20. 1–4, from *Verdcourt* 2303; 5, from *Brasnett* in *F.D.* 1506; 6, 7, from *Kerfoot* 3874. Drawn by Ann Davis.

1953, *Drummond & Hemsley* 1368 ! & just below Amani, 2 Apr. 1950, *Verdcourt* 136 !; Morogoro District: Uluguru Mts., Tegetero, 20 Mar. 1953, *Drummond & Hemsley* 1695 !

Distr. **K**3–5, 7; **T**2–4, 6, 7; ? Zaire (intermediates), Malawi and Zambia (see note)
Hab. Lowland evergreen (including rain) forest, upland evergreen forest; 500–2400 m.

Syn. *Urophyllum holstii* K. Schum. in P.O.A. C: 379 (1895)
Pauridiantha holstii (K. Schum.) Bremek. in E.J. 71: 212 (1940); T.T.C.L.: 508 (1949); I.T.U., ed. 2: 354 (1952), pro parte; Brenan in Mem. N.Y. Bot. Gard. 8: 449 (1954); K.T.S.: 453 (1961); F.F.N.R.: 414 (1962)

Note. It is with regret that I have altered the status of the well-known *P. holstii*, but it seems essential that the very close relationship between the Cameroun, E. African and Madagascan plants should be made evident. Subsp. *holstii* is very variable and some E. African specimens differ more from the type of subsp. *holstii* in some characters than do subsp. *lyallii* (Bak.) Verdc. and subsp. *paucinervis*, e.g. some Uluguru Mts. specimens have the inflorescences very densely hairy. Material frequently identified as *P. pyramidata* (K. Krause) Bremek. (type: Cameroun, Lomie, *Mildbraed* 5216 (B, holo.)) in Zambia and Malawi, usually with more compact inflorescences, is not distinguishable from material of subsp. *holstii* from S. Tanzania. It really is not feasible to suggest that W. African *P. pyramidata* is conspecific with *P. paucinervis* since there are distinct foliage differences. The S. tropical African material is, however, intermediate in many respects and might perhaps be treated as a further race of *P. paucinervis* for convenience. *Richards* 6774 (Rungwe Mt., 24 Oct. 1956, 2700 m.) is atypical in leaf-shape and indumentum, the undersurfaces being pubescent all over and the corollas are also slightly puberulous. It may represent a distinct variant. Another specimen *Procter* 1540 (Iringa District, Image Mt., Nov. 1959, 1950 m.) is from the general area of *P. bridelioides* Verdc. which it resembles in its wider sepals and less acuminate leaf-blades but differs in its glabrous corollas and pubescent styles. There is clearly a tendency to speciate in isolated montane forests.

subsp. **butaguensis** (*De Wild.*) *Verdc.* in *K.B.* 30: 283 (1975). Type: Zaire, Butagu Valley, *Bequaert* 3917 (BR, holo.)

Very similar to subsp. *holstii* but with calyx-lobes distinctly shorter and more triangular, ± 0·25–1 mm. long; stipules sometimes shorter and wider. All the material is described as a shrub.

Uganda. Toro District: S. Ruwenzori, Kijomba Ridge, Oct. 1940, *Eggeling* 4098 !; Ankole District: Kalinzu Forest, Feb. 1949, *Dale* 694 !; Kigezi District: Rubuguli, Apr. 1948, *Purseglove* 2695 !
Tanzania. Kigoma District: Kabogo, *Kyoto Univ. Exped.* 302 !
Distr. **U**2; **T**4; E. Zaire, Burundi
Hab. Evergreen forest and derived hillside scrub; 1350–2300 m.

Syn. *Urophyllum butaguense* De Wild., Pl. Bequaert. 3: 207 (1925)
U. lanuriense De Wild., Pl. Bequaert. 3: 212 (1925). Type: Zaire, Lanuri, *Bequaert* 4433 (BR, holo.)
U. butaguense De Wild. var. *exserto-stylosa* De Wild., Pl. Bequaert. 3: 207 (1925). Type: Zaire, Butagu valley, *Bequaert* 3916 (BR, holo.)
Pauridiantha butaguensis (De Wild.) Bremek. in E.J. 71: 214 (1940)
P. butaguensis (De Wild.) Bremek. var. *exserto-stylosa* (De Wild.) Bremek. in E.J. 71: 214 (1940)

Note. Bremekamp keeps up the var. *exserto-stylosa* because of thicker indumentum on the stem and larger leaves; he was, of course, well aware that the species is heterostylous. This subspecies is known from three forests where *P. dewevrei* (De Wild. & Th. Dur.) Bremek. also occurs, namely Kasatora, Kayonza and Ishasha. It would be interesting to investigate the absolute distinctness of these two taxa in the field. It seems unlikely to me that *P. claessensii* Bremek. (type: Zaire, Lower Congo, Goya, *Claessens* 1279 (BR, holo.)) and *P. bequaertii* (De Wild.) Bremek. (type: Zaire, Lubutu, *Bequaert* 6758 (BR, holo.)) should be distinguished from subsp. *butaguensis*, in fact the latter is scarcely distinguishable from subsp. *holstii* in many areas.

Distr. (of species as a whole). As above with addition of Nigeria, Cameroun, Fernando Po (subsp. *paucinervis*) and Madagascar (subsp. *lyallii*), and possibly intermediates in Malawi and Zambia

2. **P. bridelioides** *Verdc.* in K.B. 30: 284, fig. 6 (1975). Type: Tanzania, Iringa District, Dabaga Forest Reserve, *Carmichael* 24 (K, holo.!, EA, iso.!)

Shrub to about 1·5–3 m. tall; stems covered with whitish hairs, later glabrescent. Leaf-blades elliptic, 1·2–7·3(–9·5) cm. long, 0·9–3·1(–5·3) cm. wide, obtuse to subacute at the apex, broadly cuneate to rounded at the base, usually yellow-green, mat to very glossy above, glabrous above save for a few hairs on the midnerve, glabrous beneath save for sparse hairs on the main nerves; lateral nerves 6–8 on each side, very prominent beneath, impressed above and the upper leaf-surface often very bullate; petioles 3–5 mm. long, pubescent; stipules ovate-triangular, 4–8(–10·2) mm. long, 1·5–1·8 mm. wide, pubescent. Flowers heterostylous, in condensed axillary (and ? terminal) cymes ± 7 mm. long; peduncles almost obsolete; pedicels 0–1 mm. long; bracts ± 2 mm. long; all pubescent. Calyx glabrescent or pubescent; tube subglobose or campanulate, 1–1·5 mm. long; limb-tube 0·2–0·4 mm. long; lobes ovate to triangular-lanceolate, 0·9–2·5 mm. long. Corolla slightly pubescent outside, greenish white; tube 3·8–4·2 mm. long; lobes narrowly triangular, 1·7–3·5 mm. long, 1–1·5 mm. wide. Stamens exserted about ½ the length of corolla-lobes in short-styled flowers. Disc depressed, glabrous. Style minutely papillate, 2·3 mm. long in short-styled flowers, 5·5 mm. in long-styled flowers; stigma subcapitate, obscurely 2-lobed in short-styled flowers, with 2 linear lobes in long-styled flowers, 0·7–0·8 mm. long. Berries subglobose, 5 mm. in diameter. Seeds yellow-brown, ellipsoid, 8–9 mm. long, 5–6 mm. wide, strongly ridged and pitted.

TANZANIA. Buha District: 16 km. N. of Kasulu, Jan. 1955, *Procter* 346!; Iringa District: Sao Hill, Dec. 1963, *Procter* 2454!; Njombe, 18 Dec. 1931, *Lynes* V. 11!

DISTR. T4, 7; not known elsewhere

HAB. Evergreen forest, riverine forest in degraded *Combretum* bushland, thicket; 1200–1800 m.

NOTE. A sterile specimen from Dabaga (*Geilinger* 2003!) is clearly a sucker shoot and bears larger shortly acuminate leaves of the dimensions included in parenthesis in the description.

3. **P. symplocoides** (*S. Moore*) *Bremek.* in E.J. 71: 212 (1940). Type: Rhodesia, Melsetter District, Mt. Pene, *Swynnerton* 1278 (BM, holo.!, K, iso.!)

Shrub or small tree, usually much-branched, 1·8–9 m. tall; stems glabrous or nearly so. Leaf-blades lanceolate, 4·5–13·3 cm. long, 0·9–2·8 cm. wide, narrowly acuminate at the apex, cuneate at the base, glabrous above save for a few short hairs on the narrowly impressed midnerve, entirely glabrous or with a few hairs on the venation beneath; lateral nerves ± 10–12 on each side, prominent beneath; petioles 2–8 mm. long, pubescent; stipules triangular-lanceolate, 1·5–5 mm. long, 0·5–2 mm. wide, pubescent. Flowers heterostylous, 1–3 in each axil with no or only a short common peduncle 0–2·5 mm. long; pedicels 1·5–4·5 mm. long; bracts in a whorl of 2 long and 2 short, 1·5–2·5 mm. long, pubescent. Calyx pubescent; tube 1·6–1·7 mm. long; limb-tube 0–0·2 mm.; lobes lanceolate-subulate 1–4·1 mm. long. Corolla white, yellowish white or cream, glabrous outside; tube 4–4·3 mm. long; lobes triangular-lanceolate, 2–4·9 mm. long, 1·1–2 mm. wide. Anthers about ¾-exserted in short-styled flowers, just included in long-styled flowers. Disc depressed, glabrous. Style glabrous, 5·8 mm. long in long-styled flowers, 1·5 mm. long in short-styled flowers; stigma 0·9–1·2 mm. long, the lobes quite distinct, linear. Berries subglobose, 3–3·7 mm. long, 4–5 mm. across, glabrous. Seeds typically orange-brown, ovoid, 1·2 mm. long, 0·95 mm. wide.

var. ?

Leaf-blades more broadly lanceolate, up to 2·8 cm. wide, more abruptly cuneate at the base. Seeds probably smaller.

TANZANIA. Morogoro District: Uluguru Mts., Kitundu, 27 Oct. 1934, *E. M. Bruce* 55!
DISTR. **T6**; not known elsewhere
HAB. Evergreen forest; 1290 m.

NOTE. *P. symplocoides* occurs in Malawi, Mozambique and Rhodesia. The almost sterile specimen cited above as a possible variant bears only a few unripe and possibly one ripe fruit. It much resembles the true *P. symplocoides* in its 1-flowered inflorescences, glabrous stem and general appearance but the leaf-blades, although narrow, are not as narrow as in the typical material, the calyx-lobes, fruit and seeds (observed as bumps in the pericarp) are all smaller and until flowering material is available the identity of the Uluguru plant will remain dubious.

4. **P. dewevrei** (*De Wild. & Th. Dur.*) *Bremek.* in E.J. 71: 215 (1940); Hallé in Fl. Gabon 12, Rubiacées: 249, t. 51/8–14 (1966). Types: Zaire, Lower Congo and Kasai R., Wabundu, *Dewèvre* 1130 & Kimuensa, *Dewèvre* 503 & Dembo, *Gillet* (BR, syn.)

Small tree or shrub 2–8 m. tall; stems densely velvety hairy. Leaf-blades oblong-elliptic to narrowly elliptic-lanceolate, 7·5–20 cm. long, 1·8–6·8 cm. wide, acuminate at the apex, broadly to narrowly cuneate at the base, hairy on the main nerves above and either glabrescent or with scattered hairs on the rest of the surface, pubescent to velvety hairy beneath; lateral nerves 9–12 on each side; petioles 0·3–1·2 cm. long, densely hairy; stipules triangular-lanceolate, 0·6–1·3 cm. long, 3–4·5 mm. wide, hairy, deciduous. Inflorescences axillary, short, the cymes 1–1·5 cm. long; peduncles 0–1·3 cm. long; pedicels 1–3 mm. long; bracts 1·5 mm. long; all pubescent. Flowers heterostylous. Calyx pubescent or mostly ± glabrous; tube subglobose, 1 mm. long; limb ± 0·3 mm. long, very shallowly undulate-lobed, or with 4–6 broadly triangular lobes, 0·5–1 mm. long. Corolla bright yellow, greenish brown or greenish yellow, glabrous or puberulous; tube 1·8–3 mm. long; lobes ovate-oblong, 1·5–2 mm. long, 0·8–1·2 mm. wide. Stamens just exserted in long-styled flowers, filaments exserted 0·5 mm. in short-styled flowers. Disc hemispherical, grooved, glabrous. Style glabrous or papillate just below the stigma, 2·6 mm. long in long-styled flowers, 1 mm. long in short-styled flowers; stigma-lobes 0·2–1 mm. long. Berries yellow or reddish, subglobose, 6 mm. in diameter. Seeds reddish brown, ovoid, 1·2–1·5 mm. long, strongly ridged and pitted.

UGANDA. Kigezi District: Kayonza Forest, Oct. 1940, *Eggeling* 4201! & Kasatora Forest, *St. Clair-Thompson* 2509! & Ishasha Gorge, May 1950, *Purseglove* 3439!
DISTR. **U2**; Cameroun, Gabon, Central African Republic and Zaire
HAB. Evergreen forest; 1800–2400 m.

SYN. *Urophyllum dewevrei* De Wild. & Th. Dur. in Ann. Mus. Congo, Bot., ser. 2, 1 (2): 30 (1900)

5. **P. callicarpoides** (*Hiern*) *Bremek.* in E.J. 71: 216 (1940); I.T.U., ed. 2: 354 (1952); Hallé, Fl. Gabon, 12 Rubiacées: 258, t. 52 (1966). Type: Rio Muni, Muni R., 1°N., *Mann* 1826 (K, holo.!, P, iso.)

Shrub or small tree 4–15 m. tall, the branches concentrated towards the apex of the main stem; youngest parts of shoots hairy, very soon glabrous or in Zaire material persistently pubescent. Leaf-blades narrowly oblong or obovate-oblong, 15–40 cm. long, 3·7–12 cm. wide, acuminate at the apex, cuneate to shallowly emarginate at the base, sometimes bullate, at first shortly pubescent above and beneath particularly on the venation, at length practically glabrous save for a few hairs on the midnerve particularly above or in Zaire material often scabrid yellowish pubescent to densely velvety all over; lateral nerves ± 15–30 on each side; petioles 0·4–1·3 cm. long; stipules ovate-triangular, 1·1–2·3 cm. long, 0·9–1·9 cm. wide, glabrous to pubescent outside, deciduous, the scar hairy within. Inflorescences axillary, often trichotomous from the extreme base, peduncles or apparent peduncles 0·5–2·5 cm. long, the bracts borne towards the apex, 2–6 mm. long, 1–1·5 mm. wide; secondary peduncles 5 mm. long; pedicels 1–5 mm. long; all parts of inflorescence rather sparsely to very densely yellowish pubescent. Calyx hemispherical, the tube 1–1·5 mm. long, the limb 1·1–1·6 mm. long, glabrous or pubescent at the base, the lobes no more than slight undulations on the limb edge. Buds grooved and ribbed when dry. Corolla glabrous outside, whitish or pale green with blue hairs in the throat; tube 1·6–3·2 mm. long; lobes ovate-triangular, 2·2–3 mm. long, 1·3–1·7 mm. wide. Stamens almost as long as the lobes in short-styled flowers, just exserted in long-styled flowers. Disc subglobose, pubescent to tomentose, the central stigma-pit surrounded by 10 small pits visible only when hairs have worn off. Style densely hairy, 0·8–1·7 mm. long in short-styled flowers, 2–6 mm. long in long-styled flowers; stigma 0·7–1·25 mm. long. Berries green, subglobose, ± 6 mm. in diameter. Seeds dark brown, ovoid, 0·8–1 mm. long, deeply wrinkled and pitted.

UGANDA. Kigezi District: Kayonza, Ishasha Gorge, Mar. 1947, *Purseglove* 2393! & same locality, May 1950, *Purseglove* 3440! & 6 km. SW. of Kirima, 21 Sept. 1969, *Lye* 4217!

DISTR. **U2**; Zaire, Cameroun, Rio Muni and Gabon

HAB. Evergreen forest; 1050–1650 m.

SYN. *Urophyllum callicarpoides* Hiern in F.T.A. 3: 72 (1877)

NOTE. The Uganda material is much less hairy than the typical W. African plant, but there is a good deal of variation in the density of the indumentum.

6. **P. viridiflora** (*Hiern*) *Hepper* in K.B. 13: 405 (1959) & in F.W.T.A., ed. 2, 2: 168 (1963); Hallé, Fl. Gabon, 12 Rubiacées: 237 (1966). Type: Zaire, Monbuttu, Munsa, *Schweinfurth* (K, holo.!)

Shrub or small tree 3–7·5(–12) m. tall; young stems densely finely yellowish pubescent, becoming glabrescent. Leaves yellow-green (as in aluminium-accumulating plants); blades deeply furrowed, elliptic-oblong, (7–)9·6–15·5(–23·5) cm. long, 2–5·6(–8·4) cm. wide, acuminate at the apex, cuneate at the base, pubescent on main nerves particularly beneath or almost glabrous on both surfaces; lateral nerves 16–20 on each side; petiole 0·5–1·5 cm. long; stipules ovate-triangular or ovate, 0·5–1·8 cm. long, 0·3–1·2 cm. wide, acute to acuminate at the apex, adpressed pubescent, eventually deciduous. Flowers in extensive terminal and axillary inflorescences (the axillary ones actually often branches bearing one pair of leaves), 8–25 cm. long and wide; peduncles 3–8·5 cm. long; secondary peduncles 2–4 cm. long; pedicels 0·5–5 mm. long; whorls of bracts absent, but sometimes a few scattered lanceolate bracts ± 5 mm. long; all parts of inflorescence densely adpressed yellowish tomentose-pubescent. Calyx campanulate, pubescent, the tube 0·5–1 mm. long, the limb-tube 0·6–1 mm. long, the shallow teeth up to ± 0·3 mm. long. Corolla pubescent outside, green or greenish white; tube 1·4–2·7 mm. long; lobes triangular, 0·9–2·1 mm. long, 0·6–1·4 mm. wide. Stamens almost as long as the corolla-lobes in short-styled flowers, almost completely included in long-styled flowers. Disc apically pubescent. Style hairy, 2–2·8 mm. long in long-styled flowers, 0·8–0·9 mm. long in short-styled flowers; stigma 0·5–1·1 mm. long. Berries greenish, subglobose, 5–7 mm. in diameter, crowned by the cupular calyx-limb. Seeds brownish red, ellipsoid, ± 1 mm. long, strongly wrinkled and pitted.

UGANDA. Masaka District: Sese Is., Bukasa, Masekera, 26 Feb. 1933, *A. S. Thomas* 883! & Sese Is., Bugala I., Kalangalo, July 1945, *Purseglove* 1722! & Minziro Forest, July 1938, *Eggeling* 3752!
TANZANIA. Bukoba District: Munene, Sept.–Oct. 1935, *Gillman* 614! & Rubare Forest Reserve, Sept. 1958, *Procter* 1027!
DISTR. **U4**; **T1**; S. Nigeria, Cameroun, Zaire, Burundi and Central African Republic
HAB. Understory of evergreen forest, forest edges and secondary scrub; 1140–1400 m.

SYN. *Urophyllum viridiflorum* Hiern in F.T.A. 3: 74 (1877); F.W.T.A. 2: 104 (1931)
U. gilletii De Wild. & Th. Dur. in Bull. Herb. Boiss., sér. 2, 1: 26 (1900). Type: Zaire, Kisantu, *Gillet* 402 (BR, holo.)
Pamplethantha viridiflora (Hiern) Bremek. in E.J. 71: 217 (1940); I.T.U., ed. 2: 353 (1952); Verdc. in K.B. 11: 452 (1957)
P. gilletii (De Wild. & Th. Dur.) Bremek. in E.J. 71: 218 (1940)

10. RHIPIDANTHA

Bremek. in E.J. 71: 222 (1940)

Shrub; branchlets puberulous at first but later glabrescent. Leaves opposite, petiolate, the petioles laterally compressed and channelled; blades acuminate, the midnerve channelled; primary and secondary nerve-axils bearing acarodomatia, venation densely reticulate; stipules ovate-triangular, recurved at the margins, densely pubescent inside and with large sparse colleters. Flowers hermaphrodite, possibly slightly heterostylous,* 5–6-merous; inflorescences axillary, usually 4, superposed in a vertical plane in a fan-like manner, confluent in a basal pulvinus, pedunculate, the peduncles involucellate about the middle, cymose or trichotomously corymbose, 7–9-flowered; bracts and bracteoles absent. Calyx cupular, 5–6-dentate. Corolla shortly salver-shaped; tube with throat densely hairy. Stamens inserted in the throat in short-styled flowers; anthers oblong, dorsifixed, basally bilobed, the connective briefly apiculate. Disc cushion-shaped, minutely white papillate, impressed at the margins by the decurrent parts of the filaments at the base of the tube, apically lobed. Ovary 4–5-locular, with 4–5 axile placentas apically cut by a false septum which is broadly cordate; ovules numerous; style absent; stigma-lobes 4 or 5, terete, erect, short. Fruit not seen.

Bremekamp created this genus for a plant collected in the Uluguru Mts. Since he saw the type it has been destroyed in the war, but other specimens, which from the description must be congeneric, have been collected from the same area. They do not entirely agree with either K. Schumann's original description or Bremekamp's generic description but I am fairly certain only one species is involved. Further material is required to solve the problem. It is more likely to have been overlooked than to be genuinely rare. The worth of the genus must also await the collection of further material.

R. chlorantha (*K. Schum.*) *Bremek.* in E.J. 71: 222 (1940); T.T.C.L.: 528 (1949). Type: Tanzania, Uluguru Mts., Ngamba, *Stuhlmann* 8883 (B, holo. †)

Shrub or tree 2–15 m. tall; stems glabrous or with very fine pubescence just above the nodes. Leaf-blades oblong to oblong-lanceolate or broadly elliptic, 8·5–21 cm. long, 2·5–11 cm. wide, produced into a narrow obtuse acumen at the apex, cuneate at the base, finely adpressed hairy beneath and on the midrib above in very young leaves, but later glabrous or with very sparse hairs on midrib beneath; lateral nerves 10–15 on each side; petiole 1–2 cm. long (the 1–2 mm. in K. Schumann's description must be an error), glabrous or puberulous; stipules ovate or narrowly triangular, cucullate, 4·5–9 mm. long, 2·5–3 mm. wide, ± pubescent or glabrous, glandular within, the ± numerous colleters 0·5 mm. long, eventually deciduous. Inflorescences

* Bremekamp had assumed that the only specimen he saw was a short-styled one.

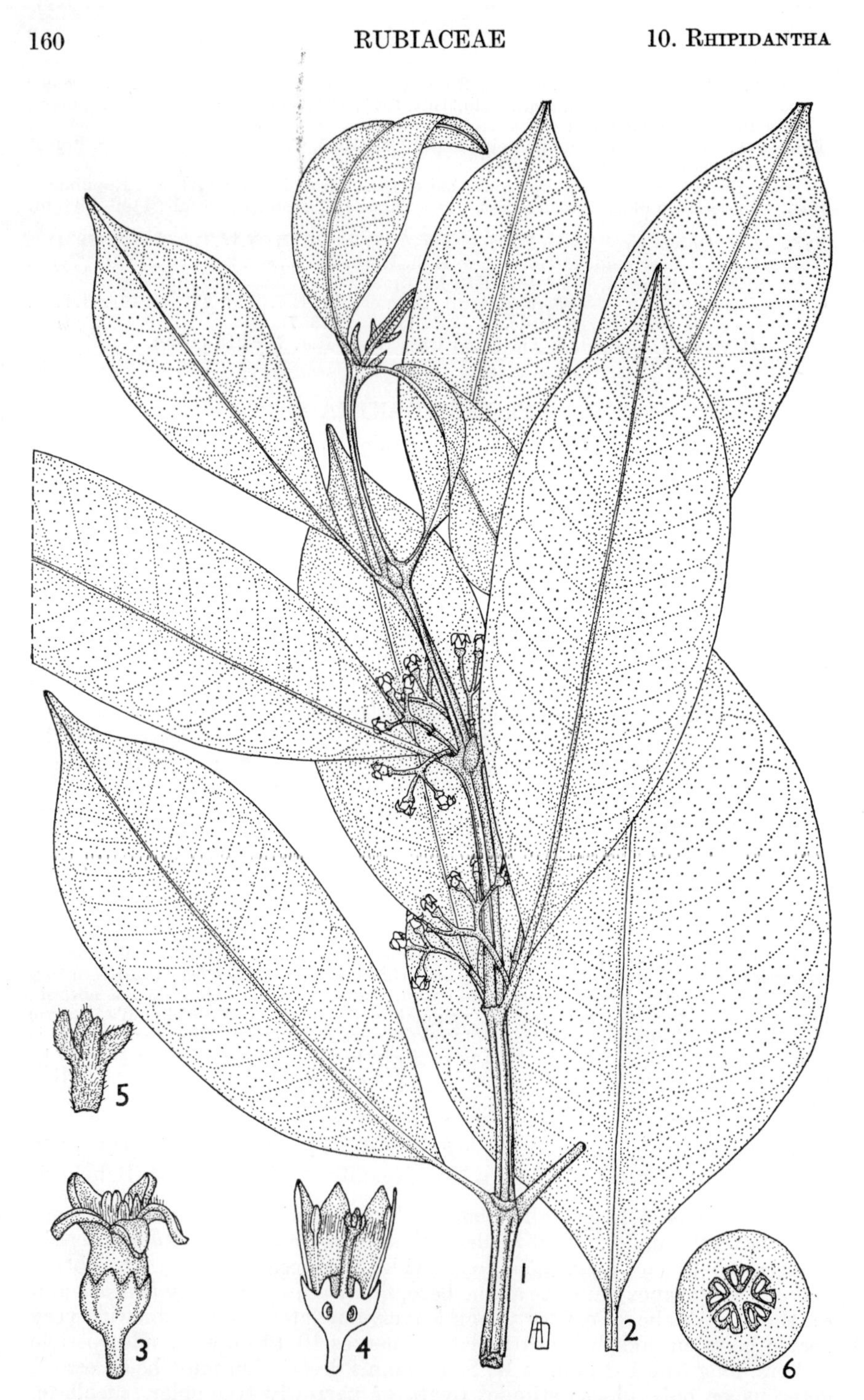

FIG. 15. *RHIPIDANTHA CHLORANTHA*—**1,** flowering branch, × ½; **2,** leaf, × ⅜; **3,** flower, × 4; **4,** longitudinal section of flower, × 4; **5,** tip of style and stigma-lobes, × 6; **6,** transverse section of ovary, × 6. 1, 3–6, from *Wallace* 490; 2, from *Paulo* 52. Drawn by Ann Davis.

axillary, sometimes drying greenish blue, 1–3(–4 *fide* Schumann), up to 2·5 cm. long from each pubescent axillary cushion; individual peduncles 0·5–1·5 cm. long, pubescent below, glabrous above, with an involucre of puberulous components comprising 2 narrowly deltoid reduced stipules 1·7–2 mm. long, ± 1 mm. wide at the base, and 2 linear reduced leaves up to 2·5 mm. long, 0·5 mm. wide, about half-way up; pedicels 0–6 mm. long, pubescent or glabrescent. Calyx-tube hemispherical, 0·5–1 mm. long; limb-tube 0·7–1·5 mm. long, repand-dentate, the teeth scarcely 0·5 mm. long, puberulous. Corolla white or very pale green; tube 1·6–3 mm. long, the throat densely hairy; lobes triangular to ovate, 2–2·3 mm. long, 1·5 mm. wide. Anthers subsessile, included in the tube, and style scarcely 0·6 mm. long, with 5 minute stigma-lobes (*fide* Schumann) or, in material seen, anthers with short filaments 1 mm. long, just exserted; stigma-lobes 4–5, ± sessile. Disc domed, annular, grooved and microscopically papillate. Fruit and seeds not seen. Fig. 15.

TANZANIA. Morogoro District: Uluguru Mts., Mwere valley, Bondwa/Mwere col, 26 Sept. 1970, *Harris et al.* 5124! & Bunduki Forest Reserve, Mar. 1955, *Paulo* 52! & N. slope of Bondwa, 24 Oct. 1972, *Pócs* 6804/B & without exact locality but presumably Uluguru Mts., 25 Nov. 1932, *Wallace* 490!

DISTR. **T6**; not known elsewhere

HAB. Evergreen forest; 1450–1740 m.

SYN. *Urophyllum chloranthum* K. Schum. in E.J. 28: 58 (1899) & 28: 488 (1900)

NOTE. It is not entirely certain that the material cited above is identical with *Stuhlmann* 8883 but I have assumed that this represents a short-styled form which would explain some of these inconsistencies. The petiole length given by K. Schumann is probably an error. He gives the corolla-tube length as 5 mm. and the lobes 2·5 mm. On the other hand it is just possible that there is more than one taxon concerned.

Tribe 5. CRATERISPERMEAE

One genus only . . . . 11. **Craterispermum**

11. CRATERISPERMUM

Benth. in Hook., Niger Fl.: 411 (1849); Hook. f. in G.P. 2: 112 (1873); Verdc. in K.B. 28: 433 (1974)

Glabrous trees or shrubs mostly with yellow-green foliage. Leaves opposite, petiolate, the blades mostly oblong or elliptic, often coriaceous; venation mostly closely reticulate; stipules intrapetiolar, broad, connate to form a tube, made up of 2 triangular parts joined by thinner tissue, undivided, persistent or deciduous. Flowers hermaphrodite, heterostylous, 5-merous, in small subcapitate or somewhat elongated occasionally 2-branched cymes, the peduncles short or less often long and slender, strongly compressed, axillary or more usually supra-axillary; bracteoles present. Calyx-tube obconic or turbinate; limb cupular, truncate, sinuate or shortly 5-dentate, persistent. Corolla salver-shaped or somewhat funnel-shaped with short or elongated tube and densely hairy or less often glabrous throat. Stamens either included in the throat or exserted; anthers linear-oblong, dorsifixed. Disc annular, thick. Ovary 2-locular; ovules solitary in each locule, pendulous from the apex; style filiform; stigma divided into 2 linear papillate branches or fusiform, bifid. Fruit subglobose, pea-like, sessile or pedicellate, 1(–2)-locular, 1-seeded; endocarp chartaceous. Seeds pendulous, hemispherical or almost bowl-shaped, dorsally convex, ventrally deeply excavated; albumen fleshy; embryo small with a superior radicle.

A small genus with 15–20 species widespread in tropical Africa and also in the Seychelles and Madagascar. Only 2 species occur in the Flora area.

NOTE. *C. orientale* K. Schum. is *Canthium crassum* Hiern.

Peduncles slender, 1·5–2·7 cm. long; inflorescences 1–2(–several)-flowered, sometimes bifid, the bracts and bracteoles well spaced; corolla-tube glabrous inside . . . . . 1. *C. longipedunculatum*
Peduncles thicker, more compressed, 0·2–1 cm. long; inflorescences several-flowered capitate cymes, the bracts and bracteoles congested; corolla-tube hairy inside . . . . 2. *C. schweinfurthii*

1. **C. longipedunculatum** *Verdc.* in K.B. 14: 350 (1960) & in K.B. 28: 434 (1974). Type: Tanzania, Morogoro District, Nguru Mts., Turiani, Ruhamba [Koluhamba], *Semsei* 1153 (EA, holo.!, K, iso.!)

Small slender glabrous tree about 8 m. tall with sparsely branched smooth green stems; bark grey, smooth. Leaf-blades elliptic-oblong, 4·5–12 cm. long, 1·7–4·5 cm. wide, with an acute acumen 0·6–1·5 cm. long, cuneate at the base, green and shining above, paler beneath, the margin minutely undulate; petiole 0·7–1·3 cm. long; stipules ovate-oblong, 4–5 mm. long, 2·5–5 mm. wide, acute or obtuse. Inflorescences supra-axillary, placed ± 4 cm. above the nodes, 2–4·5 cm. long, 1–several-flowered, dichotomous or not branched; peduncles ± flattened, 1·5–2·7 cm. long; bracts minute, 0·5 mm. long, acute, not congested. Calyx-tube urceolate, 3 mm. long and wide; lobes deltoid, 0·5–1 mm. long and wide, acute. Corolla white; tube 6–8 mm. long, 1·7–2·2 mm. wide, dilated at the apex to 3·5 mm. wide, glabrous outside and inside; lobes oblong, 5 mm. long, 1·5–2 mm. wide, the acute apex inflexed. Stamens included in tube in long-styled flowers; short-styled flowers not seen. Style 5–7 mm. long in long-styled flowers; stigma bifid, lobes oblong, 2 mm. long. Fruit (immature) urceolate-globose, 6·5 mm. long, 5·5 mm. across, narrowed to the apex and crowned with the 2 mm. long persistent calyx.

TANZANIA. Morogoro District: Nguru Mts., Turiani, Ruhamba [Koluhamba], Nov. 1953, *Semsei* 1443! & Ruhamba Peak, Apr. 1953, *Drummond & Hemsley* 1993!
DISTR. **T6**; not known elsewhere
HAB. Lower tree storey in evergreen forest; 1200 m.

NOTE. This is presumably heterostylous but no short-styled plants have been seen.

2. **C. schweinfurthii** *Hiern* in F.T.A. 3: 162 (1877); K. Schum. in P.O.A. C: 386 (1895); F.P.S. 2: 433 (1952); Verdc. in K.B. 28: 434 (1974). Type: Sudan, Equatoria [Niamniam], Khor Bodo, *Schweinfurth* 2935 (K, holo.!)

Shrub or small to medium tree 1·8–15 m. tall, glabrous; bark greyish white, warty with old swollen nodes. Leaf-blades elliptic, oblong, obovate or oblanceolate, usually yellow-green when dry, (5–)7–17 cm. long, 2–7·3 cm. wide, ± obtuse to distinctly shortly acuminate at the apex, cuneate at the base, often ± coriaceous; venation closely reticulate; petiole 1–1·7 cm. long; stipules 2·5–5 mm. long, the thicker deltoid part often with a few stiff hairs at its apex. Inflorescences supra-axillary (sometimes only slightly so), compact and subcapitate, several-flowered; peduncles mostly stout, compressed, 2–10 mm. long, thickened apically; bracts and bracteoles triangular, keeled, ± 1·5 mm. long, acuminate, very congested. Calyx-tube 0·7–1·5 mm. long; limb 0·9–1·4 mm. long, slightly toothed, the teeth 0·3–0·5 mm. long. Corolla white, sometimes tinged pink in bud, sweetly scented; tube 3·5–5·6 mm. long, densely hairy inside at the throat; lobes oblong-lanceolate to oblong-ovate, 3–5·8 mm. long, hairy inside at least at the base. Anthers with

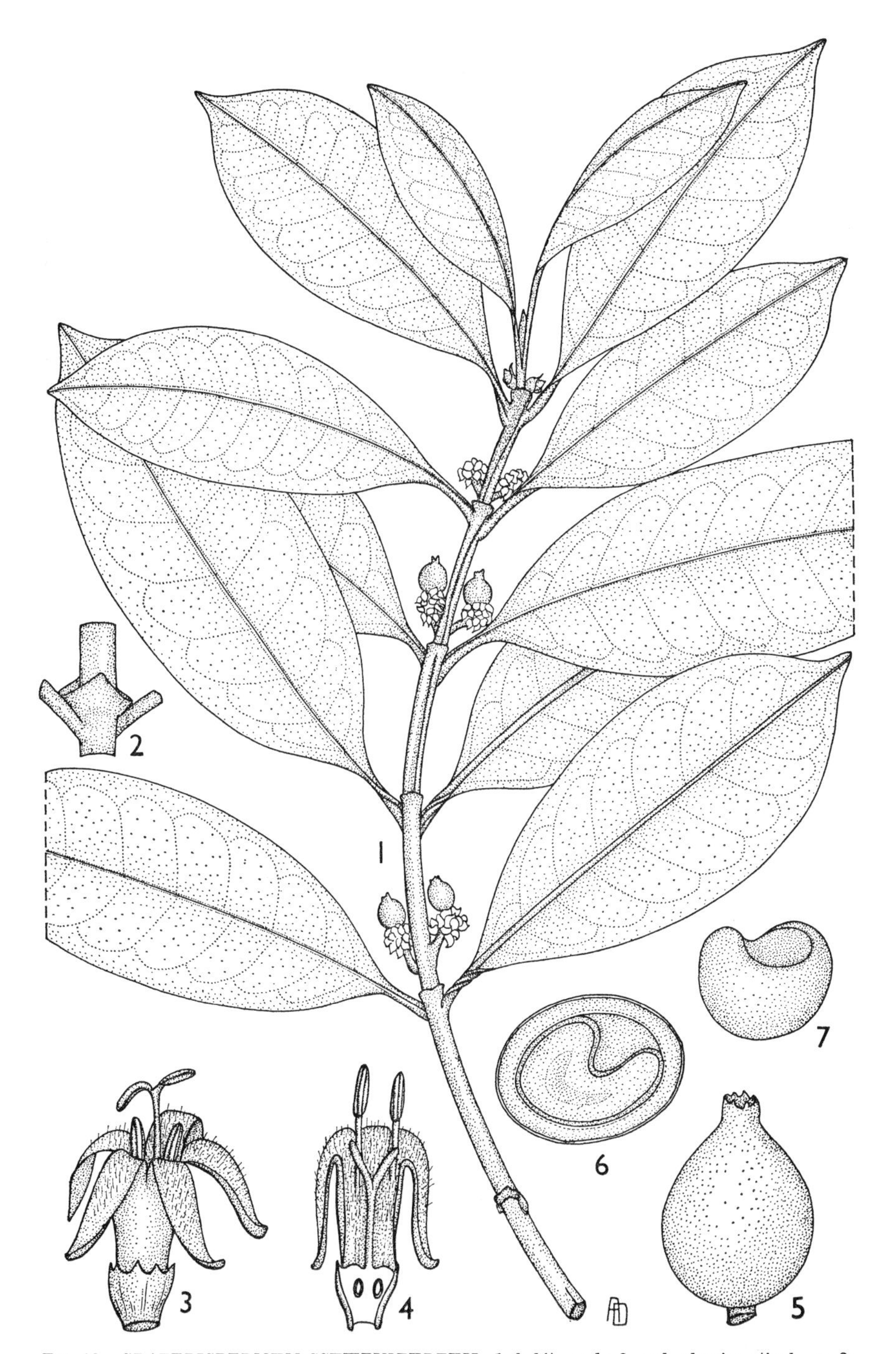

FIG. 16. *CRATERISPERMUM SCHWEINFURTHII*—**1,** habit, × ½; **2,** node showing stipule, × 2; **3,** long-styled flower, × 4; **4,** longitudinal section of short-styled flower, × 4; **5,** fruit, × 4; **6,** transverse section of fruit, × 4; **7,** seed, × 4. 1, 2, from *Ritchie* 1; 3, from *Haarer* 2170; 4, from *Lye* 3452; 5–7, from *Purseglove* 1265. Drawn by Ann Davis.

tips just exserted in long-styled flowers, completely exserted and reaching to or nearly to the tips of the corolla-lobes in short-styled flowers. Style 7–7·5 mm. long in long-styled flowers, well exserted, 2·6–4 mm. long in short-styled flowers, included; stigma-lobes linear-clavate, 1·3–2·5 mm. long. Fruit brown when dry but described as green or black, subglobose or ellipsoid, 5–6 (–7 in spirit material) mm. long and wide, sessile. Seeds dark brown, shining, bowl-shaped, longest dimension 3·4 mm., shorter dimension 2 mm., the depression deep and rounded. Fig. 16, p. 163.

UGANDA. W. Nile District: Koboko, Feb. 1934, *Eggeling* 1527 !; Bunyoro District: Bujenje, Feb. 1943, *Purseglove* 1265 !; Mengo District: Entebbe Bay, Kyiwaga Forest, 22 Sept. 1949, *Dawkins* 393 !

KENYA. N. Kavirondo District: Kakamega Forest, 3 Dec. 1962, *Ormiston* in *E.A.H.* 343/62 ! & same locality, *Holyoak* in *F.D.* 2780 ! & same locality, June 1961, *Lucas* 280 !

TANZANIA. Bukoba District: Bugandika, Sept. 1931, *Haarer* 2170 !; Ulanga District: Lukoga Forest Reserve, 4 Nov. 1961, *Semsei* 3387 !; Rungwe District: Masukulu [Mwasukulu], 18 Nov. 1912, *Stolz* 1688 !; Pemba I., Ngezi Forest, 18 Feb. 1929, *Greenway* 1482 !

DISTR. **U**1–4; **K**5; **T**1, 4, 6–8; **P**; Nigeria, Cameroun, Central African Republic, Zaire, Burundi, Sudan, Ethiopia, Mozambique, Malawi, Zambia, Rhodesia (see note) and Angola

HAB. Evergreen fringing forest of lakes and streams, swamp forest and drier evergreen forest, also in thickets, etc.; (0–)1050–1500 m.

SYN. *C. sp.* sensu T.T.C.L.: 493 (1949)
[*C. laurinum* sensu I.T.U., ed. 2: 341 (1952); Brenan in Mem. N.Y. Bot. Gard. 8: 453 (1954); K.T.S.: 437 (1961); F.F.N.R.: 405 (1962), *non* (Poir.) Benth.]

NOTE. The type of *C. schweinfurthii* is actually not " typical " of the species (although other *Schweinfurth* specimens are) and has a peduncle about 10 mm. long; specimens from Pemba have peduncles 9 mm. long. Besides the Pemba specimens there is another low altitude one (*Mgaza* 785—Bagamoyo District, Bana Forest Reserve, 29 Oct. 1965) but there seems to be no reason to separate them. The apparent absence from **T**3, Usambaras, is rather remarkable. *Jefford & Newbould* 1701 (Mpanda District, Mahali Mts., Utahya, 21 Aug. 1958) has one inflorescence with the bracts drawn out into an acumen 5–6 mm. long on each side but the other inflorescences are typical. *C. cerinanthum* Hiern, a W. African species with mostly longer slender peduncles, more slender often divided inflorescences and less coriaceous, less reticulate often more acuminate leaves, might be no more than a subspecies. Some Angolan specimens are difficult to place. *C. laurinum* (Poir.) Benth., with which the East African plant has almost invariably been identified, differs in having the fruits distinctly shortly pedicellate; also the peduncles are longer, the inflorescence often branched, the leaves blunter and the calyx-limb mostly truncate. Only one sheet from the hundreds referred to *C. schweinfurthii* has had distinct pedicels, namely *Mavi* 835 (Rhodesia, Melsetter, Lusitu R., 8 Jan. 1969), but it can scarcely be referred to *C. laurinum.*

Tribe 6. **KNOXIEAE**

Slender annual herb; flowers 4-merous, with corolla-tube 3 mm. long; ovary 2-locular; style-arms 2; fruit breaking off to leave a small cup formed of the persistent woody flanged pedicel . . . 12. **Paraknoxia**

Perennial herbs; flowers mainly 5-merous, with corolla-tube exceeding 3 mm. in length; ovary 2–5-locular; style-arms 2–5; fruit not leaving a cup-like remnant after falling . . . . . 13. **Pentanisia**

12. **PARAKNOXIA**

Bremek. in B.J.B.B. 22: 77 (1952)

Pentanisia Harv. subgen. *Micropentanisia* Verdc. in B.J.B.B. 22: 262 (1952)

Annual herbs. Leaves paired, shortly petiolate; stipules small, with several deltoid segments from a short base. Flowers small, hermaphrodite,

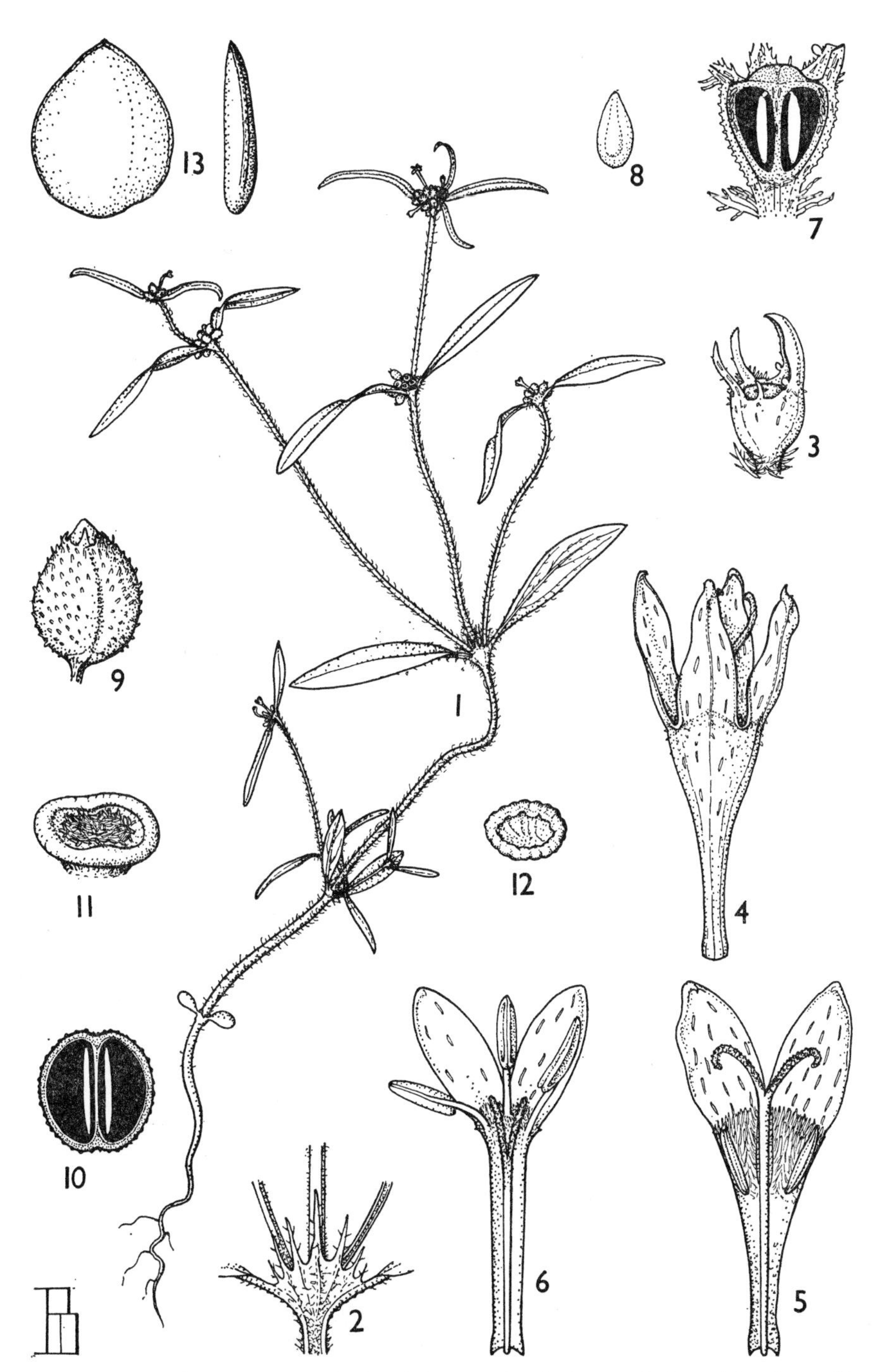

FIG. 17. *PARAKNOXIA PARVIFLORA*—**1,** habit, × 1; **2,** node showing stipule, × 4; **3,** calyx, × 14; **4,** corolla of long-styled flower, × 10; **5,** longitudinal section of long-styled flower, × 10; **6,** longitudinal section of short-styled flower, × 10; **7,** longitudinal section through ovary, × 20; **8,** ovule, × 20; **9,** fruit, × 14; **10,** transverse section of fruit, × 14; **11, 12,** pedicel-cup after fall of fruit (11 with rhaphides still present), × 14; **13,** seed, two views, × 20. 1, 10, 11, from *Polhill & Paulo* 1986; 2, from *Maitland* 1310; 3–8, from *Lewis* 5967; 9, from *Whyte*; 12, 13, from *Lugard* 99. Drawn by Diane Bridson.

dimorphic, some flowers with anthers included and style exserted and others completely vice versa, in small sessile terminal heads. Calyx-tube short; lobes small, 3–4, sometimes 1 enlarged and rest minute, or all minute. Corolla-tube narrowly funnel-shaped; lobes 3–4, oblong; throat hairy. Ovary 2-locular; ovules solitary in each locule, pendulous; style filiform; stigma bifid, the lobes filiform. Fruit ovoid, indehiscent, but longitudinally grooved and where the fruit breaks off a little cup is left being the persistent woody flanged pedicel; between this flange and the base of the fruit are masses of rhaphides. Seeds narrowly ellipsoid or ovate in outline, strongly compressed.

A monotypic genus occurring in eastern and central Africa. I formerly considered it best retained in *Pentanisia*, but agree that there are sufficient characters to separate it from that genus.

P. parviflora (*Verdc.*) *Bremek.* in K.B. 8: 439 (1953); U.K.W.F.: 405, fig. (1974). Type: Uganda, Teso District, Serere, *Chandler* 672 (EA, holo. !, BM, K, iso. !)

Erect herb 4–38 cm. tall; stems slender, often branched, hairy. Leaf-blades narrowly elliptic, lanceolate, linear or linear-oblanceolate or ovate-elliptic, 1·3–4·2 cm. long, 0·2–1(–1·7) cm. wide, ± acute at the apex, cuneate at the base, covered with short hairs; petioles up to 3 mm. long; stipules with ± 3–7 deltoid segments 1–2 mm. long from a pale chestnut-coloured base 1–1·5 mm. long. Inflorescences ± 2–4 mm. wide, supported by the apical leaf pair; there are usually 2 axillary flowering branchlets which overtop the true terminal one. Calyx-tube 1 mm. long, covered with small scaly hairs; foliaceous lobe 0·8 mm. long, 0·5 mm. wide or absent, the others very minute, crowned with hairs. Corolla white or tinged bluish mauve; tube ± 1·75–4 mm. long, dilated in long-styled flowers to 1–1·5 mm. wide for the apical third, glabrous outside; lobes ± 1·3–1·5 mm. long, 0·4–0·9 mm. wide, hairy outside. Style 3·3 mm. long in long-styled flowers, tomentose; stigma-lobes 0·6–0·8 mm. long, papillate. Fruit 1·3–1·75 mm. long, 1·25 mm. wide, somewhat acute at the apex, densely covered with scaly hairs. Seeds chestnut-coloured, 0·8–0·9 mm. long. Fig. 17, p. 165.

UGANDA. Teso District: Serere, Dec. 1931, *Chandler* 61 ! & same locality, July 1926, *Maitland* 1310 !; Mengo District: Kipayo Estate, Dec. 1913, *Dummer* 557 !
KENYA. Trans-Nzoia District: ENE. slope of Mt. Elgon, 23 Sept. 1962, *Lewis* 5967 ! & Kitale Grassland Research Station, 2 Oct. 1959, *Verdcourt* 2454 !; N. Kavirondo District: Kitosh area, 25 July 1951, *Greenway & Doughty* 8523 !
TANZANIA. Mbeya District: 261 km. S. of Iringa, between Igawa and Chimala on Great North Road, 2 Apr. 1962, *Polhill & Paulo* 1986 !; Iringa, just E. of College of National Education, 30 Mar. 1972, *Pedersen* 893 !; Songea District: about 5 km. E. of Songea, Unangwa Hill, 22 Mar. 1956, *Milne-Redhead & Taylor* 9281 !
DISTR. **U**2–4; **K**3, 5; **T**1, 4, 7, 8; Central African Republic, Zaire, Malawi, Rhodesia and Angola
HAB. Grassland, bushland, old cultivations, seasonal swamps, often on shallow damp soil on rocks; 840–2280 m.

SYN. *Pentanisia parviflora* Verdc. in K.B. 6: 383 (Jan. 1952) & in B.J.B.B. 22: 263, fig. 31/H, I (1952)
Paraknoxia ruziziensis Bremek. in B.J.B.B. 22: 77 (June 1952). Type: Zaire, Ruzizi plain, *Germain* 6684 (BR, holo. !, EA, iso.)
P. parviflora (Verdc.) Verdc. in B.J.B.B. 22: 265 (1952) (nomen eventuale)

NOTE. *Chandler* 794 (Uganda, Teso District, Serere, July 1931) is a remarkably robust form with stouter stems and very much wider leaf-blades up to 4 × 1·7 cm.

13. PENTANISIA

Harv. in Hook., Lond. Journ. Bot. 1: 21 (1842); Verdc. in B.J.B.B. 22: 233 (1952)

Perennial herbs or subshrubs, with glabrous or hairy erect, procumbent or decumbent stems. Leaves variable, small to moderate, linear to round, paired or rarely in whorls of 3, sessile or shortly petiolate; stipules with base connate with the petiole, apically fimbriate. Flowers mostly blue, small to medium-sized, hermaphrodite, dimorphic, some flowers with anthers included and style exserted and others completely vice versa, usually in few- to many-flowered terminal capitate inflorescences, frequently becoming more elongate in fruit, often spicate. Calyx-tube ovoid or ± rectangular; lobes mostly 5, 1–3 enlarged and often foliaceous, the rest small or even obsolete. Corolla-tube narrowly cylindrical, with an apical cylindrical dilation in long-styled flowers but narrowly funnel-shaped in short-styled flowers; lobes mostly 5, ovate to oblong; throat densely hairy. Ovary 2–5-locular, each locule with 1 pendulous ovule attached close to the apex; style filiform, the stigma divided into 2–5 filiform lobes corresponding with the number of locules in the ovary. Fruit dry, globular, slightly lobed, ovoid or compressed obcordate, indehiscent or tardily dehiscent into mericarps, or subglobose and rather succulent. Seeds small, compressed.

A genus of 15* species confined to tropical Africa and Madagascar; 8 are known from the Flora area. Species 1 and 2 belong to subgen. *Pentanisia* section *Pentanisia*, 3 and 4 to subgen. *Pentanisia* section *Axillares* Verdc., species 5 and 6 to subgen. *Holocarpa* (Bak.) Verdc. and species 7 and 8 to subgen. *Ouranogyne* Verdc.

Ovary 2-locular, very unusually abnormally 3-locular:
  Stems erect or procumbent, caespitose from a woody rootstock; main inflorescences terminal, although axillary ones are often present as well:
    Foliaceous calyx-lobes 0·3–1·2 cm. long; fruit black, compressed obcordate, with woody walls . . . . 1. *P. prunelloides*
    Foliaceous calyx-lobes 1–2·5(–3·5) mm. long; fruit ovoid, with thin walls . . . . 2. *P. schweinfurthii*
  Stems long, weak and straggling; rootstocks not distinctly woody; flowers in 2 axillary pedunculate inflorescences at the terminal node and often at the lower nodes also:
    Inflorescences present only at the apical nodes, spicate when in fruit but capitate during the flowering period; distinctly foetid when crushed . . . . 3. *P. foetida*
    Inflorescences spicate, present at most nodes; not known to be foetid when crushed . . . 4. *P. monticola*
Ovary 3–5-locular:
  Fruit black (rarely brown outside Flora area), globose, succulent; leaves either shorter, ovate-oblong or glabrous; stems often caespitose; indumentum of corolla-tube longer:
    Leaves narrowly elliptic, elliptic-oblong or lanceolate; stems mostly erect . . . . . *P. sykesii*

* The transference of *P. parviflora* Verdc. to a separate genus *Paraknoxia* Bremek. is upheld in this work—see genus No. 12.

Leaves mostly ovate or ovate-oblong; stems mostly decumbent in Flora area . . . 6. *P. arenaria*

Fruit chestnut-brown, globose, hard, usually shallowly lobed; leaves lanceolate, 1–10 cm. long, usually hairy; stems not caespitose; indumentum of corolla-tube short and stubby:

Corolla-tube 0·8–2·3 cm. long; foliaceous calyx-lobe 0·8–1·5(–2) mm. wide . . . . 7. *P. ouranogyne*

Corolla-tube 2·7–4 cm. long; foliaceous calyx-lobe 1–2·5 mm. wide . . . . 8. *P. longituba*

1. **P. prunelloides** (*Eckl. & Zeyh.*) *Walp.*, Repert. 2: 941 (1843); O. Kuntze, Rev. Gen. Pl. 3: 122 (1898); Schinz, Viert. Nat. Ges. Zürich 68: 437 (1923); Verdc. in B.J.B.B. 22: 248 (1952). Type: South Africa, Cape Province, Phillipstown, Mt. Winter and Katrivier Mts., *Ecklon & Zeyher* 2301 (K, PRE, W, iso.!)

Erect or semi-prostrate to completely prostrate herb 7–60 cm. tall, with mostly numerous hairy or glabrous stems from a thick woody rootstock up to 7 cm. wide. Leaf-blades oblong-elliptic or ovate to almost round, 1·3–8·5 cm. long, 0·2–3·5 cm. wide, acute or subobtuse at the apex, rounded to subcordate at the base, mostly pubescent to densely villous, less often glabrous; petiole obsolete; stipules with ± 5 narrow segments, 3–7 mm. long, 0·3–2·5 mm. wide, from a short base to 3 mm. long. Inflorescence capitate, ± villous, branched, or spicate below, up to 7 cm. long, 4 cm. wide; peduncle 3·5–34 cm. long, often densely hairy. Calyx-tube ± 1 mm. long, 0·8–1 mm. wide, pubescent; lobes unequal, 1–2 larger and foliaceous, 0·3–1·2 cm. long, 0·5–1·5(–2) mm. wide, rest small. Corolla-tube 1·1–1·8 cm. long, 1–1·5(–2) mm. wide at the apex, pubescent outside; lobes 2·5–5 mm. long, 1–2·5 mm. wide; throat densely hairy. Style in long-styled flowers exserted 2–3 mm.; stigma 2-fid, the lobes filiform, 1–1·5 mm. long. Fruit obcordate, laterally compressed, black, 2·5–4·5 mm. long, 2–3·3 mm. wide, composed of 2 indehiscent cocci which eventually separate, pubescent. Seeds brown, almost round in outline, 2·5 mm. long, 2 mm. wide, thin.

SYN. *Declieuxia prunelloides* Eckl. & Zeyh., Enum. Pl. Afr. Austr.: 363 (1837)
Diotocarpus prunelloides (Eckl. & Zeyh.) Hochst. in Flora 26: 71 (1843)

subsp. **latifolia** (*Hochst.*) *Verdc.* in B.J.B.B. 22: 250 (1952). Type: South Africa, Natal, Tafelberge, *Krauss* 249 (K, W, iso.!)

Leaves broadly ovate to round, or less often oblong, 1·6–8.5 cm. long, 1·1–3·5 cm. broad, usually densely hairy; ratio of leaf-blade length to breadth 0·8–2·8.

TANZANIA. Ufipa District: Sumbawanga to Mbisi, Jan. 1950, *Bullock* 2354! & Chala Mt., 10 Dec. 1956, *Richards* 7202! & Sumbawanga, 29 Nov. 1954, *Richards* 3412!
DISTR. T4; Malawi, Zambia, NE. South Africa
HAB. Grassland with scattered trees, rocky ground, etc.; 1800–2100 m.

SYN. *Declieuxia latifolia* Hochst. in Flora 26: 70 (1843)
Pentanisia prunelloides (Eckl. & Zeyh.) Walp. var. *latifolia* (Hochst.) Walp., Repert. 2: 941 (1843)
P. variabilis Harv. var. *latifolia* (Hochst.) Sond. in Fl. Cap. 3: 24 (1865)
P. longisepala K. Krause in E.J. 39: 532 (1907). Type: South Africa, Cape Province, Pondoland, Klein Kraa, *Bachmann* 1304 (B, holo. †)

NOTE. Subsp. *prunelloides*, with narrow leaves, occurs in South Africa and Mozambique.

2. **P. schweinfurthii** *Hiern* in F.T.A. 3: 131 (1877); Verdc. in B.J.B.B. 22: 254 (1952); U.K.W.F.: 405, fig. (1974). Type: Sudan, Dar Fertit, S. from the Gudjo, *Schweinfurth* 8 (K, holo.!, BM, W, iso.!)

Perennial pyrophytic herb, 3·8–24 cm. tall, from a woody rootstock; stems up to 25 from each root (about 3–5 from each apical branch of the rootstock), glabrous or bifariously hairy. Leaf-blades very variable, the lower round, elliptic or elliptic-obovate, the upper linear to round, 0·3–5·5 cm. long, 0·2–1·8(–2) cm. wide, acute to obtuse at the apex, cuneate at the base, mostly drying yellowish green, glabrous or rarely shortly hairy; petiole up to 2 mm. long; stipules with 2–4 deltoid lobes or 1 trifid lobe, 1–4·5(–7) mm. long from a base 1–3 mm. long. Inflorescences capitate, 1–2·5 cm. long, 0·3–1·3 cm. wide or branched and spike-like, 2–3·5 cm. long; peduncle up to 7·4 cm. long. Calyx-tube squarish, 0·6–1(–1·5) mm. long, 0·5–1(–1·5) mm. wide, glabrous or sometimes covered with bristly white hairs; lobes unequal, the longest 1–3·5 mm. long, 0·3–0·8 mm. wide, the rest minute. Corolla bright blue, white, pale lilac or purple; tube 0·6–1·3 cm. long, 1·2–2(–3) cm. wide at the apex, glabrous or pubescent with short white hairs; throat densely hairy; lobes ovate-oblong to ovate-lanceolate, 2–6 mm. long, 1–2 mm. wide. Style exserted 2–4 mm. in long-styled flowers; stigma-lobes linear, 0·5–1·5(–2) mm. long. Fruiting inflorescence spicate, up to 4 cm. long; fruits ovoid, broadest at the middle, 1·5–2·5 mm. long, 1·5–2 mm. wide, glabrous or covered with white pubescence, 2-locular, the locules often unequal, thin-walled, not ribbed, borne on ledges on the rhachis. Seeds yellow-brown, broadly elliptic in outline, concavo-convex, 2 mm. long, 1·5 mm. wide, 0·6 mm. thick, finely marked with brown.

UGANDA. W. Nile District: Koboko, July 1940, *Purseglove* 978! & Ayivu, 7 Jan. 1964, *Oakley* 2; " Elgon District ", *Evan James*! (possibly in Kenya)
KENYA. Trans-Nzoia District: S. Cherangani, 15 Feb. 1958, *Symes* 278! & Kitale, 18 Apr. 1953, *Bogdan* 3719!; N. Kavirondo District: Kakamega, Feb. 1944, *Carroll* H2!
TANZANIA. Ufipa District: on track by Katete village, 14 Dec. 1956, *Richards* 7273!; Mbeya District: Mbozi, 30 Aug. 1933, *Greenway* 3645!; Songea District: Matengo Hills, Mpapa, 1 Oct. 1956, *Semsei* 2496!
DISTR. **U**1, ?3; **K**2, 3, 5; **T**1, 4, 7, 8; widespread in tropical Africa from Nigeria and Sudan to Zaire, south to Angola and Rhodesia
HAB. Grassland, grassland with scattered trees, also *Brachystegia* and similar woodlands; always in areas subject to burning; 840–2250 m.

SYN. *P. rhodesiana* S. Moore in J.B. 40: 252 (1902). Type: Rhodesia, Salisbury, *Rand* 575 (BM, holo.!)
P. sericocarpa S. Moore in J.B. 40: 251 (1902). Type: Rhodesia, Salisbury, *Rand* 619 (BM, holo.!)
P. crassifolia K. Krause in E.J. 39: 531 (1907). Type: Rhodesia, Salisbury, Norton, *Engler* 3022 (B, holo. †)
[*P. variabilis* sensu auctt., *non* Harv.]

NOTE. Var. *puberula* Verdc. (B.J.B.B. 22: 258 (1952); type: Malawi, 25·6 km. from Fort Hill towards Katumbi Camp, *B. D. Burtt* 6095 (K, holo.!)) is a minor variant scarcely worth retaining; *Richards* 20576 (Tanzania, Namwele, Itala Hills, 23 Oct 1965) is this variety. *Richards* 2382 (Tanzania, Ufipa District, Sumbawanga, 28 Nov. 1954) shows occasional 3-fid stigmas.

3. **P. foetida** *Verdc.* in K.B. 6: 381 (1952) & in B.J.B.B. 22.: 259, fig. 32/C–E (1952); U.K.W.F.: 405 (1974). Type: Kenya, Limuru, Tigoni Dam, *Rayner* 421 (EA, holo.!, BM, BR, FI, LISC, K, P, U, iso.!)

Perennial herb, branched at the base, with several subprostrate hairy stems, up to 60 cm. long, which are erect at their apices. Most parts of the plant smell quite strongly of carbon disulphide when crushed or extracted with alcohol. Leaf-blades ovate to ovate-lanceolate, 3–8 cm. long, 1·5–2·9 cm. wide, acute at the apex, cuneate at the base, pubescent on both surfaces; petioles 0·4–1·2 cm. long; stipules with 5–7 subequal segments 2–5·5(–9) mm. long, from a base 2–4 mm. long. Inflorescences axillary, the flowers at first arranged in compact heads, but later extending and becoming spikes up to

5 cm. long; peduncles 0·8–7 cm. long; pedicels 0·3 mm. long. Flowers lilac, white, pinkish red or purple, dimorphic. Calyx-tube ovoid, 1–1·5 mm. long, 0·75–1 mm. wide, glabrescent; lobes 5, 1 foliaceous, 4–6 mm. long, 1–2 mm. wide, ciliate, the rest minute, hairy, ± 0·5 mm. long, Corolla-tube 5–9·5 mm. long, ± 1 mm. wide at the base, 1·5–2 mm. wide at the apex, pubescent outside, hairy inside the throat; lobes 5–6, oblong-lanceolate,(1·5–)2·5–3 mm. long, ± 1–1·5 mm. wide, glabrous inside, pubescent outside, the apices somewhat inflexed. Style in long-styled flowers exserted 1–3 mm.; stigma bifid, the lobes filiform, 1·5 mm. long, Fruit ovoid, brownish black, 2·5 mm. long, 2·5–3·5 mm. across, somewhat narrowed at the apex, divided into 2 indehiscent cocci between which there are numerous rhaphides. Seeds brown, elliptic, thin, 1·2 mm. long, 0·5 mm. wide.

KENYA. Kiambu District: Limuru, 16 Mar. 1933, *Mainwaring* in *Napier* 2582!; Meru District: NW. Mt. Kenya, Marimba Forest, 14 Oct. 1960, *Verdcourt & Polhill* 2992! & Nyambeni Hills Tea Estate, 8 Oct. 1960, *Polhill & Verdcourt* 260!
TANZANIA. Moshi District: Kilimanjaro, Rongai, 25 Dec. 1932, *Geilinger* 4948!
DISTR. **K**4; **T**2; not known elsewhere
HAB. Grassland, forest edges and clearings, bracken and other scrubland; 1830–2300 m.

4. **P. monticola** (*K. Krause*) *Verdc.* in K.B. 7: 362 (1952) & in B.J.B.B. 22: 260, fig. 31/J–K (1952) Type: Tanzania, Bundali Mts. and near Rungwe, *Stolz* 1286 (B, holo. †, K, U, W, iso.!)

Perennial erect weak herb 0·9–1·2 m. tall, with striate stems dichotomously branched at many of the nodes, glabrescent save for nodes and young shoots. Leaf-blades lanceolate to ovate-lanceolate, 4·6–10 cm. long, 1·1–3·3 cm. wide, acute at the apex, rounded to cuneate at the base, with long adpressed pubescence above and on the nerves beneath; petiole 1–4 (–7) mm. long; stipules with 5–6 setae, often reflexed when older, 4–9 mm. long from a base 1–2 mm. long. Flowers pinkish white or very pale purple, dimorphic, in slender axillary spikes, 1–5 cm. long, 2 from most nodes even the lowest, usually those where dichotomous branching occurs; peduncle 1–9·5 cm. long. Calyx-tube ovoid, 1 mm. long, 0·8–1 mm. wide; largest lobes 1·5–2(–3) mm. long, 0·5(–1) mm. wide, rest minute. Corolla-tube 6–8(–10) mm. long, ± 0·5 mm. wide at the base, 1·2–2 mm. wide at the apex, hairy inside; lobes 2 mm. long, 0·9 mm. wide. Style in long-styled flowers exserted 2–2·2 mm.; stigma bifid, the lobes filiform ± 1 mm. long. Fruit globose-reniform, cordate at the base, 1·5–2 mm. tall, 1·8–2 mm. across, 2-locular, each coccus oblique when separated, the interface densely covered with rhaphides, thin-walled. Seeds yellow-brown, ellipsoid, 1·4 mm. long, 0·8 mm. wide.

TANZANIA. Ufipa District: Sumbawanga, Chapota, 7 Mar. 1957, *Richards* 8539! & same area, Malonje Farm, 14 Mar. 1957, *Richards* 8719! & same area, near Zambian border, Kito Mt., 21 Apr. 1961, *Richards* 15044!
DISTR. **T**1 (see note), 4, 7; Zambia, Malawi
HAB. Coarse grassland, bushland, woodland edges, evergreen forest; 1600–2100 m.

SYN. *Otomeria monticola* K. Krause in E.J. 57: 26 (1920)

NOTE. *Tanner* 4603 (Tanzania, Ngara District, 10 Dec. 1959) almost certainly is no more than a variant of this species, differing in its hairier stems, longer calyx-lobes (up to 4 mm.) and longer corolla-lobes (up to 3 mm. long).

5. **P. sykesii** *Hutch.* in K.B. 1906: 248 (1906); Verdc. in B.J.B.B. 22: 266, fig. 32/H (1952). Type: Zambia, Batoka Plateau, *Sykes* in *Herb. Allen* 225 (K, holo.!, SRGH, iso.!)

Perennial herb with 5–20 erect, spreading or prostrate stems 8–30(–40) cm. tall from a woody rootstock ± 3 cm. in diameter, or sometimes forming mats 1·2 m. in diameter; stems narrowly bifariously hairy or scaly puberulous. Leaves paired or sometimes appearing verticillate due to short axillary branchlets, the lowest ovate, the rest lanceolate, narrowly elliptic, elliptic-oblong or almost oblanceolate, 0·65–5·9 cm. long, 0·2–1·5 cm. wide, obscurely to distinctly acute at the apex, cuneate at the base, glabrous or at most puberulous on the lateral nerves beneath, often drying yellow-green; petiole up to 2 mm. long, adnate to the stipular sheath; stipules with 3 flat linear or rarely spathulate lobes, 0·2–1·2 cm. long from a base 1·5–4 mm. long. Inflorescence dense, capitate or spicate, sometimes branched, 1·5–3 cm. long and wide; peduncle 2·5–10·5 cm. long. Flowers blue or rarely white, dimorphic. Calyx-tube blackish, quadrate, 1–1·2 mm. long and wide, glabrous; lobes unequal, the 1–2 foliaceous ones lanceolate, ovate-lanceolate or oblong-lanceolate, 0·3–1·03 cm. long, 0·9–3 mm. wide, the rest small, ± 1 mm. long, all glabrous. Corolla-tube (0·7–)1·05–1·7 cm. long, 0·3–1·2 mm. wide at the base, 1–2·5 mm. wide at the throat for a distance of 2–5 mm. in long-styled flowers, the dilation sometimes constricted at the throat so its orifice is narrow, glabrous to scaly puberulous outside; lobes broadly elliptic, ovate-oblong or oblong, (3–)4–6 mm. long, 1–3·5 mm. wide, hairy outside; throat densely hairy. Ovary and fruit 3–5-locular; style in long-styled flowers exserted 2·5–4 mm.; stigma-lobes 2–5, filiform, 0·5–1 mm. long. Fruit black, globose, ± fleshy and succulent, 4·5–8 mm. long, 3·1–5·5 mm. across, glabrous, narrowed at the apex, very shallowly lobed, crowned by the persistent calyx-lobes; walls thickish and woody, with regularly spaced channels just within the calyx-tube layer. Seeds yellow, ovoid, 2·5 mm. long, 1·5 mm. wide, laterally much compressed, with a reddish brown pattern.

subsp. **sykesii**; Verdc. in B.J.B.B. 22: 266, fig. 32/H (1952)

Lateral nerves 4–6 pairs, prominent on both surfaces.

TANZANIA. Ufipa District: new Sumbawanga road to Rukwa Escarpment, 5 Feb. 1962, *Richards* 15991! & on slopes of Kito Mt., 13 Sept. 1956, *Richards* 6182!
DISTR. **T4**; Zambia, Malawi
HAB. Coarse roadside grassland and dry burnt rocky ground; 1500 m.

NOTE. Subsp. *otomerioides* Verdc. occurs in Rhodesia and the N. Transvaal.
In the Ufipa District of Tanzania, *P. sykesii* and *P. arenaria* merge together, e.g. *Bullock* 1996 (Chapota, 3 Dec. 1949) has both the ovate broad leaves of *P. arenaria* and narrower leaves of *P. sykesii*. Over much of their range, however, these species are distinct and it is practical to maintain them separate. A specimen from Zambia, *Astle* 1436, has buff fruits 9 mm. long, 8 mm. wide, with thick spongy walls and angular seeds 4 mm. long, 1·7 mm. wide.

6. **P. arenaria** (*Hiern*) *Verdc.* in K.B. 7: 361 (1952) & in B.J.B.B. 22: 270, fig. 32/B, G (1952). Type: Angola, Pungo Andongo, near Lombe and Candumba, *Welwitsch* 5312 (LISU, holo. !, BM, COI, iso. !)

Perennial herb with caespitose, erect, decumbent or prostrate hairy stems 7–45 cm. long. Leaves paired or in whorls of 3; blades ovate, ovate-oblong or ovate-lanceolate, 1·8–5·5(–8·5) cm. long, 0·4–2·5 cm. wide, acute at the apex, ± rounded at the base, hairy or glabrescent, dark green; petiole obsolete or up to 3 mm. long; stipules with 5 setae up to 0·8–1·2 cm. long from a base 3 mm. long. Inflorescences capitate, simple or branched, 1·5–2(–4) cm. long, 1–2(–3) cm. wide; peduncle 0–9 cm. long. Flowers purplish violet, dimorphic. Calyx-tube quadrate, 1·2–1·3 mm. long, 1–2 mm. wide, hairy; lobes unequal, the foliaceous ones ovate to linear-lanceolate, 2–7 mm. long, 0·8–3 mm. wide, rest small. Corolla-tube 0·65–1·5(–1·7) cm. long, 0·5–1 mm. wide at the base, 1·5–2 mm. wide at the apex, densely pubescent outside; throat densely hairy within; lobes 4–5, round to oblong, 3–6·5 mm. long, 1–4 mm. wide, pubescent outside. Ovary and fruit 4–5-locular; style exserted 1·5–2 mm. in long-styled flowers; stigma 3–5-lobed, the lobes filiform. Fruit black, globose, probably succulent, 5 mm. long, 3–4 mm. across, hairy, somewhat narrowed to the apex, a little furrowed; walls woody and containing channels. Ripe seeds not seen.

TANZANIA. Ufipa District: Sumbawanga, Malonje, 27 Nov. 1954, *Richards* 2361! & Malonje Plateau, Mmemya Mt., 27 Oct. 1965, *Richards* 20628! & Chapota, 3 Dec. 1949, *Bullock* 1996!

DISTR. **T4**; Angola

HAB. Upland grassland, burnt rocky hillsides; 1950–2250 m.

SYN. *Pentacarpaea arenaria* Hiern, Cat. Afr. Pl. Welw. 2: 439 (1898)
Pentanisia pentagyne K. Schum. in E.J. 33: 350 (1903). Type: Angola, Malange, *Mechow* 279 (B, lecto. †, K, W, iso. !)

NOTE. As has been explained above this species grades with *P. sykesii*, but at the other end of its range it merges with *P. renifolia* Verdc. (type: Zaire, Lubumbashi [Elisabethville], *Quarré* 4808 (BR, holo. !)). Until more is understood about the variation in these three it would serve no useful purpose to unite them since they are for the most part easily distinguishable.

7. **P. ouranogyne** *S. Moore* in J.B. 18: 4 (1880); Jex-Blake, Gard. E. Afr., ed. 3, t. 10/2 (1949); Verdc. in B.J.B.B. 22: 274, figs. 31/A–E, 32/F (1952); U.K.W.F.: 405, fig. (1974). Type: Kenya, Kitui, *Hildebrandt* 2754 (BM, holo. !, K, W, iso. !)

Perennial branched or unbranched herb 3·5–60 cm. tall; stems usually with shaggy spreading white hairs, glabrescent beneath, pale yellow-brown to chestnut; main rootstock usually slender, rarely very woody. Leaf-blades lanceolate to linear-lanceolate, (1–)3–10 cm. long, (0·25–)0·4–1·8(–2·7) cm. wide, acute at the apex, cuneate at the base, mostly with bristly hairs on the venation on both surfaces, more rarely densely hairy or quite glabrous; petiole up to 1 cm. long, adnate to the stipular sheath; stipules ± membranous, pale chestnut or whitish with pale brown streaks, sheathing the stem, with ± 7 bristly hairy setae (2–)5–9·5 mm. long from a base 3–8 mm. long. Inflorescence a dense terminal head, sometimes branched, 1·3–4·5 cm. wide, not elongating to any extent; peduncle 0–16 cm. long. Flowers bright blue, rarely pink, often drying with a coppery tinge. Calyx with long hairs; tube 1–2 mm. long, 1–1·8 mm. wide; foliaceous lobes lanceolate, 3–10 mm. long, 0·8–1·5(–2) mm. wide, pubescent, the rest setiform ± 1–3 mm. long. Corolla-tube 0·7–2·3 cm. long, 0·75–1 mm. wide at the base, 1–2(–3) mm. wide at the apex, covered with very characteristic short papilla-like subglobular hairs; lobes (2–)3–5·5 mm. long, 0·8–2(–2·5) mm. wide, mostly hairy outside. Ovary and fruit 2–5-locular (predominantly 3–4); style exserted 0–3(–4) mm. in long-styled flowers; stigma 2–5-lobed, 1–2 mm. long, the lobes filiform. Fruit reddish brown, often slightly shining, globose, woody, 2·5–3·5(–5) mm. long, 2·5–4(–5) mm. across, pubescent with long hairs, broadest at the base, distinctly furrowed, the walls very thick and woody round the locules but with patches of honeycomb-like tissue just beneath the ridges. Seeds pale brown, elliptic, thin, 3 mm. long, 1·5 mm. wide, 0·2 mm. thick. Fig. 18.

UGANDA. Acholi District: Madi, Opei, Apr. 1963, *Purseglove* 1366!; Karamoja District: near Nabilatuk, Moruangaberu, 9 July 1956, *Dyson-Hudson* 19!; Mbale District: Bugishu, Cheptui, 10 Nov. 1933, *Tothill* 2235!

KENYA. Turkana District: Naitamaiong [Naitamajong] Hill, June 1934, *Champion* 326!; Kiambu/Machakos District: R. Athi at Fourteen Falls, 26 May 1963, *Verdcourt* 3631!; Teita District: Manyani, Mudanda Rock, *Hucks* 16!

TANZANIA. Shinyanga District: on road to Uduhe [Uduhi], Uchunga, 22 Jan. 1936, *B. D. Burtt* 5506!; Arusha District: Ngare Nanyuki road, 15 Dec. 1968, *Richards* 23360!; Iringa District: Great North Road, Ibumu, 16 Dec. 1961, *Richards* 15663!

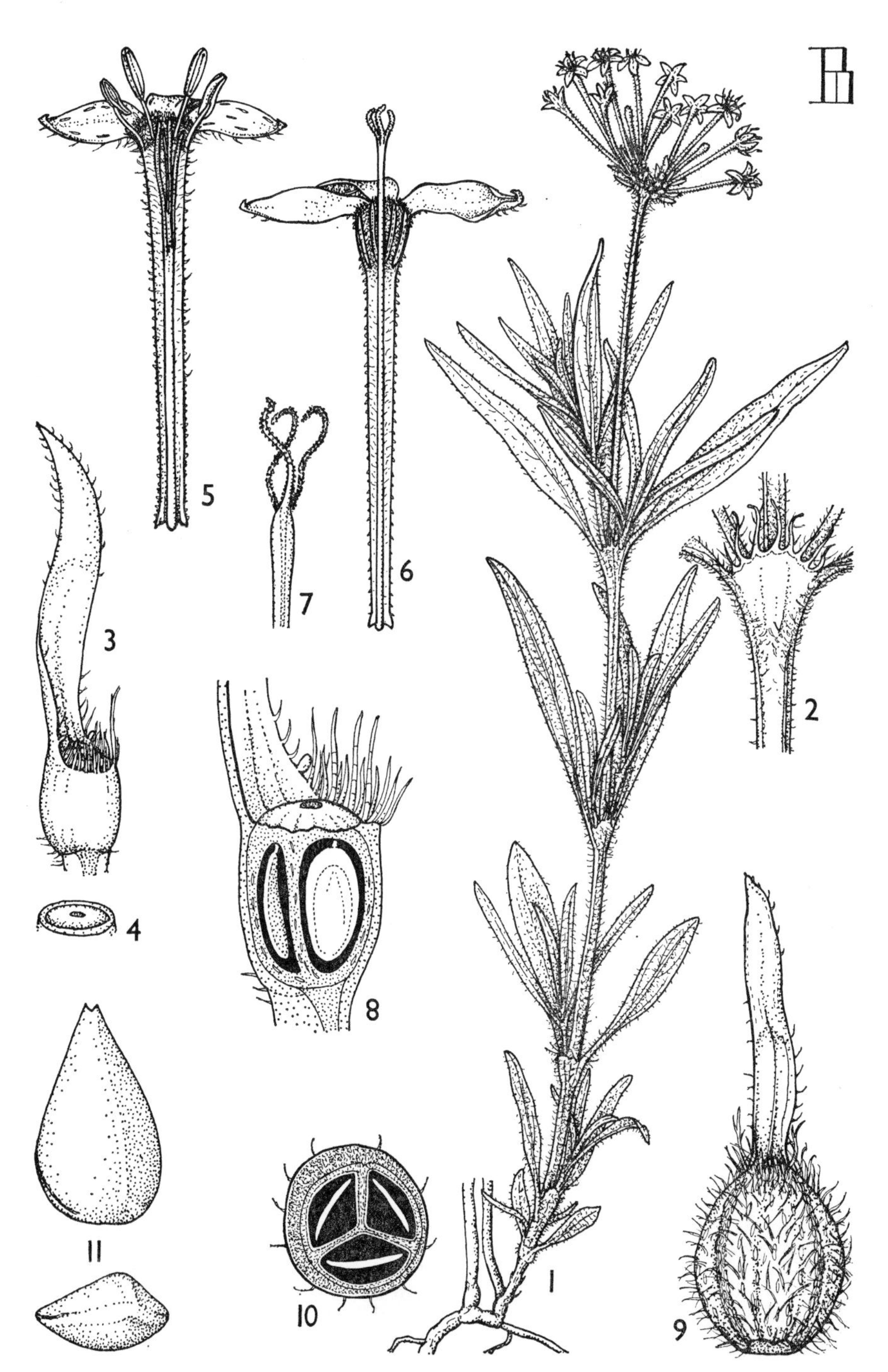

FIG. 18. *PENTANISIA OURANOGYNE*—**1,** habit, × ⅔; **2,** node showing stipule, × 2; **3,** calyx, × 10; **4,** disc, × 10; **5,** longitudinal section of short-styled flower, × 3; **6,** same of long-styled flower, × 3; **7,** tip of style and stigma-lobes, × 6; **8,** longitudinal section of ovary, × 20; **9,** fruit, × 6; **10,** transverse section of fruit, × 6; **11,** seed, two views, × 10. 1, 5, from *Richards* 23360; 2, from *Dyson Hudson* 19; 3, 4, 6–8, 11, from *Lewis* 5989; 9, 10, from *Symes* 573A. Drawn by Diane Bridson.

DISTR. **U**1, 3; **K**1–7; **T**1–3, 5–7; Somali Republic, Ethiopia
HAB. Grassland, grassland with scattered trees, bushland, *Acacia* woodland, often seasonally burnt, also in cultivations; 550–2415 m.

SYN. *P. ouranogyne* S. Moore var. *glabrifolia* Cufod. in Nuov. Giorn. Bot. Ital., n.s. 55: 87 (1948). Types: Ethiopia, El Dire, *Corradi* 2731, 2732 & 2733 (FI, syn.)

NOTE. Several specimens from **T**2 are entirely glabrous, e.g. Moshi, 12 Oct. 1925, *Haarer*.

8. **P. longituba** (*Franch.*) *Oliv.* in F. J. James, Unknown Horn of Africa, App.: 319 (1888); Verdc. in B.J.B.B. 22: 277 (1952). Type: Somali Republic, without exact locality, *Révoil* (P, holo.!)

Erect herb, branched and sometimes woody at the base, often with white papery bark, 16–60 cm. tall, very similar to the previous species but with larger flowers; rootstock often practically unbranched. Leaves paired but sometimes appearing pseudoverticillate due to arrested axillary branchlets; blades lanceolate or oblong-lanceolate to linear-lanceolate, (2·5–)3·7–7·5 cm. long, 0·35–1·45(–2·3) cm. wide (sometimes appearing narrower when the margins are revolute), acute at the apex, cuneate at the base, glabrescent or with short white bristly hairs above and beneath; petiole 0–0·7 mm. long; stipules with ± 7 setae, 0·8–4 mm. long from a base 2–5 mm. long. Inflorescence a dense terminal head 0·7–2·3 cm. wide, not elongating to any extent; peduncle 0·4–5·8(–7·8) cm. long. Flowers white or pale blue. Calyx pubescent; tube 2–3(–4) mm. long, 1·5–2·5 mm. wide; the foliaceous lobe lanceolate, 0·6–2 cm. long, 1–2·5 mm. wide, the rest reduced to tufts of hairs. Corolla-tube 2·5–4·3 cm. long, 1·5–2·5(–3) mm. wide at the apex, 1·5 mm. wide at the base, mostly with stubby indumentum similar to *P. ouranogyne*; lobes oblong-elliptic, (3–)4·5–7·5 mm. long, (1–)1·5–3 mm. wide. Ovary and fruit 3–5-locular; style exserted 2–4 mm. in long-styled flowers; stigma 3–5-fid, the lobes filiform, 1–2 mm. long. Fruit very similar to that of *P. ouranogyne*, very pale brown, globose, 4·5–5·5 mm. long, 4–6 mm. across, the walls thick and woody. Seeds as in *P. ouranogyne*.

KENYA. Northern Frontier Province: Furroli, 12 Sept. 1952, *Gillett* 13806! & Marsabit, June 1959, *T. Adamson* 16!
DISTR. **K**1; Ethiopia and Somali Republic
HAB. Open desert to *Commiphora*-*Acacia* bushland on lava plains; 960–1050 m.

SYN. *Knoxia* (*Pentanisia*) *longituba* Franch. in Révoil, Faune et Flore des Pays Çomalis: 32 (1882)
Pentanisia pentasiana Mattei in Boll. Ort. Bot. Palermo 7: 189 (1908); Cufod. in Phyton 1: 145 (1949). Type: Somali Republic (S.), Goscia, *Macaluso* G.134 (PAL, holo.!)

NOTE. There is no doubt that this is very closely related to *P. ouranogyne*, differing in little more than the longer corolla, wider foliaceous calyx-lobes and sometimes in flower colour. It would probably be best to treat it as a subspecies. Gillett on his field notes to 4947 and 4948 (collected in Somali Republic (N.), Dobo Pass, Feb. 1933) has emphasised the problem; 4948 is noted as larger all round than 4947 and paler in flower colour but " likely not distinct, intermediates occur in size and colour ". In the major part of its range, however, *P. ouranogyne* is remarkably constant in corolla size and colour and I prefer to maintain the two distinct with the suggestion that the intermediates are hybrids, which probably occur wherever the two species overlap.

Tribe 7. PAEDERIEAE

One genus only . . . . 14. **Paederia**

14. PAEDERIA

L., Mant. Pl.: 7, 52 (1767), *nom. conserv.*
Hondbessen Adans., Fam. Pl. 2: 158 (1763)
Lecontea A. Rich. in DC., Prodr. 4: 470 (1830) & Mém. Fam. Rub.: 115 (1830) & Mém. Soc. Hist. Nat. Paris 5: 195 (1834)
Siphomeris Bojer, Hort. Maurit.: 170 (1837), *nom. superfl.**

Straggling or climbing usually foetid-smelling shrubs, with mostly slender flexuous stems, more rarely suberect. Leaves mostly petiolate, opposite or in whorls of 3; stipules intrapetiolar, ovate to narrowly triangular, not divided into lobes or sometimes bifid where forming bracts at base of inflorescence, deciduous. Flowers mostly rather small, usually pedicellate, rarely subsessile, not heterostylous, in axillary or terminal branched cymes or panicles, or in fascicles. Calyx-lobes 4–5, triangular or subulate, persistent. Corolla-tube cylindrical, campanulate or narrowly funnel-shaped, the throat glabrous or hairy, the base often split for a short way into 5 parts separated by dense hairs giving a characteristic appearance; lobes 4–5, short, narrow or broad, induplicate-valvate, often hairy and sometimes margined. Stamens 4–5, inserted in the corolla-tube, the anthers either included or exserted, often at unequal heights; filaments mostly very short. Disc usually hemispherical. Ovary 2–3-locular, with a solitary erect ovule in each locule; styles filiform, free or joined; stigma-lobes 2–3, filiform, included or exserted. Fruit globose or compressed-ovoid or -ellipsoid, with a thin brittle usually shiny epicarp, splitting into 2–3 round or elliptic 1-seeded pyrenes which are dorsally compressed, marginally winged or not, membranous to coriaceous, and at length often pendulous from a detached rib-like external main nerve. Seeds the same shape as the pyrenes and intimately associated with them, dorsally very compressed, with fleshy albumen and large embryo bearing cordate leafy cotyledons.

A genus of about 50 species in the tropics of both hemispheres. The S. American species have sometimes been considered to form a separate genus *Lygodisodea* Ruiz & Pavon (e.g. by Hook. f. in G.P. 2: 134 (1873)) but this is I believe unjustified since despite the range of floral structure the genus *Paederia* forms a compact natural entity. The same argument applies to the segregation of *Lecontea* A. Rich. (*Siphomeris* Bojer). Only 2 species occur wild in the Flora area, but *P. foetida* L. has been cultivated in Tanzania (Lushoto District, Amani Nursery, 16 Feb. 1949, *Greenway* 8322 !) where it is " inclined to be a troublesome weed ". So far as is known it has not occured as an escape outside this nursery.

Leaves long petiolate, petioles 0·4–11 cm. long; lower blades, at least, ovate, mostly rounded to distinctly cordate at the base; corolla tubular:
- Leaf-blades glabrous beneath save for hairs in the axils of the main nerves; calyx-lobes short, ovate, ± 1 mm. long; anthers and stigmas included . . . . *P. foetida* (see above)
- Leaf-blades densely velvety beneath; calyx-lobes subulate, curved, 3–7 mm. long; anthers and stigmas exserted . . . . 1. *P. bojerana*

Leaves shortly petiolate, petioles 0·1–1 cm. long; blades round, ovate, elliptic or lanceolate, mostly cuneate, less often rounded at the base; corolla funnel-shaped or narrowly campanulate; calyx-lobes ± 2(–4) mm. long; anthers and stigmas included . . . . 2. *P. pospischilii*

* If *Lecontea* A. Rich. is considered not to be a homonym of *Lecontia* Torrey (1826).

1. **P. bojerana** (*A. Rich.*) *Drake* in Grandidier, Hist. Madag. 36, Hist. Nat. Pl., t. 412/a (1897). Type: Madagascar, *Bojer* (P, holo.)

Evil-smelling climbing shrub to 3·6 m. long; stems pubescent to velvety hairy with brownish or purplish hairs when young, later ridged and pubescent. Leaves opposite or in whorls of 3; blades elliptic, oblong or ovate, 2–10(–18·3) cm. long, 0·9–6·3(–13) cm. wide, acuminate at the apex, rounded to cordate at the base, pubescent with curled adpressed hairs above, densely velvety-tomentose beneath or pubescent only on the nerves; petiole 0·4–11 cm. long, pubescent to velvety pubescent; stipules ovate-oblong to triangular-acuminate, 4–10 mm. long, 2–4 mm. wide, sometimes slightly divided at the extreme tip. Inflorescences condensed axillary panicles or sometimes terminal at the ends of axillary branches and, in some cases where the leaves are suppressed, forming a compound raceme of panicles up to 21 cm. long in the fruiting state; peduncle 1–4(–10·5) cm. long, brownish or purplish velvety pubescent; pedicels 0·5–0·9 mm. long. Calyx-tube ovoid, 1·5–2 mm. long, densely pubescent; lobes 5, subulate, 3–7 mm. long, curved at the tips, densely pubescent. Corolla white or greenish yellow, ± pubescent or glabrescent outside; tube 0·75–1·2 cm. long; lobes 5, oblong-lanceolate, 2·5–6 mm. long, 1–2·2 mm. wide, hairy inside at least near throat, margined. Anthers exserted. Style 8 mm. long; stigma-lobes 6·5 mm. long. Fruit elliptic in plan, very compressed, 1·1–1·4 cm. long, 0·85–1·2 cm. wide, ± 1·5–2 mm. thick; pericarp straw-coloured, pubescent; pyrenes elliptic, 0·9–1·2 cm. long, 7–9 mm. wide, with many rhaphides on the surface, the marginal wing 1·1–3(–5 at end) mm. wide; nerves prominent on the outer surface, the main detached rib 1·1 cm. long. Seed black, 4–6·5 mm. long, 4–5·5 mm. wide.

SYN. *P. lingun* Sweet, Hort. Brit., App.; 487 (1827), *nomen nudum.*, presumably based on a specimen grown in England from material sent by Bojer from Mauritius*
Lecontea bojerana A. Rich. in DC., Prodr. 4: 470 (1830) & Mém. Fam. Rub.: 115, t. 10/1 bis (1830) & in Mém. Soc. Hist. Nat. Paris 5: 195, t. 20/1 bis (1834)
Siphomeris lingun (Sweet) Bojer, Hort. Maurit.: 170 (1837), *nom. superfl.*

subsp. **foetens** (*Hiern*) *Verdc.* in K.B. 30: 285 (1975). Types: Mozambique, N. of Sena, N'Keza and by the R. Shire about the cataracts, *Kirk* (K, syn. !) & Zambesiland, *Stewart* (BM, syn. !)

Leaf-blades with thicker more velvety indumentum beneath; calyx-lobes mostly coarser; corolla larger with larger lobes, the throat and base of the lobes inside mostly much hairier than in subsp. *bojerana*.

TANZANIA. Morogoro District: near Ruvu, June 1930, *Haarer* 1887 !; Tunduru District: 96 km. from Masasi, 19 Mar. 1963, *Richards* 17949 !; Lindi District: Nachingwea, 15 June 1952, *Anderson* 777 !
DISTR. **T**6–8; Mozambique, Malawi, Zambia, Rhodesia and South Africa (NE. Transvaal)
HAB. Coarse grassland, grassland with scrub or scattered trees, mixed dense woodland and *Chlorophora*, *Albizia* forest; 180–810 m.

SYN. *Siphomeris foetens* Hiern in F.T.A. 3: 229 (1877)
Paederia foetens (Hiern) K. Schum. in E. & P. Pf. 4(4): 125 (1891) & in P.O.A. C: 393 (1895)

NOTE. Subsp. *bojerana* occurs in Madagascar, Mauritius, and the Comoro Is.
It seemed phytogeographically important to bring out the relationships of these two taxa by treating them as subspecies of one species. *P. grevei* Drake has leaf-blades with a dense velvety indumentum and is very similar to *P. bojerana* subsp. *foetens* but has much longer flowers.

* A specimen from Mauritius collected by Bojer (K !) is labelled *Siphomeris lingun*.

2. **P. pospischilii** *K. Schum.* in E.J. 23: 469 (1897); U.K.W.F.: 407 (1974). Type: Kenya, N. of Taveta, on plain at foot of Kilimanjaro, *Pospischil* (B, holo. †)

Woody evil-smelling climber or more rarely suberect herb 0·3–9 m. long or tall; stems at first purplish, densely pubescent, later glabrous, pale, usually lenticellate and somewhat corky. Leaf-blades round, ovate, elliptic or lanceolate, 0·8–6·5 cm. long, 0·2–2·7 wide, rounded to sharply acute at the apex or rarely slightly emarginate, narrowly cuneate to rounded at the base, densely to sparsely pubescent all over or at length glabrescent; petiole 1–6(–10) mm. long; stipules ovate or triangular, 1–2·5 mm. long. Flowers 1–5 in axillary fascicles; pedicels (3–)5–8(–10) mm. long, densely pubescent save near the base of the calyx, bearing small bracts. Calyx-tube ovoid, 1–2 mm. long, glabrous; lobes 5, narrowly triangular, 1·5–2·2(–4) mm. long, very slightly joined at the base, densely pubescent. Corolla white or greenish white with a reddish purple centre, sparsely to densely hairy outside; tube narrowly funnel-shaped, 0·6–1·6 cm. long, the limb 0·3–1·1 cm. long, somewhat like that of a *Convolvulus* with thicker midpetaline areas hairy outside and thin glabrous intermediate areas, but 5-lobed, the lobes ovate, 3·6–8 mm. long, 5–6 mm. wide, densely hairy round the margins; throat densely hairy with coloured hairs. Styles joined for 0·5 mm. or free, together with stigma 1–1·2 cm. long, included. Fruit compressed-ellipsoid, 1–1·1 cm. long, 0·7–0·8 mm. wide, 2–3 mm. thick; pericarp straw-coloured, glabrous; pyrenes round or elliptic, practically flat, 9–10 mm. long, 7·5–9 mm. wide, densely covered with white surface rhaphides, the marginal wing 0·75–1·5 mm. wide, on the outside with the nerves prominent on the surface, the main detached rib 8–9 mm. long. Seeds black, 8 mm. long, 6 mm. wide. Fig. 19, p. 178.

Kenya. Northern Frontier Province: Dandu, 10 Apr. 1952, *Gillett* 12756!; Machakos District: 16 km. from Mtito Andei on road to Voi, 21 Dec. 1954, *Verdcourt* 1172!; Teita District: Tsavo National Park East, 14·4 km. Voi Gate–Sobo road, 20 Dec. 1966, *Greenway & Kanuri* 12801!

Distr. **K**1, 4, 7; Ethiopia (Ogaden), Somali Republic (S.)

Hab. Deciduous woodland, bushland and thicket, particularly of *Acacia, Commiphora, Euphorbia, Delonix, Sterculia, Grewia* and *Terminalia orbicularis*, also in open grassland with scattered bushes and grassy clearings in thicket; 450–750(–1500) m.

Note. This very characteristic plant is nevertheless very variable, particularly in leaf shape and flower size. Some small suberect plants have very narrowly lanceolate leaves and the corollas can vary in size by over 100% in one population. The Ethiopian and Somaliland material has ± glabrous stems and leaves.

Tribe 8. HEDYOTIDEAE

1. Seeds winged; climbing plant with small white corolla 4–8 mm. long . . . 15. **Danais**
   Seeds not winged; not climbing or if so then corolla scarlet, 1·8–3·2 cm. long 2
2. Calyx-limb spreading, eccentric, entire or shallowly lobed, venose, accrescent, up to 2·8 cm. wide in fruit . . . 19. **Carphalea**
   Calyx-limb not as above, usually with 4–5 distinct lobes or teeth 3
3. Ovary, placentas and fruit linear, elongated; forest floor herb with white funnel-shaped corolla 5·5–8 mm. long (**T**3) . 20. **Dolichometra**
   Ovary, placentas and fruit never linear 4
4. Flowers mostly 4-merous; leaf-blades often narrow and uninerved or with lateral nerves obscure 5

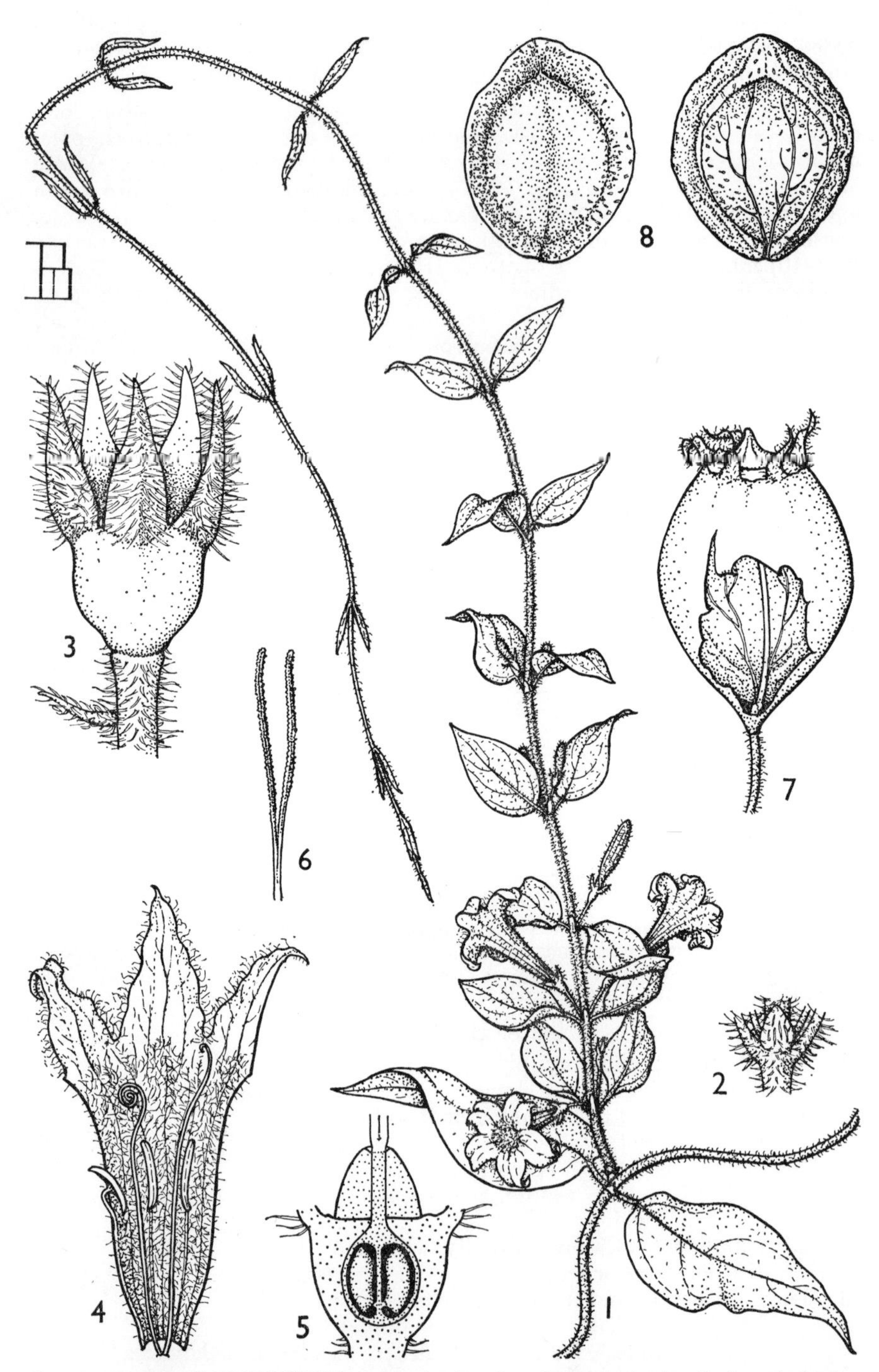

FIG. 19. *PAEDERIA POSPISCHILII*—**1,** flowering branch, × ⅔; **2,** node showing stipule, × 1; **3,** calyx × 10; **4,** longitudinal section of flower, × 3; **5,** longitudinal section of disc and ovary, × 10; **6,** style, × 2; **7,** fruit, with part of wall cut away, × 3; **8,** pyrene, two views, × 3. 1, from *Polhill & Paulo* 951; 2–6, from *Verdcourt* 1172; 7, 8, from *Mathenge* 90. Drawn by Diane Bridson.

Flowers mostly 5-merous; leaf-blades often broad and with very obvious lateral and tertiary venation 15

5. Corolla-lobes induplicate-valvate, covered on the inside with clavate hairs . . 24. **Pentanopsis**

Corolla-lobes valvate; smooth or papillate inside 6

6. Anthers and stigmas included, the latter always overtopped by the former; corolla-tube narrowly cylindrical . . 22. **Kohautia**

Anthers and/or stigmas exserted or if both included then anthers almost always overtopped by the stigmas; corolla-tube cylindrical or funnel-shaped 7

7. Corolla-tube cylindrical, at least 2 cm. long; anthers included and style exserted; flowers not heterostylous . . . 23. **Conostomium**

Corolla-tube cylindrical or more often funnel-shaped, always less than 2 cm. long, sometimes with included anthers and exserted style but in that case plant mostly heterostylous or if not corolla small* 8

8. Capsule opening both septicidally and loculicidally 9

Capsule opening only loculicidally 13

9. Beak of capsule as long as or longer than the rest of the capsule 10

Beak of the capsule shorter than the rest of the capsule 11

10. Fragile annual herb, with capsule of a very characteristic shape (fig. 32 p. 249), emarginate at the base, the beak much exceeding the rest of the capsule; stipules triangular with 2 lobes . . 25. **Mitrasacmopsis**

Robust subshrubby herb, mostly drying blackish and with rather coriaceous leaves; beak as long as the rest of the capsule; stipules 3–7-fimbriate . . 26. **Hedythyrsus**

11. Robust shrubby herb; stipules with a single deltoid lobe, sometimes ± lacerated but not fimbriate; style and stamens long-exserted (**T6**, Uluguru Mts.). . . 27. **Pseudonesohedyotis**

Herbs; stipules mostly divided into distinct fimbriae; flowers heterostylous or isostylous but both style and stamens not long-exserted 12

12. Corolla-tube bearded inside; style not shortly bifid at the apex (*apart* from the division into stigma-lobes); stigma-lobes subglobose or ovoid . . . 28. **Agathisanthemum**

* A monstrous form of *Kohautia virgata*, with green flowers on pedicels 0·4–2·3 cm. long with the corolla-tube 0·9–1·5 mm. long and anthers and style exserted, is mentioned on p. 235. If the plant being keyed has these three characters this possibility should be borne in mind.

Corolla-tube glabrous inside; style shortly bifid at the apex as well as divided into ellipsoid stigma-lobes (**K**4) . . . 29. **Dibrachionostylus**

13. Rush-like plants with linear or filiform leaves; stipule-sheath tubular, truncate or with 2 minute teeth; seeds dorsiventrally flattened . . . . . 30. **Amphiasma**

Plants not rush-like even if leaves linear; if seeds dorsiventrally flattened then leaves not linear 14

14. Capsule with thick woody wall and a solid beak, tardily dehiscent . . . 32. **Lelya**

Capsule with a horny wall, with or without a beak but never solid, early dehiscent . . . . 33. **Oldenlandia**

15. Leaves uninerved, mostly fairly small; decumbent plant of wet places with small white or blue flowers in very lax elongated axillary cymes . . . 31. **Pentodon**

Leaves larger, pinnately nerved; flowers mostly terminal. 16

16. Creeping herbs or mat-forming herb of the forest floor, often rooting at the nodes 17

Erect herbs or subshrubs 19

17. Calyx-lobes conspicuously spathulate at their apices; corolla ± 3 cm. long . . 18. **Chamaepentas**

Calyx-lobes not or much less spathulate; corolla-tube under 2 cm. long 18

18. Inflorescence ± 1–2(–3)-flowered, usually only one corolla opening at a time; flowers sessile in the axils of the leaves . 21. **Parapentas**

Inflorescences ± several-flowered, cymose, not sessile . . . . . . 16. **Pentas** (in part)

19. Flowering inflorescences capitate or lax much-branched complicated cymes; individual branches sometimes becoming spicate in fruit; plant never climbing; fruits subglobose or obtriangular; corolla-lobes narrower, not or scarcely connate at the base . . . . 16. **Pentas**

Flowering inflorescences capitate, later elongating into a long simple " spike ", rarely with axillary spikes from the upper axils, and frequently with solitary flowers at the lower nodes, or if not elongating into a spike, then plant climbing; fruits oblong; corolla-lobes more rounded, joined shortly at the base to form an annulus bearing hairs around the orifice of the tube (in subgenera of the Flora area) . . . . 17. **Otomeria**

15. DANAIS

Vent., Tabl. 2: 548 (1799)

Climbing or erect shrubs, sometimes quite extensive lianes. Leaves opposite or in whorls of 3 or 4, petiolate; stipules interpetiolar, ovate-triangular, acute. Flowers small, dimorphic, in axillary and terminal cymes, often running together to form fairly extensive panicles. Calyx-tube subglobose; lobes short, subequal, persistent. Corolla funnel-shaped or hypocrateriform; throat hairy, particularly in the case of long-styled flowers; lobes 5–6, valvate. Stamens 5–6, included in long-styled flowers, exserted in short-styled flowers. Ovary 2-locular, each locule with many ovules; style filiform, divided into 2 filiform stigmata. Capsule splitting loculicidally; seeds inserted parallel to the surface of the placenta, surrounded by a lacerate or erose wing.

A genus of about 40 species, with the exception of the following confined to the Mascarene and Comoro Is. and Madagascar. The presence of rhaphides in the tissues, occurrence of complete heterostyly and the detailed morphology of the flowers support the removal of this genus to the *Hedyotideae* rather than its retention in the *Cinchoneae*. I fully agree with Bremekamp, Verh. K. Nederl. Akad. Wet., Afd. Natuurk., ser. 2, 48 (2): 14–15 (1952), in this decision despite the winged seeds.

D. xanthorrhoea (*K. Schum.*) *Bremek.* in K.B. 3: 190 (1948). Type: Tanzania, Usambara Mts., ? Mlalo [Malo], *Buchwald* 438 (B, holo. †, K, iso. !)

Climbing herb with branched stems 6–30 m. long, pubescent or puberulous when young but soon glabrescent. Leaf-blades obovate, elliptic or obovate-oblong, 3–12 cm. long, 1–6 cm. wide, shortly acuminate at the apex, cuneate at the base, glabrous; petioles 0·7–1·5 cm. long; stipules broadly triangular or acuminate, 1·5–2 mm. long, ± glabrous. Flowers in terminal and axillary inflorescences, sometimes forming ample panicles 2·2–6 cm. long, glabrous to pubescent; peduncles 0·4–3·5 cm. long; pedicels up to 3 mm. long; bracts subulate, 2–3·5 mm. long. Calyx-tube subglobose, 1 mm. long, pubescent or glabrous; lobes deltoid or ovate, 0·5–1 mm. long, up to 0·5 mm. wide, the margins and midrib ciliolate. Corolla at first yellow or greenish-white but later orange-red or tube white and lobes bright vermilion; tube 2·5–6 mm. long, 1 mm. wide at the base, 1·7 mm. wide at the apex, glabrous outside, densely yellow pilose in the throat, particularly in long-styled flowers; lobes lanceolate, 2·5–4 mm. long, 1–1·2 mm. wide, thickened at the apices, glabrous. Stamens included in long-styled flowers but long-exserted in short-styled flowers. Style 7–8 mm. long in long-styled flowers and 3 mm. long in short-styled flowers, the stigma-arms filiform, 2–3(–4) mm. long. Capsule subglobose, 3 mm. in diameter, glabrous or puberulous, the valves sulcate. Seeds brown, 0·5–1 mm. in diameter, with a broad irregularly lacerate circumferential wing. Fig. 20, p. 182.

TANZANIA. Lushoto District: Amani, 31 Dec. 1928, *Greenway* 1079 ! & between Ngua and Kwamkoro, 19 July 1950, *Verdcourt & Greenway* 292 !; Morogoro District: Uluguru Mts., gorge below the Hululu Falls, Mgeta R., 15 Mar. 1953, *Drummond & Hemsley* 1584 !

DISTR. **T3**, 6; not known elsewhere

HAB. Rain-forest; (700–)900–1600 m.

SYN. *Urophyllum xanthorrhoeum* K. Schum. in E.J. 28: 58 (1899)
Pentas ? xanthorrhoea (K. Schum.) Bremek. in E.J. 71: 202 (1940); T.T.C.L.: 518 (1949)

NOTE. As Bremekamp points out in K.B. 3: 190 (1948) this comes very close to several Madagascan species particularly *D. rhamniphylla* Bak. and the indumentum differences mentioned by him do not hold good; nevertheless there seem to be small

FIG. 20. *DANAIS XANTHORRHOEA*—**1,** flowering branch, × 1; **2,** long-styled flower, × 3; **3,** corolla opened out to show stamens, × 3; **4,** stamen, front and back view, × 6; **5,** pistil, × 3; **6,** inflorescence of short-styled flowers, × 1; **7,** short-styled flower, × 3; **8,** pistil of same, × 3; **9,** transverse section of ovary, × 6; **10,** capsule, × 3; **11,** capsule, dehisced, × 3; **12,** seed, × 6; **13,** seed, with wing removed, × 6. 1–5, from *Greenway* 3337; 6–9, from *Greenway* 1079; 10–13, from *Koritschoner* 724. Drawn by Miss D. R. Thompson.

floral differences, e.g. lengths of corolla-lobes and stigmata. Until the genus is revised the East African species is maintained at specific rank, but subspecific rank may be the final solution.

16. PENTAS

Benth. in Bot. Mag. 70, t. 4086 (1844); Verdc. in B.J.B.B. 23: 237–371 (1953)

Mostly perennial (rarely biennial) herbs or shrubs, with erect or straggling stems from a fibrous or woody rootstock. Leaves paired or in whorls of 3–5; stipules divided into 2–many filiform colleter-tipped segments. Flowers small to very large, hermaphrodite, mono-, di- or tri-morphic, mostly in much-branched terminal complicated cymose inflorescences, the individual branches often becoming spicate in fruit. Calyx-tube ovoid or globose, sometimes with a free annular part at the top; lobes usually 5, either equal or unequal, 1–3 being larger than the others or sometimes foliaceous. Corolla-tube shortly cylindrical to narrowly tubular, 2–40 times as long as wide, hairy in the throat; lobes ovate or oblong. In monomorphic flowers stamens enclosed in an abrupt apical dilation of the tube and style exserted; in dimorphic flowers the tube is gradually dilated at the apex in short-styled flowers and abruptly dilated in long-styled flowers; in rare cases, trimorphism is shown and the third form has both stamens and style included in the tube. Ovary bilocular, with numerous ovules in each attached to placentas affixed to the septum. Capsule obtriangular or ovoid, ribbed, beaked, opening at the apex, the beak splitting into 4 valves; capsule sometimes separating into 2 cocci. Seeds minute, brownish, irregularly globose or tetrahedral, with reticulate testa. Rhaphides are plentiful in most of the tissues.

A genus of about 40 species, widely distributed throughout tropical Africa from W. Africa and Somali Republic to Angola and Natal, also in tropical Arabia, Madagascar and the Comoro Is.

Key to subgenera of Pentas

Calyx with 1–2 lobes enlarged into a white stipitate membranous lamina in the majority of flowers* . . . . 1. **Phyllopentas** (p. 184)

Calyx-lobes often foliaceous but then green and never enlarged into a stipitate lamina:

- Calyx-lobes flat, foliaceous or deltoid, nearly always 1–3 enlarged and the rest much smaller:
  - Enlarged calyx-lobes distinctly spathulate, narrowed at the base; corolla-tube 3·5–4·5 cm. long; inflorescences composed of a few 3-flowered cymes. . 4. **Chamaepentadoides** (p. 193)
  - Enlarged calyx-lobes not or scarcely spathulate; corolla-tube 0·4–9 cm. long, but mostly short; inflorescences many-flowered . . . . 6. **Pentas** (p. 199)
- Calyx-lobes subulate, narrowly spathulate or linear, subequal:
  - Leaves in whorls of 3–5 or if paired then plant a short pyrophyte ± 16 cm. tall . 5. **Longiflora** (p. 194)
  - Leaves paired; plant never a pyrophyte:
    - Corolla-tube 2·5–16 cm. long, the leaves being linear-lanceolate in the smallest flowered species:

* Except in some unusual variants of 2, *P. schumanniana*.

Calyx-lobes mostly over 1 cm. long or if shorter then corolla-lobes 1·5–2 cm. long and 0·5–1 cm. wide . . 3. **Megapentas** (p. 189)

Calyx-lobes under 1 cm. long (mostly under 0·7 cm. long); corolla-lobes up to 1·2 cm. long and 2·5 mm. wide:

Stipular setae noticeably capitellate; petioles up to 1 cm. long; corolla-tube 3·75 cm. long, gradually widening from base to throat; inflorescence ± 15-flowered, lax . 4. **Chamaepentadoides** (p. 193)

Stipular setae obscurely capitellate; petioles short; corolla-tube abruptly expanded just below the throat; inflorescence often many-flowered . . . 5. **Longiflora** (p. 194)

Corolla-tube 0·4–2·8 cm. long, in the largest-flowered species the leaves being large and oblong:

Decumbent forest floor herb . . 3. *P. ulugurica*

Shrubs or tall herbs, with the indumentum usually drying a rufous colour; calyx-lobes equal, subulate; leaves with numerous lateral nerves and a characteristic reticulate venation beneath* . . . 2. **Vignaldiopsis** (p. 187)

Subgen. 1. **Phyllopentas**

Verdc. in B.J.B.B. 23: 254 (1953)

Shrubs or subshrubs. Calyx-lobes subequal or 1 often enlarged into an ovate petaloid stipitate lamina. Flowers dimorphic.

Corolla-tube ± 1·6 cm. long . . . . 1. *P. ionolaena*
Corolla-tube ± 5 mm. long . . . . 2. *P. schumanniana*

1. **P. ionolaena** *K. Schum.* in E.J. 28: 487 (1900); T.T.C.L.: 517 (1949); Verdc. in B.J.B.B. 23: 254, fig. 31/C, D & E (1953). Types: Tanzania, Uluguru Mts., Lukwangule Plateau, *Goetze* 253 (B, holo. †); Morogoro, *E. M. Bruce* 181 (K, neo.!, EA, isoneo.!)

Straggling or shrubby herb 0·9–3·5 m. tall, with stems and peduncles covered with crisped violet or ferruginous hairs. Leaf-blades elliptic to oblong-ovate, 3·5–15 cm. long, 3·7—6·5 cm. wide, acuminate at the apex, rounded to cuneate at the base, hairy above and on the venation beneath; petiole 2–3 cm. long; stipules with 4–7 filiform pubescent capitellate setae 6–10 mm. long. Inflorescences lax, trichotomous, ± 9–12 cm. long and wide, sessile or pedunculate; bracts filiform, ± 1 cm. long, scattered on the branches. Calyx-lobes very unequal, the foliaceous one white, elliptic to rounded-ovate, 0·7–1·3 cm. long, 3·5–9·3 mm. wide, mucronate or acuminate, membranous and markedly venose, pubescent on the nerves, with the stipe

* If calyx-lobes seem approximately equal, but are not subulate and other characters do not agree, see subgen. 6, *Pentas*, p. 199.

FIG. 21. *PENTAS IONOLAENA*—**1,** flowering branch, × 1; **2,** leaf, × 1; **3,** calyx, × 3; **4,** long-styled flower, with calyx removed, × 3; **5,** short-styled flower, with calyx removed, × 3. 1, 3–5, from *Wallace* 211; 2, from *E. M. Bruce* 196. Drawn by Olive Milne-Redhead.

5 mm. long; other lobes filiform, 6–7 mm. long, deltoid basally and slightly connate above the disc, coarsely reticulate, hairy. Corolla white to mauve or pale violet; tube 1·2–1·6 cm. long, dilated at the apex for 3–4 mm. in long-styled flowers; lobes 2–4 mm. long, pubescent outside, especially at the apex. Style exserted for ± 4 mm. in long-styled flowers. Fruit 2·5 mm. long, 3 mm. wide, ribbed, the beak a little raised and the connate parts of the calyx-lobes patent, forming a horizontal nervose flange; dehiscence entirely apical. Seeds ± 0·5 mm. long. Fig. 21, p. 185.

TANZANIA. Morogoro District: Uluguru Mts., Chenzema [Kienzema]–Lukwangule Plateau, 26 Aug. 1951, *Greenway & Eggeling* 8685! & Lukwangule Plateau, above Chenzema Mission, 13 Mar. 1953, *Drummond & Hemsley* 1525! & Morningside, Jan. 1953, *Eggeling* 6469!
DISTR. T6; not known elsewhere
HAB. Moist and dry evergreen forest and derived scrub; 1200–2200 m.

SYN. [*P. schumanniana* sensu auctt., *non* K. Krause]

NOTE. *P. ionolaena* subsp. *madagascariensis* Verdc. is probably best considered a separate species.

2. **P. schumanniana** *K. Krause* in E.J. 39: 521 (1907); T.T.C.L.: 518 (1949); Verdc. in K.B. 30: 287 (1975). Type: Tanzania, Songea District, near Kwa Amakita, *Busse* 929 (B, holo. †, EA, iso.!)

Shrubby herb 1·5–2·7 m. tall, closely resembling the last species; stems with rusty indumentum on the young parts. Leaf-blades ovate or ovate-oblong, 3·5–17 cm. long, 1·5–8·4 cm. wide, acuminate at the apex, narrowly and often unequally cuneate at the base, pubescent or with scattered hairs above and on the nerves beneath; petiole 1–4·7 cm. long; stipules with ± 5 setae up to 0·7–1·2 cm. long. Inflorescences terminal together with axillary ones, small and compact when flowering, 4–11 cm. wide, enlarging in fruit to 20 cm. long, 16 cm. wide; peduncles 0·7–7 cm. long. Calyx-lobes very unequal, the foliaceous one white tinged green and green veined, elliptic to rounded-ovate, 2–6 mm. long, 1–3 mm. wide, mucronate or acuminate, membranous and markedly venose, pubescent on the nerves, with the stipe 1·5–2 mm. long (rarely none or very few of the calyces with a lobe developed into a lamina); other lobes filiform, 1·5–2·5 mm. long. Corolla lilac or white; tube ± 5–6 mm. long, dilated at the apex for 2 mm. in long-styled flowers; lobes 2–3 mm. long, pubescent outside, especially at the apex. Style exserted for ± 2–3 mm. in long-styled flowers. Fruit 2 mm. long, 3 mm. wide, the base of the persistent calyx-lobes with a net-veined pattern. Seeds ± 0·5 mm. long.

TANZANIA. Njombe District: Livingstone Mts., Madunda Mission, 2 Feb. 1961, *Richards* 14208! & Mwakete, 16 Jan. 1957, *Richards* 7830!; Songea District: Matengo Hills, Liwiri Kiteza, 5 Mar. 1956, *Milne-Redhead & Taylor* 9030!
DISTR. T7, 8; Malawi
HAB. Bushland, forest edges, sometimes on rocky outcrops; 1300–2400 m.

SYN. *P. ionolaena* K. Schum. subsp. *schumanniana* (K. Krause) Verdc. in B.J.B.B. 23: 256, fig. 31/F (1953)

NOTE. Krause's description actually fits the Uluguru species *P. ionolaena*, but the praserved isotype shows that the flowers are half the size he states. Occasionally practically none of the calyx-lobes is expanded (e.g. in *Richards* 7802, Tanzania, Njombe District, Mwakete, 15 Jan. 1957). *Robertson* 1015 (Njombe District, Igeri, 19 Feb. 1968) and *Paget-Wilkes* 362 (Iringa District, Mufindi, 10 Feb. 1969) are very similar indeed, but have no calyx-lobes at all expanded. The status of these plants needs further field investigation.

Intermediate species

P. ulugurica does not fit well into either subgen. *Phyllopentas* or subgen. *Vignaldiopsis*, but links the two groups.

3. **P. ulugurica** (*Verdc.*) *Hepper* in K.B. 14: 254 (1960). Type: Tanzania, NW. Uluguru Mts., *Schlieben* 2730 (B, holo. !, BR, iso. !)

Spreading decumbent perennial herb of forest floor, with ascending stems to 20 cm. high, pubescent with crisped ferruginous hairs. Leaf-blades ovate to elliptic, 1·3–6(–11) cm. long, 0·6–4·4(–6·5) cm. wide, acute to slightly acuminate at the apex, attenuately cuneate at the base, pubescent on both surfaces, the hairs on the midrib beneath often ferruginous; petioles 1–3·2 cm. long, ferruginous pubescent; stipules with 3–4 filiform segments up to 5 mm. long. Inflorescences terminal, trichotomous, the cymes few-flowered; peduncle up to 4 cm. long; flowers white, dimorphic. Calyx-tube ± 1 mm. long, 1·2 mm. wide; lobes subequal, subulate with a triangular base, 1·1–5 mm. long, 1 mm. wide at the base, pilose outside. Short-styled flowers: corolla-tube slightly funnel-shaped, 4·2–5 mm. long, glabrescent outside, pilose inside; lobes oblong, 4–5 mm. long, 1–1·8 mm. wide; stamens well exserted; style 3 mm. long, with stigmata 2 mm. long. Long-styled flowers: corolla-tube 4·5–5·5 mm. long, broadened at the throat to 2 mm., ± glabrous outside, glabrous inside below the middle and densely hairy above, but hairs not protruding from the throat; lobes oblong, 3–4·5 mm. long, 1·2 mm. wide; stamens included; style exserted, 6·2 mm. long; stigmata 1·7–2 mm. long. Capsule shortly obconic, 3 mm. long and wide, glabrous or pubescent, prominently costate, the beak slightly raised; calyx-lobes persistent.

TANZANIA. Morogoro District: Uluguru Mts., Tegetero, 20 Mar. 1953, *Drummond & Hemsley* 1727! & Morogoro, 17 Nov. 1935, *E. M. Bruce* 146! & Bondwa, 7 Sept. 1969, *Harris & Pócs* in *Harris* 3290!
DISTR. **T6**; not known elsewhere
HAB. Ground layer of rain-forest; 1000–1800 m.

SYN. *Tapinopentas ulugurica* Verdc. in B.J.B.B. 23: 61 (1953)

NOTE. When I described this I was uncertain about the generic placing but am convinced Hepper in K.B. 14: 253 (1954) is correct to disband *Tapinopentas* Bremek. and transfer the species elsewhere. The present species fits fairly well as an aberrant species linking the subgenera *Vignaldiopsis* and *Phyllopentas*. The capsule structure is exactly that of *Phyllopentas*. In ecology and habit it comes close to *Dolichometra leucantha* K. Schum., but that has a totally different ovary structure.

Subgen. 2. Vignaldiopsis

Verdc. in B.J.B.B. 23: 261 (1953)

Shrubs or subshrubs covered with ferruginous hairs (at least when dry). Leaves with up to 22 lateral nerves on each side and venation reticulate beneath. Calyx-lobes subulate, subequal, never foliaceous. Flowers dimorphic.

Corolla-tube 0·5–1·4 cm. long . . . . 4. *P. schimperana*
Corolla-tube 1·5–2·8 cm. long . . . . 5. *P. elata*

4. **P. schimperana** (*A. Rich.*) *Vatke* in Linnaea 40: 192 (1876); Hiern in F.T.A. 3: 45 (1877); K. Krause in N.B.G.B. 10: 602 (1929); F.P.N.A. 2: 327 (1947); T.T.C.L.: 518 (1949); Verdc. in B.J.B.B. 23: 263, fig. 31/G, H & I (1953); Hepper in F.W.T.A., ed. 2, 2: 216 (1963); U.K.W.F.: 404 (1974). Type: Ethiopia, Shire [Chiré], *Quartin Dillon* 126 (P, lecto. !, K, isolecto. !)

Shrub or woody herb (0·3–)1·3–2·7(–5) m. tall, with wrinkled blackish or purplish-black woody stems, at first densely rusty pubescent (at least when dry*), later glabrescent. Leaf-blades ovate, ovate-lanceolate or ovate-oblong, 6–21 cm. long, 1·8–8 cm. wide, acute to acuminate at the apex, cuneate to round at the base, adpressed rusty hairy above and on the characteristically reticulate (but not prominent) venation beneath; petiole 0–1·5 cm. long, velvety rusty pubescent; stipules large, with ± 6–10 linear-lanceolate brown hairy setae 0·2–1·5 cm. long from a variable base 3–8 mm. long, 4–8 mm. wide; these bases connate and forming persistent cups at the nodes on old stems. Inflorescences branched, 2·5–15 cm. wide; peduncles up to 1·5 cm. long, velvety pubescent. Calyx-tube hairy, 1·5–3 mm. long; lobes subequal, linear, 0·3–1·2 cm. long, 1 mm. wide, hairy outside, either $\frac{1}{3}$–$\frac{1}{2}$ the length of the corolla-tube or approximately equalling it. Corolla white, often tinged pink, or entirely pinkish, glabrous or pubescent; in long-styled flowers, tube funnel-shaped 0·5–1·3 cm. long, dilated at the apex for 2 mm.; in short-styled flowers the tube is cylindrical, 0·5–1·4 cm. long; lobes 3–7 mm. long, 0·75–2·5 mm. wide, the tips thickened and inflexed. Style exserted for 0·5–5 mm. long in long-styled flowers. Fruit 4–6 mm. long and wide, the persistent calyx-lobes often reflexed, costate, valvular, opening ± 1–2 mm. wide, or rarely separating loculicidally into 2 cocci.

subsp. **schimperana**; Verdc. in B.J.B.B. 23: 262 (1953)

Calyx lobes mostly approximately equalling the corolla-tube

UGANDA. Karamoja District: Mt. Morongole, 11 Nov. 1939, *A. S. Thomas* 3280!; Kigezi District: Bukimbiri, Oct. 1947, *Purseglove* 2494!; Mbale District: Bugisu, Bulago, 9 Dec. 1938, *A. S. Thomas* 2585!
KENYA. Northern Frontier Province: Mt. Nyiru, 5 Jan. 1959, *Newbould* 3467!; W. Suk District: N. Cherangani Hills, Sept. 1965, *Tweedie* 3120!; Elgeyo Escarpment, June 1932, *Gardner* in *F.D.* 2852!
TANZANIA. Mpanda District: Kungwe Mt., 11 Sept. 1959, *Harley* 9592!; Iringa District: Dabaga Highlands, Idewa Forest Reserve, 20 Feb. 1962, *Polhill & Paulo* 1552!; Njombe District: valley below Milo Mission, 30 Jan. 1961, *Richards* 14058!
DISTR. **U**1–4; **K**1–3, 5; **T**4, 7; Ethiopia, Zaire, Burundi, Malawi and Zambia
HAB. Grassland, bushland, bamboo and evergreen forest in upland areas, usually on volcanic soils or even fairly recent lava; 1450–3000 m.

SYN. *Vignaldia schimperana* A. Rich., Tent. Fl. Abyss. 1: 359 (1847)
Pentas schimperi Engl., P.O.A. A: 92 (1895), *nom. superfl.*, based on *Vignaldia schimperana*
P. thomsonii Scott Elliot in J.L.S. 32: 435 (1896). Type: Kenya, Laikipia, *Thomson* (K, lecto.!, BM, isolecto.!)
Neurocarpaea thomsonii (Scott Elliot) S. Moore in J.L.S. 37: 157 (1905)

NOTE. Subsp. *occidentalis* (Hook. f.) Verdc. occurs in Fernando Po, S. Tomé, Cameroun and also in the Ituri Forest of Zaire; records from S. Arabia (Robyns, F.P.N.A. 2: 327 (1947)) and South West Africa (Verdc. in B.J.B.B. 23: 268 (1953)) are almost certainly based on wrongly localised material.

5. **P. elata** *K. Schum.* in P O A. C: 377 (1895); Verdc. in B.J.B.B. 23: 262 (1953). Type: Tanzania, Moshi District, Himo, *Volkens* 1822 (B, holo. †, BM, iso.!)

Shrub 3–5 m. tall, very closely related to the last species; stems sparsely pubescent with grey or rusty hairs. Leaf-blades elliptic, 6–18 cm. long, 2–6·5 cm. wide, shortly acuminate at the apex, cuneate at the base, sparsely hairy; petioles 0·5–1·5 cm. long; stipules with ± 8 filiform segments 3–5 mm. long. Inflorescences axillary and terminal, 4–6 cm. wide; flowers sweet-scented; peduncles up to 3 cm. long. Calyx-tube 1·6–2 mm. long, pubescent; lobes subequal, subulate or slightly spathulate at the apices, ± 6(–9) mm. long. Corolla white, 1·5–2·8 cm. long; lobes elliptic, 5–6 mm. long, 1·5–2 mm. wide. Capsule 5 mm. long and wide.

* Often pallid in living material.

TANZANIA. Arusha District: Mt. Meru, Engare Nanyuki R. gorge, 23 Dec. 1967, *Richards* 21794! & Jekukumia to Engare Nanyuki R. gorge, 6 Nov. 1969, *Richards* 24623!; " Kilimandscharo und Meru ", Nov. 1901, *Uhlig* 1064!
DISTR. **T2**; not known elsewhere
HAB. Woodland, steep rocky slopes; 1950–2800 m.

NOTE. Bearing in mind the variation in corolla-tube length allowed within a single species in subgen. *Pentas*, it could be argued that the above should be treated as a subspecies of *P. schimperana*, but its marked geographical isolation and the lack of variation displayed by *P. schimperana* suggest it is best left as a distinct species.

Subgen. 3. **Megapentas**

Verdc. in B.J.B.B. 23: 269 (1953)

Subsucculent herbs or subshrubs. Leaves opposite, often large, petiolate or sessile. Calyx-lobes subequal, linear-deltoid or narrowly spathulate, never leafy. Flowers mostly large, with corolla-tube up to 16 cm. long, often dimorphic. Style tomentose.

Plant pubescent or hairy:
- Leaves subsessile, with blade 9·5–23 × 3·5–8 cm.; corolla-tube (6·6–)13–16 cm. long; lobes 2–2·5 × 0·6–1·2 cm.; calyx with limb-tube (i.e. free part) in fruit 0·75 mm. tall . . . . 6. *P. longituba*
- Leaves petiolate; petioles 0·2–1·4 cm. long; blade 2·8–20 × 1·5–8·8 cm.; corolla-tube 6·5–14·2 cm. long; lobes 0·9–2·1 × 0·4–0·8 cm.; calyx with limb-tube in fruit 2–5 mm. tall:
  - Calyx-lobes mostly narrowly spathulate; corolla-tube 6·5–12·5 cm. long; leaves with petioles 0·4–1·4 cm. long; blade 2·8–12 × 1·5–3·4 cm.. . . . . 7. *P. pseudomagnifica*
  - Calyx-lobes narrowly linear; corolla-tube 7·5–14·2 cm. long; leaves with petioles 0·2–1 cm. long; blade 5–20 × 1·6–8·8 cm. . . . . 8. *P. nobilis*

Plant glabrous save for style and throat of corolla-tube . . . . 9. *P. graniticola*

6. **P. longituba** *K. Schum.* in P.O.A. C: 377 (1895); T.T.C.L.: 516 (1949); Verdc. in B.J.B.B. 23: 269, fig. 32/F (1953). Types: Tanzania, Usambara,* *Holst* 418 (B, holo. †, K, photo.!); Kilosa District, Mt. Luembai, *B. D. Burtt* 4555 (K, neo.!, EA, isoneo.!).

Branched subsucculent subshrub ± 0·8–2 m. tall, the stems thick, blackish below and hard wrinkled and very finely reticulate when dry; pubescent all over with crisped hairs. Leaf-blades elliptic, narrowly ovate or elliptic-lanceolate, 9·5–23 cm. long, 3·5–8 cm. wide, acute at the apex, cuneate at the base, pubescent above and with longer matted hairs beneath particularly on the midrib; petiole very wide, scarcely 2 mm. long; stipules sheathing, connate, with 3–9 hairy filiform lobes 0·3–1·2 cm. long and bearing distinct colleters 1·5 mm. long, from a wide base 1·5–2·5 mm. long. Inflorescences terminal, lax, trichotomous, the puberulous branches up to 5 cm. long and bearing 3–6-flowered cymes. Flowers not dimorphic, white or pale mauve,

* See note at end of species.

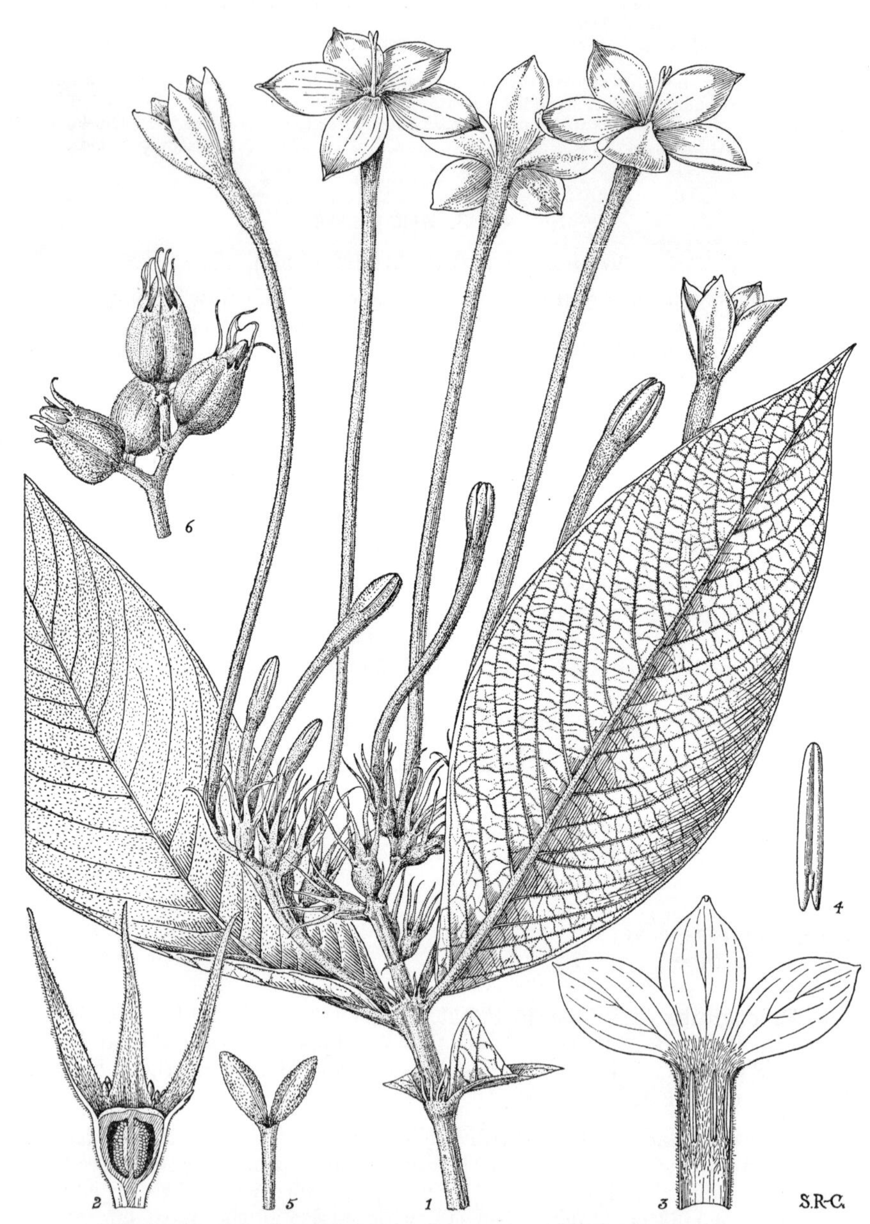

FIG. 22. *PENTAS LONGITUBA*—**1,** flowering branch, × $\frac{2}{3}$; **2,** longitudinal section of ovary and calyx, × 2; **3,** longitudinal section of upper part of corolla, × 1; **4,** stamen, × $2\frac{1}{2}$; **5,** upper part of style with stigma-lobes, × $2\frac{1}{2}$; **6,** part of infructescence, × 1. Drawn by Stella Ross-Craig. Reproduced by permission of the Bentham-Moxon Trust.

subsessile. Calyx-tube campanulate, 5–6 mm. long, 3–4 mm. wide, softly pubescent; lobes linear, 1–1·9 cm. long, 0·5–1·5 mm. wide at the base, with 1–3 stipitate glands 0·5–2 mm. long in each sinus. Corolla-tube 6·6–16 cm. long (at least some flowers with tubes 13·5 cm. long always present), 2–3·5 mm. wide, densely pubescent; throat hairy and tube villous inside for apical 2 cm.; lobes obovate, 2–2·5 cm. long, (0·6–)1–1·2 cm. wide, subacute. Stamens included. Style exserted 1–1·9 cm., scaly-tomentose above; stigma bilobed, 3 mm. long. Capsule ovoid, 1–1·5 cm. long, 7–9·5 mm. across, rounded at the base, ribbed, grooved between the locules, dehiscing by woody valves and separating into cocci; beak pointed and compressed, 4·3 mm. tall; calyx with limb-tube 0·75 mm. high, the lobes persistent. Fig. 22.

TANZANIA. Morogoro District: N. Nguru Mts., above Manyangu Forest, N. of Liwale R., 2 Apr. 1953, *Drummond & Hemsley* 1989! & Uluguru Mts., Morogoro, "The Window", 27 Feb. 1935, *E. M. Bruce* 863! & above Morogoro, near Morningside, Mar. 1954, *Eggeling* 6811!

DISTR. **T3** (see note), **6**; not known elsewhere

HAB. Crevices in granite precipices and other rocky places on mountain slopes; 900–1650 m.

SYN. *P. magnifica* Bullock in Hook., Ic. Pl. 33, t. 3265 (1935). Type: Tanzania, Kilosa District, Mt. Luembai, *B. D. Burtt* 4555 (K, holo.!, EA, iso.!)
P. longituba K. Schum. var. *magnifica* (Bullock) Bullock & M. R. F. Taylor in K.B. 1940: 57 (1940); T.T.C.L.: 516 (1949)

NOTE. No further material has been seen from the Usambara Mts., which is curious considering the striking appearance of the plant.

7. **P. pseudomagnifica** *M. R. F. Taylor* in K.B. 1940: 56 (1940); T.T.C.L.: 517 (1949); Verdc. in B.J.B.B. 23: 271, fig. 32/A (1953). Type: Tanzania, Uluguru Mts., Lupanga Peak, *B. D. Burtt* 4979 (K, holo.!, EA, iso.!)

Shrubby subsucculent herb 0·6–2 m. tall; stems thick and branched at the base, becoming woody, covered with short hairs. Leaf-blades ovate to ovate-lanceolate, 2·8–12 cm. long, 1·5–3·4 cm. wide, acute at the apex, cuneate or rounded and then cuneate at the base, shortly hairy with bristly hairs above and on the nerves beneath; petiole 0·4–1·4 cm. long; stipules with 5–7 hairy filiform setae bearing conspicuous colleters, 0·6–1·2 cm. long from a short base. Inflorescences branched and often with axillary inflorescences in 1–3 axils below the apex, each a pedunculate 2–10-flowered cyme; in small plants there may be only 13 flowers; peduncles 0·6–8·5 cm. long; pedicels 1–6 mm. long. Flowers white, sometimes tinged reddish, not dimorphic. Calyx-tube urceolate, 4–5·5 mm. long, 3–5 mm. wide, shortly adpressed hairy; lobes linear to spathulate, subequal, 0·7–1·7 cm. long, 1·5–6·7 mm. wide, or some ligulate, 4 mm. long, 0·5 mm. wide, with bristly hairs. Corolla-tube 6·5–12·5 cm. long, always some long examples in every inflorescence, 1–2·5 mm. wide, dilated at the apex to 3–4 mm. for a distance of 1–1·6 cm.; lobes ovate, 0·9–1·8 cm. long, 4–8 mm. wide, sparsely hairy outside, and with throat-hairs extending over the base inside. Stamens included. Style exserted 0·4–1·2 cm., stigma 2–3·5 mm. long. Fruit oblong, 5–7 mm. tall, 7–9·5 mm. across; veined limb-tube of calyx forming a rim 2·5 mm. tall; valves 4–5 mm. tall; intravalvular opening 5 mm. long and wide. Seeds angular-ovoid.

TANZANIA. Morogoro District: Lukwangule Plateau, above Chenzema Mission, 13 Mar. 1953, *Drummond & Hemsley* 1558! & Lukwangule, 30 Jan. 1935, *E. M. Bruce* 701! & Morogoro, *Rounce* 590!

DISTR. **T6**; not known elsewhere

HAB. Grassland, scrub, upland evergreen forest, often in rocky places; 2060–3000 m.

SYN. [*P. longituba* sensu K. Schum. in E.J. 28: 486 (1900), *non* K. Schum. (1895)]

8. **P. nobilis** *S. Moore* in J.B. 46: 37 (1908); T.T.C.L.: 516 (1949); Verdc. in B.J.B.B. 23: 274 (1953) & in Fl. Pl. Afr., t. 1690 (1974). Type: Rhodesia, Mazoe, Iron Mask Hill, *Eyles* 248 (BM, lecto. !, SRGH, isolecto. !)

Shrub or herb 0·6–1·2(–2·4) m. tall, the stems woody, ridged, grey-black, the epidermis ridged or rucked up, pubescent all over, often drying orange-ferruginous. Leaf-blades ovate, ovate-elliptic or elliptic-oblong, 5–20 cm. long, 1·6–8·8 cm. wide, acute at the apex, rounded or abruptly and then gradually cuneate, subscabrid above with very minute bristly hairs and on the costa and nerves beneath; petiole 0·2–1 cm. long; stipules with 3–5 subulate subequal setae 0·6–1·4(–2) cm. long from a short base, with conspicuous colleters. Inflorescence small and ± 6-flowered or larger, lax, up to 25 cm. long, 15 cm. wide. Flowers whitish, sometimes tinged reddish, sweetly scented, showing limited dimorphism. Calyx shortly hairy; tube 5–7 mm. long; lobes subulate or minutely spathulate, equal, 0·6–2·4 cm. long, 0·5–2 mm. wide, if a trifle spathulate then apical portion acute, 1·7 mm. long, 1·5 mm. wide. Corolla-tube pubescent, 7·5–14·2 cm. long, 1–3·5 mm. wide at base, dilated to 3–7 mm. for apical 1·2–2 cm. long; lobes ovate-oblong, 1·2–2·1 cm. long, 3–7 mm. wide, acute; throat densely hirsute. Flowers with style exserted 0–1·7 cm. and anthers included or with both included, but none with exserted anthers. Fruiting inflorescence large, lax, with peduncle 10–12 cm. long. Fruits straw-coloured, oblong, subglobose or dorsally compressed, 1–1·7 cm. long, 0·9–1·4 cm. wide, 8–9 mm. thick, with a very nervose frill formed by the limb-tube of the calyx extending above the disc for 2·5–8 mm., minutely pubescent, the bases of the calyx-lobes persistent.

TANZANIA. Ufipa District: near R. Kalambo, above Kalambo Falls, 29 Mar. 1955, *Exell, Mendonça & Wild* 1313 ! & Mmemya Mt., 20 Feb. 1951, *Bullock* 3719 !; Njombe District: Livingstone Mts., Apr. 1899, *Goetze* 852 !

DISTR. **T4**, 7; Zaire, Zambia and Rhodesia

HAB. On rocky hills in rock crevices, etc. in *Brachystegia* woodland; 1450–2450 m.

SYN. [*P. longituba* sensu De Wild. & Th. Dur. in B.S.B.B. 37: 117 (1898); Engl. in E.J. 30: 412 (1901)*, *non* K. Schum.]

P. nobilis S. Moore var. *grandifolia* Verdc. in B.J.B.B. 23: 275 (1953). Type: Tanzania, Njombe District, Livingstone Mts., *Goetze* 852 (K, holo. !)

NOTE. The large-leaved variety scarcely needs a name.

9. **P. graniticola** *E. A. Bruce* in K.B. 1933: 146 (1933); T.T.C.L.: 516 (1949); Verdc. in B.J.B.B. 23: 276, fig. 32/G & J (1953). Type: Tanzania, Mwanza, Speke Gulf, *B. D. Burtt* 2497 (K, holo. !)

Succulent, sometimes subshrubby herb 0·6–1·5 m. tall, entirely glabrous save for the throat of the corolla and the style; stems often drying reddish or straw-coloured, woody. Leaf-blades elliptic-lanceolate, 11–15 cm. long, 2·8–6 cm. wide, acute or slightly acuminate at the apex, cuneate at the base; petiole 0–1 cm. long; stipules with 3–6 setae 0·5–1·5 cm. long arising from a narrow base and bearing rather conspicuous colleters. Inflorescences much branched, loosely cymose; pedicels and secondary peduncles 0·2–3·5 cm. long, the flowers often looking as if solitary, sometimes subtended by linear-lanceolate leaves 5·2 cm. long, 2·5 mm. wide. Flowers white, sweetly scented, showing limited dimorphism. Calyx-tube yellowish, 5–6 mm. long; lobes subequal, linear or subulate, 1–2·8 cm. long, 1–2 mm. wide at their base. Corolla-tube 7–12 cm. long, 1·5–2·5 mm. wide at the base, dilating to 2·5–3·5 for an apical distance of 8–10 mm.; lobes oblong-lanceolate, 1–1·3 cm. long, 2–3·5 mm. wide, papillate internally; throat hairy with long hairs for

* The *Goetze* 85 cited is undoubtedly an error for *Goetze* 852.

2 cm. Stamens always included. Style exserted or included, minutely scaly-tomentose. Fruit straw-coloured or brownish, ellipsoidal, 0·9–1·3 cm. long, 8–9 mm. wide, 6 mm. thick, the limb-tube of the calyx 2–3(–4) mm. tall; beak little raised and intravalvular opening small.

TANZANIA. Mwanza District: Mbarika, Kamata I., 21 Mar. 1952, *Tanner* 594!; Maswa District: Shanwa, Komali Hill, June 1935, *B. D. Burtt* 5136!; Tabora District: Nyembe Bolungwa, July 1914, *Hammerstein* 5!; Dodoma District: 9 km. W. of Bagamoyo, Rungwa Forest Reserve, Sulangi, 26 Jan. 1969, *Gilbert* 5279!
DISTR. T1, 4, 5; not known elsewhere
HAB. In cracks of granite rocks on exposed hill tops; 1125–1350 m.

Subgen. 4. **Chamaepentadoides**

Verdc. in B.J.B.B. 23: 277 (1953)

Subshrubs. Calyx-lobes equal and subulate or subequal, 2–3 spathulate at the apices. Flowers not dimorphic, the style always exserted.

10. **P. hindsioides** *K. Schum.* in E.J. 34: 330 (1904), as "*hindoioides*";* T.T.C.L.: 517 (1949); Verdc. in B.J.B.B. 23: 277 (1953); U.K.W.F.: 404 (1974). Types: Tanzania, W. Usambara Mts., Sakare–Manka, *Engler* 1056 (B, holo. †); Pare District, Usangi, *Haarer* 1362 (EA, neo.!, K, isoneo.!)

Woody herb 0·5–1·8 m. tall; stems erect or trailing, with crisped whitish or purple-brown hairs above, glabrescent below. Leaf-blades ovate-lanceolate, elliptic or narrowly rhomboid, 3·3–11 cm. long, 1·2–4·5 cm. wide, acute or somewhat acuminate at the apex, cuneate at the base, glabrous to shortly adpressed hairy above and densely hairy beneath on the midrib; petiole 0·5–2·3 cm. long; stipules with 2–5 linear setae up to 2–5 mm. long, the colleters conspicuous, orange or white, 0·5 mm. long. Inflorescences ± 15-flowered, consisting of terminal cymes, each with 2–5 pedicellate flowers subumbellately arranged; pedicels 1·5–3(–7) mm. long. Flowers white or tinged faint purplish, sweet smelling. Calyx-tube 3–4 mm. long, 2–3·5 mm. wide, glabrous to densely hairy; lobes either linear-spathulate to oblanceolate, 0·8–2 cm. long (rarely only 2·5–6 mm. long) with the spathulate portion ± 0·6–1·3 cm. long, 2–5 mm. wide, or all subequal, subulate, 3–4 mm. long, 0·5 mm. wide, or a mixture of both, all densely hairy. Corolla-tube 2·7–4·5(–5·5) cm. long, 1–3 mm. wide basally, dilating to 2–3·8 mm. for an apical 8 mm.; lobes elliptic or ovate-oblong to oblong-lanceolate, 0·6–1·2 cm. long, 2–5 mm. wide, glabrous to densely hairy outside and with the throat-hairs extending over the extreme bases. Stamens totally included. Style exserted for 0–3 mm., minutely tomentose; stigma 1·7–3 mm. long. Fruit 6·5–10 mm. long, 5–9 mm. wide, glabrous to hairy. Seeds very numerous, ± 1 mm. long.

KEY TO INFRASPECIFIC VARIANTS

Some of the calyx-lobes distinctly spathulate at their apices:
 Leaves velvety hairy, mostly small, up to 6(–9) cm. long . . . . . . a. var. **hindsioides**
 Leaves glabrescent or glabrous, larger:
  Internodes 3–9 cm. long; corolla-tube up to 4·5 cm. long; leaves glabrescent or glabrous . . . . c. var. **glabrescens**

* It is obvious that K. Schumann meant to name this species after the genus *Hindsia*; the published spelling *hindoioides*, which has been followed by all subsequent authors, must be due either to a slip of the pen or to a printer's error.

Internodes 0·8–2·5 cm. long, mostly ± 1 cm.; corolla-tube up to 5·5 cm. long; leaves glabrous or almost so . . . d. var. **parensis**
Calyx-lobes mostly subulate, subequal . . b. var. **williamsii**

a. var. **hindsioides**; Verdc. in B.J.B.B. 23: 277 (1953)

Calyx-lobes mostly very spathulate at the tips. Leaves small, up to 6(–9) cm. long, velvety hairy. Internodes 2·5–10 cm. long.

KENYA. Teita District: Ngangao Hill, July 1937, *Dale* in *F.D.* 3813! & same locality, 9 Feb. 1966, *Gillett, Burtt & Osborn* 17151! & summit of Mt. Kasigau, 1 June 1969, *Gillett* 18770!
TANZANIA. Pare District: Usangi, May 1928, *Haarer* 1362! & Usangi, Kamwala, Aug. 1925, *Haarer* H.703/25!; Lushoto District: W. Usambara Mts., Shagayu Forest, May 1953, *Procter* 223!
DISTR. **K**4 (*fide* U.K.W.F.), 7; **T**3; not known elsewhere
HAB. Upland bushland, forest margins, bracken, etc. on basement complex; 1250–1950 m.

b. var. **williamsii** *Verdc.* in B.J.B.B. 23: 279 (1953). Type: Tanzania, Lushoto District, Kishiuwi Hill, *G. R. Williams* 66 (EA, holo.!, K, iso.!)

Calyx-lobes subequal, subulate, not spathulate. Leaves larger, up to 11 cm. long, hairy. Internodes 0·8–12 cm. long.

TANZANIA. Lushoto District: W. Usambara Mts., W. Shagayu [Shagai] Forest, 15 May 1953, *Drummond & Hemsley* 2540! & Lushoto–Gare path, 2 June 1961, *Mgaza* 400! & Magamba-Shume Forest Reserve, June 1951, *Eggeling* 6145B!
DISTR. **T**3; not known elsewhere
HAB. Bracken scrub, steep open rocky hillsides, abandoned native cultivations; 1350–2000 m.

NOTE. *Semsei* 2868 (Lushoto District, Lushoto–Gare path, 21 June 1959) shows characters of both var. *hindsioides* and var. *williamsii*.

c. var. **glabrescens** *Verdc.* in B.J.B.B. 23: 281, fig. 32/E (1953). Type: Tanzania, W. Usambara Mts., Makuyuni area, *Koritschoner* 631 (EA, holo.!, K, iso.!)

Calyx with some lobes distinctly spathulate and others linear. Leaves up to 10 cm. long, glabrescent or glabrous. Internodes 3–9 cm. long.

TANZANIA. Lushoto District: W. Usambara Mts., Makuyuni area, June 1935, *Koritschoner* 631! & Lushoto–Mombo road, 2·4 km. SW. of Gare turnoff, 16 June 1953, *Drummond & Hemsley* 2934!
DISTR. **T**3; not known elsewhere
HAB. Unknown, apart from cliff face; 400–1300 m.

d. var. **parensis** *Verdc.* in B.J.B.B. 23: 281 (1953). Type: Tanzania, S. Pare Mts., Shengena, *Peter* 0.III.53 [55651] (B, holo.!, K, iso.!)

Calyx with some lobes spathulate and rest subulate. Leaves up to 9 cm. long, quite glabrous or with few hairs beneath. Internodes mostly very short, 0·8–2·8 cm. long, sometimes as many as 10 nodes in a 10 cm. length of stem.

TANZANIA. Pare District: S. Pare Mts., Shengena [Schengena], Feb. 1915, *Peter* 0.III.53!
DISTR. **T**3; not known elsewhere
HAB. Unknown; 2220 m.

Subgen. 5. **Longiflora**

Verdc. in B.J.B.B. 23: 281 (1953)

Erect, often woody herbs. Leaves opposite or often in whorls of 3–5. Calyx-lobes subequal, linear-deltoid. Flowers not dimorphic, the style always exserted.

Plant a tall herb, 0·3–2 m. tall; corolla-lobes 0·3–1·3 cm. long, 1·5–4 mm. wide:
Leaves usually paired; corolla-tube 2–4·2 cm. long, 0·5–1·5 mm. wide at the base . . 11. *P. longiflora*

Leaves in whorls of 3–5 (in Flora area); corolla-tube 3–13 cm. long, 1·5–3·3 mm. wide at the base . . . . 12. *P. decora*
Plant a short pyrophyte 13–30 cm. tall; corolla-tube 6·5–13 cm. long; lobes elliptic, 1·2–2 cm. long, 5–10 mm. wide . . . . 13. *P. lindenioides*

11. **P. longiflora** *Oliv.* in Trans. Linn. Soc., ser. 2, 2: 335 (1887); F.P.N.A. 2: 328 (1947); T.T.C.L.: 517 (1949); Verdc. in B.J.B.B. 23: 282, fig. 32/D & I (1953); F.P.U., ed. 2: 160 (1972); U.K.W.F.: 404, fig. (1974). Type: Tanzania, Kilimanjaro, *Johnston* (K, holo.!, BM, iso.!)

Shrubby herb to 2 m. tall, with 2–3 main stems from a woody rootstock; stems glabrous to densely covered with whitish or rusty-orange hairs. Leaves paired or in whorls of 3; leaf-blades lanceolate or rarely ovate-lanceolate, 5–15 cm. long, 0·7–3 cm. wide, acute at the apex, narrowed at the base, glabrous to velvety or pubescent, often ferruginous; sessile or petiole 0·5–1 cm. long; stipules with 3–7 linear setae 0·1–1·3 cm. long from a base 0·5–2·5 mm. long. Flowers white or bluish white or tinged purplish. Inflorescences 20–100-flowered, up to 14 cm. wide; primary peduncles up to 15 cm. long; bracteoles filamentous, up to 3 mm. long. Calyx-tube glabrescent to velvety, often ferruginous, ± 1·5–2 mm. long and wide; lobes subequal, linear, 3–7·5 mm. long. Corolla-tube 2–4·2 cm. long, 0·5–1·5 mm. wide at the base, dilated at the apex to 1·5–2·5 mm. wide for a distance of 3–5·5 mm., pubescent outside; lobes oblong or elliptic, 3–6 mm. long, 1·5–2·9 mm. wide, acute or rather blunt; throat hairy. Stamens entirely included. Style exserted 1–6 mm., tomentose with white scaly papillae. Fruit oblong, depressed, 3–6·5 mm. tall, 4–7·5 mm. wide, ribbed, shortly pubescent with orange-brown hairs. Fig. 23, p. 196.

UGANDA. Kigezi District: Kisoro, lava plains, May 1951, *Purseglove* 3628!; Mbale District: Sebei, 9·7 km. Kapchorwa–Kaburoron [Kabururoni], 12 Oct. 1952, *G. H. Wood* 416!; Masaka District: Kalisizo, Aug. 1945, *Purseglove* 1769!
KENYA. Nandi Escarpment, Chemuse (? Chemase), 25 July 1951, *Greenway & Doughty* 8533!; Nakuru District: 8–16 km. from Nakuru on road to Eldoret, 7 Dec. 1956, *Verdcourt* 1617!; N. Kavirondo District: W. Kakamega Forest Reserve, 9 July 1960, *Paulo* 518!
TANZANIA. Arusha District: Ngongongare, 3 Aug. 1951, *Greenway & Hughes* 8569!; Lushoto District: Mtai–Malindi road, near Kidologwai, 19 May 1953, *Drummond & Hemsley* 2659! & Sangerawe, 14 Oct. 1936, *Greenway* 4675!
DISTR. U2–4; K2–6, 7 (Kasigau); T2–4, 7; Zaire, Burundi, Rwanda and Malawi
HAB. Grassland, bushland, *Myrica*-bracken scrub, thicket and forest edges, sometimes in damp places, usually on volcanic soils; 1050–2450 m.

SYN. *P. longiflora* Oliv. var. *nyassana* Scott Elliot in J.L.S. 32: 433 (1896). Type: Malawi, *Buchanan* 475 (K, holo.!, US, iso.!)
Neurocarpaea longiflora (Oliv.) S. Moore in J.L.S. 37: 157 (1905)
Pentas longiflora Oliv. forma *glabrescens* Verdc. in B.J.B.B. 23: 286 (1953). Type: Kenya, S. Kavirondo District, Kisii, *Napier* in *C.M.* 10183 (EA, holo.!)

NOTE. There are two sheets in the Kew Herbarium, *Butler* 20 & 23, said to be from localities on the Kenya coast and from altitudes of 120–180 m. From the known habits of the species these are highly improbable localities which I am not prepared to accept.

12. **P. decora** *S. Moore* in J.B. 48: 219 (1910); Verdc. in K.B. 7: 362 (1952) & in B.J.B.B. 23: 287 (1953); U.K.W.F.: 404, fig. (1974). Type: Zaire, Luendarides, *Kassner* 2419 (BM, holo.!, K, iso.!)

Herb 0·3–1·5 m. tall; stems usually single, mostly somewhat woody, glabrous to sparsely pubescent or densely hairy with short hairs. Leaves in whorls of 3–5 or rarely opposite; leaf-blades narrowly to broadly elliptic,

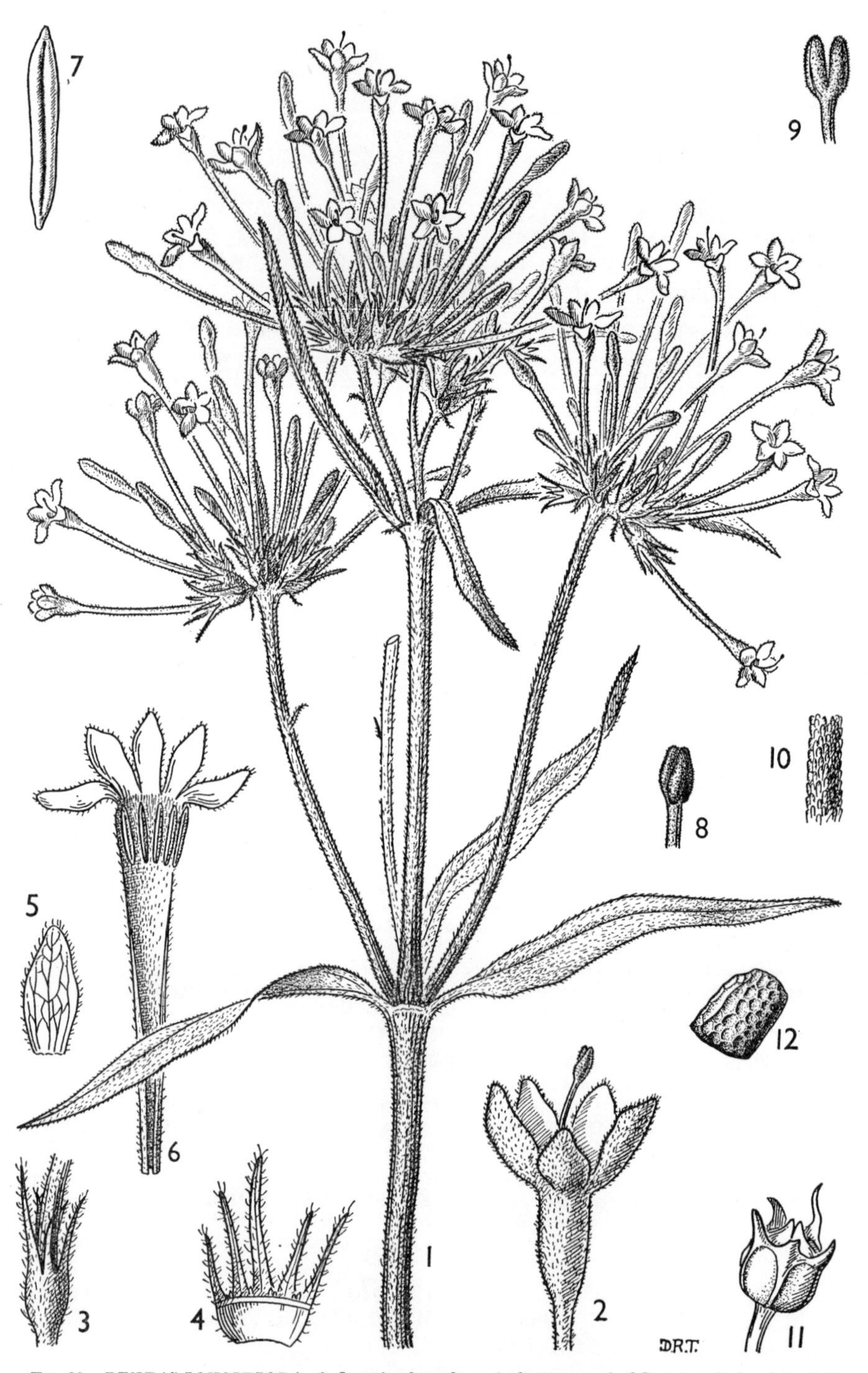

FIG. 23. *PENTAS LONGIFLORA*—**1,** flowering branch, × 1; **2,** upper part of flower, × 3; **3,** calyx, × 3; **4,** calyx, opened out, × 3; **5,** corolla-lobe, × 3; **6,** corolla opened out to show stamens, × 2; **7,** stamen, × 6; **8,** tip of style and stigma-lobes when young, × 6; **9,** same when older, × 6; **10,** detail of style, × 18; **11,** fruit, × 2; **12,** seed, × 24. All from *E. Brown* 134. Drawn by Miss D. R. Thompson.

elliptic-lanceolate, lanceolate or narrowly ovate, 3·9–13 cm. long, 0·9–5·6 cm. wide, subacute or rounded at the apex, rounded to cuneate at the base, glabrous to densely velvety; petiole obsolete; stipules with 1–5 ciliate setae, 0·5–8(–10) mm. long from a short base; often one deltoid lobe with a colleter on either side of it between the leaves in plants with 4–5 leaves in a whorl. Inflorescence consisting of 1 terminal and 2–4 axillary ones amalgamated, up to 12 cm. wide, each cluster with ± 30 flowers; peduncles 0–7·5 cm. long. Calyx glabrous to hairy; tube 2–5 mm. long, 2–4 mm. wide; lobes deltoid to filiform, 3–11 mm. long, 1–1·5 mm. wide at base, subulate at the apex, ciliate. Flowers white, sweet-scented. Corolla-tube 3–13 cm. long, 1·5–3·3 mm. wide at the base, scarcely dilated at the apex to 2·5–3·5 mm. wide for a distance of 1·2–1·7 cm., shortly hairy outside; throat densely hairy with long hairs; lobes sometimes yellowish, linear-oblong to elliptic-lanceolate, (0·5–)0·8–1(–1·8) cm. long, 1·5–4·5 mm. wide, acute, hairy outside. Stamens entirely included. Style exserted (0–)0·3–1·8 cm., densely tomentose with white papillae; stigma-lobes blackish, elliptic, 1–3 mm. long, thickened. Fruit obovoid-obtriangular, 1–1·5 cm. long, 0·7–1·1 cm. wide, prominently ribbed, pubescent; calyx-lobes reflexed, the limb-tube nervose, 3·5–4 mm. tall, exceeding the 2 mm. long beak. Seeds brown, ovoid, 1–1·5 mm. long, conspicuously reticulate.

KEY TO INFRASPECIFIC VARIANTS

Leaves mostly glabrous; corolla-tube up to 12 cm. long; individual inflorescence components ± 30-flowered:
- Capsule robust, conic, strongly ribbed . . var. **decora**
- Capsule ellipsoidal, not ribbed . . . var. **lasiocarpa**

Leaves glabrescent to velutinous; corolla-tube often only 5–6 cm. long; inflorescence usually many-flowered . . . . . . var. **triangularis**

var. **decora**; Verdc. in B.J.B.B. 23: 287 (1953)

Leaves mostly glabrous, less often sparsely pubescent or rarely hairy. Inflorescence of 1–5 clusters, each ± 30-flowered. Corolla-tube 3–12 cm. long. Capsule obovoid-obtriangular, strongly ribbed.

UGANDA. Acholi District: Imatong Mts., Apr. 1938, *Eggeling* 3566!; Karamoja District: Mt. Morongole, July 1965, *J. Wilson* 1663!; Kigezi District: Ruhinda, Jan. 1951, *Purseglove* 3536!
KENYA. W. Suk District: Kapenguria, July 1961, *Lucas* 200!; Trans-Nzoia District: Kitale Grassland Research Station, 27 July 1951, *G. R. Williams* 281!; N. Kavirondo District: Kakamega Forest Station, 17 Sept. 1949, *Maas Geesteranus* 6273!
TANZANIA. Biharamulo District: 45 km. from Biharamulo on road to Kahama, Rugongo, 24 Nov. 1962, *Verdcourt* 3450!; Ufipa District: near Mmemya Mt., 15 Feb. 1951, *Bullock* 3695!; Iringa District: Sao Hill, Feb. 1959, *Watermeyer* 30!
DISTR. **U**1, 2; **K**2, 3, 5; **T**1, 2, 4, 7; Zaire, Sudan, Zambia, Malawi, Angola
HAB. Grassland, bushland and *Combretum*, etc. woodland; 1060–2400 m.

SYN. *P. verticillata* Scott Elliot in J.L.S. 32: 431 (1896), *nomen nudum*
P. homblei De Wild. in F.R. 13: 109 (1914). Type: Zaire, Katanga, Kisangwe [Shisangwe], *Homblé* 94 (BR, holo.!)
[*P. longituba* sensu K. Krause in N.B.G.B. 10: 603 (1929), *non* K. Schum.]

var. **lasiocarpa** *Verdc.* in B.J.B.B. 23: 293 (1953). Type: Tanzania, Kigoma District, Ujiji, Machazo–Mkuti, *Peter* 46221 (B, holo.!)

Leaves quite glabrous. Inflorescence as in typical variety, but corolla-tube 7–7·5 cm. long. Capsule oblong-elliptic, 7 mm. long, 6·5 mm. wide; ribs extremely obscure.

TANZANIA. Kigoma District: Ujiji, Machazo–Mkuti, Nov. 1926, *Peter* 46221! & Ujiji, Mkuti R., Msosi, Feb. 1926, *Peter* 37187!
DISTR. **T**4; not known elsewhere
HAB. *Brachystegia* woodland; 960 m.

NOTE. As I pointed out in 1953, from the fruit alone this variant would appear to be exceedingly distinct but flowering material from the same locality is identical with var. *decora*. There is still a possibility that the type of var. *lasiocarpa* might be an abnormality.

var. **triangularis** (*De Wild.*) *Verdc.* in B.J.B.B. 23: 291 (1953); Hepper in F.W.T.A., ed. 2, 2: 215 (1963). Type: Zaire, Katanga, Lualaba Kraal, *Homblé* 934 (BR, holo.!)

Leaves pubescent to densely velvety, rarely glabrous. Inflorescence usually many-flowered. Corolla-tube 3·3–9·5 cm. long, but usually shorter and narrower than in the typical variety. Capsule as in typical variety.

UGANDA. Mengo District: Entebbe, Dec. 1930, *Hansford* 1884! & Kivuvu and Kipayo, June 1915, *Dummer* 43! & 51 km. from Kampala on Masaka road, July 1937, *Chandler & Hancock* 1777!

KENYA. Trans-Nzoia District: near Kitale, Aug. 1954, *Tweedie* 1189! & Elgon, Oct.-Nov. 1930, *Lugard* 9!; Nandi District: Sibu, *James*!

TANZANIA. Mbulu District: Great North Road, Pienaars Heights, 8 Jan. 1962, *Polhill & Paulo* 1102! & Ufiome Mt., 1 Jan. 1930, *B. D. Burtt* 2355!; Njombe, 28 Dec. 1931, *Lynes* C.61!

DISTR. **U**2, 4; **K**3, ? 5; **T**2, 4, 7; Nigeria, Central African Republic, Zaire and Ethiopia, Malawi (intermediate with *P. lindenioides*)

HAB. Grassland, grassland with scattered trees, *Brachystegia* woodland; 1300–2100 m.

SYN. *P. verticillata* Scott Elliot var. *pubescens* S. Moore in J.B. 48: 220 (1910), *nom. inval.* Type: Uganda, Mengo District, Mabira Forest, *E. Brown* 470 (BM, holo.!)

P. triangularis De Wild. in F.R. 13: 139 (1914)

P. globifera Hutch. in K.B. 1921: 374, fig. 5 (1921); Hutch. & Dalz., F.W.T.A. 2: 129 (1931). Type: Nigeria, *Lely* 386 (K, holo.!)

NOTE. (on species as a whole). It is possible that this species should be called *P. liebrechtsiana* De Wild., the type of which (Zaire, Katanga, Lukafu, *Verdick* 391 (BR, holo.!)) has very slender corolla-tubes and is intermediate between *P. decora* and *P. longiflora* (see Verdc. in B.J.B.B. 23: 287 (1953)). Two specimens from W. Tanzania are virtually indistinguishable from *P. liebrechtsiana*, e.g. *Pirozynski* 236 (Buha District, Gombe Stream Reserve, Kakombe Valley, 10 Jan. 1964). They may also be equivalent to *P. decora* var. *lasiocarpa*. Further material is needed from the Katanga to show variation before the name *decora* is given up.

13. **P. lindenioides** (*S. Moore*) *Verdc.* in K.B. 30: 344 (1975). Type: Malawi, Nyika Plateau, *Henderson* (BM, holo.!)

Pyrophytic herb 13–20(–30) cm. tall, but otherwise very similar to *P. decora*; stems simple, erect, hairy. Leaves opposite or in whorls of 3; leaf-blades elliptic, 3·2–11 cm. long, 1·4–3 cm. wide, ± acute at the apex, rounded or cuneate at the base, hairy; petioles 0–3 mm. long; stipules small, with 1–5 deltoid segments 4 mm. long. Inflorescences terminal, dense, 4–12-flowered; peduncles 1–1·5 cm. long. Flowers white, greenish or blue. Calyx-tube 2–4 mm. long, hairy; lobes deltoid to narrowly triangular, subequal, 4–6·5 mm. long. Corolla-tube 6·5–13 cm. long, 1–3 mm. wide, hairy outside; throat densely hairy, the hairs spreading over the inside of the lobes; lobes elliptic or oblong-obovate, 1·2–2 cm. long, 5–10 mm. wide, hairy outside. Stamens entirely included. Style exserted for 4–8(–12) mm., minutely tomentose; stigma-lobes 2 mm. long. Capsule not seen.

TANZANIA. Mbeya District: Kitulo [Elton] Plateau, 29 Nov. 1963, *Richards* 18437!; Rungwe District: Livingstone Forest Reserve, Dec. 1959, *Procter* 1613! & Upper Kiwira Fishing Camp, between Poroto and Rungwe Mts., 29 Nov. 1958, *Napper* 1142!; Njombe District: Poroto ridge road 35 km. E. of junction with Mbeya–Tukuyu road, 11 Nov. 1966, *Gillett* 17739!

DISTR. **T**7; Malawi, Zambia (Nyika)

HAB. Upland grassland subject to burning; 1980–2820 m.

SYN. *Heinsia lindenioides* S. Moore in J.L.S. 37: 301 (1906)

Pentas geophila Verdc. in B.J.B.B. 23: 293 (1953). Type: Tanzania, Njombe District, Livingstone Mts., Kitulo [Elton] Plateau, *Davies* E.14 (K, holo.!)

NOTE. This could be treated as a local subspecies of *P. decora* S. Moore since intermediate plants have been recorded, e.g. Tanzania, Mbeya Mt., 13 May 1956, *Milne-Redhead & Taylor* 10235 !; in fact plants very like *P. decora*, but with longer corolla-tubes and larger corolla-lobes, i.e. differing from *P. geophila* only in stature, are common in **T**7. The correct answer may be to recognise a larger-flowered variant of *P. decora* which may have a small pyrophytic ecotype. This must be decided by studying populations in the field.

Subgen. 6. **Pentas**

Verdc. in B.J.B.B. 23: 294 (1953)

Herbs or subshrubs. Leaves usually paired, in whorls of 3–4 only in two species. Calyx-lobes ± unequal, not subulate, 1–3 foliaceous, enlarged, elliptic, lanceolate, oblong or deltoid. Flowers small to large, the corolla-tube 0·4–9 cm. long, usually dimorphic, white or brightly coloured.

Corolla-tube up to 4 cm. long, usually much shorter:
- Flowers bright vermilion-scarlet; leaves with a very fine characteristic indumentum beneath; stem indumentum mostly ferruginous when dry; capsule ovoid-oblong, a little contracted above:
  - Leaves up to 6·5 cm. wide, rarely under 3 cm. and then lateral nerves 7–10 on each side and indumentum usually velvety . . . . . . 14. *P. bussei*
  - Leaves nearly always under 2 cm. wide, rarely up to 3 cm. and then lateral nerves 3–6 on each side and indumentum sparse . . . . . . 15. *P. parvifolia*
- Flowers white, mauve, blue or pink, only rarely red and then of a deeper crimson shade; leaf indumentum coarser and capsule obtriangular:
  - Flowers white, not dimorphic*, style always exserted; leaves thin in texture, with lower petioles 1–4(–7) cm. long; corolla-tube 7–9·5(–11) mm. long . . . 16. *P. micrantha*
  - Flowers white or coloured, dimorphic:
    - Corolla-tube usually under 10 mm. long, rarely up to 11 mm.:
      - Leaves in whorls of 3–4; herb 16–20(–30) cm. tall; plant of upland grassland in **U**1 . . . . . 19. *P. purseglovei*
      - Leaves paired; mostly taller herbs or shrubs:
        - Flowers and fruits in laxer cymose inflorescences; corolla-lobes devoid of hairs on the inner face or if present (species 21) then corolla white or pale:
          - Taller herbs with 1–3 stems from a taproot; corolla-tube usually 4–10 mm. long:

* This is not an easy character; from single specimens it may be impossible to tell unless the anthers are exserted, but it is a valuable character often easily observed in populations in the field.

Corolla reddish or purple, more rarely white; tube usually 8–10 mm. long . . . 17. *P. zanzibarica*
Corolla white or blue; tube 4–5 mm. long . . . 18. *P. pubiflora*
Pyrophytic herb with numerous short stems from a thick woody rootstock; corolla ± white, the tube 4–4·5 mm. long . 21. *P. arvensis*
Flowers and fruits in very dense globular heads, or if rather laxer then throat hairs extending up over the inner surface of the corolla-lobes which are often erect; flowers indigo or deep mauve . . . . . . 20. *P. purpurea*
Corolla-tube 1–4 cm. long . . . 22. *P. lanceolata*
Corolla-tube 3–9 (mostly ± 7) cm. long (Kenya, Mt. Suswa) . . . . 23. *P. suswaensis*

Section *Coccineae*

Verdc. in B.J.B.B. 23: 296 (1953)

Shrubs or subshrubs with woody erect or scrambling stems. Corolla distinctly vermilion-scarlet. Flowers dimorphic. Capsule eventually dividing into 2 cocci and showing little apical dehiscence. Species 14, 15.

This section has distinct affinities with the genus *Otomeria* (see Verdc. in B.J.B.B. 23: 246 and fig. 30) and is well characterised.

14. **P. bussei** *K. Krause* in E.J. 43: 134 (1909); T.T.C.L.: 517 (1949); Verdc. in B.J.B.B. 23: 297, fig. 33/C, E–H (1953). Type: Tanzania, Lindi District, Rondo [Mwera] Plateau, *Busse* 2628 (B, holo. †, BR, EA, K, iso. !)

Erect or somewhat scrambling shrub or herb 0·6–4 m. tall; stems with sparse to dense white or brown hairs above, usually drying ferruginous or yellowish, sometimes densely velvety. Leaf-blades ovate-lanceolate or ovate-oblong, 3·5–15 cm. long, 1·7–6·5 cm. wide, acute or acuminate at the apex, cuneate at the base, mostly very discolorous, sparsely shortly hairy above, pubescent to velvety beneath with very fine short white hairs; petiole up to 2 cm. long; stipules with 3–9 linear setae 0·4–1·5 cm. long from a triangular base 1–6 mm. long. Inflorescences terminal and axillary, dense or lax, up to 8 cm. wide, many-flowered; peduncles up to 4 cm. long. Calyx-tube ± 1·5 mm. long and wide, glabrous or velvety; lobes foliaceous, very unequal; 1–3 much enlarged, narrowly lanceolate, 0·5–1·8 cm. long, 1–4 mm. wide, acute, 3-nerved, pubescent, the rest small, linear, deltoid or lanceolate, 1·5–7 mm. long, 0·3–1·3 mm. wide. Flowers vermilion-scarlet. Long-styled flowers with corolla-tube 0·7–2 cm. long, 0·5–1·5 mm. wide at the base, dilated at the apex to 1·5–3 mm. wide for an apical cylindrical portion of 2·5–5(–7) mm., glabrous to hairy outside, with white hairs inside save at the throat where they are scarlet; lobes narrowly oblong, elliptic or oblong-lanceolate, (2·5–)3–10(–12) mm. long, 1–4·5 mm. wide; stamens entirely included; style usually bluish purple, exserted 1–9 mm., the stigmas scarlet or purple, 1·5–4·5 mm. long. Short-styled flowers very similar; corolla-tube 0·7–1·7 cm. long, 0·5 mm. wide at the base, gradually expanding to the throat which is 1–2·5 mm. wide; exserted anthers usually blue-purple. Fruiting inflorescences with individual branches ± spicate. Capsules oblong to obovoid, 3–6 mm. long, 2·25–3·5 mm. wide, 10-ribbed, crowned with the persistent calyx-lobes; beak only slightly raised.

KENYA. Kwale District: Buda Mafisini Forest, 16 Aug. 1953, *Drummond & Hemsley* 3813!; Kilifi District: Mida, Sept. 1929, *Graham* in *F.D.* 2065! & Mtwapa creek, 3 June 1934, *Napier* 3361 in *C.M.* 6286!

TANZANIA. Tanga District: Bomalandani, 10 Aug. 1953, *Drummond & Hemsley* 3687!; Buha District: Kasakela Reserve, 17 Nov. 1962, *Verdcourt* 3338!; Kilosa District: Ilonga, 25 May 1968, *Renvoize* 2348!; Zanzibar I., Nungwi, 10 July 1950, *Oxtoby* 28! & same locality, 10 Apr. 1963, *Faulkner* 3175!

DISTR. **K**2, 4 (intermediates with sp. 15), 7; **T**1–8; **Z**; Somali Republic (S.), Zaire, Burundi, Zambia and Malawi; also cultivated in Trinidad, New Guinea and formerly at Kew

HAB. Grassland, bushland and woodland (including *Brachystegia*, etc.), dry evergreen forest, also on coral rocks by sea, rocky slopes, disturbed places; 0–1800 m.

SYN. [*P. klotzschii* sensu Vatke in Oest. Bot. Zeitschr. 25: 231 (1875), quoad *Hildebrandt* 1124]
P. coccinea Stapf in Bot. Mag. 149, t. 9005 (1924). Type: Tanzania, Amani, a plant cultivated at Kew from seeds sent by *Rogers* (K, holo.!)
P. flammea Chiov., Fl. Somala 2: 231 (1932). Type: Somali Republic (S.), Upper Juba, Baddada, *Senni* 395 (FI, holo.!)

NOTE. There are some nomenclatural difficulties concerned with this species. The possibility that it should be called *P. zanzibarica* (Klotzsch) Vatke has been discussed in detail by me (B.J.B.B. 23: 300 (1953)). The evidence for this is a fragment of type preserved at Kew, but I think some confusion may have occurred and refuse to upset the nomenclature on inadequate evidence. The name *zanzibarica* has been used consistently for another species for a very long while and the utmost confusion would result from any changes. In B.J.B.B. 23: 302 (1953) I described three forms of *P. bussei*: forma *brevituba* Verdc. (type: Tanzania, Rufiji, *Musk* 114 (EA, holo.!)), recorded from Kenya and Tanzania, and distinguished by having a corolla-tube 7–10 mm. long; forma *minor* Verdc. (type: Tanzania, W. Usambara Mts., near Makuyuni, Gomba [Ngomba], *Zimmermann* in *Peter* 0.III.213 [55818] (B, holo.!)), recorded from Tanzania and Zaire, and an arbitrary way of disposing of certain plants with leaf-blades 4·8–9 × 1·3–6 cm., which are intermediate between this and the next species; forma *glabra* Verdc. (type: Tanzania, Uluguru Mts., above Morogoro, SW. of Silesian [Schlesien] Mission Station, *Peter* 32165 (B, holo.!, K, iso.!)), known only from Tanzania, with glabrous foliage. Some specimens said to have been collected in Angola by J. Gossweiler in 1907 must I think be wrongly labelled.

15. **P. parvifolia** *Hiern* in J.L.S. 16: 262, t. 7 (1877); T.T.C.L.: 518 (1949); Verdc. in B.J.B.B. 23: 303 (1953); U.K.W.F.: 404, fig. (1974). Type: Kenya, Mombasa, *Hildebrandt* 1994 (BM, holo.!, K, W, iso.!)

Subshrub 0·6–2·5 m. tall, with erect or straggling often irregularly branched stems but not a true liane; in straggling forms the branches are frequently very unequally developed on opposite sides of a node; stems pale or purplish brown, woody, pubescent with white or ferruginous hairs above, glabrous below and the epidermis often peeling. Leaf-blades elliptic-lanceolate, ovate-lanceolate or oblong-elliptic, 1·2–9(–10·5) cm. long, 0·3–0·8(–2) cm. wide, acute at the apex, cuneate at the base, glabrescent to pubescent above, with fine white pubescence beneath; petiole 0–6 mm. long; stipules with 5–9 setae up to 6 mm. long from a brown base up to 4 mm. long. Inflorescence terminal, combined with axillary ones from the apical node, mostly 2–3 cm. wide, often only few-flowered; in one form the branches are distinctly spicate. Flowers scarlet, rarely pink and white, mostly dimorphic, but sometimes practically trimorphic. Calyx-tube obovoid, ± 1·5 mm. long and wide, glabrous to velutinous; lobes 5–6, unequal, the largest 0·3–1·2 cm. long, 0·7–2(–6) mm. wide, the rest short, 1·5 mm. long. Short-styled flowers; corolla-tube 0·7–1·8(–2·2) cm. long, 0·7 mm. wide, gradually dilated to 1–2 mm. at the apex; lobes oblong to linear-oblong, 2·5–10 mm. long, 1–3·3 mm.

wide, glabrous or pubescent outside; stamens with anthers well exserted or rarely almost included; throat densely hairy. Long-styled flowers similar; corolla-tube enlarged cylindrically at apex to 1·5–5 mm. for a distance of 2–4 mm.; style exserted 1–6 mm.; stigma 2–2·5 mm. long; stamens completely included. Capsules as in *P. bussei.*

forma **parvifolia**; Verdc. in B.J.B.B. 23: 303 (1953)

Fruiting inflorescences compact, the branches not markedly spicate.

UGANDA. Karamoja District: Moroto Mt., 5 Oct. 1952, *Verdcourt* 767 ! & same locality, 4 Sept. 1956, *Bally & Hardy* 10717 ! & same locality, June 1942, *Dale* 251 !
KENYA. Northern Frontier Province: Mathews Range, Wamba, 28 Nov. 1958, *Newbould* 2916 !; Machakos District: Makindu, 10 Apr. 1902, *Kassner* 562 !; Teita District: S. of Voi, Sagala Hill, 11 Dec. 1961, *Polhill & Paulo* 956 !
TANZANIA. Pare District: NW. spur of N. Pare Mts. above Kifaru Estate, 19 May 1968, *Bigger* 1851 !; Uzaramo District: Dar es Salaam, Wazo Hill, 25 Apr. 1968, *Batty* 36 !; Lindi District: Rondo Plateau, Feb. 1951, *Eggeling* 6039 !; Zanzibar I., Ras Nungwi, 29 Apr. 1952, *Tidbury* in *Williams* 169 ! (intermediate)
DISTR. **U**1; **K**1–4, 6, 7; **T**2, 3, 6–8; **Z** (intermediate); S. Ethiopia (Moyale)
HAB. Grassland with scattered trees, thickets, dry evergreen forest, *Brachystegia* woodland, bushland, rocky hills; 0–2400 m.

SYN. *P. mombassana* Oliv. in Trans. Linn. Soc., ser. 2, 2: 335 (1887); Britten in *op. cit.* 4: 16 (1894), *nomen nudum*

NOTE. I originally separated a forma *intermedia* (B.J.B.B. 23: 305 (1953); type: Tanzania, E. Usambara Mts., Sigi, *Verdcourt* 200 (EA, holo. !, K, iso. !)), with leaf-blades 1–2 cm. wide, but it seems hardly worth separating. There are on *Scheffler* 4 (Kenya, Machakos District, Kibwezi, Dec. 1905) leaf-blades up to 6·4 × 3 cm., but these have only 5 lateral nerves. It seems better placed here than in *P. bussei.* There are intermediates between the two species but they are undoubtedly to be maintained as distinct species.

forma **spicata** *Verdc.* in B.J.B.B. 23: 306 (1953). Type: Kenya, Chyulu foothills, *Bally* 735 (EA, holo. !)

Fruiting and sometimes flowering inflorescences with the branches distinctly spicate.

KENYA. Kiambu District: Fourteen Falls, near where Donyo Sabuk road crosses R. Athi, 2 Jan. 1960, *Verdcourt* 2606 !; Machakos District: Tsavo National Park, Mtito Andei–Mzima Springs road, 17 Jan. 1961, *Greenway* 9755 !; Masai District: Garabani Hill, 26 Mar. 1940, *V. G. L. van Someren* 217 !
DISTR. **K**4, 6, 7; not known elsewhere
HAB. Seashore scrub, grassland with scattered trees, and open bushland; 0–1710 m.

Section *Monomorphi*

Verdc. in B.J.B.B. 23: 307 (1953)

Biennial or perennial herbs with often thin leaves. Flowers small, mostly white, not dimorphic; style always exserted. Species 16.

16. **P. micrantha** *Bak.* in J.L.S. 21: 408 (1885); Verdc. in B.J.B.B. 23: 307 (1953). Type: Madagascar, Tanala Forest, *Baron* 3292 (K, holo. !)

Herb 45–90 cm. tall, ? annual, biennial or a short-lived perennial (but habit of typical race not certain), stems hairy when young. Leaves ovate to ovate-oblong or elliptic, 5·5–14·5 cm. long, 1·5–6·3 cm. wide, acute at the apex, cuneate at the base, thin and membranous, sparsely or moderately pubescent above and on the venation beneath; petiole 0·3–3·5(–7) cm. long; stipules with 4–7 setae up to 6 mm. long from a base 0·5–3·5 mm. long, hairy. Inflorescence a small lax cluster ± 1·5 cm. wide or trichotomous; individual cymes lax, few-flowered, 2–3 cm. wide; flowers white or lavender-blue; peduncles 0–6·5 cm.; pedicels ± 1·5 cm. long. Calyx-tube 1–1·5 mm. long,

1–1·5 mm. wide, hairy; lobes unequal, the longest lanceolate, (3–)4–8(–10) mm. long, 1–3·5 mm. wide, rest 1–2·3 mm. long, 0·25–0·5 mm. wide. Buds characteristically clavate. Corolla-tube 6·5–10(–11) mm. long, 0·8–1·5 mm. wide at the base, abruptly expanded at the apex to 1·5–2·5 mm. for 1·5–2 mm., the resulting dilated portion being urceolate, all hairy outside; lobes ovate, 1·8–3 mm. long, 1–1·75 mm. wide, acute, hairy on the main nerve outside. Stamens completely included. Style glabrous, exserted 0–2·5 mm.; lobes of stigma 0·75–1·5(–2) mm., mostly practically immersed in the dense throat hairs, sometimes the lower parts immersed in the corolla-tube. Capsules obtriangular or oblong-ovoid, 2·5–5 mm. long and wide, prominently ribbed, glabrous or pubescent. Seeds minute, 0·5 mm. long, bluntly angular.

subsp. **wyliei** (*N.E. Br.*) *Verdc.* in B.J.B.B. 23: 308, fig. 33/I (1953) & in K.B. 12: 355 (1957) & in K.B. 30: 287 (1975). Type: South Africa, Natal, Ungoya, *Wood* 7590 (K, holo. !, BM, BOL, NH, PRE, US, iso. !)

Lower petioles mostly ± 1–4(–7) cm. long. Corolla-tube 7–9·5(–11) mm. long.

TANZANIA. Lushoto District: Amani–Muheza road, Sigi, 19 Apr. 1968, *Renvoize* 1593 !; Uzaramo District: Pugu Hills, July 1958, *Tweedie* 1643 !; Rufiji District: Mafia I., Kirongwe, 25 Aug. 1937, *Greenway* 5160 !; Zanzibar I., 27 km. on Chwaka road, 28 Dec. 1961, *Faulkner* 2965 !; Pemba I., Pandani, 13 Dec. 1930, *Greenway* 2723 !

DISTR. **T**3, 6, 8; **Z**; **P**; Mozambique, Zambia and South Africa (Natal)

HAB. Bushland, woodland and forest, sometimes on rock outcrops and in seepage areas, and in cultivations; 0–750 m.

SYN. *P. wyliei* N.E. Br. in K.B. 1901: 123 (1901); Wood, Natal Plants 4, t. 344 (? 1904)
P. extensa K. Krause in Z.A.E. 2: 311 (1911), excl. specim. ex Kivu, *nomen nudum*
P. zanzibarica (Klotzsch) Vatke var. *membranacea* Verdc. in B.J.B.B. 23: 323 (1953), quoad typum solum. Type: Tanzania, Lindi District, Rondo Plateau, *Eggeling* 6040 (EA, holo. !, K, iso. !)
P. zanzibarica (Klotzsch) Vatke var. *pembensis* Verdc. in B.J.B.B. 23: 324 (1953). Type: Tanzania, Pemba I., Ngezi Forest, *Vaughan* 608 (EA, holo. !, K, iso. !)

NOTE. It appears I was misled by small differences in corolla shape, etc. when I described the two varieties of *P. zanzibarica* mentioned in synonymy above. They do not seem to be distinguishable from *P. micrantha* subsp. *wyliei*. The typical subspecies occurs in Madagascar.

Section *Pentas*

Verdc. in B.J.B.B. 23: 319 (1953)

Herbs or subshrubs. Flowers white or brightly coloured, mostly lilac to indigo, markedly heterostylous, always with forms with either style or anthers strongly exserted and occasionally with a third form with both included. Species 17–23.

17. **P. zanzibarica** (*Klotzsch*) *Vatke* in Oest. Bot. Zeitschr. 25: 232 (1875); T.T.C.L.: 518 (1949); Verdc. in B.J.B.B. 23: 319, fig. 33/A, B (1953); U.K.W.F.: 405 (1974). Types: Zanzibar, *Peters* (B, holo. †); Zanzibar, *Hildebrandt* 1128 (W, neo. !, BM, K, isoneo. !)

Herb or shrubby herb or rarely a shrub 0·3–2·6 m. tall, with 1–2(–6) stems from a somewhat woody rootstock; stems greenish or purple tinged, mostly strict and unbranched, often densely hairy above. Leaf-blades lanceolate to ovate or elliptic, (2·8–)4–14·5 cm. long, (0·5–)1·4–5·8 cm. wide, acute at the apex, cuneate at the base, hairy on both surfaces; petiole 0–1 cm. long; stipules with ± 7 setae 4–9(–14) mm. long from a short base. Inflorescences lax or somewhat globose, terminal and axillary, 2–6·5 cm. across; peduncles 0–15 cm. long; bracts 6 cm. long. Flowers white, pink or mostly lilac, bluish mauve or in one variety bright crimson-red. Calyx-tube hairy, 1–1·3 mm. long, 1–1·5 mm. wide; lobes unequal, 1–9 mm. long, 0·5–1·5(–2·5) mm. wide. Long-styled flowers; corolla-tube (4–)5–9(–11) mm. long, hairy; lobes oblong-elliptic, 1·5–5·5(–6·5) mm. long, 0·8–2·7(–3) mm. wide; style exserted 0·5–5·5 mm.; stigma-lobes 1·5–2·5 mm. long; stamens completely included; throat densely hairy. Short-styled flowers very similar, the anthers exserted 2–3 mm. and style and stigma completely enclosed. Capsule pubescent, (2–)3–4(–5·5) mm. long and wide, with beak 1–2 mm. tall.

KEY TO INFRASPECIFIC VARIANTS

Corolla usually white; tube 5–9 mm. long; leaves not particularly thin and membranous; W. Uganda . . . . b. var. **intermedia**
Corolla normally coloured, if white then only as albinos in a coloured population:
  Mostly subshrubby, with bright red-crimson flowers; W. Uganda . . . . . e. var. **rubra**
  Mostly herbaceous, with blue, lilac, mauve or pink flowers, rarely red* or whitish:
    Leaves 4–11·5 × 1·4–3·7 cm.; corolla variously coloured . . . . . a. var. **zanzibarica**
    Leaves 11–14·5 × 5·2–5·8 cm.:
      Corolla-tube about 6 mm. long (Tanzania, Moshi District) . . . . . c. var. **latifolia**
      Corolla-tube up to 10 mm. long; leaf-blades thin (Elgon) . . . . . d. var. **tenuifolia**

a. var. **zanzibarica**; Verdc. in B.J.B B. 23: 319, fig. 33/A, B (1953)

Herb or shrubby herb ± 0·6–1·8 m. tall. Leaves mostly ovate lanceolate, 4–11·5 cm. long, 1·4–3·7 cm. wide. Corolla blue, lilac, mauve or rarely whitish or red; tube (4–)6–9(–10) mm. long.

UGANDA. Ankole District: Mbarara, 17 Oct. 1925, *Maitland* 841!; Kigezi District: Lake Bunyonyi, Bufundi, Dec. 1938, *Chandler & Hancock* 2604!; Mengo District: 160 km. NW. of Kampala on the Mubende road, June 1915, *E. Brown* in *Dummer* 2660!
KENYA. Naivasha District: Hells Gate Valley, 12 Sept. 1964, *Richards* 19165!; Masai District: Olekaitororr Escarpment, 32 km. from Narok on Nairobi road, 14 July 1962, *Glover & Samuel* 3115!; Kwale District: Mrima Hill Forest edge, 4 Sept. 1957, *Verdcourt* 1867!
TANZANIA. Arusha District: Ngurdoto National Park, 10 Apr. 1965, *Richards* 20151!; Tanga District: Muheza [Muhesa], Bombwera, 18 July 1969, *Faulkner* 4246!; Morogoro District: Uluguru Mts., 21 Oct. 1935, *E. M. Bruce* 12!; Zanzibar I., Mkokotoni, 1 Apr. 1960, *Faulkner* 2524!
DISTR. **U**2, 4; **K**1, 3, ? 4, 5–7; **T**1–3, 5–8; **Z**; also in E. Zaire
HAB. Open grassland, grassland with scattered trees, forest glades and edges, cultivation, etc., sometimes on lava soils; 0–2600 m.

SYN. *Pentanisia zanzibarica* Klotzsch in Peters, Reise Mossamb., Bot. 1: 286 (1861)
Pentas stolzii K. Schum. & K. Krause in E.J. 39: 522 (1907), pro parte quoad *Uhlig* 270!
[*P. purpurea* sensu auctt., *non* Oliv.]

NOTE. The above concept certainly covers more than one minor race. A good deal of evidence could be listed for considering the common coastal race (e.g. Kenya, Kwale District, Shimba Hills, Makadara, *R. M. Graham* in *F.D.* 1692) worthy of formal distinction from the race occurring on upland volcanic soils (e.g. Kenya, Masai District, 48 km. along main road to Narok, 16 June 1956, *Verdcourt* 1500), but there are so many intermediate variants that the pattern is obscured. In my original

* See note at foot of this page.

monograph I placed anything with truly red flowers in var. *rubra*, but this is not correct, true var. *rubra* is shrubby and has a distinctive distribution. Odd specimens of var. *zanzibarica* occur which, according to their collectors, have truly red flowers. Most specimens from the Teita Hills (e.g. *Murray* 20, ? Wemgha, 6 Nov. 1965) have more ovate leaf-blades.

b. var. **intermedia** *Verdc.* in B.J.B.B. 23: 324 (1953); F.P.U., ed. 2: 162 (1972). Type: Uganda, Kigezi District, Kackwehano Farm, *A. S. Thomas* 3783 (EA, holo. !, K, ENT, iso. !)

Herb 0·6–1·3 m. tall. Leaf-blades 2·8–3·5 cm. long, 0·8 cm. wide. Corolla white; tube 5–9 (mostly about 5) mm. long.

UGANDA. Ankole District: Kichwamba, 26 June 1945, *A. S. Thomas* 4172 !; Kigezi District: Kackwekano Farm, May 1951, *Purseglove* 3615 ! & same locality, 3 July 1945, *A. S. Thomas* 4224 !
DISTR. **U**2; Rwanda, Burundi
HAB. Grassland, rocky hillsides; 1200–2100 m.

SYN. *P. extensa* K. Krause in Z.A.E. 2: 311 (1911), pro parte, *nomen nudum*

NOTE. The two specimens reported in my original revision from **K**3 do not appear to belong to this variety, but are more or less typical var. *zanzibarica*.

c. var. **latifolia** *Verdc.* in B.J.B.B. 23: 326 (1953). Type: Tanzania, Moshi District, Lyamungu, *Wallace* 983 (EA, holo. !)

Herb 0·6 m. tall. Leaf-blades 12 cm. long, 5·5 cm. wide. Corolla reddish pink; tube 6 mm. long.

TANZANIA. Moshi District: Lyamungu, Coffee Research Station, Sept. 1941, *Wallace* 983 ! & near Moshi, Mweka, 15 June 1968, *Harris* 1898 !
DISTR. **T**2; not known elsewhere
HAB. Clearings in evergreen forest; 1400–1440 m.

d. var. **tenuifolia** *Verdc.* in K.B. 30: 288 (1975). Type: Kenya, Elgon, Endebess, *Webster* in *Herb. Amani* 9644 (EA, holo. !)

Leaf-blades ovate to ovate-elliptic or elliptic, up to 12 cm. long, 5·6 cm. wide, thin; petiole up to 2·5 cm. long. Corolla probably pink; tube up to 10 mm. long.

KENYA. Trans-Nzoia District: Elgon, Endebess, *Webster* in *Herb. Amani* 9644 !
DISTR. **K**3; not known elsewhere
HAB. Not known, presumably evergreen forest; ± 1800 m.

SYN. *P. zanzibarica* (Klotzsch) Vatke var. *membranacea* sensu Verdc. in B.J.B.B. 23: 324 (1953), typo excluso

e. var. **rubra** *Verdc.* in B.J.B.B. 23: 322 (1953). Type: a specimen cultivated in Kenya, Nairobi, *Greenway* 8467 (EA, holo. !, K, iso. !)

Herb or shrub 0·5–2·6 m. tall. Leaves often bullate, blades 8–13(–17) cm. long, 2–5·8(–6·5) cm. wide. Corolla red or crimson-red, less often bright reddish pink, frequently with dense purple-red hairs but often glabrescent; tube 5–10 mm. long. Buds sometimes drying with a marked collar around the throat.

UGANDA. Toro District: Ruwenzori, Bwera, *Osmaston* 3747 !; Kigezi District: Mabungo, 20 Dec. 1933, *A. S. Thomas* 1089 ! & Bufumbira, Kisoro, May 1951, *Purseglove* 3627 !
DISTR. **U**2; E. Zaire, also cultivated in Kenya
HAB. Grassland and open bushland, mostly on lava plains; 1500–2700 m.

SYN. [*P. zanzibarica* sensu Robyns, F.P.N.A. 2: 330 (1947), excl. *Fries* 1668, *non* (Klotzsch) Vatke]
[*P. lanceolata* sensu Robyns, F.P.N.A. 2: 329 (1947); Jex-Blake, Gard. E. Afr., ed. 3, t. 17/6 (1949), *non* (Forssk.) Deflers]
[*P. coccinea* sensu auctt., *non* Stapf]

NOTE. As explained in the note after var. *zanzibarica*, my original conception of var. *rubra* included some red-flowered forms of var. *zanzibarica*. Var. *rubra* has been much cultivated in and around Nairobi under the name *P. coccinea* (i.e. *P. bussei*), but although the flowers are somewhat similar to that in colour, var. *rubra* is distinctly deeper crimson in shade and not so vermilion-scarlet; also the indumentum is much coarser and the capsules totally different in shape.

NOTE. (on species as a whole). The nomenclature of this species has already been mentioned under *P. bussei*, see note on p. 201. There is practically no dividing line between this species and the short-flowered variant of *P. lanceolata* (Forssk.) Deflers, but the number of intermediates occurring is sufficiently small to make joining them totally impractical.

Other variants occur in Malawi and Zambia.

18. **P. pubiflora** *S. Moore* in J.L.S. 38 : 254 (1908) ; F.P.N.A. 2 : 331 (1947) ; Verdc. in B.J.B.B. 23 : 327, fig. 35/B, C (1953) ; Hepper in F.W.T.A., ed. 2, 2 : 216 (1963) ; U.K.W.F. : 405, fig. (1974); Verdc. in K.B. 31 : 186 (1976). Type : Uganda, Ruwenzori, *Wollaston* (BM, holo. !)

Herb or subshrub 0·6–1·5(–3) m. tall ; stems erect or somewhat decumbent, woody below, hairy or glabrescent. Leaf-blades ovate to lanceolate, (3–)9·5–14·5 cm. long, (1·2–)3–5 cm. wide, acute at the apex, cuneate at the base, pubescent ; petiole 0–0·6(–3) cm. long ; stipules with ± (5–)7–9 filiform setae 0·5–1·2 cm. long from a short base. Inflorescences of terminal and axillary components, usually wide and corymbose, up to 12 cm. wide (smaller, 2–5·5 cm. wide, in the W. African race) ; peduncles ± 2–4·5 cm. long. Flowers white, rarely tinged pale blue or pinkish, or in W. African race blue. Calyx-tube 1·5 mm. long ; lobes unequal, 1 foliaceous, ovate 2–6·5 mm. long, 0·75–2 mm. wide, the rest deltoid or triangular, 1–2 mm. long. Long-styled flowers ; corolla-tube funnel-shaped, 4–5 mm. long, 1–1·25 mm. wide below, dilated above to 1·75 mm. ; lobes oblong-lanceolate, 2–3 mm. long, 1–1·25 mm. wide ; stamens entirely included ; style exserted (1–)2–3·5 mm. Short-styled flowers : similar, tube up to 5·5 mm. long, 2 mm. wide at the throat ; lobes 1·5–2 mm. long ; anthers exserted 0·5–1·25 mm. Capsule 2–4·5 mm. long, 2–3 mm. wide, hairy, ribbed ; beak 1 mm. tall.

subsp. **pubiflora** ; Verdc. in B.J.B.B. 23 : 327, fig. 35/B, C (1953)

Stems erect. Leaf-blades 10–13 cm. long, 3·5–4·5 cm. wide. Inflorescence up to 12 cm. wide. Flowers white tinged pale blue or pinkish. Capsule 4·5 mm. long, 3 mm. wide.

UGANDA. Kigezi District : Bukimbiri, May 1950, *Purseglove* 3388 ! & Kinaba Gap, Dec. 1938, *Chandler & Hancock* 2603 ! ; Mbale District : Bugisu, Sipi, 31 Aug. 1932, *A. S. Thomas* 417 !

KENYA. Trans-Nzoia District : NE. Elgon, Sept. 1954, *Tweedie* 1267 ! ; Nakuru District : Dundori, 9 July 1958, *Verdcourt* 2206 ! ; Kericho District : SW. Mau Forest, Sambret Catchment, 19 Sept. 1962, *Kerfoot* 4339 !

DISTR. **U**2, 3 ; **K**2, 3, 5 ; Zaire

HAB. Grassland, scrub, dry riverine forest and forest edges, sometimes on lava slopes ; 1230–2550 m.

SYN. *P. pubiflora* S. Moore var. *longistyla* S. Moore in J.L.S. 38 : 255 (1908). Type : Uganda, Ruwenzori, *Wollaston* (BM, holo. !)

NOTE. A specimen from Tanzania, Mpanda District, Kabwe R., just S. of Pasagulu, 7 Aug. 1973, *Harley* 9184, and another from Buha District, Gombe Stream Reserve, Mkenke Valley, Feb. 1972, *Parnell* 2287, belong to a glabrescent variant of *P. pubiflora* with corolla-tube slightly larger, about 6 mm. long ; similar variants occur in Burundi, Nigeria and Cameroun. Formerly I confused them with another taxon under the name *P. pubiflora* subsp. *bamendensis* but the type of this proves to be the same as *P. ledermannii* K. Krause which I mistakenly synonymised with *P. schimperana* (A. Rich.) Vatke subsp. *occidentalis* (Hook. f.) Verdc. More study is needed before a name can be given to this glabrescent variant of *P. pubiflora*.

19. **P. purseglovei** *Verdc.* in B.J.B.B. 23 : 329 (1953). Type : Uganda, Imatong Mts., Langia, *Purseglove* 1421 (EA, holo. !, K, iso. !)

Perennial herb up to 16–30 cm. tall ; stems simple, glabrescent or with 2 lines of hairs, presumably several from a woody rootstock. Leaves in whorls of 3–4 ; blades ovate to ovate-lanceolate, 2–6 cm. long, 1–2·2 cm. wide, acute at the apex, cuneate at the base, sparsely hairy above, densely so on the main nerves beneath or glabrescent save for marginal ciliae; petioles 0–1 mm. long; stipules small with 3 subulate lobes up to 4 mm. long. Flowers bluish green or dark purple, in small dense terminal inflorescences ± 3 cm. across; peduncles ± 5 mm. long. Calyx-tube ± glabrous, obconic, ± 1·5 mm. long; lobes glabrous, narrowly triangular to lanceolate, the largest 2·5–3·5(–6) mm. long, 1–1·5 mm. wide, others intermediate or minute (1 mm. long, 0·3 mm. wide). Corolla-tube 5 mm. long, widened at the apex to 1·5–2·5 mm., glabrescent outside, throat very densely hairy; lobes oblong-elliptic, 2·5–3 mm. long, 1 mm. wide. Long-styled flowers: stamens completely included; style exserted 1·5 mm.; stigma bifid, the lobes 1·5 mm. long. Short-styled flowers with anthers exserted 2 mm. Capsules not seen.

UGANDA. Acholi District: Imatong Mts., Langia, Apr. 1943, *Purseglove* 1421!
DISTR. U1; S. Sudan
HAB. Upland grassland; 2790 m.

NOTE. This is known only from two specimens, one from Uganda and one from the Sudan.

20. **P. purpurea** *Oliv.* in Trans. Linn. Soc. 29: 83 (1873); Verdc. in K.B. 7: 363 (1952) & B.J.B.B. 23: 330 (1953); Hepper in F.W.T.A., ed. 2, 2: 215 (1963). Type: Tanzania, Biharamulo District, Usui, *Grant* 140 (K, holo.!)

Herb 0·4–1·3 m. tall, mostly ± 30–45 cm., with 1–4 hairy or glabrescent mostly unbranched stems from a woody rootstock. Leaves paired; blades elliptic-, ovate- or oblong-lanceolate or elliptic, (3–)4·5–13·5 cm. long, 1–4·5 cm. wide, ± acute at the apex, rounded or cuneate at the base, pubescent above and on the venation beneath where the hairs have a bristly appearance; petiole obsolete or up to 7 mm. long; stipules with 1–10 setae, 1–9(–12) mm. long from a short base. Flowers mostly deep purple-violet or indigo but sometimes apparently paler, e.g. said to be lavender (*fide Emson* 355). Inflorescence a small dense head 1·3–3(–4) cm. in diameter, scarcely enlarging in fruit, or sometimes laxer up to 6·5 cm. wide, sessile or peduncle 2–20 cm. long. Calyx-tube 1–2 mm. long and wide; lobes oblong, deltoid or minute, 1·5–8·5 mm. long, 0·5–2 mm. wide, strigosely ciliate. Long-styled flowers: corolla-tube 4–8(–9) mm. long, often split longitudinally at the base along the filament-sutures, cylindrical and scarcely dilated; lobes linear-oblong, 1·25–4 mm. long, 0·5–1·25 mm. wide, tips inflexed and hairy, and with the long matted throat hairs extending over the lower third of the interior surface; anthers either with tips 0·5 mm. below the throat-orifice or situated nearly in the middle of the tube. Style exserted 1·5–2·5(–4) mm., the stigma-lobes 1–1·5 mm. long. Short-styled flowers very similar but anthers exserted 1·5 mm. (often hidden by the throat hairs). Capsule 3–4 mm. long, 3 mm. wide, with a rounded triangular beak 1·5–2 mm. tall.

subsp. **purpurea**; Verdc. in B.J.B.B. 23: 330 (1953)

Anthers in long-styled flowers with tips ± 0·5 mm. below the throat.

TANZANIA. Biharamulo District: 45 km. from Biharamulo on Kahama road, Rugongo, 24 Nov. 1962, *Verdcourt* 3454!; Buha District: 90 km. on road from Kigoma to Kasulu, near Kasulu, 16 Nov. 1962, *Verdcourt* 3329!; Iringa District: Dabaga Highlands, Kilolo, 9 Feb. 1962, *Polhill & Paulo* 1400!; Songea District: by R. Nakawali, 9 Mar. 1956, *Milne-Redhead & Taylor* 9076!
DISTR. T1, 4, 7, 8; ? Guinée, Nigeria, Cameroun, Zaire, Malawi, Mozambique and Rhodesia
HAB. Grassland, woodland (including *Brachystegia*), bracken scrub; 1335–2100 m.

SYN. *P. stolzii* K. Schum. & K. Krause in E.J. 39: 522 (1907). Type: Tanzania, Songea District, upper Ngaka [Magaka] valley, *Busse* 937 (B, syn. †, EA, lecto.!)
[*P. zanzibarica* sensu auctt., *non* (Klotzsch) Vatke]

NOTE. Tall plants 0·8–1·1 m. tall were separated off in my original monograph as var. *buchananii*, but although extremes are distinctive there is no real reason for maintaining it. The validity of subsp. *mechowiana* (K. Schum.) Verdc., on the other hand, distinguished by having the anthers inserted lower in the tube (in long styled flowers), needs further investigation. It has not been recorded from the Flora area.

DISTR. (of species as a whole). As above with Angola

21. **P. arvensis** *Hiern* in F.T.A. 3: 47 (1877); Verdc. in B.J.B.B. 23: 335, fig. 35/E, F (1953); Hepper in F.W.T.A., ed. 2, 2: 215 (1963); U.K.W.F.: 404 (1974). Type: Sudan, Mittuland, Derago, *Schweinfurth* 2775 (K, holo. !, BM, W, iso. !)

Pyrophytic herb, with 4–9 unbranched herbaceous stems (15–)30–50 cm. tall from a very woody rootstock ± 1 cm. wide, ± hairy in 2 lines, often showing signs of burning. Leaves paired or rarely in whorls of 3; blades elliptic to lanceolate, 2·5–8 cm. long, 0·4–2·6 cm. wide, bluntly acute or obtuse at the apex, cuneate at the base, with adpressed crisped hairs on both surfaces; petiole obsolete or up to 2 mm. long; stipules with 3–5 linear or deltoid lobes 3–4(–9) mm. long from a short base. Inflorescences trichotomous, few–many-flowered, laxly globose, each cluster 2–2·5 cm. wide. Flowers white or less often pinkish. Calyx-tube 1–1·25 mm. long and wide; lobes oblong, the largest up to 3 mm. long, 0·5–1 mm. wide, the rest smaller, or minute. Long-styled flowers: corolla-tube 4·3–5(–6) mm. long, widened above to 1·5–2 mm. at the apex; lobes oblong or elliptic, 1·75–3 mm. long, 0·75–1·5 mm. wide, hairy over the basal half inside; stamens entirely included; style exserted 1 mm., stigma thickened, the lobes 1 mm. long. Short-styled flowers similar: corolla-tube 4 mm. long; lobes 2·5–3 mm. long, 1·5 mm. wide; throat densely hairy; anthers exserted 2 mm. Fruiting inflorescence 1·5–3 cm. wide; peduncle 6·5–10 cm. long. Capsules subglobose, 2·5–4 mm. long, 2·5–3·5 mm. wide; beak 1·5 mm. tall.

UGANDA. Acholi District, *Dawe* 855!; "Elgon District", 1905, *James*! (possibly from Kenya)
KENYA. N. Kavirondo District: Kakamega Forest, Jan. 1944, *Carroll* H8!; Kavirondo, *Scott Elliot* 7025!; Nyanza basin, *Battiscombe* 653!
DISTR. **U**1, 3; **K**5; Nigeria, Central African Republic, Sudan
HAB. Grassland subject to regular burning; 1200–1530 m.

SYN. *Neurocarpaea arvensis* (Hiern) Hiern, Cat. Afr. Pl. Welw. 2: 438 (1898)

NOTE. Hepper (K.B. 14: 254 (1960)) has suggested that *Dawe* 855 might be *P. nervosa* Hepper, but I believe I am correct in referring it to the present species.

22. **P. lanceolata** (*Forssk.*) *Deflers*, Voy. Yemen: 142 (1889); Verdc. in K.B. 6: 377 (1951) & in B.J.B.B. 23: 339, fig. 35/D, G (1953); F.P.U., ed. 2: 160, fig. 103 (1972); U.K.W.F.: 404, fig. (1974). Type: Yemen, Hadie Mts., *Forsskål* (C, holo. !, BM, iso. !)

Herb or subshrub with erect or straggling mostly woody stems 0·5–1·3 m. tall, hairy. Leaf-blades ovate, lanceolate, ovate-lanceolate or elliptic, 3–13 cm. long, 1–6 cm. wide, acute at the apex, cuneate at the base, pubescent to densely velvety on both surfaces; petioles 0–5 cm. long; stipules with 3–9(–14) setae, 2–9 mm. long, bearing small colleters, from a short base. Inflorescence with terminal and axillary components combined into a single cluster. Calyx-tube hairy, 1–3 mm. long, 1·5–2·5 mm. wide; lobes very unequal, the largest lanceolate, 0·5–1·3 cm. long, 0·5–3 mm. wide, the smallest 1–3 mm. long. Flowers often trimorphic, either with style exserted and anthers included, anthers exserted and style included or both included. Long-styled flowers: corolla-tube (1–)2·3–3·9 cm. long, dilated at the apex to 3(–6) mm. wide for a distance of 4–8 mm., hairy or glabrous outside; lobes oblong-ovate to elliptic, 0·3–1 cm. long, (1–)1·5–4·5 mm. wide; throat hairy within; anthers completely included; style exserted 1·5–5·5 mm.; stigma 2–5 mm. long. Short-styled flowers very similar, corolla-lobes 4·5–8 mm. long; anthers exserted 2·5–4 mm.; style and stigma usually completely enclosed or rarely tips of stigma-lobes exserted 2·5 mm. Flowers with both style and stigma and anthers included: tube 2–4 cm. long; lobes ovate-oblong 0·5–1·1 cm. long, 2·5–3·3 mm. wide; anthers sometimes with tips exserted 0·25 mm. but usually included; style and stigma always included. Fruit obtriangular, 4–6 mm. tall and wide; beak 1–2 mm. tall.

NOTE. The division of *P. lanceolata* into infraspecific variants is difficult; in my original monograph I divided it into 13, one of which has since been raised to specific rank, and many of the others are not recognised here or merely mentioned in notes. It is not possible to produce a key to these variants which will work efficiently and they are best named from geographical considerations; nevertheless it is not feasible to ignore the variants completely since extremes are very distinct. W. H. Lewis (in Ann. Missouri Bot. Gard. 52: 195 (1965)) has carried out cytological work on the species and found that var. *lanceolata*, var. *leucaster* and var. *oncostipula* are diploid, whereas var. *nemorosa* is tetraploid. After quoting my statement " var. *nemorosa* is similar to var. *oncostipula* but distinct in the field by virtue of its larger flowers " he states " provided that future studies from more individuals substantiate the tetraploidy of var. *nemorosa* then I believe the taxon should be recognised at the specific level ". Against this view the flowers of var. *lanceolata* are much the largest. I cannot believe that specific rank is indicated for any of the variants I recognised, although it must be admitted that some specimens of var. *nemorosa* have distinctively narrow long petiolate leaves. There are many doubtless who would prefer not to recognise more than two variants and this they may do by calling the long-flowered specimens subsp. *lanceolata* and the short-flowered ones subsp. *quartiniana*.

KEY TO INFRASPECIFIC VARIANTS

Corolla predominantly white, more rarely tinged lilac or pink; tube 2–4 cm. long (subsp. *lanceolata*) . . . . a. var. **lanceolata**

Corolla usually pink, lilac, mauve, magenta, etc., rarely white; tube 1–2·2 cm. long (subsp. *quartiniana*):

- Leaves mostly ovate-lanceolate or ovate, rarely lanceolate in some Kenya variants and then on mostly well-developed petioles:
  - Inflorescences ± congested in flower and fruit:
    - Erect herbs or subshrubs; stems very densely hairy; leaves hairy, mostly with very short petioles; flowers usually very showy (Kenya Highlands, Uganda and NW. Tanzania) . b. var. **leucaster**
    - Scrambling to erect herbs or subshrubs; stems sparsely to densely hairy; leaves pubescent to densely hairy, with short to long petioles; flowers mostly less showy (Tanzania) . . d. var. **oncostipula**
  - Inflorescences with branches lax and spicate in fruit:
    - Leaves smaller, with 6–11 lateral nerves on each side; stipular setae 3–9; corolla-lobes 6·5–8·5 mm. long . . c. var. **nemorosa**

Leaves larger, with 11–15 lateral nerves on each side; stipular setae 10–14; corolla-lobes 3 mm. long (E. Usambara Mts.) . . . . . f. var. **usambarica**
Leaves mostly lanceolate, subsessile; strictly erect herb with dense inflorescences (**T4**) . e. var. **angustifolia**

subsp. **lanceolata**; Verdc. in B.J.B.B. 23: 339, fig. 35/D, G (1953)

Flowers larger, sometimes trimorphic; corolla predominantly white or tinged pink, with tube 2–4 cm. long.

a. var. **lanceolata**

UGANDA. Acholi District: Rom Mt., June 1930, *Liebenberg* 305!; Karamoja District: Napak, June 1950, *Eggeling* 5920! & Mt. Moroto, June 1963, *Tweedie* 26551!

KENYA. Northern Frontier Province: Mt. Kulal, 27 July 1958, *Verdcourt* 2262!; Naivasha District: 16 km. E. of Naivasha, 9 Sept. 1962, *Lewis* 5928!; Nairobi, *Battiscombe* 428!

TANZANIA. Masai District: Ololmoti Volcano, at Oldonyo Wass Camp, 16 Sept. 1932, *B. D. Burtt* 4347!; Arusha District: S. slope of Mt. Meru, 1 Nov. 1959, *Greenway* 9606!; Moshi District: Kilimanjaro, Bismarck Hill to Marangu, 1 Mar. 1934, *Greenway* 3899!

DISTR. U1; K1–4, 6; T2; also in Arabia, Ethiopia, Sudan and Afars Issas

HAB. Grassland, bushland, thicket, also in evergreen forest, sometimes on rocky cliffs and escarpments; 1440–3000 m.

SYN. *Ophiorrhiza lanceolata* Forssk., Fl. Aegypt.-Arab.: 42 (1775)
Manettia lanceolata (Forssk.) Vahl, Symb. Bot. 1: 12 (1790)
Neurocarpaea lanceolata (Forssk.) R. Br. in Salt, Voy. Abyss. App. 4: 64 (1814); Britten in J.B. 35: 129 (1897), excl. syn.
Vignaldia quartiniana A. Rich. emend Schweinf. var. *grandiflora* Schweinf., Beitr. Fl. Aeth.: 140 (1867). Type: Ethiopia, Bellaka, 1854, *Schimper* 338 (B, holo. †)
Virecta lanceolata (Forssk.) Baill., Hist. Pl. 7: 380 (1879)
Pentas ainsworthii Scott Elliot in J.L.S. 32: 433 (1896). Type: Kenya, Machakos/Kitui Districts, Ukamba, *Scott Elliot* 6437 (K, holo.!, BM, iso.!)
P. schweinfurthii Scott Elliot in J.L.S. 32: 432 (1896). Type: Yemen, Menacha, *Schweinfurth* 1370 (BM, holo.!)
[*P. longiflora* sensu Cufod. in Phyton 1: 139 (1949), *non* Oliv.]
P. lanceolata (Forssk.) Deflers var. *lanceolata* forma *velutina* Verdc. in B.J.B.B. 23: 344 (1953). Type: Tanzania, Meru, near Ngongongare, *Peter* 0.I.62 (B, holo.!)
P. lanceolata (Forssk.) Deflers var. *membranacea* Verdc. in B.J.B.B. 23: 344 (1953). Type: Tanzania, Arusha, Temi R., *Lindeman* 817 (EA, holo.!)
[*P. carnea* sensu auctt., *non* Benth. sensu stricto]

NOTE. The nomenclatural confusion concerning this species has been explained in K.B. 6: 377 (1952). The names *Mussaenda luteola* Delile, *Vignaudia luteola* (Delile) Schweinf. and *Pseudomussaenda lanceolata* (Forssk.) Wernham are all nomenclatural synonyms of *Pentas lanceolata*, the same type being cited in synonymy; the plant to which they have always been attached is now known as *Pseudomussaenda flava* Verdc. The variants I described in my original monograph are not worth retaining—they correspond to variants with large thin glabrescent leaves and smaller thicker densely velvety leaves.

subsp. **quartiniana** (*A. Rich.*) *Verdc.* in B.J.B.B. 23: 364 (1953). Type: Ethiopia, Maiguiga, *Quartin & Dillon* 6 (P, holo.!, S, W, iso.!)

Flowers smaller, usually dimorphic; corolla predominantly coloured pink, mauve or blue, rarely white, with tube (1·3–)1·4–1·6(–2·15) cm. long.

SYN. *Vignaldia quartiniana* A. Rich., Tent. Fl. Abyss. 1: 357 (1847)
Pentas quartiniana (A. Rich.) Oliv. in Trans. Linn. Soc. 29: 82, t. 46 (1873)
P. verruculosa Chiov. in Atti Reale Accad. Ital., Mem. Cl. Sci. Mat. Nat. 11: 35 (1940). Type: Ethiopia, Shoa, Sciassamanna, *Senni* 2258 (FI, holo.!)

b. var. **leucaster** (*K. Krause*) *Verdc.* in B.J.B.B. 23: 347, 34/C (1953). Type: Rwanda, Lake Mohasi, *Mildbraed* 444 (B, holo. †, BR, iso. (fragment)!)

Erect herb to ± 1 m.; stems mostly very densely covered with spreading hairs. Leaves ovate, mostly densely hairy; petiole very short. Flowers numerous in dense heads, which do not become spicate in fruit, often rather showy; corolla-tube 1·4–2 cm. long; lobes 4–6·5 × 1·2–3·75 mm.

UGANDA. Kigezi District: Ruhinda, Jan. 1951, *Purseglove* 3537!; Mbale District: 9·6 km. N. of Busia, Buteba [Butiba], Oluchor Hill, 4 May 1951, *G. H. Wood* 187!; Mubende District: near Kasambya, 15 May 1957, *Griffiths* 11!
KENYA. W. Suk District: Kapenguria, 13 May 1932, *Napier* 1902!; Trans-Nzoia District: Hoey's Bridge, 15 Aug. 1963, *Heriz-Smith & Paulo* 926!; Kavirondo District: Kakamega Forest, 9 Dec. 1956, *Verdcourt* 1640!
TANZANIA. Ngara District: Bugufi, Jan. 1936, *Chambers* 32! & Nyamyaga [Nyamiaga], Mukagezi, 6 Jan. 1961, *Tanner* 5591B!
DISTR. **U**1–4; **K**1–3, 4 (intermediates) 5, ?6; **T**1; E. Zaire, Rwanda, Sudan and Ethiopia
HAB. Grassland, open bushland, grassland with scattered trees and also in cultivations; 1080–2400 m.

SYN. *P. leucaster* K. Krause in Z.A.E. 2: 312 (1911)
P. coerulea Chiov. in Atti Reale Accad. Ital., Mem. Cl. Sci. Mat. Nat. 11: 34 (1940). Type: Ethiopia, Galla Sidama, Saio, near Taber, *Giordano* 2452 (FI, holo.!)

c. var. **nemorosa** (*Chiov.*) *Verdc.* in B.J.B.B. 23: 349 (1953). Type: Kenya, Aberdare Mts., Tusu, *Balbo* 678 (TOM, lecto.!)

Usually erect or somewhat straggling herb up to 1·5 m. tall; stems sparsely to fairly densely hairy. Leaves ovate or frequently ovate-lanceolate to elliptic-lanceolate, glabrescent to hairy; petiole short to quite long. Flowers in few–many-flowered heads which usually become quite open and spicate in fruit, often quite showy; corolla-tube 1·3–2·2 cm. long; lobes 5·5–8·5 × 2·5–4 mm.

KENYA. Northern Frontier Province: Mathews Range, Dunyos, *J. Bally* 28 in *Bally* 3628!; Fort Hall District: Thika, gorge near Blue Posts Hotel, 19 Feb. 1953, *Drummond & Hemsley* 1223!; Nairobi District: Nairobi–Kiambu road, 3 Dec. 1950, *Verdcourt* 392!
TANZANIA. Moshi District: 8 km. S. of Moshi, by R. Njoro, 3 Nov. 1955, *Milne-Redhead & Taylor* 7206! & 7207!
DISTR. **K**1, 3, 4, 6; **T**2; Ethiopia
HAB. Edges of paths and glades in mostly secondary evergreen forest, also in grassy places; 1380–2300 m.

SYN. *P. parvifolia* Hiern var. *nemorosa* Chiov. in Lav. Ist. Bot. R. Univ. Modena 6: 52 (1935)

NOTE. At least some specimens of this differ cytologically from other varieties of *P. lanceolata*, see general note on p. 209.

d. var. **oncostipula** (*K. Schum.*) *Verdc.* in B.J.B.B. 23: 351 (1953). Types: Tanzania, W. Usambara Mts., Sakare, *Engler* 973 (B, holo. †); W. Usambara Mts., Balangai–Garaya [Ngaraya], *Peter* 15976 (B, neo.!)

Erect or often very long straggling herb up to 3 m.; stems sparsely to densely hairy. Leaves mostly ovate or elliptic-ovate, with short to rather long petioles. Flowers numerous in dense heads, which do not become spicate in fruit, less showy; corolla-tube 1–2 cm. long; lobes 3–7 × 1–3 mm.

TANZANIA. Moshi District: Kilimanjaro, 20 Dec. 1924, *T. W. Lewis* 259!; Lushoto District: W. Usambara Mts., Mkuzi, 19 Aug. 1950, *Verdcourt* 320!; Morogoro District: Uluguru Mts., Tanana, 24 Jan. 1935, *E. M. Bruce* 628!
DISTR. **T**2, 3, 6, 7; ? **Z** (see note); not known elsewhere
HAB. Evergreen forest clearings, secondary bushland, bracken-*Myrica* zone, streamsides, roadsides and abandoned cultivations; 765–2010 m.

SYN. *Pentanisia nervosa* Klotzsch in Peters, Reise Mossamb. Bot. 1: 287 (1861). Type: Zanzibar I., *Peters* (B, holo. †, K, fragment!)
P. oncostipula K. Schum. in E. J. 34: 329 (1904)!

NOTE. The fragment of *P. nervosa* preserved at Kew, which was sent to Dr. O. Stapf from Berlin by Diels and Krause, has the corolla-tube about 1·4–1·5 cm. long and is definitely not *P. zanzibarica*. Nothing has been seen from Zanzibar I. which matches it, neither did Peters visit any area where var. *oncostipula* grows. Nevertheless the fragment does appear to belong to this variety.

A form with the leaves densely velvety has been separated off as forma *velutina* Verdc. in B.J.B.B. 23: 352 (1953)—type: Tanzania, Lushoto District, Shume, World's View, *Greenway* 8462 (EA, holo. !, K, iso. !); extremes of this are distinctive and have been seen from several places in the W. Usambara Mts. and also from **T7**, Mufindi.

Var. *alba* Verdc. (in B.J.B.B. 23: 349 (1953); type: Tanzania, Lushoto District, Amani, *Verdcourt & Greenway* 279 (EA, holo. !, K, iso. !)) is based on a cultivated plant with white flowers and a woody based stem up to 2·5 cm. in diameter. Its origin is not known but it is probably derived from var. *oncostipula* or var. *leucaster*.

e. var. **angustifolia** *Verdc.* in K.B. 30: 288 (1975). Type: Tanzania, Ufipa District, near Mmemya Mt., *Bullock* 3665 (K, holo. !, EA, iso. !)

Strictly erect herb ± 1 m. tall; stems rather densely hairy. Leaves distinctly narrowly to broadly lanceolate (save at base). Flowers numerous in dense heads which probably do not become spicate in fruit, less showy; corolla-tube 1·4–1·9 cm. long, the lobes 4–6 × 1·5–2 mm.

TANZANIA. Ufipa District: near Mmemya Mt., 12 Feb. 1951, *Bullock* 3665 ! & Kasamvu, 21 Mar. 1950, *Bullock* 2699 ! & road to Rukwa, Ilemba Gap, 12 Mar. 1959, *Richards* 11175 !
DISTR. **T4**; not known elsewhere
HAB. Grassland and woodland; 1500–2100 m.

f. var. **usambarica** *Verdc.* in B.J.B.B. 23: 351 (1953). Type: Tanzania, E. Usambara Mts., Ngwelo [Nguelo] on Ngambo track, *Zimmermann* in *Herb. Amani* 7877 (EA, holo. !)

? Erect herb. Leaves ovate or elliptic, with short to long petioles, mostly large, up to 6 cm. wide and with 11–15 lateral veins on each side; stipules with 10–14 filiform setae, 5–8 mm. long, more developed than in other variants. Inflorescence 3·5–4·5 cm. broad in flower, expanding to 11 cm. in fruit, the individual branches being spicate and lax. Flowers not showy; corolla 1–1·5 cm. long; lobes 3 × 1·2 mm.

TANZANIA. Lushoto District: E. Usambara Mts., Derema to Ngambo, May 1917, *Peter* 20216 ! & Monga to Sangerawe, Dec. 1914, *Peter* 0.III.11 ! & Nguelo to Derema, May 1918, *Peter* 23411 !
DISTR. **T3** (E. Usambara Mts.); not known elsewhere
HAB. Presumably in clearings in rain-forest; 850–950 m.

NOTE. This variety forms a link between *P. lanceolata* and *P. zanzibarica*. Curiously no material appears to have been collected recently.

23. **P. suswaensis** *Verdc.* in Kirkia 5: 274 (1966); U.K.W.F.: 404 (1974). Type: Kenya, Mt. Suswa, *Verdcourt* 710 (EA, holo. !)

Subshrub or woody herb 0·9–1·2 m. tall; stems densely pubescent. Leaves opposite or in whorls of 3; blades elliptic, 3–6 cm. long, 1·5–3·5 cm. wide, subacute or obtuse at the apex, cuneate at the base, densely covered with short spreading rather rigid hairs; petioles 2–6 mm. long; stipules with 5 setae, 2–5 mm. long, from a 1 mm. long base. Inflorescence usually no more than 10-flowered. Calyx-tube 2 mm. long, densely pubescent, strongly ribbed; lobes unequal, oblong, elliptic or subspathulate, 3–7 mm. long, 1·2–4 mm. wide, rather obtuse at the apex, the apical parts often blackish or bluish inside in living plants in fruit, the main nerves parallel and conspicuous. Corolla white or greenish cream; tube (3–)5·5–9 cm. long (usually ± 7 cm.), 3 mm. wide, pubescent outside; lobes 4–5, oblong, 6–10 mm. long, 2–2·5 mm. wide. Style exserted 1 cm. in long-styled flowers, tomentellous. Capsule oblong, 6–8 mm. long, 5–6·5 mm. wide, strongly ribbed, densely pubescent.

KENYA. Naivasha District: Mt. Suswa, northern part of the Caldera, 26 Feb. 1964, *Verdcourt & Glover* 3983 ! & same locality, in Caldera near edge of moat, 23 Mar. 1963, *Glover & Bally* 3534 !; Masai District: Mt. Suswa, outer western slopes of the mountain, Aug. 1952, *Verdcourt* 710 !
DISTR. **K3**, 6; not known elsewhere
HAB. Bushland on rocky volcanic ground; 1650–2100 m.

SYN. *P. lanceolata* (Forssk.) Deflers subsp. *longituba* Verdc. in B.J.B.B. 23: 344 (1953) & in K.B. 17: 500 (1964)

NOTE. The completely dimorphic flowers distinguish this from other long-flowered species in subgenera *Megapentas* and *Chamaepentadoides*, apart from differences in the calyx and capsule. This species is certainly a specialised derivative of *P. lanceolata* and it is interesting to see that a specimen of the latter from Uganda, Karamoja District, June 1950, *Eggeling* 5920 !, shows some characters approaching *P. suswaensis*, particularly the calyx-lobes and few-flowered inflorescences.

17. OTOMERIA

Benth. in Hook., Niger Fl.: 405 (1849); Verdc. in B.J.B.B. 23: 5–34, 249 (1953)

Annual or perennial erect, subprostrate or twining herbs with mostly hairy stems. Leaves paired; stipules with base divided into several narrow segments. Flowers small and white or rather large and coloured, hermaphrodite, monomorphic or dimorphic, in cymose heads, which in fruit develop into a long simple spike with the fruits geminately arranged and a solitary remotely placed flower in the axil at the base of the spike (save in subgen. *Volubiles* where the cluster remains dense). Calyx-tube ovoid or elongate-oblong; lobes unequal, 5, 1–3 foliaceous and larger than the rest, alternating with small colleters. Corolla-tube long and narrow, with a markedly ovoid-oblong apical dilation in long-styled forms; throat densely hairy; lobes in small-flowered species elliptic, but in large-flowered species broader, ovate to orbicular, narrowing to the base where the lobes are often connate for a short distance. Stamens completely exserted in short-styled forms. Style exserted in long-styled forms, the stigma bifid with filiform lobes; anthers completely included in the apical dilation of the corolla tube. Capsule oblong, compressed, ribbed, opening by apical valves and also frequently splitting longitudinally. Seeds small, reticulate.

A small genus of 8 species widely distributed in tropical Africa, 4 of which occur in the Flora area.

As has been mentioned in my revision the circumscription of this genus is very unsatisfactory and forms a reticulate pattern with *Pentas*; it could be combined with that genus but the result would be no more satisfactory. Subgen. *Otomeria*, with small white flowers, is very closely related to *Pentas* subgen. *Pentas* sect. *Monomorphi* Verdc., whereas subgen. *Neotomeria* Verdc. is closely related to *Pentas* subgen. *Pentas* sect. *Coccineae* Verdc. Nevertheless these two subgenera of *Otomeria* are united by their inflorescence and capsule structure. It seems practical to retain the classification adopted in my revision.

Plants not climbing; fruiting inflorescence a spike (subgen. *Neotomeria* Verdc.):
- Flowers bright pink, rose or usually scarlet, not dimorphic, the style always exserted; plant of marshy areas . . . . 1. *O. elatior*
- Flowers pale pinkish, dimorphic; plants of dry areas:
  - Leaves 2·5–5·5 cm. wide, with venation raised and reticulate beneath; corolla white and pink . . . . . 2. *O. madiensis*
  - Leaves 0·8–2 cm. wide; corolla flesh-coloured with a blue eye. . . . . 3. *O. oculata*

Plant climbing; fruiting inflorescence capitate (subgen. *Volubiles* Verdc.) . . . . 4. *O. volubilis*

1. **O. elatior** (*DC.*) *Verdc.* in B.J.B.B. 23: 18, fig. 3/A–D (1953); Hepper in F.W.T.A., ed. 2, 2: 214 (1963); Hallé, Fl. Gabon, 12. Rubiacées: 117 (1966); F.P.U., ed. 2: 160 (1972); U.K.W.F.: 405 (1974). Type: Angola, *da Silva* (P, holo.!)

Erect or rarely straggling herb 0·35–3 m. tall, with single unbranched or sparsely branched glabrescent pubescent or hairy stem. Leaf-blades ovate-elliptic or rarely linear, 1·5–9·5 cm. long, 0·7–3·2 cm. wide, ± acute at the apex, rounded or cuneate at the base, pubescent to densely hairy on both surfaces, or glabrous above and pubescent beneath on the nerves, or altogether glabrous save for a very minute pubescence near the edges above and on the venation beneath; petiole obsolete or 1–6 mm. long; stipules with 1–3 flat linear setae 1–5 mm. long, 0·2–1 mm. wide, and also 2–6 short setae or sessile colleters from a short base. Flowering inflorescence 1–6·5 cm. long, becoming 4–37 cm. long in fruit, with a peduncle 0–30 cm. long; solitary nodal flower with pedicel up to 2 cm. long. Calyx-tube ± 2 mm. long, glabrous to hirsute; lobes very unequal, 1–3 foliaceous, lanceolate, 0·5–2·4 cm. long, 1·2–4·8 mm. wide, the rest 1–4 mm. long, 0·5–1·5 mm. wide. Corolla-tube 1·7–2·7 cm. long, sparsely hairy or glabrous below, 0·25–2·3 mm. wide below, dilated above to 1·25–3 mm. wide for a distance of 3–5 mm., the dilation urceolate and usually rounded at the base, externally hairy with long multicellular crimson or purple hairs; lobes ovate, orbicular or elliptic-spathulate, 0·5–1·8 cm. long, 0·25–1 cm. wide, mucronate from the emarginate tip, connate at the base for 0·5–1·5 mm.; throat orifice 0·5–1·5 mm. in diameter; throat densely hairy, the apical hairs crimson and spreading over the orifice and connate parts of the corolla-lobes. Anthers completely included. Style exserted 0–5 mm.; stigmas elliptic, 0·5–1 mm. long. Fruit oblong, compressed, chestnut or purple-brown, 0·6–1·2 cm. long, 4–6 mm. wide, 2·5–5 mm. thick (rarely smaller, 5 × 3·5 mm.), strongly ribbed, pubescent or hairy; dehiscence apical and accompanied by some longitudinal splitting. Seeds angular, ± 0·7 mm. long.

UGANDA. Bunyoro District: Budongo Forest, 30 Nov. 1938, *Loveridge* 151!; Teso District: Serere, Mar. 1932, *Chandler* 649!; Masaka District: Lake Nabugabo, SW. side, 7 Oct. 1953, *Drummond & Hemsley* 4671!

KENYA. Uasin Gishu District: Burnt Forest, *Webster* in *E.A.H.* 8834!; N. Kavirondo District: Mumias, 29 July 1913, *Battiscombe* 668!; Masai District: Lolgorien, Sept. 1933, *Napier* 2916!

TANZANIA. Near Bukoba, Aug. 1931, *Haarer* 2129!; Mbeya District: Mbosi Circular Road, 12 Jan. 1961, *Richards* 13881!; Songea District: Matengo Hills, about 2·5 km. E. of Ndengo, 4 Mar. 1956, *Milne-Redhead & Taylor* 8971!

DISTR. **U**1–4; **K**3, 5, 6; **T**1, 4, 7, 8; W. Africa from Mali to Cameroun and Angola, Central African Republic, Sudan to Mozambique and Rhodesia

HAB. Always in swampy places, inundated grassland and permanent swamps; 1030–1590 m.

SYN. *Sipanea elatior* DC., Prodr. 4: 415 (1830); A. Rich. in Mém. Soc. Hist. Nat. Paris 5: 276 (1834)
Pentas elatior (DC.) Walp., Repert. 6: 57 (1846)
Otomeria dilatata Hiern in F.T.A. 3: 50 (1877). Type: Nigeria, N. Nupe, *Barter* 1237 (K, holo.!)
[*O. madiensis* sensu Hiern in F.T.A. 3: 50 (1877), pro parte quoad *Schweinfurth* 3565, *non* Oliv.]

NOTE. In the SW. of Tanzania and adjoining part of Zambia a variant occurs with a more showy corolla, the lobes being 1·6–1·8 cm. long and 1 cm. wide, narrowed at the base to 2·5 mm. In my revision this was formerly recognized as forma *speciosa* (Bak.) Verdc. (in B.J.B.B. 23: 23 (1953); *Pentas speciosa* Bak. in K.B. 1895: 67 (1895); *Otomeria speciosa* (Bak.) Scott Elliot in J.L.S. 32: 437 (1896); type: Zambia, Fwambo, *Carson* (K, holo.!)). It seems scarcely worth maintaining despite the distinctiveness of extreme forms, e.g. Ufipa District, Lake Rukwa to Lake Tanganyika, *Nutt*!

2. **O. madiensis** *Oliv.* in Trans. Linn. Soc. 29: 83, t. 47 (1873); Hiern in F.T.A. 3: 50 (1877), pro parte; Verdc. in B.J.B.B. 23: 25 (1953). Type: Uganda, W. Nile District, Madi, *Grant* 691 (K, holo.!)

Erect herb 0·45–1·2 m. tall, with pubescent stems. Leaf-blades oblong-elliptic, 8–13 cm. long, 2·6–6 cm. wide, acute at the apex, narrowly cuneate at the base, very sparsely pubescent above, shortly bristly hairy on the venation beneath; lateral nerves 9–12 on each side, prominent beneath, the tertiary venation ± prominent and reticulate; petiole 0–2 mm. long; stipules with ± 3 setae up to 6 mm. long. Flowers dimorphic, in a spike (6–)29–35 cm. long; peduncle 5–12 cm. long. Calyx-tube 1–5 mm. long; lobes unequal, the foliaceous ones oblong, 1–1·4 cm. long, 1–2 mm. wide, 3-nerved, bristly pubescent. Corolla-tube red or white, 1·7–2·2 cm. long, dilated at the apex up to 3 mm. for 4 mm.; lobes white flushed pink or with 3 red lines, elliptic-oblong, 7–8 mm. long, 2·3–4 mm. wide, acute at the apex, narrowed to the base, pubescent outside on the central portion; throat-orifice purple or pink; style exserted 1·5–2·5 mm. in long-styled flowers; stigma-lobes up to 2 mm. long; in short-styled flowers the corolla-tube is more gradually dilated. Fruits obovoid-oblong, 5–6 mm. long, 3–4 mm. wide, ribbed.

UGANDA. W. Nile District: Koboko, May 1938, *Hazel* 473!; Bunyoro District: top of Lake Albert escarpment near the Sonso R., May 1941, *Eggeling* 4307! & Chiope, 4 Jan. 1906, *Dawe* 853!
DISTR. **U**1, 2; Sudan and NE. Zaire
HAB. Open woodland; ± 1000 m.

SYN. *O. blommaertii* De Wild., Pl. Bequaert. 5: 429 (1932). Type: NE. Zaire, *Blommaert* 259 (BR, holo.!)

NOTE. *O. madiensis* and *O. oculata* are closely related and a few specimens are more or less intermediate, e.g. *Bagshawe* 1598! (Uganda, Bunyoro, Victoria Nile, near Fajao, 4 May 1907).

3. **O. oculata** *S. Moore* in J.B. 18: 4 (1880); Verdc. in B.J.B.B. 23: 25 (1953); U.K.W.F.: 405 (1974). Type: Kenya, Kitui, *Hildebrandt* 2756 (BM, holo.!, K, W, iso.!)

Erect herb or subshrubby herb, 0·3–0·6 m. tall, with shortly adpressed hairy stems. Leaf-blades lanceolate or narrowly rhomboid, 2·5–6·5(–8·5) cm. long, 0·7–2·3 cm. wide, ± acute at the apex, narrowly cuneate at the base, shortly hairy above and with longer adpressed bristly hairs beneath; petiole obsolete or up to 8 mm. long; stipules with 1–3 flat segments 1·5–3 mm. long. Inflorescence capitate at first but soon becoming a spike 5–32 cm. long, with flowers rather widely spaced at the base; peduncles 4·5–10 cm. long; flowers dimorphic. Calyx-tube blackish, 1–2 mm. long, glabrous or bristly; larger lobes linear, 0·5–1·5 cm. long, 0·5–1·3 mm. wide. Corolla-tube white, reddish or pink, 1·8–3·2 cm. long, dilated at the apex in long-styled flowers to 1·2–2·25 mm. for a distance of 4–5·5 mm.; lobes cream, pink or yellowish to greenish flesh-coloured, with a crimson or purple pubescent eye, narrowly to broadly elliptic, 0·5–1 cm. long, 2–5·5 mm. wide, slightly connate at the base; style exserted 1–2·5 mm., stigma-lobes 1–2 mm. long; corolla-tube in short-styled flowers gradually widened at the apex. Fruit obtriangular-oblong, ribbed, 3·5–6 mm. long, 2·5–4·5 mm. wide, pubescent. Fig. 24, p. 216.

UGANDA. Karamoja District: near Kunyao, *J. Wilson* 917!
KENYA. Northern Frontier Province: Sololo, 2 Aug. 1952, *Gillett* 13673!; Lake Baringo, Main I., 14 July 1956, *J. G. Williams* in *E.A.H.* 11097!; Fort Hall District: Mabaloni Rocks, 12 Dec. 1952, *Verdcourt & Bally* 841!
DISTR. **U**1; **K**1, 3, 4; S. Ethiopia
HAB. Bushland, woodland, often in rock-clefts, also on lava flows; 540–1650 m.

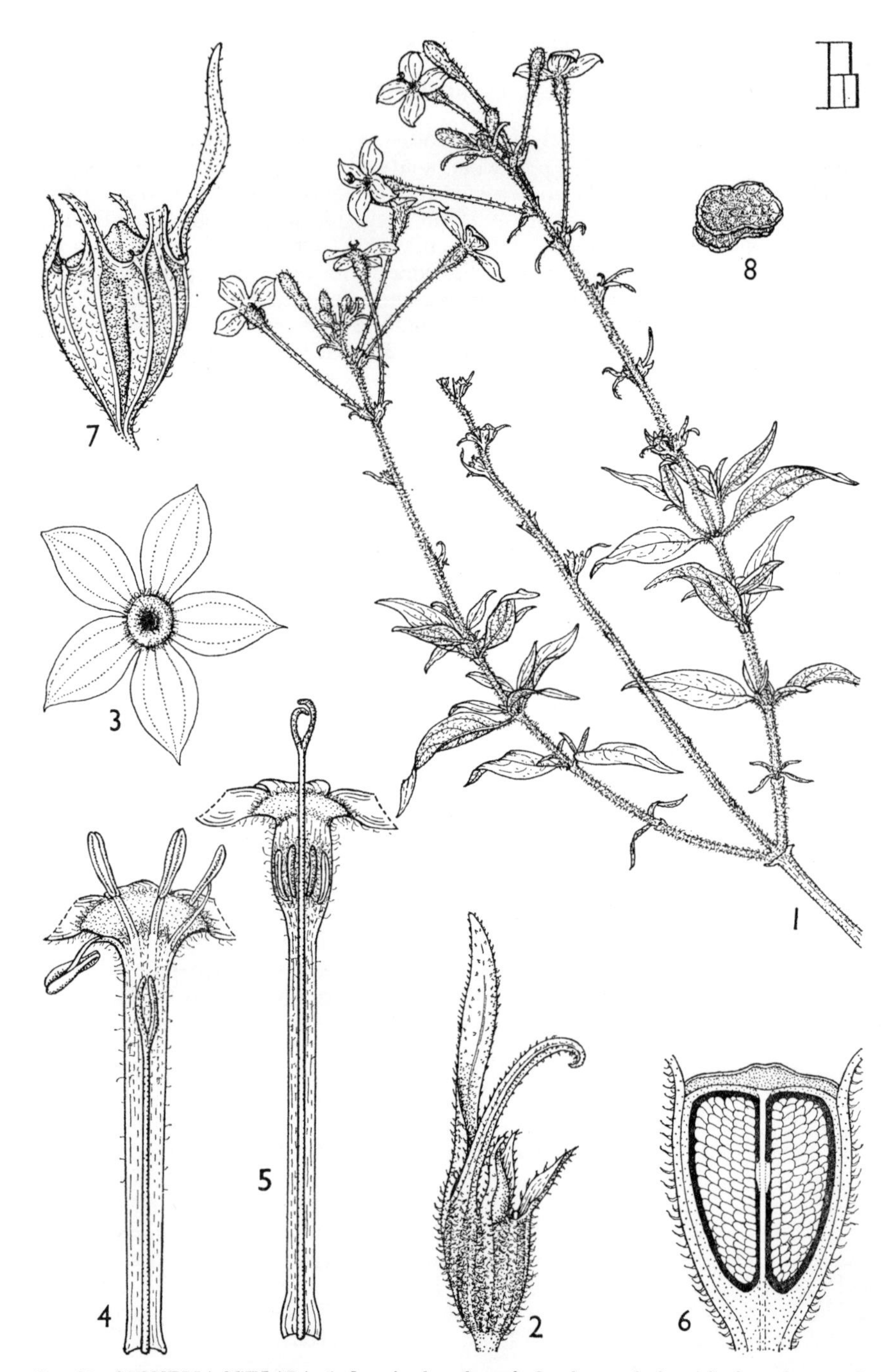

FIG. 24. *OTOMERIA OCULATA*—**1,** flowering branch, × $\frac{2}{3}$; **2,** calyx, × 6; **3,** corolla from above, × 2; **4,** longitudinal section of short-styled flower, × 3; **5,** same of long-styled flower, × 3; **6,** longitudinal section of ovary, × 12; **7,** fruit, × 6; **8,** seed, × 20. 1, from *J. G. Williams* in *E.A.H.* 11097; 2, 3, 6, from *Gillett* 13673; 4, from *Verdcourt & Bally* 841; 5, from *Bally* 8373; 7, 8, from *Greenway* 9174. Drawn by Diane Bridson.

4. **O. volubilis** (*K. Schum.*) *Verdc.* in K.B. 7: 361 (1952) & in B.J.B.B. 23: 26, fig. 3/E (1953); Hepper in F.W.T.A., ed. 2, 2: 215 (1963); Hallé, Fl. Gabon, 12 Rubiacées: 118, t. 22 (1966). Type: Cameroun, W. of Lake Barombi [Barombi-ba-Mbu], *Preuss* 471 (B, lecto. †, BM, K, isolecto. !)

Scandent shrub up to 3–5 m. long, rooting at the lower nodes and between them; roots very fine, up to 20 cm. long; stems with 2 lines of short adpressed hairs above, glabrous below. Leaf-blades ovate-lanceolate to elliptic-lanceolate, 3·6–10 cm. long, 1·1–4 cm. wide, acute to slightly acuminate at the apex, abruptly cuneate at the base, glabrous or very minutely pubescent near the midrib above, with minute adpressed hairs on the nerves beneath; petiole 0·7–2·3(–3·5) cm. long; stipules with 3–5 setae up to 3–4 mm. long from a short base. Flowers not dimorphic, in congested head-like inflorescences 2–3 cm. wide, which do not become spike-like in the fruiting stage. Calyx-tube oblong, 2·5–2·7 mm. long, 0·9–1·3 mm. wide, tomentose; lobes unequal, 1–2 larger, elliptic-lanceolate, 6–10(–12) mm. long, 1·3–3(–5) mm. wide, acute, 3-nerved, 3–4 smaller, deltoid, linear or falcate, 2·5–7 mm. long, 0·75–1·5 mm. wide, with sessile colleters between the lobes. Corolla crimson or bright carmine; tube 1·8–2·8(–3·2) cm. long, 0·5–1 mm. wide below, widening apically to ± 1·5 mm. for a distance of 2–3 mm., pubescent with short hairs; lobes elliptic-oblong, 5–9 mm. long, 2·5–5 mm. wide, mucronate at the tips which are inflexed and have long hairs externally, main part of lobes shortly hairy outside; throat hairy and hairs spreading on to the extreme base of the slightly connate corolla-lobes. Stamens entirely included. Style exserted 0·5–5 mm.; stigmas lanceolate, 0·8–1·2 mm. long. Fruits oblong to ellipsoid-oblong, 7–10 mm. long, 3–5 mm. wide, glabrous, ribbed. Seeds brown, angular, ± 0·5 mm. long.

UGANDA. Bunyoro District: Bugoma, Nkwaki, 28 Nov. 1905, *Dawe* 735!; ? Entebbe 2 May 1910, *Dawe* 986! (locality doubtful)
DISTR. U2, ? 4; Nigeria, Cameroun, Gabon, Central African Republic, Zaire
HAB. Evergreen forest; 1250 m.

SYN. *Pentas volubilis* K. Schum. in E.J. 23: 421 (1897)
P. dewevrei De Wild. & Th. Dur. in Compte Rendu Soc. Bot. Belge 38: 198 (1900). Type: Zaire, Mutembe, Stanley Falls, *Dewèvre* 1089 (BR, holo. !)
Otomeria batesii Wernham in J.B. 54: 226 (1916). Type: Cameroun, *Bates* (BM, holo. !)*

18. CHAMAEPENTAS

Bremek. in Verh. K. Nederl. Akad. Wet., Afd. Natuurk., ser. 2, 48 (2): 46 (1952)

Procumbent perennial herbs. Leaves paired, petiolate, well developed and with the lateral venation clearly evident; stipule-sheath divided into several subulate colleter-tipped lobes. Flowers white, medium-sized, hermaphrodite, probably dimorphic, mostly in 3-flowered terminal inflorescences, less often reduced to 1–2, ebracteolate. Calyx-tube campanulate; lobes 5, unequal, distinctly spathulate. Corolla-tube long, narrowly tubular, the throat densely hairy within; lobes 5, ovate-oblong. Stamens included in long-styled forms, the upper part of the corolla-tube widened at the point of inclusion. Ovary 2-locular, each locule with numerous ovules on peltate placentas; style exserted, minutely tomentellous; stigma bilobed, the lobes cylindrical. Fruit a capsule with thinly bony walls, quite markedly beaked, septicidally and loculicidally dehiscent at the apex. Seeds numerous, brown, angular, pitted.

* In my original revision *Bates* 418 & 985 were given as syntypes, but this is quite erroneous.

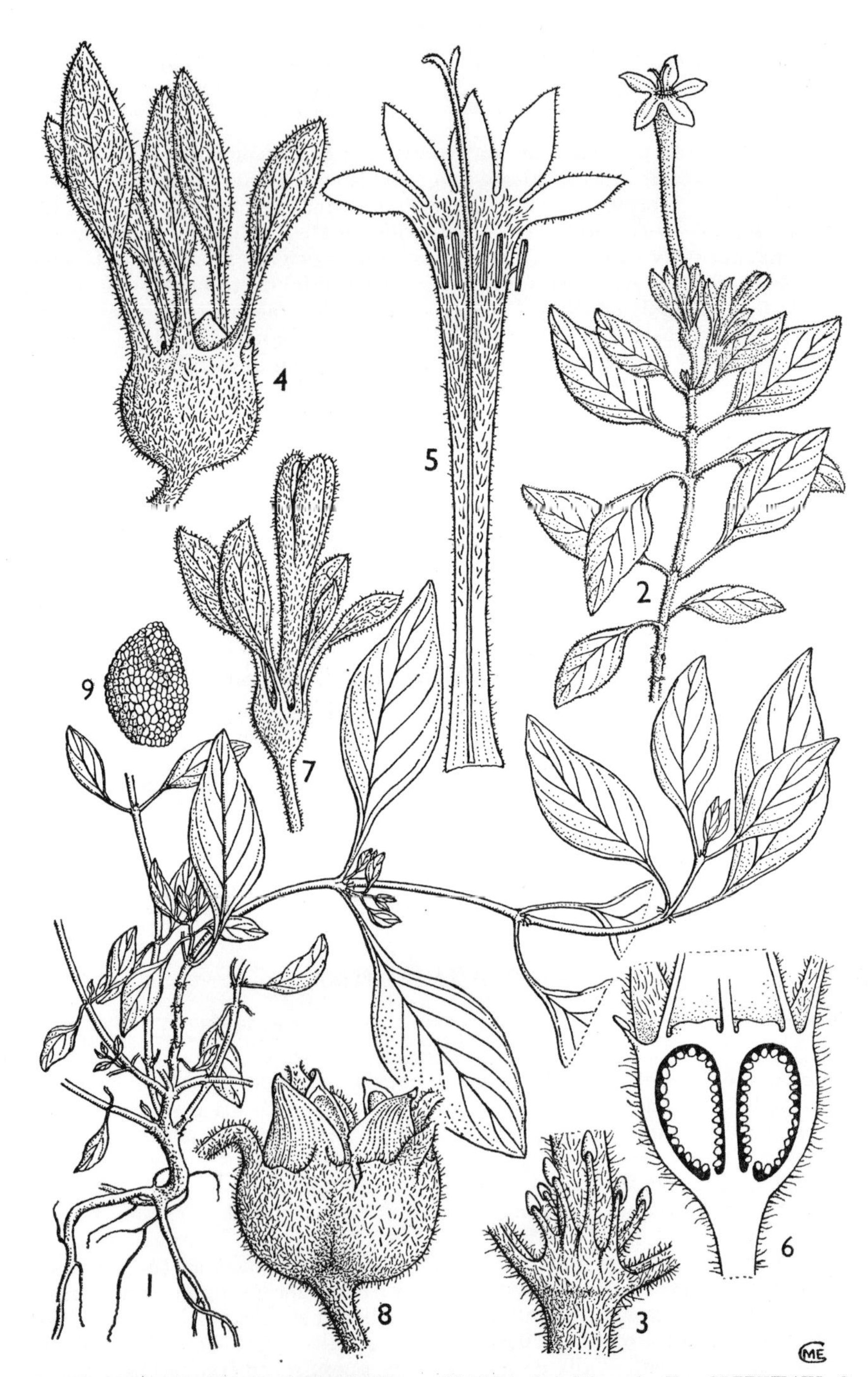

FIG. 25. *CHAMAEPENTAS GREENWAYI* var. *GLABRA*—**1**, habit, × $\frac{2}{3}$. Var. *GREENWAYI*—**2**, flowering shoot, × $\frac{2}{3}$; **3**, node showing stipule, × 10; **4**, flower bud, × 2; **5**, corolla, opened out, and style, × 2; **6**, longitudinal section of ovary, × 8; **7**, young fruit, × 3; **8**, dehisced capsule, × 4; **9**, seed, × 24. All from *Greenway* 6570 (both varieties). Drawn by Mrs M. E. Church.

A monotypic genus, the single species scarcely generically separable from *Pentas hindsioides* K. Schum. The typical variety of the latter has spathulate calyx-lobes very similar to *Chamaepentas* but var. *williamsii* Verdc. has subulate lobes. It might be better to merge the two genera.

C. greenwayi *Bremek.* in Verh. K. Nederl. Akad. Wet., Afd. Natuurk., ser. 2, 48 (2) : 47 (1952). Type : Tanzania, S. Pare Mts., Vidani Mt., *Greenway* 6570 in part (K, holo. !, EA, iso. !)

Stems up to 30 cm. long, sparsely to densely covered with spreading hairs. Leaf-blades elliptic, 1·2–4·5 cm. long, 0·5–3·2 cm. wide, bluntly acute at the apex, cuneate at the base, sparsely to densely hairy on both surfaces or practically glabrous save for the margins; 3–6 lateral nerves clearly visible on each side; petiole 3–9 mm. long, pubescent or densely hairy; stipule sheath 1 mm. long, divided into 3–5 lobes 0·5–1·7 mm. long, the colleters distinctly capitate. Pedicels 1·5–6 mm. long, glabrescent to hairy. Calyx-tube 1·8–2·8 mm. long, glabrous or hairy; lobes of varying sizes, the smallest 7–10 mm. long, 1·5–2 mm. wide, intermediates 1·1–1·3 cm. long, 2·5–3·5 mm. wide, and largest 1·3–1·7 cm. long, 3·8–4·5 mm. wide, densely pubescent or glabrescent save for the margins. Corolla white; tube 4–4·5 cm. long, glabrous or pubescent, 1·5–2·5 mm. wide; lobes ovate-oblong, 0·8–1 cm. long, 3–4·5 mm. wide, glabrous inside. Style exserted 3 mm. in long-styled flowers; stigma-lobes 3 mm. long. Capsule oblong, ± 5 mm. tall, 8 mm. wide, glabrescent or pubescent, the beak ± 2·5–3 mm. tall.

var. **greenwayi**; Bremek. in Verh. K. Nederl. Akad. Wet., Afd. Natuurk., ser. 2, 48 (2): 47 (1952)

Stems, leaves, calyx and corolla densely hairy or pubescent outside. Fig. 25/2–9.

TANZANIA. Pare District: S. Pare Mts., Vidani Mt., 7 July 1942 *Greenway* 6570 in part !
DISTR. **T3**; not known elsewhere
HAB. In rock crevices in shade of great boulders on mountain peak; 1650 m.

var. **glabra** *Bremek.* in Verh. K. Nederl. Akad. Wet., Afd. Natuurk., ser. 2, 48 (2): 48 (1952). Type: Tanzania, S. Pare Mts., Vidani Mt., *Greenway* 6570 in part (K, holo. !, EA, iso. !)

Stems, leaves and calyx glabrescent or with only sparse hairs or only ciliate on the margins; corolla glabrous outside. Fig. 25/1.

TANZANIA. Pare District: S. Pare Mts., Vidani Mt., 7 July 1942, *Greenway* 6570 in part !
DISTR. **T3**; not known elsewhere
HAB. Growing mixed with typical variety

NOTE. Since only one gathering of this species is known, the characters involved seem quite adequate to maintain the variety described by Bremekamp. Study of more adequate populations might reveal intermediates. Also further information on heterostyly in this species is needed.

19. CARPHALEA

Juss., Gen.: 198 (1769); Homolle in Bull. Soc. Bot. Fr. 58: 616 (1937); Verdc. in K.B. 28: 423 (1974)

Dirichletia Klotzsch in Monatsber. K. Preuss. Akad. Wiss. 1853: 494 (1853) & in Peters, Reise Mossamb., Bot. 1: 292, t. 47, 48 (1862)

Small shrubs with erect branched stems. Leaves paired or in whorls of 3, petiolate. Stipules with (1–)3–5 linear or filiform colleter-tipped setae from a short base. Flowers medium-sized, hermaphrodite, dimorphic, some flowers with anthers included and style exserted and others completely vice versa, in rather dense few–many-flowered terminal inflorescences. Calyx-tube narrowly obconic or turbinate, ribbed; limb variously deeply 4–5-lobed or, in sect. *Dirichletia*, eccentrically elliptic, the tube ± placed at one of the foci, sometimes shallowly 3–4-lobed.* Corolla-tube very narrowly cylindrical; lobes 4–5, oblong or ovate; throat densely hairy. Ovary 2–3-locular, each locule with a slender basal placenta bearing ± (3–)4–6 ovules; style filiform, the stigma divided into 2(–4) filiform lobes. Fruit obconic, sometimes curved, of bony texture, strongly ribbed, the ribs running out into the strongly nervose accrescent calyx-limb or lobes, apparently always indehiscent, 1–2-seeded. Seeds narrowly oblong-obconic.

A small genus of about 15 species confined to eastern and central tropical Africa, Socotra and Madagascar. The 3 tropical African species belong to the section *Dirichletia* (Klotzsch) Verdc. and are very poorly defined.

C. glaucescens (*Hiern*) *Verdc.* in K.B. 28: 424 (1974). Type: Somali Republic (S.), Tola R., *Kirk* (K, holo. !)

Erect or scrambling often much-branched shrub (or even described as a small tree) 0·6–3·5 m. tall; stems pubescent at first, later glabrous. Leaf-blades elliptic, elliptic-lanceolate or obovate-elliptic, 1–8·5 cm. long, 0·4–2·1(–2·7) cm. wide, ± obtuse to distinctly acute at the apex, cuneate at the base, glabrous to mostly shortly scabrid pubescent above and beneath, and with longer hairs on the nerves beneath; petioles obsolete to 4(–6) mm. long; stipules with mostly 3–5 setae 1–5 mm. long from a base 0·5–3 mm. long. Flowers in rather dense inflorescences up to 5 cm. across, said by some collectors to be sweet-scented; peduncles 0·1–1·5 cm. long; pedicels 1·5–5(–9) mm. long. Calyx-tube obconic to cylindrical, 1·5–3 mm. long, usually densely hairy, less often glabrous; limb white, greenish, pale mauve to pink, basically eccentrically elliptic, 0·5–1·9(–2) cm. long, 0·5–1·5 cm. wide in flowering state, becoming accrescent and venose in fruiting state, 1·3–2·8 cm. long, 0·85–2·3 cm. wide, sometimes shallowly obtusely 2–4-lobed, glabrous to shortly pubescent. Corolla white or pale pink, sometimes the tube tinged purple at the base (? sometimes blue); tube 1·2–3(–3·5) cm. long, usually densely hairy outside, less often glabrous; lobes ovate, 3·5–4·5(–6) mm. long, 1·5–2(–3) mm. wide, glabrous to hairy outside and sometimes inside also; throat densely hairy. Style exserted 0–6 mm.; stigma-lobes 1–1·25(–2·25) mm. long. Fruit obconic, often curved, 4–8(–10) mm. long, 2–6 mm. wide, strongly ribbed, pubescent. Seeds pale with walls of testa-cells brown, obconic-fusiform, 2·3–2·7(–4) mm. long, 1–1·1 mm. wide.

subsp. **glaucescens**

Leaves more elliptic, usually not all narrowly lanceolate. Fig. 26.

KENYA. Northern Frontier Province: Ndoto Mts., Latakwen, 31 Dec. 1958, *Newbould* 3496!; Machakos District: 237 km. from Mombasa on Nairobi road, 15 Apr. 1960, *Verdcourt & Polhill* 2690!; Teita District: Voi–Mwatate, Nov. 1955, *Ossent* 134!

TANZANIA. Moshi District: near Lake Chala, Mar. 1958, *Bally* 12009! & Kilimanjaro, 600–900 m., *Johnston*!; Pare District: N. Pare Mts., NW. Spur, above Kitari Estate, 19 May 1968, *Bigger* 1842!

DISTR. **K**1, 4, ? 6, 7; **T**2, 3; Somali Republic (S.) and Ethiopia

HAB. Deciduous woodland, bushland and thicket, sometimes on rock outcrops; (70–)200–900(–1500) m.

SYN. *Dirichletia glaucescens* Hiern in F.T.A. 3: 51 (1877); K.T.S.: 438 (1961)
D. asperula K. Schum. in P.O.A. C: 378 (1895). Type: Kenya, Teita District, Ndi Mt., *Hildebrandt* 2595 (B, holo. †, BM, K, iso. !)
D. ellenbeckii K. Schum. in E.J. 33: 336 (1903). Types: Ethiopia, Gobelle valley, *Ellenbeck* 1053A & Arussi-Galla [Arrosi-balla], Buchar, *Ellenbeck* 2013 &

* Rarely 2 contiguous flowers have their calyx-tubes fused and then 2 corollas appear to emerge from one calyx bearing 2–3 lobes.

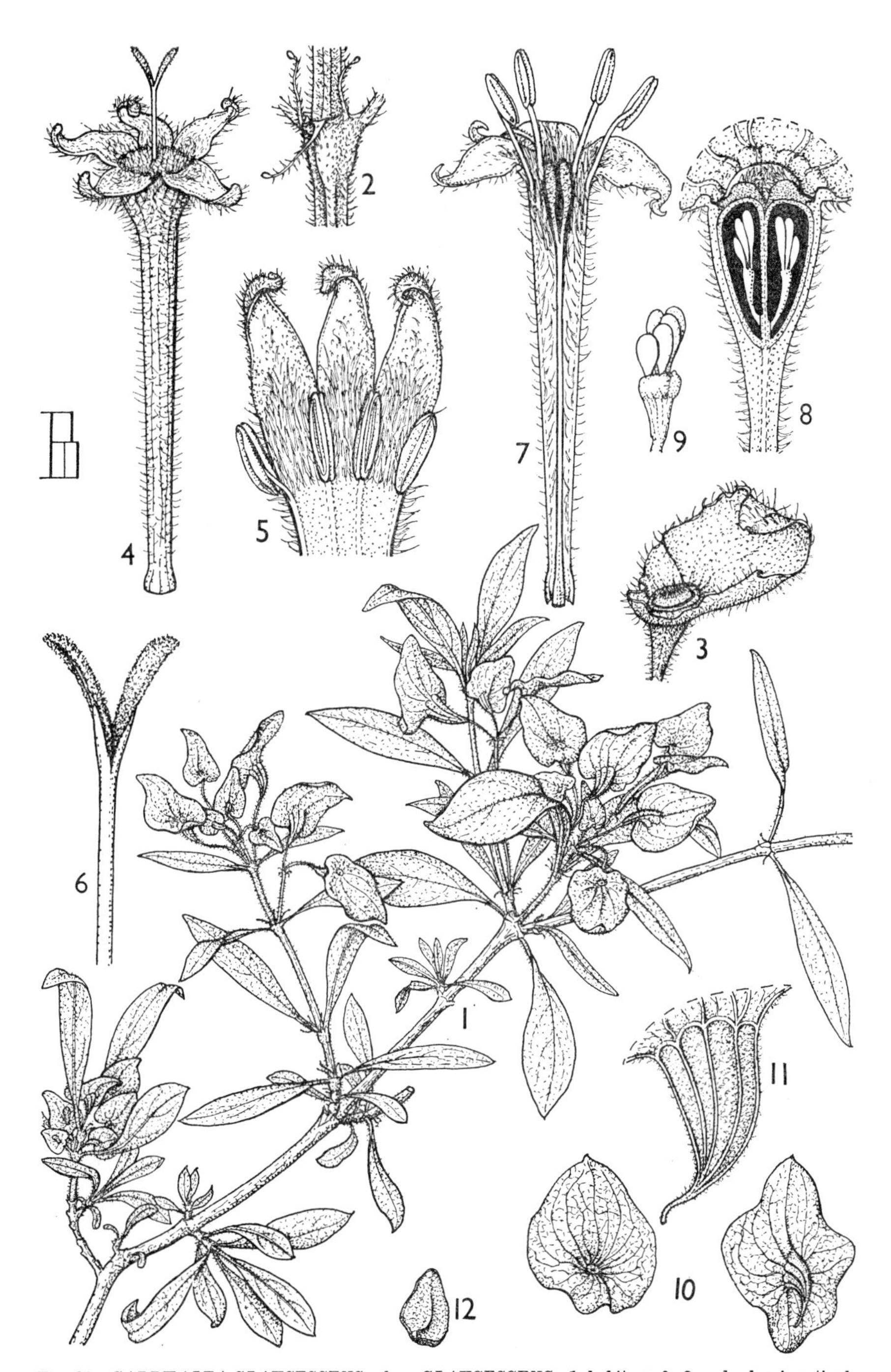

FIG. 26. *CARPHALEA GLAUCESCENS* subsp. *GLAUCESCENS*—**1,** habit, × $\frac{2}{3}$; **2,** node showing stipule, × 4; **3,** calyx, × 3; **4,** long-styled flower, × 3; **5,** longitudinal section of upper part of same, × 6; **6,** tip of style and stigma-lobes, × 6; **7,** short-styled flower, × 3; **8,** longitudinal section of ovary, × 10; **9,** detail of placenta and ovules, × 10; **10,** fruit, front and back views, × 1; **11,** fruit, with calyx-limb removed, × 4; **12,** seed, × 20. 1, from *Tweedie* 4018; 2, from *Ossent* 163; 3, 7–9, from *Verdcourt & Polhill* 2690; 4–6, from *Verdcourt* 1170; 10–12, from *Verdcourt* 2069. Drawn by Diane Bridson.

Kenya, Northern Frontier Province, Boran, Karro Guddi, *Ellenbeck* 2169a (all B, syn. †)
D. sp. sensu K.T.S.: 438 (1961)

NOTE. In the Central and Coastal Provinces of Kenya and the Northern Province of Tanzania the corolla-tube is uniformly small, with a few exceptions, e.g. *Greenway* 10465! (Kenya, Teita District, 32 km. W. of Voi on Nairobi road, 14 Jan. 1962) which has the corolla up to 2·5 cm. long and leaf-blades up to 7 × 2·6 cm.—possibly just the product of a very wet season in that locality. In the north specimens with longer corolla-tubes are much more frequent, but distributed erratically, and I have assumed that those with the longest seen are equivalent to *D. ellenbeckii*, which could be retained as a variety if desired.

subsp. **angustifolia** *Verdc.* in K.B. 28: 425 (1974). Type: Tanzania, 72 km. from Iringa on Dodoma road, Nyangolo Scarp, *Verdcourt* 3078 (EA, holo. !, K, iso. !)

Leaves mostly narrowly lanceolate, 3·6–8·2 cm. long, 0·7–2·7 cm. wide, narrowly acute at the apex.

TANZANIA. Mpwapwa District: Kongwa Ranch, 16 Feb. 1966, *Leippert* 6270!; Iringa District: Lukose R., *Goetze* 482! & Ruaha National Park, Kinyantupa [Chinaputa] Escarpment, 11 Jan. 1966, *Richards* 20953!
DISTR. T5, 7; not known elsewhere
HAB. Deciduous bushland and woodland, mainly of *Combretum*, *Acacia*, *Commiphora*, *Sterculia*, *Strychnos*, *Lannea*, *Sclerocarya*, etc., also *Brachystegia* woodland; 700–1350 m.

SYN. [*Dirichletia pubescens* sensu T.T.C.L.: 494 (1949), *non* Klotzsch]

NOTE. *Carphalea pubescens* (Klotzsch) Verdc. is widespread in S. tropical Africa and when compared with the majority of specimens of *C. glaucescens* is clearly distinguishable by the longer flowers and larger acuminate leaves with closer less arcuate venation. The S. Tanzania populations are however very much closer to *C. pubescens* than the material from Kenya particularly that from Mpwapwa District. Typical material of *C. glaucescens* subsp. *angustifolia* has the corolla-tube 1·5–1·8 cm. long, but *Hornby* 182! (Mpwapwa, 18 Feb. 1930) has it just over 3 cm. long. This specimen could equally well be referred to *C. pubescens*. See also note after subsp. *glaucescens*.

20. DOLICHOMETRA

K. Schum. in E.J. 34: 331 (1904)

Perennial procumbent herb with ascending flowering stems. Leaves paired, petiolate; stipules ± triangular, basally joining the petioles to form a very short sheath, apically very shortly lobed. Flowers small, hermaphrodite, dimorphic but showing rather limited heterostyly, in few-flowered terminal and axillary mostly pedunculate cymes. Calyx-tube linear-fusiform; lobes 5, linear-oblong or slightly spathulate, equal or very nearly so, with sessile colleters between the lobes. Corolla-tube narrowly funnel-shaped; lobes 5, narrowly triangular- or oblong-lanceolate. Stamens included, the filaments long or short. Ovary 2-locular; ovules ± 18*, attached uniseriately to elongate-fusiform centrally attached placentas; style included, long or short; stigma bifid, the lobes cylindrical-filiform, just exserted in the long-styled form. Fruit linear-fusiform, indehiscent. Rhaphides very abundant in all tissues.

A monotypic genus apparently restricted to the E. Usambara Mts. in NE. Tanzania.

D. leucantha *K. Schum.* in E.J. 34: 331 (1904); Verdc. in K.B. 12: 354 (1957) & in B.J.B.B. 28: 266 (1958). Types: Tanzania, E. Usambara Mts., Amani, *Engler* 613 & 722 (B, syn. †); Amani, Mt. Bomole, *Verdcourt* 55 (EA, neo. !, K, isoneo. !)

* The " 4–6 " of the original description appears to be erroneous.

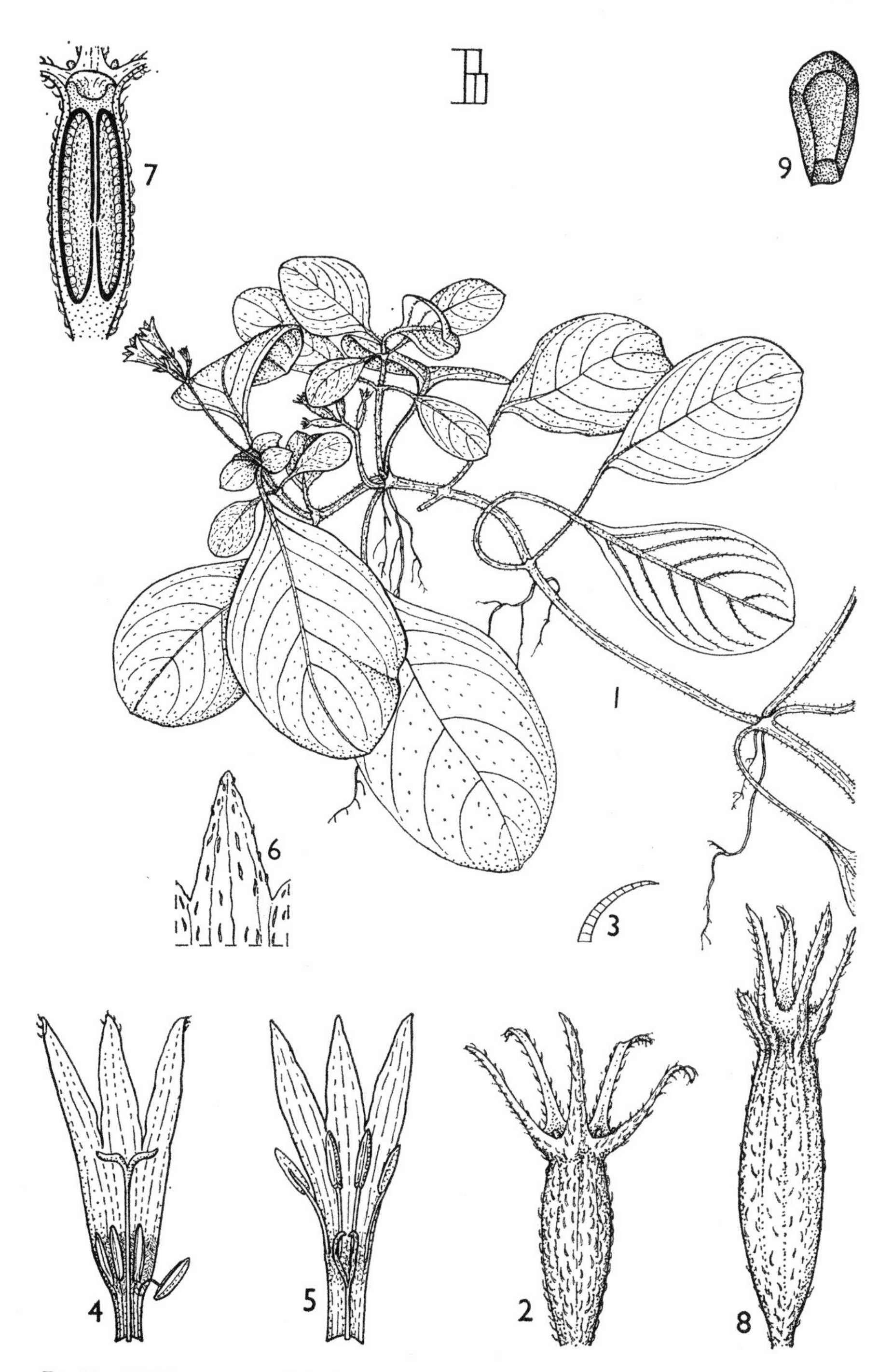

FIG. 27. *DOLICHOMETRA LEUCANTHA*—**1,** habit, × $\frac{2}{3}$; **2,** calyx, × 8; **3,** multicellular hair from calyx, × 30; **4,** longitudinal section of long-styled flower, × 5; **5,** same of short-styled flower, × 5; **6,** corolla-lobe, from outside, × 6; **7,** longitudinal section of ovary, × 10; **8,** fruit, × 5; **9,** seed, × 20. 1–3, 5, 7, from *Drummond & Hemsley* 3449; 4, 8, 9, from *Tanner* 2487; 6, from *Greenway* 4731. Drawn by Diane Bridson.

Prostrate, perennial, creeping herb rooting at the nodes; stems slightly succulent, 10–45 cm. long, pubescent with short adpressed hairs. Leaf-blades elliptic to ovate-elliptic, 1·7–8·5(–9) cm. long, 0·7–4·7 cm. wide, obtuse or ± acute at the apex, narrowly cuneate at the base, sparsely to rather densely pubescent with short hairs above and beneath, particularly on the nerves; petiole (0·2–)1–2·5(–5) cm. long; stipules ± 3 mm. long, basal 1 mm. connate to the petioles, the apical 3–5 lobes 0·5–1 mm. long. Inflorescences 6–10-flowered, pubescent; peduncles up to 4·5 cm. long; pedicels obsolete or ± 1 mm. long; at some nodes, apart from the long-peduncled several-flowered inflorescence there is present in the opposite axil an additional 1–2-flowered inflorescence subsessile or on a much shorter peduncle (appearing as a long pedicel in single-flowered inflorescences). Calyx-tube 3–4 mm. long, pubescent; lobes 1·5–3 mm. long, 0·4–0·8 mm. wide. Corolla white; tube 5·5–8 mm. long, 1–1·5 mm. wide at the base, 3–5 mm. wide at the throat; lobes 2·5–4·5 mm. long, 1·5–2 mm. wide at the base; both tube and lobes shortly pubescent outside. Filaments ± 1 mm. long in long-styled flowers, the anthers immersed in a part of the corolla-tube bearing dense long hairs; in short-styled flowers the filaments are ± 3·5 mm. long but the tube is not densely hairy inside; anthers ± 2 mm. long in both. Style 5–6 mm. long and pubescent, with stigma-lobes 1·5 mm. long in long-styled flowers; in short-styled flowers the style is 1·5 mm. long and the stigma-lobes 2 mm. long. Fruits 6–7 mm. long, 1·5 mm. wide, pubescent, ± 30-seeded. Seeds brown particularly on the angles, very angular, polygonal, often resembling slices through a prism, the 2 large faces triangular, the 3 narrow ones oblong, reticulate, longest dimension ± 0·7 mm. Fig. 27, p. 223.

TANZANIA. Lushoto District: Amani, track to Mt. Bomole, 20 Apr. 1968, *Renvoize & Abdallah* 1616! & Amani, by Sigi R., 11 Nov. 1936, *Greenway* 4731! & Amani–Kwamkoro road, about 1·6 km. SW. of Amani, 25 July 1953, *Drummond & Hemsley* 3449! & Mt. Bomole, 15 Apr. 1921, *Soleman* in *Herb. Amani* 5956!

DISTR. **T3**; not known elsewhere

HAB. Characteristic mat-forming herb in the lowest herb layer of rain-forest, particularly on paths with *Oplismenus* and *Pseudechinolaena*; 700–950 m.

NOTE. This species appears to be strictly endemic to the E. Usambara Mts., save for its occurrence on Mt. Mlinga, the highest point of a small range situated between the E. Usambaras and the east coast (based on a visual record by the author in 1950).

21. PARAPENTAS

Bremek. in Verh. K. Nederl. Akad. Wet., Afd. Natuurk., ser. 2, 48 (2): 50 (1952); Verdc. in B.J.B.B. 23: 53 (1953)

Procumbent perennial forest floor herbs. Leaves paired, petiolate, well developed and with the lateral venation clearly evident; stipule-sheath divided into several subulate colleter-tipped lobes. Flowers white or lilac, hermaphrodite, isostylous or heterostylous, terminal or pseudo-axillary, sessile, usually solitary or sometimes paired. Calyx-tube campanulate; lobes 5, ± equal, linear or linear-oblong. Corolla-tube narrowly tubular, glabrous or sparsely to densely pilose within throat; lobes 5, elliptic, minutely papillate inside. Stamens and style exserted in isostylous species, in normally heterostylous species the stamens well-exserted in one form and included in the other but in *P. setigera* (Hiern) Verdc. (*P. gabonica* Bremek.) either both stamens and style are included or the anthers are exserted and the style included. Ovary 2-locular with peltate placentas and numerous ovules. Fruit an obconic or globose capsule, sometimes compressed beneath, not beaked, loculicidally dehiscent or, in *P. setigera,* dividing into 4 valves. Seeds numerous, brown, angular, pitted.

A small genus of 3 or 4 species restricted to Africa and possibly an undescribed one in Madagascar.

Flowers distinctly heterostylous, the throat of the corolla rather densely hairy . . . 1. *P. battiscombei*
Flowers isostylous, both stamens and style exserted; throat of corolla glabrous or only sparsely hairy . . . . . . 2. *P. silvatica*

1. **P. battiscombei** *Verdc.* in B.J.B.B. 23 : 54 (1953) ; U.K.W.F. : 398 (1974). Type : Kenya, E. Mt. Kenya Forests, *Battiscombe* 695 (EA, holo. !, K, iso. !)

Stems 15–30 cm. long, often forming carpets and rooting at the nodes, the flowering shoots ± erect or ascending for 5–10 cm., rather sparsely to densely adpressed pubescent. Leaf-blades ovate or elliptic-ovate, 0·6–4·5 cm. long, 0·3–2·4 cm. wide, rather bluntly acute, cuneate to rounded or rounded and then attenuately cuneate at the base, with scattered curly hairs above and on margins but mostly confined to the venation beneath; petiole 0·3–2 cm. long, pubescent; stipule-sheath 1–1·5 mm. long, divided into 3–5 filiform fimbriae 1–2(–3·8) mm. long, with scarcely evident colleters. Pedicels very short, attaining 2 mm. in fruit, pubescent. Calyx-tube subglobose, ± 1 mm. long and wide, hairy; lobes ± foliaceous, unequal or subequal, oblong to narrowly spathulate, (2–)3–6 mm. long, (0·4–)1–2·2 mm. wide, with scattered hairs as on the leaves. Corolla white, pale blue or lilac, sometimes with a darker median stripe on the lobes; tube 1·4–2·3 cm. long, 1–2·2 mm. wide, very sparsely hairy outside, densely hairy within the throat; lobes oblong-elliptic, 5–8 mm. long, 1·6–4·5 mm. wide, glabrous inside. Style exserted 2–5 mm. in long-styled flowers, 1·4 cm. long in short-styled flowers, glabrous; stigma-lobes filiform 1–2·5 mm. long. Capsule brown, oblong-ovoid, 3–4 mm. long, 2·5–3 mm. wide, only slightly beaked.

KENYA. Meru District: Nyambeni Hills, where R. Thangatha meets the south circular track to Maua, 10 Oct. 1960, *Polhill & Verdcourt* 283 !; Embu District: SE. Mt. Kenya Forest, near Thuchi and S. Mara Rivers, 18 Sept. 1951, *D. Davis* 27 !; Teita District: Ngangao Forest, 6 Feb. 1953, *Bally* 8768 !
DISTR. **K**4, 7; not known elsewhere
HAB. Forest floor herb, particularly in clearings of upland evergreen forest; 1410–2100 m.

NOTE. Bremekamp had proposed to describe *Bally* 8768 as a new species of *Chamaepentas* before he had his attention drawn to the fact that it was conspecific with the above. He considered that the lack of bracteoles and hairy corolla throat should exclude it from *Parapentas*. The habit is however that of *Parapentas*, the capsule is scarcely if at all rostrate, the calyx-lobes are not so characteristically spathulate as in *Chamaepentas* where they are actually more or less stipitate and, more important perhaps, the style lacks the dense tomentellous indumentum which is characteristic of *Chamaepentas* and some sections of *Pentas*.

2. **P. silvatica** (*K. Schum.*) *Bremek.* in Verh. K. Nederl. Akad. Wet., Afd. Natuurk., ser. 2, 48 (2): 52 (1952); Verdc. in B.J.B.B. 23: 56 (1953) & in K.B. 30: 289 (1975). Type: Tanzania, W. Usambara Mts., Mlalo, *Holst* 511 (B, holo. †)*

Stem straggling or prostrate, (0·15–)0·5–1 m. long, rooting at the nodes, the flowering shoots ascending to 10 cm., the younger parts with dense violet or ferruginous adpressed hairs, the older parts more sparsely pubescent. Leaf-

* For discussion of *Buchwald* 632, chosen as a neotype by Bremekamp, see note at the end of subspecies *silvatica*.

blades ovate or elliptic-ovate, 0·6–8 cm. long, 0·5–4·8 cm. wide, obtuse to acute at the apex, very shortly mucronulate, gradually cuneate or rounded then narrowly cuneate at the base, sparsely to fairly densely adpressed pubescent above and beneath; petiole 0·3–4 cm. long, adpressed pubescent; stipule-sheath 1–2 mm. long, divided into 3–5 fimbriae, 2–5 mm. long; colleters mostly fairly evident. Flowers 1–3, axillary and terminal, opening one at a time, sessile, isostylous, both stamens and style exserted. Calyx-tube subglobose, 1·2–1·5 mm. long, pubescent; lobes linear-oblong or slightly spathulate, (1–)1·5–6 mm. long, 0·4–1·25 mm. wide, pubescent, the base sometimes triangularly widened in fruit and veined. Corolla white or bluish lilac or blue; tube 0·8–1·7 cm. long, pubescent or sometimes glabrescent outside; throat glabrous or somewhat hairy; lobes elliptic-oblong, (1·7–)4–7 mm. long, 0·6–2 mm. wide. Stamens exserted; filaments 2–4 mm. long. Style slightly exserted to 2 mm. or only as long as the tube; stigma well exserted, 1·6–3 mm. long. Capsule oblong or obconic, 2–3·5 mm. long, 2·5–3 mm. wide, ribbed, speckled with dark resin patches, sparsely hairy; fruiting pedicels up to about 1·5 mm. long.

subsp. **silvatica**

Leaf-blades narrower, 0·6–3·3(–4) cm. long, 0·5–1·8(–2·5) cm. wide; petioles ± 1 cm. long. Floral parts near the lower of the limits stated.

TANZANIA. Lushoto District: 6·4 km. NE. of Lushoto, Mkuzi, 10 Apr. 1953, *Drummond & Hemsley* 2064! & Mkuzi–Kinguelo road, 25 Dec. 1950, *Greenway* 8483!; Ulanga District: 35 km. S. of Mahenge, Sali, 20 Mar. 1932, *Schlieben* 1925!
DISTR. **T**3, 6; not known elsewhere
HAB. Floor of upland evergreen forest; 1000–1650 m.

SYN. *Oldenlandia silvatica* K. Schum. in P.O.A. C: 376 (1895)
O. procurrens K. Schum. in E.J. 34: 329 (1904). Type: Tanzania, W. Usambara Mts., Mlalo, Shagayu Forest, *Engler* 1403 (B, holo. †)
Virecta obscura K. Schum. in E.J. 34: 331 (1904). Type: Tanzania, W. Usambara Mts., Sakare, *Engler* 984 (B, holo. †)
Parapentas parviflora Bremek. in Verh. K. Nederl. Akad. Wet., Afd. Natuurk., ser. 2, 48 (2): 53 (1952). Type: Tanzania, Pare Mts., Hotulwa, *Luchman* 21 (EA, holo.!)
P. procurrens (K. Schum.) Verdc. in B.J.B.B. 23: 55 (1953)

NOTE. I hinted in B.J.B.B. 23: 57 (1953) that I thought Bremekamp had applied the name *P. silvatica* strictly to the East Usambara plant in error; on geographical grounds it is clearly to be referred to the West Usambara plant, i.e. what Bremekamp called *P. parviflora* and I called *P. procurrens*. K. Schumann's description is short but his mention of petiole to 0·8 cm. is clearly in agreement with this. Bremekamp's selection of *Buchwald* 632, the Kew sheet of which approaches the East Usambara subspecies, but the British Museum sheet of which has small leaf-blades 2·5 × 1·2 cm., as neotype is best set aside; unfortunately this specimen is unlocalised. There is a good deal of overlap between the populations concerned and subspecific status seems appropriate.

subsp. **latifolia** Verdc. in K.B. 30: 289 (1975). Type: Tanzania, E. Usambara Mts., Mlinga Mt., *Drummond & Hemsley* 1445 (K, holo.!, EA, iso.)

Leaf-blades wider, up to 8 cm. long and 4·8 cm. wide; petioles up to 4 cm. long. Floral parts near the upper of the limits stated. Fig. 28.

TANZANIA. Lushoto District: Amani, Mt. Bomole, 8 May 1950, *Verdcourt* 181!; Tanga District: Mt. Mlinga, 17 Aug. 1950, *Verdcourt* 309!; Morogoro District: S. Nguru Forest Reserve, Liwale R., Manyangu [Minangya], 21 Aug. 1951, *Greenway & Farquahar* 8635!
DISTR. **T**3, 6; not known elsewhere
HAB. Floor of rain-forest; 900–1080(–1450) m.

SYN. [*Parapentas silvatica* sensu Bremek. in Verh. K. Nederl. Akad. Wet., Afd. Natuurk., ser. 2, 48 (2): 53 (1953) & Verdc. in B.J.B.B. 23: 56 (1953), *non* (K. Schum.) Bremek. sensu stricto]

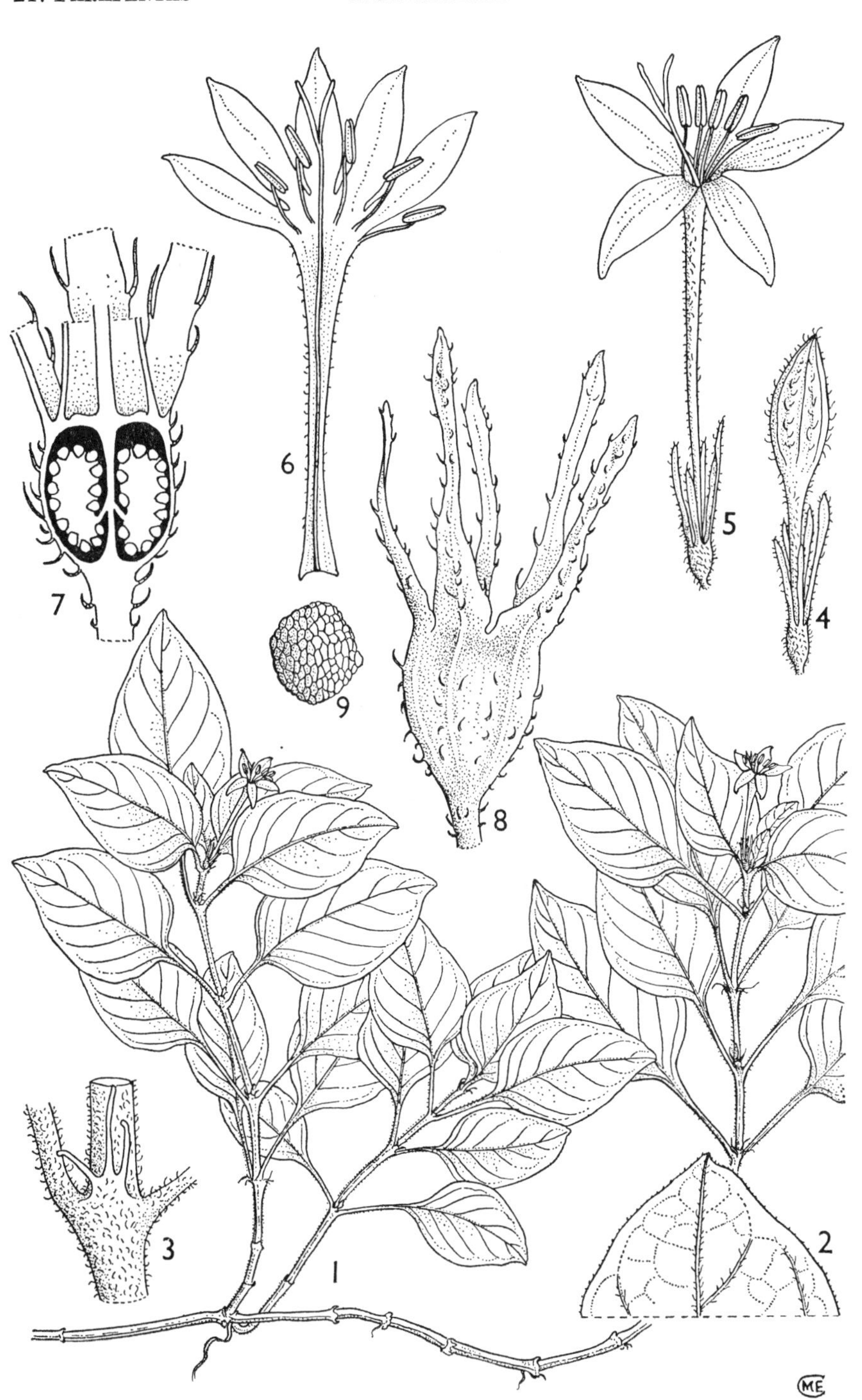

FIG. 28. *PARAPENTAS SILVATICA* subsp. *LATIFOLIA*—**1,** habit, × ⅔; **2,** lower surface of leaf apex, × 4; **3,** node showing stipule, × 4; **4,** flower bud, × 4: **5,** flower, × 3; **6,** corolla, opened out, and style, × 3; **7,** longitudinal section of ovary, × 16; **8,** fruit, × 6; **9,** seed, × 18. 1–3, from *Drummond & Hemsley* 1445; 4–7, from *Greenway & Farquhar* 8635; 8, 9, from *Don Carlos* 5965. Drawn by Mrs M. E. Church.

22. **KOHAUTIA**

Cham. & Schlecht. in Linnaea 4: 156 (1829); Bremek. in Verh. K. Nederl. Akad. Wet., Afd. Natuurk., ser. 2, 48 (2): 56 (1952)

Annual or perennial herbs or rarely subshrubs, mostly erect, often much branched. Leaves sessile, opposite, nearly always linear and mostly 1-nerved, less often penninerved or 3–5-nerved from the base; stipules with 2–several fimbriae or rarely reduced to a simple lobe. Flowers small or medium-sized, never heterostylous, in panicles, corymbs, occasionally in heads, more rarely in pairs, triads or even solitary. Calyx-lobes 4 (rarely 5), small, equal, subulate to ovate-triangular or triangular. Corolla-tube narrowly cylindrical; lobes 4 (rarely 5); throat glabrous or pilose inside. Stamens always included in the upper swollen part of the corolla-tube or only the anther-tips exserted. Ovary 2-locular; ovules numerous, immersed in fleshy peltate placentas; style always included, the stigma not divided, cylindrical, or divided into 2 filiform lobes mostly with the tips well below the base of the anthers or just reaching them. Capsule globose or ellipsoid, scarcely beaked, loculicidally splitting at the apex. Seeds numerous, angular, mostly not becoming viscid when moistened, somewhat alveolate.

A genus of about 60 species occurring throughout Africa and also in Madagascar and parts of tropical Asia. The structure of the flowers supports Bremekamp's claim that this is undoubtedly a genus distinct from *Oldenlandia*; it is also fairly easy with practice to separate them by habit alone. He divides the genus into subgen. *Pachystigma*, with two series, *Barbatae* and *Imberbae*, and subgen. *Kohautia* (*Eukohautia*), with two series *Diurnae* and *Kohautia* (*Noctiflorae*). The classification is illustrated in the form of a key on p. 230.

Stigma cylindrical or ovoid, not divided into lobes:
 Corolla-throat bearded with flattened hairs:
  Decumbent much-branched herb, with many ascending flowering shoots (rarely a small annual), the leaves rather concentrated at the base, pubescent above; peduncles very well developed, 4–12 cm. long, much longer than the preceding internodes (**K1**, 7) . . . 2. *K. prolixipes*
  More erect herbs, with more scattered leaves separated by longer internodes; peduncles, if as long, then not much longer than the preceding internodes:
   Corolla-limb mostly scarlet or bright orange- or vermilion-red, the lobes 4·5–9 mm. long, 3–5 mm. wide; plant glabrescent or scabridulous-papillate:
    Widened part of corolla-tube not constricted above, more funnel-shaped, the constrictions at the base of the widened part often nodular inside; calyx-lobes (0·5–)3–4(–5·5) mm. long; adnate part of filaments below anthers glabrous; more robust plants up to 70 cm. tall, the pedicels rarely attaining 3 cm. in length after flowering; coastal plant . 1. *K. obtusiloba*
    Widened part of corolla-tube constricted both above and below; calyx-lobes

1·2–1·8 mm. long; adnate part of filaments below the anther with a few hairs; more slender shorter plants, mostly about 25 cm. tall, the pedicels often attaining 4 cm. after flowering; only known in Flora area from Kigoma District of **T4** . . . . . . 3. *K. microcala*

Corolla-limb mostly lilac, blue, mauve or white, the lobes 1·9–4·5 mm. long, 1·3–4 mm. wide; plant glabrous to distinctly hairy; calyx-lobes 1–7 mm. long; widened part of corolla-tube constricted above and below; adnate part of filaments with a few hairs . 4. *K. longifolia*

Corolla-throat not bearded within; lobes pale rose, lilac, red or white, 1·3–1·8 mm. long, 0·8–1·2 mm. wide . . . . . 5. *K. virgata*

Stigma divided into 2 filiform lobes:

Corolla-lobes very distinctly mucronate, giving the buds a distinctive apiculum 1–2 mm. long; corolla red, crimson-pink, rose or rarely white, the lobes 0·45–1 cm. long, 3–7 mm. wide; flowers in corymbs . . 7. *K. grandiflora*

Corolla-lobes not so distinctly mucronate; flowers smaller and not in corymbs:

Pedicels all very slender, 1–2 cm. long, eventually sometimes attaining 4 cm. in fruiting stage; calyx-lobes narrowly triangular, 1·3 mm. long; capsule subglobose (rare plant in **T4**) . . . 8. *K. confusa*

Pedicels mostly short or very short, 0·5–7 mm. long, the majority very short but sometimes a few (and these often 1-flowered branchlets simulating pedicels) 0·7–2·6 cm. long:

Corolla scarlet-pink, pink or crimson-purple; lobes 1·5–4·5(–7·5) mm. long, 0·8–2·3(–4) mm. wide; stems scabridulous with small white papillae; capsule distinctly oblong . . 6. *K. coccinea*

Corolla white, yellowish, blue or lilac, rarely truly red; capsule more globose:

Flowers mostly borne singly along the branchlets of the inflorescence; corolla-tube 1·1–1·4 cm. long; stems glabrous to scabridulous pubescent 9. *K. caespitosa*

Flowers mostly borne in pairs along the branchlets of the inflorescence, both subsessile or one sessile and the other subsessile; stems scabridulous or very distinctly papillate:

Corolla usually white, yellow or buff, tube 0·5–1·5 cm. long; leaves narrowly lanceolate to elliptic-

lanceolate, 2–10 mm. wide; stigma not reaching the anthers . 10. *K. lasiocarpa*
Corolla often blue or purple, tube 2·5–4·7 mm. long; leaves linear to narrowly elliptic, 1–3 mm. wide; stigma touching the anthers . . . . . . 11. *K. aspera*

SYNOPSIS OF CLASSIFICATION

Style ending in a single ovoid or cylindrical stigma; corolla-limb usually red or purple . subgen. **Pachystigma**
Corolla-throat ± densely bearded . . . series *Barbatae* (species 1–4)
Corolla-throat glabrous inside . . . series *Imberbae* (species 5)
Style ending in 2 filiform stigmas; corolla white or coloured . . . . subgen. **Kohautia**
Corolla-limb red, blue or purple . . . series *Diurnae* (species 6–8)
Corolla white, yellowish or yellow-brown, less often pale blue or lilac (in species 11) series *Kohautia* (species 9–12)

1. **K. obtusiloba** (*Hiern*) *Bremek*. in Verh. K. Nederl. Akad. Wet., Afd. Natuurk., ser. 2, 48 (2): 66 (1952); Jex-Blake, Gard. E. Afr., ed. 4, t. 4/4 (1957). Types: Tanzania, Bagamoyo District, Kingani R., *Kirk* & Mozambique, *Forbes* (both K, syn.!)

Annual or probably sometimes perennial erect herb 10–70 cm. tall, branched from the base; stems 4-ribbed, glabrous or the ribs scabridulous. Leaf-blades linear to linear-lanceolate, 1·5–6·8 cm. long, 1–3·5(–6) mm. wide, acute at the apex, narrowed to the base, scabridulous, particularly above and on the margins; stipule-sheath 1·2–2 mm. long, divided into 4 very unequal fimbriae, (0·5–)2·2–5 mm. long. Inflorescences lax, made up of 5–7-flowered terminal cymes augmented by 3-flowered cymes from the upper axils; peduncle (1·7–)5–9·3 cm. long; pedicels 0·25–1·5 cm. long, extending to 3·3 cm. long after flowering. Calyx-tube subglobose, 1·5 mm. in diameter, glabrous; lobes linear-lanceolate, (0·5–)3–4(–5·5) mm. long, ± 0·8 mm. wide, keeled, scabridulous on the margins. Corolla with pale green tube (? always) and scarlet or vermilion limb; tube 0·8–1·2 cm. long, 0·7 mm. wide, the apical part 2–3 mm. long, 1·5–2·5 mm. wide, densely hairy within the throat; lobes broadly elliptic, 4·5–9 mm. long, 3–5 mm. wide, bluntly acute, joined for up to 1·5 mm. at the base. Style 3·5 mm. long; stigma 1·8 mm. long. Capsule almost hemispherical, 2–2·5(–4) mm. tall, 4–4·8 mm. wide, glabrous. Seeds blackish, subglobose or angular, ± 0·4 mm. long.

KENYA. Mombasa District: Changamwe, 1 June 1934, *Napier* in *C.M.* 6291! & near Mombasa, Feb. 1876, *Hildebrandt* 1968!; Kilifi, on road to Mombasa, 9 July 1945, *Jeffery* 250!
TANZANIA. Tanga District: 8 km. SE. of Ngomeni, 29 July 1953, *Drummond & Hemsley* 3536! & Kange Estate, 10 Nov. 1951, *Faulkner* 810!; Bagamoyo District: just N. of Bagamoyo, mouth of Kingani R., 27 Feb. 1961, *Verdcourt* 3098A!
DISTR. **K**7; **T**3, 6; Mozambique
HAB. Coarse grassland, particularly in swampy areas on black cotton soil, e.g. "mbugas", also as a weed; 0–270 m.

SYN. ? *Oldenlandia zanguebariae* Lour., Fl. Cochinch.: 78 (1790). Type: "Zanzibar", *Loureiro* (not traced; may belong to next species)
O. obtusiloba Hiern in F.T.A. 3: 56 (1877); K. Schum. in P.O.A. C: 376 (1895)

NOTE. I would agree with Bremekamp that a *Johnston* specimen at Kew stated to be from Kilimanjaro 5000 ft. (1500 m.) is probably wrongly localised. I believe some generalised expedition labels were put on to specimens actually found during the earlier part of the expedition, i.e. nearer the coast. The distinguishing characters given by Bremekamp to separate this species from *K. longifolia* seem to be valid, i.e. the corolla-tube is funnel-shaped or at least not constricted above the dilated part in *K. obtusiloba*; also in this species the adnate parts of the filaments just below the anthers are glabrous, not hairy and the throat is more densely hairy. In *K. longifolia* there are dents at the top of the inflated part of the corolla-tube containing the anthers corresponding to intrusions inside; in *K. obtusiloba* there are intrusions only at the base and these are often distinctly nodular. The application of the Loureiro name is not certain; it could equally well be applied to the next species.

2. **K. prolixipes** (*S. Moore*) *Bremek.* in Verh. K. Nederl. Akad. Wet., Afd. Natuurk., ser. 2, 48 (2): 67 (1952). Type: Kenya, Kwale District, near Avisana, Daruma, *Kassner* 442 (BM, holo. !, K, iso. !)

Perennial decumbent herb (9–)15–35 cm. tall, much branched from a woody base and with many ascending flowering branches; stems rather densely shortly pubescent. Leaf-blades narrowly lanceolate to narrowly elliptic-lanceolate, 1·2–3·1 cm. long, 2–5(–7) mm. wide, acute at the apex, ± rounded at the base, densely shortly pubescent above and on the midvein beneath; stipule-sheath 1–3 mm. long, divided into (1–)2–5 fimbriae, 1·2–3 mm. long. Inflorescences often 5-flowered, long-stalked; peduncle 4–12 cm. long, glabrous; pedicels 0·3–2 cm. long. Calyx-tube subglobose, 1·2 mm. in diameter, glabrous; lobes oblong-lanceolate, (1·2–)1·8–2·4 mm. long, glabrous or sparsely ciliate, keeled. Corolla magenta; tube 5–6 mm. long, 0·8 mm. wide, the widened part 2–2·5 mm. long, 1–1·3 mm. wide, the throat shortly hairy within; lobes oblong, 3·2–5·5 mm. long, (1·3–)1·8–2·5 mm. wide, bluntly acute, joined for up to 1 mm. at the base. Style 1 mm. long; stigma 1·2 mm. long. Capsule ± hemispherical, 2·2–3 mm. tall, 3–3·2 mm. wide. Seeds (immature) blackish, trigonous, ± 0·4 mm. long.

KENYA. Northern Frontier Province: Mt. Nyiru, June 1936, *Jex-Blake* 16 !; Teita District: Sagala, 4 Feb. 1953, *Bally* 8715 ! & Sagala Hills, 1 Jan. 1971, *Faden & Evens* 71/43 !; Kilifi District: Mariakani, *R. M. Graham* in *F.D.* 1720 !

DISTR. K1, 7; not known elsewhere

HAB. Rocky places, bushland, grassland; 120–2400 m.

SYN. *Oldenlandia prolixipes* S. Moore in J.B. 43: 351 (1905) & 45: 115 (1907)
O. pedunculata K. Schum. & K. Krause in E.J. 39: 519 (1907), *nom. superfl.*, also based on *Kassner* 442 (B, holo. †, BM, K, iso. !)

NOTE. The altitude 2400 m. for *Jex-Blake* 16 seems very high for a plant which is mostly restricted to the coastal province; it may be that Lady Jex-Blake was merely giving the altitude of the mountain and that her specimen actually came from the lower slopes. *Bally* 8715 is from an altitude of 1050 m. *C. R. Field* 56 (Kenya, Galana Ranch, Dakabuka, 21 Sept. 1972) is a small erect annual up to 17 cm. tall, presumably a plant flowering in its first year. The fine pubescence on the upper leaf surface and inflorescence show it cannot be separated from *K. prolixipes*.

3. **K. microcala** *Bremek.* in Verh. K. Nederl. Akad. Wet., Afd. Natuurk., ser. 2, 48 (2): 73 (1952). Type: Zambia, near Kalungwishi R., *Walter* 5 (K, holo. !)

Erect annual herb with unbranched or mostly branched slender yellowish green papillate-puberulous stems, 20–25(–30) cm. tall. Leaf-blades narrowly linear, filiform or narrowly lanceolate, 1–3·5 cm. long, 0·5–3 mm. wide, blunt at the apex, passing unnarrowed into the stipule-sheath at the base, shortly scabridulous above, ± glabrous beneath; stipule-sheath 1–1·5 mm. long, divided into 2–5 subulate or filiform fimbriae 1·5–3·5(–4) mm. long. Inflorescences very lax, cymose; peduncles up to 3 cm. long; pedicels slender,

1–2 cm. long, elongating to 4 cm., slightly papillate. Calyx-tube ellipsoid, 1 mm. long, often papillate; lobes triangular-lanceolate, 1·2–1·8 mm. long, keeled, scabridulous at margins and on the costa, not accrescent in fruit. Corolla red, violet or orange-red; tube 5–6 mm. long, glabrous outside or slightly scabridulous near the throat, the apical dilated part 2–2·2 mm. long; throat sparsely hairy within; lobes ± round or broadly elliptic, 4·5–5 mm. long, 4·2–4·6 mm. wide, very shortly acuminate. Style 1·5 mm. long; stigma 0·7 mm. long. Capsule depressed subglobose, 2·2 mm. tall, 3·2 mm. wide. Seeds olive-brown, ellipsoid-angular, 0·4 mm. long, 0·2 mm. wide, strongly reticulate.

TANZANIA. Kigoma District: Ujiji area, mountain W. above Kigoma, 12 Feb. 1926, *Peter* 36773! & same locality and date, *Peter* 36757b! & Kigoma to Machazo [Machaso], 18 Feb. 1926, *Peter* 37007! & Machazo [Machaso] to Mkuti R., 19 Feb. 1926, *Peter* 37114!
DISTR. T4; Mozambique, Malawi and Zambia
HAB. Woodland; 800–960 m.

NOTE. Bremekamp distinguishes this from many of its allies on the structure of the stipules which he claims bear only 2 short fimbriae but even some of those specimens he cites have more. It also keys out in his key in a couplet where the leaves should be linear or filiform but the material cited above has narrowly lanceolate leaves and such wider leaves can also be seen on some of the specimens he cites.

4. **K. longifolia** *Klotzsch* in Peters, Reise Mossamb., Bot. 1: 297 (1861); Bremek. in Verh. K. Nederl. Akad. Wet., Afd. Natuurk., ser. 2, 48 (2): 68 (1952). Types: Mozambique, Sena, *Peters* (B, holo. †); neotype: Gonubi Hill, *Schlechter* 12181 (K, neo. !)

Annual or perennial erect or decumbent herb 13–70(–105) cm. tall, little to usually well branched; stems glabrous, puberulous or shortly hairy. Leaves linear to narrowly lanceolate or less often elliptic-lanceolate, 0·5–9·3 cm. long, 0·12–1·2 cm. wide, acute at the apex, narrowed to the base, glabrous, scabridulous or shortly pubescent; stipular sheath 0·5–2 mm. long, produced into 2–11 fimbriae 1–9 mm. long. Inflorescences strict to very lax, trichotomous or dichasial the branchlets mostly monochasial or flowers in triads; pedicels 0·3–4·5 cm. long, glabrous to pubescent. Calyx-tube subglobose, 1–1·5 mm. long, glabrous to hairy; lobes narrowly triangular to lanceolate, 1–7 mm. long, ciliolate on the margins. Corolla lilac, pink, red, blue or white, the tube often greenish, narrowly cylindrical, 3·8–6·5 mm. long, the widened part 1–3 mm. long, 0·8–1·5 mm. wide, glabrous to shortly hairy outside; lobes elliptic, 1·9–4·5(–6) mm. long, 1·3–4 mm. wide, slightly joined at the base, sometimes shortly hairy outside; limb often darker at the throat. Style 1·1–1·8 mm. long; stigma 0·9–1·3 mm. long. Capsules subglobose or hemispherical, 1·2–3(–4) mm. long, 1·8–3·2 mm. wide, glabrous to shortly hairy. Seeds olive-brown, irregularly oblong-ovoid, ± 0·5 mm. long, reticulate.

KEY TO THE INFRASPECIFIC VARIANTS

Calyx-lobes under 5 mm. long, mostly 2–3 mm.:
  Stems glabrous to scabridulous or pubescent; capsule glabrous:
    Leaf-blades mostly wider and more lanceolate; corolla-lobes 3–4·5(–6) × (1·5–)2–4 mm. . . . . a. var. **longifolia**
    Leaf-blades mostly linear to linear-lanceolate; corolla-lobes 2(–3) × 1·5(–2) mm. . d. var. **effusa**
  Stems, calyx, capsule and sometimes outsides of corolla pubescent to hairy . . . b. var. **vestita**

Calyx-lobes attaining 6–7 mm. long; stems, etc. hairy as in var. *vestita* . . . . c. var. **macrocalyx**

a. var. **longifolia**

Stems glabrous or puberulous to pubescent above. Leaves mostly wider and more lanceolate. Corolla-lobes 3–4·5(–6) mm. long, (1·5–)2–4 mm. wide. Capsule glabrous.

TANZANIA. Mpanda District: without exact locality, 7 May 1958, *Jones* 85 !; Rungwe, 12 Oct. 1910, *Stolz* 330 !; Songea District: about 9·5 km. SW. of Songea, near R. Mtanda, 25 Mar. 1956, *Milne-Redhead & Taylor* 9295 !; Zanzibar I., Walezo, 16 Feb. 1930, *Vaughan* 1189 ! & Massazine, 26 June 1960, *Faulkner* 2613 !
DISTR. **T**3, 4, 6–8; **Z**; Mozambique, Malawi, Zambia, Rhodesia and South Africa (Natal)
HAB. Grassland, sometimes marshy, *Brachystegia* and other woodlands, also as a weed in cultivations; 30–1560 m.

SYN. ? *K. macrophylla* Klotzsch in Peters, Reise Mossamb., Bot. 1: 297 (1861). Type: Mozambique, Cabaceira, *Peters* (B, holo. †)
[*Oldenlandia caffra* sensu Hiern in F.T.A. 3: 58 (1877), pro parte, *non* Eckl. & Zeyh.]
O. longifolia (Klotzsch) K. Schum. in P.O.A. C: 376 (1895), *non* (Schumach.) DC., *nom. illegit.*

b. var. **vestita** *Bremek.* in Verh. K. Nederl. Akad. Wet., Afd. Natuurk., ser. 2, 48 (2): 70 (1952). Type: Zambia, 54·6 km. NE. of Livingstone, *Hutchinson & Gillett* 3496 (K, holo. !)

Similar to the typical variety but much hairier, including the calyx, fruit and often outside of corolla. Calyx-lobes under 5 mm. long, mostly 2–3 mm.

KENYA. Kwale District: Shimba Hills, 13 Apr. 1968, *Magogo & Glover* 826 !; north of Mombasa to Lamu and Witu, *Whyte* !
TANZANIA. Tanga District: Sawa, 28 Aug. 1967, *Faulkner* 4018 !; Tabora District: 11 km. on [Tabora–] Itigi road, 27 Apr. 1962, *Tallantire* 230 !; Uzaramo District: near Dar es Salaam University, 26 May 1968, *Batty* 96 !
DISTR. **K**7; **T**1, 3, 4, 6; Mozambique, Malawi, Zambia, Rhodesia
HAB. Grassland, grassy swamps, bushland, clearings in *Brachystegia* woodland and old cultivations; 0–1380 m.

c. var. **macrocalyx** *Bremek.* in Verh. K. Nederl. Akad. Wet., Afd. Natuurk., ser. 2, 48(2): 70 (1952). Type: Tanzania, Dodoma, *Derhain* (K, holo. !)

Exactly as in var. *vestita* but calyx-lobes attaining 6–7 mm. in length.

TANZANIA. Tabora District: near Tabora, Usagari, 15 Apr. 1958, *Jones* 64 ! & Urambo, June 1960, *Grundy* 23 !; Dodoma, 8 Mar. 1924, *Derhain* !
DISTR. **T**4, 5; not known elsewhere
HAB. Grassland, railway banks and as weed in tobacco; 750 m.

d. var. **effusa** (*Oliv.*) *Verdc.* in K.B. 30: 290 (1975). Type: Tanzania, Morogoro District, banks of Mgeta R., *Grant* (K, holo. !)

Stems glabrous or scabridulous. Leaves mostly linear to linear-lanceolate. Corolla-lobes ± 2(–3) mm. long and 1·5(–2) mm. wide. Capsule glabrous.

KENYA. Tana River District: 48 km. S. of Garsen, Kurawa, 5 Oct. 1961, *Polhill & Paulo* 603 !
TANZANIA. Ufipa District: Lake Rukwa, Milepa, 15 June 1936, *Lea* 45 !; Masasi District: 16 km. on Masasi–Tunduru road, 5 May 1962, *Boaler* 577 !; Lindi District: Nachingwea, 13 Apr. 1955, *Anderson* 1046 !
DISTR. **K**7; **T**4, 6, 8; Mozambique, Malawi and Zambia
HAB. Deciduous woodland, sometimes in cultivated areas and on footpaths, etc.; 15–870(–1400) m.

SYN. *Oldenlandia effusa* Oliv. in Trans. Linn. Soc. 29: 84, t. 48 (1873); Hiern in F.T.A. 3: 59 (1877), pro parte
Kohautia effusa (Oliv.) Bremek. in Verh. K. Nederl. Akad. Wet., Afd. Natuurk., ser. 2, 48(2): 72 (1952)

NOTE. I have been unable to satisfactorily separate *K. effusa* at specific level; there is great variation in the width of the leaves and size of the corolla-limb, and narrow-leaved specimens do not always have flowers as small as the dimensions given by

Bremekamp. The variability is particularly evident in the many specimens collected by Milne-Redhead & Taylor in the Songea District (**T**8). As Bremekamp points out, the type of var. *effusa* is a lax and weak specimen grown in a wet place.

NOTE (on species as a whole). Apart from the varieties mentioned above, Bremekamp also records his var. *psilogyna* from Tanzania, Dar es Salaam (*Stuhlmann* 7775 in part). Var. *psilogyna* is based on Mozambique, Moribane, *Dawe* 490 (K, holo. !) and differs from var. *vestita* in having the ovary glabrous.

5. **K. virgata** (*Willd.*) *Bremek.* in Verh. K. Nederl. Akad. Wet., Afd. Natuurk., ser. 2, 48 (2): 77 (1952); Hepper in F.W.T.A., ed. 2, 2: 209 (1963); U.K.W.F.: 398 (1974). Type: probably Ghana [Guinea], *Thonning* (B, holo., S, iso.)

Perennial (? rarely annual) erect herb, 9–60 cm. tall, with few to many rather slender strict or spreading branches from a woody base; stems glabrous, scabridulous or quite densely pubescent with short spreading hairs. Leaf-blades linear to narrowly elliptic, 1–4 cm. long, 0·5–5 mm. wide, acute and often, particularly in wider leaves, ending in a seta up to 1 mm. long, gradually narrowed at the base, glabrous to quite densely covered with spreading hairs at least above and on margins and midvein beneath; stipule-sheath 1–2 mm. long, divided into 2–6 filiform fimbriae (1–)2–6 mm. long. Inflorescences lax, with 2–several flowers at each node, some with short pedicels others long, those in the lower parts of the inflorescence actually axillary but the leaves reduced to stipules only which form hyaline bracts; peduncles 0·1–3(–6) cm. long, internodes of inflorescence 0·6–3·3 cm. long; pedicels 1–9(–18) mm. long, glabrous. Calyx-tube subglobose, ± 1 mm. long, 1·5 mm. wide, glabrous; lobes narrowly triangular, 1·2–2 mm. long, ending in a short seta, glabrous or somewhat scabridulous on the margins. Corolla white to purplish brown with lobes white, pale rose, red or lilac (sometimes only at the apex outside); tube 3·5–5·5 mm. long, the widened part 1–1·3 mm. long, 0·9–1·2 mm. wide, glabrous outside and throat not bearded within; lobes ovate-elliptic, 0·9–1·9 mm. long, 0·8–1·2 mm. wide, ± acute. Style 2·5–3·5 mm. long; stigma-lobes 1·2–1·4 mm. long. Capsule transversely oblong or subglobose, 1–2 mm. long, 1·5–3 mm. wide, glabrous; fruiting pedicel up to 1·3 cm. long. Seeds dark brown, subconic to irregularly ovoid, ± 0·3 mm. long.

UGANDA. Teso District: Serere, Mar. 1933, *Chandler* 1129 !
KENYA. Kiambu District: Ruiru, 5 Aug. 1930, *Napier* 380 !; Kwale District: 22·4 km. S. of Mombasa, Twiga, 14 Oct. 1962, *Verdcourt* 3286 !; Kilifi District: Malindi area, May 1959, *Rawlins* 614 ! & 659 !
TANZANIA. Pangani District: 29 km. SW. of Pangani, 17 Mar. 1950, *Verdcourt* 116 !; Morogoro District: Morogoro–Dakawa road, 14·4 km. N. of Morogoro, 25 Mar. 1953, *Drummond & Hemsley* 1784 !; Rufiji District: S. bank of Rufiji R., Utete, 2 Dec. 1955, *Milne-Redhead & Taylor* 7467 !; Zanzibar I., Walezo Ridge, 9 Nov. 1930, *Vaughan* 1672 ! & Marahubi, 17 May 1935, *Vaughan* 2229 !
DISTR. **U**3; **K**4, 7; **T**2, 3, 6; **Z**; widespread in tropical Africa from Guinée to South West Africa, Zaire and Sudan to South Africa, also in Comoro Is. and Madagascar
HAB. Open grassy places in coastal bushland, deciduous woodland, secondary grassland, rocky places, waste ground, roadsides and in old cultivations and rice fields; 0–1080(–1620) m.

SYN. *Hedyotis virgata* Willd., Sp. Pl. 1: 567 (1797)
Oldenlandia virgata (Willd.) DC., Prodr. 4: 426 (1830); Hiern in F.T.A. 3: 59 (1877); F.W.T.A. 2: 132 (1931)
O. caffra Eckl. & Zeyh., Enum. Pl. Afr. Austr.: 360 (1837). Type: South Africa, Uitenhage, Zwartkops R., *Zeyher* 2705 (K, ? iso. !)—see Bremek., loc. cit.: 79 (1952)
Kohautia parviflora Benth. in Hook., Niger Fl.: 403 (1849). Type: Ghana, Accra, *Vogel* (K, holo. !)
Oldenlandia parviflora (Benth.) Oliv. in Trans. Linn. Soc. 29: 84 (1873); Hiern in F.T.A. 3: 60 (1877), quoad typum solum

Kohautia virgata (Willd.) Bremek. var. *oblanceolata* Bremek. in Verh. K. Nederl. Akad. Wet., Afd. Natuurk., ser. 2, 48(2): 80 (1952). Type: South Africa, Transvaal, Kruger National Park, Ikukusa, *Cholmondeley* (PRE, holo.)

NOTE. Specimens with broader leaves ending in setae are to be found scattered throughout the distribution of this species and since intermediates occur I have not considered the variety worth retaining.

A plant from Tanzania (Uzaramo District, Dar es Salaam, 8 May 1970, *Harris* 4155 !) is a monstrous form of this species with many of the floral organs leaf-like; there are however a few normal flowers and fruits on some specimens. The Kew sheet showed no normal flowers and was at first thought to be an intergeneric hybrid. The pollen proved to be 95% infertile; there are a few large fat 5-furrowed grains but the rest are deformed, collapsed and tiny.

6. **K. coccinea** *Royle*, Ill. Himal.: 241, t. 53/1 (1835); Bremek. in Verh. K. Nederl. Akad. Wet., Afd. Natuurk., ser. 2, 48 (2): 82 (1952); Hepper in F.W.T.A., ed. 2, 2: 210 (1963); U.K.W.F.: 398, fig. (1974). Type: India, Himalaya, Budraj, *Royle* (LIV, holo.)

Annual erect, unbranched or sparsely branched herb 7–45(–70) cm. tall, the stems scabridulous with small white papillae. Leaves linear or linear-lanceolate, 1–7·7 cm. long, 1–8 mm. wide, acute at the apex, narrowed at the base, ± smooth or sometimes scabridulous above; stipular sheath 1–3 mm. long, produced into 2–6 filiform fimbriae 1·5–3(–5) mm. long. Inflorescences spike-like, raceme-like or in well-grown specimens distinctly monochasial or dichasial, rarely reduced to 2 flowers; peduncles 0·8–4 cm. long; pedicels or apparent pedicels (sometimes 1-flowered branchlets simulate pedicellate flowers) 0·5–7(–15) mm. long, densely scabridulous. Calyx-tube ellipsoid, ± 1·2–1·5 mm. long, densely papillate; lobes linear, 1·8–5·2(–7) mm. long, ± 0·25 mm. wide, scabridulous. Corolla variously coloured, the limb usually pink, scarlet-pink or crimson-purple but paler beneath, the venation often purplish black or dark red, the tube usually greenish with a red tinge or sometimes yellowish, mostly white at the throat (one note says some specimens can have all white flowers but this may be an error, likewise the one which states pale blue); corolla scabridulous or rarely glabrous outside; tube often bent, narrowly cylindrical, 2–6 mm. long, the widened part 1·2–3·7 mm. long, 0·7–1·3 mm. wide; lobes (3–)4(–5), oblong-elliptic, 1·5–4·5(–7·5) mm. long, 0·5–2·3(–4) mm. wide, subacute. Style 1·2–1·5 mm. long; stigmata 1–1·2 mm. long. Capsule oblong-ellipsoid, 3–5·5 mm. tall, 2·5–4·5 mm. wide, the beak slightly raised, usually scabridulous. Seeds dark brown, irregularly angular, 0·3 mm. long.

UGANDA. W. Nile District: Paidha, 20 Aug. 1953, *Chancellor* 178 !; Kigezi District: Ruzumbura, Apr. 1939, *Purseglove* 678 !; Mbale District: Sebei, Kyosoweri, 16 Oct. 1955, *Norman* 295 !

KENYA. Trans-Nzoia District: S. Cherangani, 14 Sept. 1957, *Symes* 208 !; Nairobi District: Thika Roadhouse, 1 July 1951, *Verdcourt* 530 !; Masai District: N. slopes of Kilimanjaro, above Laitokitok, 14 Feb. 1933, *C. G. Rogers* 413 !

TANZANIA. Mbulu District: Mbulumbul, 23 June 1944, *Greenway* 6912 !; Ufipa District: Namanyere, 4 Apr. 1950, *Bullock* 2820 !; Songea District: near Kitai, by R. Rovuma, 17 Apr. 1956, *Milne-Redhead & Taylor* 9669 !; Zanzibar, May 1960, *Tallantire* 146 !

DISTR. **U**1–4; **K**1–6; **T**1, 2, 4, 7, 8; **Z**; Senegal, N. Nigeria, Zaire, Ethiopia, Malawi, Zambia, Rhodesia and Mozambique, also in N. India

HAB. Grassland, including seasonally wet areas, grassland with scattered trees, woodland edges, gravel paths, waste ground, shallow soil on rock outcrops and also as a weed in cultivated fields, by roadsides, etc.; 880–2400 m.

SYN. *Hedyotis abyssinica* A. Rich., Tent. Fl. Abyss. 1: 363 (1847). Type: Ethiopia, Adua, *Schimper* 1902 (P, holo., BM, K, MO, iso. !)

Oldenlandia abyssinica (A. Rich.) Hiern in F.T.A. 3: 57 (1877); K. Schum. in P.O.A. C: 376 (1905); F.W.T.A. 2: 131 (1931)

O. coccinea (Royle) Hook. f. in Fl. Brit. India 3: 69 (1880)

O. macrodonta Bak. in K.B. 1895: 67 (1895). Type: Zambia, Lake Tanganyika, Fwambo, *Carson* 107 (K, holo. !)

Pentas modesta Bak. in K.B. 1895: 290 (1895). Type: Zambia, Mwero, Kalungwizi R., *Carson* 33 (K, holo.!)

NOTE. Specimens growing side by side can vary greatly in flower size and colour, e.g. *Verdcourt* 2462, 2463 (Kenya, Trans-Nzoia District, Kitale Grassland Research Station, 2 Oct. 1959); in the former the corolla has the tube greenish, the inside of the lobes scarlet but the outside pale; in the latter the corolla is smaller with the tube tinged dark purplish brown and the limb pink. The flowers can vary in size by over 100%, the shorter unbranched plants often having the smallest flowers.

7. **K. grandiflora** *DC.*, Prodr. 4: 430 (1830); Bremek. in Verh. K. Nederl. Akad. Wet., Afd. Natuurk., ser. 2, 48 (2): 88 (1952); Hepper in F.W.T.A., ed. 2, 2: 210, fig. 244 (1963). Types: Senegal, without locality, *Bacle* & Kounoun, *Perrottet & Leprieur* (G-DC, syn.)

Annual erect sparsely branched or unbranched herb 10–80 cm. tall, the stems glabrous or sparsely scabridulous with minute white papillae. Leaves linear to linear-lanceolate, 2–10·5 cm. long, 0·7–9 mm. wide, tapering, acute or apiculate at the apex, narrowed to the base, minutely scabridulous above particularly near the margins, otherwise glabrous; stipule-sheath 1–1·5(–2) mm. long, adnate to the leaf-base, produced into (1–)2 closely adpressed fimbriae 0·5–3 mm. long. Inflorescences corymbose or paniculate, often joined to form 2 wide corymbs up to 7 cm. wide; peduncle 0–7 cm. long; pedicels mostly very short, 0(–5) mm. long, finely scabridulous-papillate. Calyx-tube ovoid, 1–1·5 mm. long, papillate; lobes subulate, 1–2 mm. long, scabridulous. Corolla red, crimson-pink or less often rose or white, glabrous or very slightly scabridulous outside; tube narrowly cylindrical, 1·2–1·7 cm. long, the widened part 4–5 mm. long, 0·9–1·5 mm. wide; lobes elliptic, 0·45–1·1 cm. long, 2·5–7 mm. wide, acute and distinctly mucronate, the terminal mucro up to 1–2 mm. long, very distinct in the bud. Style 2·5–4·5 mm. long; stigma-lobes 1·5 mm. long. Capsule subglobose, often compressed or depressed, 2–4·5 mm. long, 3–4(–5) mm. in diameter, glabrous; beak noticeably raised at dehiscence. Seeds dark brown, angular, subconic, 0·5–0·8 mm. long, margined. Fig. 29.

UGANDA. Karamoja District: 6·4 km. NE. of Iriri, 29 Sept. 1962, *Lewis* 5997! & 1 km. from Karita, 10 Oct. 1964, *Leippert* 5089! & Lochoi (? Lochowai), 24 May 1940, *A. S. Thomas* 3538!

DISTR. U1; W. Africa from Senegal to Cameroun, Central African Republic and Sudan

HAB. Grassland and grassland with scattered *Acacia* etc., usually on seasonally inundated black clay soils, less often on stony soil; 1050–1400 m.

SYN. *Oldenlandia grandiflora* (DC.) Hiern in F.T.A. 3: 57 (1877), pro parte; K. Schum. in P.O.A. C: 376 (1895), pro parte

8. **K. confusa** (*Hutch. & Dalz.*) *Bremek.* in Verh. K. Nederl. Akad. Wet., Afd. Natuurk., ser. 2, 48 (2): 89 (1952). Type: Senegal, near Galam, *Heudelot* 220 (K, lecto.!, P, isolecto.)

Erect annual herb, with branched glabrous or very slightly scabridulous stems, 35–70 cm. tall. Leaves linear, 3–6 cm. long, 0·8–1·5 mm. wide, narrowly acute at the apex, narrowed to the base, glabrous; stipular sheath 1–1·5 mm. long, with 2 filiform setae 0·7–1·5 mm. long. Inflorescences very lax panicles; pedicels very graceful, 1–1·5 cm. long, lengthening up to 2–4 cm. long in fruit, glabrous. Calyx-tube ellipsoid, 0·8 mm. long, papillose; lobes narrowly triangular, 1·3 mm. long, slightly scabridulous at the margins. Corolla said to be red; tube 4·5 mm. long, minutely papillate, the widened part 1·5 mm. long, 0·5 mm. wide, glabrous inside; lobes 1·3 mm. long, 0·7 mm. wide. Style 1·3 mm. long; stigma-lobes 1·7 mm. long. Capsule subglobose, 3 mm. in diameter, glabrous. Seeds chestnut, angular, 0·3 mm. long.

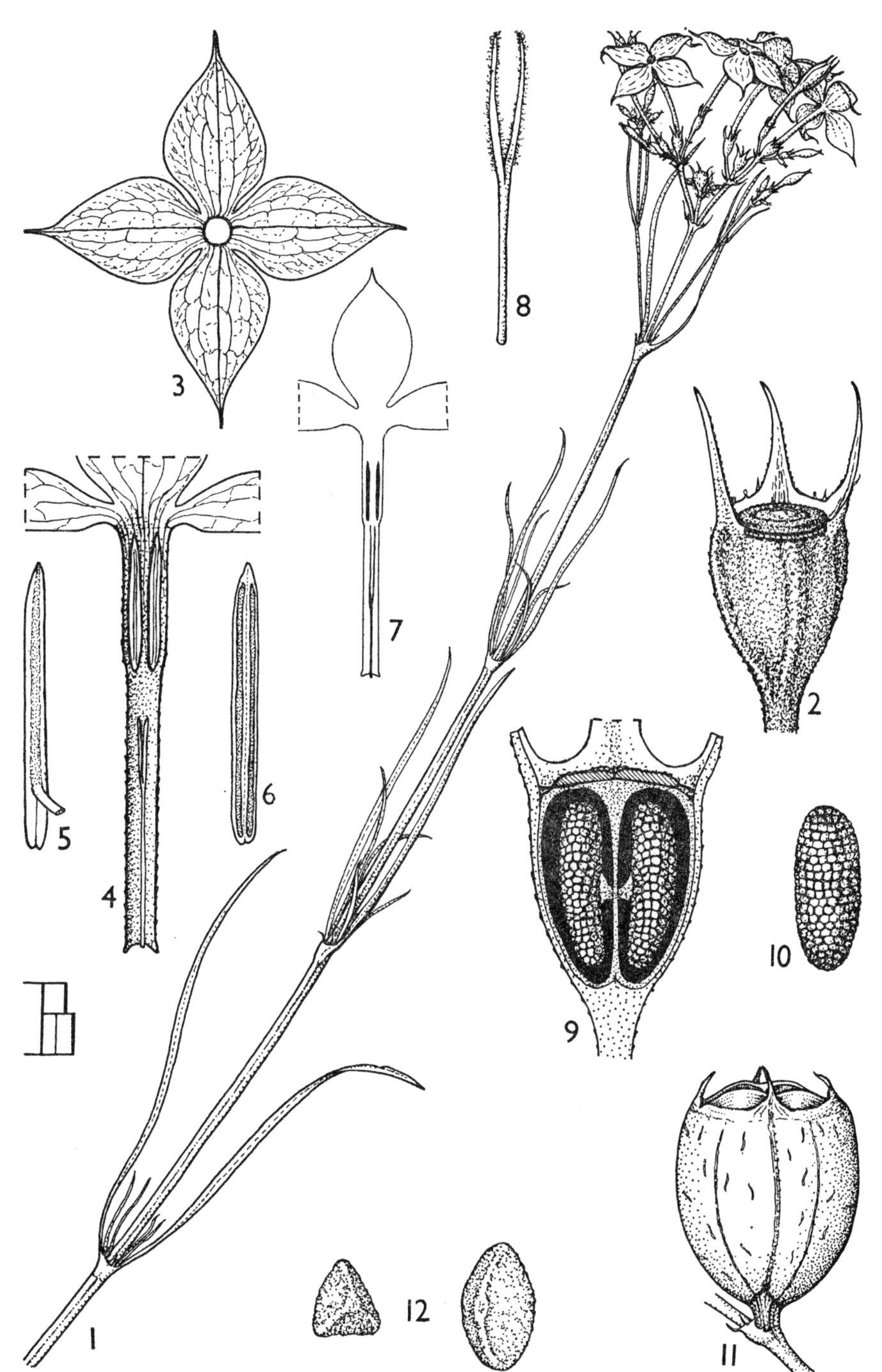

FIG. 29. *KOHAUTIA GRANDIFLORA*—**1,** flowering branch, × ⅔; **2,** calyx, with one lobe removed, × 10; **3,** corolla from above, × 2; **4,** longitudinal section of flower, × 4; **5, 6,** stamen, back and front views, × 8; **7,** longitudinal section of another flower, showing variation, × 2; **8,** style and stigma-lobes, × 4; **9,** longitudinal section of ovary, × 14; **10,** ovules on one placenta, × 14; **11,** capsule, × 6; **12,** seed, two views, × 20. 1, 3–6, from *J. Wilson* 363; 2, 7–10, from *Lewis* 5997; 11, from *Norman* 241; 12, *from Tweedie* 770. Drawn by Diane Bridson.

TANZANIA. Near S. end of Lake Tanganyika, *Cameron* !*
DISTR. **T4**; Senegal, Guinée, Malawi

SYN. [*Kohautia stricta* sensu DC., Prodr. 4: 430 (1830), pro parte, *non* (Smith) DC. sensu stricto]
[*Oldenlandia effusa* sensu Hiern in F.T.A. 3: 59 (1877), *non* Oliv.]
O. confusa Hutch. & Dalz., F.W.T.A. 2: 131 (1931)

9. **K. caespitosa** *Schnizl.* in Flora 25, Beibl. 1: 145 (1842); Bremek. in Verh. K. Nederl. Akad. Wet., Afd. Natuurk., ser. 2, 48 (2): 104 (1952); U.K.W.F.: 398 (1974). Type: Sudan, Kordofan, Arasch Cool, *Kotschy* 138 (W, holo., K, iso.!)

Annual, perennial or sometimes subshrubby herb, usually erect, 10–90 cm. tall, branched at the base with several to many caespitose glabrous to densely scabridulous pubescent stems; rootstock containing a red dye. Leaves linear to linear-lanceolate or narrowly elliptic, 0·5–8 cm. long, 0·3–5(–12) mm. wide, acute to narrowly apiculate at the apex, narrowed to the base, glabrescent to densely scabridulous-pubescent or -papillose; stipular sheath 1–8 mm. long, produced into 2–5 filiform fimbriae 0·5–3 mm. long. Inflorescences often trichotomous or dichasial, the branches often again similarly branched, the flowers usually solitary at the nodes; peduncles 0–8 cm. long; pedicels almost obsolete or up to 1·1(–2·4) cm. long (though often in these cases the stalk is a reduced branchlet). Calyx-tube ovoid, ± 1 mm. long, verruculose, scabridulous or glabrous; lobes lanceolate to triangular, 0·5–2 mm. long, glabrous or with indistinctly scabridulous margins, the teeth often with secondary elements between and sometimes developed into quite distinct filaments. Flowers scented; corolla white, grey, buff, yellowish, pink or lilac, the tube often purplish and the lobes ochraceous, nearly always pale inside, glabrous or papillate outside; tube narrowly cylindrical, (0·8–)1·15–1·4 cm. long, the widened part 1·5–3 mm. long, 0·5(–1·5) mm. wide; lobes linear-oblong to narrowly elliptic, 2–6 mm. long, 0·6–1·2(–1·7) mm. wide. Style 1·5–4 mm. long, in one variety just touching the anthers. Capsule subglobose or ovoid, 1·5–4 mm. tall, 2–3(–5·5) mm. wide, verruculose, hispidulous or glabrous, the beak slightly raised at dehiscence. Seeds pale olive-brown, angular-subconic, ± 0·3 mm. long, reticulate.

KEY TO INFRASPECIFIC VARIANTS

Ovary and capsule verruculose or shortly hispid:
  Ovary and capsule ± verruculose . . . a. var. **caespitosa**
  Ovary and capsule shortly hispid . . . c. var. **delagoensis**
Ovary and capsule ± glabrous:
  Elements between the calyx-teeth not so well developed; corolla smaller and more graceful, the lobes 0·5–0·7 mm. wide . b. var. **amaniensis**
  Elements between the calyx-teeth well developed, ± 0·5 mm. long; corolla more robust, the lobes 1–1·8 mm. wide . . d. var. **kitaliensis**

a. var. **caespitosa**; Bremek. in Verh. K. Nederl. Akad. Wet., Afd. Natuurk., ser. 2, 48(2): 106 (1952)

Ovary and capsule verruculose; corolla typically papillose outside.

KENYA. Northern Frontier Province: 80 km. SW. of Wajir, 10 Dec. 1971, *Bally & Smith* 14494!
DISTR. **K1**; Sudan, Somali Republic, Ethiopia (Eritrea) and Arabia
HAB. Open *Acacia, Commiphora* bushland; 200 m.

* Probably Tanzania, possibly Zambia or Zaire, see Gillett in K.B. 14: 319 (1960).

SYN. *Hedyotis caespitosa* (Schnizl.) Walp., Repert. 6: 56 (1846)

NOTE. The cited specimen is not very verruculose and scarcely typical, but seems best placed here.

b. var. **amaniensis** (*K. Krause*) *Bremek.* in Verh. K. Nederl. Akad. Wet., Afd. Natuurk., ser. 2, 48(2): 108 (1952). Type: Tanzania, Tanga District, Moa, Mtotohovu, *Braun* in *Herb. Amani* 1383 (B, holo. †, EA, iso. !)

Ovary and capsule glabrous or practically so. Corolla glabrous outside.

UGANDA. Karamoja District: base of Mt. Debasien, between Amaler and Napienyenya, Jan. 1936, *Eggeling* 2541 ! & Kangole, 22 May 1940, *A. S. Thomas* 3466 !
KENYA. Northern Frontier Province: Mathews Range, Ngeng, 10 Dec. 1958, *Newbould* 3162 !; N. Frontier/Turkana District: base of Turkwell Gorge, Aug. 1962, *J. Wilson* 1274 !; Masai District: Ol Orgesailie, 11 Aug. 1951, *Verdcourt* 580 !; Tana River District: Kurawa, 24 Sept. 1961, *Polhill & Paulo* 555 !
TANZANIA. Arusha District: Arusha–Nairobi road, about 32 km. from Arusha, Ol'donyo Sambu, 28 Dec. 1961, *Greenway* 10407 !; Lushoto District: Mombo–Lushoto road, 1 Aug. 1960, *Semsei* 3070 !; Kondoa District: Bubu R., delta at confluence with Kondoa R., 12 Dec. 1927, *B. D. Burtt* 825 !
DISTR. **U**1; **K**1, 2, 3 (Baringo), 4, 6, 7; **T**1–3, 5, 8; Ethiopia, Somali Republic, also some intermediates in Rhodesia
HAB. Grassland especially in semi-desert areas, *Acacia-Commiphora* and similar bushland and scrub, sandy plains, flats, dry river beds, dunes, lava outcrops, etc., also roadsides, cultivations and waste places; 0–1890 m.

SYN. *Oldenlandia amaniensis* K. Krause in E.J. 43: 129 (1909)
O. sagensis Cufod. in Nuov. Giorn. Bot. Ital. n.s. 55: 84 (1948). Types: numerous syntypes from Ethiopia, Galla-Sidamo and Kenya, mainly *Corradi* (FI, syn.)
Kohautia caespitosa Schnizl. var. *hispidula* Bremek. in K.B. 8: 439 (1953). Type: Kenya, Northern Frontier Province, El Wak, *Gillett* 13341 (K, holo. !, EA, iso.)

NOTE. A specimen from **T**8 (about 6·5 km. W. of Songea, 30 Mar. 1956, *Milne-Redhead & Taylor* 9430) has the ovary very slightly verruculose. *K. caespitosa* var. *hispidula* has the stems and leaves distinctly pubescent with short white hairs but there are many specimens from N. Kenya showing this tendency and I have not kept it up.

c. var. **delagoensis** (*Schinz*) *Bremek.* in Verh. K. Nederl. Akad. Wet., Afd. Natuurk., ser. 2, 48(2): 107 (1952). Type: Mozambique, Delagoa Bay, Rikatla, *Junod* 203 (Z, holo.)

Ovary and capsule shortly hispid. Corolla mostly glabrous outside.

TANZANIA. Just N. of Iringa, 14 July 1956, *Milne-Redhead & Taylor* 11141 !
DISTR. **T**7; ? Zaire, ? Burundi, Ethiopia, Mozambique, Rhodesia, South Africa and South West Africa
HAB. Secondary grassland on sand; 1620 m.

SYN. *Oldenlandia delagoensis* Schinz in Mém. Herb. Boiss. 10: 64 (1900)

NOTE. The corolla in the specimen cited above is not entirely glabrous. A specimen from northern Kenya, *Bally & Smith* 14548 (5 km. NE. of El Wak, 11 Dec. 1971, from open *Acacia* bushland on limestone with scattered *Delonix*, 380 m.) is technically var. *delagoensis*, but is slender and very different from the specimen from **T**7; it has fine spreading pubescence on the calyx-tube and capsule and papilla-like hairs on the stem. It is closer to var. *amaniensis* in habit.

d. var. **kitaliensis** *Verdc.* in K.B. 30: 290 (1975). Type: Kenya, Kitale, *Tweedie* 2595 (K, holo. !)

Elements between the calyx-teeth mostly well developed, ± 0·5 mm. long; calyx-teeth shorter and more triangular. Inflorescence with very few of the flowers subsessile, others with pedicels up to 2 cm. long; corolla more robust, the widened part of tube up to 1·5 mm. wide.

KENYA. Trans-Nzoia District: Kitale, 5 Mar. 1953, *Bogdan* 3676 ! & Kitale, Milimani, Apr. 1969, *Tweedie* 3629 ! & Kitale, Apr. 1963, *Tweedie* 2595 !
DISTR. **K**3; not known elsewhere
HAB. Moist grassland with scattered *Combretum*; 1830–1860 m.

NOTE. The four available specimens have a distinctly different facies from var. *amaniensis*, which is almost unknown from the Kenya highland area. The differences are less easy to describe. A rather poor specimen " from Nandi to Mumias " collected by A. Whyte has been annotated as var. *amaniensis* by Bremekamp. *Brodhurst Hill* 108 ! (Kenya, Kipkarren, Aug. 1931) does seem to be var. *amaniensis* with short narrow corolla-lobes.

10. **K. lasiocarpa** *Klotzsch* in Peters, Reise Mossamb., Bot. 1: 296 (1861); Bremek. in Verh. K. Nederl. Akad. Wet., Afd. Natuurk., ser. 2, 48 (2): 110 (1952). Type: Mozambique, Sena, *Peters* (B, holo. †)

Erect annual or perennial herb 15–55(–100) cm. tall, much branched, the stems sometimes quite woody at the base, scabridulous with white papillae or tomentellous. Leaves narrowly lanceolate to elliptic-lanceolate, 1·5–6 cm. long, 1·5–10·5 mm. wide, acute at the apex and cuneate at the base, scabridulous above and on the midnerve beneath; stipular sheath produced into a triangular lobe 1–2(–3) mm. long with (1–)2(–7) fimbriae 1–5 mm. long and several small teeth on either side. Inflorescences usually extensive and much branched, trichotomous or dichasial, the branches often again branched, the flowers ± subsessile, nearly always in pairs at the nodes or pedicels (or apparent pedicels) up to 8(–13) mm. long. Calyx-tube subglobose, 1–1·2 mm. in diameter, verruculose to scabrid-pubescent; lobes lanceolate, 0·8–2 mm. long, scabridulous on the margins and main nerve. Corolla white, buff or yellowish, rarely greenish or blue; tube filiform-cylindrical, 0·5–1·5 cm. long, the widened part 0·8–2·8 mm. long, 0·5–1·2 mm. in diameter; lobes elliptic, (1–)1·3–4·5 mm. long, 0·5–1·3 mm. wide. Style 2·5–5 mm. long; stigma-lobes 1·5–2 mm. long, in some variants just reaching the anthers. Capsule subglobose, 2–3·3 mm. tall, 2·5–3·5(–4) mm. wide, verruculose or scabrid-pubescent, the beak slightly raised at dehiscence. Seeds pale to dark brown, irregularly ellipsoid or angular, ± 0·5 mm. long, granular.

KEY TO INFRASPECIFIC VARIANTS

Ovary shortly stiffly hairy to scabridulous-pubescent; capsule glabrescent to setulose or scabridulous-pubescent:
- Corolla-tube (0·8–)1·2–1·5 cm. long:
  - Corolla-lobes 2·5–4·5 mm. long . . . a. var. **lasiocarpa**
  - Corolla-lobes 1·3–2·2 mm. long . . . b. var. **breviloba**
- Corolla-tube (4–)5–6·5(–7) mm. long; lobes 1–2 mm. long; ovary and capsule more densely scabrid-pubescent . . . c. var. **subverticillata**

Ovary papillate-tuberculate; capsule glabrescent or obscurely papillate; corolla-tube (3·5–)5–6·5 mm. long; lobes ± 1 mm. long . . . d var. **eritreensis**

a. var. **lasiocarpa**; Bremek. in Verh. K. Nederl. Akad. Wet., Afd. Natuurk., ser. 2, 48 (2): 111 (1952)

Corolla-tube (0·8–)1·2–1·5 cm. long; lobes 2·5–4·5 mm. long; ovary shortly stiffly hairy and capsule sparsely setulose or glabrescent.

TANZANIA. Kondoa District: Kandaga scarp, 20 Jan. 1928, *B. D. Burtt* 1220 !; Iringa District: Mazombe area, Aug. 1936, *Ward* U.85 !
DISTR. T5, 7; Mozambique, Zambia, Rhodesia, Angola and South West Africa
HAB. Not adequately known, sometimes on cultivated ground; 1440 m.

SYN. *Oldenlandia lasiocarpa* (Klotzsch) Hiern in F.T.A. 3: 55 (1877)

b. var. **breviloba** *Bremek.* in Verh. K. Nederl. Akad. Wet., Afd. Natuurk., ser. 2, 48(2): 112 (1952). Type: Rhodesia, Trelawney, *Jack* 184 (K, holo. !)—specimen annotated as holotype but not cited as such

Corolla-tube, etc. similar to var. *lasiocarpa*, but corolla-lobes shorter, 1·3–2·2 mm. long.

TANZANIA. Mwanza District: Mbarika, Buhunda, 6 May 1952, *Tanner* 739!; Ufipa District: Rukwa Valley, Milepa, 19 Mar. 1947, *Pielou* 152!; Chunya District: Rukwa Valley, Udinde Village, path to Kikamba R., 11 Aug. 1968, *Sanane* 247!
DISTR. **T1**, 4, 7; Mozambique, Zambia, Rhodesia, Angola and South Africa
HAB. Open deciduous woodland and also cultivations, on fine sandy ground; 780–1200 m.

c. var. **subverticillata** (*K. Schum.*) *Bremek.* in Verh. K. Nederl. Akad. Wet., Afd. Natuurk., ser. 2, 48(2): 112 (1952). Type: Angola, Sopola, *Welwitsch* 5321 (B, holo. †, K, LISC, iso.!)

Corolla-tube (4–)5–6·5(–7) mm. long; lobes ± 1–2 mm. long; ovary and capsule shortly densely scabridulous-pubescent.

TANZANIA. Ngara District: Bushubi, Mu Rusagamba, 14 Feb. 1961, *Tanner* 5761!; Tabora, 20 Jan. 1926, *Peter* 35187!
DISTR. **T1**, 4; Mozambique, Malawi, Rhodesia, Angola, South West Africa and South Africa (Transvaal)
HAB. Sandy plain; 1200–1350 m.

SYN. *Oldenlandia subverticillata* K. Schum. in E.J. 23: 419 (1897); Hiern in Cat. Afr. Pl. Welw. 2: 444 (1898)

d. var. **eritreensis** *Bremek.* in Verh. K. Nederl. Akad. Wet., Afd. Natuurk., ser. 2, 48(2): 113 (1952). Type: Ethiopia, Eritrea, Beni-Amer, Carajai, *Pappi* 6384 (FI, holo.)

Corolla-tube (3·5–)5–6·5 mm. long; lobes ± 1 mm. long; ovary papillate-tuberculate; capsule glabrescent or obscurely papillate.

TANZANIA. Kigoma, 8 Mar. 1964, *Pirozynski* 552!; Morogoro District: 30 km. W. of Morogoro, Mkata Ranch, 9 Mar. 1974, *Wingfield* 2636!; Iringa District: Ruaha National Park, above Mwagusi R., 10 Feb. 1966, *Richards* 21268!
DISTR. **T4**, 6, 7; Ethiopia (Eritrea)
HAB. Grassland on heavy clay, also on sandy lake beaches; 420–840 m.

NOTE. It is not certain if any of the above specimens is correctly named. The Kigoma specimen has been referred to *K. aspera* on account of the small corolla (3·5 mm. long), but the general facies particularly of the foliage is that of *K. lasiocarpa*. *Wingfield* 2173 and 1936 (Pare District, Korogwe–Same road from just past Makanya Sisal Estate, 20 Oct. 1972 and 2 km. past the Mkomazi turn-off, 5 Mar. 1972, respectively) are perhaps both forms of *K. lasiocarpa* with glabrous ovary and narrow leaf-blades but the material is scarcely adequate for a decision. It is a variable species and Bremekamp's varieties are not very satisfactory. Study of populations in the field may show they are better disregarded.

11. **K. aspera** (*Roth*) *Bremek.* in Verh. K. Nederl. Akad. Wet., Afd. Natuurk., ser. 2, 48(2): 113 (1952); Lebrun in Mitt. Bot. Staats., München 10: 444, fig. 5 (1971); U.K.W.F.: 398 (1974). Type: "India orientali", *Heyne* (whereabouts uncertain)

Annual sparsely to densely branched herb 13–40 cm. tall, the stems scabrid with white papillae. Leaves linear to narrowly elliptic, 1·5–3(–4·7) cm. long, 0·5–4 mm. wide, acute at the apex, narrowed to the base, glabrous above, papillate on midvein beneath and on the margins; stipule-sheath 0·7–1·5 mm. long, divided into 2 fimbriae 0·5–2 mm. long. Inflorescences lax, the flowers mostly in pairs well spaced on the rhachis, entirely monochasial or branched at the base, trichotomous or dichasial; peduncle 1–4(–8) cm. long; pedicels 0·1–1·3(–2·6) cm. long; all parts papillate like the stems. Calyx-tube ellipsoid, 1–1·2 mm. long, papillate; lobes linear-lanceolate, 0·7–2 mm. long, glabrous or minutely scabridulous. Corolla white, brownish, greenish, pale blue or purple; tube narrowly cylindrical, sometimes curved, 2·5–4·7 mm. long, the widened part 0·7–2 mm. long, 0·5–0·8 mm. wide, mostly sparsely papillate outside with downwardly directed hairs inside near the base of the widened part; lobes oblong, 1–1·2 mm. long, 0·5–0·9 mm. wide. Style ± 0·6 mm. long; stigma-lobes 1·2–1·4 mm. long. Capsule oblong-subglobose, 2·5–3(–4) mm. tall, 2·5–4 mm. wide, ± smooth or papillate, the dehisced beak slightly raised. Seeds pale brown, angular, 0·4–0·6 mm. long.

UGANDA. Karamoja District: Kangole, Aug. 1957, *J. Wilson* 388! & Kokumongole, 28 May 1939, *A. S. Thomas* 2846! & Mt. Komolo, July 1930, *Liebenberg* 386!
KENYA. NE. Elgon, May 1960, *Tweedie* 2008!; Nairobi District: 19·2 km. E. of Nairobi, Njiro Farm, 21 June 1951, *Bogdan* 3081!; Embu District: 29 km. SSW. of Embu, 20 Feb. 1957, *Bogdan* 4437!
TANZANIA. Mbulu District: Yaida valley, Makomero, 17 Jan. 1970, *Richards* 25162!; Kondoa District: Kolo, 12 Jan. 1962, *Polhill & Paulo* 1150!; Mbeya District: Rujewa, 1 Apr. 1956, *Anderson* 1114!
DISTR. **U**1; **K**1, 3, 4, 6; **T**1–5, 7; Senegal, Mauritania, Mali, Niger, Arabia, Ethiopia, Sudan, Botswana, South Africa (Transvaal, Orange Free State) and South West Africa, also in Cape Verde Is., India, Australia
HAB. Grassland, open bushland and thicket, also overgrazed wet ground and as a weed in arable fields, particularly on heavy black clay soils; 750–1950 m.

SYN. *Hedyotis aspera* Roth, Nov. Pl. Sp. Ind. Or.: 94 (1821)
Oldenlandia aspera (Roth) DC., Prodr. 4: 428 (1830)
Hedyotis (*Kohautia*) *strumosa* A. Rich., Tent. Fl. Abyss. 1: 364 (1847). Type: Sudan, Kordofan, Abu-Gerad, *Kotschy* 46 (P, holo., MO, iso.)
Oldenlandia strumosa (A. Rich.) Hiern in F.T.A. 3: 58 (1877); K. Schum. in P.O.A. C: 376 (1895)

12. **K. sp. A**

Annual herb 15–30 cm. tall; stems densely branched above, slightly subscabrid-papillate. Leaves pseudoverticillate, linear, 1–1·7 cm. long, 1·2 mm. wide, acute at the apex, ± glabrous; stipule-sheath 0·5 mm. long, with 3 setae 0·5–1·5 mm. long. Inflorescences terminal and axillary, lax; peduncle 4–8 cm. long; pedicels 2–8 mm. long, papillose, lengthening to 1·4 cm. Calyx-tube ellipsoid to oblong, 1·8 mm. long and wide, densely papillate; lobes triangular-lanceolate, 1 mm. long, ± ciliate. Corolla colour not known, perhaps greenish; tube 2·8 mm. long; lobes elliptic, 1 mm. long. Anthers enclosed in the top of the tube. Style 2 mm. long, the stigmatic arms 1 mm. long, tips just exserted. Capsule subglobose, 2·5 mm. long, 2·8 mm. wide, laterally subcompressed, obscurely papillate, with persistent calyx-lobes and low beak, both ± 0·5 mm. long.

TANZANIA. Masai District: NE. Lake Manyara, 26 May 1962, *Dingle* 239!
HAB. On soda flats; 975 m.

NOTE. This had been named *K. lasiocarpa*, but the inflorescence fits neither *K. aspera* nor *K. lasiocarpa*.

23. CONOSTOMIUM

Cufod. in Nuov. Giorn. Bot. Ital. n.s. 55: 85 (1948); Bremek. in Verh. K. Nederl. Akad. Wet., Afd. Natuurk., ser. 2, 48 (2): 125 (1952)

Oldenlandia L. sect. *Conostomium* Stapf in J.L.S. 37: 517 (1906)

Perennial herbs or branched shrublets mostly with erect stems. Leaves sessile, linear to lanceolate, the lateral nerves rather obscure; stipules with the truncate sheath bearing hairs and colleters or 2–several teeth. Flowers small to fairly large, never truly heterostylous, either in terminal corymbs, spike-like inflorescences or solitary or in pairs or fascicles in the axils of the leaves. Calyx-lobes 4, linear-subulate. Corolla salver-shaped with a long narrowly cylindrical tube; lobes 4; throat and lobes within glabrous or sparsely pilose. Stamens always included in the tube or only the tips of the anthers exserted. Ovary 2-locular; ovules numerous, immersed in the peltate placentas; style usually exserted or rarely included, in which case the stigmas do not reach the bottoms of the anthers; stigma-lobes oblong or linear-oblong, densely papillate. Capsule ovoid or subglobose, crowned by the persistent calyx-lobes and with a usually well developed beak, loculicidally dehiscent at the apex. Seeds numerous, angular, smooth.

A small genus of 9 species confined to the eastern and central parts of tropical Africa and also in S. Africa. Bremekamp divides the genus into 3 subgenera 2 of which occur in the Flora area and are mentioned in the key below.

Corolla-tube large and conspicuous, 9–15 cm. long; flowers sessile in the axils of most of the leaves and forming a spike-like inflorescence; leaves with lateral nerves visible (subgen. *Conostomium*) . . . . . 1. *C. quadrangulare*
Corolla-tube much smaller, 2–5·7 cm. long; flowers pedicellate* in the axils of the upper leaves; leaves narrower with lateral nerves scarcely visible (subgen. *Beckia* Bremek.) . 2. *C. kenyense*

1. **C. quadrangulare** (*Rendle*) *Cufod.* in Nuov. Giorn. Bot. Ital., n.s. 55: 85 (1948); Bremek. in Verh. K. Nederl. Akad. Wet., Afd. Natuurk., ser. 2, 48(2): 134 (1952); U.K.W.F.: 398, fig. (1974). Type: Ethiopia, Lake Stephanie, *Donaldson-Smith* (BM, holo.!)

Perennial (or annual *fide Tweedie* 2477) herb (13–)30–60 cm. tall, with 1–several unbranched or sparsely branched glabrous 4-ribbed stems from a woody root. Leaf-blades linear-lanceolate to lanceolate, 1·8–6·8 cm. long, (0·2–)0·8–1·9 cm. wide, narrowly tapering acute at the apex, narrowed to subcordate at the base, glabrous save for some hairs above near the base; stipule-sheath truncate, 0·5–1·5 mm. long, fringed with hairs and obscure colleters. Flowers axillary, sessile, forming a leafy spike-like inflorescence (1·5–)10–25(–40) cm. long. Calyx-tube ellipsoid, 3–4 mm. long, glabrous or ciliate on the ribs; lobes 0·5–1·3 cm. long. Corolla white or pinkish or tube and lobes often greenish outside or tinged crimson; tube (5–)9–15 cm. long, widened to 2–6 mm. at the apex at part where stamens are included, this widened part pilose outside and inside at the point of insertion of the stamens; lobes ovate or triangular-ovate, 1–2 cm. long, 2–8 mm. wide, long-acuminate at the apex; buds very distinctly apiculate. Style exserted (1–)3–8(–22) mm.; stigma-lobes 3–5·5 mm. long. Capsule ovoid, 5–7 mm. long, 4–6 mm. wide, 8-ribbed, glabrous, crowned by the persistent calyx-lobes, quite conspicuously beaked, the beak acute, compressed laterally, 2–3·5 mm. long; a pedicel up to 2 mm. long may be apparent in the fruiting stage. Seeds chestnut brown, somewhat shining, trigonous-ellipsoid, ± 0·6–0·8 mm. long. Fig. 30, p. 244.

UGANDA. Karamoja District: without precise locality, 3 Nov. 1933, *H. B. Johnston* 635! & at base of Mt. Debasien, Napianyenya [Napyenenya], *Eggeling* 2598! & Oropoi valley, July 1930, *Liebenberg* 155!
KENYA. Northern Frontier Province: Samburu District, Kichich, 14 Dec. 1958, *Newbould* 3335!; W. Suk District: second camp from Morobus [Moribus] Pass, May 1932, *Napier* 2067!; Meru District: Isiolo, Dec. 1957, *J. Adamson* 692!
DISTR. **U**1, 3; **K**1, 2, ? 3, 4, 7; Sudan, Ethiopia
HAB. Deciduous bushland, grassland with scattered trees, always in dry places, mostly in sandy areas by dry rivers, rocky ledges, stony overgrazed land, etc.; 60–1350(–1800) m.

* *C. longitubum* (Beck) Cufod., another species in this subgenus, has sessile flowers; it has so far been recorded only from Ethiopia and Somalia.

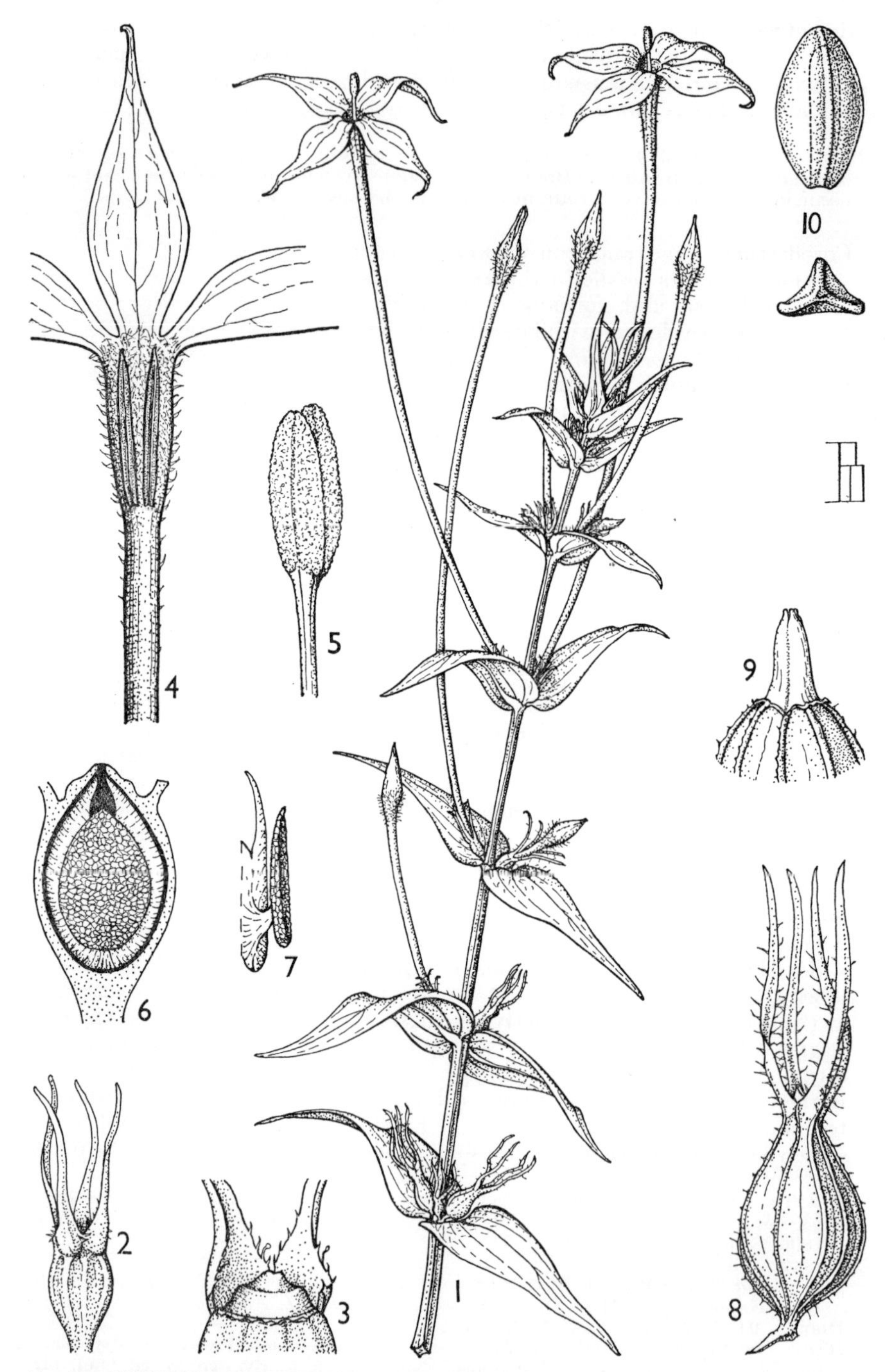

FIG. 30. *CONOSTOMIUM QUADRANGULARE*—**1,** flowering branch, × $\frac{2}{3}$; **2,** calyx, × 3; **3,** part of calyx, with lobes cut away to show disc, × 4; **4,** longitudinal section of upper part of corolla, × 2; **5,** tip of style and stigma-lobes, × 4; **6,** longitudinal section of ovary (one of two locules shown), × 4; **7,** placenta and part of septum, × 4; **8,** capsule, × 3; **9,** tip of capsule with calyx removed to show beak, × 4; **10,** seed, two views, × 30. 1, from *Popov* 1550; 2–7, from *Napier* 2067; 8, 9, from *Kirrika* 132; 10, from *Hemming* 305. Drawn by Diane Bridson.

SYN. *Pentas quadrangularis* Rendle in J.B. 34: 127 (1896)
Neurocarpaea quadrangularis (Rendle) Rendle in J.B. 36: 29 (1898)
Oldenlandia megistosiphon K. Schum. in E.J. 28: 56 (1900). Type: Kenya, without locality [Ost.-Afrika], *Fischer* 234 (B, holo. †)
Otomeria ? heterophylla K. Schum. in E.J. 33: 336 (1903). Type: S. Ethiopia*, Girma, near Mt. Robé along the R. Daua, *Riva* 159 (B, holo. †)
Oldenlandia dolichantha Stapf in J.L.S. 37: 518 (1906) & in Bot. Mag. 133, t. 8165 (1907). Type: Uganda/Sudan, Nile Province, *Dawe* 945 (K, holo. !)

NOTE. Only one field label mentions a slight scent; it is probably scented at night. Further information is needed on this point. A sheet from NE. Aberdares, *Dawson* 549 is the basis for the **K3** record and for the altitude limits of 2100 m. (7000 ft.). It seems dubious if this species could really occur at this height.

2. **C. kenyense** *Bremek.* in Verh. K. Nederl. Akad. Wet., Afd. Natuurk.' ser. 2, 48(2): 132 (1952); U.K.W.F.: 398 (1974). Type: Kenya, Northern Frontier Province, Lorogi [Leroki], *D. G. B. Leakey* 35 in *F.D.* 3466 (K, holo. !, EA, iso. !)

Perennial herb or shrub 0·15–1·5 m. tall, with mostly much branched 4-ribbed stems, usually woody at the base, at first scabridulous or shortly scabrid-pubescent, soon glabrescent, or entirely glabrous. Leaf-blades narrowly linear to linear-lanceolate, 0·6–6(–7) cm. long, 0·2–7 mm. wide, acute at the apex, narrowed to the base, ± glabrous or scabridulous to scabrid-pubescent, the margins often inrolled; stipular sheath 0·5–2 mm. long bearing 0–5 small teeth 0·5–1·2 mm. long on its truncate margin. Flowers axillary, pedicellate, forming terminal raceme-like inflorescences at the apices of the branchlets which bear reduced and less conspicuous leaves; pedicels straight or slightly to quite distinctly S-curved, 0·2–1 cm. long shortly pubescent. Calyx-tube ovoid, 1·2–2 mm., glabrescent to densely scabrid-pubescent with tubercle-based hairs; lobes lanceolate, 1–6 mm. long. Corolla white or cream, often tinged blue, purple or green outside, sweetly scented; tube 2–5·2(–5·7) cm. long, glabrous outside or shortly scabrid-pubescent, at the widest part 2–3 mm. wide; lobes oblong-lanceolate, (0·4–)0·8–1·5 cm. long, (1·1–)1·7–5 mm. wide, glabrous to shortly scabrid-pubescent outside, ± acuminate but buds not so distinctly apiculate as in last species. Style either included with stigma-tips up to 2 cm. below the orifice of the throat or slightly longer than the tube; stigma-lobes (2–)3·5–8 mm. long. Capsule subglobose, 1·5–3 mm. tall, (2–)2·5–4·2 mm. wide, glabrous or sparsely scabridulous, very distinctly beaked, the beak compressed, 1–3 mm. tall, dehiscing into 4 valves. Seeds yellow-brown, narrowly ellipsoid, trigonous, 0·8 mm. long.

KENYA. Northern Frontier Province: Mathews Range, Mathews Peak [Ol Doinyo Lengiyo], 20 Dec. 1958, *Newbould* 3516 !; Fort Hall District: Yatta Plateau,Maboloni Rock, 7 Dec. 1952, *Bally* 8374 !; Meru District: Isiolo, Dec. 1957, *J. Adamson* 685 !
DISTR. **K1**, 2, 4; S. Ethiopia
HAB. Deciduous bushland, dry grassland, rocky hillsides, crevices in rocks; (150–)600–1800 m.

SYN. *C. kenyense* Bremek. var. *subglabrum* Bremek. in Verh. K. Nederl. Akad. Wet., Afd. Natuurk., ser. 2, 48(2): 132 (1952). Type: Ethiopia, Galla-Sidamo, near Dande, *Corradi* 2705 (FI, holo.)
C. camptopodum Bremek. in Verh. K. Nederl. Akad. Wet., Afd. Natuurk., ser. 2 48(2): 133 (1952). Type: Kenya, Turkana District, Nataparin [" Wateparin "]' *Champion* 136 (K, holo. !),
C. microcarpum Bremek. in Verh. K. Nederl. Akad. Wet., Afd. Natuurk., ser. 2, 48(2): 133 (1952). Type: Kenya, edge of Turkana Desert, *Pole-Evans & Erens* 1558 (K, holo. !)
C. floribundum Agnew, U.K.W.F.: 398 (1974), *nom. nud.*

* According to an itinerary prepared by Mr. J. B. Gillett, Girma is in Kenya but Mt. Robé is in Ethiopia.

NOTE. The indumentum is too variable to maintain the var. *subglabrum*. The plant varies in leaf size and in stature, some of those growing in rock crevices being very small. The S-shaped pedicel used by Bremekamp to characterise *C. camptopodum* appears in other specimens clearly belonging to *C. kenyense*. *C. camptopodum* and *C. microcarpum* are based on rather scrappy depauperate specimens. There do not seem to be any reliable specific characters. Another problem emphasised by Bremekamp is the presence of short-styled flowers in the subgenus *Beckia*; he knew these to occur in *C. longitubum* (Beck) Cufod. and *C. brevirostrum* Bremek. and all the known specimens of C. *hispidulum* Bremek. and *C. microcarpum* Bremek. Stapf had suggested that such flowers were brachystylous but Bremekamp thought them to be teratological. They also occur in *C. kenyense* and it is clear that although true heterostyly does not occur the styles can be of two lengths. Only field studies will establish the true nature of this phenomenon.

24. PENTANOPSIS

Rendle in J.B. 36 : 28 (1898) ; Bremek. in Verh. K. Akad. Wet., Afd. Natuurk., ser. 2, 48(2) : 139 (1952)

Small shrub with much branched stems. Leaves paired but sometimes appearing verticillate due to the presence of short undeveloped shoots; stipules sheathing the stem, adnate to the petioles, the apical margin fimbriate. Flowers medium-sized, dimorphic, solitary or in fascicles of 3 terminating short undeveloped shoots. Calyx-tube ovoid; lobes 4, linear-lanceolate. Corolla with long slender tube, narrowly funnel-shaped towards the apex, glabrous outside, sparsely hairy inside; lobes 4, oblong-elliptic to lanceolate, induplicate at the base in bud, densely covered all over the inside with blunt stubby hairs. Ovary 2-locular; ovules numerous on peltate placentas; style with bifid stigma just exserted in long-styled form, much shorter in the short-styled form; anthers not exserted in either form but at different levels in the tube or only the tips of the anthers exserted in the short-styled form. Capsule obovoid or ellipsoid, loculicidally dehiscent, only indistinctly beaked. Seeds brown, flattened.

A monotypic genus confined to NE. tropical Africa and only just occurring in the Flora area. It differs from *Conostomium* subgenus *Beckia* in the fimbriate stipules, the corolla-lobes covered with hairs inside and induplicate below in bud, the capsule less rostrate, the pollen grains with long colpi and in having the testa walls smooth and not punctate. It has a fairly distinctive facies and little would be gained in uniting the two.

P. fragrans *Rendle* in J.B. 36: 29 (1898); Bremek. in Verh. K. Nederl. Akad. Wet., Afd. Natuurk., ser. 2, 48(2) : 139 (1952). Type : Somali Republic (N.), Wagga Mt., *Lort Phillips* (BM, holo. !)

Small shrub (0·8–)1·3–2·1 m. tall; stems erect or rarely scrambling, glabrous or nearly so, the bark often peeling. Leaf-blades linear-lanceolate to elliptic-lanceolate, 1–2·8 cm. long, 1–5 mm. wide, acute and mucronulate at the apex, narrower to the base, usually revolute, glabrescent to fairly densely scabrid with short hairs, the margins similarly with very short hairs, or sometimes the leaves with short pubescence beneath but ± glabrous above, 1-nerved; sessile on to the stipule-sheath which is 2–3 mm. long, fimbriae 2–3, narrowly deltoid to deltoid, up to 0·5 mm. long. Pedicels 1–4 mm. long. Calyx-tube 2–2·5 mm. long, glabrous to sparsely scabrid; lobes broadly linear, 0·25–1·1 cm. long, 1–1·2 mm. wide, acute, mostly ciliate with very short hairs. Corolla white or tinged with purple on tube, with carnation-like scent; tube 2–3·7 cm. long; lobes (0·5–)1–1·5 cm. long, (1·5–)4–8 mm. wide, acute. Anthers 1–2 mm. long in long-styled flowers, 2·1–3 mm. long in short-styled flowers. Style 1·6–2·2 cm. long in short-styled flowers and 2·1–3 cm. long in long-styled flowers; stigma-lobes 2·2–7 mm. long. Capsule 4·5–7 mm. long, 4–5

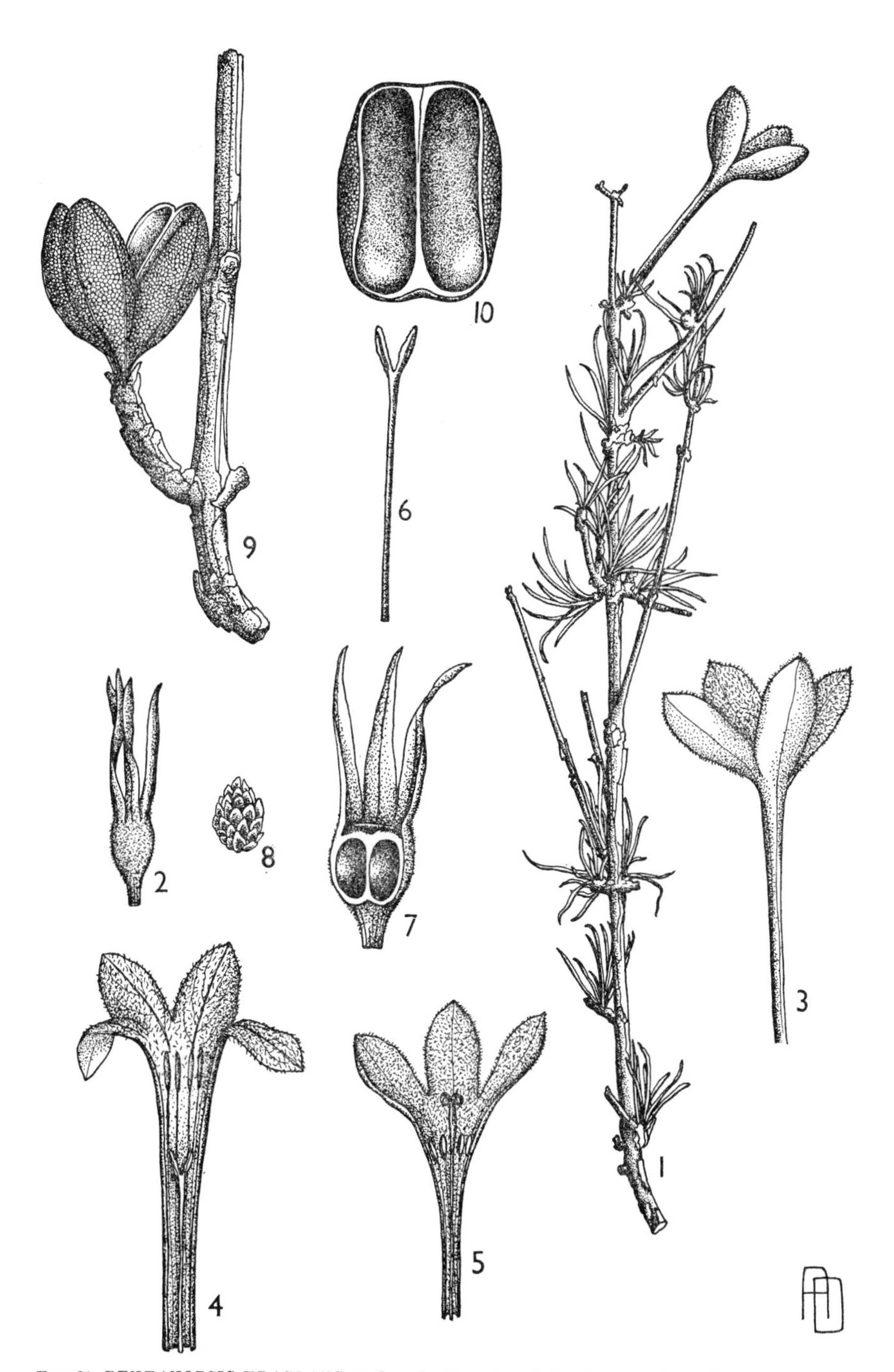

FIG. 31. *PENTANOPSIS FRAGRANS*—**1,** flowering branch, × ⅔; **2,** calyx, × 4; **3,** corolla, × 1⅓; **4,** longitudinal section of short-styled flower, × 1⅓; **5,** same of long-styled flower, × 1⅓; **6,** style, × 2; **7,** longitudinal section of ovary, × 6; **8,** placenta, showing ovules, × 10; **9,** fruiting branchlet, × 4; **10,** valve of capsule, × 6. 1, from *Hemming* 1726; 2–8, from *Ellis* 138; 9, 10, from *Hemming* 4774 .Drawn by Ann Davis.

mm. wide, glabrous to sparsely very shortly puberulous, shiny inside. Seeds elliptic, flat, 1·5 mm. long, 1 mm. wide. Fig. 31, p. 247.

KENYA. Northern Frontier Province: Dandu, 14 Apr. 1952, *Gillett* 12789!
DISTR. **K1**; Ethiopia (Ogaden), Somali Republic
HAB. *Commiphora-Acacia* bushland on " black cotton soil "; 750 m.

SYN. *Conostomium squarrosum* Bremek. in K.B. 11: 169 (1956). Type: Ethiopia, Ogaden, between Gorrahi and Wardere, *Ellis* 138 (K, holo.!)

NOTE. Despite the redescription of this species in the genus *Conostomium* Bremekamp (*in litt.*) has stated that this was a slip due to forgetting about *Pentanopsis* and that he still considers the two genera should be maintained distinct.

25. MITRASACMOPSIS

Jovet in Archiv. Mus. Nat. Hist. Paris, sér. 6, 12: 589 (1935)

Diotocranus Bremek. in Verh. K. Nederl. Akad. Wet., Afd. Natuurk., ser. 2, 48(2): 148 (1952)

Annual herb with short erect branched stems. Leaves paired, narrowly (rarely broadly) elliptic or elliptic-lanceolate; stipules with 2–4 fimbriae from a short base. Flowers small, not dimorphic, in small, usually few-flowered terminal dichasial inflorescences. Calyx-tube ovoid; lobes 4, triangular-lanceolate. Corolla-tube very short; lobes ovate, throat hairy. Ovary 2-locular, each locule with few ovules; style not exserted; stigma bifid, the lobes linear. Capsule of very characteristic shape, both loculicidal and septicidal, cordate and bilobed at the base, with the fertile part produced into a very distinct compressed obtuse central beak. Seeds few, compressed subglobose, bluntly angular, distinctly reticulately pitted.

A monotypic genus occurring in tropical Africa and Madagascar.

M. quadrivalvis *Jovet* in Archiv. Mus. Nat. Hist. Paris, sér. 6, 12: 590, figs. 1–14 (1935); Verdc. in K.B. 30: 291 (1975). Types: Madagascar, Mania, *Perrier de la Bâthie* 12544 & W. Itremo, Mts. Analamamy, *Perrier de la Bâthie* 12475 & Tsaratanana, *Perrier de la Bâthie* 16104 & 16104 bis & Sambirano, Monongarivo, *Perrier de la Bâthie* 3802 (all P, syn.)

Herb 8–40 cm. tall, with glabrous to sparsely hairy square stems. Leaf-blades 0·8–3·2 cm. long, 0·1–1·1 cm. wide, acute or subacute but mucronulate at the apex, cuneate at the base, margin often narrowly revolute, glabrescent to scabridulous above, glabrous beneath or the midvein and margins ciliate; petiole obsolete or very short; stipular fimbriae up to 1·5–2 mm. long from a base 0·2–1·5 mm. long, hairy. Pedicels obsolete or up to 3 mm. long in the fruiting stage. Calyx puberulous to densely scabrid pubescent; tube shallow, 0·5 mm. long or less, 1–1·1 mm. wide; lobes 0·5–0·8 mm. long, 0·3 mm. wide. Corolla white, pink or pale mauve, tube 0·6–1 mm. long, widest at the base; lobes ovate, 0·4–0·7 mm. long, 0·4 mm. wide, papillate inside. Style 0·3–0·6 mm. long; stigmas linear, 0·2–0·3 mm. long. Capsule 2 mm. wide, beak 1 mm. long and wide, puberulous to densely scabrid pubescent. Seeds brown, ± 20 per capsule, ± 0·3 mm. long. Fig. 32.

TANZANIA. Near Bukoba, Sept. 1931, *Haarer* 2182!; Kigoma District: Kabogo Mts., 5 Apr. 1963, *Kyoto Univ. Exped.* 419!
DISTR. **T1, 4**; Zaire, Burundi, Zambia and Angola, also in Madagascar
HAB. Floor of *Brachystegia spiciformis* open woodland, also shallow soil on rock ledges, etc.; 950–1200 m.

SYN. *Diotocranus lebrunii* Bremek. in Verh. K. Nederl. Akad. Wet., Afd. Natuurk., ser. 2, 48(2): 148 (1952); Tennant in K.B. 22: 438 (1968). Type: Zaire, Kwango District, Mt. Sörensen, *Lebrun* 155 (BR, holo.)

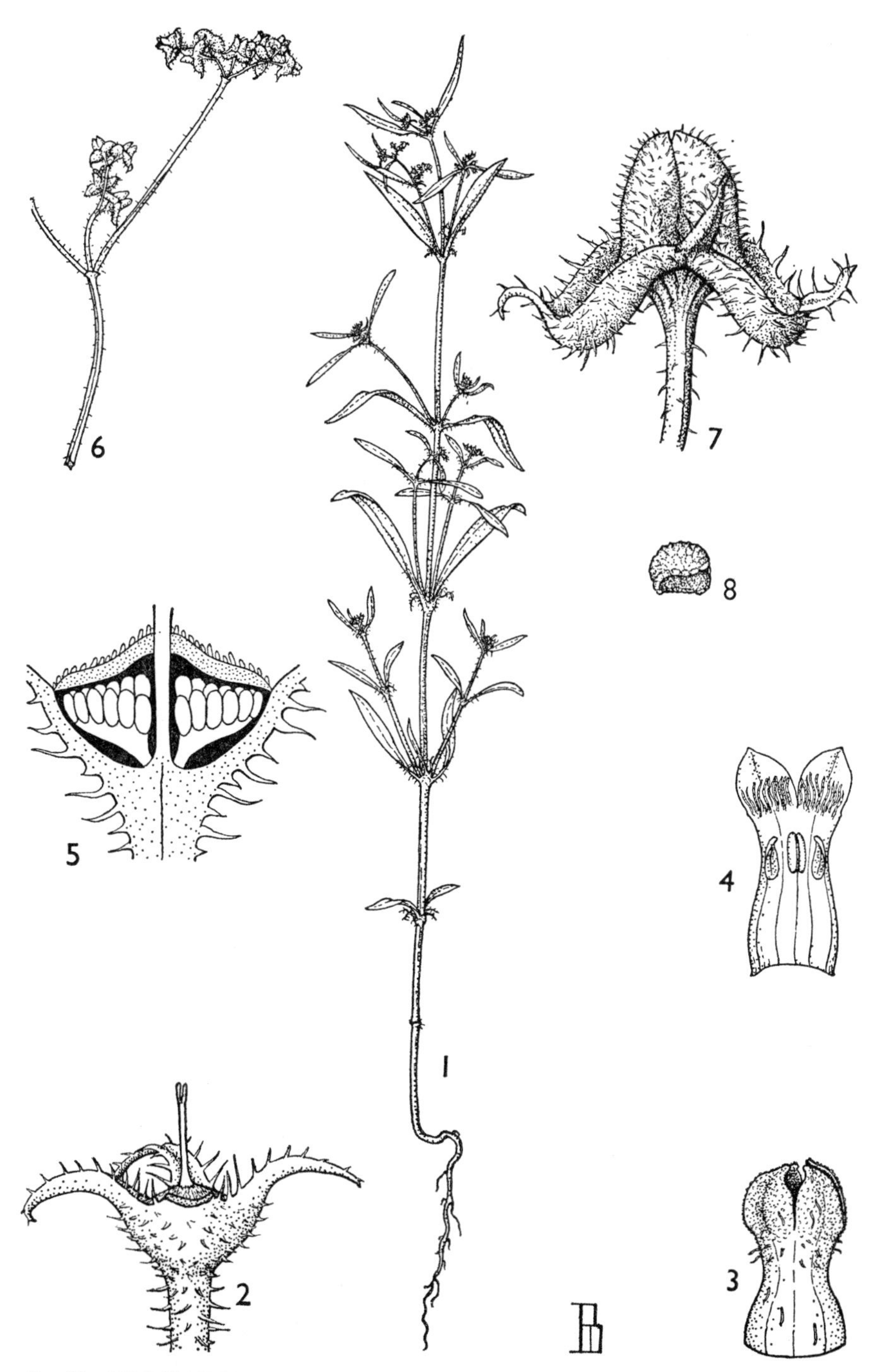

FIG. 32. *MITRASACMOPSIS QUADRIVALVIS*—**1,** habit, × $\frac{2}{3}$; **2,** calyx, with one lobe removed, and style, × 20; **3,** corolla, × 20; **4,** longitudinal section of corolla, × 20; **5,** longitudinal section of ovary, × 30; **6,** infructescences, × 2; **7,** capsule, × 14; **8,** seed, × 30. 1, from *Richards* 5121; 2, 4–6, from *Haarer* 2182; 3, from *E. A. Robinson* 4999; 7, 8, from *Richards* 14972. Drawn by Diane Bridson.

D. lebrunii Bremek. var. *sparsipilus* Bremek. in Verh. K. Nederl. Akad. Wet., Afd. Natuurk., ser. 2, 48(2): 149 (1952). Type: Tanzania, near Bukoba, *Haarer* 2182 (K, holo. !, EA, iso. !)

NOTE. Bremekamp distinguishes the variety by having the stems sparsely pilose with spreading white hairs, margins of leaf-blades and costa with long hairs and the ovary and capsule densely pubescent. Now that very much more material is available, mainly from Zambia, there seems to be no point in formally distinguishing it since the indumentum proves to be very variable. Jovet attributed his genus to the Loganiaceae but there is no doubt that it is the same as Bremekamp's more recently described *Diotocranus*. Bremekamp himself suggested this possibility in Acta Botanica Neerlandica 15: 7 (1966).

26. HEDYTHYRSUS

Bremek. in Verh. K. Nederl. Akad. Wet., Afd. Natuurk., ser. 2, 48(2): 149 (1952)

Small erect shrubs, invariably turning blackish on drying and with distinctly discolorous leaves. Leaves very shortly petiolate, elliptic to lanceolate, rather thick, closely placed, the lateral nerves invisible or very obscure; stipule sheath usually hairy, divided into several subulate fimbriae. Flowers heterostylous, small, in many-flowered dense terminal corymbs or panicles. Calyx-lobes 4, triangular or lanceolate, often with fimbriae between. Corolla shortly subcylindrical, slightly widened above; lobes 4; throat glabrous or sparsely hairy inside. Stamens well exserted in short-styled flowers. Ovary 2-locular; ovules few on peltate placentas; style filiform; stigma-lobes subglobose. Capsule depressed hemispherical, produced into a conical beak which splits loculicidally and septicidally into 4 diverging valves. Seeds few, much compressed, elliptic or oblong, sometimes subangular and often slightly winged at both ends or all round, reticulate.

A genus of 2 very closely allied species confined to upland areas in tropical Africa. It might be advisable to treat these taxa as subspecies since the differences are rather trivial.

H. thamnoideus (*K. Schum.*) *Bremek.* in Verh. K. Nederl. Akad. Wet., Afd. Natuurk., ser. 2, 48(2): 151 (1952). Types: Tanzania, Uluguru Mts., Lukwangule, *Stuhlmann* 9167, 9187 & 9193 (B, syn. †)

Subshrub 0·2–2 m. tall; stems drying blackish, ± glabrous or with only a few lines of hairs on young parts; internodes short. Leaf-blades narrowly elliptic to lanceolate, discolourous, green above (blackish green when dry), whitish or silvery beneath, 0·8–2(–2·5) cm. long, 2·5–8 mm. wide, acute at the apex, cuneate at the base, revolute at the margins, glabrous or with a few hairs on the margins and main nerve beneath; petiole short, scarcely 1(–1·5) mm. long; stipule-sheath hairy, 1–3 mm. long, with 3–7 fimbriae 1–4 mm. long, often ciliate. Inflorescences (1–)1·5–2·5 cm. wide; peduncle 0–1·5 cm. long, sometimes bifariously hairy; pedicels 0–1 mm. long. Calyx-tube ovoid, 0·5–1·5 mm. long, glabrous, pubescent or with a few hairs at top between the lobes; lobes 1·3–2·5 mm. long, glabrous or with some hairs on margin at the base. Corolla white, pink or pale lilac; tube 1·8–4 mm. long, longest in the short-styled flowers, glabrous or hairy outside; lobes oblong-ovate, 1·5–3 mm. long, 1–1·3 mm. wide. Style 1·7–2·8 mm. long in short-styled flowers, 3–5·8 mm. long in long-styled flowers; stigma-lobes 0·2–0·5 mm. long. Capsule 1–1·2 mm. tall excluding the 1–1·6 mm. long beak, 1·8–2 mm. wide. Seeds chestnut brown, 1–1·2 mm. long. Fig. 33.

TANZANIA. Morogoro District: Uluguru Mts., Lupanga Peak, 26 Dec. 1931, *B. D. Burtt* 3477 !; Mbeya District: Kitulo [Elton] Plateau, 29 Nov. 1963, *Richards* 18439 !;

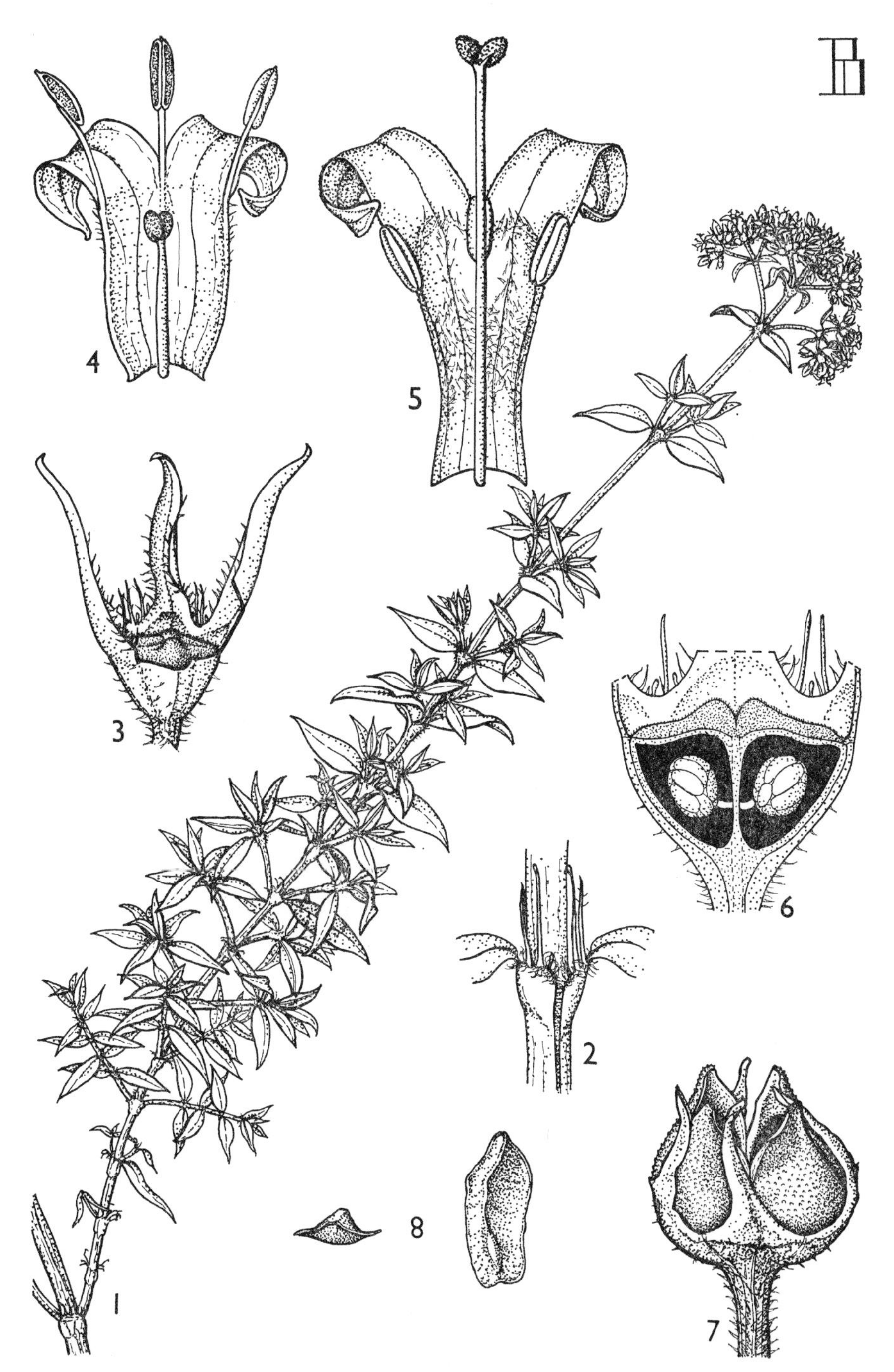

FIG. 33. *HEDYTHYRSUS THAMNOIDEUS*—**1,** flowering branch, × $\frac{2}{3}$; **2,** node showing stipules, × 6; **3,** calyx, with one lobe removed, × 10; **4,** longitudinal section of short-styled flower, × 10; **5,** same of long-styled flower, × 10; **6,** longitudinal section of ovary, × 20; **7,** capsule, × 12; **8,** seed, two views, × 20. 1–4, 6, from *Rounce* 614; 5, from *Richards* 7782; 7, 8, from *Greenway & Eggeling* 8687. Drawn by Diane Bridson.

Rungwe District: NW. of Rungwe Mt., edge of slope above the lower Kiwira R., 24 Oct. 1947, *Brenan & Greenway* 8195!

DISTR. T6, 7; Zaire, Rwanda, Burundi, Malawi and NE. Zambia

HAB. Stunted montane forest, grassland, grassland with scrub of *Pteridium*, *Hypericum*, *Philippia*, *Helichrysum*, etc., subalpine moorland, often on rocky hills and in rock crevices; 1450–2880 m.

SYN. *Oldenlandia thamnoidea* K. Schum. in E.J. 28: 56 (1899) & in E.J. 28: 485 (1900)

NOTE. *H. spermacocinus* (K. Schum.) Bremek. differs in the wider more diffuse inflorescences containing more flowers, the calyx-lobes not exceeding the beak in fruit and the larger leaves; it occurs in Angola and Zambia. A field study in Isoka District, Zambia, an area where the two overlap, is needed. As mentioned above, subspecific rank may prove more suitable, but no difficulty is experienced in naming specimens although only a relatively few sheets of *H. spermacocinus* have been seen.

27. PSEUDONESOHEDYOTIS

Tennant in K.B. 19: 277 (1965)

Shrub with several ? erect stems* or sometimes a liane. Leaves paired, shortly petiolate, uninerved; stipules with a single deltoid lobe from a short sheathing base on young stems; later these lobes are deciduous leaving only the truncate sheath. Flowers hermaphrodite, not dimorphic, small, in composite fairly lax terminal inflorescences plus inflorescences from the upper axils, mostly many-flowered. Calyx-tube very short, campanulate; lobes 4, equal, lanceolate-triangular. Corolla-tube very shortly cylindrical; lobes 4, oblong-lanceolate, just over twice as long as the tube. Stamens 4, exserted further than the corolla-lobes. Ovary 2-locular, each locule with many (18–20) ovules on a placenta affixed close to the base; style filiform, exserted for about the same distance as the stamens; stigma small, capitate, slightly bifid. Fruit capsular, with a low conical beak, loculicidal from the apex of the beak and septicidal below, crowned by the persistent sepals.

A monotypic genus confined to the Uluguru Mts. in Tanzania; allied to *Nesohedyotis* (Hook. f.) Bremek. and *Hedythyrsus* Bremek. If the segregate genera of Bremekamp are to be accepted then this genus takes its place alongside them. If *Oldenlandia* and the segregate genera related to it are merged with *Hedyotis* L. (see p. 269) then *Pseudonesohedyotis* will also have to be sunk into it.

P. bremekampii *Tennant* in K.B. 19: 278 (1965). Type: Tanzania, Uluguru Mts., Mkambaku Mt., *Schlieben* 3600 (PRE, holo., K, photo.!)

Shrub about 1 m. across; stems at first shortly bristly pubescent, soon glabrous, pale and somewhat corky, rather thickened and leafless at the nodes beneath. Leaf-blades lanceolate, 2–4 cm. long, 0·6–1·3 cm. wide, acute to acuminate at the apex, cuneate at the base, the margins slightly revolute, glabrous or pubescent at margins beneath; petiole up to 4 mm. long, often margined with bristly hairs; stipules with sheath ± 1 mm. long, lobe 3–5 mm. long, 2–2·5 mm. wide at base, glabrous or pubescent, the rhaphides often very obvious on the surface. Inflorescences up to 5–6 cm. across; pedicels up to 3 mm. long, pubescent. Calyx-tube 0·8 mm. long, ± pubescent; lobes 1·7–2·3 mm. long, 0·75–1 mm. wide, glabrous. Corolla white or pinkish; tube 1·4 mm. long; lobes 3·2 mm. long, 1·2 mm. wide, minutely apiculate. Filaments 4 mm. long. Style 6 mm. long; stigma 0·2 mm. long. Capsule 1·8 mm. long, 2 mm. wide. Seeds few, black, discoid, 0·65 mm. in diameter, 0·2 mm. thick, convex dorsally, flattened ventrally. Fig. 34.

* The habit is not evident; the stems are perhaps always straggling.

FIG. 34. *PSEUDONESOHEDYOTIS BREMEKAMPII*—**1,** flowering branch, × ⅜; **2,** node showing stipules, × 6; **3,** part of inflorescence, × 4; **4,** flower bud, × 6; **5,** flower, × 6; **6,** corolla, opened out, × 6; **7,** flower with corolla removed, × 8; **8,** longitudinal section of ovary, × 16. 1, 2, 4, from *E. M. Bruce* 700; 3, 5–8, from *Schlieben* 3379. Drawn by Mrs M. E. Church.

TANZANIA. Morogoro District: Uluguru Mts., Lukwangule Peak, 30 Jan. 1935, *Bruce* 700! & Uluguru Mts., 4 Feb. 1933, *Schlieben* (probably 3379!)* & Uluguru Mts., Mkambaku Peak, 26 Feb. 1933, *Schlieben* 3600
DISTR. T6; not known elsewhere
HAB. Upland evergreen forest; 1370–2650 m.

28. AGATHISANTHEMUM

Klotzsch in Peters, Reise Mossamb., Bot. 1: 294 (1861); Bremek. in Verh. K. Nederl. Akad. Wet., Afd. Natuurk., ser. 2, 48(2): 152 (1952); Lewis in Ann. Missouri Bot. Gard. 52: 182 (1965)

Perennial herbs or small shrubs with simple or branched stems. Leaves sessile or nearly so, narrowly lanceolate to ovate or oblong; lateral nerves mostly strong; stipule-sheath produced into a 3–13 fimbriated lobe. Flowers usually heterostylous, small, in many-flowered cymes or corymbs or in subglobose heads. Calyx-lobes 4, lanceolate to ovate-triangular, keeled. Corolla shortly tubular, densely hairy at the throat; lobes 4, often with some additional filamentous lobes between. Ovary 2(–3)-locular; ovules numerous on peltate placentas; style filiform, shortly hairy, usually undivided but sometimes bifid at the apex; stigma-lobes subglobose or ovoid. Disc farinose or very shortly pubescent. Capsule subglobose, produced into a conical beak which opens both loculicidally and septicidally. Seeds numerous, angular.

A small genus of 5–6 ill-defined species confined to tropical Africa and the Comoro Is., seemingly closely allied to *Hedyotis* and possibly best considered a section of that genus. Pollen aperture morphology indicates that it may be nearer to the American *Hedyotis* L. subgen. *Edrisia* (Raf.) Lewis rather than to the Asiatic section *Diplophragma* Wight & Arn. as Bremekamp suggested.

Calyx-lobes ± 2 (rarely up to 3) mm. long; inflorescences mostly lax but sometimes flowers in collections of dense subcapitulate clusters, always distinctly heterostylous; disc shortly pubescent; mostly coastal areas . . . 1. *A. bojeri*
Calyx-lobes exceeding 2 mm. and up to 9 mm. long; inflorescences not lax, the flowers in capitula, either heterostylous or not; disc mostly finely papillate; mostly inland areas:
 Flowers usually heterostylous, in corymbs of small capitula 0·7–2·2(–3·2) cm. in diameter; calyx-lobes 2·5–3·5(–4·6) mm. long . . . . 2. *A. quadricostatum*
 Flowers not distinctly heterostylous, both anthers and style exserted; capitula very often large, 1·1–4 cm. in diameter; calyx-lobes 3·3–9 mm. long . . . . . 3. *A. globosum*

1. **A. bojeri** *Klotzsch* in Peters, Reise Mossamb., Bot. 1: 294 (1861); Bremek. in Verh. K. Nederl. Akad. Wet., Afd. Natuurk., ser. 2, 48(2): 154 (1952). Types: Zanzibar, *Bojer* (B, syn. †, K, isosyn. !) & *Peters* (B, syn. †)

Branched herb or subshrub 0·3–1·5 m. tall; young stems densely pubescent, becoming glabrescent with age and often the brown bark peeling off. Leaves subsessile, pale; blades narrowly lanceolate to lanceolate, narrowly elliptic or ovate-oblong, 1·2–5·8 cm. long, 0·15–1(–2·1) cm. wide, subacute to acute and sometimes apiculate at the apex, cuneate at the base, glabrous or slightly

* Tennant, loc. cit.: 279 (1965) explains the confusion of numbers.

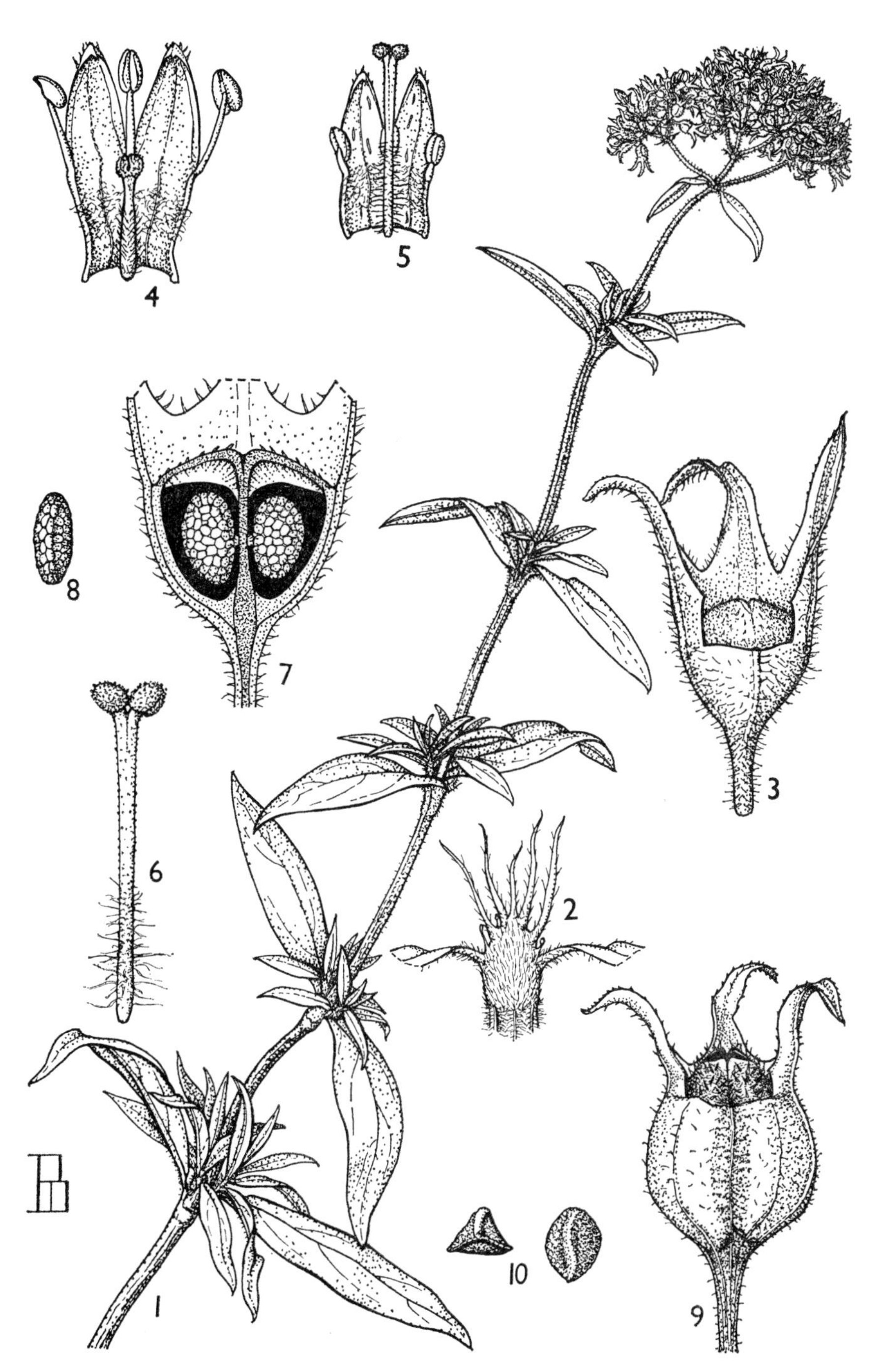

Fig. 35. *AGATHISANTHEMUM BOJERI* var. *BOJERI*—**1,** habit, × $\frac{2}{3}$; **2,** node showing stipule, × 4; **3,** calyx, with one lobe removed, × 10; **4,** longitudinal section of short-styled flower, × 8; **5,** same of long-styled flower, × 8; **6,** style and stigma-lobes, × 14; **7,** longitudinal section of ovary, × 16; **8,** ovules on one placenta, × 16; **9,** capsule, with one calyx-lobe removed, × 10; **10,** seed, two views, × 20. 1, 9, 10, from *Tweedie* 1243; 2, from *Faulkner* 1205; 3, 4, 6–8, from *Tweedie* 1216; 5, from *Rawlins* 731. Drawn by Diane Bridson.

scabridulous above, pubescent on the nerves beneath; petiole 0–1·2 mm. long; leaves often appearing whorled owing to presence of short undeveloped leafy shoots present at practically every node; stipule-sheath pubescent, 0·5–4·5 mm. long with (2–)5 fimbriae 1–5·8 mm. long. Inflorescences laxly corymbose, 0·9–7 cm. across or sometimes quite dense subcapitulate clusters; peduncles 0·2–4 cm. long; pedicels 0–1·5(–3) mm. long; flowers distinctly heterostylous. Calyx-tube 0·9–1·7 mm. long, glabrous or pubescent, the limb-tube 0·2–0·8 mm. long; lobes lanceolate, 1·4–2(–3·2) mm. long (4·6 in an atypical isostylous specimen). Corolla white, creamy white or greenish white; tube 1–2·5 mm. long, glabrous or very sparsely hairy outside; lobes ovate-oblong, 1·5–2·7 mm. long, 0·65–1·8 mm. wide. Disc very shortly pubescent. Style 0·7–2 mm. long in short-styled flowers, 3–3·5(–6 *fide* Bremekamp) mm. long in long-styled flowers, usually filiform and undivided but sometimes bifid at the apex; stigma-lobes 0·15–0·5 mm. long. Capsule (1–)1·5–2 mm. long, (1·3–)1·8–2·5 mm. wide. Seeds black, ovoid-trigonous, 0·32 mm. long, reticulate.

var. **bojeri**

Calyx pubescent. Fig. 35, p. 255.

KENYA. Kwale District: Cha Shimba forest, 1 Feb. 1953, *Drummond & Hemsley* 1088!; Mombasa, English Point, 26 May 1934, *Napier* 3479 in *C.M.* 6290!; Kilifi District: Malindi, Oct. 1951, *Tweedie* 1042!

TANZANIA. Lushoto District: Korogwe area, Magunga estate, 9 May 1953, *Faulkner* 1205!; Bagamoyo District: Kidomole, Apr. 1964, *Semsei* 3837!; Uzaramo District: 9·5 km. S. of Dar es Salaam, Mbagala Mission, 30 Nov. 1953, *Milne-Redhead & Taylor* 7453!; Zanzibar I., Wanda area, 24 Jan. 1929, *Greenway* 1159! & Kizimbani, 25 Mar. 1960, *Faulkner* 2515!; Pemba I., Mkoani, 9 Aug. 1929, *Vaughan* 478!

DISTR. **K**7; **T**3, 4, 6, 8; **Z**; **P**; Somali Republic (N.), Mozambique, Rhodesia, South Africa, Comoro Is. (Mayotte) and Madagascar

HAB. Grassland, forest edges, bushland, open woodland and also as a weed in cultivations; particularly in coastal areas; 0–1200 m.

SYN. *Oldenlandia bojeri* (Klotzsch) Hiern in F.T.A. 3: 53 (1877); T.T.C.L.: 507 (1949); U.O.P.Z.: 384, fig. (1949)
Agathisanthemum bojeri Klotzsch subsp. *australe* Bremek. in Verh. K. Nederl. Akad. Wet., Afd. Natuurk., ser. 2, 48(2): 156 (1952). Type: South Africa, Transvaal, Komatipoort, *Schlechter* 11864 (K, holo.!)

var. **glabriflorum** *Bremek.* in Verh. K. Nederl. Akad. Wet., Afd. Natuurk., ser. 2, 48(2): 156 (1952). Type: Mozambique, Pungue valley, *Vasse* 280 (P, holo.)

Calyx glabrous.

TANZANIA. Rungwe District: Masoko Crater, 25 Sept. 1932, *Geilinger* 2654!
DISTR. **T**7; Mozambique
HAB. Probably grassland; 1500–2000 m.

SYN. *A. bojeri* Klotzsch subsp. *australe* Bremek. var. *glabriflorum* Bremek. in Verh. K. Nederl. Akad. Wet., Afd. Natuurk., ser. 2, 48(2): 157 (1952). Type: Mozambique, Beira railway, *Cecil* 10 (K, holo.!)

NOTE. The Geilinger specimen has calyx-lobes 2·6(–3) mm. long but the disc is distinctly pubescent. This is a very variable species particularly in leaf-shape, laxity of inflorescence and sizes of the floral parts. I have not however been able to maintain Bremekamp's division into a northern and southern subspecies. The shape and length of the leaf-blades—the characters he uses—do not seem to correlate with geography in the way he suggests.

2. **A. quadricostatum** *Bremek.* in Verh. K. Nederl. Akad. Wet., Afd. Natuurk., ser. 2, 48(2): 160 (1952). Type: Zaire, Katanga, Kasiki, *de Witte* 409 (BR, holo.)

Erect herb with 1–several mostly sparsely branched glabrous to pubescent 4-ribbed stems 0·23–1·2 m. tall from a woody rootstock; bark eventually peeling from the older stems. Leaves subsessile; blades narrowly ovate to lanceolate, 1–7·5 cm. long, 0·2–2(–3) cm. wide, acute at the apex, cuneate at the base, glabrescent to scabridulous-pubescent on both surfaces or only margins and nerves beneath scabridulous; leaves mostly appearing whorled owing to the presence of short undeveloped leafy shoots at practically every node; stipule-sheath pubescent, 1–4(–7) mm. long with 3–9(–15) fimbriae 3–5(–10) mm. long. Flowers mostly heterostylous, in dense clusters arranged in corymbs or panicles, each cluster 0·7–2·2(–3·2) cm. wide; apparent peduncles 0–4·4 cm. long; pedicels 0·5–3 mm. long. Calyx-tube 0·9–1·5 mm. long, glabrous to densely pubescent; limb-tube 0·4–0·8 mm. long; lobes 2·5–3·5(–4·6) mm. long, glabrous save for ciliate margins, or densely pubescent. Corolla white, greenish or tinged purple, glabrous outside; tube 1·6–2 mm. long; lobes triangular-lanceolate, 1·7–2 mm. long, 0·8 mm. wide. Disc minutely papillate or sometimes with hairs. Style 1·4 mm. long in short-styled flowers, 2·6–4 mm. long in long-styled flowers. Capsule glabrous or pubescent, compressed-ellipsoid, 1·5–3 mm. tall, 1·6–2·8 mm. wide, grooved between the loculi, the beak ± 1 mm. long, crowned by the persistent calyx-lobes. Seeds blackish, ovoid-trigonous, 0·5 mm. long, reticulate.

TANZANIA. Buha District: Kasakela Reserve, Melinda stream, 19 Nov. 1962, *Verdcourt* 3377!; Mpanda District: Kungwe Mt., Selimweguru, 24 July 1959, *Newbould & Harley* 4606!; Tunduru District: 96 km. from Masasi, 19 Mar. 1963, *Richards* 17950!

DISTR. **T**1, 4, 7, 8; Zaire, Burundi, Mozambique, Malawi, Zambia, Rhodesia and South Africa, Transvaal (*fide* Bremekamp)

HAB. Grassland, bracken scrub, secondary *Brachystegia* and other woodland; 500–1950 m.

SYN. ? *A. petersii* Klotzsch in Peters, Reise Mossamb., Bot., 1: 295 (1861). Type: Mozambique, Querimba, *Peters* (B, holo. †)

A. quadricostatum Bremek. var. *pubescens* Bremek. in Verh. K. Nederl. Akad. Wet., Afd. Natuurk., ser. 2, 48(2): 161 (1952). Type: Malawi, Zomba, *Purves* 114 (K, holo. !)

NOTE. *A. bojeri*, with its usually rather open inflorescence of heterostylous flowers, short calyx-lobes, pubescent disc and mostly coastal distribution, is clearly distinct from *A. globosum* with isostylous flowers in very compact globose heads, longer calyx-lobes, finely papillate disc and mostly upland distribution. The " species " described above is in many ways intermediate between these two and, although undoubtedly not just direct hybrids since the two do not now occur together, it is very likely that it has arisen by hybridisation in the past. In Kenya and Uganda where the two could not have been in contact during any time relevant to the problem, no intermediate populations of any kind exist but in Tanzania, particularly in the west where it is very likely that *A. bojeri* could have occurred, it may have been eliminated through hybridisation with *A. globosum* to give rise to the ill-defined " species " here recognised for convenience. Heterostyly is not at all constant e.g. in *Milne-Redhead & Taylor* 8491 (Tanzania, 12 km. W. of Songea, by Kimarampaka stream) the inflorescence is as open as in many specimens of *A. bojeri*, the calyx-lobes are 4·6 mm. long, the disc has some hairs and the flowers have both anthers and style well exserted; it clearly has characters of both *A. bojeri* and *A. globosum* but the general appearance is that of *A. quadricostatum*. Another intermediate specimen is *Lunan* H 86/46 (Tanzania, E. of Tabora, Igalula), a poor specimen made the type of a supposed new species, *A. assimile*, by Bremekamp. This has an open inflorescence, calyx-lobes 4 mm. long, the disc distinctly hairy and the flowers with both anthers and style exserted. He distinguished it from *A. bojeri* by having the leaf-nerves all springing from the lower half of the leaf-blade (not true so far as can be seen from the specimen), the stipule-sheath, calyx-lobes and capsule all longer. Further it is distinguished from *A. angolense* Bremek. by not having a cork layer which peels, although in the type specimen the bark is clearly peeling off; in the diagnosis it is among other things distinguished from *A. quadricostatum* by its hairy disc and in not having a 4-angled stem, but the first character is variable in *A. quadricostatum* and the type of *A. assimile* has some parts of the stem very distinctly 4-ribbed. The genus is much commended to local workers in East Africa for field and experimental studies; only by such means will the confusion be resolved.

3. **A. globosum** (*A. Rich.*) *Bremek.* in Verh. K. Nederl. Akad. Wet., Afd. Natuurk., ser. 2, 48(2): 161 (1952); U.K.W.F.: 401, fig. (1974). Type: Ethiopia, Kouaietha, *Quartin Dillon* (P, syn.) & Tecli, *Schimper* 512 (P, syn., BM !, K !, MO, iso.)

Rather strictly erect herb (11–)25–60(–90) cm. tall, with (1–)several usually sparsely branched stems from a woody rootstock; branches densely pubescent with short spreading hairs; whole plant often drying the peculiar yellow-green of an aluminium accumulator. Leaves subsessile, elliptic, oblong-elliptic or oblong-lanceolate, (1·9–)2·5–7·5 cm. long, 0·6–2·5 cm. wide, acute at the apex, cuneate at the base, glabrescent to pubescent or scabrid on both surfaces but always some hairs on the margins; petiole less than 1 mm. long; aborted leafy shoots are present in most axils; stipule-sheath triangular, 2–9 mm. long, bearing (0–)3–5 linear fimbriae 4–9 mm. long or sometimes produced as a practically unfimbriated triangular lobe. Flowers not heterostylous, anthers and stigmas touching in bud, (3–)4(–5)-merous, in dense subglobose inflorescences 1–1·4 cm. in diameter, sessile, supported by a pair of leaves, terminal and also terminating slender axillary branches from the upper part of the stems; pedicels ± 1 mm. long. Calyx-tube 1·2–2·1 mm. long, glabrous or pubescent; limb-tube 0·8–1·8 mm. long; lobes linear-lanceolate, 3·3–7(–9) mm. long, shortly ciliate on margins and usually costa or densely pubescent. Corolla variable in colour, white, cream, pale yellow, blue, lilac, pink, purple or mauve, the lobes sometimes coloured at least outside and the tube whitish, glabrous, puberulous or minutely papillate; tube (1·8–)3–4 mm. long; lobes triangular, 2·2–4·1 mm. long, 1–1·5 mm. wide. Disc minutely papillate. Anthers exserted, often dark. Style white to yellow, 3·7–6·5 mm. long, exserted; stigma-lobes 0·4–0·7 mm. long. Capsule compressed-ellipsoid, 2·4 mm. tall, 2·4 mm. wide, grooved between the loculi, the beak ± 0·8 mm. long, crowned by the persistent calyx-lobes, mostly pubescent. Seeds dark purplish brown, ovoid-trigonous, 0·6 mm. long, 0·5 mm. wide.

UGANDA. S. Ankole, Nov. 1956, *Chenery* in *E. A. H.* H138/56–7 !; Kigezi District: Kambuga, May 1947, *Purseglove* 2420 !; Mengo District: 1·9 km. Kiwoko to Tweyanze, 23 Apr. 1956, *Langdale-Brown* 2070 !

KENYA. W. Suk District: 9·6 km. SW. of Kapenguria and 0·5 km. N. of Keringet, 26 Sept. 1962, *Lewis* 5987 !; Trans-Nzoia District: Kitale, 19 Sept. 1961, *Verdcourt* 3212 !; N. Kavirondo District: Kakamega Forest Station, 17 Sept. 1949, *Maas Geesteranus* 6275 !

TANZANIA. Biharamulo, 15 Nov. 1962, *Verdcourt* 3310 !; Ufipa District: 14·4 km. from Zambian border, Chilo Village, 2 Jan. 1968, *Richards* 22849 !; Njombe District: Madehani, 26 Jan. 1914, *Stolz* 2468 !

DISTR. **U**2, 4; **K**2, 3, 5; **T**1, 4, 6, 7; Zaire, Gabon, Burundi, Ethiopia, Malawi, Mozambique, Zambia, Rhodesia and Angola

HAB. Grassland, grassland with scattered trees, *Brachystegia* woodland and derived formations, wasteland, roadsides, etc., often on stony or rocky ground; 900–2400 m.

SYN. *Hedyotis globosa* A. Rich., Tent. Fl. Abyss. 1: 360 (1847)
Oldenlandia globosa (A. Rich.) Hiern in F.T.A. 3: 54 (1877) & in Cat. Afr. Pl. Welw. 2: 440 (1898)
Agathisanthemum globosum (A. Rich.) Bremek. var. *subglabrum* Bremek. in Verh. K. Nederl. Akad. Wet., Afd. Natuurk., ser. 2, 48(2): 163 (1952). Type: Tanzania, Njombe District, Msima Stock Farm, *Emson* 357 (K, holo. !, EA, iso.)

NOTE. The material now available shows that this species varies considerably in the degree of indumentum on the foliage. The exsertion of the anthers also varies considerably. The absence of true heterostyly needs checking in the field. A good deal of what Bremekamp has called *A. globosum* for Mozambique and elsewhere is not distinguishable from other material labelled *A. quadricostatum* (see note on p. 257).

29. DIBRACHIONOSTYLUS

Bremek. Verh. K. Nederl. Akad. Wet., Afd. Natuurk., ser. 2, 48(2): 163 (1952)

Perennial herb with woody rhizomatous rootstock; stems 4-ribbed or almost winged, mostly much branched. Leaves sessile, linear or linear-lanceolate, the lateral nerves rather obscure; stipules with several fimbriae from a short sheath. Flowers small, markedly heterostylous, in many-flowered terminal rather dense corymbiform inflorescences. Calyx-lobes 4, small, narrowly triangular, separated by colleters. Corolla-tube almost cylindrical, slightly narrowed to the base, short, glabrous inside; lobes 4, oblong-ovate. Ovary 2-locular; ovules numerous on peltate placentas; style shortly divided into 2 branches; stigmas ellipsoid. Capsule subglobose, shortly beaked, apically loculicidally and septicidally dehiscent. Seeds numerous, angular.

A monotypic genus restricted to East Africa, closely related to *Hedythyrsus* and *Agathisanthemum*. Lewis, in Ann. Missouri Bot. Gard. 52: 186, 202 (1965), would not separate this genus from *Oldenlandia*.

D. kaessneri (*S. Moore*) *Bremek.* in Verh. K. Nederl. Akad. Wet., Afd. Natuurk., ser. 2, 48(2): 164 (1952); U.K.W.F.: 401 (1974). Type: Kenya, Nairobi, *Kassner* 957 (BM, holo. !*, K !, MO, iso.)

Herb 15–40 cm. tall; stems glabrous. Leaf-blades 1·5–5·4 cm. long, 2–8 mm. wide, acute or rather blunt, very shortly mucronate, cuneate at the base, glabrous; stipular sheath 1–3 mm. long, divided into 5–7 fimbriae (1–)1·5–3(–5) mm. long. Component parts of inflorescence subglobose, 0·5–1·5 cm. diameter, expanding in fruit to 2·5 cm.; peduncles 1·8–6 cm. long; pedicels 0·5–1 mm. long. Calyx glabrous; tube ovoid, 0·8–1 mm. long; lobes 0·8–1 mm. long. Corolla more or less white, mauve or lilac; tube 1·1–2 mm. long, often slightly longer in short-styled flowers; lobes 1·8–2 mm. 1·2–1·4(–1·8) mm. wide, obtuse, but shortly mucronulate. Stamens in short-styled form longer than the lobes, and just about level with the throat in long-styled form. Style 3·7–4 mm. long in long-styled form excluding the 0·25–0·8 mm. long branches, 1·5 mm. long in short-styled form excluding the 0·25–0·6 mm. long branches; stigmas 0·4–0·5 mm. long. Capsule 1·2–1·8 mm. tall, 1·5–2·2 mm. wide, glabrous, the beak up to 0·5 mm. tall. Seeds pale yellow-brown, sharply trigonous, ± 0·35 mm. long, reticulate. Fig. 36, p. 260.

KENYA. Fort Hall/Kiambu Districts: Thika, 30 June 1947, *Bogdan* 848 !; Nairobi District: Nairobi National Park, near main entrance, 23 May 1961, *Verdcourt & Polhill* 3145 ! & near Thika Road House, just N. of Nairobi on road to Thika, 22 Oct. 1950, *Verdcourt* 360 !

DISTR. **K4**; not known from elsewhere (the specimen cited by Bremekamp from Tanzania is from the Meru in Kenya not Mt. Meru)

HAB. Seasonally wet grassland, with or without scattered trees, particularly on murram soils, sometimes in rather rocky places; 1140–1860 m.

SYN. *Oldenlandia kaessneri* S. Moore in J.B. 43: 249 (1905), *non* K. Schum. & K. Krause (1907)

NOTE. A specimen of this plant bearing a label East Griqualand is, as Bremekamp points out, clearly wrongly labelled.

* S. Moore also mentions in parenthesis a specimen from Nairobi collected by A. Blayney Percival, but it seems reasonable not to consider the two as syntypes.

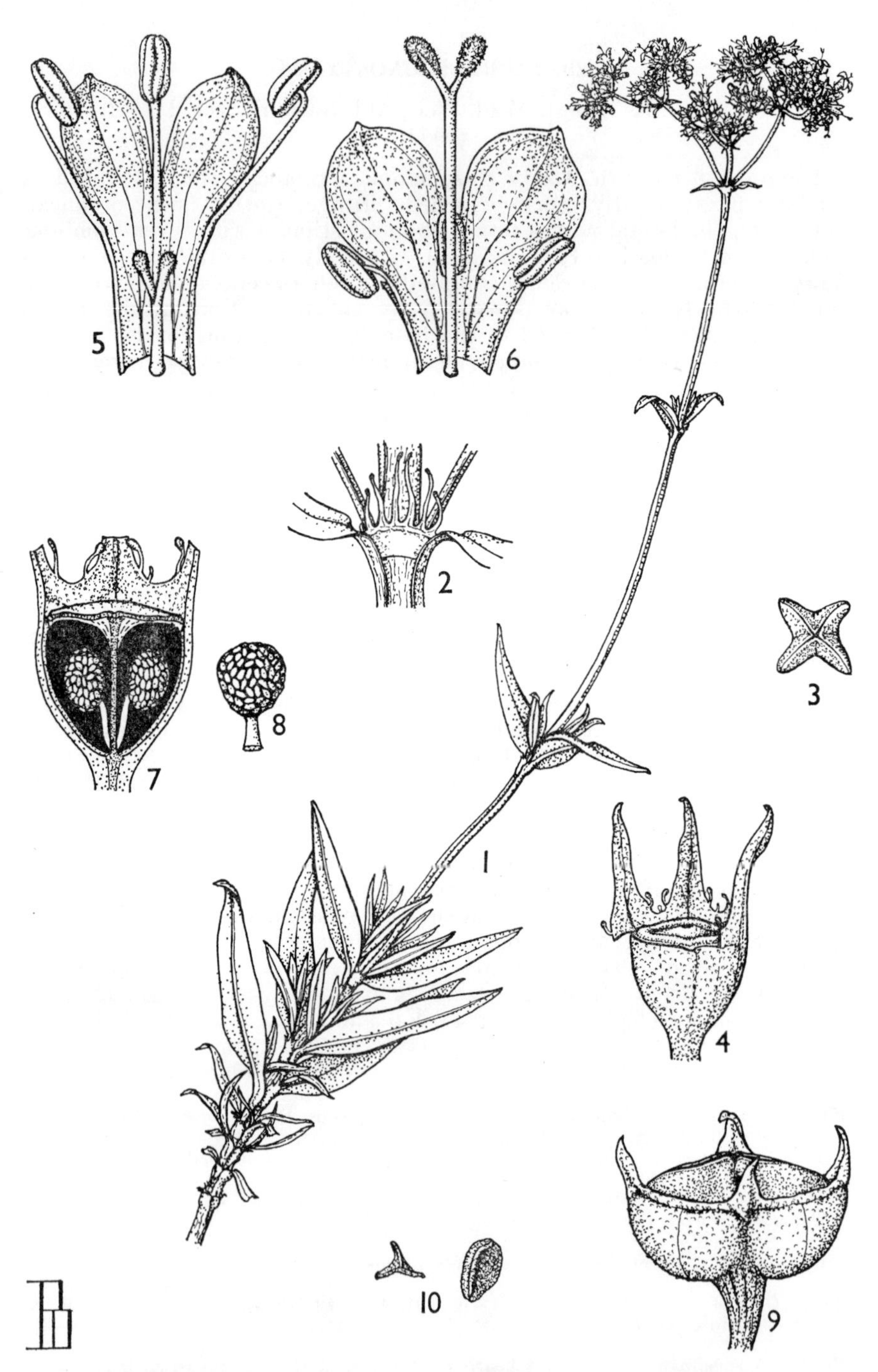

FIG. 36. *DIBRACHIONOSTYLUS KAESSNERI*—**1,** flowering branch, × ⅔; **2,** node showing stipule, × 2; **3,** flower bud from above, × 10; **4,** calyx, with one lobe removed, × 12; **5,** longitudinal section of short-styled flower, × 10; **6,** same of long-styled flower, × 10; **7,** longitudinal section of ovary, × 20; **8,** ovules on one placenta, × 20; **9,** capsule, × 12; **10,** seed, two views, × 20. 1, from *Napier* 2593; 2–4, 6–8, from *Verdcourt & Polhill* 3145; 5, from *Kabuye* 2; 9, from *Perdue & Kibuwa* 8185; 10, from *Dowson* 151. Drawn by Diane Bridson.

30. AMPHIASMA

Bremek. in Verh. K. Nederl. Akad. Wet., Afd. Natuurk., ser. 2, 48(2): 168 (1952)

Perennial herbs or subshrubs, with rather strict rush-like stems, usually glabrous or glabrescent. Leaves sessile, opposite; blades linear or filiform, rather rigid and usually erect or nearly so, acute and hard-pointed, the margins revolute; stipule-sheath mostly tubular, truncate or sometimes with 2 small cusps. Flowers 4-merous, isostylous or heterostylous, in small rather dense terminal and axillary capituliform, umbelliform or cymose inflorescences, sometimes combining to form a more ample inflorescence; bracts connate in pairs. Calyx-lobes triangular. Corolla mostly white; tube cylindrical or somewhat funnel-shaped; throat ± densely hairy; lobes ovate or ovate-oblong, hairy in lower half inside. Stamens included in the tube in long-styled flowers, exserted in isostylous and short-styled flowers; filaments glabrous; anthers dorsifixed. Ovary 2-locular; ovules fairly numerous on peltate slightly stipitate placentas affixed just below the middle of the septum; style glabrous, included in short-styled flowers, exserted in long-styled and isostylous flowers; stigma-lobes filiform, densely covered with long papillae. Disc cushion-shaped, farinose. Capsule globose, only slightly beaked, the beak dehiscing loculicidally. Seeds dark brown, not very numerous, dorsiventrally compressed, oblong, smooth.

A small genus of probably 5–6 species in south-central and SW. Africa. I am unable to distinguish some of the 8 species listed by Bremekamp. Lewis—see Ann. Missouri Bot. Gard. 52: 183, 202 (1965)—would not separate this genus from *Oldenlandia*. Only one species occurs in the Flora area.

A. luzuloides (*K. Schum.*) *Bremek.* in Verh. K. Nederl. Akad. Wet., Afd. Natuurk., ser. 2, 48(2): 170 (1952). Type: Malawi, Nyika Plateau, *Whyte* (B, holo. †, K, iso. !)

Perennial herb, with erect stems 14–40 cm. tall from a many-headed rhizome; stems strict, simple or sparsely branched, glabrous. Leaves held erect; blades linear, 1–4 cm. long, 1–2(–3) mm. wide, scabridulous above and ± ciliolate at the base of the midvein beneath; stipular sheath 2–5 mm. long, truncate or with 2 minute teeth ± 0·2 mm. long. Inflorescences 2–5, situated at the apex and at the first 2 nodes beneath, capitulate, ± 6 mm. long, ± 6 mm. wide; peduncles 2–8 mm. long; pedicels 0–1·5 mm. long; bracts leaf-like, 3–9 mm. long, 1–2 mm. wide; flowers heterostylous. Calyx-tube glabrous, 1·2 mm. long; lobes 1–1·2 mm. long, with ciliolate margins. Corolla white or tinged lavender, glabrous or papillate outside; tube slightly funnel-shaped, 2·8–4 mm. long, pilose inside; lobes 2·2 mm. long, 1·6 mm. wide, lower half pilose inside. Style 4 mm. long in long-styled flowers, 2 mm. long in short-styled flowers; stigma 1 mm. long. Capsule 2 mm. long, 2·5 mm. wide. Seeds not seen. Fig. 37, p. 262.

TANZANIA. Njombe District: Ukinga Mts., Lipanga Mt., June 1899, *Goetze* 998!
DISTR. **T7**; Malawi
HAB. Grassy slopes; 2500 m.

SYN. *Oldenlandia luzuloides* K. Schum. in E.J. 28: 55 (1899)
O. luzuloides K. Schum. in E.J. 30: 411 (1901), *nom. illegit.* Type: Tanzania, Ukinga Mt., *Goetze* 998 (B, holo. †, EA, K, iso. !)

NOTE. This species was presumably based on material at Berlin sent as a duplicate from Kew. K. Schumann curiously used the same name twice, so the second is a later homonym, but fortunately the two types belong to the same species. *A. assimile* Bremek., also described from the Nyika Plateau and supposed to have a different, less capituliform inflorescence and isostylous flowers, can be no more than a minor variant.

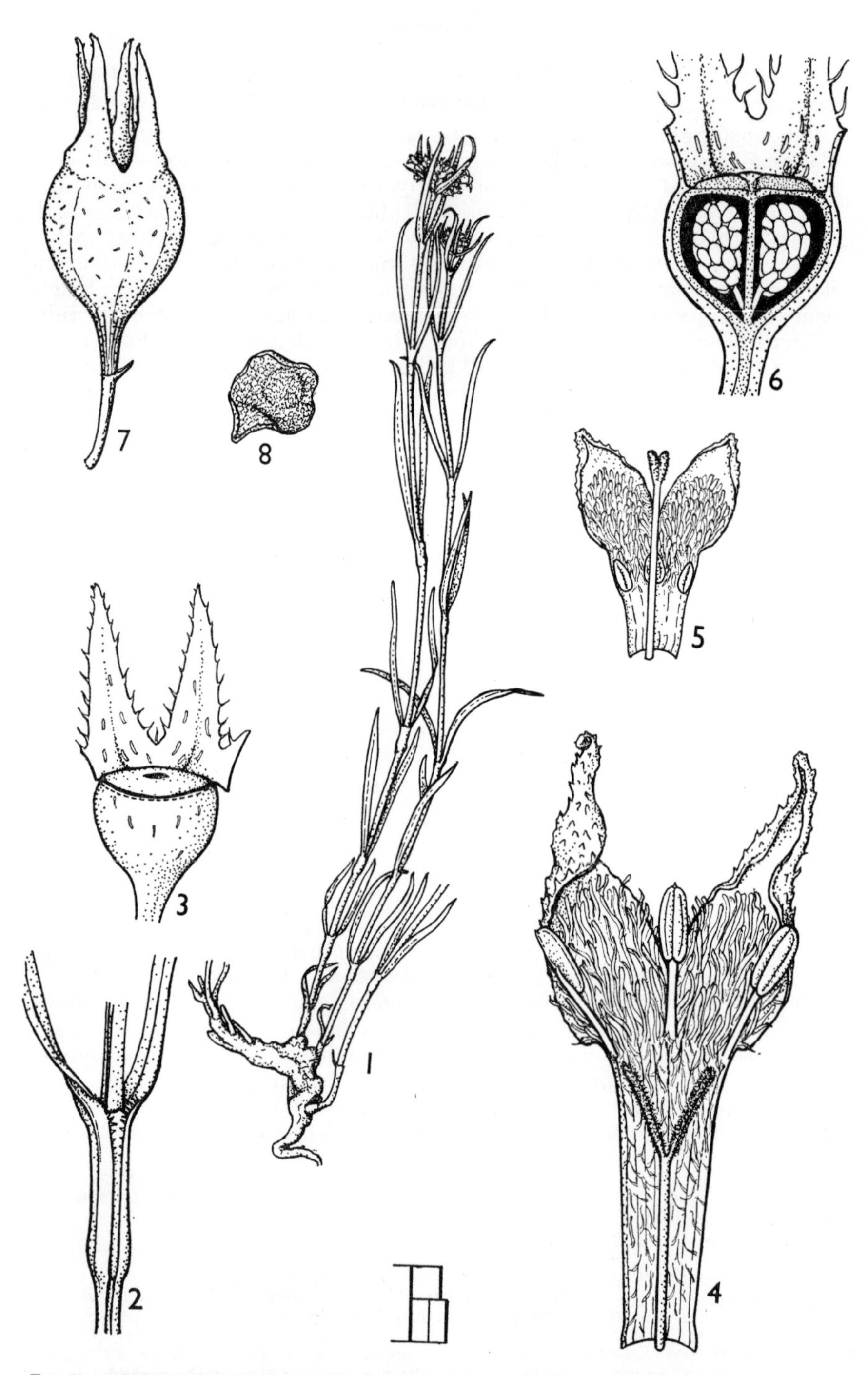

FIG. 37. *AMPHIASMA LUZULOIDES*—**1**, habit, × ⅔; **2**, node showing stipule and leaf-bases, × 4; **3**, calyx, with two lobes removed, × 14; **4**, longitudinal section of short-styled flower, × 14; **5**, same of long-styled flower, × 8; **6**, longitudinal section of ovary, × 20; **7**, capsule, × 8; **8**, seed, × 20. 1, 2, 5, 7, 8, from *Whyte* 128; 3, 4, 6, from *Pawek* 2195. Drawn by Diane Bridson.

31. PENTODON

Hochst. in Flora 27: 551 (1844); Bremek. in Verh. K. Nederl. Akad. Wet., Afd. Natuurk., ser. 2, 48(2): 175 (1952)

Glabrous subsucculent annual or short-lived perennial herbs with slender rootstocks. Leaves paired, sessile, penninerved but lateral nerves obscure; stipules with the short sheath divided into (1–)2–5 narrow fimbriae. Flowers small, hermaphrodite, dimorphic or not, in few–several-flowered very lax pedunculate terminal or apparently axillary (solitary at the nodes) inflorescences which appear to be verticillate, with 1–4 flowers from 1–4 widely spaced nodes. Calyx-tube obconic or campanulate; lobes 5, equal, very narrowly triangular. Corolla-tube narrowly funnel-shaped, throat hairy; lobes 5, ovate-triangular. Stamens exserted in short-styled forms, included in long-styled and equal-styled forms. Ovary 2-locular; placentas peltate with numerous ovules; style filiform, glabrous; stigma 2-lobed, the lobes filiform, included in short-styled forms, often exserted in equal-styled forms, exserted in long-styled forms. Fruit capsular, campanulate or oblong, the beak only slightly raised, loculicidally dehiscent. Seeds small, numerous, brown, angular.

A genus of probably only 2 species, one restricted to the Somali Republic (S.), the other widespread in Africa and extending to Arabia and Madagascar. A third species has been described from America, namely *P. halei* (Torrey & Gray) A. Gray; this occurs in the U.S.A. from Florida to Texas, Cuba and Nicaragua. Bremekamp suggests that it is conspecific with *P. pentandrus* and other writers (e.g. Correll & Johnston, Man. Vasc. Pl. Texas: 1490 (1970)) have accepted this. Clearly it cannot be specifically distinct. The material is rather uniform in facies, having small elliptic leaves and the inflorescences shorter than in most African material; curiously it is with some coastal E. African material that it can be most easily matched in the characters just mentioned; but the flowers appear to be more like those of var. *pentandrus*, although unfortunately no material has been seen which throws light on the heterostyly or otherwise of the New World material. It seems possible that the American populations are due to very few introductions of atypical material from Africa quite possibly long ago. Studies on the presence of heterostyly in these populations would be very interesting.

P. pentandrus (*Schumach. & Thonn.*) *Vatke* in Oest. Bot. Zeitschr. 25: 231 (1875); K. Schum. in P.O.A. C: 377 (1895), as *pentander*; Bremek. in Verh. K. Nederl. Akad. Wet., Afd. Natuurk., ser. 2, 48(2): 176 (1952), as *pentander*; Hepper in F.W.T.A., ed. 2, 2: 213 (1963); F.P.U., ed. 2: 160 (1972); U.K.W.F.: 401 (1974). Type: Ghana, *Thonning* (C, holo., S, iso.)

Usually annual or short-lived perennial herb, with weak decumbent, procumbent or rarely suberect, often single stems 4–90 cm. long. Leaf-blades linear-lanceolate, elliptic-lanceolate or elliptic, (1·3–)1·5–8 cm. long, (0·3–)0·45–2·5 cm. wide, subacute to sharply acute at the apex, rounded to cuneate at the base, sessile; stipule-base 0·5–3(–5) mm. long, fimbriae 0·5–3 mm. long. Inflorescences 0·7–9 cm. long; peduncles (0·3–)0·6–6·5 cm. long; pedicels 0·2–1·5 cm. long, spreading. Calyx-tube 0·5–1·5 mm. long; limb-tube 0·25–1 mm. long; lobes 0·5–1(–1·5) mm. long. Corolla white, pink, pale to deeper blue or pale mauve, very often with the limb blue and the tube pale yellow; tube (1·5–)2–4·5 mm. long, narrowly to widely funnel-shaped, widest in heterostylous forms, (1–)2–4 mm. wide; lobes ovate-triangular, 1–3 mm. long, 0·8–2 mm. wide, the throat hairs extending up over the inside. Style in long-styled, short-styled and equal-styled forms 2–3·5 mm., 1–1·5 mm. and 1 mm. long respectively; stigma-lobes in short-styled forms short, ± 0·8–1·5 mm. long, in long-styled forms longer, 1·5–2 mm. long. Capsule (2–)3–4 mm.

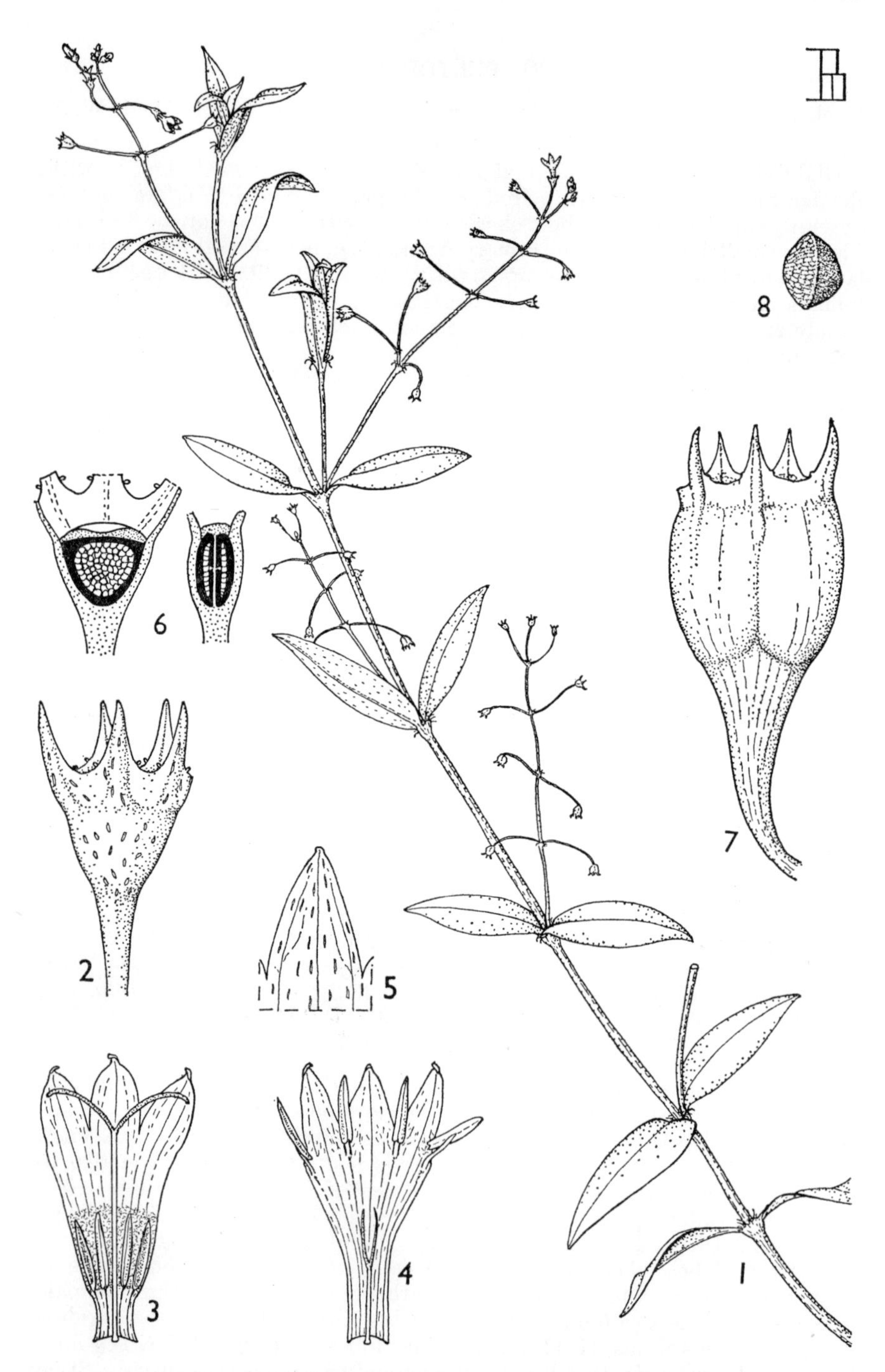

FIG. 38. *PENTODON PENTANDRUS* var. *MINOR*—**1,** habit, × $\frac{2}{3}$; **2,** calyx, × 10; **3,** longitudinal section of long-styled flower, × 6; **4,** same of short-styled flower, × 6; **5,** corolla-lobe, outer side, × 8; **6,** longitudinal sections of ovary, × 10; **7,** capsule, × 6; **8,** seed, × 20. 1, from *Milne-Redhead & Taylor* 8020; 2, 4–6, from *Rawlins*; 3, from *Tanner* 2905; 7, 8, from *Tanner* 2630. Drawn by Diane Bridson.

long, (2–)2·8–3·5 mm. wide; beak 0·5 mm. tall, crowned by the persistent calyx-lobes. Seeds black, angular, 0·3 mm. long.

var. **pentandrus**; Bremek. in Verh. K. Nederl. Akad. Wet., Afd. Natuurk., ser. 2, 48(2): 177 (1952)

Flowers isostylous or with style exserted and anthers included but never with flowers truly heterostylous; corolla usually small, the tube ± 2 mm. long, rather more cylindrical than funnel-shaped.

UGANDA. Bunyoro District: Bugamba, May 1943, *Purseglove* 1570!; Busoga District: 16 km. E. of Jinja, by margin of Lake Victoria, 3 Oct. 1952, *G. H. Wood* 467A!; Mengo District: Entebbe, Sept. 1944, *Maitland* 226!
KENYA. Northern Frontier Province: Mathews Range, Kichich, 23 Dec. 1958, *Newbould* 3563! & Mathews Range, 11 June 1959, *Kerfoot* 1144! & same area, Mandasion, 7 Dec. 1960, *Kerfoot* 2557!
TANZANIA. Arusha District: Ngurdoto Crater National Park, Longil, E. Lake, 24 Feb. 1966, *Greenway & Kanuri* 12388!; Ufipa District: Lake Tanganyika, Kipili, 1 Feb. 1950, *Bullock* 2381!; Rungwe District: Mbaka R., 13 Jan. 1912, *Stolz* 1828!
DISTR. **U**2–4; **K**1, ? 4, ? 5, ? 7; **T**1–4, 6, 7, 8; widespread in tropical Africa from Cape Verde Is., W. Africa (Senegal–Angola), Arabia, Sudan, Ethiopia and Somali Republic to Malawi and Zambia, Rhodesia, Botswana, South West Africa, Madagascar; also in U.S.A. (Florida & Texas), Cuba, Nicaragua and Brazil (see p. 263)
HAB. Swamp, lake and river margins, seasonally wet ground, muddy hollows, etc.; (?200–)450–2250 m.

SYN. *Hedyotis pentandra* Schumach. & Thonn. in K. Danske Vid. Selsk. Nat. Math. Afh. 3: 71 (1827)
Oldenlandia pentandra (Schumach. & Thonn.) DC., Prodr. 4: 427 (1830), *non* Retz.
O. macrophylla DC., Prodr. 4: 427 (1830); Hiern in F.T.A. 3: 63 (1877). Type: Gambia, Albreda, *Leprieur & Perrottet* (G-DC, holo.)
Pentodon abyssinicus Hochst. in Flora 27: 552 (1844). Type: Ethiopia, *Schimper** (?TUB, holo.)

var. **minor** *Bremek.* in Verh. K. Nederl. Akad. Wet., Afd. Natuurk., ser. 2, 48(2): 179 (1952). Type: South Africa, Natal, Durban, *Krauss* 332 (? TUB, holo., BM, K, MO, iso.!)

Flowers heterostylous, some with style included and anthers well-exserted and others exactly vice versa; corolla usually longer, the tube ± (2·5–)3–4 mm. long, usually distinctly funnel-shaped and often with a wide throat up to 4 mm. wide. Fig. 38.

UGANDA. Busoga District: Bulamogi, 11·2 km. W. of Kaliro, Namalemba A.L.G. Plantation, 27 Aug. 1952, *G. H. Wood* 332!
KENYA. Kwale District: Mrima road, a few km. from main Shimoni road, 5 Sept. 1957, *Verdcourt* 1887!; Kilifi District: Arabuko-Sokoke Forest, Mida, 3 Dec. 1961, *Polhill & Paulo* 890!; Tana River District: Kurawa, 23 Sept. 1961, *Polhill & Paulo* 538!
TANZANIA. Tanga Bay, 4 Nov. 1929, *Greenway* 1857!; Rufiji District: Utete, 4 Oct. 1954, *Anderson* 979!; Songea District: 11 km. W. of Songea, Ulamboni valley, 1 Jan. 1956, *Milne-Redhead & Taylor* 8020!; Zanzibar I., Zingwe Zingwe, 21 Jan. 1929, *Greenway* 1099! & Mahonda rice fields, 25 Sept. 1960, *Faulkner* 2720!; Pemba I., Madunga, 29 July 1929, *Vaughan* 417!
DISTR. **U**3; **K**7; **T**3, 6, 8; **Z**; **P**; Mozambique, Rhodesia and South Africa (Transvaal, Natal) and Swaziland, also in Seychelles**
HAB. As for var. *pentandrus* with addition of saline habitats near the sea; also in damp sand and on coral rocks; often on black soil; 0–1080***

SYN. *Pentodon decumbens* Hochst. in Flora 27: 552 (1844). Type: as for var. *minor*
Hedyotis pentamera Sond. in Fl. Cap. 3: 12 (1865). Type: as for var. *minor*

NOTE. There is no doubt that Bremekamp is correct in dividing this species into two taxa. Since var. *pentandrus* has long-styled forms certain specimens can be very difficult to name. Studies of heterostyly in populations of this species, particularly

* No number was mentioned in the original reference but the following gathering was obviously intended, Modat, *Schimper* 1750 (BM, K!, MO).
** Material mostly too poor to assign to a variety.
*** Highest altitude for a short-styled form; a long-styled form from Suji in the Pare Mts. from an altitude of 1500 m. is considered by Bremekamp to be var. *minor* but there is some doubt.

the inland ones, are needed. Around Mkomazi in Lushoto District, Tanzania, for example both forms have been recorded (var. *pentandrus*, Mkomazi, 23 Apr. 1934, *Greenway* 3981 ! and var. *minor*, Mkomazi R., 3·2 km. NE. of Lake Manka, 1 May 1953, *Drummond & Hemsley* 2348 !). Bremekamp also uses the size of the capsule as a distinguishing character but I have not found it constant.

32. LELYA

Bremek. in Verh. K. Nederl. Akad. Wet., Afd. Natuurk., ser. 2, 48(2): 181 (1952); W. H. Lewis in Ann. Missouri Bot. Gard. 52: 189 (1965)

Prostrate perennial herb. Leaves paired, small, shortly petiolate; stipules in lower part of stem with sheath produced into an undivided triangular lobe, in the upper part deeply bifid into narrowly triangular fimbriae or less often with 3 fimbriae. Flowers white, small, hermaphrodite, dimorphic, in terminal triads or sometimes solitary. Calyx-tube ovoid or ellipsoid; lobes 4, equal, oblong-elliptic or oblong-spathulate, rather thick, exceeding the tube, joined at the base to form a free tube. Corolla-tube shortly cylindrical or narrowly funnel-shaped, sparsely pilose inside; lobes 4, elliptic-oblong to oblong-lanceolate. Stamens completely exserted in short-styled forms. Ovary 2-locular, with peltate placentas and 8 to fairly numerous ovules; style stoutly filiform, minutely bifid at apex, the stigma-lobes short, subglobose or ellipsoid. Fruit capsular with bony walls, ellipsoid, produced into a solid beak, at length apically loculicidally dehiscent. Seeds few per locule, blackish, angular, pitted.

A monotypic genus occurring in widely separated areas of tropical Africa and scarcely separable from *Oldenlandia*, but apart from the bony fruits the pollen differs from all known species of that genus (see Lewis loc. cit.: 191 (1965) for details).

L. prostrata (*Good*) *W. H. Lewis* in Ann. Missouri Bot. Gard. 52: 189 (1965). Type: Angola, Cuanza Norte, Capijongo, Lucala, *Gossweiler* 7385 (BM, holo. !)

Perennial herb, with many stems 2·5–10 cm. long radiating from a woody rootstock, much branched at the base, sometimes rooting at nodes *fide Staples* 153; often forming small mats; branchlets usually spreading pubescent at first, later glabrous or all subglabrescent. Leaf-blades elliptic to almost linear, 0·5–1·5 cm. long, 1–10 mm. wide, acute, apiculate or somewhat obtuse at the apex, cuneate at the base, the margins often recurved, rather thick, frequently drying the curious yellow-green of an aluminium-accumulating plant, glabrous or scabridulous above, pubescent or glabrous beneath with 2–3 distinct lateral nerves; petiole 1–2·5 mm. long, united with the stipule sheath in lower half; stipule-sheath triangular-ovate, 1–2·5 mm. long; lobes 3–5, 0·8–3 mm. long. Pedicels 1–4 mm. long, mostly pubescent. Calyx-tube 0·4–1·8 mm. long, pubescent or glabrous, the limb 2·5–3·5 mm. long, the tubular part up to 0·3–1 mm. long, the lobes 1–2·5 mm. long, up to 1 mm. wide, acute or subobtuse, keeled. Corolla glabrous or sparsely hairy outside; tube 1·6–3 mm. long; lobes 2–5 mm. long, 1·2 mm. wide. Style in short-styled flowers 2 mm. long, in long-styled flowers 3–5·5 mm. long; stigma-lobes 0·4–0·6 mm. long. Capsule globose or ovoid, 2 mm. long, 2–2·3 mm. wide, with short spreading hairs or glabrescent, the beak 0·8 mm. long. Seeds round in outline, 0·7 mm. in diameter.

KEY TO INFRASPECIFIC VARIANTS

Stems short, the plants up to 20 cm. across:
  Stems, petioles and undersides of leaf-blades,
    corolla and calyx-tube ± pubescent . . var. **prostrata**

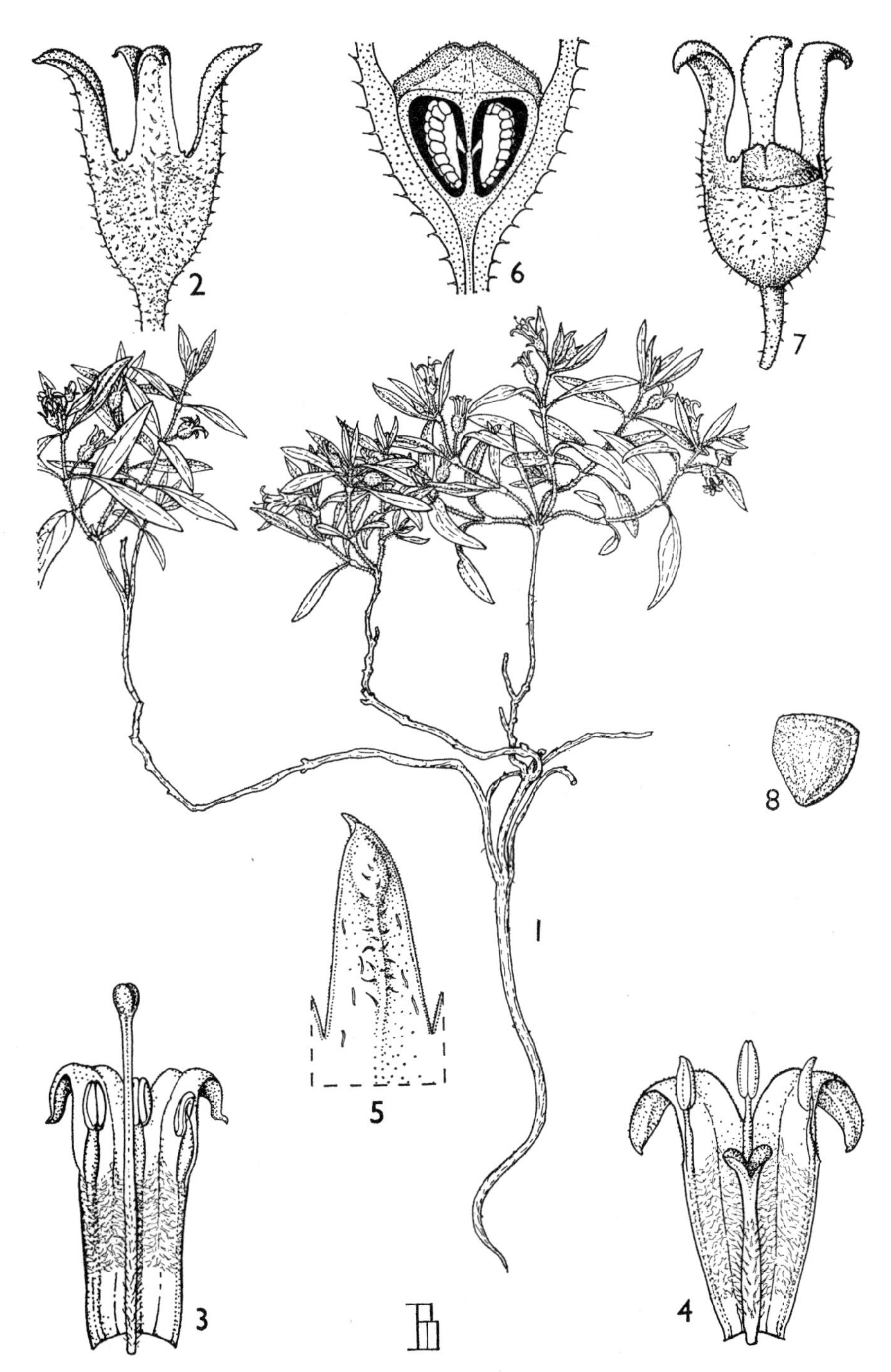

FIG. 39. *LELYA PROSTRATA* var. *PROSTRATA*—**1,** habit, × 1; **2,** calyx, × 8; **3,** longitudinal section of long-styled flower, × 8; **4,** same of short-styled flower, × 8; **5,** corolla-lobe, outer side, × 10; **6,** longitudinal section of ovary, × 14; **7,** capsule, with one calyx-lobe removed, × 6; **8,** seed, × 20. 1–3, 5–7, from *Lewis* 6067; 4, from *Robson & Angus* 221; 8, from *Salubeni* 1202. Drawn by Diane Bridson.

Stems and petioles ± glabrescent; leaf-blades, corolla and calyx-tube glabrous . . var. **angustifolia**
Stems elongated, the plants up to at least 30 cm. long . . . . var. **elongata**

var. **prostrata**

Stems and petioles densely pubescent; leaf-blades elliptic to lanceolate, pubescent beneath; corolla pubescent outside above; ovary pubescent. Fig. 39, p. 267.

TANZANIA. Iringa District: 7·2 km. NE. of John's Corner, 22 Oct. 1962, *W. H. Lewis* 6067 !; Njombe District: Njombe, Oct. 1931, *Staples* 153 ! & Lupembe road, Nov. 1928, *Haarer* 1605 !
DISTR. **T4**, 7; Nigeria, Zaire, Malawi, Zambia and Angola
HAB. Bare ground, seasonally burnt grassland; 1300–1830 m.

SYN. *Spermacoce prostrata* Good in J.B. 65, Suppl. 2: 42 (1927)
Lelya osteocarpa Bremek. in Verh. K. Nederl. Akad. Wet., Afd. Natuurk., ser. 2, 48(2): 181 (1952); Hepper, F.W.T.A., ed. 2, 2: 212 (1963). Type: N. Nigeria, Bauchi Plateau, *Lely* P96 (K, holo. !, MO, iso.)

var. **angustifolia** (*Bremek.*) *W. H. Lewis* in Ann. Missouri. Bot. Gard. 52: 189 (1965). Type: Tanzania, Iringa District, Idodi area, *Ward* 16 (K, holo. !, EA, iso. !)

Stems and petioles glabrescent; leaf-blades very narrowly elliptic to linear, glabrous; corolla and ovary glabrous

TANZANIA. Iringa District: Idodi area, Sept. 1936, *Ward* 16 ! & Mufindi, 29 Sept. 1968, *Paget-Wilkes* 173 !
DISTR. **T7**; Malawi, Zambia
HAB. Grassland; 1800 m.

SYN. *L. osteocarpa* Bremek. var. *angustifolia* Bremek. in Verh. K. Nederl. Akad. Wet., Afd. Natuurk., ser. 2, 48(2): 183 (1952)

NOTE. It is scarcely surprising that Bremekamp overlooked Good's name; the description is totally erroneous particularly with regard to the number of ovules and structure of the fruit. It is very questionable if the variety should be recognised; *Paget-Wilkes* cited above has hairy stems and many Central African specimens are intermediate.

var. **elongata** *Verdc.* in K.B. 30: 466 (1975). Type: Tanzania, Buha District, Mugunga, *Peter* 38565 (B, holo. !)

Stems elongated, the shoots over 30 cm. long. Leaf-blades glabrous. Inflorescence many-flowered.

TANZANIA. Buha District: Mugunga, 11 Mar. 1926, *Peter* 38565 !
DISTR. **T4**
HAB. Not known; 1315–1400 m.

33. OLDENLANDIA

L., Sp Pl.: 119 (1753) & Gen. Pl., ed. 5: 55 (1754); Bremek. in Verh. K. Nederl. Akad. Wet., Afd. Natuurk., ser. 2, 48(2): 183 (1952)

Annual or perennial herbs or rarely subshrubs, the stems erect or prostrate, simple or branched, or rarely cushion-herbs. Leaves opposite; stipules with 1–several fimbriae from a short base which is often adnate to the leaf-base. Flowers mostly small, hermaphrodite, heterostylous or isostylous, in terminal or axillary lax or dense inflorescences or sometimes fasciculate or solitary at the nodes. Calyx-lobes 4 (rarely more), mostly small, equal, narrowly to broadly triangular. Corolla-tube usually short, cylindrical; lobes 4; throat often hairy. Stamens enclosed or exserted. Ovary 2-locular; ovules mostly numerous on peltate placentas; style short or long, included or exserted, filiform; stigma-lobes linear to subglobose. Capsule subglobose to oblong, usually with a loculicidally dehiscent beak. Seeds mostly numerous, angular or subglobose, smooth or alveolate, often becoming viscid when moistened; testa-cells smooth to distinctly punctate or granular.

A large genus, sometimes estimated at 300 species (but probably nearer a third that number) in the tropics of both the Old and New Worlds. Many botanists consider that *Oldenlandia* L. itself (let alone the many segregate genera proposed by Bremekamp) should be merged with *Hedyotis* L., e.g. Torrey & Gray, *Hedyoteae* in Fl. N. Amer. 2(1): 37–43 (1841), Fosberg in Va. Journ. Sci. 2: 106–111 (1941) & in Castanea 19: 25–37 (1954), Lewis in Southwest. Nat. 3: 204–207 (1959) & in Rhodora 63: 216–223 (1961), Fukuoka in Tonan Ajia Kenkyu 8: 305–336 (1970), etc. Shinners in Field & Lab. 17: 166–169 (1949) and Lewis *loc. cit.* (1961) merge *Houstonia* L. (based on N. American species) with *Hedyotis*. In other papers, however, Lewis continues to use the genus *Oldenlandia* (Lewis in Grana Palynologica 5: 330–341 (1964)) and in a more extensive paper, Cytopalynological Study of African Hedyotideae (Rubiaceae), Ann. Missouri Bot. Gard. 52: 182–211 (1965), he upheld many of Bremekamp's segregate genera as well as *Oldenlandia*. If all the genera closely related to *Hedyotis* are sunk into it, then it forms an unwieldy unit covering a very wide range of structure and habit. It is true that Bremekamp's segregate African genera were based mainly on a study of the African taxa only and he himself predicted that an extension of this study on a worldwide basis would result in the necessity to erect numerous new genera. The African genera mostly have distinct facies although the technical characters are frequently trivial; no one can doubt the distinctness of genera such as *Kohautia* or *Mitrasacmopsis*, which I would certainly keep distinct even if *Oldenlandia* and *Hedyotis* were merged. For the purposes of this Flora account I have maintained nearly all Bremekamp's genera.

Bremekamp has divided the genus into a number of subgenera and a key to these will be found on pages 184–6 of his monograph. Those occurring in East Africa are listed below showing which species belong to them. Two of the subgenera are subsequent to his monograph, one being new and the other a demotion of a genus he upheld. I have not entirely followed Bremekamp's order for this account.

1. *Anotidopsis* (Hook. f.) K. Schum. (species 1–4)
2. *Orophilum* Bremek. (species 5–10)
3. *Hymenophyllum* Bremek. (species 11)
4. *Gymnosiphonium* Bremek. (species 12)
5. *Tardavelinum* Bremek. (species 13)
6. *Trichopodium* Bremek. (species 14)
7. *Octoneurum* Bremek. (species 15)
8. *Aneurum* Bremek. (species 16)
9. *Stachyanthus* Bremek. (species 17)
10. *Euryanthus* Bremek. (species 18)
11. *Eionitis* (Bremek.) Verdc. (species 19)
12. Unnamed group (species 20)
13. *Polycarpum* Bremek. (species 21–23)
14. *Cephalanthium* Bremek. (species ? 24, 25–29)
15. *Oldenlandia* (*Euoldenlandia*) (species ? 30, 31–37)

Whether the plants are heterostylous or isostylous is a useful character in this genus. The following are heterostylous, i.e. they exist in two very distinct forms, one with flowers having a short included style and anthers well-exserted, and the other with the anthers included and the style well-exserted: 3–10, 13, [14, some variants not seen in East Africa], 15, 17–29, 31 (var. *holstii*), 34 (some varieties). The following are isostylous, i.e. with the anthers and style of more or less equal lengths or at least at constant relative heights or dolichostylous with the style long but the opposite form with the anthers exserted missing: 1, 2, 11, 12, 14, 16, 30, 31 (most varieties), 32, 33, 34 (typical variety), 35–37.

Sepals 4 (to p. 278):
- Creeping perennial plant with stems producing flowering shoots bearing densely tufted congested imbricate leaves, the internodes almost suppressed; flowers crowded in the centres of the tufts, almost sessile and practically hidden (**T4**) . . . . . 20. *O. cyperoides*
- Habit quite different, the leaves not so densely tufted or, if so, then inflorescences pedunculate and obvious:

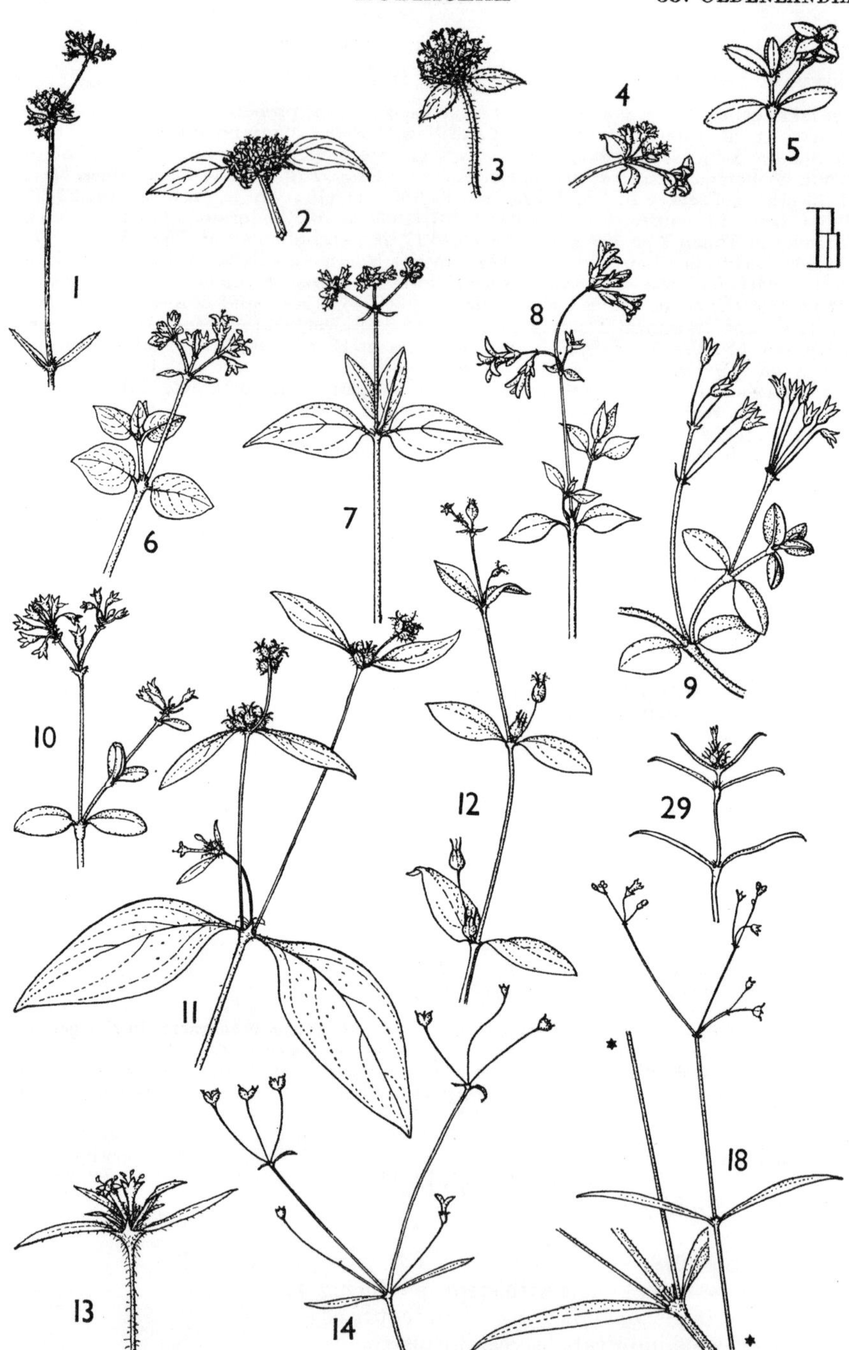

FIG. 40. *OLDENLANDIA*—portion of flowering shoot, × 1. Species numbered as in text. **1,** *O. angolensis* var. *angolensis*; **2,** *O. goreensis* var. *goreensis*; **3,** *O. bullockii*; **4,** *O. oxycoccoides*; **5,** *O. monanthos*; **6,** *O. friesiorum*; **7,** *O. johnstonii* subsp. *A*; **8,** *O. rupicola*; **9,** *O. hockii*; **10,** *O. aegialodes*; **11,** *O. pellucida* var. *pellucida*; **12,** *O. manyoniensis*; **29,** *O. nematocaulis* var. *nematocaulis*; **13,** *O. borrerioides*; **14,** *O. rosulata* var. *parviflora*; **18,** *O. patula*. 1, from *Maitland* 814; 2, from *Dawkins* 852; 3, from *Tanner* 1022; 4, from *Renvoize* 1964; 5, from *D. Davies* 24; 6, from *Batty* 791; 7, from *Faulkner* 4159; 8, from *Archbold* 691; 9, from *Harley* 9486; 10, from *Greenway* 5216; 11, from *Milne-Redhead & Taylor* 9276; 12, from *Greenway & Polhill* 11585; 13, from *Polhill & Paulo* 601; 14, from *Milne-Redhead & Taylor* 10083; 18, from *Migeod* 129; 29, from *St. Clair-Thompson* 787. Drawn by Diane Bridson.

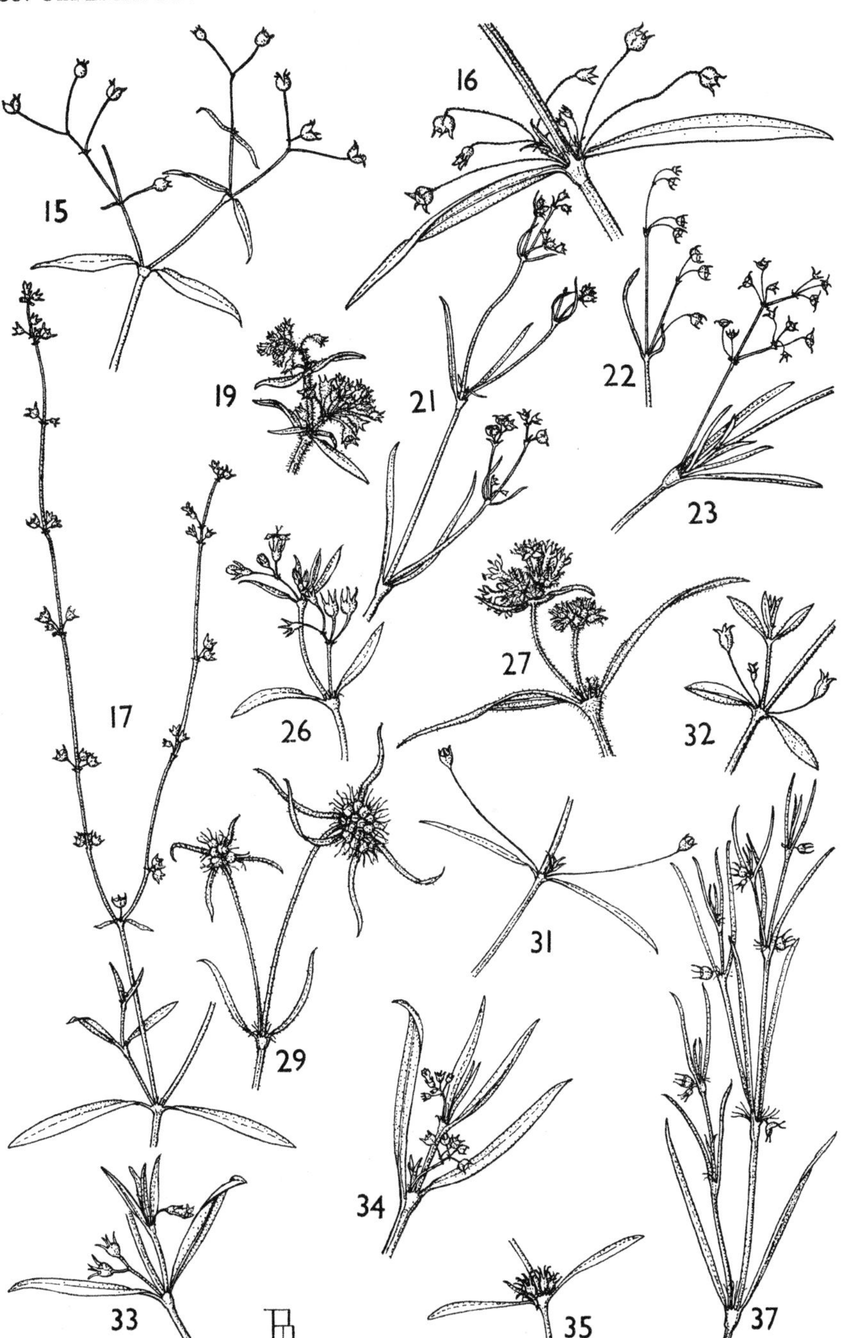

FIG. 41. *OLDENLANDIA*—portion of flowering shoot, × 1. Species numbered as in text. **15,** *O. affinis* subsp. *fugax*; **16,** *O. lancifolia* var. *scabridula*; **17,** *O. flosculosa* var. *flosculosa*; **19,** *O. richardsonioides* var. *richardsonioides*; **21,** *O. taborensis*; **22,** *O. duemmeri*; **23,** *O. microcarpa*; **26,** *O. scopulorum*; **27,** *O. wiedemannii* var. *wiedemannii*; **29,** *O. nematocaulis* var. *nematocaulis*; **31,** *O. herbacea* var. *herbacea*; **32,** *O. pumila*; **33,** *O. corymbosa* var. *linearis*; **34,** *O. fastigiata* var. *somala*; **35,** *O. capensis* var. *pleiosepala*; **37,** *O. acicularis*. 15, from *Lye* 2822; 16, from *Milne-Redhead & Taylor* 8069; 17, from *Vaughan* 1690; 19, from *Greenway & Rawlins* 9289; 21, from *Grant*; 22, from *Lye* 2884; 23, from *Richards* 13401; 26, from *Harvey* 170; 27, from *Haarer* 1247; 29, from *Tanner* 4973; 31, from *Bogdan* 1208; 32, from *Haarer* 1915; 33, from *Tweedie* 757; 34, from *Mathew* 6353; 35, from *Verdcourt* 3445; 37, from *Napier* 221. Drawn by Diane Bridson.

Inflorescences terminal or terminal and axillary, capitate to very lax (to p. 276):
Leaves broader, ovate to broadly lanceolate or elliptic, sometimes the lateral nerves very clearly evident (to p. 274):
Flowers in dense terminal and sometimes axillary clusters or, if inflorescences compound, then at least the pedicels very short, 0–1·5(–3) mm. long:
Plant not erect or if ascending then leaves and corolla not as below; inflorescences all sessile or pedunculate:
Inflorescence consisting of a panicle of small heads (fig. 40/7) . 7. *O. johnstonii*
Inflorescence simply capitate:
Corolla-tube shortly cylindrical, 0·3 mm. long; plant glabrescent or rarely pubescent . 2. *O. goreensis*
Corolla-tube narrowly funnel-shaped, 4–5 mm. long; plant densely hairy . . . 13. *O. borrerioides*
Plant erect; leaves elliptic to elliptic-lanceolate with very evident lateral nerves; corolla-tube narrowly cylindrical, 3·2–4·5 mm. long; at least terminal inflorescence pedunculate . . 11. *O. pellucida*
Flowers in rather dense to extremely lax terminal and lateral inflorescences or at least the pedicels well dedeveloped and clearly evident:
Leaves distinctly fleshy; straggling plant of the sea-shore with flowers in fairly lax terminal cymes . . . . . 10. *O. aegialodes*
Leaves not distinctly fleshy; not a plant of the sea-shore:
Inflorescence very characteristic, consisting mostly of sessile central flowers with pedicellate lateral flowers or at some lower nodes ± all the flowers sessile; plant an erect annual with narrowly cylindrical corolla (fig. 42, p. 288) . . 12. *O. manyoniensis*
Inflorescence not of this characteristic form; plant prostrate:
Plants mostly densely hairy; flowers shortly pedicellate in clusters subsessile in the axils, i.e. peduncle scarcely developed . . . 3. *O. bullockii*
Plants glabrescent or only minutely pubescent or, if

hairy, then peduncles well developed:
Flowers white or lilac, in less spreading inflorescences; leaves more ovate; capsule not so perfectly globose:
Leaves distinctly larger, 0·8–3·8 × 0·45–1·9 cm., with lateral nerves very clearly visible . . 6. *O. friesiorum*
Leaves smaller (at least in East African material), 0·3–1·2(–3·7) × 0·2–0·9 (–1·9) cm., the lateral nerves visible or invisible:
Leaves drying a pale yellow-green colour, small, 4–6·5 × 2–5 mm.; inflorescences 1–5-flowered, the pedicels 1–2(–14) mm. (fig. 40/4) . . 4. *O. oxycoccoides*
Leaves not drying a pale yellow-green colour, 0·3–1·2(–3·7) × 0·3–0·9(–1·9) cm.; inflorescence 3–several-flowered:
Leaves not fleshy, the lateral nerves invisible or often clearly visible; pedicels 0·5–2·5 (–7) mm. long, but mostly very short in East African material; peduncles 0–2 cm. long; plants usually over 10 cm. long (fig. 40/8) . . 8. *O. rupicola*
Leaves rather fleshy, the lateral nerves not visible; pedicels 0·2–1 cm. long, mostly long; peduncle up to 2·5 cm. long; plants small, usually no more than 10 cm. long (fig. 40/9) . . 9. *O. hockii*
Flowers bright blue in lax divaricate paniculate inflorescences (fig. 41/15);

leaves narrowly lanceolate, tapering acuminate; capsules perfectly globose, glabrous, with the beak ± obsolete . . . 15. *O. affinis*

Leaves narrower, linear to linear-lanceolate or narrowly elliptic, the lateral nerves mostly not visible:

Inflorescence capitate, the flowers with the pedicels scarcely evident:

More robust herbs:

Calyx-lobes shorter, 1–1·2 mm. long; plant drying a yellowish green colour and growing in boggy places, glabrous save for one variant . . . 1. *O. angolensis*

Calyx-lobes longer, 1·5–3 mm. long; plant not drying yellowish green; not growing in marshy places, mostly pubescent:

Mostly perennial tussock herbs; calyx-lobes triangular-lanceolate:

Calyx-lobes without very obvious hyaline margins . 27. *O. wiedemannii*

Calyx-lobes with very obvious hyaline margins . . 28. *O. ichthyoderma*

Slender annual herb; calyx-lobes linear-lanceolate . 29. *O. nematocaulis* (robust forms)

Minute erect filiform unbranched herb 2–6 cm. tall; corolla narrowly cylindrical, 2·6 mm. long . . 29. *O. nematocaulis*

Inflorescences laxer:

Inflorescences elongate, the flowers arranged along it in spaced fascicles of 2–3:

Inflorescence (fig. 41/17) spike-like, the pedicel 0–4 mm. long; capsule shortly beaked . . 17. *O. flosculosa*

Inflorescence (fig. 40/18) raceme-like, the pedicels 2–8 mm. long; capsule with practically no beak . . . . 18. *O. patula*

Inflorescence not of this structure:

Plant of sand-dunes with densely hairy stems and somewhat succulent leaves . . . 19. *O. richardsonioides*

Not growing on sand-dunes:

Leaves distinctly lanceolate, 1·2–6·5 × 0·1–1·6 cm.; plant a lax diffuse straggler with divaricate panicles of blue flowers (fig. 41/15); capsule exactly globose . . 15. *O. affinis*

Leaves mostly smaller; plant mostly erect or ascending:

Calyx-lobes broadly triangular, joined at the base so that there is a short tubular limb above the calyx-tube; sinuses very narrow; leaf-blades up to 3 cm. long; corolla-lobes 3–3·5 mm. long; perennial herb 24. *O. sp. A*

Calyx-lobes narrowly triangular to lanceolate, not joined at the base, the broad sinuses reaching to the top of the calyx-tube or if broader with narrow sinuses then plant a delicate annual:

Slender annual herbs with lax to very lax inflorescences:

Stems less leafy, the leaves narrower, up to 2·2 mm. wide; basal leaves rosulate, spathulate or elliptic, often persistent and very different from the stem leaves; stipule-sheath very short, ± 0·3 mm. long; pedicels very slender, 0·4–3·8 mm. long; flowers not heterostylous in East African variants (but see note on p. 290) . 14. *O. rosulata*

Stems more leafy, the leaves more elliptic, up to 3·5 mm. wide; basal leaves not very different in shape; stipule-sheath 1·5–2 mm. long, well dedeveloped; pedicels 1–8 mm. long; flowers distinctly heterostylous . . . 25. *O. uvinsae*

Mostly robuster plants with more condensed inflorescences; lowermost leaves never rosulate:

Calyx-lobes mostly short, 0·3–1·1 mm. long or, if not, then calyx often more obconic or hemispherical:

Inflorescence (fig. 41/21) more condensed, the pedicels short, 1(–2, rarely–4 in fruit) mm. long; testa cells very strongly punctate . 21. *O. taborensis*

Inflorescence (fig. 41/22, 23) laxer, the pedicels longer, 1–9 mm. long:

Testa-cells strongly punctate; Uganda . . . 22. *O. duemmeri*

Testa-cells smooth; Tanzania . . 23. *O. microcarpa*

Calyx-lobes (0·5–)1–2 mm. long; pedicels 1–7 (–16) mm. long; testa cells smooth to finely punctate; usually a tufted perennial herb; capsule oblong 26. *O. scopulorum**

Inflorescences axillary, often the flowers solitary or several at the nodes or in cymes or fascicles; sometimes the flowers appear falsely terminal due to the shortness of the young terminal shoots and their overtopping by inflorescences from the upper axils or rarely to the suppression of some of the upper leaves:

Flowers 1–several at the nodes, each solitary on its peduncle or only rarely 2–3 per peduncle due to the suppression of leaves near the apex of the shoot:

Leaves broader, ovate, elliptic, or broadly lanceolate:

Flowers heterostylous; leaves mostly elliptic-oblong, the lateral nerves usually ± visible; procumbent herb with very long shoots; capsule obconic-campanulate; common . . . . . 5. *O. monanthos*

Flowers isostylous; leaves narrowly elliptic, the lateral nerves ± invisible; capsule distinctly elongate; rare introduction . 32. *O. pumila*

Leaves narrower, linear, linear-lanceolate or narrowly lanceolate:

* A few specimens of *O. wiedemannii* with longer pedicels than usual will key here, but can usually be recognised by the longer leaves, absence of axillary inflorescences at lower nodes and denser indumentum.

Capsule distinctly broader than long, rather saccate at the base; spreading herb with long shoots . 16. *O. lancifolia*

Capsule ± globose or oblong, not at all saccate at the base:

Pedicels longer, (0·2–)0·4–2·6 cm. long, the flowers never appearing ± sessile when young:

Capsule markedly beaked; corolla narrowly tubular, 0·2–1·1 cm. long (according to variety); flowers isostylous or heterostylous; seeds developing a distinct layer of mucilage when soaked in water*; testa-cells very strongly punctate; plant usually erect . . 31. *O. herbacea*

Capsule much less beaked; corolla shortly cylindrical, 0·6–1 mm. long; flowers isostylous; testa-cells not strongly punctate; plant mostly prostrate . . . 33. *O. corymbosa* var. *caespitosa*

Pedicels mostly very short, 1–2(–4) mm. long, the flowers sometimes ± sessile at first:

Plant ± prostrate:

Calyx-lobes 0·6–1·1 mm. long . 35. *O. capensis*

Calyx-lobes 1·5 mm. long, often becoming 2·5–3 mm. long 36. *O. geminiflora*

Plant strictly erect; rare plant of rock pools in **K3** . . 37. *O. acicularis*

Flowers in 2–several-flowered inflorescences or in fascicles, at least at some nodes; at other nodes the flowers may be solitary:

Leaves elliptic-oblong, under 3·5 cm. long; flowers mostly solitary, but rarely paired or in shortly pedunculate cymes of several flowers; flowers heterostylous . . . 5. *O. monanthos*

Leaves linear to linear-lanceolate or rarely wider and then up to 7 cm. long:

At least some stipules with 2 narrow often curved calliper-like lobes; tiny erect annual 5–10 cm. tall, with erect leaves; whole plant shortly pubescent; flowers isostylous (**T**4) . . . . . 30. *O. forcipitistipula*

Stipules not of this shape; plants

* These characters will distinguish this species from *O. rosulata*, which has seeds which do not become mucilaginous when wet and smooth testa-cells; in habit the two are often very similar.

glabrous or at most slightly scabridulous or papillate:
Flowers (fig. 41/33) in pedunculate 2–several-flowered cymes, sometimes mixed with solitary flowers; flowers isostylous . 33. *O. corymbosa*
Flowers (fig. 41/34) in sessile to very shortly pedunculate fascicles, often 10–20 flowers at a node including those borne on short undeveloped axillary shoots or, if peduncles well developed, *then* flowers heterostylous . 34. *O. fastigiata*
Sepals 5–8 . . . . 35. *O. capensis* var. *pleiosepala*

1. **O. angolensis** *K. Schum.* in E.J. 23: 412 (1897); Bremek. in Verh. K. Nederl. Akad. Wet., Afd. Natuurk., ser. 2, 48(2): 195 (1952). Type: Angola, Malange, *Mechow* 379 (B, holo.†, K, iso.!)

Short-lived perennial (or ? sometimes annual) herb 15–60 cm. tall; stems erect or suberect, mostly sparsely branched, slender, glabrous (save at the nodes) or rarely hairy. Leaf-blades narrowly elliptic to linear-lanceolate, (0·3–)1–2 cm. long, 1–4·5 mm. wide, subacute at the apex, cuneate at the base, rather rigid and held erect, mostly drying a distinct yellowish green and with surface finely reticulate, glabrous but margins very slightly scabridulous or rarely hairy all over; petiole obsolete or very short; stipule-sheath ± 1 mm. long, usually with a few hairs, rarely densely hairy, produced into a 2-fid lobe 1·5–3 mm. long. Inflorescences dense, subcapitate, 5–15-flowered, 5–8 mm. in diameter, terminal and also in leaf-axil below the terminal head and also sometimes on axillary side branches from terminal and lower nodes, usually sessile but peduncles up to 5 cm. long; pedicels 0·5–1·5(–3) mm. long. Flowers not dimorphic, equal-styled. Calyx-tube subglobose, 1 mm. long, glabrous to sparsely or rarely spreading hairy; lobes deltoid, ± 1–1·2 mm. long, glabrescent or with rather long white marginal hairs or rarely more densely covered with spreading hairs. Corolla white to dull mauve; tube 0·5 mm. long; lobes ovate-lanceolate, 1–1·3 mm. long. Capsule purple, subglobose, 1·5 mm. long, 2 mm. wide; beak slightly raised, 0·5 mm. long, glabrescent to spreading hairy. Seeds black, angular, 0·4–0·5 mm. long.

var. **angolensis**

Plant glabrescent or with sparse indumentum only. Fig. 40/1, p. 270.

UGANDA. Masaka District: NW. side of Lake Nabugabo, 9 Oct. 1953, *Drummond & Hemsley* 4695! & same locality, Aug. 1935, *Chandler* 1300! & Bukakata, June 1925, *Maitland* 814!
TANZANIA. Bukoba District: Bugandika, Jan. 1932, *Haarer* 2458!; Songea District: about 1·5 km. E. of Songea, 4 Feb. 1952, *Milne-Redhead & Taylor* 8486!; Tunduru District: about 1·5 km. E. of Mawese, 19 Dec. 1955, *Milne-Redhead & Taylor* 7728!
DISTR. **U4**; **T1**, 8; Cameroun, Zaire, South-central Africa, Angola
HAB. Swamps, damp depressions with sedges, seasonally flooded grassland; 330–1200 m.

var. **hirsuta** *Verdc.* in K.B. 30: 292 (1975). Type: Tanzania, Tunduru District, Mawese, *Milne-Redhead & Taylor* 7730 (K, holo.!, EA, iso.)

Plant covered with spreading hairs.

TANZANIA. Tunduru District: about 1·5 km. E. of Mawese, 19 Dec. 1955, *Milne-Redhead & Taylor* 7730!

DISTR. **T8**; known only from the above gathering
HAB. Boggy grassland; 450 m.

2. **O. goreensis** (*DC.*) *Summerh.* in K.B. 1928: 392 (1928); Bremek. in Verh. K. Nederl. Akad. Wet., Afd. Natuurk., ser. 2, 48(2): 196 (1952); Hepper in F.W.T.A., ed. 2, 2: 211 (1963); F.P.U., ed. 2: 159 (1972); U.K.W.F.: 400 (1974). Type: Senegal, Cape Verde Peninsula, near Gorée, Kounoum, *Perrottet* 484 (P, holo.)

Annual or possibly sometimes a short-lived perennial herb with prostrate, decumbent or ascending stems (7–)10–40(–90) cm. long, branched at the base, sometimes rooting at the nodes, glabrescent to sparsely hairy. Leaf-blades elliptic to ovate, 0·5–2·5(–3·3) cm. long, 0·3–1·5(–2) cm. wide, obtuse or subacute at the apex, rounded or cuneate at the base, glabrous save for the scabridulous to ciliate margins and adpressed hairs on the midvein and some nerves; petiole 0·5–2 mm. long, glabrous or ciliate; stipule-sheath 0·7–1(–2) mm. long, produced into a 2-fid lobe 1–2 mm. long, ciliate. Inflorescences terminal, sessile, 9–25-flowered, subglobose, those on reduced side branches in lower axils appearing like axillary fascicles; pedicels 1–3 mm. long, glabrous to pilose. Flowers not dimorphic, 4-merous or rarely 5–6-merous. Calyx-tube subglobose, ± 1 mm. long, glabrous to spreading pubescent; lobes narrowly triangular, 1–1·5 mm. long, ± 0·5 mm. wide, ciliate. Corolla white or less often pink or red, glabrous to pilose outside; tube 0·3 mm. long, sparsely pilose inside; lobes ovate-triangular, 0·8–1·2 mm. long, 0·7 mm. wide. Capsule subglobose, 1·5 mm. long, 2 mm. wide, glabrescent to pilose; beak slightly raised, puberulous. Seeds black, angular, 0·4 mm. long.

var. **goreensis**; Bremek. in Verh. K. Nederl. Akad. Wet., Afd. Natuurk., ser. 2, 48(2): 197 (1952)

Stems glabrous or glabrescent. Fig. 40/2, p. 270.

UGANDA. Kigezi District: Kachwekano Farm, Dec. 1951, *Purseglove* 3725!; Busoga District: Butembe Bunya, N. end of Luvia I., 15 Jan. 1953, *G. H. Wood* 641!; Mengo District: Kampala, Kings Lake, 29 Aug. 1935, *Chandler & Hancock* 13!
KENYA. Trans-Nzoia District: Kitale, 15 Aug. 1956, *Bogdan* 4213! & Sandums Bridge, Nov. 1965, *Tweedie* 3208!; Uasin Gishu District: Kipkarren, Mar. 1932, *Brodhurst Hill* 711!
TANZANIA. Bukoba District: Kalema, Aug. 1931, *Haarer* 2114!; Lushoto District: W. Usambara Mts., Jaegertal, 18 Jan. 1966, *Archbold* 662!; Mpanda District: Kasoje, 17 July 1959, *Newbould & Harley* 4419!
DISTR. **U2–4**; **K3**; **T1–4, 7, 8**; W. Africa from Senegal to Angola, Zaire, Burundi, Sudan, Ethiopia, southern tropical Africa to South Africa (Transvaal), Madagascar, Seychelles and Mascarene Is.
HAB. Swampy places, marshy grassland, mud by ponds, drained swamps, rock crevices in wet areas, etc.; 780–1920 m.

SYN. *Hedyotis goreensis* DC., Prodr. 4: 421 (1830)
[*Oldenlandia trinervia* sensu Hiern in F.T.A. 3: 63 (1877) & in Cat. Afr. Pl. Welw. 2: 449 (1898); K. Schum. in P.O.A. C: 374 (1895), *non* Retz.]

var. **trichocaula** *Bremek.* in Verh. K. Nederl. Akad. Wet., Afd. Natuurk., ser. 2, 48(2): 198 (1952). Type: Togo, Lome, *Warnecke* 239 (K, holo.!, EA, iso.!)

Stems sparsely to fairly densely spreading hairy.

KENYA. Kwale District: Shimba Hills, 21 km. SW. of Kwale, valley S. of Mwele Mdogo Forest, 7 Feb. 1953, *Drummond & Hemsley* 1162! & Shimba Hills, Lango ya Mwagandi [Longo Magandi] area, 21 Mar. 1968, *Magogo & Glover* 378!
TANZANIA. Uzaramo District: about 16 km. W. of Dar es Salaam, 30 Nov. 1955, *Milne-Redhead & Taylor* 7510! & 16 km. SW. of Dar es Salaam, Ukonga, 3 Aug. 1968, *Harris* 2091!; Tunduru District: Kipendele, 7 Nov. 1950, *Tanner*!; Zanzibar I., Mahonda, 14 Feb. 1962, *Faulkner* 3011! & without precise locality, 7 Nov. 1873, *Hildebrandt* 1002!

DISTR. **K7**; **T3**, 6, 8; **Z**; Mozambique, W. Africa
HAB. Edges of swamps, seasonal pools and water-holes; 0–375 m.

SYN. [*O. goreensis* (DC.) Summerh. var. *goreensis* sensu Bremek. in Verh. K. Nederl. Akad. Wet., Afd. Natuurk., ser. 2, 48(2): 196 (1952), quoad *Hildebrandt* 1002 et *Boivin*]

NOTE. Bremekamp mentions that in the type of his var. *trichocaula* the flowers are mainly 5-merous whereas in var. *goreensis* they are mainly 4-merous but I have been unable to confirm this for the E. African material. Since there is some geographical basis for recognising the hairy coastal variant in E. Africa I have done so. Bremekamp himself suggested *Tanner* s.n. cited above might be var. *trichocaula*. A specimen at Kew from India (Pondicherry) must have been introduced. More difficult is the position of *O. anagallis* Bremek. (in Verh. K. Nederl. Akad. Wet., Afd. Natuurk., ser. 2, 48(2): 199 (1952); type: Tanzania, Njombe District, Upper Ruhudje, Lupembe, *Schlieben* 350 (B, holo., BM, iso. !)). Bremekamp admits that this is very close to *O. goreensis* but distinguished by the much smaller leaves, absence of adpressed hairs on the costa above and 4- rather than 3-colporate pollen grains; he gives the leaf-blade size as 4–7 × 2–5 mm. My study of the BM isotype which Bremekamp did not see shows leaf-blades up to 11 × 7 mm., mainly 3-colporate pollen grains and the flowers are probably heterostylous not isostylous as stated. It seems to be intermediate between *O. bullockii* and *O. goreensis* (DC.) Summerh. var. *trichocaula* Bremek.

3. **O. bullockii** *Bremek.* in K.B. 13: 382 (1959); U.K.W.F.: 400 (1974). Type: Kenya, first and second day's march from Mumias, *Whyte* (K, holo. !)

Annual or perennial prostrate or straggling plant, with sparsely to densely hairy stems 10–30(–50) cm. long, sometimes rooting at the nodes and occasionally forming mats or cushions. Leaf-blades ovate to elliptic-ovate, 0·35–1·7 cm. long, 0·2–1·2 cm. wide, acute to rounded at the apex, cuneate to rounded at the base, always conspicuously ciliate at the margins, the surface either glabrous or the midvein slightly hairy above and the lower surface sparsely hairy; petiole 0·5–2 mm. long, glabrous to hairy; stipule-sheath ± 0·5–1 mm. long, produced into a 2-fid lobe 1·5–2 mm. long, ciliate. Inflorescences terminal or on very short axillary shoots, sessile, mostly (3–)7–15-flowered, rather laxly subglobose, up to ± 1·3 cm. wide; pedicels 1–5 mm. long, glabrous or densely pilose. Flowers heterostylous, 4-merous. Calyx-tube subglobose, 1 mm. long, glabrous or densely spreading pilose; lobes narrowly oblong-triangular to lanceolate, 1·2–2 mm. long, 0·4–0·9 mm. wide, glabrous or with margins and midvein ciliate. Corolla white, pale mauve or pink; tube 0·3–0·6 mm. long; lobes oblong-lanceolate, (0·8–)1·4–1·8 mm. long, 0·5–0·7 mm. wide, glabrous or sparsely hairy on the midnerve outside. Style in short-styled flowers 0·3–0·4 mm. long, in long-styled flowers 1·8–2 mm. long. Capsule campanulate, 1–1·5 mm. long, 1·7–2 mm. wide, glabrous to densely spreading hairy, crowned by the persistent calyx-lobes; beak slightly raised, ± 0·5 mm. tall. Seeds black, ovoid, distinctly angular, ± 0·3 mm. long. Fig. 40/3, p. 270.

UGANDA. Bunyoro District: Kiryandongo, Mar. 1942, *Purseglove* 1326!; Teso District: Serere, Feb. 1933, *Chandler* 1086!; Masaka District: Minziro [Minzito] Hill, Oct. 1925, *Maitland* 1071! & Katera–Kyebe [Kiebbe] road, 1·6 km. from Katera, 1 Oct. 1953, *Drummond & Hemsley* 4515!
KENYA. Uasin Gishu District: Kipkarren, Sept. 1931, *Brodhurst Hill* 349!; N. Kavirondo District: first and second day's march from Mumias, *Whyte*!
TANZANIA. Bukoba District: Kashambya, Oct. 1931, *Haarer* 2331!; Ufipa District: Sumbawanga–Chapota road, 12 Mar. 1957, *Richards* 8661!; just N. of Iringa, 15 July 1956, *Milne-Redhead & Taylor* 11158!
DISTR. **U2–4**; **K3**, 5; **T1**, 4–8; Mozambique, Zaire, Burundi
HAB. Seasonally dry swamps, damp grassland, irrigation furrows, seepage zones, wet ditches, etc.; 900–1800 m.

SYN. *O. verticillata* Bremek. var. *trichocarpa* Bremek. in Verh. K. Nederl. Akad. Wet., Afd. Natuurk., ser. 2, 48(2): 200 (1952). Type: Tanzania, Bukoba District, Nyakato, *Haarer* 2060 (K, holo. !, EA, iso. !)

O. verticillata Bremek. in Verh. K. Nederl. Akad. Wet., Afd. Natuurk., ser. 2, 48(2): 199 (1952), *non* L., *nom. illegit.* Type: as for *O. bullockii* Bremek.

NOTE. The species is divided into two varieties by Bremekamp the typical one being represented only by the Whyte specimen. This actually is not quite so glabrous as he describes and several sheets from Teso District of Uganda show variation in the indumentum of the pedicel, capsules, etc. I have therefore not followed him.

4. **O. oxycoccoides** *Bremek.* in K.B. 13: 382 (1959). Type: Tanzania, Iringa District, Ngwazi [Nguazi], *Carmichael* 359 (EA, holo.!, K, iso.!)

Prostrate short-lived perennial mat-forming herb, with ± glabrous stems (2–)10–30 cm. long, rooting at the nodes. Leaf-blades ovate, 4–6·5 mm. long, 2–5 mm. wide, subacute to rounded at the apex, rounded at the base, often rather thick, shortly ciliate around the margins but otherwise glabrous; petiole 0·5–1 mm. long; stipule-sheath and lobe 0·4–1 mm. long, produced into 2 fimbriae 0·2–0·5 mm. long, ciliate. Flowers heterostylous, 1–5, terminal or axillary, in fascicles without a common peduncle; pedicels 1–2 mm. lengthening to 7(–14) mm. after flowering. Calyx-tube obconic, 0·8 mm. long, glabrous or sparsely pubescent; lobes oblong-triangular, 1 mm. long, 0·5 mm. wide. Corolla white with pinkish-mauve limb or reddish green or pink; tube 0·2 mm. long; lobes triangular, 1·5 mm. long, 1 mm. wide, glabrous. Style in short-styled flowers 0·1 mm. long, in long-styled flowers 1·2 mm. long; stigma-lobes subglobose, 0·1–0·2 mm. in diameter. Capsule 1 mm. tall, 1·7 mm. wide, ± glabrous, not beaked. Seeds dark brown. Fig. 40/4, p. 270.

TANZANIA. Iringa District: Johns Corner, 12 Mar. 1962, *Polhill & Paulo* 1729! & Mufindi, around Lake Ngwazi, 7 May 1968, *Renvoize & Carr* 1964! & Sao Hill, Mar. 1959, *Watermeyer* 47!
DISTR. **T7**; not known elsewhere
HAB. Marshes and tracks on black clay soil; 1800–1860 m.

SYN. [*O. rupicola* (Sond.) O. Kuntze var. *parvifolia* sensu Bremek. in Verh. K. Nederl. Akad. Wet., Afd. Natuurk., ser. 2, 48(2): 209 (1952), quoad *Geilinger* 2037 *non* Bremek. sensu stricto]

NOTE. This species is closely related to *O. bullockii* Bremek., but differs in its smaller blunter leaves, few-flowered inflorescences, non-beaked capsule and glabrous stems. *Richards* 17672 (Njombe District, Njombe–Songea road, near Kifanya, 28 Feb. 1963) with pilose stems is intermediate between the two. Watermeyer reports it as " a nuisance " in the garden.

5. **O. monanthos** (*A. Rich.*) *Hiern* in F.T.A. 3: 60 (1877); Engl., Hochgebirgsfl. Trop. Afr.: 396 (1892); K. Schum. in P.O.A. C: 374 (1895); Bremek. in Verh. K. Nederl. Akad. Wet., Afd. Natuurk., ser. 2, 48 (2): 199 (1952); U.K.W.F.: 400, fig. (1974). Type: Ethiopia, Enschedcap, *Schimper* 1370 (P, holo., K!, MO, iso.)

Perennial prostrate scrambling or suberect herb (2–)7–60 cm. long, the stems sometimes rooting at the nodes, glabrous to sparsely spreading hairy. Leaf-blades elliptic to elliptic-oblong or elliptic-lanceolate, 0·5–3·5 cm. long, 2–9(–12) mm. wide, acute at the apex, cuneate or somewhat rounded at the base, rather thick and often drying a dark colour above, the margins and sometimes the midvein scabrid but rest of the blade glabrous; petiole 1·2–3 mm. long, sometimes adnate to the stipule-sheath for its entire length; stipule-sheath 0·7–3 mm. long, produced into a triangular lobe bearing 4–6 fimbriae 2–2·5(–4) mm. long. Flowers mostly solitary at the nodes, sometimes paired or rarely in fascicles, 4-merous, heterostylous; peduncle 0–10(–16) mm. long; pedicels 0·3–2·6 cm. long, glabrous. Calyx-tube campanulate, 1–1·5 mm. long, glabrous; lobes ovate-triangular, 1–2 mm. long, 0·8–1·5 mm. wide, often ciliolate or ciliate. Corolla white, bluish, pink or violet; tube 1–3(–5) mm. long, the throat densely hairy; lobes oblong-lanceolate, 1·7–3·5(–4) mm. long, 1–1·8(–2) mm. wide, hairy at the base. Style 1·5 mm. long in short-styled flowers, 2–3·2 mm. long in long-styled flowers; stigma-lobes 0·3–1 mm. long. Capsule obconic-campanulate, 2 mm. long, 2·5 mm. wide, glabrous, the beak scarcely raised. Seeds dark brown, ellipsoid, 0·7 mm. long, flattened on one side, conically raised on the other, strongly reticulate. Fig. 40/5, p. 270.

UGANDA. Karamoja District: Mt. Debasien, Zebiel, 9 Jan. 1937, *A. S. Thomas* 2218!; Mbale District: Mt. Elgon, Butandiga, Jan. 1918, *Dummer* 3684! & Mt. Elgon, Bugisu, 22 Mar. 1951, *G. H. Wood* 114!

KENYA. Trans-Nzoia District: NE. Elgon, May 1948, *Tweedie* 712!; Nairobi, *Napier* 473!; Londiani, 2 Nov. 1951, *D. Davis* 24!

TANZANIA. Masai District: Ngorongoro Rest Camp, 3 Apr. 1941, *Bally* 2251!; Arusha District: Ngurdoto Crater, SE. of Leopard Point, 27 Oct. 1965, *Greenway & Kanuri* 12214!; Moshi District: Kilimanjaro, 19 Jan. 1934, *Schlieben* 4584!

DISTR. **U**1, 3; **K**1–6; **T**2; Ethiopia

HAB. Upland rain-forest and surrounding woodland, bushland and grassland; 1350–2800(–3500) m.

SYN. *Hedyotis monanthos* A. Rich. in Tent. Fl. Abyss. 1: 359 (1847)
Oldenlandia violacea K. Schum. in P.O.A. C: 374 (1895). Type: Tanzania, Kilimanjaro, Marangu, *Volkens* 848 (B, holo., BM, K, iso.!)
O. roseiflora K. Schum. & K. Krause in E.J. 39: 516 (1907). Type: Tanzania, Arusha District, Ngongongare [Engongo-Engare] and Momela [Njoro Lkatenda], *Uhlig* 452 (B, holo., EA, iso.!)

6. **O. friesiorum** *Bremek.* in Verh. K. Nederl. Akad. Wet., Afd. Natuurk., ser. 2, 48 (2): 204 (1952); U.K.W.F.: 400 (1974). Type: Kenya, Meru District, Churi R., *R.E. & T. C. E. Fries* 1848 (S, holo., K, iso.!)

Perennial procumbent carpet-forming herb with stems (6–)15–30 cm. long, sometimes rooting at the nodes, glabrous. Leaf-blades ovate or oblong-ovate, 0·8–3·8 cm. long, 0·45–1·9 cm. wide, acute or very slightly acuminate at the apex, rounded at the base, with short dense flat adpressed hairs on the margins and for a narrow belt parallel to the margin on upper surface, but otherwise glabrous save for a few hairs on the midvein above; petiole 1–4 mm. long, glabrous; stipule-sheath 0·5–2 mm. long, pubescent or glabrous, produced into a triangular lobe up to 2 mm. long bearing 5–7 filiform fimbriae 1–3 mm. long. Flowers heterostylous, in few-flowered terminal or axillary cymes, the cymes solitary or in threes; peduncles 0·15–1·4 cm. long or apparent peduncles 3 cm. long, glabrous; pedicels 1–3 mm. long, glabrous; bracts at junction hairy. Calyx-tube subglobose, 1 mm. long, glabrous; lobes narrowly triangular or oblong-lanceolate, 1·8–2 mm. long, 0·5–0·9 mm. wide, glabrous or ciliate. Corolla white, pale lilac, blue or pink; tube 2 mm. long, the throat densely hairy; lobes oblong, 2–3 mm. long, 0·8–1·2 mm. wide, hairy at the base. Style glabrous or puberulous, 1·2 mm. long in short-styled flowers, 3–4·5 mm. long in long-styled flowers; stigma-lobes 0·4–0·8 mm. long. Capsule subglobose or depressed subglobose, rather thick-walled, (2–)2·5–3 mm. tall, (2–)2·5–3·2 mm. wide, the beak scarcely raised, very rounded. Seeds black, ellipsoid, 0·8–0·9 mm. long, 0·6–0·7 mm. wide, 0·4–0·5 mm. tall, ± flattened on one side, conically raised on the other, strongly reticulate. Fig. 40/6, p. 270.

KENYA. Fort Hall District: Kimakia Forest Reserve, 12 Feb. 1965, *Gillett* 16642! & same locality, 11 Dec. 1960, *Lucas, Polhill & Verdcourt* 5!; Meru District: NW. Mt. Kenya, Marimba Forest, 14 Oct. 1960, *Verdcourt & Polhill* 2991!

TANZANIA. Mbeya District: Poroto Mts., road to Kitulo [Elton] Plateau, 18 Jan. 1961, *Richards* 13943!; Rungwe Mt., 30 Nov. 1958, *Napper* 1193!; Njombe District: Ukinga Mts., Madehani, 3 Dec. 1913, *Stolz* 2338!

DISTR. **K**4; **T**7; not known elsewhere

HAB. Upland evergreen forest clearings and edges, bushland, grassland, roadsides and gravel tips, etc.; 1800–2550 m.

7. **O. johnstonii** (*Oliv.*) *Engl.*, Hochgebirgsfl. Trop. Afr.: 397 (1892) & in P.O.A. C: 375 (1895); Bremek. in Verh. K. Nederl. Akad. Wet., Afd. Natuurk., ser. 2, 48 (2): 205 (1952); U.K.W.F.: 400 (1974). Type: Tanzania, Kilimanjaro, *Johnston* (K, holo.!)

Perennial usually completely prostrate herb, with radiating stems 0·2–1 m. long from a central fibrous rootstock and often rooting at the nodes; young branches sometimes with lines of fine short hair particularly at nodes, but otherwise practically glabrous. Leaf-blades ovate or elliptic to elliptic-lanceolate or lanceolate, 1·5–6 cm. long, 0·5–1·8(–2·2) cm. wide, acute at the apex, rounded to cuneate at the base, sparsely very shortly scabrid-puberulous on both surfaces or only on the nerves and margin; petiole 2–5 mm. long, scabrid pubescent; stipule-sheath hairy, 1–1·5 mm. long, produced into a triangular lobe 2–2·5 mm. long bearing 5–9 filiform fimbriae 1–5 mm. long. Flowers dimorphic, in axillary and terminal inflorescences, usually consisting of small heads usually in threes; common peduncles 0·6–3 cm. long; secondary peduncles 1–8 mm. long; pedicels obsolete to 0·5 mm. long. Calyx-tube obconic, ± 1 mm. long, glabrous to densely pubescent; lobes triangular-lanceolate, 2 mm. long, 0·5–1 mm. wide, usually keeled, slightly puberulous. Corolla white or slightly pinkish; tube 1·2–1·5 mm. long, densely hairy in the throat; lobes oblong, 2–2·5 mm. long, 0·8 mm. wide, hairy at the base. Style glabrous to shortly pilose, 1 mm. long in short-styled flowers, 3 mm. long in long-styled flowers. Capsule ovoid or oblong-subglobose, 1·5 mm. tall, 2 mm. wide, glabrous to puberulous. Seeds black, angular, ± 0·5 mm.

subsp. **johnstonii**

Ovary and capsule glabrous or nearly so.

KENYA. Nairobi District: Karura Forest, June 1961, *Verdcourt* 518! & Nairobi National Park, 1 Sept. 1962, *Lewis* 5906!; Masai District: Ngong, *McDonald* 824!
TANZANIA. Moshi District: Kilimanjaro, Marangu, 12 Dec. 1963, *Archbold* 3471!; Pare District: Kamwala, 12 Oct. 1925, *Haarer*!; Lindi District: Nyengedi, 19 Mar. 1935, *Schlieben* 6146!
DISTR. **K**4, 6; **T**2, 3, 7, 8; not known elsewhere
HAB. Dry evergreen forest, wooded grassland, grassland near water; 350–2000 m.

SYN. *Hedyotis johnstonii* Oliv. in H. Johnston, Kilimanjaro Exped. App.: 341 (1886) & in Trans. Linn. Soc., ser. 2, 2: 335 (1887)

subsp. **A**

Ovary and capsule with short spreading hairs. Fig. 40/7, p. 270.

KENYA. Kwale District: Shimba Hills, 14 Jan. 1964, *Verdcourt* 3917!; Mombasa I., below Fort Jesus, Aug. 1961, *Tweedie* 2203!; Kilifi District: 40 km. NW. of Malindi, Marafa, 18 Nov. 1961, *Polhill & Paulo* 785!
TANZANIA. Tanga district: Kange forest clearing, 29 Nov. 1955, *Faulkner* 1782! & Maweni, 7 June 1965, *Faulkner* 3552!; Pangani District: Bushiri, 11 May 1950, *Faulkner* 564!
DISTR. **K**7; **T**3; not known elsewhere
HAB. Evergreen forest, *Brachystegia* woodland edges, grassland with scattered trees and bushes, shady places on limestone outcrops; 0–450(–600) m.

NOTE. The slight difference in indumentum is scarcely adequate to distinguish a named taxon, but the two races are mentioned here because of their phytogeographic interest.

8. **O. rupicola** (*Sond.*) *O. Kuntze*, Rev. Gen. Pl. 1: 293 (1891); Bremek. in Verh. K. Nederl. Akad. Wet., Afd. Natuurk., ser. 2, 48 (2): 208 (1952). Type: South Africa, Natal, Tagoma, *Gerrard & McKen* 1364 (TCD, holo., BM, K, iso.!)

Perennial procumbent usually mat-forming herb, with glabrous to hairy much-branched stems often rooting at the nodes, 8–45 cm. long, or prostrate with erect flowering shoots (2·5–)7–10 cm. tall. Leaf-blades elliptic-ovate to ovate, 0·3–1·2(–3·7) cm. long, 3–9(–19) mm. wide, acute at the apex, rounded to cuneate at the base, sometimes rather thick, scabridulous to hairy above, glabrous beneath or sometimes slightly hairy near the margin; petiole 0·5–2(–5) mm. long, glabrous to sparsely hairy; stipule-sheath 0·5–1 mm. long, glabrous to sparsely hairy, with 5–7 lobes 1–3 mm. long. Flowers heterostylous in terminal triads or few- to several-flowered cymes, less often solitary; peduncle 0–1·5(–2) cm. long; pedicels 0·5–2·5(–7) mm. long, glabrous. Calyx-tube obconic, 1–1·5 mm. long, glabrous, hairy or with sparse to dense subglobose papillae, sometimes purple-tipped; lobes usually broadly triangular, less often ovate or lanceolate, 1–4 mm. long, 0·7–1 mm. wide, keeled, glabrous or sometimes hairy on the midvein and margins, free or connate at the base. Corolla white, pale blue, lilac or pink; tube 2–7(–10) mm. long, hairy in the throat; lobes triangular to oblong-lanceolate, 1·5–5·5 mm. long, 1·2–2 mm. wide, hairy inside. Style 2–3(–6) mm. long in short-styled forms, 5–8(–12) mm. long in long-styled forms; stigma-lobes 0·3–0·7 mm. long. Capsule subglobose, 2–2·5 mm. long, 2·2–3 mm. wide, glabrous, papillate or hairy, beak slightly raised. Seeds blackish, elliptic, ± 0·5(–0·8) mm. long, flat on one side, somewhat conically raised on the other, strongly reticulate. Fig. 40/8, p. 270.

var. **rupicola**; Bremek. in Verh. K. Nederl. Akad. Wet., Afd. Natuurk., ser. 2, 48(2): 209 (1952)

Stems glabrescent. Ovary glabrous to densely covered with short or subglobose papillae.

KENYA. Teita District: Yale Peak, 13 Sept. 1953, *Drummond & Hemsley* 4299! & Vuria Peak, 17 Apr. 1960, *Verdcourt & Polhill* 2722!

TANZANIA. Lushoto District: Magamba Forest Reserve, 6 Sept. 1958, *Semsei* 2801!; Morogoro District: Uluguru Mts., Lupanga Peak track, 16 Aug. 1951, *Greenway & Eggeling* 8623!; Iringa District: Image Mt., 28 Feb. 1962, *Polhill & Paulo* 1622!

DISTR. **K**7; **T**3, 5–7; Mozambique, Malawi, Rhodesia and South Africa (Natal and Transvaal)

HAB. Rock crevices and rock outcrops, etc. in exposed *Philippia-Myrsine* scrub, and upland forest; 1500–2220 m.

SYN. *Hedyotis rupicola* Sond. in Fl. Cap. 3: 12 (1865)

Oldenlandia oliverana K. Schum. in Engl., Hochgebirgsfl. Trop. Afr.: 397 (1892) & in P.O.A. C: 375 (1895). Type: Mozambique, Namuli Plateau, Mahua Country, *Last* (B, holo.†, K, iso.!)

O. junodii Schinz in Viert. Nat. Ges. Zürich 52: 431 (1907), *non* Schinz (1900), *nom. illegit.* Type: South Africa, Transvaal, Mt. Mamotsuri, *Junod* 2007 (Z, holo.)

O. rogersii S. Moore in J.B. 59: 229 (1921). Type: South Africa, Transvaal, Lydenburg, Graskop, *Rogers* 14857 (BM, holo.!)

O. rupicola (Sond.) O. Kuntze var. *psilogyna* Bremek. in Verh. K. Nederl. Akad. Wet., Afd. Natuurk., ser. 2, 48(2): 209 (1952). Type: as for *O. oliverana* K. Schum.

O. rupicola (Sond.) O. Kuntze var. *parvifolia* Bremek. in Verh. K. Nederl. Akad. Wet., Afd. Natuurk., ser. 2, 48(2): 209 (1952). Type: as for *O. junodii* Schinz (1907)

O. greenwayi Bremek. in Verh. K. Nederl. Akad. Wet., Afd. Natuurk., ser. 2, 48(2): 207 (1952). Type: Tanzania, Mpwapwa District, Kiboriani Mts., *Greenway* 2436 (K, holo.!, EA, iso.!)

NOTE. There is a good deal of variation in the indumentum of the ovary, etc. and certainly a great deal in the size of the leaves and corolla; I have not been able to retain Bremekamp's varieties cited in synonymy above; var. *hirtula* (Sond.) Bremek. which occurs in S. Africa has hairy stems, leaves, calyx-tubes and capsules, and often has much larger leaves and flowers. The species as a whole is very variable and extremes can appear markedly different. Some especially robust specimens, said to be scramblers or climbers with large leaves and flowers, occur in Malawi.

9. **O. hockii** *De Wild.* in F.R. 13: 108 (1914); Bremek. in Verh. K. Nederl. Akad. Wet., Afd. Natuurk., ser. 2, 48 (2): 203 (1952). Type: Zaire, Lubumbashi [Elisabethville), *Hock* (BR, holo.!)

Small presumably decumbent perennial herb 4–9 cm. long; stems very shortly scabridulous-pubescent. Leaves blue-green or purplish green, elliptic or elliptic-ovate, 0·8–1·3 cm. long, 5–8 mm. wide, subacute at the apex, ± rounded at the base, somewhat fleshy, revolute, glabrous or minutely scabrid above, glabrous beneath, discolorous, the rhaphides particularly evident on the lower surface; midvein evident but no lateral veins visible; petioles ± 1 mm. long; stipule-sheath ± 1 mm. long, divided into ± 3 fimbriae 0·8 mm. long. Flowers heterostylous, in several-flowered erect axillary ± dichasial cymes; peduncles (0·1–)1–2·5 cm. long; pedicels 0·2–1 cm. long, glabrous or minutely pubescent. Calyx-tube campanulate, 1·5 mm. long; lobes ovate, 1–1·8 mm. long, 0·5–0·8 mm. wide, mostly minutely scabrid-ciliate on the margins. Corolla white; tube 2·1–2·5 mm. long, hairy at the throat; lobes ovate, 1·5–1·8 mm. long, 1·3 mm. wide, apiculate, densely papillate inside; style hairy, 1·3 mm. long in short-styled flowers, 3 mm. long in long-styled flowers; stigma-lobes oblong or elliptic, 0·6–0·8 mm. long, densely papillate hairy. Ripe capsule not seen. Fig. 40/9, p. 270.

TANZANIA. Kigoma District: 104 km. S. of Kigoma, Mugombazi, 31 Aug. 1959, *Harley* 9486!; Mpanda District: Mahali Mts., Utahya, 6 Sept. 1958, *Newbould & Jefford* 2374!

DISTR. T4; Zaire

HAB. Recently burnt ground in *Brachystegia-Julbernardia* woodland and open bushland; 900–1200 m.

NOTE. It does not seem possible to separate the Tanzanian plant from *O. hockii*, but it does differ from the type of the latter in its longer peduncles and wider more obviously marginally ciliate calyx-lobes.

10. **O. aegialodes** *Bremek.* in K.B. 8: 438 (1953). Type: Tanzania, Mafia I., Kikutani, *Greenway* 5216* (K, holo.!, EA!, U, iso.)

Perennial decumbent herb with somewhat branched stems 15–40 cm. long, slightly woody below, ridged above and with small papillae on the ridges; perhaps slightly succulent. Leaves succulent; blades elliptic or elliptic-oblong, 0·6–1·1 cm. long, 3–6·5 mm. wide, subacute, obtuse or rounded at the apex, rounded or broadly cuneate at the base, markedly revolute when dry, glabrous save for dense papillae on the margins at the base; petioles ± 1 mm. long, densely papillate pubescent; stipule-sheath 1–1·5 mm. long, densely papillate pubescent, with 1–2 teeth 0·3–1 mm. long. Flowers in congested corymbs at the apices of the shoots, terminal and axillary, heterostylous; peduncles up to 1·7 cm. long; secondary peduncles ± 1·5 mm. long, the junctions with hyaline bracts ± 1 mm. long or ringed with stipule-sheaths devoid of leaves; pedicels 1·5–3·5 mm. long, widened at the apex and bearing hyaline ovate bracts ± 1 mm. long. Calyx-tube narrowly campanulate, ± 1·5 mm. long, glabrous; lobes ovate, 1·3 mm. long, 0·6 mm. wide, glabrous. Corolla greenish white, ± glabrous; tube 1 mm. long, densely hairy at the throat; lobes oblong-lanceolate, 1·3 mm. long. Style 2 mm. long in long-

* The citation 5236 given in the original reference is wrong.

styled flowers; stigma-lobes 0·4 mm. long. Capsule glabrous, 1·5 mm. long, 2 mm. wide; beak slightly raised, under 0·5 mm. high. Seeds few, dark brown, subglobose, 0·4 mm. long. Fig. 40/10, p. 270.

TANZANIA. Rufiji District: Mafia I., Kikutani, 3 Sept. 1937, *Greenway* 5216!
DISTR. T6; known only from the type gathering
HAB. With Labiatae and Acanthaceae on sand overlying recent coral near the seashore above high water mark

11. **O. pellucida** *Hiern* in Cat. Afr. Pl. Welw. 2: 448 (1898); Bremek. in Verh. K. Nederl. Akad. Wet., Afd. Natuurk., ser. 2, 48 (2): 212 (1952). Type: Angola, Huila, Empalanca, *Welwitsch* 5344 (LISU, holo., BM, K, iso.!)

Erect branched or unbranched annual herb 4–40 cm. tall; stems glabrous or scabrid-papillate on the ribs. Leaf-blades elliptic-lanceolate, 1·2–6·5 cm. long, 0·6–2·8 cm. wide, acute to acuminate at the apex, cuneate at the base, thin, sparsely very shortly strigose-pubescent above, later glabrescent, margin usually pubescent, glabrescent beneath save for the midvein; petiole 2–5(–10) mm. long, glabrous or sparsely ciliate; stipule-sheath 0·5–1 mm. long, usually produced into a lobe 1·5 mm. long bearing 2–5 fimbriae 1–3 mm. long. Flowers not heterostylous, arranged in small heads 4 mm. wide, terminal or axillary, the heads often in threes; peduncle 0·2–1·5 cm. long; pedicels 0–1·2 mm. long. Calyx-tube subglobose, 0·5 mm. long, densely covered with short spreading hairs; lobes lanceolate, 0·7–2 mm. long, 0·5 mm. wide, ciliolate on the margins and midvein. Corolla white or tube green and lobes pale lilac, sometimes with a double mauve spot on each; tube very narrowly cylindrical, 3·2–4·5 mm. long, 0·4–0·9 mm. wide; lobes broadly elliptic, 0·4–1·8 mm. long, 0·4–1·2 mm. wide. Style 1·7–5 mm. long; stigma-lobes 0·4–1·2 mm. long. Capsule ± biglobose, 1·5 mm. tall, 2·2 mm. wide, densely covered with short spreading bristly hairs. Seeds black, ellipsoid, ± 0·5 mm. long, strongly reticulate.

var. **pellucida**

Plants mostly weak and relatively unbranched 4–35 cm. tall; leaves very tenuous; calyx-lobes 0·7–1·2 mm. long; corolla-tube 1·5–3 mm. long; anthers usually very slightly exserted. Fig. 40/11, p. 270.

TANZANIA. Mpanda District: 19·2 km. N. of Kasogi, Belengi, 3 Aug. 1959, *Harley* 9137!; Ulanga District: 35 km. S. of Mahenge, Ngongo, 18 May 1932, *Schlieben* 2214!; Songea District: about 5 km. E. of Songea, Unangwa Hill, 22 Mar. 1956, *Milne-Redhead & Taylor* 9276!
DISTR. T4, 6, 8; Sierra Leone, Nigeria, Cameroun, Sudan, Rhodesia, Angola
HAB. Moist rock outcrops, damp soil by river in gallery forest; 900–1200 m.

SYN. *O. golungensis* Hiern in Cat. Afr. Pl. Welw. 2: 451 (1898); Bremek. in Verh. K. Nederl. Akad. Wet., Afd. Natuurk., ser. 2, 48(2): 212 (1952). Type: Angola, Golungo Alto, *Welwitsch* 3083 (BM, holo.)
O. pellucida Hiern var. *robustior* Bremek. in Verh. K. Nederl. Akad. Wet., Afd. Natuurk., ser. 2, 48(2): 205 (1952). Type: Tanzania, Ulanga District, 35 km. S. of Mahenge, Ngongo, *Schlieben* 2214 (BR, holo., K, iso.!)
[*O. echinulosa* sensu Hepper in F.W.T.A., ed. 2, 2: 211 (1963), *non* K. Schum. sensu stricto]

var. **echinulosa** (*K. Schum.*) *Verdc.* in K.B. 30: 292 (1975). Type: Malawi, without locality, *Buchanan* 498 (B, holo. †, K, iso.!)

Plants often more robust and branched, up to 40 cm. tall, but sometimes unbranched; leaves less tenuous; calyx-lobes up to 2 mm. long; corolla-tube up to 4·5 mm. long; anthers usually included.

TANZANIA. Ufipa District: Muse Gap, 2 June 1951, *Bullock* 3924!; Mbeya, 1 Mar. 1932, *R. M. Davies* 417!; Njombe District: Lupembe, Upper Ruhudje, Feb. 1931, *Schlieben* 204!

DISTR. **T4, 7, 8**; Zaire, Burundi, Sudan, Zambia, Rhodesia, Malawi and Angola
HAB. Woodland, edges of cultivation, boggy grassland; 1200–2400 m.

SYN. *O. echinulosa* K. Schum. in P.O.A. C: 375 (1895); Bremek. in Verh. K. Nederl. Akad. Wet., Afd. Natuurk., ser. 2, 48(2): 213 (1952)

NOTE. Although extremes of the two taxa mentioned above are very distinctive, intermediates occur; the floral characteristics mentioned by Bremekamp do not always work. There is a tendency for var. *pellucida* to occur in wild habitats and var. *echinulosa*, the more robust weedy variant, to occur in disturbed habitats. Most W. African material seems to be nearer var. *pellucida*.

12. **O. manyoniensis** *Bremek.* in K.B. 21: 244 (1967). Type: Tanzania, Dodoma District, S. of Itigi Station on the Chunya road, *Greenway & Polhill* 11585 (K, holo. !, EA, iso.)

Annual slender erect herb 10–30 cm. tall, not or sparsely branched; stems sharply 4-ribbed, glabrous or sparsely pubescent. Leaf-blades elliptic to ovate or ovate-lanceolate, 0·4–2·8 cm. long, 0·15–1·3 cm. wide, acute at the apex, cuneate at the base, thin, discolorous, glabrous save for the ciliate margins and base of the midnerve beneath or pubescent above and on midnerve beneath; petiole obsolete or ± 1 mm. long, ciliate; stipule-sheath small, up to about 0·8 mm. long, pubescent, the margins only obscurely fimbriate. Flowers isostylous, in terminal and axillary cymes; central flower of the cyme ± sessile, the laterals on pedicels 0·4–1·1 cm. long. Calyx-tube ovoid, 0·5 mm. long, glabrous or pubescent; lobes linear-lanceolate, 1–1·6 mm. long, ciliate. Corolla white; tube slender, widened at the throat, 3–3·5 mm. long, glabrous inside and out; lobes ovate, 1·2 mm. long, 0·8 mm. wide. Style ± 3 mm. long; stigma-lobes 0·8 mm. long. Capsule ovoid-globose, 2 mm. long, 2·3 mm. wide, mostly distinctly shortly spreading pubescent, crowned by the persistent calyx-lobes and produced into a distinct compressed beak ± 1 mm. long, so that whole fruit appears broad near the base and narrowed to the apex. Seeds ellipsoid, angular, longest dimension 0·65 mm., shortest dimension 0·35 mm. Fig. 40/12, p. 270, and 42, p. 288.

TANZANIA. Tabora, 10 May 1970, *Tabora Boys School* TAB 18!; Dodoma District: 15 km. S. of Itigi Station on the Chunya road, 16 Apr. 1964, *Greenway & Polhill* 11585!
DISTR. **T4, 5**; not known elsewhere
HAB. Deciduous thicket; ± 1300 m.

NOTE. Bremekamp has made this the type of a subgenus *Gymnosiphonium*

13. **O. borrerioides** *Verdc.* in K.B. 30: 292, fig. 7 (1975). Type: Kenya, Tana River District, 48 km. S. of Garsen, Kurawa, *Polhill & Paulo* 601 (K, holo. !, EA, iso.)

Annual erect herb 6–16 cm. tall, 4–many branched at the base, the branches often decumbent at the base, densely pubescent with white spreading hairs ± 1 mm. long, mostly arranged bifariously. Leaf-blades narrowly elliptic-lanceolate, 9 mm. long, 2–3·5 mm. wide, acute at the apex, merging at the base with the stipule sheath and the lower epidermal cells becoming larger and more evident in this hyaline area, discolorous, rather densely scabrid-pubescent above and on the margins and prominent midnerve beneath; petiole undeveloped; stipule-sheath and leaf-bases combined, forming a hairy cup up to ± 2 mm. tall, 6 mm. wide, produced into a lobe ± 2 mm. long bearing 2 fimbriae 1–2 mm. long; the margins of the leaf-bases bear hairs ± 1 mm. long, distinctly longer than on the leaf margins higher up. Flowers heterostylous, in small terminal and apparently lateral sessile few-flowered heads ± 6 mm. wide, supported by the terminal leaves; the terminal head is often overtopped by 2 lateral branchlets from the terminal leaf-axils;

FIG. 42. *OLDENLANDIA MANYONIENSIS*—**1,** habit, × 1; **2,** node with flower and fruits, × 6; **3,** flower, from above, × 8; **4,** flower, with corolla opened out to show fertile parts and ovary cut longitudinally, × 8; **5,** ovules on one placenta, × 50; **6,** capsule, × 8; **7,** half a placenta with some seeds displaced, × 12; **8,** seeds, × 16; **9,** detail of testa, × 100. All from *Greenway & Polhill* 11585. Drawn by Mary Grierson.

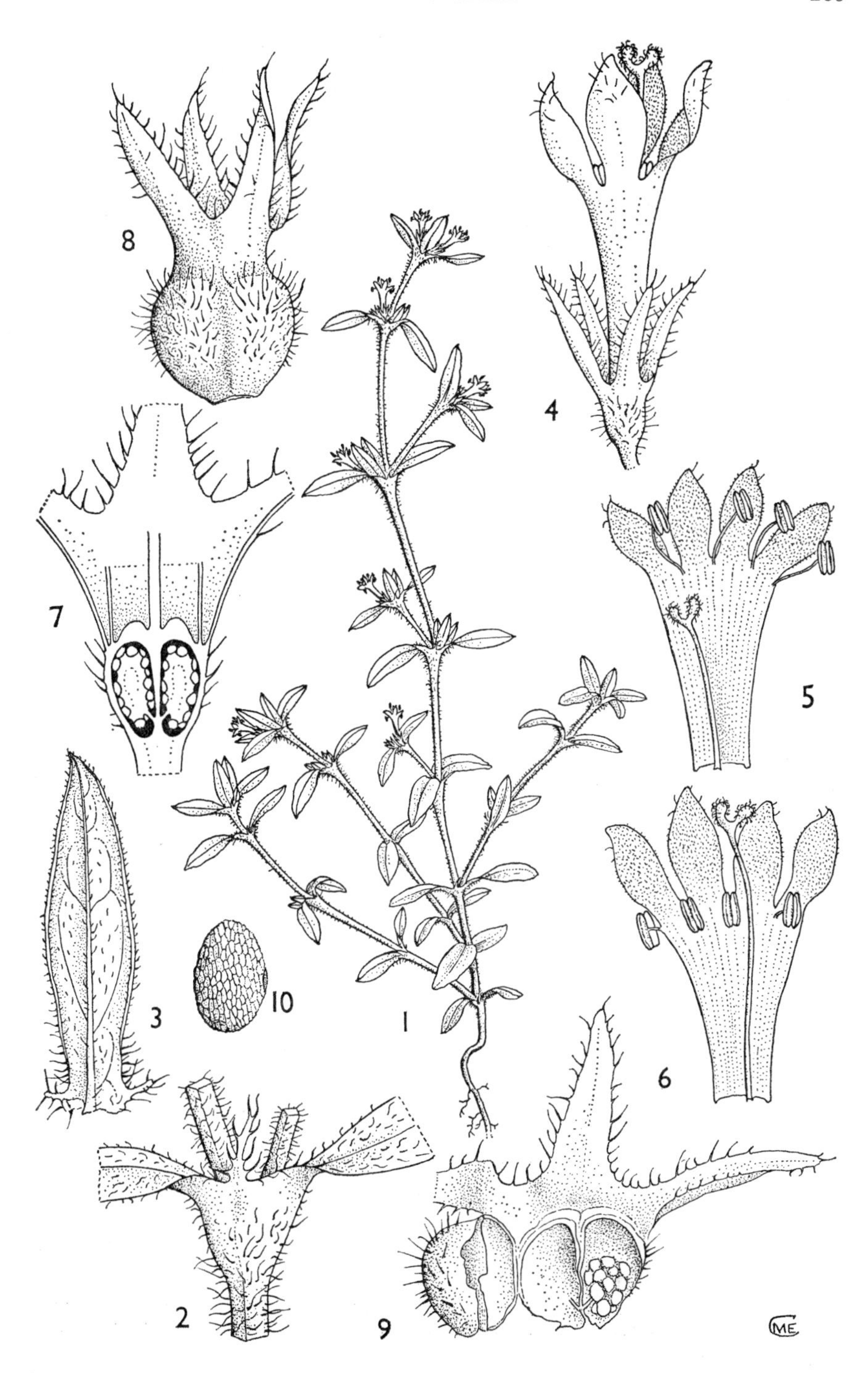

FIG. 43. *OLDENLANDIA BORRERIOIDES*—**1**, habit, × 1; **2**, node showing stipule, × 6; **3**, lower surface of leaf, × 4; **4**, long-styled flower, × 8; **5**, corolla, opened out, and style of short-styled flower, × 6; **6**, corolla, opened out, and style of long-styled flower, × 6; **7**, longitudinal section of ovary, × 18; **8**, capsule, × 10; **9**, capsule, opened out, × 10; **10**, seed, × 40. All from *Polhill & Paulo* 601. Drawn by Mrs M. E. Church.

pedicels almost obsolete. Calyx-tube obovoid, 1·5 mm. long, 1 mm. wide, covered with long spreading hairs 0·8 mm. long; lobes narrowly lanceolate, 2·5–3 mm. long, joined at the base for 0·8 mm., the margins ciliate with hairs 0·5 mm. long, tipped with 1–2 longer hairs ± 1 mm. long. Corolla white, narrowly funnel-shaped, 4–5 mm. long, glabrous inside; lobes oblong-elliptic, 2·2–3 mm. long, 0·8–1·2 mm. wide, hairy outside, papillate inside. Style 7 mm. long in long-styled flowers, 4·5 mm. long in short-styled flowers, the stigma-lobes 0·7 mm. long, 0·5 mm. wide in the former, 0·5 mm. long and wide in the latter, in both cases inclusive of the long stigmatic hairs. Unripe capsules obovoid, 1·7 mm. long, 1·8 mm. wide, hairy. Ripe seeds not seen. Fig. 40/13, p. 270, and 43, p. 289.

KENYA. Tana River District: Kurawa, 5 Oct. 1961, *Polhill & Paulo* 601!
DISTR. **K**7; not known elsewhere
HAB. Coastal bushland on black clay soil; 15 m.

14. **O. rosulata** *K. Schum.* in E.J. 23: 416 (1897); Hiern, Cat. Afr. Pl. Welw. 2: 447 (1898); Bremek. in Verh. K. Nederl. Akad. Wet., Afd. Natuurk., ser. 2, 48 (2): 225 (1952). Type: Angola, Huila, next to R. Monino, Varzens, *Welwitsch* 5320 (B, holo. †, BM, K, iso.!)

Slender erect annual herb 4–35 cm. tall, with glabrous or basally scabrid pubescent stems. Basal leaves rosulate, the blades spathulate, lanceolate or elliptic, 3–7 mm. long, 1·3–3 mm. wide, obtuse, glabrous, often eventually disappearing before maturity; stem leaves with filiform, linear or linear-lanceolate blades 0·3–3 cm. long, 0·3–2·2 mm. wide, glabrous; petioles obsolete; stipule-sheath 0·3 mm. long with 2 short teeth 0·5 mm. long. Flowers heterostylous or isostylous, terminal or axillary, either solitary or in groups of 2 or 3, in larger plants forming a paniculate inflorescence; peduncles 0·9–2·8 cm. long; pedicels very graceful, 0·4–3·8 cm. long. Calyx-tube subglobose, 0·5–0·7 mm. long, glabrous or with minute hairs; lobes broadly triangular, 0·3–0·9 mm. long. Corolla white to pale purple, sometimes with a dark spot or streaks near the throat; tube 1·2–3·6 mm. long, somewhat hairy at the throat with white or purple hairs; lobes oblong-ovate, 0·7–2 mm. long. Style 1·2 mm. long in short-styled forms and 2·9 mm. long in long-styled forms; stigma-lobes 0·6–0·8 mm. long; isostylous forms have the style of intermediate length and the stigma-lobes are at the same height as or higher than the anthers. Capsule subglobose, 1–1·8 mm. tall, 2 mm. wide, glabrous, beaked, the valves rounded and well separated after dehiscence. Seeds black, angular, ± 0·4 mm. long, reticulate.

var. **parviflora** *Bremek.* in Verh. K. Nederl. Akad. Wet., Afd. Natuurk., ser. 2, 48(2): 208 (1952). Type: Rwanda, Rushake–Bumba, *Becquet* 268 (BR, holo.)

Flowers isostylous, often small, with corolla-tube 1·2–2(–3) mm. long and lobes 0·7 mm. long. Fig. 40/14, p. 270.

UGANDA. Kigezi District: Rubuguli, Apr. 1948, *Purseglove* 2699! & Bufumbira, Nyamagana, Kamukumu Hill, 25 Feb. 1953, *Norman* 213!; Mengo District: Kampala, Kings Lake, 14 Nov. 1935, *Chandler & Hancock* 85!
TANZANIA. Iringa District: Sao Hill, Apr. 1959, *Watermeyer* 81!; Songea District: Mkuanga Hill, 11 Apr. 1956, *Milne-Redhead & Taylor* 9609! & by Lumecha Bridge, 4 May 1956, *Milne-Redhead & Taylor* 9896!
DISTR. **U**2, 4; **T**7, 8; Rwanda, Burundi and Zambia
HAB. Mainly grassland, seasonally swampy grassland, grassland with scattered trees, roadsides, ditch edges, cultivations and also rock outcrops in *Brachystegia* woodland; 450–2040 m.

NOTE. I am not at all certain of the validity of Bremekamp's varieties; certainly most of the East African material is isostylous. Some populations may however be uniformly long-styled; *Milne-Redhead & Taylor* 9007 (Tanzania, Songea District, Matengo Hills, about 1·5 km. N. of Miyau by R. Utili, 2 Mar. 1956) shows 9 such plants on one sheet and a further sheet from **T8** is similar. This may correspond to var. *rosulata*; I have not seen two forms in any collection examined from other areas—all plants have been dolichostylous. *O. rosulata* and *O. herbacea*, particularly weak forms of the latter which may even have some broader lowermost leaves (? cotyledons) preserved, have been confused. They can always be distinguished if ripe seeds are available. Those of *O. herbacea* immersed in water for an hour develop a thick coating of mucilage whereas those of *O. rosulata* do not; furthermore the testa cells of *O. herbacea* are very coarsely granulate whereas those of *O. rosulata* are smooth, the differences being so marked that they can be seen by viewing the seed under high magnification without making microscopical preparations.

Mutch 109 (Tanzania, Buha District, 0·8 km. N. of Malagarasy road bridge on Kibondo–Kigoma road, 2 Feb. 1973) is possibly a variant of *O. rosulata* related to var. *rosulata*, but differs in having the anthers in long-styled flowers (the only ones seen) situated at the base of the corolla-tube, which is 2·3 mm. long with broadly ovate-elliptic lobes 1–1·5 × 1–1·2 mm.; the style is 3 mm. long with stigmas 0·8–1·3 mm. long.

var. **littoralis** *Verdc.* in K.B. 30: 293 (1975). Type: Kenya, Kwale District, Ukunda, *Symes* 133 (EA, holo. !)

Flowers probably isostylous, the corolla shorter and broader than in other varieties, ± 1 mm. long and wide; lobes ± 0·5 mm. long; capsule transversely oblong, compressed.

KENYA. Kwale District: Ukunda, 9 Aug. 1957, *Symes* 133 !
DISTR. **K7**; not known elsewhere
HAB. Grassland in cultivated area by sea

DISTR. (of species as a whole). As above with addition of Zaire, Angola and South Africa (Transvaal and Natal)

NOTE. *Oldenlandia rosulata* K. Schum. var. *scabra* Gilli in Ann. Naturhistor. Mus. Wien 77: 23 (1973), type: Tanzania, W. slopes of Livingstone Mts., between Madunda and Lupingu, *Gilli* 412 (W, holo. !), is distinguished by having scabrid leaves, 5-colporate pollen grains 22μ in diameter, corolla-tube 2 mm. long, corolla-lobes 1 mm. long, style 1·2 mm. long and stigma 0·5 mm. long. Scabrid leaves occur in many specimens throughout the range of the species and the pollen character is of dubious value; 4- and 5-colporate grains occur in the allied *O. microcalyx* K. Schum. and in many species of *Hedyotideae* the number varies in the same plant.

15. **O. affinis** (*Roem. & Schult.*) *DC.*, Prodr. 4: 428 (1830); Bremek. in Verh. K. Nederl. Akad. Wet., Afd. Natuurk., ser. 2, 48 (2): 226 (1952); Hepper in F.W.T.A., ed. 2, 2: 212 (1963). Type: India, *Koenig* (B, holo.)

Perennial herb with erect scrambling or trailing stems 0·2–1·2 m. tall radiating from a woody rootstock; stems glabrous, with 2 raised lines on each internode below the stipules. Leaf-blades elliptic-oblong to elliptic- or linear-lanceolate, 1·2–6·5 cm. long, 0·1–1·6 cm. wide, acute to tapering acuminate at the apex, cuneate to rounded at the base, scabridulous above particularly near the margins, glabrous beneath; petiole very short or not developed; stipule-sheath 1 mm. long, with 1–5 short fimbriae or colleters 0·2–0·5 mm. long, sometimes with bristly hairs. Flowers heterostylous, in lax terminal or pseudo-axillary lax dichasia often running together to form ample panicles; pedicels 0·2–1(–1·5) cm. long (up to 3 cm. in inflorescences reduced to 1 flower). Calyx-tube subglobose, 0·6–1 mm. long, glabrous; lobes triangular, 0·5–1 mm. long, keeled, glabrous or with ciliate margins. Corolla dark blue, blue-purple to deep violet; tube 3–4·5 mm. long, hairy inside; lobes elliptic or oblong, 1·8–3·2 mm. long, 0·8–1·5 mm. wide. Style 3 mm. long in short-styled flowers, 4·5 mm. long in long-styled flowers; stigma-lobes 1–1·3 mm. long. Capsule globose, 2–2·8 mm. tall, 2·2–2·5 mm. wide, glabrous, obscurely 8-ribbed, scarcely beaked, crowned by the calyx-lobes. Seeds brown, angular, ± 0·5 mm. long, hilum borne on one of the keels, reticulate.

SYN. *Hedyotis affinis* Roem. & Schult., Syst. Veg. 3: 194 (1818)

subsp. **fugax** (*Vatke*) *Verdc.* in K.B. 30: 293 (1975). Type: Zanzibar, *Hildebrandt* 1007* (W, holo., BM, iso. !)

Plant more robust with coarser shorter pedicels, more evident bracts, rather longer calyx-lobes and a thicker walled more globose capsule. Fig. 41/15, p. 271.

UGANDA. W. Nile District: Koboko, May 1939, *Hazel* 728 !; Busoga District: Lake Victoria, Lolui I., 13 May 1964, *G. Jackson* U59 !; Masaka District: near Lake Navugabo, 19 July 1951, *Norman* 27 !
KENYA. Kwale District: Cha Simba Forest, 1 Feb. 1953, *Drummond & Hemsley* 1076 !; Kilifi District: Kibarani Forest, *G. M. Jeffery* 30 !; Lamu District: Mararani, Boni Forest, 16 Sept. 1961, *Gillespie* 364 !
TANZANIA. Tanga District: Sawa, 22 May 1957, *Faulkner* 1980 !; Morogoro District: Uluguru Mts., Tanana, Feb. 1935, *E. M. Bruce* 825 !; Njombe District: Wangingombe, 29 Mar. 1962, *Polhill & Paulo* 1940 !; Zanzibar I., Kisimbani, 25 Mar. 1960, *Faulkner* 2514 !; Pemba I., Tundaua, June 1928, *Vaughan* 346 ! & 348 !
DISTR. **U**1–4; **K**7; **T**3, 4, 6–8; **Z**; **P**; widespread in tropical Africa from Liberia to Angola, Gabon, Sudan, and southwards to Natal and Transvaal; Madagascar and Comoro Is.
HAB. Grassland, swampy ground, bushland, evergreen forest margins and clearings, grassland with scattered trees and also in cultivations; 0–1500 m.

SYN. *Hedyotis decumbens* Hochst. in Flora 27: 552 (1844); Sond. in Fl. Cap. 3: 11 (1865). Type: South Africa, Durban, *Krauss* 305 (B, holo. †, BM !, MO, iso.)
H. (*Kohautia ?*) *fugax* Vatke in Oest. Bot. Zeitschr. 25: 232 (1875)
Oldenlandia decumbens (Hochst.) Hiern in F.T.A. 3: 54 (1877); K. Schum. in P.O.A. C: 376 (1895); Hiern, Cat. Afr. Pl. Welw. 2: 442 (1898); F.W.T.A. 2: 132 (1931), *non* Spreng. 1815, *nom. illegit.*
[*O. affinis* sensu Bremek. in Verh. K. Nederl. Akad. Wet., Afd. Natuurk., ser. 2, 48(2): 226 (1952), pro parte; Hepper in F.W.T.A., ed. 2, 2: 212 (1963), *non* (Roem. & Schult.) DC. sensu stricto]

NOTE. Subsp, *affinis* occurs in India and the Malay Peninsula.

16. **O. lancifolia** (*Schumach.*) *DC.*, Prodr. 4: 425 (1830); Hiern in F.T.A. 3: 61 (1877); K. Schum. in P.O.A. C: 375 (1895); Bremek. in Verh. K. Nederl. Akad. Wet., Afd. Natuurk., ser. 2, 48(2): 230 (1952); Hepper in F.W.T.A., ed. 2, 2: 212 (1963); U.K.W.F.: 400 (1974). Type: Ghana, Valley of Aquapim, *Thonning* 210 (C, lecto., S, isolecto.)

Perennial (rarely annual) straggling or prostrate herb, often much branched near the base into almost simple stems, (5–)20–60(–90) cm. long, which sometimes root at the nodes, glabrous or in some variants scabridulous when young; sometimes forming loose mats. Leaf-blades linear to linear-lanceolate, less often elliptic or lanceolate, 1–6 cm. long, 0·2–1·2 cm. wide, acute at the apex, cuneate at the base, scabridulous above near the margins, glabrous beneath or with midvein scabridulous; petioles not developed, very short and adnate to the stipule sheath which is 1 mm. long and bears 2–5 linear fimbriae 1·5 mm. long. Flowers not heterostylous, often solitary at the nodes (pseudo-axillary) or sometimes several at the nodes but then these actually borne on very reduced axillary shoots; pedicels slender, 0·5–3 cm. long, glabrous or scabridulous. Calyx-tube bowl-shaped, 0·8 mm. tall, 1·5 mm. wide, glabrous or with scattered very short hairs; lobes triangular, 1–1·8 mm. long, acuminate at the apex, glabrous or sparsely scabridulous. Corolla white, sometimes tinged pink or purple; tube 1 mm. long, glabrous inside; lobes triangular, 1–2 mm. long. Style slightly longer than the tube; stigma-lobes 0·7–1·4 mm. long. Capsule depressed subglobose, 2·2–3 mm. tall including the 1 mm. tall beak, 3·2–5 mm. wide, grooved at the middle. Seeds pale brown, angular, 0·3–0·4 mm. long, strongly reticulate.

* 1007a at Kew is from Bagamoyo on the mainland but Vatke clearly says Insula Sansibar for 1007; 1007b at BM is also from Bagamoyo.

SYN. *Hedyotis lancifolia* Schumach. in Schumach. & Thonn., Beskr. Guin. Pl.: 72 (1827)

var. **scabridula** *Bremek.* in Verh. K. Nederl. Akad. Wet., Afd. Natuurk., ser. 2, 48(2): 232 (1952). Type: Tanzania, Njombe District, upper Ruhudje R., Lupembe, *Schlieben* 626 (K, holo. !)

Young stems, pedicels and leaf-blades scabridulous; calyx-tube often puberulous. Leaf-blades attaining maximum dimensions. Fig. 41/16, p. 271.

UGANDA. W. Nile District: Madi, Dec. 1862, *Grant* !; Mengo District: Kome I., 7 July 1951, *Norman* 16 ! & Kipango, Nov. 1913, *Dummer* 460 !
KENYA. Trans-Nzoia District: Kipkarren, *Brodhurst Hill* 287 ! & 8 km. Kitale–Endebess road, May 1969, *Napper* 2137 !; Embu, 1 Sept. 1962, *Lewis* 5909 ! (see note)
TANZANIA. Mwanza District: Bwiru, 2 Jan. 1954, *Tanner* 1527 !; Buha District: Kaberi Swamp, 10 Aug. 1950, *Bullock* 3128 !; Songea District: about 12 km. W. of Songea by the Kimarampaka stream, 7 Jan. 1956, *Milne-Redhead & Taylor* 8069 !; Zanzibar I., Mwera Swamp, 19 Aug. 1960, *Faulkner* 2694 ! & same locality, 20 Aug. 1964, *Faulkner* 3413 !; Pemba I., 5 km. on Wesha road, 30 Sept. 1929, *Vaughan* 688 !
DISTR. **U**1–4; **K**3, 4, 7; **T**1–4, 6–8; **Z**; **P**; W. Africa from Sierra Leone to Angola and Sudan, Zaire, southwards to the Transvaal
HAB. Swamps, swampy grassland, drainage ditches, lake-side marshes, sometimes actually with roots in the water; also recorded from dry ground near lakes, wet abandoned cultivations and on black soil in forest; 0–1860 m.

var. **seseensis** *Bremek.* in Verh. K. Nederl. Akad. Wet., Afd. Natuurk., ser. 2, 48(2): 232 (1952). Type: Masaka District, Sese Is., Soyi, *Maitland* 415 (K, holo. !)

Plant glabrous save for the marginal area of the leaf-blades which are scabridulous above. Leaf-blades mostly relatively shorter and broader, up to 3 cm. long, 7 mm. wide.

UGANDA. Busoga District: Butembe Bunya, N. end of Luvia Is., 15 Jan. 1953, *G.H. Wood* 643 !; Masaka District: Sese Is., Bugala, 5 June 1932, *A. S. Thomas* 73 !; Mengo District: 2·5 km. S. of Entebbe, Hippo Bay, 20 May 1951, *Norman* 4 !
DISTR. **U**3, 4; not known elsewhere
HAB. With ruderals on almost bare sand by lake shores, cultivated ground and pathsides; 1110–1200 m.

NOTE. Bremekamp has divided this species into 7 varieties which are mostly very minor in character and have been ignored by most workers (e.g. Hepper in F.W.T.A. ed. 2). There are certainly two distinguishable entities in the Flora area and I have maintained Bremekamp's names for these. His description of the pedicels of var. *seseensis* in the flowering stage as being only 0·5 mm. long is clearly an error but even if 5 mm. were meant certain sheets annotated by him have flowering pedicels 1 cm. long. There is scant morphological justification for separation this from var. *lancifolia*, but since it forms a rather isolated population it is not unreasonable to maintain the varietal name.

DISTR. (of species as a whole). Widespread in tropical Africa from Senegal to Ethiopia, thence to Transvaal and Angola; also introduced into S. America and West Indies

17. **O. flosculosa** *Hiern* in F.T.A. 3: 60 (1877); K. Schum. in P.O.A. C: 375 (1895); Bremek. in Verh. K. Nederl. Akad. Wet., Afd. Natuurk., ser. 2, 48(2): 234 (1952). Type: Zanzibar, *Hildebrandt* 1348 (K, holo. !, BM !, MO, iso.)

Perennial herb often much branched at the base with glabrous or puberulous weakly erect or straggling stems 30–50 cm. tall. Leaf-blades linear, oblong or very narrowly elliptic, 1–3·5 cm. long, 1·5–5·5 mm. wide, acute at the apex, cuneate at the base, glabrous or sparsely puberulous; petioles not developed; stipule-sheath 1–2 mm., divided into 3–5 fimbriae up to 2–2·5 mm. long. Flowers heterostylous, in 2–4(–7)-flowered glomerules, as many as 8 arranged in spicate fashion, the " spikes " either terminal or axillary up to 8 cm. long; when terminal they are arranged dichasially, the centre terminal part either a " spike " or a solitary flower overtopped by long lateral " spikes " (the " spikes " are really a series of nodes bearing few flowers from which the leaves have disappeared); pedicels 0–4 mm. long. Calyx-tube subglobose, 0·8 mm. long and wide, glabrous or puberulous; lobes ovate-triangular to narrowly triangular, 0·8–1 mm. long, glabrous or with margins and midvein ciliate. Corolla white or with lobes tinged pink beneath at the apex; tube 1·5 mm. long, hairy inside; lobes elliptic-oblong, 1·3–1·5 mm. long, papillate inside. Style 0·8 mm. long in short-styled flowers, 3 mm. long in long-styled flowers; stigma-lobes filiform, 0·5 mm. long. Capsule depressed subglobose, 1·2 mm. tall, 1·8–2 mm. wide, glabrous or puberulous, the beak slightly raised, ± 0·2 mm. tall. Seeds dark purplish brown, almost round or elliptic in outline, ± 0·3 mm. long, raised on the hilar side, flat on the other side.

var. **flosculosa**; Bremek. in Verh. K. Nederl. Akad. Wet., Afd. Natuurk., ser. 2, 48(2): 235 (1952)

Stem leaves and calyx glabrous or at most the nodes hairy and the angles of the stem with minute hairs. Fig. 41/17, p. 271.

TANZANIA. Lindi District: Tendaguru, 2 July 1924, *L. S. B. Leakey* in *Migeod* 72! & same locality, 25 Feb. 1926, *Migeod* 96!;* Zanzibar I., Tunguu road, 24 Apr. 1950, *Oxtoby* 2, 2a! & Kiungani, 2 Feb. 1929, *Greenway* 1311! & 25·6 km. along Chwaka road, 24 Apr. 1960, *Faulkner* 2536!
DISTR. **T8**; **Z**; not known elsewhere
HAB. Grassy places on sand, coral outcrops and also in open bushland; 0–90 m.

var. **hirtella** *Bremek.* in Verh. K. Nederl. Akad. Wet., Afd. Natuurk., ser. 2, 48(2): 235 (1952). Type: Tanzania, Mlayar, *Jaeger* 88 (W, holo., BM, iso.!)

Stems and calyx distinctly puberulous, the leaf-blades sparsely puberulous and calyx-lobes ciliate.

TANZANIA. Masai District: Mlayar, 11 July 1906, *Jaeger* 88!
DISTR. **T2**; not known elsewhere
HAB. Open deciduous bushland; ± 950 m.

NOTE. Bremekamp cites this number again under *O. wiedemannii* with the addition of Massai Steppe, but probably in error for 78; a specimen of *Jaeger* 78 at the BM is indeed *O. wiedemannii*.

18. **O. patula** *Bremek.* in Verh. K. Nederl. Akad. Wet., Afd. Natuurk., ser. 2, 48(2): 236 (1952). Type: Tanzania, Lindi District, Tendaguru, *Migeod* 129 (BM, holo.!)

Erect annual herb 18–35 cm. tall, with spreading 4-ribbed axillary branchlets; stems scabrid-papillose. Leaf-blades linear-lanceolate, 1·5–4·5 cm. long, 1·5–6 mm. wide, attenuate to an acute apex, cuneate at the base, scabridulous above with very short papillae or glabrous, scabridulous on the midvein beneath and on the margins; petiole not developed; stipule-sheath 0·5–1·5 mm. long, the margins with 5–7 fimbriae up to 3 mm. long. Flowers heterostylous, in terminal monochasial raceme-like inflorescences, the rhachis slender with pairs of flowers well separated, the internodes 1–2·5 cm. long; pedicels 3–8 mm. long. Calyx-tube shallowly campanulate, 0·7 mm. long, glabrous; lobes triangular, 0·6–0·8 mm. long, glabrous. Corolla white;** tube narrowly funnel-shaped, 1·5 mm. long, sparsely pilose at the throat; lobes 1·2–1·6 mm. long; style 0·7 mm. long in short-styled flowers, 2·5 mm. long in long-styled flowers. Capsule depressed subglobose, 0·9–1·2 mm. tall, 1·8 mm. wide, glabrous, the beak not raised. Seeds not seen. Fig. 40/18, p. 270.

TANZANIA. Lindi District: Tendaguru, 21 Mar. 1926, *Migeod* 129! & same locality, 17 June 1930, *Migeod* 766!
DISTR. **T8**; not known elsewhere
HAB. Dry places in wooded grassland; 225 m.

* Not cited by Bremekamp in his monograph, but specimens annotated by him at BM.
** Obtained from field note-books; not mentioned on type.

19. **O. richardsonioides** (*K. Schum.*) *Verdc.* in K.B. 28: 420 (1974). Type: Kenya, Lamu District, Kiunga, *Riva* 1650 (FI, holo.!)

Annual erect or suberect herb, branched at every node, with lowest stems as long as the height of the plant, so that plant appears dense and broadly obconical, or branched from the base and more irregular in shape, 5–10 cm. tall, 9–25 cm. wide, or not branched at base but with a few lateral branches, 5–14 cm. tall; stems pale, densely covered with short spreading white hairs, internodes long or short. Leaves sessile, somewhat succulent; blades linear- or elliptic-lanceolate to lanceolate, 0·5–2·5 cm. long, (0·8–)2·5–7·5 mm. wide, acute and apiculate at the apex, somewhat narrowed to the base, with short white hairs above and on the main nerve beneath, the margin often revolute; stipule-sheath 1–2 mm. long, with 2–5 filiform setae 1–4·5 mm. long. Flowers heterostylous, in pseudoterminal and axillary 3–9-flowered shortly pubescent or glabrescent cymes; peduncle at first obsolete but enlarging after flowering to 5 mm. long or up to 1·2 cm.; pedicels 1–5 mm. long. Calyx-tube campanulate, 0·8–1 mm. long, shortly pubescent and with very evident surface rhaphides; lobes ovate-triangular, 1–1·3 mm. long, 0·8 mm. wide, acute, shortly pubescent. Corolla white or suffused with carmine; tube 1·7–2 mm. long, glabrous outside, densely hairy at the throat inside; lobes oblong-elliptic to ovate-lanceolate, 1·5–3 mm. long, 0·8–1·1 mm. wide, hooked at the apex, papillate-pubescent at the base inside. Style 0·8–1·2 mm. long in short-styled flowers, 3–3·3 mm. long in long-styled flowers, hirtellous; stigma-lobes 0·6–0·8 mm. long or stigma scarcely lobed in short-styled forms. Capsule campanulate, 1–1·5 mm. long, 1·4 mm. wide, with short spreading hairs or glabrescent, crowned with the persistent calyx-lobes; beak slightly raised, ± 0·2 mm. tall. Seeds dark brown, irregularly ellipsoidal, ± 0·5 mm. long.

var. **richardsonioides**

Plants compact, obconic, densely branched from the base; most parts densely pubescent. Inflorescences short and dense. Fig. 41/19, p. 271.

KENYA. Lamu District: Oseni [Osine], 8 Oct. 1957, *Greenway & Rawlins* 9289! & 48 km. N. of Lamu, Mkokoni–Ashuwei [Aswe], Sept. 1956, *Rawlins* 107!
DISTR. **K7**; Somali Republic (S.)
HAB. Sand-dunes with Cyperaceae in *Hyphaene-Erythroxylum* bushland; sea-level

SYN. *Mitratheca richardsonioides* K. Schum. in E.J. 33: 335 (1903)
Eionitis psammophila (Chiov.) Bremek. var. *compacte-hirtella* Bremek. in Verh. K. Nederl. Akad. Wet., Afd. Natuurk., ser. 2, 48(2): 168 (1952). Type: Somali Republic (S.), Mogadishu, *Senni* 1337 (FI, holo.!)
Oldenlandia marginata Bremek. in K.B. 13: 383 (1959). Type: Kenya, Lamu District, Oseni, *Greenway & Rawlins* 9289 (EA, holo.!, K, iso.!)

var. **gracilis** *Verdc.* in K.B. 28: 421 (1974). Type: Kenya/Somali Republic (S.) border, Kiunga Archipelago, *Gillespie* 254 (K, holo.!)

Slender plants not branched at the base, but often with several lateral branches; most parts densely pubescent.

KENYA. Lamu District/Somali Republic (S.) border: Kiunga Archipelago, 23 Aug. 1961, *Gillespie* 254!
DISTR. **K7**; Somali Republic (S.)
HAB. On consolidated dunes of white calcareous sands; 6 m.

NOTE. Two other varieties var. *laxiflora* (Bremek.) Verdc. and var. *hirtella* (Chiov.) Verdc. occur in Somali Republic (S.). The marked variation in habit in this species is probably related to the zonation of habitats in the sand dunes, but until somebody has studied a population of this interesting species in the field I have maintained the conception of distinct varieties. If they prove to be merely ecotypic forms then they can be dispensed with. Certainly comparing the only four specimens known from the Flora area of this species, here referred to two varieties, one might suspect them of being distinct species but var. *hirtella* bridges the gap.

20. **O. cyperoides** *Verdc.* in K.B. 30: 693, fig. 1 (1976). Type: Tanzania, Kigoma District, north of Lugufu towards Kigamba, *Peter* 36688 (K, holo. !, EA !, B, iso.)

Prostrate perennial herb, with creeping stems producing flowering shoots 2–5 cm. tall and bearing densely tufted congested imbricate leaves, the internodes being suppressed; stems glabrescent. Leaves linear-subulate, 2·5–3 cm. long, 0·9 mm. wide, drawn out into a filiform apiculum at the apex, broadened at the base into the stipule-sheath, very shortly scabridulous above, glabrous beneath, the midrib beneath broad, mostly occupying almost the total width of the leaf; stipule-sheath 3 mm. long, hyaline, with 3–6 setae 1–2 mm. long. Flowers heterostylous, crowded in the centre of the tufts, the actual arrangement very obscure but probably condensed cymes; pedicels up to 1 mm. long. Calyx-tube obconic-campanulate, 1 mm. long, glabrous; lobes subulate, 3–3·5 mm. long, ciliate, the joined part of the limb ± 1 mm. long. Corolla-tube cylindrical, 5 mm. long, hairy inside; lobes oblong-elliptic to ovate, 2–2·5 mm. long, 1–1·1 mm. wide. Stamens with filaments exserted 1 mm. in short-styled flowers, minutely hairy. Style 2·9 mm. long in short-styled flowers, hairy; stigma-lobes linear, 1·1 mm. long, hairy. Capsule ovoid or subglobose, 1·5–2 mm. long, 1·5 mm. wide, glabrous, the beak rounded and scarcely raised. Seeds (immature) angular, 0·8 mm. long, 0·5 mm. wide, strongly reticulate, apparently not becoming sticky on moistening; the testa-cells prominently dotted.

TANZANIA. Buha District: Kasulu [Kassulo]–Kivumba, 24 Feb. 1926, *Peter* 46281 !; Kigoma District: Uvinsa, N. of Lugufu towards Kigamba, 10 Feb. 1926, *Peter* 36688 !

DISTR. **T4**; not known elsewhere

HAB. Bushland and grassland; 1070–1350 m.

NOTE. Bremekamp, to whom I sent a portion, thought this species might belong to subgen. *Aneurum*, but the habit is very different. It seems as if it might be related to subgen. *Polycarpum*, with the peduncles reduced, but the seeds should become slimy when moistened. More mature seeds are required; until then I am leaving its position doubtful.

21. **O. taborensis** *Bremek.* in Verh. K. Nederl. Akad. Wet., Afd. Natuurk., ser. 2, 48 (2): 237 (1952). Type: Tanzania, Tabora [Kazeh], *Grant* (K, holo. !)

Annual or possibly perennial erect herb 17–35 cm. tall, usually branched from the base into several to many stems, glabrous or scabrid-papillose on the ribs with short spreading hairs. Leaf-blades linear, 0·9–5 cm. long, (0·5–)1(–3·5) mm. wide, revolute, apiculate at the apex, sparsely scabrid-pubescent above and beneath; petiole not developed; stipule-sheath 2–3 mm. long, divided into 2–7 fimbriae 1–4 mm. long. Inflorescences terminal and axillary panicles of peduncled 2–9-flowered glomerules; peduncles 0·2–2·5 cm. long; pedicels up to 1 mm. long in bud but lengthening to 2(–4) mm. in fruit. Flowers heterostylous. Calyx-tube subglobose, ± 0·8–1 mm. long, glabrous; lobes triangular-lanceolate, 1 mm. long, sparsely ciliate on the margins. Corolla white; tube 1·4 mm. long, hairy inside; lobes oblong-elliptic, 1·3–2 mm. long, 0·7 mm. wide, papillate inside. Style in short-styled flowers 0·4 mm. long, in long-styled flowers 2·8 mm. long; stigma-lobes 0·5–0·8 mm. long. Capsule depressed subglobose, 1·2–1·5 mm. tall, 2–2·2 mm. wide, glabrous, crowned by the persistent calyx-lobes; beak slightly raised, ± 0·4 mm. tall. Fig. 41/21, p. 271.

TANZANIA. Shinyanga District: Msalala, *Hannington* !;* Nzega District: 72 km. N. of Tabora, 26 Nov. 1926, *Wallace* 42!; Tabora District: Uyogo-nyembe, 5 June 1913, *Braun* in *Herb. Amani* 5406! & Tabora, 19 Jan. 1926, *Peter* 35174!
DISTR. T1, 4, 5; not known elsewhere
HAB. Fields, open places; 1100–1200 m.

SYN. [*O. parviflora* sensu Oliv. in Trans. Linn. Soc. 29: 84 (1873); Hiern in F.T.A. 3: 60 (1877), *non* (Benth.) Oliv. sensu stricto]

NOTE. This is an unsatisfactorily circumscribed species, differing from *O. duemmeri* only in the structure of the inflorescence and shorter pedicels. It would probably be better treated as a variety of *O. duemmeri*, but insufficient evidence is available. *Jones* 40 (Tanzania, Kigoma District, Nguruka, 15 Feb. 1958, from *Brachystegia* woodland at 900 m.) has more prostrate wiry shoots and very fine filiform leaves but seems to belong to this species. *O. scopulorum* Bullock is close and Bremekamp cites *Braun* in *Herb. Amani* 5406 (see above) under both species but has annotated the Kew sheet as *O. scopulorum*; it belongs I think to *O. taborensis*. If ripe seeds are available the testa-cell sculpture differences mentioned by Bremekamp work, the cells of *O. taborensis* being quite coarsely sculptured, whereas those of *O. scopulorum* are not sculptured or only finely sculptured. *Peter* 36131 (Kigoma District, E. of Malagarasi, 2 Feb. 1926) has a tufted habit rather similar to that of *O. cyperoides*, but is probably a variant of *O. taborensis* or *O. scopulorum*; the very coarsely granulate testa cells suggest the former. The seeds, however, do not seem to become slimy when moistened, but this is a very difficult character to be certain about.

22. **O. duemmeri** *S. Moore* in J.B. 54: 250 (1916); Bremek. in Verh. K. Nederl. Akad. Wet., Afd. Natuurk., ser. 2, 48(2): 237 (1952). Type: Uganda, Bugoye, *Dummer* 2624 (BM, holo.!, K, MO, iso.!)

Short-lived perennial or sometimes annual herb 7–40 cm. tall, much branched from the extreme base with several to numerous erect, suberect or trailing branches; stems glabrous or pubescent at the base. Leaf-blades linear or linear-lanceolate, 0·7–1·5(–4) cm. long, 0·8–1·6 mm. wide, apiculate at the apex, sparsely scabrid-pubescent above, glabrous beneath save sometimes on the midrib; petiole not developed; stipule-sheath 1–2 mm. long, pubescent, divided into 2 slender fimbriae about 1–2 mm. long, sometimes with 1–2 shorter subsidiary ones. Inflorescences terminal and axillary, the terminal panicles made up of pedunculate 1–4-flowered groups, the axillary ones sometimes reduced to a single flower; peduncles 0·5–2 cm. long; pedicels 1–6(–9) mm. long. Flowers heterostylous. Calyx-tube subglobose, ± 0·6 mm. long, glabrous; lobes triangular-lanceolate, 0·4–1·1 mm. long, glabrous or sparsely ciliate on the margins with papilla-like hairs. Corolla white or faintly tinged mauve; tube 1–2 mm. long, hairy inside; lobes elliptic-oblong, 1·4–2 mm. long, 0·8 mm. wide, papillate inside. Style 0·7 mm. long in short-styled flowers, 2–3·5 mm. long in long-styled flowers; stigma-lobes 0·5–0·7 mm. long. Capsule depressed semi-globose, 1–1·6 mm. tall, 1·7–2·1 mm. wide, glabrous, crowned by the persistent calyx-lobes, beak slightly raised, ± 0·3 mm. tall. Seeds brown, ellipsoid, somewhat angular, 0·35 mm. long. Fig. 41/22, p. 271.

UGANDA. Masaka District: NW. side of Lake Nabugabo, 9 Oct. 1953, *Drummond & Hemsley* 4690! & 2–3 km. S. of Kasokero, 12 May 1969, *Lye* 2884!; Mengo District: near Entebbe, N. of Kisi, 31 Aug. 1969, *Lye* 3685!
DISTR. U4; not known elsewhere
HAB. Grassland, often seasonally wet; 1080–1170 m.

NOTE. Bremekamp states that *O. microcarpa* differs from *O. duemmeri* in the structure of the inflorescence, longer entirely glabrous leaves, slightly smaller flowers and much smaller fruits; also he states that the scabrid-papillose filaments are an unexpected feature. There is nothing important in the supposed differences in dimensions and although the leaves are undoubtedly longer they are not entirely glabrous.

* This small piece is mounted on the same sheet as the *Speke & Grant* specimen but is not mentioned in Bremekamp's account.

Moreover the filaments of *O. duemmeri* are slightly papillose. On the other hand *O. microcarpa* has quite smooth testa-cells whereas *O. duemmeri* and *O. taborensis* have the testa-cells distinctly coarsely granulate. It is curious that Bremekamp does not mention this since it is a character of which he makes much use. It seems advisable to keep the two distinct. He does give a sketch on the holotype of *O. microcarpa* showing smooth cells. *B. D. Burtt* 3625 (Tanzania, Dodoma District, Kazikazi, 20 May 1932) is a plant scarcely to be distinguished from the above and is more or less intermediate with the next species. It is taller (about 40 cm.) with 4–8 strictly erect stems from the extreme base and has rather less lax inflorescences.

23. **O. microcarpa** *Bremek.* in Verh. K. Nederl. Akad. Wet., Afd. Natuurk., ser. 2, 48(2): 238 (1952). Type: Tanzania, Shinyanga, *B. D. Burtt* 2422 (K, holo.!)

Annual herb with main stem usually much branched near the base, 20–40 cm. tall, erect or ? somewhat straggling; branchlets slender, ribbed, glabrous or sparsely scaly-pubescent. Leaf-blades linear, 1·5–4·3 cm. long, 0·5–2 mm. wide, blunt to acuminate at the apex, narrowed at the base, glabrous or sparsely shortly scabrid pubescent on the upper surface and on the midvein beneath, revolute; petioles not developed; stipule-sheath 1–2 mm. long, divided into ± 5 unequal filiform fimbriae 0·5–3 mm. long. Flowers heterostylous, in axillary and terminal corymbs and panicles, each often 7- or more flowered; peduncle slender, up to 1·8 cm.; pedicels 2–3(–6) mm. long. Calyx depressed subglobose or broadly obconic, 0·5–1 mm. long; lobes narrowly lanceolate, 0·3–1 mm. long, glabrous, keeled. Corolla white; tube 1·8 mm. long, densely hairy at the throat; lobes ovate-oblong, 1·4–1·8 mm. long, 0·8 mm. wide, papillate inside. Style in short-styled flowers 0·6 mm. long, in long-styled flowers 2·5 mm. long; stigma-lobes 0·6–0·8 mm. long. Capsule depressed subglobose or broadly obconic, 1–1·2 mm. tall, 1·7–2·2 mm. wide, the beak very slightly raised, scarcely 0·2 mm. tall. Seeds dark brown, irregularly ellipsoid, angular, ± 0·4 mm. long. Fig. 41/23, p. 271.

TANZANIA. Shinyanga, 1 Apr. 1932, *B. D. Burtt* 3745! & Shinyanga, new aerodrome, 6 May 1931, *B. D. Burtt* 2422!; Tabora District: Ugalla R., Isimbira, 25 Oct. 1960, *Richards* 13401!

DISTR. T1, 4; not known elsewhere

HAB. Marshy ground, grassy river banks, also cultivations; 1080–1200 m.

NOTE. Very close to *O. duemmeri* but differing considerably in the ornamentation of the testa cells. *Richards* 13401, cited above, is less branched and had been annotated as a new species but it agrees closely in inflorescence and testa cell characters with the Shinyanga material.

24. **O. sp. A**

Perennial herb with thin woody branched stems 15–17 cm. tall from an unknown rootstock; stems finely densely papillate. Leaves linear-lanceolate, 2–3 cm. long, 0·8–3 mm. wide, acute at the apex, narrowed at the base into the stipule-sheath, glabrous save for the slightly scabridulous-papillate margins; stipule-sheath 1·5–2 mm. long with 2–4 setae, the outer ones very short, the inner up to 1 mm. long. Flowers heterostylous, in lax ± 5-flowered cymes; peduncles 2–2·5 cm. long, glabrous; secondary peduncles 4–5 mm. long; pedicels 0·5–10 mm. long; calyx-tube glabrous, ovoid, 1·8 mm. long and wide; limb-tube 0·7 mm. long; lobes triangular, 1·5–2 mm. long, 0·9–1·2 mm. wide, rather foliaceous in appearance, equal or slightly unequal, keeled on the back, minutely scabridulous-ciliate on the margins. Corolla rose or white; tube 2·8–3·5 mm. long; lobes oblong, 3–3·5 mm. long, 1·2–1·3 mm. wide, acuminate with inflexed apex, papillate inside. Stamens exserted in short-styled flowers, the filaments exserted 0.5 mm.; style 2·2 mm. long; stigma-lobes 1 mm. long. Anthers well included in long-styled flowers, near the middle of the tube; style 4·2 mm. long, hairy at the middle; stigma-lobes 1·3 mm. long. Immature capsule obconic, 2 mm. long, 2·2 mm. wide.

TANZANIA. Buha District: Mbirira [Birira] to Kisuzi [Nisusi], 27 Feb. 1926, *Peter* 37934! & Kivumba to Muhoro, 25 Feb. 1926, *Peter* 37667!
DISTR. **T4**; not known elsewhere
HAB. Unknown; 1280–1450 m.

NOTE. The placing of this species in the genus is not certain.

25. **O. uvinsae** *Verdc.* in K.B. 30: 693 (1976). Type: Tanzania, Kigoma District, W. of Uvinsa Railway Station on R. Malagarasi, *Peter* 36338 (B, holo.!)

Annual herb with branched stems, 8–14 cm. tall; stems glabrous or sometimes slightly papillate on the angles below the nodes. Leaf-blades narrowly elliptic to lanceolate, 0·7–2 cm. long, 2–3·5 mm. wide, acute at the apex, narrowed into the stipule-base, glabrous; stipule-bases ± scarious, 1·5–2 mm. long, with 2–4 weak setae 0·5–1·7 mm. long. Flowers heterostylous, few in lax cymes; peduncles 0·5–1·5 cm. long, glabrous; pedicels 1–8 mm. long, thickened beneath the capsules, glabrous. Calyx glabrous; tube subglobose or obconic, 0·8 mm. long; lobes triangular, 0·5–1 mm. long, glabrous. Corolla pale lilac, drying pale, but with 2 blue marks at the base of the corolla-lobes; tube 1·8–2·2 mm. long; lobes elliptic, 1·5 mm. long, 0·8–1·2 mm. wide, papillate inside. Filaments exserted 0·8 mm. in short-styled flowers; style 0·8 mm. long, ± hairy above the stigma-lobes, 0·7 mm. long, hairy. Style exserted 1 mm. in long-styled flowers, the stigma-lobes 0·8 mm. long; stamens well-inserted. Capsule glabrous, obconic to subglobose, 1·5–2 mm. long and wide, the beak very low. Seeds dark, depressed-ellipsoid, umbonate on one side, 0·35 mm. long, 0·3 mm. wide, 0·12 mm. thick, strongly reticulate, apparently becoming slimy when moistened.

TANZANIA. Kigoma District: W. of Uvinsa Railway Station on R. Malagarasi, 6 Feb. 1926, *Peter* 36338!
DISTR. **T4**; known only from the type
HAB. Not known; 990 m.

26. **O. scopulorum** *Bullock* in K.B. 1932: 497 (1932); Bremek. in Verh. K. Nederl. Akad. Wet., Afd. Natuurk., ser. 2, 48(2): 239 (1952); U.K.W.F.: 401 (1974). Type: Kenya, Mt. Elgon, *Lugard* 40 (K, holo.!)

Usually a tufted perennial herb 2–25 cm. long or tall, with many prostrate, erect or decumbent stems from a branched woody rootstock, or less often an annual herb with single stem, usually much branched near the base, or rarely erect and unbranched, sometimes forming definite mats; branchlets glabrous to quite densely scabrid-pubescent. Leaf-blades linear to linear-lanceolate, 0·7–2·2(–4) cm. long, 0·7–3·5(–4) mm. wide, acute and mucronulate at the apex, cuneate at the base, glabrous save for the sparsely papillate margins or with sparse to rather dense papilla-like hairs on the upper surface; petioles not developed; stipule-sheath 1–2 mm. long, papillate, bearing 2(–3) main filiform fimbriae 1–5 mm. long and several intermediate shorter ones. Flowers heterostylous, in terminal and apparently axillary subcorymbose (2–)3–5-flowered inflorescences; peduncle slender, 0·25–2·8 cm. long; pedicels 1–7(–16) mm. long. Calyx-tube campanulate, 0·8 mm. long, glabrous; lobes triangular or lanceolate, (0·5–)1·2–2 mm. long, keeled, slightly scabrid-papillose. Corolla white, pink, lilac, mauve or pale blue, the lobes often with 2 darker mauve marks; tube (1·5–)2–3 mm. long, densely hairy in the throat; lobes ovate-oblong, (1·4–)2–3 mm. long, ± 1·5(–1·9) mm. wide, densely papillate inside. Style in short-styled flowers ± 1 mm. long, in long-styled flowers (2·6–)3–4 mm. long; stigma-lobes 0·7–2 mm. long. Capsule subglobose, 1·5–2 mm. tall, 2–2·6 mm. wide, glabrous, the beak 0·3–0·4 mm. tall. Seeds dark brown, ovoid, angular, 0·4 mm. long, strongly reticulate. Fig. 41/26, p. 271.

KENYA. Uasin Gishu District: Kapsabet, 25 June 1951, *G. R. Williams* 257!; N. Nyeri District: N. slopes of Mt. Kenya, Nanyuki–Meru road about 110 km. from Nyeri, 4 Feb. 1933, *C. G. Rogers* 398!; Masai District: Trans Mara, Narok–Lolgoris road, top of Isuria Escarpment, 15 Apr. 1961, *Glover, Gwynne & Samuel* 526B!
TANZANIA. Musoma District: about 32 km. on Musoma–Mwanza road, 13 Nov. 1962, *Verdcourt* 3297! & Serengeti, Kirawira Plain, 27 Apr. 1965, *Richards* 20297!; Tabora District: Ugalla R., 192 km. SW. of Tabora, 3 Oct. 1949, *Bally* 7525!
DISTR. **K**3–6; **T**1, 2, 4, 5; Rwanda
HAB. Grassland including seasonally wet areas, grassy clearing in *Acacia* and other woodland, gardens, roadsides and other open places in scrub, often in rock crevices, etc.; 900–2520 m.

SYN. *O. scopulorum* Bullock var. *lanceolata* Bremek. in Verh. K. Nederl. Akad. Wet., Afd. Natuurk., ser. 2, 48(2): 239 (1952). Type: Kenya, Londiani, *Lindblom* 5. 1920 (S, holo.)
O. eludens Bremek. in Verh. K. Nederl. Akad. Wet., Afd. Natuurk., ser. 2, 48(2): 252 (1952). Type: Rwanda, Kagera, *Bredo* 2257 (BR, holo.)
O. somala Bremek. var. *scabridula* Bremek. in Verh. K. Nederl. Akad. Wet., Afd. Natuurk., ser. 2, 48(2): 253 (1952). Type: Tanzania, Shinyanga, *R. D. Bax* 197 (K, holo.!)
O. filipes Bremek. in K.B. 11: 169 (1956). Type: Kenya, Londiani, *Davis* 23 (K, holo.!, BM, iso.!)

NOTE. The variety differs in having longer internodes, broader leaf-blades and very slightly longer calyx-lobes and corolla. Many specimens show equally wide leaf-blades and the characters are inadequate to separate two taxa at any rank. In my opinion *O. filipes* is only a variant of this species—whether a genetic one or merely a well grown specimen due to the habitat is not certain. It differs in having pedicels up to 1·8 cm. long and larger flowers with the corolla-lobes 3·5 × 2 mm. Some of the Tanzania material, particularly that from Musoma District, has much longer leaves than typical Kenya material, also the inflorescences are sometimes branched and the flowers more often white. It comes close to the specimens separated by Bremekamp as *O. eludens* Bremek., but despite the fact he places the two supposed species in different subgenera I see no real differences; it is more or less identical with *O. somala* Bremek. var. *scabridula* Bremek. The testa-cells of many specimens have been examined and some (including those of the type) are finely dotted but many others are devoid of any sculpture. These two types do not correlate with other differences in habit, indumentum, etc. Kenya highland material usually has coloured flowers, whereas the Tanzania material mostly has white flowers.

27. **O. wiedemannii** *K. Schum.* in E.J. 28: 57 (1899), as *wiedenmannii*;* Bremek. in Verh. K. Nederl. Akad. Wet., Afd. Natuurk., ser. 2, 48 (2): 240 (1952); U.K.W.F.: 400 (1974). Types: Tanzania, near Moshi, *Wiedemann* (B, holo. †); Moshi, *Haarer* 474 (K, neo.!, EA, isoneo.!)

Perennial (or ? sometimes annual) erect herb (5–)7·5–30(–60) cm. tall, with several to many stems from a slender often woody taproot, frequently forming tussocks; stems mostly with dense short spreading pubescence, rarely glabrous. Leaf-blades linear to linear-lanceolate, 1–6·5 cm. long, 0·8–4·5(–7) mm. wide, acute at the apex, narrowly cuneate at the base, the margins often revolute, puberulous above and on the midvein beneath save in "var. *glabricaulis*"; petiole not developed; stipule-sheath 2–4 mm. long, divided into up to 4 filiform fimbriae 2–4(–7) mm. long. Flowers heterostylous, in dense heads 0·8–1·2 cm. wide; true peduncle 0–5 mm. long but the leaves of the nodes containing the head are sometimes so reduced or hidden that the previous internode simulates a peduncle 2–3·5(–6) cm. long; pedicels usually

* Schumann names the species *O. wiedenmannii*, but cites the collector as Wiedemann; Krause in E.J. 39: 519 (1907) uses the name *O. wiedenmannii* and says collected by Wiedenmann. Bremekamp assumes a mistake, but both are possible German names.

scarcely 0·5 mm. long or not developed, rarely up to 3(–8) mm. long. Calyx-tube subglobose, 1 mm. long, 1·5 mm. wide, hairy or glabrous; lobes triangular to triangular-lanceolate, 1·5–2·5(–3) mm. long, puberulous or at least ciliate on the margins and the midvein. Corolla white, pale lilac, pale pink or bluish; tube (1·5–)2–2·8(–4) mm. long, hairy at the throat; lobes oblong-elliptic, 1·6–2·2 mm. long, 0·8–1 mm. wide, papillate inside and with a few scattered long hairs outside. Style 1·8 mm. long in short-styled flowers, 3 mm. long in long-styled flowers, hairy at the base; stigma-lobes linear, 0·4–1·2 mm. long, densely covered with rather long capitate hairs. Capsule subglobose, 1–1·8 mm. tall, 1·5–2 mm. wide, usually hairy, rarely ± glabrous, the beak slightly raised. Seeds yellow-brown, irregularly ellipsoid, 0·4 mm. long, usually raised on hilar side, flattened on opposite side.

var. **wiedemannii**

Inflorescence compact; calyx-lobes usually 1·5–2·5 mm. long; corolla-tube 2·5–3 mm. long. Fig. 41/27, p. 271.

KENYA. Northern Frontier Province: Ndoto Mts., Losirikan, 5 Jan. 1957, *Newbould* 3350!; Machakos District: bottom of Mua Hills, 30 May 1958, *Verdcourt & Napper* 2167!; Masai District: just S. of Kajiado, 15 June 1957, *Greenway* 9196!

TANZANIA. Musoma District: Central Serengeti Plains, Engare Nanyuki, 4 June 1962, *Greenway & Turner* 10695!; Mbulu District: Magugu, 7 May 1962, *Polhill & Paulo* 2379!; Lushoto District: Sunga–Manolo road, 26 May 1953, *Drummond & Hemsley* 2781!

DISTR. **K**1, 3, 4, 6, 7; **T**1–3, 5, 7; not known elsewhere

HAB. Usually rocky places in bushland and thicket, mostly on volcanic soil, also gravel screes, limestone pavements, lake shores, old cultivations and dry evergreen juniper forest; 340–1950(–2400) m.

SYN. *O. uhligii* K. Schum. & K. Krause in E.J. 39: 518 (1907). Type: Tanzania, NE side of Mt. Meru, *Uhlig* 718 (B, holo. †, EA, iso. !)

O. kaessneri K. Schum. & K. Krause in E.J. 39: 520 (1907). Type: Kenya, Machakos District, Sultan Hamud, *Kassner* 665 (B, holo. †, BM!, K!, MO, iso.)*, *non* S. Moore (1905), *nom. illegit.*

O. wiedemannii K. Schum. var. *glabricaulis* Bremek. in Verh. K. Nederl. Akad. Wet., Afd. Natuurk., ser. 2, 48(2): 241 (1952). Type: based on *O. kaessneri* K. Schum. & K. Krause

NOTE. Bremekamp gives the Kew sheet *Kassner* 653 as the type of his var. *glabricaulis*, but since he cites *O. kaessneri* in synonymy the holotype must be the specimen destroyed at Berlin. I have not kept up var. *glabricaulis*; there is considerable variation in the indumentum of this species and many specimens have glabrescent stems. Moreover the isotype definitely has some traces of indumentum. A specimen from **T4** (Buha District, Matende, 15 Aug. 1950, *Bullock* 3168) is very atypical in having some pedicels up to 6 mm. long but this may have been due to unusual burning, the plant appearing to have been burnt after the young shoots had appeared. *Peter* 55730 (Tanzania, Lushoto District, Lake Manka near Mkomazi, 6 June 1915) consists of 7 small annuals with smaller inflorescences than normal, shorter calyx-lobes and less indumentum, but I believe they are *O. wiedemannii* flowering during the first year.

var. **laxiflora** *Verdc.* in K.B. 30: 293 (1976). Type: Tanzania, Ruaha National Park, *A. Bjørnstad* 2558 (O, holo. !, EA, iso. !)

Inflorescences laxer; calyx-lobes 2·8–3·2 mm. long; corolla-tube 2 mm. long.

TANZANIA. Iringa District: Ruaha National Park, track to Ikuka Flats, 23 km. N. of Msembe, 15 Feb. 1973, *A. Bjørnstad* 2558!

DISTR. **T**7; not known elsewhere

HAB. Marginal *Brachystegia* woodland; 1130 m.

NOTE. *Napier* 954 (Kenya, Teita District, Voi, Mwatate, 6 May 1931) has some pedicels to 8 mm. and comes near to this variety.

* The K, BM and MO sheets are numbered 653 and it is very probable that the number cited in E.J. 39 is erroneous.

28. **O. ichthyoderma** *Cufod.* in Nuov. Giorn. Bot. Ital., n.s. 55: 83 (1948); Bremek. in Verh. K. Nederl. Akad. Wet., Afd. Natuurk., ser. 2, 48 (2): 242 (1952). Ethiopia, Sagan Omo, Lake Rudolf, Elolo, *Corradi* 2754 (FI, lecto.!)

Perennial herb 15–25 cm. tall, with usually several stems from a thin woody rootstock; stems simple or sparsely branched, ± 4-angled, scabrid-papillate. Leaf-blades linear-lanceolate, 1·5–5(–7) cm. long, 1–3·5 mm. wide, acute and apiculate at the apex, narrowed to the base, sessile, scabrid with short papilla-like hairs above and on the main nerve beneath; stipule-sheath 3–5 mm. long, ± scabrid, with 5–7 setae 2–5 mm. long. Flowers heterostylous, in condensed sessile terminal heads up to 1·1 cm. wide; pedicels 1–2 mm. long. Calyx-tube turbinate, 1·4 mm. long, very slightly scabrid; lobes ovate-lanceolate, 3–3·3 mm. long, 1·4 mm. wide, joined at the base for ± 0·4–1 mm., green in the middle but with broad hyaline margins 0·75 mm. wide, scabrid on the margins and midribs. Corolla white; tube 3–3·3 mm. long, hairy at the throat; lobes oblong-elliptic, 1·5–1·8 mm. long, 1·2 mm. wide, uncinate at the apex, with longer papilla-like hairs inside. Filaments exserted ± 0·3–0·7 mm. in short-styled flowers, the style 1·7–2 mm. long, pilose; stigma-lobes 0·5–0·8 mm. long. Anthers included in long-styled flowers, the style exserted 1·8 mm., the stigma-lobes 0·5 mm. long. Capsule 1·7–2 mm. long, 2·2 mm. wide, ± glabrous.

KENYA. Northern Frontier Province: Wamba–Archers Post road, S. of Mt. Lolokwi [Ololokwe], 22 Dec. 1972, *Bally & Smith* 14748!; Meru District: Meru National Park, ridge between Mughwango North and South swamp, 21 May 1972, *Ament & Magogo* 328!; Tana River District: 1·6 km. N. of SKT 15, 21 Dec. 1964, *Gillett* 16518!
DISTR. **K**1, 4, 7; Somali Republic (N.) and Ethiopia
HAB. *Acacia-Commiphora* open bushland; 100–950 m.

NOTE. It seems unlikely that *O. capitata* Bremek. (in Verh. K. Nederl. Akad. Wet., Afd. Natuurk., ser. 2, 48: 241 (1952); type: Somali Republic (N.), near Erigavo, El Sheikh, *Peck* 216 (EA, holo.!, K, iso.!)) is any more than a very minor variant of *O. ichthyoderma*. The stipule fimbriae are certainly much shorter, the habit is somewhat smaller and there are slight differences in the calyx.

29. **O. nematocaulis** *Bremek.* in Verh. K. Nederl. Akad. Wet., Afd. Natuurk., ser. 2, 48 (2): 222 (1952); Verdc. in K.B. 30: 295, fig. 8 (1975). Type: Tanzania, Rungwe District, between Tukuyu and Mporoto, Mbeye, *St. Clair-Thompson* 787 (K, holo.!)

Slender strict annual herb 2–20 cm. tall, with either unbranched or single branched stems with some of the branchlets basal and almost as long as or overtopping the main stem; stems angled, densely to sparsely scabridulous with short papilla-like hairs, particularly on the angles. Leaf-blades linear to linear-lanceolate, 0·25–2·2 cm. long, 0·1–1·5 mm. wide, acute at the apex, scarcely narrowed at the base, scabridulous-pubescent or glabrous above and scabridulous on margins and midrib beneath, revolute; petiole not developed; stipule-sheath 0·7–2 mm. long, divided into (1–)2–5 filiform pubescent setae 0·5–3·5 mm. long. Flowers heterostylous, in sessile mostly subglobose 1–many-flowered inflorescences supported by the 2 terminal leaves or sometimes the apical internodes suppressed and lower axillary inflorescences joined with the terminal one and apparently supported by a whorl of 4 leaves, up to 4–9 mm. in diameter in fruit but minute specimens may bear only 1 fruit; pedicels often purplish, 0·2–1·5 mm. long, sometimes papillate. Calyx-tube ovoid or subglobose, 0·6–0·8 mm. long, scabridulous; lobes often purplish, linear-lanceolate, 1–3 mm. long, mostly 0·1–0·2 mm. wide, probably attaining the maximum length only in fruit, the margins scabridulous-ciliate. Corolla white, pink, bluish white or pale mauve; tube slender, 2–3·3

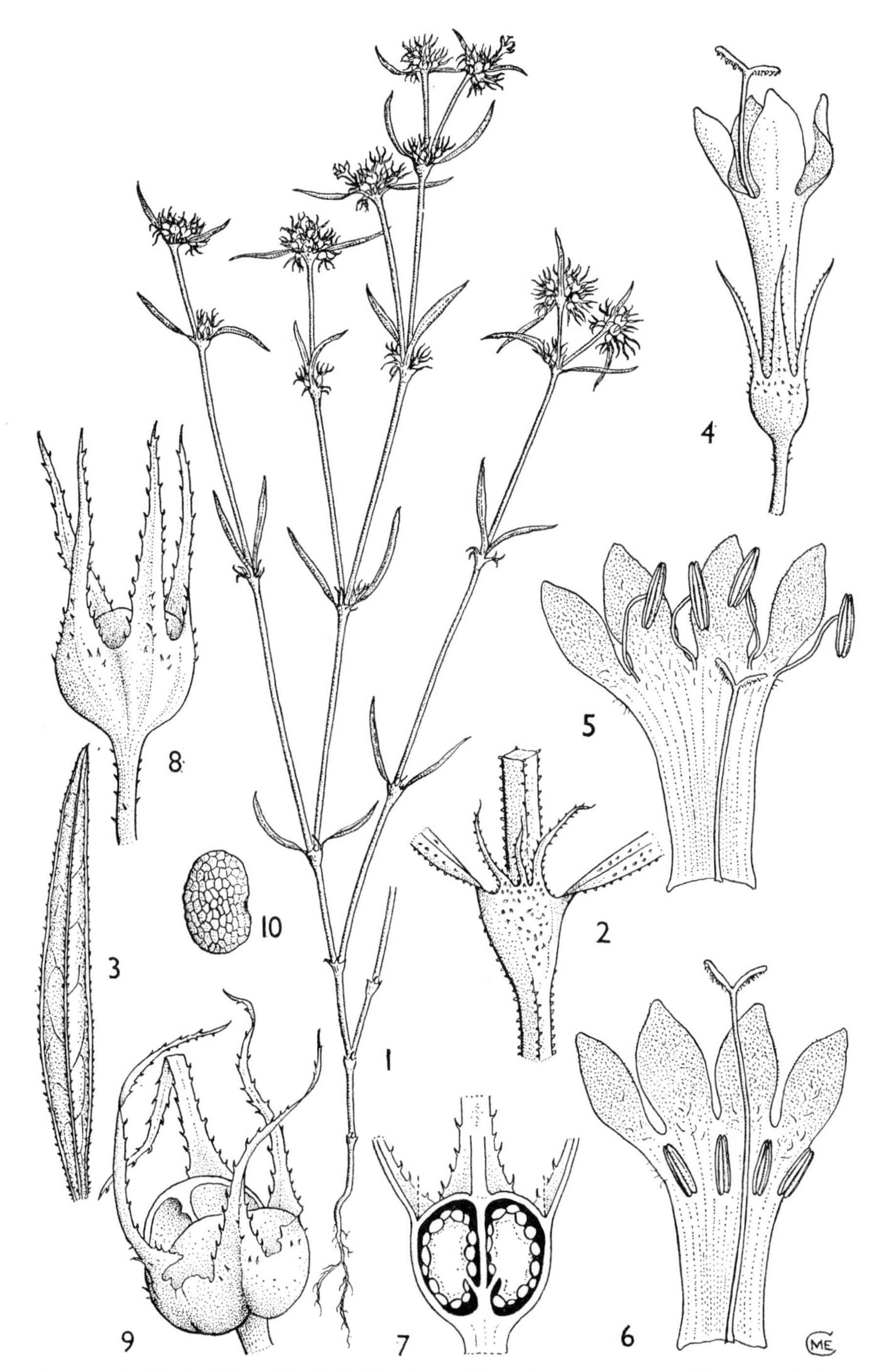

FIG. 44. *OLDENLANDIA NEMATOCAULIS*—**1,** habit, × 1; **2,** node showing stipule, × 6; **3,** lower surface of leaf, × 4; **4,** long-styled flower, × 10; **5,** corolla opened out, and style of short-styled flower, × 12; **6,** corolla opened out, and style of long-styled flower, × 12; **7,** longitudinal section of ovary, × 20; **8,** capsule, × 12; **9,** dehisced capsule, × 12; **10,** seed, × 40. All from *Tanner* 4973. Drawn by Mrs M. E. Church.

mm. long, widened at the throat, upper half hairy inside in short-styled forms, throat densely barbate in long-styled forms; lobes oblong-elliptic, 1–1·5 mm. long, 0·7–0·9 mm. wide. Filaments exserted 1 mm. in short-styled flowers, included in the throat in long-styled flowers. Style 2 mm. long in short-styled flowers, the stigma-lobes 0·8 mm. long, 3·3–4·3 mm. long in long-styled flowers, the stigma-lobes 0·3–1 mm. long. Capsule subglobose to transversely oblate, 0·9–1·8 mm. tall (including the beak), 1·2–1·8 mm. wide, papillate, slightly scabridulous or glabrous, the beak very rounded, scarcely raised or ± 0·5 mm. tall, the persistent calyx-lobes slightly decurrent. Seeds brown, ellipsoid, 0·25–0·35 mm. long, reticulate, the testa-cells granular.

var. **nematocaulis**

Flowers with pedicels 0·2–1·5 mm. long. Figs. 40/29, p. 270, 41/29, p. 271, and 44, p. 303.

TANZANIA. Ngara District: Bushubi, Keza, 15 May 1960, *Tanner* 4973!; Buha District: Mbirira [Birira]–Kisuzi [Nisusi], 27 Feb. 1926, *Peter* 37864!; Kigoma District: Uvinsa, N. of Lugufu towards Kigamba, Mwagao, 10 Feb. 1926, *Peter* 36724b!; Rungwe District: Mbeya to Tukuyu road, Usafwa, Isongole, 7 May 1975, *Hepper, Field & Mhoro* 5361!

DISTR. T1, 4, 7; Burundi, N. Zambia

HAB. Bushland and grassland on ironstone, damp grassland, thin soil on pumice lava rubble and in rock crevices; 835–1860 m.

NOTE. I at first considered that *Tanner* 4973, a robust fruiting specimen, represented a new species and did not associate it with Bremekamp's species. Examination of 4 gatherings made by Peter show that intermediates occur. As with so many ephemerals the stature varies enormously according to conditions. Bremekamp describes the flowers of this species as isostylous, but the style is well exserted in the type although all of the several specimens seem to be of the same form.

var. **pedicellata** *Verdc.* in K.B. 30: 296 (1975). Type: Tanzania, near Malagarasi R. bridge on Kibondo–Kasulu road, *Mutch* 66 (EA, holo.!)

Flowers with pedicels attaining 2–4 mm.

TANZANIA. Buha District: 0·2 km. downstream from Malagarasi R. bridge on Kibondo–Kasulu road, 9 Mar. 1972, *Mutch* 86! & 0·4 km. on same road from same point, 24 Feb. 1972, *Mutch* 66!

DISTR. T4; not known elsewhere

HAB. Bare ground, and grassy places on lava soil; about 1500 m.

30. **O. forcipitistipula** *Verdc.* in K.B. 30: 695, fig. 2 (1976). Type: Tanzania, Kigoma District, SW. of Malagarasi Railway Station, *Peter* 35841 (K, holo.!, B, iso.)

Slender annual herb 5–10 cm. tall, mostly unbranched or with a few lateral branches; stems angular, mostly densely pubescent particularly on the angles. Leaves held erect, the blades linear, 0·7–1·7 cm. long, 0·5–1·2 mm. wide, apiculate at the apex, passing unnarrowed into the stipule-base, pubescent on upper surface and on midrib beneath; stipule-base 1–1·2 mm. long, pubescent, the limb various, sometimes with a rectangular lobe 1 mm. long bearing 2 fine subulate setae 2·2 mm. long or less often with several additional setae, the outer pair often recurved and horn-like and sometimes with a single seta from an ovate lobe or merely a triangular lobe 1·3–2·6 mm. long. Flowers isostylous, in 2–5-flowered cymes; peduncles 3·5–6·5 mm. long, pubescent; pedicels 1–3 mm. long, pubescent. Calyx pubescent; tube hemispherical, 0·8 mm. long, 1 mm. wide; lobes lanceolate, 0·9 mm. long. Corolla white; tube 1–1·6 mm. long, sparsely hairy inside below the middle; lobes ovate, 1–1·2 mm. long, 0·8 mm. wide, papillate inside. Tips of anthers just exserted; style 1·5 mm. long but well exserted in the case of short corolla-tubes, with literally 2–3 hairs; stigma-lobes 0·5 mm. long, papillate. Capsule transversely rounded-oblong, 1·1 mm. long, 1·7 mm. wide, sparsely hairy. Seeds pale chestnut coloured, angular-ovoid, 0·35 mm. long, 2·6 mm. wide, 2 mm. thick, reticulate, the testa becoming slimy when moistened.

TANZANIA. Kigoma District: SW. of Malagarasi Railway Station, 30 Jan. 1926, *Peter* 35841 ! & 35842c !
DISTR. **T4**; not known elsewhere
HAB. Rocky hills; 1062–1100 m.

31. **O. herbacea** (*L.*) *Roxb.*, Hort. Bengal.: 11 (1814) & Fl. Indica, ed. Carey & Wall. 1: 445 (1820); DC., Prodr. 4: 425 (1830); Bremek. in Verh. K. Nederl. Akad. Wet., Afd. Natuurk., ser. 2, 48 (2): 244 (1952); Hepper in F.W.T.A., ed. 2, 2: 212 (1963); F.P.U., ed. 2: 160, fig. 101 (1972); U.K.W.F.: 401 (1974). Type: Ceylon, *Hermann* 4.19 (BM, holo. !)

Annual or perennial erect, decumbent or spreading branched herb 7–60 cm. tall, with glabrous 4-ribbed stems. Leaf-blades linear to linear-lanceolate, 0·6–5·5 cm. long, 1–3·5(–4) mm. wide, acute at the apex, cuneate at the base, glabrous or with a few setae at the margins; petiole not developed; stipule-sheath short, rarely exceeding 0·5 mm., truncate, with a few setae ± 0·3 mm. long but not fimbriate. Flowers usually isostylous but in one variety markedly heterostylous, solitary or paired at the nodes; pedicels graceful, spreading (0·3–)0·8–3 cm. long. Calyx-tube ovoid, 0·5–1 mm. long, glabrous, papillate or shortly hairy; lobes narrowly triangular, 0·5–2·5 mm. long, scabridulous on the margins. Corolla white, lilac or mauve or tube green and lobes mauve with purple marks; tube cylindrical, 0·2–1·1 cm. long; lobes ovate, 1–3 mm. long. Stigma-lobes filiform, 0·7–0·9 mm. long. Capsule pale straw-coloured, subglobose, 2·2–5 mm. long, 1·5–2 mm. wide, crowned by the blackish calyx-lobes, glabrous, papillate or shortly hairy; beak 0·8–1 mm. long. Seeds brown, ovoid to ellipsoid, angular, 0·2–0·4 mm. long, reticulate.

KEY TO INFRASPECIFIC VARIANTS

Corolla-tube short, 2–3·7(–4) mm. long; flowers not heterostylous, the anthers never well exserted:
- Calyx-tube and capsule glabrous . . . . . var. **herbacea**
- Calyx-tube and capsule papillate or shortly hairy . var. **papillosa**

Corolla-tube longer, (3·5–)7–9(–11) mm. long:
- Flowers distinctly heterostylous, the anthers well exserted in short-styled plants; mostly perennial decumbent herbs with less slender corolla-tube . . . . var. **holstii**
- Flowers not distinctly heterostylous, the anthers either just included or only with the tips protruding from the throat; mostly erect annuals with very slender corolla-tube . . . var. **goetzei**

var. **herbacea**; Bremek. in Verh. K. Nederl. Akad. Wet., Afd. Natuurk., ser. 2, 48(2): 245 (1952)

Mostly erect annual herbs; flowers small, isostylous, with corolla-tube 2–3·7 mm. long; ovary and capsule glabrous. Fig. 41/31, p. 271.

UGANDA. W. Nile District: Koboko, May 1938, *Hazel* 471 !; Teso District: Serere, 26 Oct. 1955, *Langdale-Brown* 1616 !; Masaka District: Sese Is., Bugala I., 8 Oct. 1958, *Symes* 476 !
KENYA. Trans-Nzoia District: 12·8 km. from Kitale on road to Cherangani Hills, 12 Aug. 1961, *Symes* 760 !; Machakos District: Mbooni Hills, 9 Oct. 1947, *Bogdan* 1208 !; Kwale District: Mariakani, 17 Sept. 1961, *Polhill & Paulo* 485 !

TANZANIA. Musoma District: Majita, Bwasi, 23 Mar. 1959, *Tanner* 4088 !; Rungwe Mt., 13 Sept. 1932, *Geilinger* 2200 !; Newala, 9 Apr. 1959, *Hay* 52 !; Pemba I., Mtangatwani, 30 Aug. 1929, *Vaughan* 618 !

DISTR. **U**1–4; **K**3–7; **T**1–4, 6–8; **P**; widespread in tropical and southern Africa; also in Asia

HAB. Grassland, thicket, bushland and wooded grassland, often on rock outcrops; also in old cultivations; 0–2160 m.

SYN. *Hedyotis herbacea* L., Sp. Pl.: 102 (1753)
Oldenlandia heynii G. Don., Gen. Syst. 3: 531 (1834); Oliv. in Trans. Linn. Soc. 29: 84 (1873); Hiern in F.T.A. 3: 59 (1877), pro parte; K. Schum. in P.O.A. C: 376 (1895). Type: India, *Heyne* (K-WALL. 867, ? iso. !)
Hedyotis dichotoma A. Rich., Tent. Fl. Abyss. 1: 361 (1847), *non* Roth., *nom. illegit.* Type: Ethiopia, Shire (Chiré), *Quartin Dillon* (P, holo.)
H. heynii (G. Don) Sond. in Fl. Cap. 3: 10 (1865)

NOTE. Bremekamp has described a var. *flaccida* (in Verh. K. Nederl. Akad. Wet., Afd. Natuurk., ser. 2, 48(2): 248 (1952)); no type is cited but it is probably meant to be based on *Hedyotis trichopoda* A. Rich., Tent. Fl. Abyss. 1: 361 (1847), type: Ethiopia, Chiré, *Quartin Dillon & Petit* (P, holo.). It has leaf-blades up to 2 mm. wide rather than 1 mm. and although extremes can be very marked, e.g. Uganda, Busoga District, Lolui I., 18 June 1953, *G. H. Wood* 798 ! variants with longer wider leaves occur sporadically throughout the range of the typical plant and every intermediate occurs; it does not seem worth recognizing.

var. **papillosa** (*Chiov.*) *Bremek.* in Verh. K. Nederl. Akad. Wet., Afd. Natuurk., ser. 2, 48(2): 249 (1952). Type: Somali Republic (S.), Bur Meldac, *Paoli* 703 (FI, holo. !)

Mostly erect annual herbs; flowers small, isostylous, with corolla similar to that of var. *herbacea*; ovary and capsule distinctly papillose or densely shortly hairy.

UGANDA. W. Nile District: summit of Mt. Wati [Eti], 19 July 1953, *Chancellor* 7 in part !

KENYA. Northern Frontier District: Dandu, 9 May 1952, *Gillett* 13140 !; Masai District: Emali, 10 Mar. 1940, *V. van Someren* 82 ! & SE. of Sultan Hamud, Iltoroto Hill, 26 Feb. 1969, *Napper* 1907a !

TANZANIA. Shinyanga District: Old Shinyanga, 7 Apr. 1952, *Welch* 151 !; Mbulu District: Pienaars Heights, 4 May 1963, *Polhill & Paulo* 2327 !; Dodoma District: 20·5 km. E. of Itigi Station, 11 Apr. 1964, *Greenway & Polhill* 11511a !

DISTR. **U**1; **K**1, 6; **T**1–8; Rwanda, Somali Republic (S.), Zambia

HAB. Grassland, bushland, open *Brachystegia* woodland; (0–)300–1650 m.

SYN. *O. dichotoma* A. Rich. var. *papillosa* Chiov. in Result. Sci. Miss. Stef.–Paoli, Coll. Bot.: 89 (1916)

var. **holstii** (*K. Schum.*) *Bremek.* in Verh. K. Nederl. Akad. Wet., Afd. Natuurk., ser. 2, 48(2): 249 (1952). Type: Tanzania, Lushoto District, Kwa Mshusa [Mschuza], *Holst* 8942 (B, holo. †, BM, iso. !)

Mostly decumbent perennial herbs; flowers larger, heterostylous, the anthers long exserted in brachystylous form, the corolla-tube (3·5–)7–9 mm. long; ovary and capsule glabrous.

KENYA. Northern Frontier Province: Moyale, 6 July 1952, *Gillett* 13515 !; Teita District: Ndara Mts., Feb. 1871, *Hildebrandt* 2435 ! & Wesu Hospital, May 1955, *Ossent* 87 !

TANZANIA. Lushoto District: E. Usambara Mts., Ndola, 24 May 1950, *Verdcourt & Greenway* 219 !; Ufipa District: Nsanga Mt., Malonje Plateau, 13 Mar. 1959, *Richards* 12107 !; Njombe District: 11 km. S. of Njombe, by R. Hagafilo, 8 July 1956, *Milne-Redhead & Taylor* 10779 !

DISTR. **K**1, 7; **T**2–4, 6, 7; Ethiopia, Somali Republic (N.)

HAB. Montane scrub, *Philippia* heath, grassland, open woodland, often on rock outcrops; 600–2190 m.

SYN. *O. holstii* K. Schum. in P.O.A. C: 376 (1895) & in E.J. 30: 411 (1901), excl. *Goetze* 840

NOTE. Extremes of this variant are very distinctive and it is difficult to believe that it is not specifically distinct from *O. herbacea*, but there is considerable variation in the length of the corolla-tube. Some specimens with the corolla-tube only 2 mm. long are heterostylous, e.g. *Lewis* 5931 (Kenya, Teita Hills, SE. slope of Mt. Vuria, 13 Sept. 1962).

var. **goetzei** *Bremek.* in Verh. K. Nederl. Akad. Wet., Afd. Natuurk., ser. 2, 48(2): 249 (1952). Type: Tanzania, Tukuyu [Langenburg], *Goetze* 840 (K, holo. !, EA, iso. !)

Annual or perennial mostly erect herb; flowers larger, isostylous or at least the anthers never long-exserted, the corolla-tube 0·7–1·1 cm. long; ovary and capsule glabrous.

TANZANIA. Songea District: about 6 km. N. of Songea, Chandamara Hill, 23 Mar. 1956, *Milne-Redhead & Taylor* 9287 ! & 5 km. E. of Songea, Unangwa Hill, 22 Mar. 1956, *Milne-Redhead & Taylor* 9274 !; Tunduru District: just E. of Songea District boundary, 5 May 1956, *Milne-Redhead & Taylor* 10559 !
DISTR. **T**7, 8; Zambia, Malawi, Rhodesia
HAB. Secondary grassland, *Brachystegia* woodland, often amongst rock outcrops; 855–1200 m.

SYN. [*O. holstii* sensu K. Schum. in E.J. 30: 411 (1901) quoad *Goetze* 840, *non* K. Schum. (1895)]

NOTE. Apart from the floral character given by Bremekamp there are certainly distinct differences in habit and corolla shape; it is more erect than var. *holstii* with narrow leaves and drying a paler colour; the corolla-tube is more slender. More evidence is needed about heterostyly; it is quite probable that Bremekamp is correct in assuming it is of a different character from that found in var. *holstii*. The style varies in length considerably, being included to well exserted, but the anthers are either just included or have their tips just protruding from the throat.

32. **O. pumila** (*L.f.*) *DC.*, Prodr. 4: 425 (1830); Bremek. in Verh. K. Nederl. Akad. Wet., Afd. Natuurk., ser. 2, 48 (2): 250 (1952). Type: India, Tranquebar, *Koenig* (not found)

Annual herb, much branched at the base; stems mostly prostrate, 3–17 cm. long, scabridulous on the ribs. Leaf-blades elliptic to elliptic-lanceolate or linear-lanceolate, 0·5–1·6(–3) cm. long, 1–6 mm. wide, acute at the apex, narrowed to the base, minutely sparsely scabrid-setulose above, distinctly scabridulous on the margins, ± glabrous beneath; petioles not developed; stipule-sheath 0·5–2 mm. long, produced at the middle, with 3–5 unequal fimbriae up to 0·8(–1·2) mm. long. Flowers not heterostylous, usually solitary at the nodes, rarely apparently in terminal pedunculate 2–3-flowered cymes, or in axillary 2-flowered cymes; peduncle 3–9 mm. long; pedicels ± 2–3 mm. long, lengthening to 1·1(–1·9) cm. in fruit, sometimes minutely scabridulous. Calyx-tube ellipsoid to subglobose, ± 1 mm. long; lobes triangular, 1–1·2(–3) mm. long, 0·5–1 mm. wide, minutely setulose on the margins. Corolla white; tube 1 mm. long; lobes ovate, 0·6–0·7 mm. long and wide. Style 0·8–1 mm. long, hairy; stigma-lobes 0·7 mm. long. Capsule oblong-ovoid, usually distinctly taller than wide, ± 2–3 mm. long, 1·5–2·2 (–2·8) mm. wide, beak 0·2–0·7 mm. tall. Seeds brown, obtusely broadly conic or ellipsoid, ± 0·3 mm. long, strongly reticulate. Fig. 41/32, p. 271.

UGANDA. Toro District: Rwenzoli National Park, NW. shore of Lake Edward, 4 Oct. 1962, *Lewis* 6009 ! & same area, Kayanja, 5 Oct. 1968, *Lock* 68/222 !; Ankole District: Lake Kyasanduka, 19 Nov. 1967, *Lock* 67/133 !
TANZANIA. Morogoro, May 1930, *Haarer* 1915 !
DISTR. **U**2; **T**6; an Asian species (India, Indo-China, W. part of Malesia) introduced into Africa and also found as a weed in Jamaica
HAB. Dried black cotton soil along lake shores (but unknown for Tanzania specimen); 480–990 m.

SYN. *Hedyotis pumila* L. f., Suppl. Pl.: 119 (1781)
Oldenlandia crystallina Roxb., Fl. Indica, ed. Carey & Wall. 1: 443 (1820); Hook. f., Fl. Brit. India 3: 65 (1880); S. Moore in Fawcett & Rendle, Fl. Jam. 7: 34 (1936), *nom. superfl.* based on *Hedyotis pumila* L.f.

NOTE. The testa-cells are usually coarsely granulate in this species, but quite a number of seeds taken from *Haarer* 1915 had smooth testa-cells; possibly the granulation is not evident until the seed is quite mature or there may be variation in this character. *Lock* 68/222 cited above also has the cells not punctate. There are, however, difficulties; several specimens from Uganda determined by Bremekamp as *O. caespitosa* (Benth.) Hiern var. *major* Bremek. (e.g. Teso District, Serere, Dec. 1931, *Chandler* 128) are very similar in habit to *O. pumila* but most of the flowers are in pedunculate 2-flowered cymes; the testa-cells moreover are granulate in *some* of these specimens. Despite Bremekamp's description of *O. pumila* as having flowers nearly all solitary, Indian material quite often has a number of the flowers in 2-flowered cymes. In the W. Indies and Asia *O. pumila* and *O. corymbosa* are abundantly distinct but in Africa some variants of *O. corymbosa* seem scarcely distinguishable.

33. **O. corymbosa** *L.*, Sp. Pl.: 119 (1753); Hiern in F.T.A. 3: 62 (1877), pro parte; K. Schum. in P.O.A. C: 375 (1895), pro parte; F.W.T.A. 2: 132 (1931), pro parte; Bremek. in Verh. K. Nederl. Akad. Wet., Afd. Natuurk., ser. 2, 48 (2): 254 (1952); Hepper in F.W.T.A., ed. 2, 2: 211 (1963); Lewis in Grana Palynologica 5: 330 (1964) & in Ann. Missouri Bot. Gard. 53: 257 (1966); Hallé, Fl. Gabon, 12 Rubiacées: 99, fig. 17/3 (1966); U.K.W.F.: 401, fig. as "*linearis*" (1974). Type: based on *Oldenlandia humilis hyssopifolia* Plumier, Nova Plantarum Americanarum Genera: 42, t. 36 (1703); probably a specimen from Martinique collected by Plumier* now no longer in existence

Annual herb, sparsely to very densely branched near the base; stems prostrate to ± erect, 1·5–30 cm. long, ridged, glabrous or scabridulous or pubescent on the ribs. Leaf-blades linear to narrowly elliptic, 0·6–3·5(–5·5) cm. long, 0·5–7 mm. wide, acute and apiculate at the apex, narrowed to the base, glabrous to sparsely scabridulous above and on margins and also beneath, particularly on the main nerve; petioles not developed; stipule-sheath 0·5–2(–3) mm. long, produced at the middle with (2–)3–5 unequal fimbriae, 0·5–1(–2·5) mm. long. Flowers not heterostylous, variously arranged, either 1–several single flowers in the axils or in 2–5(–6)-flowered pedunculate umbel-like inflorescences, both kinds present on one branch or even at one node, the peduncles and pedicels mostly long and slender but rarely the flowers are fasciculate; peduncles (0–)0·5–1·8(–2·3) cm. long; pedicels (1·8–)3–6(–13) mm. long. Calyx-tube ellipsoid, 0·7–1 mm. long; lobes triangular, 0·5–1·8 mm. long, setulose on the margins. Corolla white or tinged blue, pink or purple or with 2 pink stripes on each lobe; tube 0·6–1 mm. long; lobes ovate to oblong, 0·5–1·2 mm. long. Style 0·5–1·5 mm. long. Capsule ovoid, 1·2–2·2 mm. tall, (1–)1·8–2·8 mm. wide, the beak scarcely raised. Seeds pale brown, ellipsoid or very obtusely depressed conic, ± 0·3 mm. long, reticulate.

KEY TO INFRASPECIFIC VARIANTS

Inflorescences nearly all 2–5(–6)-flowered:
- Leaves mostly elliptic, up to 7 mm. wide . . . var. **corymbosa**
- Leaves mostly linear, 1–1·5(–3) mm. wide; plants more strictly erect with leaves held more erect than in other variants:
  - Plants mostly over 10 cm. tall . . . . var. **linearis**
  - Plants very small, mostly 1·5–7 cm. tall . . var. **nana**

Inflorescences mostly all 1-flowered . . . . var. **caespitosa**

var. **corymbosa** (typical form)

Plants erect or prostrate. Leaf-blades mostly narrowly elliptic. Inflorescences nearly all 3–5(–6)-flowered. Style glabrous.

KENYA. Kisumu-Londiani District: Lake Victoria, Nanga, Aug. 1934, *Turner* in *C.M.* 6714!; Kilifi District: Malindi, 10 Aug. 1949, *Bogdan* 2565!; Lamu District: 88 km. NE. of Lamu, 23 July 1961, *Gillespie* 22!

* See Plumier, ed. J. Burman, Plantarum Americanarum . . . t. CCXII/I (1755–60) and Urban in F.R. Beih. 5: 87 (1920).

TANZANIA. Ufipa District: Rukwa Valley, Milepa, 21 Feb. 1947, *Pielou* 108!; Uzaramo District: Dar es Salaam, 26 Feb. 1971, *Wingfield* 1165!
DISTR. **K**5, 7; **T**4–6; widespread in tropical Africa and also as a weed throughout most of the world; probably native only in Africa and India
HAB. Open woodland, sandy edges of seasonal watercourses in dry scrub, as a weed in sandy places; 0–1110 m.

SYN. *Hedyotis corymbosa* (L.) Lam., Tab. Encycl. 1: 272 (1792)

NOTE. *G. H. Wood* 649 (Uganda, Busoga District, Bukoli, Lugala Landing, on shore of Lake Victoria, 26 Mar. 1953, growing in sand at edge of the lake shore, 1125 m.) is a curious plant having a tufted probably perennial (despite collector's "annual") habit, 4–5-flowered inflorescences with peduncle 1 cm. long, pedicels 2–3 mm. long and rather long calyx-lobes, thus resembling *O. scopulorum* in habit but growing in the wrong habitat and differing in the smaller isostylous flowers. At present I leave it as an anomalous form of *O. corymbosa* var. *corymbosa* not unlike *Turner* in *C.M.* 6714 cited above. *Newbould* 3385 (Kenya, Northern Frontier District, Ndoto Mts., Latakwen, 2 Jan. 1959) is intermediate between this and *O. fastigiata* Bremek. and in the W. Indies such intermediates are common. *O. fastigiata* has been maintained since it has become a ± fixed and easily recognisable entity in East Africa but it is very likely not a sound biological species; it would probably easily interbreed with all variants of *O. corymbosa*. *Polhill & Paulo* 2060 (Tanzania, Iringa District, 8 km. S. of point where the Great North Road cuts the Great Ruaha River, 18 Apr. 1962) is a similar plant. *Gillespie* 22 cited above shows great variation in the inflorescence; mostly the flowers are solitary on long peduncles and often up to 3 at one node or the peduncles are very short (a fascicle) or long (a cyme) bearing several flowers.
Lewis (*loc. cit.*) shows that there are three cytotypes, a diploid, a tetraploid and a hexaploid based on x = 9, all three being found in Africa. Bremekamp's suggestion of an African origin for the species is supported by cytopalynological evidence.

var. **corymbosa** (usual African form)

Plants erect or prostrate. Leaf-blades mostly narrowly elliptic. Inflorescences mostly 2(–3)-flowered but also few to many solitary flowers. Style glabrous or with a few hairs.

UGANDA. Lango District: 4·8 km. S. of Aboki [Aboke], 30 Sept. 1962, *W. H. Lewis* 6001!; Toro District: eastern slope of Ruwenzori, near Nyakalengija, 3 Oct. 1962, *W. H. Lewis* 6006!; Mengo District: 6·4 km. NE. of Entebbe, 7 Oct. 1962, *W. H. Lewis* 6019!
KENYA. Northern Frontier Province: Moyale, 14 Oct. 1952, *Gillett* 14041!; Embu District: Emberre, 29 Sept. 1932, *M. D. Graham* 2252!
TANZANIA. Lushoto District: 4·8 km. NE. of Lushoto, Magamba–Mkuzi road, 18 Apr. 1953, *Drummond & Hemsley* 2130!; Tabora, 21 Mar. 1958, *Jones* 61A!; Mpanda District: Rukwa Plain, Sonta, 26 Oct. 1963, *Richards* 18279!; Zanzibar I., Masingini Ridge, 1 Feb. 1929, *Greenway* 1275! & Mazizini [Massazine], 3 Feb. 1960, *Faulkner* 2482!
DISTR. **U**1–4; **K**1–4, 7; **T**2–4, 6, 7; **Z**; widespread throughout Africa, with similar forms in other parts of the world
HAB. Long and short grassland, bushland, montane scrub, shallow soil on rocks, sandy river ridges, furrows and dry ponds on black cotton soil, cultivated and disturbed ground; 0–2280 m.

SYN. [*O. caespitosa* (Benth.) Hiern var. *major* sensu Bremek. in Verh. K. Nederl. Akad. Wet., Afd. Natuurk., ser. 2, 48(2): 263 (1952), pro parte]

NOTE. Several Zanzibar specimens are intermediate with *O. fastigiata* and exactly similar to W. Indian intermediates of the same type. All variants of *O. corymbosa* have the potential of producing such intermediates.

var. **linearis** (*DC.*) *Verdc.* in K.B. 30: 296 (1975). Type: Senegal, Bay of St. Louis, *Perrottet* (G, holo., P, iso., K, photo.!)

Plants mostly strictly erect. Leaves held more erect than in other variants, blades usually linear (only 1–1·5(–3) mm. wide) and often longer. Inflorescences 1–2(–3)-flowered. Style glabrous. Capsules usually smaller than in var. *corymbosa*. Fig. 41/33, p. 271.

UGANDA. Karamoja District: near Nabilatuk, 9 Aug. 1956, *Dyson-Hudson* 100! & Pian, Moruangaberu [Emoruagaberru], 21 July 1958, *Dyson-Hudson* 471!; Teso District: Serere, Sept. 1931, *Chandler* 950!

KENYA. Turkana District: 19 km. N. of Kacheliba, 7 Oct. 1964, *Leippert* 5037!; Trans-Nzoia District: ENE. slope of Mt. Elgon, 24 Sept. 1962, *W. H. Lewis* 5977! & Kitale, 12 Sept. 1956, *Bogdan* 4275 in part!
TANZANIA. Mwanza District: Urima, Liyoma, 4 May 1954, *Tanner* 712!; Moshi, June 1927, *Haarer* 327!; Dodoma District: Itigi thicket, Itigi–Singida road, 26 Mar. 1965, *Richards* 19880!
DISTR. **U**1, 3; **K**2–4, 7; **T**1–8; Senegal, Sudan, Ethiopia, Burundi, Zaire, Zambia and Rhodesia
HAB. Short grassland, bushland, hard pans, shallow soil over rocks, also in roadside ditches and cultivations; 0–2280 m.

SYN. *O. linearis* DC., Prodr. 4: 425 (1830); Bremek. in Verh. K. Nederl. Akad. Wet., Afd. Natuurk., ser. 2, 48(2): 258 (1952); Hepper in F.W.T.A., ed. 2, 2: 211 (1963)

NOTE. This has nominally usually been maintained as a separate species from *O. corymbosa* and typical specimens are readily recognizable; there is, however, little difference except in the width of the leaves and many intermediates occur which are difficult to assign. On the coast intermediates with *O. fastigiata* occur, e.g. *Faulkner* 3621 (Tanzania, Tanga District, Sawa, 18 Aug. 1965) consists of specimens of normal var. *linearis* and intermediates with *O. fastigiata*. Bremekamp has synomymized *O. subtilis* S. Moore (in J.B. 43: 249 (1905); type: Kenya, Kitui District, Galunka, *Kassner* 781 (BM, holo.!, K!, MO, iso.)) with *O. linearis* but it is very different in habit, although this may be due to habitat differences—the slender specimens on the type sheets of *O. subtilis* clearly grew in very thin soil in rocky places. Moreover, in Kenya true var. *linearis* is an upland plant and no other material has been seen from **K**7. I prefer to consider it either a state of var. *caespitosa* or perhaps deserving of varietal rank.

var. **nana** (*Bremek.*) *Verdc.* in K.B. 30: 296 (1975). Type: Uganda, Mbale District, Bugisu, Bulago, *A. S. Thomas* 320 (K, holo.!)

Similar to var. *linearis* but differing in the small stature, only 1·5–7 cm. tall.

UGANDA. Kigezi District: without further locality, 12 Jan. 1933, *Rogers & Gardner* 351! & Kigezi, Mabungo Rest Camp, *Eggeling* 918!; Mbale District: Mt. Elgon, Kapchorwa, 7 Sept. 1954, *Lind* 245!
KENYA. Trans-Nzoia District: 12·8 km. from Kitale on the Cherangani road, 12 Aug. 1961, *Symes* 754! & ENE. slopes of Mt. Elgon, 23 Sept. 1962, *W. H. Lewis* 5970!
TANZANIA. Mpanda District: Kasoje, 17 July 1959, *Newbould & Harley* 4415! (showing range in size to larger specimens)
DISTR. **U**2, 3; **K**2, 3; **T**4; not known elsewhere
HAB. Short grassland, thin turf over rocks, paddocks, pathsides, etc.; 780–2280 m.

SYN. *O. linearis* DC. var. *nana* Bremek. in Verh. K. Nederl. Akad. Wet., Afd. Natuurk., ser. 2, 48(2): 259 (1952)

NOTE. Bremekamp would separate this from small specimens of broad-leaved variants of *O. corymbosa*, e.g. Uganda, Mengo District, Kampala, Makindye Hill, 9 Oct. 1962, *W. H. Lewis* 6026!, but gradual merging occurs in all directions. Var. *nana* may just represent specimens growing in thin soil but there is a geographical basis for recognition and many large specimens of *O. corymbosa* are to be found in similar habitats. Breeding experiments would soon establish if the variety was well founded. Var. *nana* also merges directly with var. *caespitosa*.

var. **caespitosa** (*Benth.*) *Verdc.* in K.B. 30: 298 (1975). Type: Liberia, Cape Palmas, *Vogel* 51 (K, holo.!)

Typically a small herb 2–30 cm. tall with several prostrate stems, but sometimes erect and unbranched. Leaf-blades typically small and narrowly elliptic, 0·7–2(–2·8) cm. long, 1·5–3(–4·5) mm. wide. Flowers typically all solitary at the nodes, but often also a few in 2-flowered cymes or sometimes, where reduced branchlets are leafless, simulating 2–3-flowered cymes. Style often sparsely hairy at the middle.*

UGANDA. Toro District: Nyakasura School, 20 May 1932, *Shillito* 103!; Mbale District: Bugisu, Bulucheke, 22 Feb. 1950, *L. M. Forbes* 227!; Mengo District: Entebbe, 13 Sept. 1961, *Rose* 199!

* Bremekamp uses this as a character to separate *O. caespitosa*, which he maintains as a species, from *O. capensis*, etc. On the type sheet, however, he has added style glabrous in pencil against his drawing.

KENYA. Northern Frontier Province: Moyale, 3 July 1952, *Gillett* 13476!; W. Suk District: 48 km. from Kitale on Kitale–Lodwar road, 4 Oct. 1952, *Verdcourt* 736!; Nairobi District: 17 Sept. 1962, *W. H. Lewis* 5948!

TANZANIA. Lushoto District: Korogwe, 29 May 1969, *Archbold* 1010!; Kilosa District: Mikumi National Park, Mkata Plain, 1 May 1968, *Renvoize* 1847!; Songea District: near Mshangano Fish Ponds, 21 Mar. 1956, *Milne-Redhead & Taylor* 9265!; Pemba I., Chake Chake, 23 Aug. 1929, *Vaughan* 575!

DISTR. **U**2–4; **K**1–5, 7; **T**1–3, ? 4, 6–8; **P**; widespread from Cape Verde Is., N. and W. Africa to the Somali Republic, Natal and Angola, Madagascar, Mascarene Is. and Middle East

HAB. Open grassland, degraded montane scrub, shallow seasonally wet crevices on rock outcrops, and as a weed of cultivated ground; 0–2300 m.

SYN. *O. herbacea* (L.) Roxb. var. *caespitosa* Benth. in Hook., Niger Fl.: 403 (1849)
O. caespitosa (Benth.) Hiern in F.T.A. 3: 61 (1877); Bremek. in Verh. K. Nederl. Akad. Wet., Afd. Natuurk., ser. 2, 48(2): 262 (1952); Hepper in F.W.T.A., ed. 2, 2: 212 (1963)
O. corymbosa L. var. *subpedunculata* O. Kuntze, Rev. Gen. Pl. 3: 121 (1893). Type: Mozambique, without locality, *O. Kuntze* (K, iso.!)
O. caespitosa (Benth.) Hiern var. *subpedunculata* (O. Kuntze) Bremek. in Verh. K. Nederl. Akad. Wet., Afd. Natuurk., ser. 2, 48(2): 263 (1952)
O. caespitosa (Benth.) Hiern var. *major* Bremek. in Verh. K. Nederl. Akad. Wet., Afd. Natuurk., ser. 2, 48(2): 263 (1952). Type: South Africa, Natal, Durban, *Gueinzius* 133 (P, holo.)

NOTE. Bremekamp maintains this as a distinct species and extremes of var. *corymbosa* and var. *caespitosa* are indeed very different. There are, however, very many specimens difficult to place. Presumably Bremekamp would include anything here which had a hairy style whether or not the flowers were solitary or paired. Certainly many of the specimens he has determined as *O. caespitosa* var. *major* are not specifically distinct from others determined as *O. corymbosa*. Moreover, many specimens of var. *major* from Uganda have the testa cells punctate just as in *O. pumila*. Bremekamp admits that *O. caespitosa* is difficult to define. At the other extreme of its range of variation var. *caespitosa* is difficult to distinguish from *O. capensis*. Here again Bremekamp uses the hairy versus glabrous style character; *Archbold* 837 (Tanzania, Dar es Salaam, 11 July 1966) and *Renvoize & Abdallah* 1847 (Tanzania, Kilosa District, Mikumi National Park, 1 May 1968) are examples of the many specimens difficult to place. *Newbould & Jefford* 2376, pro parte (Tanzania, Kigoma District, Mahali Mts., 6 Sept. 1958) is a small annual plant intermediate in characters between *O. fastigiata* and *O. corymbosa* var. *caespitosa*. Lewis (1964) shows that *O. corymbosa* and *O. caespitosa* are conspecific.

34. **O. fastigiata** *Bremek.* in Verh. K. Nederl. Akad. Wet., Afd. Natuurk., ser. 2, 48(2): 260 (1952); U.K.W.F.: 401 (1974). Type: Tanzania, Ulanga District, Mahenge, Mbangala, *Schlieben* 1797 (BR, holo., BM, K, iso.!)

Annual or perennial herb with stems mostly branched at the base, erect, straggling or ascending, (6–)15–60 cm. long, ribbed, glabrous or slightly to rarely densely scabridulous on the ribs. Leaf-blades linear to linear-lanceolate, 1·2–4·5(–6) cm. long, 1–5(–8) mm. wide, acute at the apex, narrowed to the base, entirely glabrous or with very small obscure papilla-like hairs above, particularly when very young, and sometimes on margins and main nerve beneath; petioles not developed; stipule-sheath 1–2 mm. long, scabrid-papillose, bearing 3–5 fimbriae of varying lengths, 0·5–2·5 mm. long. Flowers isostylous or heterostylous, in 3–7(–10)-flowered sessile or pedunculate axillary cymes or fascicles; peduncle 0–3 cm. long, usually very much shorter than the subtending leaf, typically $\frac{1}{4}$ the length; pedicels 1·5–4 mm. long, glabrous or rarely scabridulous. Calyx-tube subglobose, 0·5–0·8 mm. long, glabrous; lobes narrowly triangular, 1–2 mm. long, scabridulous on the margins, keeled. Corolla white or tinged blue or lilac-pink; tube 1–1·6 mm. long, the throat densely hairy; lobes ovate-oblong, 0·7–2 mm. long, 0·6 mm. wide, papillate inside. Style in short-styled flowers 0·6 mm. long, in long-styled flowers 2·8 mm. long, in isostylous flowers 0·6–1·2 mm. long; stigma-lobes 0·3–1 mm. long. Capsule depressed globose, 1–2 mm. tall, 1·5–2·5 mm.

wide, glabrous, the beak scarcely raised. Seeds pale brown, bluntly conical, angular, ± 0·2 mm. long, strongly reticulate.

KEY TO THE INFRASPECIFIC VARIANTS

Inflorescences sessile or shortly pedunculate, the peduncules up to 2 (rarely 10) mm. long; flowers not heterostylous . . . . var. **fastigiata**
Inflorescences distinctly pedunculate, the peduncles up to 3 cm. long; flowers heterostylous:
  Leaf-blades not so thin, up to 6 × 0·5 cm.; inflorescences not so lax, the pedicels up to 2 mm. long . . . . . var. **somala**
  Leaf-blades very thin, up to 7 × 1·2 cm.; inflorescences lax, the pedicels up to 5 mm. long . . . . var. **pseudopentodon**

var. **fastigiata**

Inflorescence mostly sessile or only shortly pedunculate, the peduncles 2 mm. long, rarely longer, up to 1 cm. Flowers not heterostylous, mostly smaller, the tube of the corolla ± 1 mm. long.

KENYA. Kitui District: Kibwezi–Kitui road, at 11·2 km. after crossing the Athi R., 22 Apr. 1969, *Napper & Kanuri* 2039 !; Masai District: R. Ewaso [Eusso] Nyiro, W. of Lake Magadi, 21 Jan. 1951, *Verdcourt, Greenway & Eggeling* 420 !; Kwale District: Mrima Hill, 4 Sept. 1957, *Verdcourt* 1870 !
TANZANIA. Mbulu District: Lake Manyara National Park, Chem Chem R., 22 June 1965, *Greenway & Kanuri* 11887 !; Lushoto District: Mkomazi R., on road between Mkomazi and Buiko, 30 Apr. 1953, *Drummond & Hemsley* 2311 !; Morogoro District: 32 km. NE. of Morogoro, Songa [Mabatini], 14 Sept. 1960, *Paulo* 786 !; Zanzibar I., without locality, Sept. 1873, *Hildebrandt* 908 !
DISTR. **K**1, 2, 4, 6, 7; **T**2, 3, 5–8; **Z**; Zaire, Sudan, Ethiopia, Somali Republic, Mozambique and Malawi
HAB. Grassland, thickets and open *Acacia* woodland, usually in open or temporary habitats, e.g. sandy roadsides, murram pans, waterhole edges, saline flats, also sand banks and rock crevices in rivers, and as a weed in cultivations; 0–1350(–1800) m.

SYN. *O. fastigiata* Bremek. var. *longifolia* Bremek. in Verh. K. Nederl. Akad. Wet., Afd. Natuurk., ser. 2, 48(2): 261 (1952). Type: Tanzania, Moshi District, Gwari, *Haarer* 1449 (K, holo. !, EA, iso. !)

NOTE. *Drummond & Hemsley* 2311, cited above, is a small erect annual growing on saline flats, with narrow leaves and tightly contracted inflorescences; material cultivated at Kew has long straggling stems, much looser inflorescences and much wider leaves, up to 7 mm. wide rather than 1·5 mm. This admirably shows the plasticity of this group and how little reliance can be placed on characters derived from the habit. The wild form would have been referred by Bremekamp to var. *fastigiata* and the cultivated ones to var. *longifolia*.

var. **somala** (*Bremek.*) *Verdc.* in K.B. 30: 299 (1975). Type: Somali Republic (S.), Bulo Aran, *Bisi* 134 (FI, holo. !)

Inflorescences distinctly pedunculate. Flowers distintly hererostylous, often larger, the tube of the corolla up to 1·6 mm. long. Fig. 41/34, p. 271.

KENYA. Northern Frontier Province: S. Turkana, 17·6 km. S. of Kangetet, Ayangyangi Swamp, 23 May 1970, *Mathew* 6353 !; Tana River District: Bura, 17 Mar. 1963, *Thairu* 77 !; Lamu District: Witu, *Thomas* 23 !
TANZANIA. Masai District: track to Kitumbeine and Lisingita area, 9 Jan. 1969, *Richards* 23702 ! & same locality 2 Jan. 1969, *Richards* 23599 !; Pare, *Uhlig* 894 !
DISTR. **K**1, ? 2, 4, 7; **T**2, 3; Ethiopia and Somali Republic (S.)
HAB. Grassland areas in *Acacia* bushland, by seasonal waterholes, etc.; 60–1220 m.

SYN. *O. somala* Bremek. in Verh. K. Nederl. Akad. Wet., Afd. Natuurk., ser. 2, 48(2): 251 (1952), excl. var. *scabridula*

NOTE. *D. Wood* 1380 (Kenya, Kitui District, 33 km. from Galana R. Camp on Dakadima road, 15 Nov. 1967) has the inflorescences not much shorter than the leaves at some nodes but is undoubtedly this species.

var. **pseudopentodon** *Verdc.* in K.B. 30: 299 (1975). Type: Kenya, 1 km. S. of Garsen, *Gillett & Kibuwa* 19909 (K, holo.!, EA, iso.)

Weak straggling herb. Leaf-blades very thin, up to 7 cm. long, 1·2 cm. wide. Inflorescences lax; peduncles up to 3 cm. long; secondary peduncles up to 1·4 cm. long; pedicels up to 5 mm. long. Flowers heterostylous (short-styled only seen).

KENYA. Tana River District: 1 km. S. of Garsen, W. of Tana R., 15 July 1972, *Gillett & Kibuwa* 19909!
DISTR. **K**7; not known elsewhere
HAB. Open damp grassland; 15 m.

NOTE. Perhaps just an abnormal seasonal form of var. *somala* which occurs in the area. Could be confused with *Pentodon* if not examined carefully.

35. **O. capensis** *L.f.*, Suppl. Pl.: 127 (1781); DC., Prodr. 4: 424 (1830); Hiern in F.T.A. 3: 62 (1877); K. Schum. in P.O.A. C: 375 (1895); Bremek. in Verh. K. Nederl. Akad. Wet., Afd. Natuurk., ser. 2, 48(2): 265 (1952); Hepper in F.W.T.A., ed. 2, 2: 211 (1963). Type: South Africa, Cape of Good Hope, *Thunberg* (S, holo.)

Prostrate or ascending annual herb (2–)8–22 cm. tall or long, sometimes forming mats; stems branched, glabrescent or with minutely setulose ribs. Leaf-blades linear to narrowly elliptic, 0·7–3·4 cm. long, 0·5–4·5 mm. wide, subobtuse to usually acute at the apex, cuneate at the base, scabrid papillose above, on the inrolled margins and often on the main nerve beneath but mostly glabrous beneath; petiole absent or apparent petiole 1–2 mm. long; stipule-sheath 1·2–2 mm. long, with 3–7 long and short fimbriae up to 1·5 mm. long. Flowers not heterostylous, sometimes solitary at the nodes or few in subsessile axillary fascicles; peduncles obsolete or up to 2 mm. long; pedicels 1–3 mm. long. Calyx-tube obconic to campanulate, ± 1 mm. long, glabrous or sparsely pubescent; lobes 4–8, triangular or linear-subulate, 0·6–1·1 mm. long, 0·2–0·3 mm. wide, margins setulose-ciliolate. Corolla white, deep lilac, dull reddish, pink or lilac; tube cylindrical, 0·6–0·9 mm. long; lobes 4, 0·5–0·6 mm. long. Anthers included. Style 0·2 mm. long, glabrous; stigma-lobes 0·2 mm. long, adjacent to the anthers. Capsule obconic, 1·75–2 mm. long and wide, glabrous or shortly pubescent, the beak straight, slightly raised, 0·3–0·4 mm. tall. Seeds brown, ellipsoid-angular or ± cone-shaped with flat-base, 0·3 mm. long.

var. **capensis**; Bremek. in Verh. K. Nederl. Akad. Wet., Afd. Natuurk., ser. 2, 48(2): 266 (1952); Hepper in F.W.T.A., ed. 2, 2: 211 (1963)

Calyx-lobes 4.

UGANDA. Masaka District: Katera, 16 Sept. 1961, *Rose* 10023! & Sese Is., Bugala I., 12 Oct. 1958, *Symes* 508!
KENYA. Elgeyo Escarpment, *Harger*!; Thika District: Ruiru, 6 Sept. 1962, *W. H. Lewis* 5908! (shows tendency to *O. caespitosa* and *O. fastigiata*)
TANZANIA. Tanga District: 6·4 km. N. Tanga, 18 Oct. 1962, *W. H. Lewis* 6059!; Rungwe District: Kyimbila, Kabasa, 10 Sept. 1910, *Stolz* 254!; Songea District: 12 km. E. of Songea by Nonganonga stream, 27 Dec. 1955, *Milne-Redhead & Taylor* 7760!
DISTR. **U**4; **K**3, 4; **T**2–4, 6–8; more or less throughout Africa from Morocco and Egypt to the Cape Peninsula, also in Madagascar and Yugoslavia
HAB. Grassy places, damp sandy ground, cultivated ground; 30–2400 m.

var. **pleiosepala** *Bremek.* in Verh. K. Nederl Akad. Wet. Afd. Natuurk., ser. 2, 48 (2): 267 (1952). Type: Russia, Azerbaydzhan, Lenkoran, *Hohenacker* (LE, lecto., K, isolecto.!)

Calyx-lobes 5–8. Fig. 41/35, 271.

UGANDA. Ankole District: Mbarara, Ruizi Gorge, 17 Sept. 1959, *Tallantire* 3!
TANZANIA. Buha District: 64 km. from Kibondo on the Kasulu road, Malagarasi Ferry, 24 Nov. 1963, *Verdcourt* 3445!; Arusha District: Arusha National Park, Lake Kawanga ya Matheo, 4 Nov. 1969, *Richards* 24587B/1!; Mpanda District: S. end of Mahali Peninsula, Kalya, 13 Aug. 1959, *Harley* 9416!;
DISTR. **U**2; **T**1, 2, 4, 7; similar to typical variety; also in Transcaucasia and Iran, Madagascar
HAB. Swampy ground in *Brachystegia* woodland, damp sandy seasonal river beds, paddy fields on black marshy soil, short grassland; 600–1440 m.

SYN. *Karamyschewia hedyotoides* Fisch. & Mey. in Bull. Soc. Nat. Mosc. 1838: 266 (1838). Type as for var. *pleiosepala*
Theyodis octodon A. Rich., Tent. Fl. Abyss. 1: 364 (1847). Type: Ethiopia, Shire [Chiré], *Quartin Dillon* (P, holo.)
Oldenlandia hedyotoides (Fisch. & Mey.) Boiss., Fl. Orient. 3: 11 (1875); Hiern in F.T.A. 3: 64 (1877)

NOTE. *Huxley* 128 (Tanzania, Moshi District, Machame Central Girl's School, 22 Feb. 1955) is a distinctive hairy-capsuled variant of *O. capensis* showing an approach to *O. fastigiata*, and some of the inflorescences have quite distinct peduncles. *Greenway & Kanuri* 12348 (Tanzania, Arusha District, Ngurdoto Crater National Park, Ngongongare, 19 Feb. 1966) is a similar plant but rather near to *O. corymbosa* var. *caespitosa*. It is possible these two variants should be given names under either *O. capensis* or *O. corymbosa*.

36. **O. geminiflora** (*Sond.*) *O. Kuntze*, Rev. Gen. Pl. 1: 292 (1891); Bremek. in Verh. K. Nederl. Akad. Wet., Afd. Natuurk., ser. 2, 48: 268 (1952), *non* K. Schum. (1900). Type: South Africa, Transvaal, N. side of Magaliesberg, *Zeyher* 756 (? LY, holo., K, iso.!)

Perennial branched herb 12 cm. tall; stems ribbed, glabrescent or pubescent with flattened white hairs. Leaf-blades lanceolate to linear-lanceolate, 1·6–3 cm. long, 1·8–6 mm. wide, acute at the apex, ± narrowed to the base, pubescent; stipule-sheath whitish, 1–1·5 mm. long, with 2–5 short setae 0·5 mm. long. Flowers 1–3 together at most axils, the pedicels very short at first, later becoming 2–4 mm. long. Calyx-tube ovoid, 1·2 mm. long; lobes narrowly oblong-lanceolate, (0·8–)1·5 mm. long, becoming 2·5–3 mm. long, 0·6–0·8 mm. wide, sparsely shortly ciliate. Corolla white, 1 mm. long; lobes elliptic-ovate, 1 mm. long, 0·5 mm. wide. Anthers enclosed in the tube. Style and capitate stigma ± 0·8 mm. long.

TANZANIA. Buha District: Mbirira [Birira] to Lake Manyoni, 26 Feb. 1926, *Peter* 46298!; Iringa District: Ruaha National Park, 5 km. NE. of Msembe, Great Ruaha R., 14 Dec. 1971, *Bjørnstad* 1176!
DISTR. **T**4, 7; South Africa and ? Rhodesia
HAB. With short-lived annuals on dry sand bank; 790–1300 m.

SYN. *Hedyotis geminiflora* Sond. in Linnaea 23: 51 (1850) & in Fl. Cap. 3: 10 (1865)

NOTE. *O. geminiflora* is known only from two old South Africa specimens and a specimen from Rhodesia and these are more robust plants than the Tanzanian specimens, with longer coarser leaves. The true status of these specimens is doubtful—it could be looked on as a form of *O. capensis* with abnormally long sepals or a northern variant of *O. geminiflora* which is itself dubiously distinct. It is pointless to discuss this without much more material.

37. **O. acicularis** *Bremek.* in Verh. K. Nederl. Akad. Wet., Afd. Natuurk., ser. 2, 48(2): 269 (1952); U.K.W.F.: 401 (1974). Type: Kenya, Uasin Gishu District, Kipkarren, *Brodhurst Hill* 221 (K, holo.!, EA, iso.!)

Annual slender erect herb; stem sparsely branched, ± 15 cm. tall, with 4 papillose-pubescent ribs. Leaf-blades linear, 1·2–3 cm. long, 0·5–0·8 mm. wide, mucronulate, margins inrolled, shortly scabridulous-papillose-pubescent above and on the main nerve beneath; petiole obsolete; stipule-sheath 1·5–3 mm. long, produced into 5 slender setae 1–4 mm. long. Flowers isostylous, solitary at the nodes or in pairs, pedicels 1–1·5(–4) mm. long. Calyx-tube campanulate, 0·8 mm. long, with few short hairs; lobes linear-lanceolate, 1·2–1·8 mm. long, the margins and main nerve scabrid-pubescent. Corolla white; tube ± 1·2 mm. long; lobes ovate, 1·2 mm. long; throat hairy. Style slightly shorter than the stigma-lobes. Capsule campanulate, laterally compressed, 1·5 mm. long, 1·8 mm. wide, glabrous, crowned by the persistent calyx-lobes; beak slightly raised, 0·2 mm. tall; seeds brown, angular, ± 2·2 mm. long. Fig. 41/37, p. 271.

KENYA. Uasin Gishu District: Kipkarren, *Brodhurst Hill* 221!
DISTR. **K**3; known only from the type gathering
HAB. Growing in rock pools

NOTE. More material of this is needed; when better understood it may prove to be a variant of the very variable ephemeral *O. nematocaulis* but there appear to be differences in the inflorescences and flowers. The supposed technical differences could all be illusory.

Tribe 9. ANTHOSPERMEAE

Calyx-lobes unequal, 1 or more enlarged; flowers hermaphrodite; stigmas ± smooth; corolla-tube exceedingly fine, almost thread-like in some species . . . . 34. **Otiophora**
Calyx-lobes equal; flowers unisexual or polygamous, sometimes showing dimorphism in style length; stigmas feathery; corolla-tube narrow . . . . 35. **Anthospermum**

34. OTIOPHORA

Zucc. in Abh. Akad. Muench. 1: 315 (1832); Verdc. in J.L.S. 53: 383–412 (1950) & in Garcia de Orta, Ser. Bot. 1 (1–2): 25 (1973)

Annual or perennial herbs or subshrubs, with erect, straggling or procumbent stems. Leaves very variable, mostly small, linear to round, paired or in whorls of 3 or pseudo-verticillate, sessile or shortly petiolate; stipules connate with the petiole to form a sheath, divided into ± linear segments. Flowers small, hermaphrodite, not dimorphic, paired, arranged in short to long spikes or dense heads or combinations of both. Calyx-tube ovoid or angular; lobes 5, 1–3 foliaceous, much larger than the rest and resembling the leaves of the species concerned, or 1–2 foliaceous, 1–2 smaller foliaceous and the remainder in each case reduced to minute teeth or bristles. Corolla-tube mostly filiform, less often narrowly funnel-shaped; lobes 4–5, ovate to narrowly lanceolate; throat glabrous or in one species densely hairy. Stamens 4–5, well exserted, the filaments glabrous or less often hairy or with hairy projections. Ovary bilocular, the ovules solitary in each locule, erect, attached to the partition very close to the base; style glabrous, filiform, well exserted; stigma bifid (rarely 3-fid), the lobes filiform. Fruit ovate, subglobose or oblong, usually splitting into 2 (rarely 3) cocci one of which is crowned by the persistent foliaceous calyx-lobe or lobes. Seeds black, granulate, subglobose, ovoid or oblong, depressed ventrally, convex dorsally, sometimes slightly keeled mid-dorsally and with a round, oval or linear depression surrounding the hilum corresponding to the shape of the seed.

A small genus of about 20 species confined to tropical and S. Africa and Madagascar, the 2 species occurring in the latter also present on the mainland.

Filaments hairy at the base; leaves linear to linear-lanceolate, pseudoverticillate and appearing to be in whorls of 6–10; plant of wet places . . . . 3. *O. pycnostachys*

Filaments glabrous; leaves mostly wider, not pseudoverticillate; not usually growing in wet places:

Mature inflorescence a lax spike; plant nearly always an erect annual or short-lived perennial with single stems; corolla-tube 3·5–6 mm. long . . . . . . 1. *O. scabra*

Mature inflorescence a head or sometimes if in short dense spikes up to 3 cm. long in fruiting stage then plant straggling or with numerous stems from a woody rootstock:

Leaves subcoriaceous, ± sessile, glabrous or ciliate, very closely placed at the apices of the shoots, the internodes 1–4 mm. long; stems leafless below, covered with persistent stipule-sheaths . . . 2. *O. pycnoclada*

Leaves not subcoriaceous, often hairy but never so densely conferted:

Corolla-tube very filiform 0·7–1·2 cm. long, usually with narrow lobes 5 mm. long; calyx-tube subglobose, glabrous; seeds subglobose; leaves very variable, 3–4 cm. long; plant erect, caespitose . . . . . 5. *O. caerulea*

Corolla mostly shorter or if not then plant a long straggler and without the other characters combined:

Small erect or procumbent herb; corolla-tube 3·5–6 mm. long; seeds ovoid, oblong or subglobose:

Leaves ovate-lanceolate, linear-lanceolate or elliptic, 1·3–6·5 mm. wide:

Calyx-tube densely hairy; leaves 0·7–2 cm. long; plant usually erect:

Young stems bifariously pubescent; leaf-blades mostly shorter and narrower; seeds oblong . . . . 4. *O. stolzii*

Young stems densely hairy; leaf-blades mostly longer and wider; seeds subglobose (but unknown in one variety occurring in the Flora area) 8. *O. parviflora*

Calyx-tube glabrous; leaves 0·3–1·2 cm. long; plant probably procumbent. . . . . 6. *O. sp. A*

Leaves ± ovate, acute, broadest at the base or middle, 1–1·9 × 0·6–1·1 cm.; plant suberect or procumbent . . . . 7. *O. villicaulis*

Straggling plant with usually 2–3 long stems; corolla-tube 0·7–1·2 cm. long; seeds tapering-elongate . 9. *O. pauciflora*

1. **O. scabra** *Zucc.* in Abh. Akad. Muench. 1: 315 (1832); Drake, Hist. Nat. de Plantes 4, t. 412 (1897); K. Schum. in E. & P. Pf. IV. 4: 133, fig. 42/F (1891); Verdc. in J.L.S. 53: 391, fig. 5, 3/h & 4/s (1950). Type: Madagascar, Emirna, *Bojer* (M, holo. photo. !, K, P, W, iso. !)

Annual or rarely perennial herb or undershrub 0·15–1 m. tall; stems slender, erect or rarely procumbent, much-branched, covered with dense white scaly hairs above but glabrescent below or rarely ± glabrous. Leaves opposite or appearing verticillate owing to the presence of very abbreviated axillary shoots; blades lanceolate to linear-lanceolate, 1–7·5 cm. long, 0·2–1(–2) cm. wide, acute at the apex, cuneate at the base, somewhat revolute at the margins, glabrous to densely covered with adpressed scaly white hairs particularly above; petioles almost or quite obsolete; stipules with 3–4 fimbriae ± 1·5–3 mm. long. Flowers geminate in terminal spikes 1·2–15 cm. long (up to 26·5 cm. in fruit); pedicels obsolete. Calyx-tube ovoid, 1 mm. long, densely covered with spreading bristly hairs; lobes 5, 1 foliaceous, lanceolate, 3–5 mm. long, 1–2 mm. wide, with entire revolute margins usually covered with white bristly hairs, rest of lobes small or minute, unequal, densely hairy. Corolla white or usually red, lilac, mauve or blue; buds with limb abruptly expanded into an ovoid head; corolla-tube very narrowly filiform, (2–)3·5–6 mm. long, 0·2 mm. wide, glabrous; throat glabrous; lobes 4–5, lanceolate, 2–2·5 mm. long, 0·5–0·75 mm. broad, usually hairy outside. Stamens 4–5, exserted for up to ± 2 mm.; filaments glabrous. Style filiform, the stigma bilobed, lobes filiform, 1–1·5 mm. long, style and stigma together exserted for a distance equalling the corolla-lobes. Fruit pale yellow-brown to dark purplish brown, ovoid, 2 mm. long, 1·2–1·5 mm. wide, densely covered with white hairs. Seeds black, ovoid, 1·4 mm. long, 0·9 mm. wide, ventrally flattened, granulated, slightly keeled dorsally and with a narrow depression surrounding the hilum.

var. **scabra**

Stems ± erect, hairy; leaves usually hairy. Fig. 45, p. 318.

TANZANIA. Bukoba District: near Nyakato, Aug. 1931, *Haarer* 2133 !; Iringa District: Iheme, 23 Feb. 1962, *Polhill & Paulo* 1588 !; Songea District: about 11 km. W. of Tunduru District boundary on Songea–Tunduru road, 4 June 1956, *Milne-Redhead & Taylor* 10492 !

DISTR. T1, 4, 6–8; Zaire, Burundi, Malawi, Zambia, Rhodesia, Angola and Madagascar

HAB. Grassland, bushland, grassland with scattered trees in rocky places, scrub and woodland of *Uapaca, Brachystegia, Isoberlinia*, etc., also in cultivations; 880–1725 m.

SYN. *Pentanisia spicata* S. Moore in J.B. 46: 38 & 76 (1908). Type: Rhodesia, Mazoe, Iron Mask Hill, *Eyles* 522 (BM, holo. !, S, iso. !)

NOTE. In my original revision some Angolan specimens were separated as var. *glabra* Verdc. (in J.L.S. 53: 395 (1950)). A specimen from E. Africa may be referred to this (Tanzania, Songea District, about 8 km. SE. of Mbamba Bay, 6 Apr. 1956, *Milne-Redhead & Taylor* 9483 !), but the variety is possibly not worth retaining.

2. **O. pycnoclada** *K. Schum.* in E.J. 30: 415 (1901); Verdc. in J.L.S. 53: 401 fig. 3/d (1950) & in Garcia de Orta, Sér. Bot. 1(1–2): 29 (1973). Type: Tanzania, Ukinga Mts., Kipengere Ridge, *Goetze* 967 (B, holo. †, BM, K, iso. !*)

* Goetze sheets at BM & K are numbered 965—they are almost certainly isotypes and one of the numbers is an error, probably the published one.

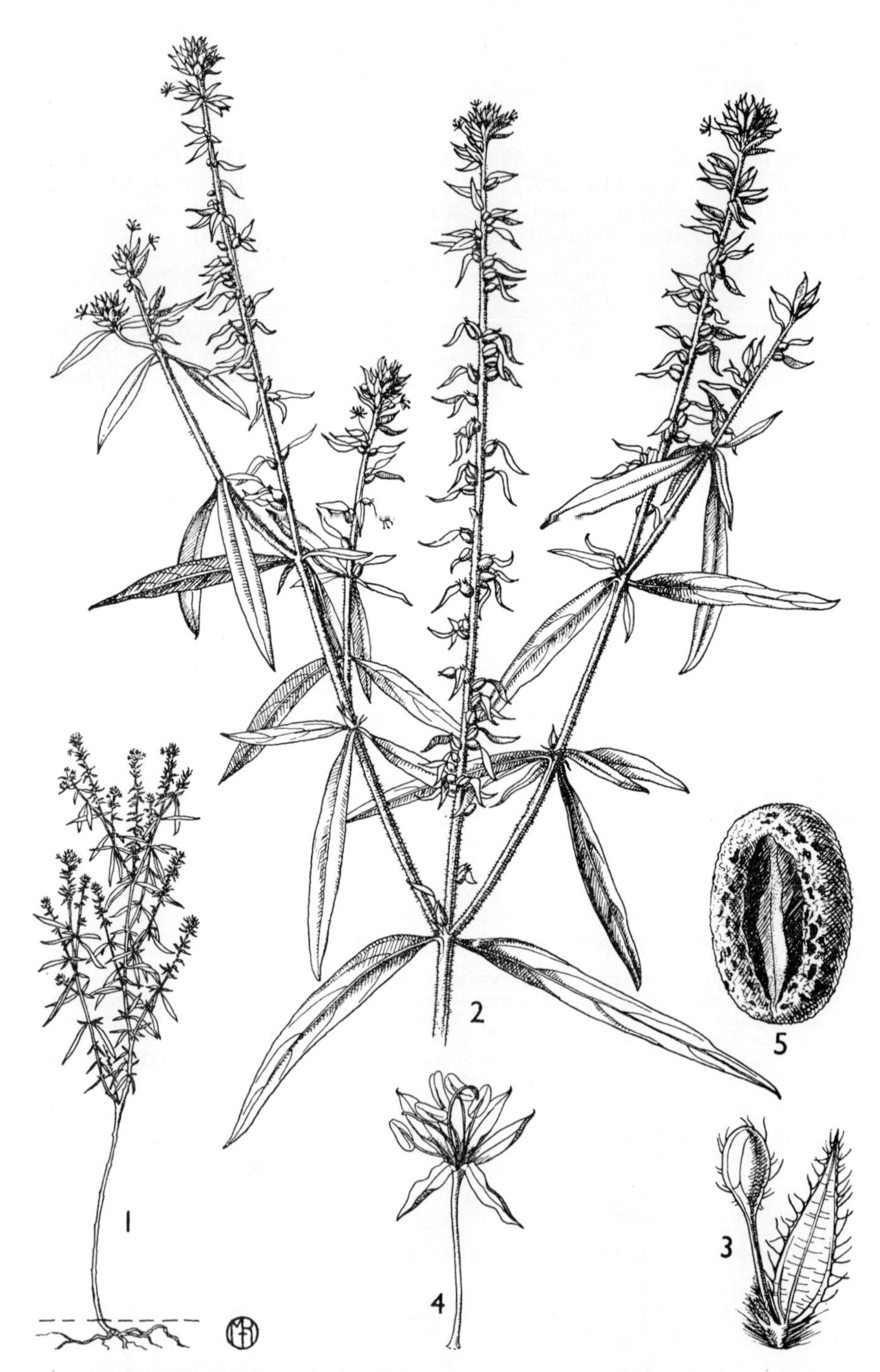

FIG. 45. *OTIOPHORA SCABRA* var. *SCABRA*—**1,** habit, × $\frac{1}{6}$; **2,** upper portion of plant, × 1; **3,** flower bud, × 7; **4,** flower, with calyx removed, × 7; **5,** seed, × 20. All from *Schlieben* 1951. Drawn by Olive Milne-Redhead.

Perennial subshrub 30–60 cm. tall, probably much branched from a woody base; stems grey-black with flaking epidermis, glabrous or pubescent in patches, leafless at the base, densely and closely covered with small sheaths formed at the nodes by persistent bases of stipules and leaves; internodes very short, mostly (1–)2–4 mm. long. Leaves rather coriaceous, densely conferted; leaf-blades ovate-oblong to oblong-lanceolate, 0·6–1 cm. long, 2–5 mm. wide, subacute at the apex, rounded or cuneate at the base, glabrous or the margins shortly ciliate; petioles obsolete or scarcely 0·5 mm. long; stipules with sheathing base ± pubescent, fimbriae setiform, 2–3 mm. long. Flowers geminate in small terminal heads, sessile. Calyx ellipsoid, 1–2 mm. long, 1 mm. wide, pilose with long white hairs or glabrescent; lobes 5, either 1 foliaceous and 4 small, or 2 foliaceous and 3 small, with several colleters between the small lobes; foliaceous lobes similar in texture to the leaves, lanceolate, 5–6 mm. long, 1·5 mm. wide, acute, glabrous or shortly ciliate; other lobes setiform, 1·5–2·5 mm. Corolla pale pink or violet; buds with limb abruptly expanded into a subcylindrical head quite distinct from the filiform tube, the limb portion covered with long hairs particularly on the nerves; tube very slender, 6 mm. long, 0·2 mm. wide, glabrous; throat glabrous; lobes 5, lanceolate, 3·5–5 mm. long, 1 mm. wide, covered with long hairs outside. Stamens 5, exserted for a distance ± equalling the corolla-lobes; filaments glabrous. Style filiform, stigma bilobed, lobes filiform 2–3·5 mm. long, the style and stigma together exserted the same distance as the stamens. Immature fruit ellipsoid, 2·5 mm. long. Immature seeds oblong, 2 mm. long, 1 mm. wide.

TANZANIA. Njombe District: Njombe–Milo road, 28 Jan. 1961, *Richards* 14029! & Ukinga Mts., Kipengere Ridge, May 1899, *Goetze* 965!*
DISTR. **T7**; not known elsewhere
HAB. Rocky places on dry hillsides; 1950–2700 m.

3. **O. pycnostachys** *K. Schum.* in P.O.A. C: 393 (1895); Verdc. in J.L.S. 53: 402, fig. 4/G, M, Q (1950). Type**: Tanzania, *Stuhlmann* (B, holo. †, K, iso.!)

Perennial herb or subshrub 0·15–1 m. tall, with several erect or straggling branched stems from a woody base; stems with white hairs in 2 narrow lines on the upper parts, glabrous below. Leaves pseudoverticillate and appearing to be in whorls of 6–10, an effect due to non-developed leafy branches in the axils of the 2 main stem leaves; blades lanceolate, 1–5 cm. long, 1–8 mm. wide, acute, somewhat coriaceous, the margins often slightly revolute, entirely glabrous or with a few short hairs on the margins and midvein above, sessile; stipules pubescent, fimbriate into ± 5 segments (1–)1·5–4 mm. long, from a broad saccate persistent base. Flowers in heads or very compact subcylindrical spikes (1·5–)2–3 cm. long and sometimes with a few lax flowers beneath the main part of the inflorescence; sometimes lateral inflorescences arise in the axils of the leaves of the uppermost node these sometimes appearing to coalesce with the main inflorescence to form a single compact head. Calyx-tube elongate-oblong, 1·5 mm. long, ± 1 mm. wide, densely hairy; lobes 5, 1 foliaceous, linear-lanceolate, 3–5 mm. long, 0·5–1 mm. wide, acute, hairy, particularly on the margins, rest small, setiform but often somewhat dilated at the apex, hairy. Corolla white, pinkish or crimson; buds with limb very abruptly expanded into a subcylindrical head, the base of which forms a right-angle with the filiform tube; tube 5–8 mm. long, 0·25 mm. wide; lobes lanceolate, 3–5·5 mm. long, 1 mm. wide, with scattered hairs at the

* See footnote on page 317.
** No information given in original reference—probably omitted in error. There is every reason to believe the *Stuhlmann* sheet at Kew written up in K. Schumann's handwriting is an isotype and certainly nothing to prove that it is not.

apex outside. Stamens 5, exserted for 3·5–5 mm.; filaments filiform, with ± 1 mm. close to the base somewhat dilated and covered with long white hairs. Style tenuously filiform, the stigma bifid, lobes filiform, ± 1 mm. long, style and stigma together exserted 3–4 mm. Fruit yellowish brown, elongate-ovoid, 3–4 mm. long, 1·5–2 mm. wide, slightly to densely hairy. Seeds brownish, oblong, 2 mm. long, 0·75 mm. wide, often distinctly square-cut at both ends, granulate, somewhat keeled for practically the whole length of the dorsal face.

UGANDA. Masaka District: Lake Nabugabo, Oct. 1932, *Eggeling* 582 ! & same locality, June 1937, *Chandler* 1708 ! & E. side of Lake Kayanja, 7 Mar. 1971, *Lye* 5909 !
TANZANIA. Bukoba District: near Kaagyia, Aug. 1931, *Haarer* 2106 !; Iringa District: Kibengu, 17 Feb. 1962, *Polhill & Paulo* 1515 !; Songea District: Matengo Hills, valley of R. Halau about 3 km. SE. of Miyau, 12 Jan. 1956, *Milne-Redhead & Taylor* 8227 !
DISTR. **U4**; **T1**, **7**, **8**; Zaire, Zambia
HAB. Bogs, swamps and other more or less permanently inundated habitats; 1125–1920 m.

SYN. *O. caespitosa* S. Moore in J.B. 63: 146 (1925). Type: Zambia, Luwingo, *Jelf* 5 (BM, holo. !)

4. **O. stolzii** (*Verdc.*) *Verdc.* in K.B. 14: 351 (1960). Type: Tanzania, Njombe District, Madehani, *Stolz* 2467 (K, holo. !, BM, EA, iso. !)

Perennial tufted or cushion herb 10–22 cm. tall, with up to ± 25 simple or branched erect or suberect mostly bifariously pubescent stems from a woody rootstock. Leaves paired; blades narrowly elliptic or linear- to elliptic-lanceolate, 0·7–2 cm. long, 1·3–8 mm. wide, subacute or obtuse at the apex, rounded or cuneate at the base, usually with revolute margins, pubescent with short bristly hairs above and with longer hairs on the midvein above and beneath or practically glabrous on both surfaces; petiole obsolete or up to 1 mm. long; stipules with 3–4 subulate fimbriae 0·5–2 mm. long from a short base. Flowers in small terminal heads ± 1 cm. long and wide which in the fruiting stage elongate to 1–3 cm., or very few at the apices of the shoots. Calyx-tube ovoid, 0·8 mm. long, densely pilose with white hairs; lobes 4–5, 1–2 foliaceous, 2–6 mm. long, 0·5–2 mm. wide, similar in shape and indumentum to the leaf-blades, rest setiform, up to 0·8 mm. long. Corolla rose, pale blue, white, lilac or pink; buds with limb abruptly expanded into an ovoid head, glabrous or with a few bristly hairs at the apex; tube narrowly filiform, (3–)4·5–5·5 mm. long, 0·2 mm. wide, glabrous; lobes 4–5, lanceolate, 2·2–2·8 mm. long, 0·8–1 mm. wide, glabrous or with a few hairs outside at the apex. Stamens 4–5, exserted for usually just over half the length of the corolla-lobes; filaments glabrous. Style filiform, stigma bifid (rarely trifid), lobes filiform 1·5 mm. long; style and stigma together exserted for about the length of the corolla lobes. Fruits elongate-ovoid, 2·2–2·5 mm. long, 1–1·5 mm. wide, spreading hairy. Seeds brown, oblong, 1·5–1·9 mm. long, 0·9–1·1 mm. wide, granulate, slightly keeled dorsally and with an oblong depression around the hilum.

TANZANIA. Ufipa District: Mbisi, 2 Apr. 1950, *Bullock* 2791 !; Njombe District: Bundali, Namulapi, 3 Mar. 1914, *Stolz* 2574 ! & Kipengere Mts., 9 Jan. 1957, *Richards* 7612 !
DISTR. **T4**, **7**; Malawi, Zambia (Nyika)
HAB. Grassland in rocky places in rock crevices and also recorded from a "muddy bank"; 2000–2520(–2700) m.

SYN. *O. parviflora* Verdc. var. *stolzii* Verdc. in J.L.S. 53: 407 (1950) & in B.J.B.B. 23: 63 (1953)
[*O. parviflora* sensu Verdc. in J.L.S. 53: 407 (1950), quoad *Stolz* 2574 & *Thomson*]

NOTE. The seeds seen are not quite mature. It is difficult to be sure how many stems there are from a mature plant since all the specimens are probably parts of cushions. *Thomson* s.n. ('N. of Lake Nyasa') has the stems, foliage and flowers more hairy. Further material is needed to elucidate the relationship of this species to *O. multicaulis* Verdc. (type: Zambia, SW. of Dobeka Bridge, *Milne-Redhead* 2739 (K, holo. !)), particularly ripe fruits.

5. **O. caerulea** (*Hiern*) *Bullock* in K.B. 1933: 471 (1933); Verdc. in J.L.S. 53: 404, fig. 3/B, C, fig. 4/E, R (1950) & in Garcia de Orta, Sér. Bot. 1(1–2): 29 (1973). Type: Angola, Huila, Quilenques, R. Monino, *Welwitsch* 5314 (LISU, holo., BM, K, iso. !)

Perennial herb 11·5–45 cm. tall, with 3–5 erect shoots from each part of a divided woody rootstock; stems with 2 narrow rows of short bristly hairs on opposite sides or hairy all over. Leaves paired or occasionally appearing pseudoverticillate if short axillary shoots remain undeveloped; blades linear, lanceolate or elliptic-lanceolate, (0·7–)1·7–5 cm. long, (0·1–)0·2–1·1(–1·7) cm. wide, acute at the apex, cuneate at the base, often drying a characteristic yellowish tinge, glabrous, sessile; stipules ± pubescent, with (1–)2–3 lanceolate or subulate segments (if 1 then with short lateral lobes) 1–5 mm. long from a short broad base, which is adnate to the leaf-bases, and sometimes with small lobes between the main lobes. Flowers in compact heads 1·2(–2·5) cm. long, 1·3–3(–4) cm. wide, terminating the main and axillary branches and branchlets. Calyx-tube ovoid or subglobose, ± 1 mm. long and wide, glabrous; lobes 5, either 1 large foliaceous lobe and 4 short setiform ones, 2 large foliaceous lobes and 3 short setiform ones, or 2 longer foliaceous lobes, 1 subfoliaceous lobe and 2 short setiform ones; foliaceous lobes linear to lanceolate, 0·5–2 mm. wide, acute, glabrous or somewhat ciliate; setiform lobes 1·5–3 mm. long, somewhat bristly. Corolla blue to mauve or mauve-pink; buds with limb abruptly expanded into an ovoid or subcylindrical head, glabrous or slightly hairy; tube narrowly filiform, (0·5–)0·7–1·2 cm. long, 0·25 mm. wide; lobes lanceolate, (4–)5–6 mm. long, 0·75–1 mm. wide, glabrous or slightly hairy outside. Stamens 5, ± equalling the corolla-lobes; filaments glabrous. Style filiform, stigma bifid, lobes filiform, up to 3 mm. long; style and stigma together exserted 4–6 mm. Fruits subglobose, 1·5–3 mm. long and wide, glabrous. Seeds dark brownish, subglobose, ± 1·5 mm. in diameter, granulate, with a slight dorsal keel and a circular depression surrounding the hilum.

TANZANIA. Buha District: 16 km. from Manyovu on Kasulu road, 22 Nov. 1962, *Verdcourt* 3423 !; Iringa District: Kibengu, 14 Feb. 1962, *Polhill & Paulo* 1482 !; Njombe, 12 July 1956, *Milne-Redhead & Taylor* 11123 !

DISTR. T1, 4, 7; Zaire, Burundi, Malawi, Zambia and Angola

HAB. Open and wooded grassland; 1500–2700 m.

SYN. *Pentanisia caerulea* Hiern, Cat. Afr. Pl. Welw. 2: 471 (1898)
Otiophora pulchella De Wild. in Ann. Mus. Congo, Bot. Sér. 4 (1): 230(1903). Type: Zaire, Katanga, *Verdick* 320 (BR, holo. !)

NOTE. This species shows extraordinary variation in the width and shape of the leaf-blades which vary from broadly lanceolate (e.g. Iringa District, Sao Hill, 17 Dec. 1961, *Richards* 15709 !) to almost linear (e.g. Ufipa District, Sumbawanga, Chapota, 7 Mar. 1957, *Richards* 8546 !). Examined side by side it is difficult to avoid the conclusion that these two specimens belong to different species but intermediates are not infrequent. One specimen has been seen from Zambia (Mwinilunga, 10 Nov. 1962, *W. H. Lewis* 6207) which is abnormally hairy with long hairs on the stems, leaves, calyx-tubes and lobes.

6. **O. sp. A**

Perennial cushion herb, with several branches each much branched from the base, from a narrow woody rootstock; stems probably suberect, 5–10 cm. long, with 2 narrow lines of short pubescence. Leaves paired; blades narrowly elliptic, 0·3–1·2 cm. long, 1·5–3·2 mm. wide, obtuse or subacute at the apex, cuneate at the base, the margins sometimes revolute, glabrous; petiole obsolete to ± 0·5 mm. long; stipules with 3 subulate fimbriae 0·5–1 mm. long from a short base. Flowers few in small terminal clusters and also some in the axils beneath. Calyx-tube ovoid, ± 1 mm. long, glabrous; lobes 4–5, 1–2 foliaceous, elliptic, 2–3 mm. long, 0·5–1 mm. wide, glabrous, the rest setiform, 0·5–1·5 mm. long, sometimes very sparsely hairy. Corolla lilac, glabrous; buds with limb abruptly expanded into an oblong-ovoid head, glabrous; tube filiform, 4–5 mm. long; lobes 4–5, oblong-elliptic, 3 mm. long, 0·8 mm. wide. Stamens 4–5, exserted for almost the same length as the corolla-lobes; filaments glabrous. Style filiform, stigma bifid, the lobes filiform, 1·5 mm. long, exserted for about the same length as the stamens. Fruits ovoid, 2 mm. long, 1·5 mm. wide, glabrous. Ripe seeds not seen.

TANZANIA. Njombe District: Mwakete, 16 Jan. 1957, *Richards* 7817!
DISTR. **T7**; not known elsewhere
HAB. Short coarse hillside grassland on gritty soil; 2400 m.

NOTE. This is closely allied to *O. multicaulis* Verdc. but differs in branching, habit and the glabrous flowers; until more material in ripe fruit is available it is not certain if it is a variant of that species or an undescribed species.

7. **O. villicaulis** *Mildbr.* in N.B.G.B. 15: 637 (1941): Verdc. in J.L.S. 53: 411 (1950). Type: Tanzania, Songea District, Matengo Hills, Ugano, *Zerny* 378 (B, holo. †)

Small perennial herb 3·5–15 cm. tall, with up to 15–20 procumbent or subprocumbent densely hairy often branched stems spreading from a narrow ± woody rootstock. Leaves paired; blades mostly ovate but sometimes ovate-lanceolate to round, 1–2 cm. long, 0·6–1·5 cm. wide, obtuse or acute at the apex, broadest near the base, widely cuneate, glabrescent to densely covered with long hairs, particularly beneath; sessile or petioles very short; stipules small, fimbriate into 3–5 segments 0·8–2 mm. long, hairy. Flowers in dense terminal heads 0·8–1·5 cm. wide; pedicels very short. Calyx-tube ± 1 mm. long, covered with long hairs; lobes 4–5, 1–2 foliaceous and ± ovate, 1·5–5·5 mm. long, 0·5–1·5(–2) mm. wide, glabrescent to hairy, the rest minute, pilose. Corolla lilac or bluish lilac; buds with limb abruptly expanded into an ovoid head; corolla-tube filiform, (4–)5–6 mm. long, 0·25 mm. wide; lobes 4–5, ovate-lanceolate to lanceolate, (1·5–)2–3 mm. long, 1 mm. wide, glabrous outside or less often with a few hairs (not in Flora area). Stamens 4–5, exserted ± 2 mm.; filaments glabrous. Style filiform, the stigma bilobed, lobes filiform, 1·25–2 mm. long, style and stigma together exserted for ± 3·5 mm. Fruit pale brown, ovoid to subglobose, 2 mm. long, 1·5 mm. wide, pilose. Seeds dark reddish brown, depressed ovoid or subglobose, 1–1·3 mm. long, 0·8–1 mm. wide, somewhat keeled dorsally, with an elliptic to round depression around the hilum.

var. **villicaulis**

Buds glabrous; corolla-tube mostly longer.

TANZANIA. Njombe District: Ukinga Mts., Madehani, 26 Jan. 1914, *Stolz* 2466!; Songea District: Matengo Hills, Miyau, 10 Jan. 1956, *Milne-Redhead & Taylor* 8168! & same locality, 29 Feb. 1956, *Milne-Redhead & Taylor* 8168A!
DISTR. **T7**, 8; Zaire, Zambia, Malawi, ? Angola
HAB. Bare patches in secondary grassland; 1620–2000 m.

SYN. *O. latifolia* Verdc. in J.L.S. 53: 406, fig. 4/C (1950). Type: Zaire, Lubumbashi, near Likasi Road, *Hirschberg* 70 (K, holo.!)

NOTE. When I wrote my original monograph I thought *O. latifolia* might well prove to be the same as *O. villicaulis* but had seen no material from near the type locality of the latter; it is clear I was over-cautious and the two are undoubtedly conspecific.

A minor variant, described as *O. latifolia* var. *bamendensis* Verdc., occurs in Cameroun; this might be best considered a subspecies but the distinguishing characters are far less constant than I previously thought and it might be best not to recognize it at all. The Zambian variant, var. *villosa* (Verdc.) Verdc., has even denser longer indumentum.

8. **O. parviflora** *Verdc.* in J.L.S. 53: 407 (1950). Type: Zambia, just NE. of Mufulira, *Cruse* 161 (K, holo. !)

Perennial tufted herb with 3–20 simple or sparsely branched stems 6·5–20 cm. tall, densely covered with long spreading hairs or with shorter bifarious hairs; rootstock slender (? but never annual as first reported by me). Leaves paired; blades elliptic or the lower almost ovate, 0·7–2·6 cm. long, 0·4–1 cm. wide, subacute at the apex, cuneate at the base, usually with revolute margins, glabrous above or with a few hairs on the base of the midrib or shortly pilose, hairy beneath, sometimes only near and on the lower part of the midrib; petiole 1–2 mm. long, glabrous or densely hairy; stipules with 3 subulate fimbriae 0·5–1 mm. long from a short base. Flowers in small terminal heads ± 1 cm. across. Calyx-tube subglobose, 1 mm. in diameter, densely covered with white hairs; lobes 5 or all but foliaceous ones obsolete; foliaceous lobes 1–2, often with 1–2 smaller ones added, 2·8–6·5 mm. long, 1–2·3 mm. wide, hairy or glabrescent, rest ± obsolete. Corolla white, lilac-blue, red or mauve; buds with limb abruptly expanded into an ovoid head, glabrous or hairy; tube narrowly filiform, 3·5–6 mm. long, 0·2 mm. wide, glabrous or with scattered hairs; lobes 4–5, oblong-elliptic, 2·5 mm. long, 0·6 mm. wide, glabrous. Stamens 4–5, exserted 2·2 mm.; filaments glabrous. Style filiform, exserted 0·8 mm.; lobes filiform, 1·8 mm. long. Capsule subglobose, 2 mm. in diameter, hairy. Seeds subglobose, 1·25 mm. long, 1 mm. wide, concave dorsally, depressed ventrally, reticulate and ± keeled, subacute at the apex, ± truncate at the base.

var. **parviflora**

Leaf-blades more densely hairy. Corolla white, lilac-blue or red, the lobes and buds with scattered long hairs outside.

TANZANIA. Buha District: Mbirira [Birira]–Kisuzi [Nisusi], 27 Feb. 1926, *Peter* 46307 !
DISTR. **T4**; Zaire, Zambia
HAB. ? Woodland; ± 1300 m.

NOTE. The single specmen cited above has the leaf-blades (other than the margins and midribs) less hairy than in typical material from Zambia. *Thomson* s.n. cited on p. 321 under *O. stolzii* might possibly belong here; in the absence of ripe fruits it is difficult to be certain of the identification.

var. **iringensis** *Verdc.* in J.L.S. 53: 408 (1950). Type: Tanzania, near Njombe, Msima Stock Farm, *Emson* 372 (K, holo. !, EA, iso. !)

Leaf-blades ± glabrous or with few hairs on the midrib above, with denser hairs on and near the mibrib beneath. Corolla glabrous, red or mauve.

TANZANIA. Mbeya District: Ruaha National Park, 25 km, W. of Magangwe Ranger Post, Isunkaviola Mt., 27 Dec. 1972, *Bjørnstad* 2456 !; Njombe District: Msima Stock Farm, *Emson* 372 !
DISTR. **T7**; Malawi
HAB. Hillside grassland and *Brachystegia* woodland; 1780 m.

NOTE. No ripe fruits have been seen of var. *iringensis*; until they are available its position will be doubtful.

9. **O. pauciflora** *Bak.* in J.L.S. 20: 170 (1883); Verdc. in J.L.S. 53: 408, fig. 4/F, 4/0 (1950). Type: Madagascar, Tananarive, Imerina, *Lyall* 305 (K, holo. !)

Perennial herb with 2–3 usually straggling stems 0·6–1 m. long bearing many axillary branchlets, arising from a woody rootstock; stems hairy when young, the hairs mostly in 2 lines, later glabrescent and the epidermis often flaking off. Leaf-blades ovate to lanceolate, 0·5–3·5 cm. long, 0·3–1·2 cm. wide, acute at the apex, rounded or broadly cuneate at the base, often discolorous, glabrescent or with short hairs above and mostly on the venation beneath, usually somewhat revolute; petiole obsolete or 1–4 mm. long; stipules with 3–5 setiform segments 2–4 mm. long from a short base. Flowers in 2–10-flowered terminal heads which may elongate into short spikes after flowering, usually with a few axillary flowers near the tops of the shoots as well. Calyx-tube oblong, 1 mm. long, glabrous or hairy; lobes 5, either 2 foliaceous and 3 small, 1 foliaceous and 4 small, or 3 small and 1 large and 1 medium foliaceous or all small; foliaceous lobes ovate, 4–5 mm. long, 1·5–2·5 mm. wide, acute at the apex, glabrescent or sparsely hairy and usually ciliate round margins; setiform lobes 2–3 mm. long, hairy. Corolla white, lilac or pale pink; buds with limb expanded into an ovate or subcylindrical head, usually rather hairy particularly at the apex; tube filiform, 0·6–1·2 cm. long, sometimes hairy; lobes 5, lanceolate, 3·5–5 mm. long, 0·5–1 mm. wide, usually with scattered hairs externally. Stamens 5, rather shorter than the lobes. Style and stigma exserted for 3·5–5 mm. ; stigma bilobed, lobes filiform, 1–2·5 mm. long. Fruit brown, elongate-oblong, 4 mm. long, 2 mm. wide, 1 mm. thick, hairy. Seeds black, elongate-oblong, truncate at both ends or elongate-tapering, 1·5–3 mm. long, 0·5–1 mm. wide, dorsally with a central keel, reticulate, ventrally with an elongate depression surrounding the hilum.

subsp. **burttii** (*Milne-Redh.*) *Verdc.* in Garcia de Orta, Sér. Bot. 1(1–2): 29 (1973). Type: Zaire, Virunga Mts., between Mts. Sabinio and Vissoke, Mashiga Pass, *B. D. Burtt* 2993 (K, holo. !, EA, iso.!)

Leaf-blades mostly more lanceolate and almost sessile. Inflorescence with more flowers than in typical subspecies. Foliaceous calyx-lobes less ovate. Fruits smaller, 2·5–3 mm. long. Seeds rather smaller, 1·5–2 mm. long and distinctly tapering.

UGANDA. Kigezi District: Lake Mutanda, Mushongero, 1 Feb. 1939, *Loveridge* 474! & Bukimbiri, Bufumbira, Oct. 1947, *Purseglove* 2502! & Luhizha, Kinaba, Mar. 1947, *Purseglove* 2364!
TANZANIA. Iringa District: Mufindi, 29 Apr. 1969, *Paget-Wilkes* 499!
DISTR. **U**2; **T**7; E. Zaire, Rwanda and Burundi
HAB. Short grassland and hillside scrub; 1800–2400 m.

SYN. *O. burttii* Milne-Redh. in B.J.B.B. 18: 103 (1946)
O. pauciflora Bak. var. *burttii* (Milne-Redh.) Verdc. in J.L.S. 53: 411 (1950)

DISTR. (of species as a whole). As above with the addition of Madagascar

35. **ANTHOSPERMUM**

L., Sp. Pl.: 1058 (1753) & Gen. Pl., ed. 5: 479 (1754)

Herbs or small shrubs with glabrous to densely hairy stems. Leaves opposite or verticillate, in whorls of 3(–6), the blades mostly narrow, more rarely ovate or oblong; stipule-sheath adnate to the petioles at base with mostly 1 or less often 2–3 narrow lobes. Flowers mostly small or very small, dioecious, polygamous or hermaphrodite, axillary, sessile or rarely in panicles; bracts minute, 2–3 at base of the ovary. Calyx-tube ovoid, ellipsoid or obovoid, the limb mostly minute, 4–5-toothed or 2–4-lobed, the lobes sometimes ovate. Corolla in ♂ flowers with tube campanulate, funnel-shaped or cylindrical, lobes 4(–5), linear to elliptic-lanceolate, longer or shorter than the tube, often revolute; stamens exserted, the filaments variously inserted at base or the throat of tube; styles 2, or joined at base, often short in ♂ flowers but exserted in ⚥ ones; in ♀ flowers the corolla-tube is often very minute with 2–4 reduced erect lobes; styles 2, or joined at the base, the stigmas long and feathery hairy; in certain species they are sometimes very long in certain specimens and form a tangled skein falling over the foliage. Ovary 2-locular, each locule containing a single erect ovule. Fruit separating septicidally into 2 cocci, each compressed ellipsoid, ventrally plane or grooved, dorsally convex, indehiscent or sometimes dehiscing ventrally. Seeds conforming in shape to the cocci.

A genus of about 40 species, the majority confined to South Africa but with several species in East and South-central Africa and also a few in Madagascar; one reaches Arabia. There has clearly been a northward migration from South Africa. The species are difficult and the following account is recognized to be rather unsatisfactory.

Stems straggling, trailing or ascending from a usually perennial rootstock; leaf-blades lanceolate to ovate-lanceolate, 0·2–1·6 cm. wide; fruits pulley-block shaped bearing no traces of calyx-lobes, nearly always quite glabrous (save in one variety known from a single specimen) . . . 1. *A. herbaceum*

Stems erect; annual herbs to perennial shrubs:

Annual herbs with ± strict, unbranched to sparsely branched stems:

Leaf-blades smaller, up to 2·2 cm. long, 1·2 mm. wide, glabrous or pubescent beneath; herb 15–70 cm. tall; fruit without traces of persistent calyx-lobes . . . . . 2. *A. ternatum*

Leaf-blades larger, up to 4·7 cm. long, 0·7 cm. wide, mostly densely hairy. . . . . 3. *A. randii*

Shrubs with sparsely to very densely branched stems:

Leaf-blades mostly exceeding 1·3 cm. in length; much less ericoid in appearance:

Leaf-blades revolute, densely hairy:

Main venation not distinctly white beneath; mostly herbaceous . . . . . 3. *A. randii*

Main venation distinctly white beneath; shrub with woody stems . . . . . 4. *A. rosmarinus*

Leaf-blades quite flat, not revolute, ± ciliate but not densely hairy; nervation not white beneath; fruits with calyx-lobes persistent; shrub 0·35–1·8 m. tall . . . . . 5. *A. welwitschii*

Leaf-blades 0·2–1·3 cm. long, 1–3 mm. wide, revolute; ericoid shrubs with densely packed whorls of leaves and mostly with numerous side-branches; fruits with persistent calyx-lobes:

Leaf-blades glabrous, sparsely ciliate or hairy on the midrib beneath; stipules with (1–)3–5(–7) setae; shrub 0·45–4·5 m. tall . . 7. *A. usambarense*

Leaf-blades mostly densely covered with spreading hairs; stipules typically with 1 seta; shrubs 0·6–2·4 m. tall . . . . . 6. *A. whyteanum*

1. **A. herbaceum** *L.f.*, Suppl. Pl.: 440 (1781); Murray, Syst. Veg., ed. 14: 919 (1784); Brenan in Mem. N.Y. Bot. Gard. 8: 455 (1954); U.K.W.F.: 407, fig. (1974). Type: South Africa, Cape of Good Hope, *Thunberg* (LINN 1233·5, holo.!, ? UPS, iso.)

Straggling, trailing or ascending perennial or perhaps sometimes annual herb 0·025–1·5(–2) m. long or tall, usually much branched at the base and usually with a ± woody tap-root; stems slender, glabrous to densely covered with short scaly hairs or with longer hairs on the young parts, reddish or purplish brown, the epidermis often peeling from the older stems, usually branched. Leaf-blades lanceolate to ovate-lanceolate, 0·5–5·5 cm. long, 0·2–1·6 cm. wide, acute at the apex, cuneate to rounded at the base, discolorous, with small reddish spots on both surfaces, shortly scaly pubescent above and on the margins and main nerves beneath or ± glabrescent on both surfaces; petioles 1–6 mm. long; stipules with 3–5 linear fimbriae 1·5–5·5 mm. long from a base ± 2 mm. long. Flowers several, axillary, ± sessile or pedicels under 0·5 mm. long, hermaphrodite or unisexual, monoecious, dioecious or polygamous and also to some extent dimorphic; bracts lanceolate, 1–4 mm. long, 0·8–1 mm. wide, ciliate. Calyx-tube oblong, 1·2 mm. long, 1 mm. wide, smooth or muriculate, emarginate at the apex; lobes not developed. Corolla white, cream, yellow or greenish; tube narrowly cylindrical or reduced, often hairy above, 2–4 mm. long in ♂ and ⚥ flowers, 0·5–0·6 or 3–3·5 mm. long in ♀ flowers; lobes in ♂ and ⚥ flowers elliptic, 2–4 mm. long, 0·5–1 mm. wide, with long hairs outside particularly near the apex; lobes in ♀ flowers either very small, 0·5 mm. long, or normal-sized, 2·6 mm. long, 0·8 mm. wide. Stamens well exserted in ♂ and ⚥ flowers, the filaments projecting beyond the tube for 1–2·2 mm., absent in ♀ flowers. Styles joined at base for 0·2–0·5 mm., the free feathery or papillate parts very variable in length, rudimentary and about 0·6–1·5 mm. long in ♂ flowers, 3·5–9 mm. long in ⚥ and ♀ flowers or up to 1·7 cm. long and tangled with the stems and foliage in what may be termed " super " ♀ forms, either exceeding the anthers or slightly shorter than them in ⚥ flowers. Fruits straw-coloured, oblong or somewhat ovoid, 2-coccous, with a median longitudinal groove so that the general shape is like that of a pulley block, 2·1–3 mm. long, 1·5–1·8 mm. wide, smooth to densely minutely muriculate or rarely densely white hairy, each coccus with 2 ventral grooves separated by a raised ± median rib; the fruit sits on a headrest-shaped structure acute at each end, which is presumably a combination of minute pedicel and 2 small bracteoles and after the fruit has fallen this straightens out to form a thickened elliptic persistent organ ± 1·5 mm. long, 1 mm. wide, acute at both ends. Seeds ellipsoid, 1·8–2·8 mm. long, 1·2–1·6 mm. wide, strongly compressed, ventrally excavated, soft and with a membranous brown shining finely reticulate testa, at length separating from the cocci which appear to be indehiscent.

var. **herbaceum**

Fruits glabrous, or minutely muriculate. Fig. 46.

UGANDA. Ankole District: Igara, Kyamahungu, May 1937, *Purseglove* 687!; Kigezi District: Maziba, Dec. 1944, *Puresglove* 1607!; Mbale District: Bulago, 27 Aug. 1932, *A. S. Thomas* 317!

KENYA. NE. Elgon, Mar. 1960, *Tweedie* 1997!; Kiambu District: Muguga, 19 Sept. 1952, *Verdcourt* 728!; Teita Hills, Yale Peak, 13 Sept. 1953, *Drummond & Hemsley* 4302!

TANZANIA. Lushoto District: W. Usambara Mts, SW Shagai Forest, Kwegoka, 24 May 1953, *Drummond & Hemsley* 2717!; Iringa District: Dabaga Highlands, Kilolo, 10 Feb. 1962, *Polhill & Paulo* 1414!; Songea District: Matengo Hills, 3 km. NE. of Mpapa, by R. Luhekea, 25 May 1956, *Milne-Redhead & Taylor* 10452!

DISTR. **U**2, 3; **K**1–7; **T**1–3, 6–8; Arabia, Ethiopia, E. Zaire, Rwanda, Mozambique, Malawi, Zambia, Rhodesia and South Africa

HAB. Forest edges, woodland, scrub, thicket, rocky grassland, heathland, also often as a weed by roadsides and in cultivations; 900–2590(–3240) m.

SYN. *A. lanceolatum* Thunb., Prodr. Pl. Cap.: 32 (1794); De Wild., Pl. Bequaert. 2: 301 (1923); F.P.N.A. 2: 371 (1947). Type: none cited, presumably a Thunberg specimen from the Cape

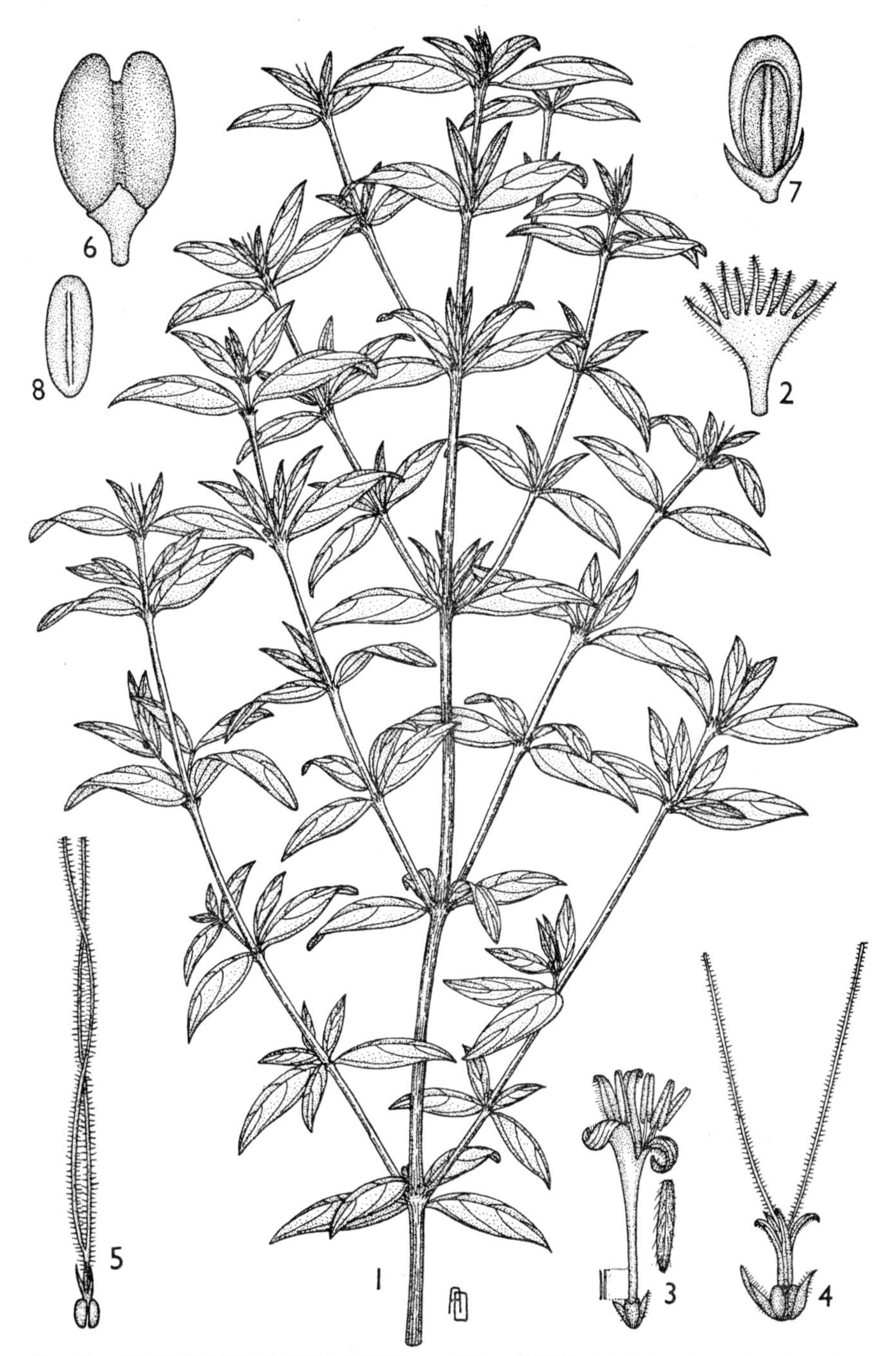

FIG. 46. *ANTHOSPERMUM HERBACEUM* var. *HERBACEUM*—**1,** habit, × $\frac{2}{3}$; **2,** stipules, × 2; **3,** male flower, × 4, showing position of rudimentary style-arms and detail of a corolla-lobe; **4,** female flower, × 4; **5,** super female flower, × 4; **6,** fruit, × 8; **7,** side view of fruit cut longitudinally, × 8; **8,** seed, × 8. 1, 2, 6–8, from *Milne-Redhead & Taylor* 10452; 3, from *Polhill & Paulo* 1414; 4, from *Milne-Redhead & Taylor* 9399; 5, from *Rounce* 422. Drawn by Ann Davis.

A. muriculatum A. Rich., Tent. Fl. Abyss. 1: 345 (1847); Hiern in F.T.A. 3: 229 (1877). Types: Ethiopia, Wodjerat, *Petit* (P, syn.) & Koubi, *Schimper* 732 (P, syn., K, isosyn. !)

A. mildbraedii K. Krause in Z.A.E. 2: 341 (1911). Type: Rwanda, W. of Lake Mohasi, *Mildbraed* 497 (B, holo. †, BR, fragment !)

NOTE. The specimens with long tangled styles, e.g. *Rounce* 422 (Tanzania, Morogoro District, Nguru Mts., 15 July 1935) are extraordinary but presumably the species is in the process of becoming adapted to wind pollination. It is very similar in habit to *Otiophora pauciflora* Bak. and easily confused, but readily distinguishable since, in that species, the fruit is crowned by the persistent calyx-lobe, not smooth and quite devoid of any trace of the calyx-lobes as in *A. herbaceum*. Study in the field is badly needed of the various sexual forms of the flowers. *Archbold* mentions in the note to her 1182 (Tanzania, Lushoto, 16 Jan. 1970) that there seems to be three kinds of flowers—(a) short tube with 4 exserted stamens, (b) short tube with 2 exserted hairy stigmas, and (c) longer tube with 2 enclosed stigmas and stamens with lost anthers or staminodes. I have not seen ⚥ or ♀ flowers with the stigmas included nor any specimens with the anthers included. Field observations are needed to check if the smaller ♂ flowers ever set seed, if the plant is ever truly dioecious and if flowers exist with the stamens included.

var. **villosicarpum** *Verdc.* in K.B. 30: 299 (1975). Type: Kenya, Northern Frontier Province, Furroli, *Gillett* 13946 (K, holo. !)

Fruits densely covered with spreading white hairs. Leaf-blades narrowly elliptic-lanceolate, ± 1 cm. long, 3–4 mm. wide, with bristly white hairs, particularly beneath. Stems densely spreading pubescent.

KENYA. Northern Frontier Province: Furroli, 20 Sept. 1952, *Gillett* 13946 !
DISTR. **K1**; not known elsewhere
HAB. *Olea*, *Juniperus* scrub on quartzite ridge; 1800 m.

NOTE. This had been annotated as a possible new species but, despite the very distinctive indumentum of the fruits, I think it is only a well-marked variety, although a final decision must await more material.

2. **A. ternatum** *Hiern*, Cat. Afr. Pl. Welw. 2: 499 (1898). Types: Angola, Huila, near Eme, *Welwitsch* 5339 (LISU, syn., BM, K, isosyn. !) & Panda Forests about Eme, *Welwitsch* 5340 (LISU, syn., K, isosyn. !)

Erect unbranched or sparsely branched annual herb 15–70 cm. tall; stems reddish, densely covered with short white spreading pubescence but eventually glabrescent beneath. Leaves in closely placed whorls of 2–3 but appearing densely verticillate due to reduced shoots in the axils, turning reddish; blades linear, 1–2·2 cm. long, 0·8–1·2 mm. wide, sharply pointed at the apex, narrowed to the base, glabrous save for the almost obsolete petioles; upper surface reticulate, the cells often obvious at ×10; stipule-sheath 0·5 mm. long with 1 seta 2 mm. long, pubescent. Flowers several in the axils of the upper 5–22 verticils of short leaves, the whole inflorescence 3–12 cm. long; pedicels very short; flowers unisexual or sometimes ⚥. Male: calyx-tube ovoid, 1·5 mm. long, hairy around the rim but lobes ± obsolete; corolla cream; tube narrow, 1·8 mm. long; lobes elliptic-lanceolate, 2 mm. long, 1 mm. wide, with a few scattered hairs outside; anthers exserted 1·4 mm., ± 1–1·2 mm. of filaments showing; styles joined for 0·3 mm., lobes 1 mm. long. Female: calyx similar to ♂, lacking real lobes; corolla cream; tube small, 0·5–0·8 mm. long; lobes 0·5–1 mm. long, 0·2 mm. wide; styles joined for 1 mm., arms 0·6–1·3 cm. long, pubescent. Hermaphrodite: calyx-tube similar, hairy at apex; corolla greenish, tube ± 3 mm. long; lobes narrowly elliptic-lanceolate, 2·5 mm. long, 0·8 mm. wide, hairy outside; filaments exserted 2 mm.; styles joined at base for 0·8 mm., free arms 6·2 mm. long. Fruits pubescent or glabrous save at apex, the cocci 2 mm. long, 0·8 mm. wide, not crowned by the calyx-limb.

TANZANIA. Njombe District: about 4 km. N. of Njombe District Boundary on Songea road, 6 July 1953, *Milne-Redhead & Taylor* 10760!; Songea District: about 44 km. N. of R. Hanga bridge on Njombe road, 30 June 1956, *Milne-Redhead & Taylor* 10934!; Tunduru District: just E. of Songea District Boundary, 5 June 1956, *Milne-Redhead & Taylor* 10602!
DISTR. T4, 5, 7, 8; Malawi, Zambia and Angola
HAB. *Brachystegia* woodland, often on poor eroded slopes, rocky hillsides, etc.; 880–1695 m.

NOTE. The material from East Africa has more coarsely pubescent stems than the Angolan types and much of the material from south-central Africa is similar but the differences are scarcely of varietal significance. What may be a ± hairy-leaved variant occurs in Mozambique.

3. **A. randii** *S. Moore* in J.B. 40: 253 (1902). Type: Rhodesia, Salisbury, *Rand* 475 (BM, holo.!)

Annual or shrubby herb 0·7–2 m. tall; stems mostly sparsely branched, at first densely covered with white spreading hairs but finally glabrescent and with a somewhat peeling epidermis. Leaves in whorls of 3–4 or at least some leaves opposite and also with short undeveloped axillary leafy shoots giving a verticillate appearance; blades linear-lanceolate to linear-oblong, 1–3·5 (–4·7) cm. long, (1–)2·5–3(–7) mm. wide, acute or subacute at the apex, narrowed to the base, revolute, ± discolorous, pubescent or scabrid above and more densely so beneath, particularly on the main nerves which are not markedly whitish; petioles obsolete; stipules with mostly 1 seta 3–5 mm. long from a base 0·5–1 mm. long. Flowers in ± sessile axillary several-flowered clusters, ⚥ or unisexual and monoecious or dioecious, 4-merous (⚥) or 5-merous (♀); pedicels 0–0·8 mm. long. Male flowers: calyx-tube 0·8–1 mm. long, the lobes minute; corolla yellow or yellowish green, pubescent outside; tube 1·5–3 mm. long; lobes elliptic, 2·6 mm. long, 1 mm. wide; filaments exserted 2–2·5 mm.; rudimentary styles 0·5 mm. long, almost free. Female flowers with corolla-tube 2·3 mm. long, lobes 1·5 mm. long; style 4·7–15 mm. long, joined and glabrous for 0·7–1·5 mm., the branches papillose. Fruit glabrous or pubescent, the cocci 2–3 mm. long, 1–1·2 mm. wide, not crowned with the calyx-limb.

TANZANIA. Mpanda District: below Kungwe Mt., 6 Sept. 1959, *Harley* 9529!; Ufipa District: Mbisi Forest, 13 Mar. 1957, *Richards* 8666!; Kondoa District: Kolo, 2 Feb. 1928, *B. D. Burtt* 1303!
DISTR. T4, 5; Zaire, Zambia, Rhodesia and Angola
HAB. Rocky hills and bracken hillsides; 1650–2400 m.

NOTE. I at first considered this should be united with *A. rosmarinus*, the two being kept varietally distinct; *Richards* 8666 I had referred to *A. rosmarinus* but I now believe it is merely a large-leaved *A. randii*. Further material of *A. rosmarinus* is needed.

4. **A. rosmarinus** *K. Schum.* in E.J. 30: 417 (1901); T.T.C.L.: 483 (1949). Type: Tanzania, Njombe District, Kinga Mts., Mwigi [Muigi] Mt., *Goetze* 1010 (B, holo. †, K, iso.!)

Shrub 2 m. tall; stems basally woody, densely covered with white spreading hairs but finally glabrescent and with somewhat peeling epidermis; internodes mostly rather short, 0·5–2·3 cm. long. Leaves opposite and also with short undeveloped axillary leafy shoots giving a verticillate appearance; blades linear-lanceolate to linear-oblong, (1–)1·3–5·5 cm. long, (0·3–)0·55–1 cm. wide, acute or subacute at the apex, narrowed to the base, strongly revolute, ± discolorous, pubescent above and more densely so beneath particularly on the main nerves which are distinctly whitish; petioles obsolete; stipules with 3 setae 3–6 mm. long from a base 0·5–1 mm. long. Flowers in sessile axillary several-flowered clusters, unisexual and dioecious.

Calyx-tube 1 mm. long; lobes very short. Corolla yellow or yellowish green, pubescent outside; tube in ♂ flowers 1·5–1·8 mm. long; lobes oblong-elliptic, 2·5 mm. long, 1 mm. wide; filaments 2·5–3 mm. long; anthers 2·2 mm. long; styles reduced to 2 free arms ± 0·8 mm. long. Fruits not known.

TANZANIA. Njombe District: Kinga Mts., Mwigi [Muigi] Mt., June 1899, *Goetze* 1010!
DISTR. **T7**; not known elsewhere
HAB. Dry slopes; 2100 m.

NOTE. See note after species 3, *A. randii* above. K. Schumann's measurement of the corolla-tube as 4 mm. long and lobes as 1 mm. long seem to be erroneous.

5. **A. welwitschii** *Hiern*, Cat. Afr. Pl. Welw. 2: 500 (1898). Type: Angola, Huila, Panda forests, near Eme, *Welwitsch* 5335 (LISU, holo., BM, iso.!)

Annual or short-lived perennial herb, subshrub or shrub 0·35–1·8 m. tall, with woody stems and often numerous short lateral branchlets, reddish brown, glabrous to densely spreading pubescent above, glabrescent towards the base where the epidermis peels off in papery strips. Leaves opposite, often appearing pseudoverticillate due to numerous abbreviated shoots; blades elliptic to oblong-oblanceolate or very narrowly oblanceolate to elliptic-lanceolate, 0·75–4 cm. long, 0·8–5 mm. wide, acute or very shortly acuminate at the apex, narrowed to the base, glabrous or with some obscure scaly marginal hairs, often ± shining and reticulate above; petiole obsolete or only ± 1 mm. long; stipule-sheath 1·5 mm. long, hairy, with 1–5 setae, 0·7–4·5 mm. long. In some specimens the internodes are very short giving the plant a very leafy appearance. Flowers axillary, usually rather numerous at each node in the axils of the main leaves and also of the numerous short leaves of the abbreviated shoots, the total inflorescence attaining up to 20 cm., with numerous flowering nodes on the main stem and the short lateral branchlets, ± sessile or pedicels ± 1 mm. long. Male: calyx-tube obconic, 0·5 mm. long; lobes unequal, oblong to triangular or lanceolate, 0·5–1·5 mm. long; corolla whitish or yellowish green, sometimes dotted with purple; tube 1 mm. long; lobes oblong-elliptic, 3–3·5 mm. long, 1–1·2 mm. wide; filaments 3 mm. long, anthers 2·2 mm. long; rudimentary stigma ± 1 mm. long. Hermaphrodite flowers: calyx-lobes triangular, 1 mm. long, apiculate; corolla-tube 2·2 mm. long, lobes 1 mm. long, 0·8 mm. wide, a few hairs at the apex outside; anthers 1·2 mm. long, exserted 0·7 mm.; styles 1·9 mm. long, joined for 0·6 mm. Female: calyx-tube oblong or ellipsoid, 0·5–2 mm. long; lobes triangular or ovate to elliptic-lanceolate, 0·3–1·8 mm. long, up to 0·8 mm. wide, acute; corolla-tube ellipsoid, 0·5–1 mm. long, lobes oblong or elliptic-lanceolate, 0·5 mm. long, 0·25 mm. wide; styles feathery, 0·55–1 cm. long, joined at the base for 1·5–3 mm. Fruits glabrous, 2 mm. long, covered with the persistent calyx-teeth.

UGANDA. Karamoja District: Mt. Kadam [Debasien], 30 May 1939, *A.S. Thomas* 2934!
KENYA. Naivasha District: near Njorowa Gorge, Hobley's Volcano, 9 July 1961, *Verdcourt* 3201!; Masai District: top of Siyabei Gorge, about 16 km. from Narok, 15 July 1962, *Glover & Samuel* 3199! & near same locality, 7 Feb. 1949, *Bally* 6593!
TANZANIA. Mbeya Mt., Apr. 1959, *Procter* 1188! & 16 Apr. 1960, *Kerfoot* 1685! & near P.C.'s house, just N. of Peak, 13 May 1956, *Milne-Redhead & Taylor* 10226! & 10228!
DISTR. **U1**; **K1**, 3, 4, 6; **T2**, ? 3, 7; Zambia, Malawi, Angola and possibly in the Transvaal
HAB. Grassland, thicket, *Tarchonanthus* scrub, often on rocky ground or rough lava soils; 1650–3000 m.

SYN. *A. cliffortioides* K. Schum. in E.J. 30: 416 (1901); T.T.C.L. 482 (1949). Type: Tanzania, Mbeya Mt., *Goetze* 1082 (B, holo.†, BM, K, iso.!)
A. sp. sensu K.T.S.: 425 (1961)

NOTE. Hedberg, A.V.P.: 176, 328 (1957), has sunk *A. cliffortioides* under *A. usambarense* but I am certain this is not correct. I am not, however, at all satisfied with the treatment above. Only one sheet, *Milne-Redhead & Taylor* 10775 (8 km. S. of Njombe, 8 July 1956), had been named *A. welwitschii* and it differs from the other material in the much more developed short lateral side branches; in general facies, however, there is great similarity. As interpreted above the species shows very great variability in the sizes of the floral parts and in the condensation of the nodes but this is a feature of the genus. It is possible that *A. ammannioides* S. Moore is not specifically distinct.

6. **A. whyteanum** *Britten* in Trans. Linn. Soc., ser. 2, 4: 16 (1894); T.T.C.L.: 483 (1949); Brenan in Mem. N.Y. Bot. Gard. 8: 455 (1954). Type: Malawi, Mlanje Mt., *Whyte* 48 (BM, holo.!)

A much-branched shrub or subshrub (0·25–)0·6–2·4 m. tall, similar in habit to the next species in most respects; stems densely covered with much coarser spreading hairs but ultimately glabrescent and the epidermis often peeling on the older stems. Leaves verticillate in whorls of 3 and also with short axillary shoots giving an appearance of more; blades elliptic or narrowly ovate-lanceolate to linear-lanceolate, 0·3–1·3 cm. long, 1–3 mm. wide, apiculate, narrowed to the base, strongly revolute, mostly densely covered with spreading hairs; stipules with a single seta 2–3 mm. long from a base 0·5 mm. long; petioles obsolete. Flowers in sessile axillary clusters, unisexual and dioecious, apparently single between a pair of bracts and a pair of bracteoles. Calyx-tube obconic, 1·5 mm. long, hairy, the hairs upwardly directed and apical ones 0·5 mm. long beyond the tip of the calyx, hiding the triangular 0·3 mm. long lobes. Corolla greenish white, white, pink or yellowish, sometimes marked dull purple; tube 2 mm. long in ♂ flowers, 0·5 mm. long in ♀ flowers; lobes oblong-lanceolate or oblong-elliptic in ♂ flowers, 2·8 mm. long, 0·8–1·1 mm. wide, oblong-lanceolate and smaller in ♀ flowers, 0·6–0·7 mm. long, 0·2 mm. wide. Filaments exserted 2 mm. in ♂ flowers. Styles rudimentary, free, ± 2 mm. long in ♂ flowers, 8·5 mm. long and joined at the base for 2 mm. in ♀. Fruit hairy or glabrous.

TANZANIA. Songea District: Matengo Hills, Lupembe Hill, 10 Jan. 1956, *S. Paulo* in *Milne-Redhead & Taylor* 8177! & same locality, 3 Jan. 1936, *Zerny* 269!
DISTR. T8; Mozambique, Malawi, Zambia and Rhodesia
HAB. Amongst rocks in upland grassland; 1860 m.

SYN. *A. albohirtum* Mildbr. in N.B.G.B. 15: 637 (1941). Type: Tanzania, Matengo Highlands, Lupembe Mt., *Zerny* 269 (W, holo.!)

NOTE. Published and unpublished records of this species from East Africa probably all refer to *A. usambarense*; I have seen only the gatherings cited above and it must be presumed at the extreme north of its range in this locality. In Rhodesia the species is usually a small subshrub from a very woody rootstock. *Brenan & Greenway* 8235 (Tanzania, Mbeya–Sao Hill road, 29 Oct. 1947) has the habit of *A. whyteanum* and stipules with a single seta, but the stems have the indumentum of *A. usambarense* and the leaves are glabrous; moveover the fruits are not crowned by the persistent calyx-lobes.

7. **A. usambarense** *K. Schum.* in E.J. 28: 112 (1899); F.P.N.A. 2: 372, t. 37 (1947); T.T.C.L.: 483 (1949); Brenan in Mem. N.Y. Bot. Gard. 8: 455 (1954); A.V.P.: 176, 328 (1957); K.T.S.: 425 (1961). Types: Tanzania, W. Usambara Mts., *Holst* 420 (B, syn. †) & Kwai, *Eick* (B, syn. †)

A much-branched shrub 0·45–4·5 m. tall, with dense heath-like foliage, the whorls of leaves running together at the apices of the shoots; stems covered with dense spreading short grey hairs when young, but later glabrescent or sometimes ± persistently hairy, the often reddish epidermis peeling off. Leaves opposite or verticillate in whorls of 4–6 and also pseudo-verticillate due to short undeveloped shoots, forming sometimes big subglobose fasciculate clusters; blades narrowly oblong-elliptic to linear-lanceolate, 0·2–1·5 (–3·5) cm. long, 2–5 mm. wide, but often appearing narrow (± 0·5 mm. wide) due to strong revolution of the margins, apiculate at the apex, narrowed to the base, glabrous, sparsely ciliate or hairy on midrib beneath; stipules with (1–)3–5(–7) setae 0·6–3(–5) mm. long from a base 1 mm. long; petioles obsolete. Flowers in sessile axillary clusters, unisexual and dioecious. Calyx-tube conic, 0·8 mm. long, lobes in ♀ flowers 2 small and 2 larger, 0·2–0·5 mm. long. Corolla greenish white or yellowish, often tinged maroon in bud or entirely pinkish in ♀ flowers; in ♂ flowers tube 0·8 mm. long, lobes elliptic, 2·7 mm. long, 1 mm. wide; in ♀ flowers much reduced, the tube 0·4 mm. long with firmer midpetaline areas and the lobes short, oblong, 0·2 mm. long. Stamens with filaments ± 1 mm. long, the anthers well exserted, absent in ♀ flowers. Styles cream, feathery, 4·5–6·5 mm. long, joined for ± 0·5–2·7 mm. of their length. Fruits oblong-ellipsoid, 1·5–2 mm. long, 0·7–1 mm. wide, crowned with the small calyx-lobes.

UGANDA. Karamoja District: Mt Morongole, June 1942, *Dale* U. 259!; Mbale District: Mt. Elgon, Madangi, Nov. 1939, *Dale* U.48! & Mt. Elgon, crater, Jan. 1918, *Dummer* 3348!

KENYA. Naivasha District: Aberdare National Park, Fort Jerusalem, 30 July 1960, *Polhill* 229!; Nanyuki District: Mt Kenya, Sirimon Track at 2550 m., 1 Oct. 1964, *Gillett* 16258!; Masai District: W. side of Nasompolai [Enesambulai] valley, 25 July 1970, *Greenway & Kanuri* 14535!

TANZANIA. Kilimanjaro, near Horombo [Peter's] Hut, 22 Feb. 1934, *Greenway* 3756!; Lushoto District: at head of Kwai valley, Matondwe Hill, 28 Feb. 1953, *Drummond & Hemsley* 1351!; Iringa District: Dabaga Highlands, Kibengu, 17 Feb. 1962, *Polhill & Paulo* 1518!; Rungwe District: NW. of Rungwe Mt., Kiwira R., lower fishing camp, 24 Oct. 1947, *Brenan & Greenway* 8193!

DISTR. **U**1–3; **K**1 (Nyiru), 2–6; **T**2, 3, 7, 8; Sudan, Zaire, Malawi, Rhodesia

HAB. Typical heath zone of many mountains, exposed upland grassland, also forest margins and clearings, rocky outcrops; 1300–4050(–4500 fide *Grote* 3903) m.

SYN. *A. leuconeuron* K. Schum. in E.J. 30: 517 (1901); T.T.C.L.: 483 (1949). Type: Tanzania, Mbeya Mt., *Goetze* 1081 (B, holo.†, BM, BR, K, P, iso.!)

A. aberdaricum K. Krause in N.B.G.B. 10: 609 (1929). Types: Kenya, Aberdares, Kinangop Summit, *Fries & Fries* 2574 (B, syn.†, BR, K, S, UPS, isosyn.!) & same area, Satima summit, *Fries & Fries* 2465 (B, syn.†, K, UPS, isosyn.!)

A. uwembae Gilli in Ann. Naturhist Mus. Wien 77: 18, fig. 1 (1973). Type: Tanzania, Njombe District, Uwemba, *Gilli* 399 (W, holo.!)

NOTE. Very closely related to some S. African species, notably to *A. ciliare* L.

8. **A. prittwitzii** *K. Schum. & K. Krause* in E.J. 39: 570 (1907); T.T.C.L.: 483 (1949). Type: Tanzania, "Nordlich Nyassaland", Lager Kidoko, *Prittwitz & Gaffron* 57 (B, holo. †)

Ericoid subshrub, densely branched from the base, up to 50 cm. tall; taproot stout; branches ± erect, sparsely puberulous above, glabrous beneath, with reddish brown epidermis. Leaves 12–16-verticillate;* blades linear-lanceolate, 0·8–1·2 cm. long, 2–3 mm. wide, acuminate at the apex, cuneate at the base, margin slightly reflexed, ± pilose on both surfaces; stipules connate with the petioles to form a truncate cupulate sheath, 3 mm. long, sparsely puberulous. Flowers minute and inconspicuous, shortly pedicellate (pedicels 1–1·5 mm. long). Hermaphrodite: calyx-tube glabrous, ovoid, 0·5 mm. long; limb obsoletely 4-toothed; corolla glabrous, 3–3·5 mm. long; tube funnel-shaped; lobes 4, ovate-oblong, acute, longer than the tube. Stamens inserted on the throat; anthers ± 1 mm. long, exserted. Style 4 mm. long, equalling the corolla-lobes, divided into 2 filiform papillose stigmas. Fruit 2 mm. long, dividing into 2 cocci.

TANZANIA. ? District: Lager Kidoko, 28 Aug. 1901, *Prittwitz & Gaffron* 57

DISTR. **T**7 (fide Gillett)

HAB. Unknown

* Presumably including abbreviated axillary shoots, i.e. pseudoverticillate.

Note. Its authors compare this with *A. usambarense** but distinguish it by its larger broader leaf-blades in more separated whorls. I have seen no material nor have I been able to identify it with any known species. The habit and leaves could fit *A. whyteanum* but the stipules do not agree unless the description is very faulty.

Tribe 10. SPERMACOCEAE

Ovary usually 2-locular; stigmas 2 or 1 capitate stigma; fruit with 2 cocci, capsular or circumscissile:
- Fruit indehiscent, with 2 cocci or a capsule splitting longitudinally:
  - Succulent creeping plant of the seashore, with imbricated leaves joined by quite broad sheathing stipules with very short processes; stems rooting at the nodes; fruits indehiscent; seeds not lobed . . . . . . **36. Hydrophylax**
  - Plant not a littoral succulent or, if somewhat so (*Diodia* subgen. *Pleiaulax*), then leaves not imbricated, stipules with longer processes, stems not rooting at the nodes, fruits dividing into cocci and seeds lobed:
    - Fruit with 2 cocci separating but ± indehiscent . **37. Diodia****
    - Fruit capsular with 2 valves, either dehiscing from base to apex (sect. *Arbulocarpus*) or in the reverse direction (sect. *Borreria*) or with 2 cocci, 1 or both ± dehiscent . . . **38. Spermacoce**
- Fruit circumscissile about its middle, the top lifting off like a lid; flowers minute, clustered in sessile globose inflorescences at the nodes; seeds with a dorsal ± X-shaped groove . . . **39. Mitracarpus**

Ovary 3-locular; stigmas 3; fruit with 3 cocci; straggling procumbent herbs . . . . . **40. Richardia**

36. HYDROPHYLAX

L.f., Suppl. Pl.: 126 (1781)

Prostrate subsucculent perennial herbs confined to the littoral zone. Leaves sessile, ovate, oblong or elliptic, closely imbricated on short flowering shoots; stipule-sheath cupular, adnate to the petioles, with 1–2 short setae or up to 8 very small processes. Flowers solitary, axillary, hermaphrodite, isostylous or heterostylous. Calyx-tube oblong-ovoid, 4-angled; limb irregularly lobed or with 4 persistent ovate-lanceolate acute lobes. Corolla somewhat fleshy; tube broadly funnel-shaped, hairy inside for upper half; lobes 4, valvate, ovate to elliptic-lanceolate. Filaments filiform; anthers linear, exserted. Disc fleshy. Ovary 2-locular; ovules solitary in each locule, attached to the centre of the septum; style filiform, pubescent; stigma-lobes linear, exserted. Fruit indehiscent, small or relatively large, ellipsoid to oblong-ovoid, compressed, often curved, corky, 4-angled, keeled between the angles, or 8-keeled, 1–2-locular, 1–2-seeded. Seeds linear-oblong,

* Die Pflanze is nahe verwandt mit *A. usambarense* K. Sch., aber durch grössere vor allen Dingen breitere Blätter sowie besonders in den oberen Teilen der Stengel zeimlich entfernt stehende Blattuvirle unterschieden.

** Since the division into *Spermacoce* and *Diodia* is not based on very reliable characters, the key to the species of *Spermacoce* also includes those of *Diodia*.

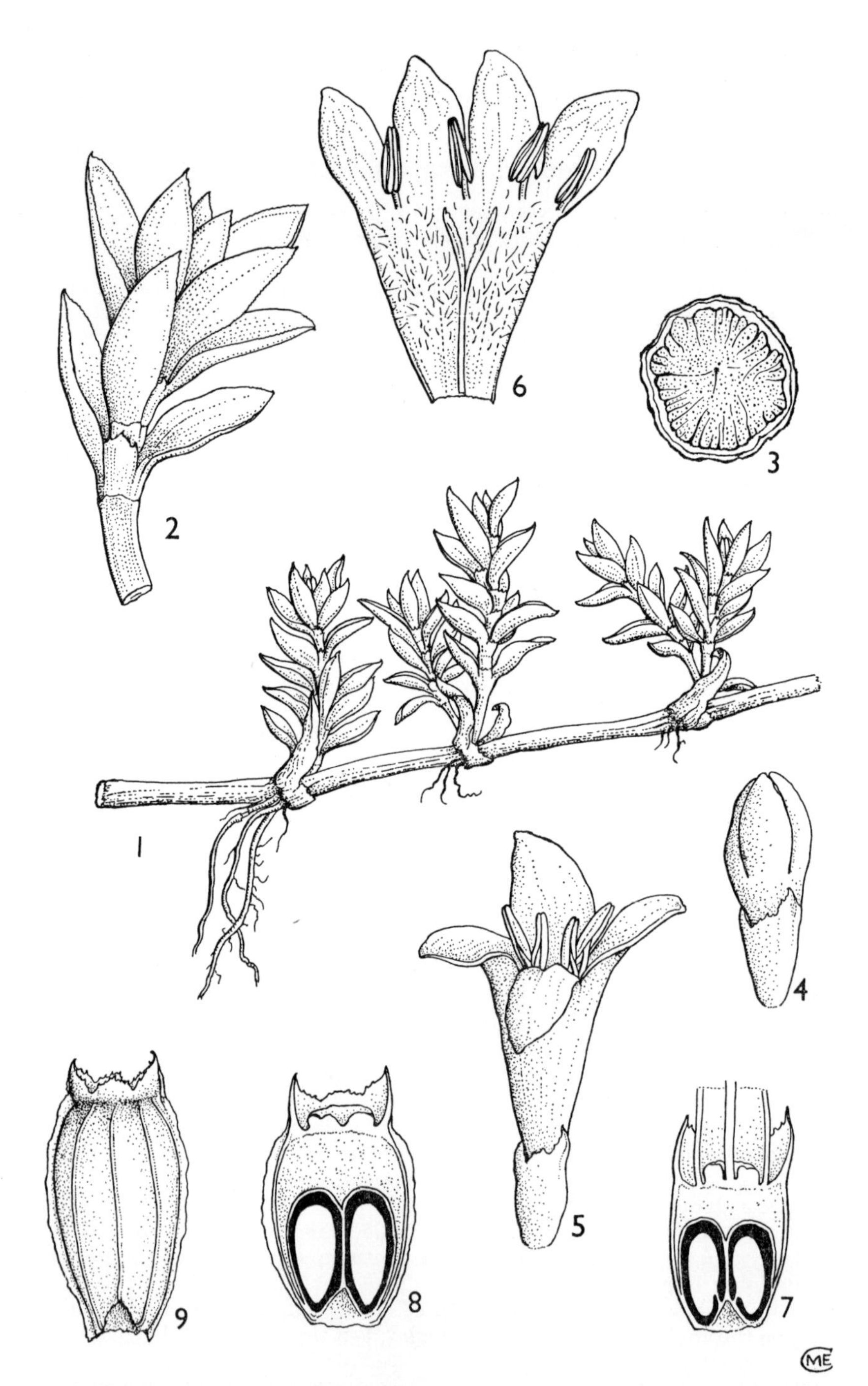

FIG. 47. *HYDROPHYLAX MADAGASCARIENSIS*—**1,** habit, × 1; **2,** branchlet, × 2; **3,** transverse section of stem, × 1; **4,** flower bud, × 6; **5,** flower, × 6; **6,** corolla, opened out, and style, × 6; **7,** longitudinal section of ovary, × 10; **8,** same of young fruit, × 10; **9,** older immature fruit, × 10. 1, 2, 4–9, from *Frazier* 1124; 3, from *Perrier* 18181. Drawn by Mrs M. E. Church.

dorsally convex, ventrally longitudinally deeply 2-sulcate when viewed in transverse section but grooves covered over externally; testa minutely granular; endosperm cartilaginous.

A genus related to *Diodia* and distributed from Thailand, India, Ceylon and Madagascar to Tanzania, Mozambique, Natal and E. Cape Province, with 3 species, 1 of which has very recently been discovered in the Flora area. I am not at all certain if all plants of the genus exhibit heterostyly but Medley Wood (in litt. on *Wood* 11683) specifically mentions that of 410 flowers of *H. carnosa* Sond. 180 were short-styled and 230 long-styled.

H. madagascariensis *Roem. & Schult.*, Syst. Veg. 3: 527 (1818); Denis in Rev. Gén. Bot. 31: 118, t. 1 (1919). Type: E. Madagascar, *Commerson* (B-WILLD 2674, holo.*)

Perennial herb spreading over an area of about 0·5 sq. m., with long creeping stems 30 cm. to several m. long, bearing short flowering shoots with densely imbricated leaves; rhizomes up to 2·5 cm. in diameter (in Madagascar); main stems with peeling straw-coloured epidermis when dry, glabrous. Leaves imbricated, probably somewhat longitudinally folded in life, narrowly elliptic to elliptic-lanceolate, 0·7–1·5 cm. long, 2·5–4 mm. wide, apiculate at the apex, scarcely narrowed at the base, the margins minutely serrulate with small papillae; stipule-sheath 3–4 mm. long, truncate, with 1–4 small teeth or irregularly lacerate. Calyx-tube ovoid, 2 mm. long, 1·5 mm. wide, 8-ribbed, 2 ribs stronger than the others; limb 0·75–1·5 mm. tall, irregularly undulate, slightly lacerate. Corolla white; tube narrowly funnel-shaped, 3·5–4·5 mm. long; lobes elliptic-lanceolate, 2·5–3·5 mm. long, 1–1·4 mm. wide. Anthers 1·5–2 mm. long, included or exserted save for extreme base. Style 2–2·5 mm. long; stigma-lobes ± unequal, very slender, 1·3–2 mm. long. Fruit compressed ellipsoid, 3·3 mm. long, 2 mm. wide, ribbed. Ripe seeds not seen. Fig. 47.

TANZANIA. Pangani District: S. of Mto Kama, 12 km., south of Mkwaja, 13 Aug. 1974, *Frazier* 1124!
DISTR. **T3**; Madagascar
HAB. Seaward edge of beach platform on ± bare quartz sand; sea-level

NOTE. *Hydrophylax carnosa* Sond., known from South Africa (E. Cape Province, Natal) and S. Mozambique, seems to differ in foliage. I would hesitate to suggest it was a synonym of *H. madagascariensis* without considerably more material of the latter for comparison.

37. DIODIA

L., Sp. Pl.: 104 (1753) & Gen. Pl., ed. 5: 45 (1754)

Annual or perennial erect or mostly decumbent or prostrate herbs; stems often 4-angled. Leaves opposite or sometimes appearing whorled due to the presence of reduced axillary shoots, ± sessile or shortly petiolate; blades linear to ovate; stipules with bases united to the petiole, divided into fimbriae. Flowers mostly small, not heterostylous, in small axillary clusters or, in one group of American species, terminal in spikes or capitula. Calyx-tube ellipsoid, ovoid or obconic; lobes 2–4 (rarely 5–6), sometimes with some small accessory teeth, ± persistent. Corolla-tube ± funnel-shaped, glabrous or hairy at the throat; lobes 4 (rarely 5–6), valvate. Stamens 4 (rarely 5–6), exserted, the filaments inserted at the throat. Ovary 2(rarely–3–4)-locular; ovules solitary in each cell, attached to the middle of the septum; style

* The small specimen in the Willdenow Herbarium which is labelled *Hydrophylax*, Madagascar, *Commerson*, may, be the holotype unwritten up; it is obviously at any rate a portion of the original gathering.

filiform exserted; stigma 2-lobed or ± capitate. Fruit of 2 (rarely 3–4) indehiscent cocci. Seeds oblong, dorsally convex, ventrally longitudinally grooved, rarely with a transverse groove or in one subgenus (*Pleiaulax* Verdc.) distinctly lobed.

A genus of some 30–50 species, mostly in the New World. The circumscription of the genera *Diodia*, *Borreria*, *Spermacoce* and several satellite genera has been a very difficult matter indeed. If only the African species were involved it would seem reasonable to treat all as synonyms of *Spermacoce* but a survey of the New World material shows that *Diodia* is quite distinctive and best kept separate. *D. aulacosperma* belongs to subgenus *Pleiaulax* Verdc. characterized by its lobed seeds. It does not seem quite worthy of generic rank since traces of such lobing can be seen in other species of *Spermacoce* and *Diodia* in the New World.

Plant mostly of higher elevations; corolla-tube 1·8 mm. long;* seeds not lobed . . . 1. *D. sarmentosa*
Strand plant; corolla-tube (3–)4–7 mm. long; seeds distinctly and characteristically lobed (see fig. 48/7, p. 338) . . . . . 2. *D. aulacosperma*

1. **D. sarmentosa** *Sw.*, Prodr. Veg. Ind. Occ.: 30 (1788) & Fl. Ind. Occ. 1: 231 (1797); S. Moore in Fl. Jam. 7: 117, fig. 35 (1936); Adams, Fl. Pl. Jam.: 731 (1972). Type: Jamaica, *Swartz* (ubi?)

Straggling, scrambling or procumbent herb 1–3·6 m. long, often with many lateral branches from the main stem; stems 4-angular, pubescent on the angles but at length glabrous. Leaf-blades often rather yellowish green, elliptic, 1·8–6·3 cm. long, 0·7–2·8 cm. wide, acute at the apex, narrowed to the base, scabrid above with dense very short to longer tubercle-based hairs, pubescent beneath; petiole 1–5 mm. long; stipule-bases 1–2 mm. long with lines of hairs, bearing 5–7 setae 1–7 mm. long. Flowers usually few in axillary clusters at most nodes, the inflorescences up to 1·2 cm. in diameter in fruiting stage. Calyx-tube glabrous, obconic, 1·5–2 mm. long; lobes 4, often unequal, oblong-lanceolate or narrowly triangular, 1·5–3 mm. long, 0·8 mm. wide, ciliate. Corolla mauve or white; tube glabrous, funnel-shaped, 1·8 mm. long; lobes triangular, 1 mm. long, 1 mm. wide, with a few hairs outside. Filaments exserted 0·5 mm. Style exserted 1·5 mm., minutely papillate. Cocci ½-oblong-ellipsoid, 3·5–5 mm. long, 2·5 mm. wide, 1·2 mm. thick or sometimes more globose, quite definitely not readily dehiscent. Seeds dark blackish red, compressed ellipsoid, 2–4 mm. long, 1·5 mm. wide, 0·8 mm. thick, with a broad ventral groove, finely rugulose. Fig. 48/8–10, p. 338.

UGANDA. Mengo District: Buvuma, Namunyoro [Namunyolo], Mar. 1925, *Maitland* 1079! & Entebbe, Sept. 1922, *Maitland* 205! & Lolui I., 5 May 1965, *G. Jackson* 35565!
KENYA. Kwale District: Shimba Hills, 14·5 km. SW. of Kwale, Pengo Forest, 12 Feb. 1953, *Drummond & Hemsley* 1205!
TANZANIA. Lushoto District: Amani, 22 Jan. 1950, *Verdcourt* 53!; Rufiji District; Mafia I., Kilindoni, 10 Sept. 1937, *Greenway* 5242!; Mpanda District: Kungwe Mt., Kasoje, 16 July 1959, *Newbould* 4377!; Zanzibar I., without precise locality, Sept. 1873, *Kirk*! & Sept. 1873, *Hildebrandt* 1139!; Pemba I., 9·6 km. on road between Chake and Weti, 20 Jan. 1933, *Vaughan* 2060!
DISTR. **U**2, 4; **K**7; **T**3, 4, 6; **Z**; **P**: widespread in tropical Africa, also in tropical Asia, America and the Mascarenes
HAB. Evergreen forest, rocky places, lake shores, coconut plantations, etc.; 0–1150 m.

SYN. *D. pilosa* Schumach. & Thonn., Beskr. Guin. Pl.: 76 (1827). Type: Ghana, Akwapim, *Thonning* 212 (C, holo., P–JU, iso.)
Spermacoce pilosa (Schumach. & Thonn.) DC., Prodr. 4: 553 (1830); Hiern in F.T.A. 3: 235 (1877); F.W.T.A. 2: 135 (1931)

* If corolla-tube under 1 mm. long, see *Spermacoce*, particularly *S. tenuior* and *S. confusa*.

Diodia breviseta Benth. in Hook., Niger Fl.: 424 (1849); Hiern in F.T.A. 3: 231 (1877). Type: Fernando Po, *Vogel* (K, holo.!)
[*D. scandens* sensu auctt., e.g. Hepper, F.W.T.A., ed. 2, 2: 216, fig. 245 (1963), *non* Sw.]

Note. There has been a tendency to consider *D. scandens* as the plant occurring in Africa and *D. sarmentosa* as a synonym. Swartz was much too careful a botanist to describe the same species under two different names on the same page; in any case an authentic specimen of *D. scandens* collected in Hispaniola by Swartz exists in the BM.

2. **D. aulacosperma** *K. Schum.* in P.O.A. C: 394 (1895); Cufod., E.P.A.: 1021 (1965); Verdc. in K.B. 30: 300, t. 42 (1975). Type: Kenya, Lamu I., *Hildebrandt* 1903 (B, holo. †, K, iso.!)

Annual, erect, decumbent or ± prostrate, unbranched to much-branched, usually semi-succulent herb 3–35(–90) cm. long or tall; stems distinctly 4-angular or ± terete, glabrous, glabrescent or covered with long white spreading hairs or upper parts glabrous and lower hairy. Leaves usually distinctly subsucculent; blades oblong, elliptic, elliptic-lanceolate or even almost round, 0·8–5 cm. long, 0·3–2·2 cm. wide, ± acute to very rounded at the apex, broad to very distinctly narrowed to the base, margins mostly ± thickened, pubescent to hairy or scabrid on both surfaces or glabrescent save for the distinctly scabrid margin; petiole 1–3(–7) mm. long; stipule-sheath 2–3 mm. long, usually hairy, with (3–)5–8 setae 1–6 mm. long. Flowers (1–)4–10 in dense axillary clusters up to 1·6 cm. wide in fruit; pedicels obsolete or up to 0·5 mm. long. Calyx-tube turbinate, hairy, 1·5–2 mm. long; lobes 4(–6), usually unequal, triangular, oblong-triangular, oblong or rounded-ovate, (1–)1·5–4·5 mm. long, 0·5–2 mm. wide, acute or obtuse. Corolla white, pink, blue or mauve; tube funnel-shaped to campanulate, (3–)4–7 mm. long, up to 4(–6) mm. wide, upper half hairy outside; lobes ovate-triangular, 2–4·5 mm. long, 1–3 mm. wide, hairy outside. Filaments exserted 1–2 mm. Style exserted 3–4 mm., papillate; stigma 0·6–1 mm. wide. Fruits oblong-ovoid, 2·5–4 mm. long, 2 mm. wide, usually covered with ± inflated blunt hairs but sometimes scabrid or glabrescent, splitting into 2 indehiscent cocci. Cocci oblong-ellipsoid, 2·5–4 mm. long, 1·5–2 mm. wide, 1·2 mm. thick, the seed-ribs showing through the walls faintly. Seeds very characteristic, oblong, 2·4 mm. long, 1·6 mm. wide, 1·1 mm. thick, dorsal surface with 2 longitudinal grooves and 10–12 deep lateral fissures on each side making the surface resemble a sternum and ribs; ventral surface similar but with only the main groove of the raphe; the actual seed surface is also pitted.

var. **aulacosperma**

Leaf-blades more succulent, rounded, oblong or elliptic, not so markedly narrowed to the base; lateral nerves less evident or not visible. Fig. 48/1–7, p. 338.

Kenya. Kilifi District: near Malindi, 2 Jan. 1954, *Davis* 91! & Watamu, Oct. 1965, *Tweedie* 3163!; Lamu I., 28 Jan. 1958, *Verdcourt* 2115!
Tanzania. Pangani District: Mkwaja–Buyuni, 28 Nov. 1915, *Peter* 55832!; Zanzibar I., Marahubi, 3 Apr. 1952, *R. O. Williams* 156! & same locality, 2 Jan. 1963, *Faulkner* 3146!
Distr. **K7**; **T3**; **Z**; Somali Republic (S.)
Hab. A strand plant, often quite near high-tide mark, in grassland and bushland, usually on sand, sometimes a weed of cultivation; 0–30 m.

Syn. [*Diodia breviseta* sensu Hiern in F.T.A. 3: 231 (1877), quoad spec. *Hildebrandt* cit., *non* Benth.]
Hypodematium somalense Chiov., Result. Sci. Miss. Stef.-Paoli, Coll. Bot.; 99 (1919). Type: Somali Republic (S.), Bender Suguma, *Paoli* 267, 277 & Giumbo, *Paoli* 241, 254 & Mogadishu, *Paoli* 124 (all FI syn!)
Diodia physotricha Chiov. in Agric. Colon. 20: 45 (1926) & in Atti Ist. Bot. Univ. Pavia, ser. 4, 7: 135 (1936); Cufod., E.P.A.: 1021 (1965). Type: Somali Republic (S.), Kismayu, *Gorini* 43 (FI, holo.)

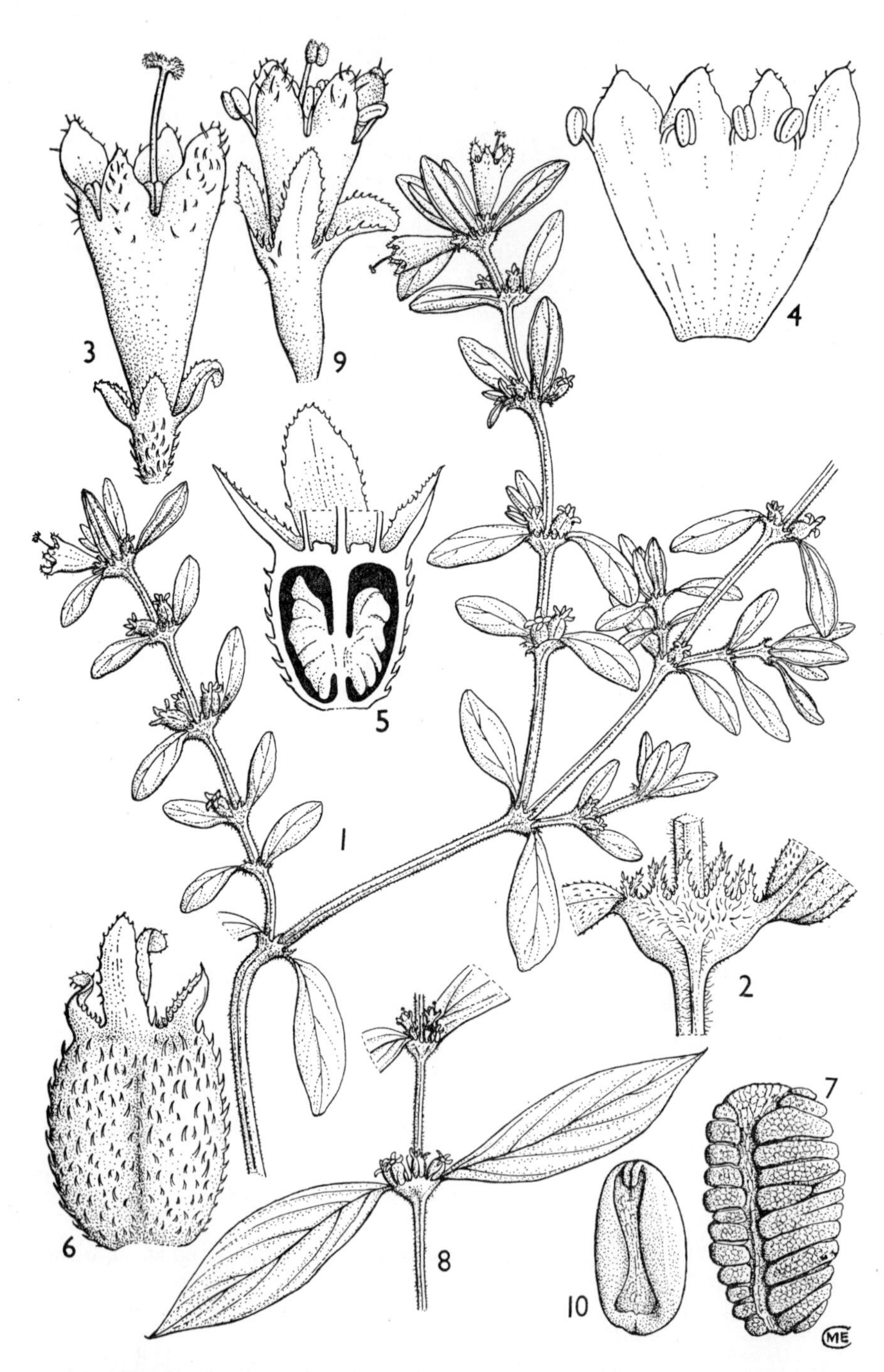

FIG. 48. *DIODIA AULACOSPERMA* var. *AULACOSPERMA*—**1,** habit, × 1; **2,** node showing stipule, × 4; **3,** flower, × 4; **4,** corolla, opened out, × 3; **5,** longitudinal section of ovary, × 8; **6,** fruit, × 8; **7,** seed, × 12. *D. SARMENTOSA*—**8,** part of flowering branchlet, × 1; **9,** flower, × 10; **10,** seed, × 12. 1–7, from *Schlieben* 12117; 8–10, from *Verdcourt* 53. Drawn by Mrs M. E. Church.

D. physotricha Chiov. var. *cyclophylla* Chiov. in Atti Ist. Bot. Univ. Pavia, ser. 4, 7: 135, t. 6 (1936). Type: Somali Republic (S.), S. of Merca, *Ciferri* 93 (FI, holo., K, iso. !)
Arbulocarpus somalensis (Chiov.) Cufod., E.P.A.: 1023 (1965)

NOTE. *Gillespie* 83 (Kenya, Lamu District, 88 km. NE. of Lamu, Kiunga Point, 29 July 1961) is undoubtedly almost identical with the type specimens of *Hypodematium somalense*, but I do not think this is worth keeping up even as a variety. *Verdcourt* 2115 cited above contains elements bridging the short-leaved and long-leaved specimens; long-leaved specimens with erect hairy stems could be referred to a variety based on this name if considered necessary.

var. **angustata** *Verdc.* in K.B. 30: 300 (1975). Type: Tanzania, Tanga District, Sawa, *Faulkner* 1666 (K, holo. !, EA, iso.)

Leaf-blades less succulent, ovate to elliptic, markedly contracted to the base; blade up to 4·5 × 2·2 cm. and petiole up to 7 mm. long; lateral nerves much more obvious. Stems usually less hairy and corolla mostly smaller, the tube 3–4 mm. long.

KENYA. Mombasa District: Likoni, 30 May 1934, *Napier* 3321 in *C.M.* 6276 !; Tana River District: S. of Garsen, Karawa, June 1959, *Rawlins* 779 !; Lamu District: 88 km. NE. of Lamu, Kiunga Point, 23 July 1961, *Gillespie* 28 !
TANZANIA. Tanga District: Sawa, 29 July 1967, *Faulkner* 3984 ! & 12·8 km. S. of Moa, Bomandani, 5 Aug. 1953, *Drummond & Hemsley* 3652 !; Pangani, 14 Sept. 1950, *Faulkner* 713 !
DISTR. **K7**; **T3**; Socotra
HAB. As for var. *aulacosperma*; 0–5 m.

SYN. [*Spermacoce hispida* sensu Balf. f. in Trans. Roy. Soc. Edin. 31: 119 (1888), *non* L.]

38. SPERMACOCE

L., Sp. Pl.: 102 (1753) & Gen. Pl., ed. 5: 44 (1754)

Borreria G. F. W. Mey., Prim. Fl. Esseq.: 79, t. 1 (1818)
Octodon Schumach. & Thonn., Beskr. Guin. Pl.: 74 (1827)
Hypodematium A. Rich., Tent. Fl. Abyss. 1: 348 (1848), *non* Kunze (1833)
Dichrospermum Bremek. in B.J.B.B. 22: 75 (1952)
Arbulocarpus Tennant in K.B. 12: 386 (1958)

Annual or perennial herbs or small subshrubs, with glabrous, pubescent, hispid or scabrid very often 4-angled prostrate to erect stems. Leaves opposite or falsely whorled, sessile or petiolate, the petioles often united with the stipule-sheath, which is mostly divided into 1–many ± filiform fimbriae. Flowers mostly small or occasionally medium-sized, hermaphrodite, not heterostylous (except in sect. *Galianthe* (Griseb.) K. Schum. which does not occur in Africa), sessile, mostly in axillary ± globose often very many-flowered clusters or less often in terminal capitula, supported by 1–2(–more) pairs of leaves forming bracts, or 1–few in the axils (some extra-African species have extensive terminal inflorescences); sometimes the axillary nodal clusters run together to form a spike-like inflorescence the leaf pairs forming scattered or congested bracts; the flowers in the globose clusters are frequently intermixed with numerous ± scarious filiform bracteoles. Calyx-tube obovoid, turbinate or obconic; lobes 2–4(–8), mostly triangular, oblong or lanceolate, often ± persistent, sometimes with intermediate denticles. Corolla funnel-shaped or salver-shaped, the tube sometimes very slender; throat glabrous to hairy; lobes (3–)4, mostly spreading, valvate. Stamens 4, the filaments inserted in the tube or at the throat, the linear to oblong anthers included or mostly exserted. Ovary 2-locular, the amphitropous ovules solitary in each locule, attached to the middle of the septum; style filiform, mostly exserted; stigma capitellate or with 2 short lobes. Fruit mostly a 2-valved capsule dehiscing from the apex downwards with the septum

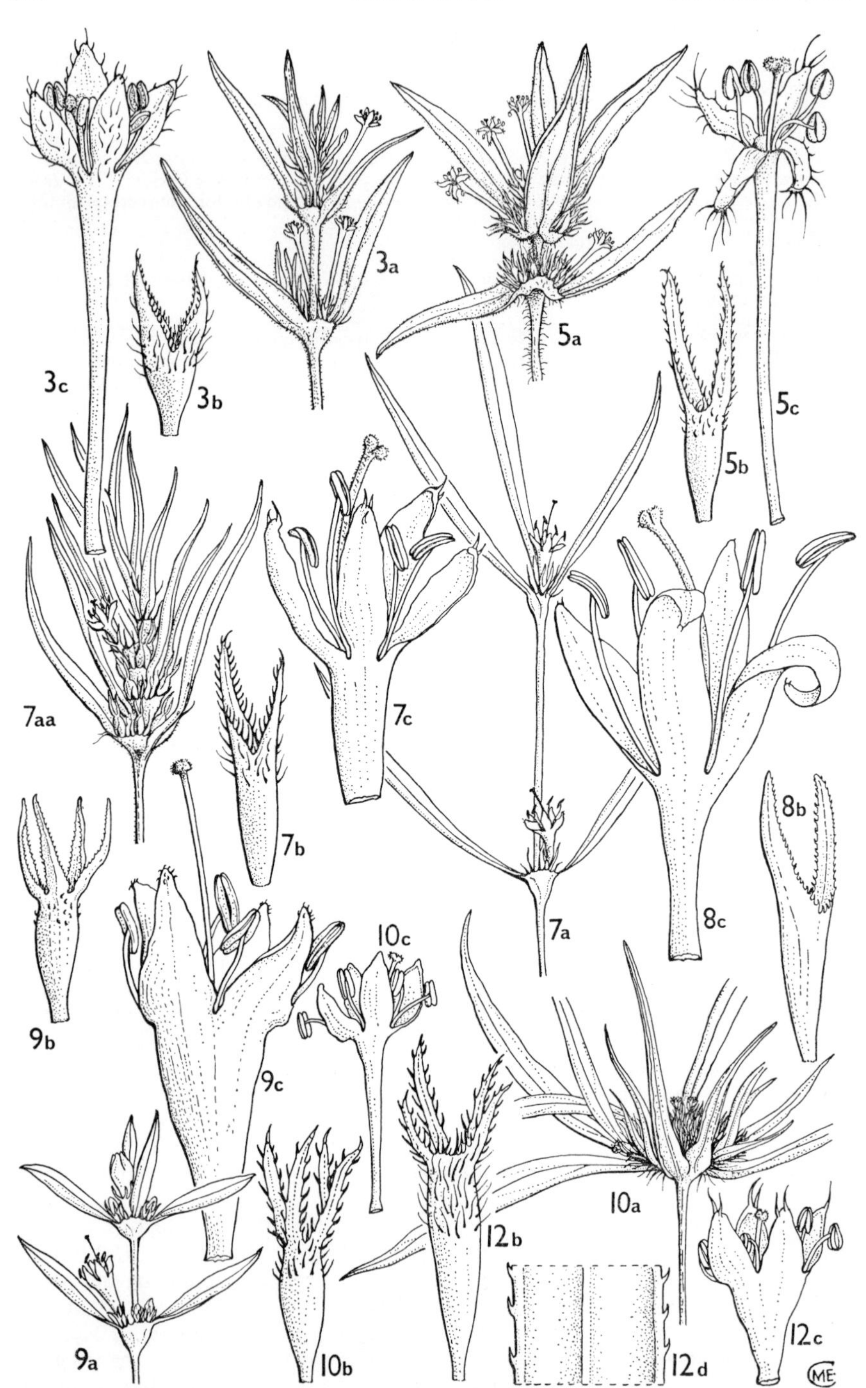

FIG. 49. *SPERMACOCE*—inflorescence (a), calyx (b), corolla (c) and detail of lower leaf surface (d). Species numbered as in text. **3,** *S. filituba*, a × 1, b, c × 6; **5,** *S. arvensis*, a × 1, b, c × 6; **7,** *S. subvulgata* var. *subvulgata*, a, aa × 1, b, c × 6; **8,** *S. taylorii*, b, c × 3; **9,** *S. latituba*, a × 1, b, c × 4; **10,** *S. chaetocephala*, a × 1, b, c × 12; **12,** *S. radiata*, b, c × 12, d × 12. 3, from *Faulkner* 4767; 5, from *Greenway & Polhill* 11486; 7a, from *Greenway & Kanuri* 14142; 7aa, b, c, from *Greenway & Polhill* 11461; 8, from *Milne-Redhead & Taylor* 8821; 9, from *Milne-Redhead & Taylor* 9482; 10, from *Bally* 10817; 12, from *Hazel* 683. Drawn by Mrs M. E. Church.

disappearing (in sect. *Borreria* (G. F. W. Mey.) Verdc.) or sometimes 2-coccous, one dehiscent but the other remaining ± closed (sect. *Spermacoce*) or in a few species (sect. *Arbulocarpus* (Tennant) Verdc.) the capsule splitting from the base upwards but valves remaining attached by the calyx-limb which is not split across, the whole falling off like a lid, a persistent septum being left behind. Seeds oblong, ellipsoid or ovoid, usually shining brown with a thin often clearly reticulate testa, ventrally grooved; albumen horny or fleshy.

A large genus of worldwide distribution in the tropics and subtropics, with some 150–250 species according to various estimates, mainly American but with many species in Africa and some also in most tropical and subtropical areas; many have become weeds of cultivation. Some of the species are very similar in facies to members of the *Oldenlandia* group of genera but can always be absolutely distinguished by the solitary ovules and discoid pluricolpate pollen grains.

It has been accepted practice for a hundred years or so to split up this genus, but although the technical characters of the fruit dehiscence sound admirable on paper there is no associated habit facies and it is not possible without additional knowledge to assign flowering material to the genera concerned. I have therefore followed Hooker in G.P. 2: 145 (1873) in his circumscription of the genus with the addition of *Octodon* (which he kept separate) to the synomymy as suggested by Hepper (K.B. 14: 260 (1960)). Bremekamp (Rec. Trav. Bot. Néerland. 31: 305 (1934)) who has been responsible for much generic segregation in this family states his position as follows. "The differences between *Diodia*, *Spermacoce* and *Borreria* are also very small and hardly of sufficient importance to justify their separation. In contradistinction with the difference between *Diodia* and *Hemidiodia* which is not only taxonomically of little value but also difficult to see the differences between *Diodia*, *Spermacoce* and *Borreria* are at least easily recognizable. For this admittedly purely opportunistic reason I have retained these genera." I agree entirely but I must report that Steyermark (*in litt.*), although agreeing with much of this argument, thinks the technical characters are adequate and maintains the genera, thus avoiding a great many name changes since very many S. American species have been described in *Borreria*. As far as the African species are concerned I would also unite *Spermacoce* with *Diodia* and the very close relationship can be seen by comparing *Diodia sarmentosa* with *Spermacoce princeae*, but a survey of the New World material of *Diodia* has convinced me this would not be a wise course—certainly the type species of *Diodia*, *D. virginiana* L. differs widely in its fruit structure.

I have found no satisfactory logical way to arrange the species, in fact this will be impossible until a thorough revision is made on a world basis. K. Schumann's classification of *Borreria* (E. & P. Pf. IV.4: 143 (1891)) as follows, is of no value for arranging the African species.

Sect. *Borreria*
Series *Tenellae* K. Schum.
Subseries *Virides* K. Schum. (including *Octodon*)–*S. laevis*, *S. ocimoides*, *S. pusilla* and *S. dibrachiata* belong here
Subseries *Glaucae* K. Schum.
Subseries *Radiatae* K. Schum.—*S. radiata* and *S. chaetocephala* belong here
Series *Latifoliae* K. Schum. *S. latifolia* belongs here
Sect. *Trachiphyllum* K. Schum.
Sect. *Galianthe* (Griseb.) K. Schum.

For convenience the species of *Diodia* have also been included in the following key. Many of the species keyed out on capsule and seed characters early in the key are keyed out again on floral characters later, so if fruits are not present these couplets may be passed over, starting at dichotomy 8.

1. Capsule-valves almost circular, held ± at 180° to each other 2
   Capsule-valves oblong or fruit not dehiscing 3
2. Calyx-lobes 4; valves of capsule 2·5–3 mm. wide; seeds and flowers unknown . 27. *S. sp. D*
   Calyx-lobes 2; valves of capsule 5 mm. wide; seeds with a marked cream-coloured appendage . . . . . . 28. *S. congensis*
3. Capsule splitting from the base, both valves lifting off, joined at the apices by the

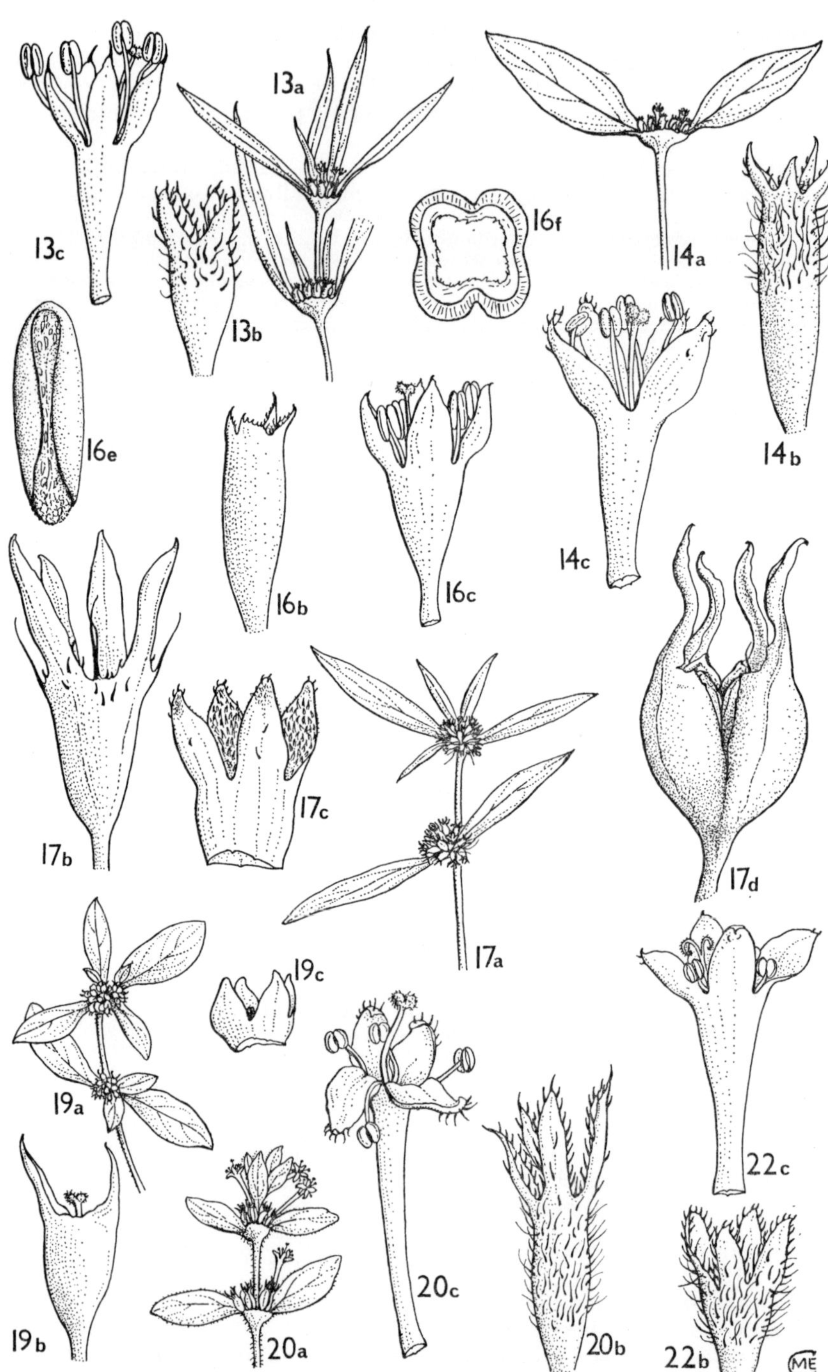

FIG. 50. *SPERMACOCE*—inflorescence (a), calyx (b), corolla (c), capsule (d), seed (e), transverse section of stem (f). Species numbered as in text. **13,** *S. pusilla*, a × 1, b, c × 16; **14,** *S. laevis*, a × 1, b, c × 16; **16,** *S. quadrisulcata*, b, c, e × 12, f × 6; **17,** *S. natalensis*, a × 1, b, c, d × 12; **19,** *S. ocymoides*, a × 1, b × 16, c × 20; **20,** *S. hispida*, a × 1, b, c × 8; **22,** *S. latifolia*, b, c × 8. 13, from *Norman* 94; 14, from *Archbold* 1104; 16, from *Milne-Redhead & Taylor* 9082; 17, from *Milne-Redhead & Taylor* 8761; 19, from *Symes* 512; 20a, from *Mwasumbi* 10113; 20b, c, from *Harris* 5764; 22, from *Lye* 2824. Drawn by Mrs M. E. Church.

persistent calyx; annual with funnel-shaped corolla-tube 5–8·5 mm. long; inflorescences mostly terminal (sect. *Arbulocarpus*) . . . . . . 26. *S. sphaerostigma*
Capsule splitting from the apex to the base into 2 separate valves or fruit dividing into indehiscent or tardily dehiscent cocci 4
4. Seeds of very characteristic shape, lobed in a " rib-cage " manner (fig. 48/7, p. 338) . *Diodia aulacosperma*, p. 337
Seeds not lobed 5
5. Fruit dividing into ± indehiscent cocci; stems long and straggling or prostrate, much branched; corolla-tube 1·8 mm. long . . . . *Diodia sarmentosa*, p. 336
Fruit capsular, with 2 valves 6
6. Seeds provided with a very distinct cream-coloured basal conical appendage (elaeosome), fig. 51/28e, 29e 7
Seeds without an appendage (a small whitish aril may be present) or fruits lacking 8
7. Corolla-tube 6–11 mm. long . . . . 28. *S. congensis*
Corolla-tube 3 mm. long . . . . . 29. *S. milnei*
8. Perennial herb with ± 20–100 decumbent to suberect stems from a woody usually multi-headed rootstock . . . . 18. *S. minutiflora*
Annual or if perennial then with only a few stems from a small stock 9
9. Flowers in very dense terminal heads supported by 1 pair of leaves, mostly without any additional flowers at the nodes beneath or rarely at one node beneath; corolla usually blue or violet, but sometimes pale. 10
Flowers in all or at least several of the upper nodes, but in some cases these run together to form a spicate or even capitate collection of inflorescences, which are easily distinguished from the terminal heads mentioned above by the numerous supporting leaves, which appear to radiate due to suppression of the nodes 13
10. Corolla-tube funnel-shaped, 5–8·5 mm. long; plant annual, with white, blue or purple flowers . . . . 26. *S. sphaerostigma*
Corolla-tube very narrow, or if funnel-shaped then plant perennial with a woody rootstock 11
11. Calyx-lobes 4; corolla-tube 1–2·1 cm. long . 30. *S. dibrachiata*
Calyx-lobes 2; corolla-tube 5·5–7(–11) mm. long (Ufipa District) 12
12. Corolla-lobes hairy on the upper half or only at the tips; plant prostrate, suberect or

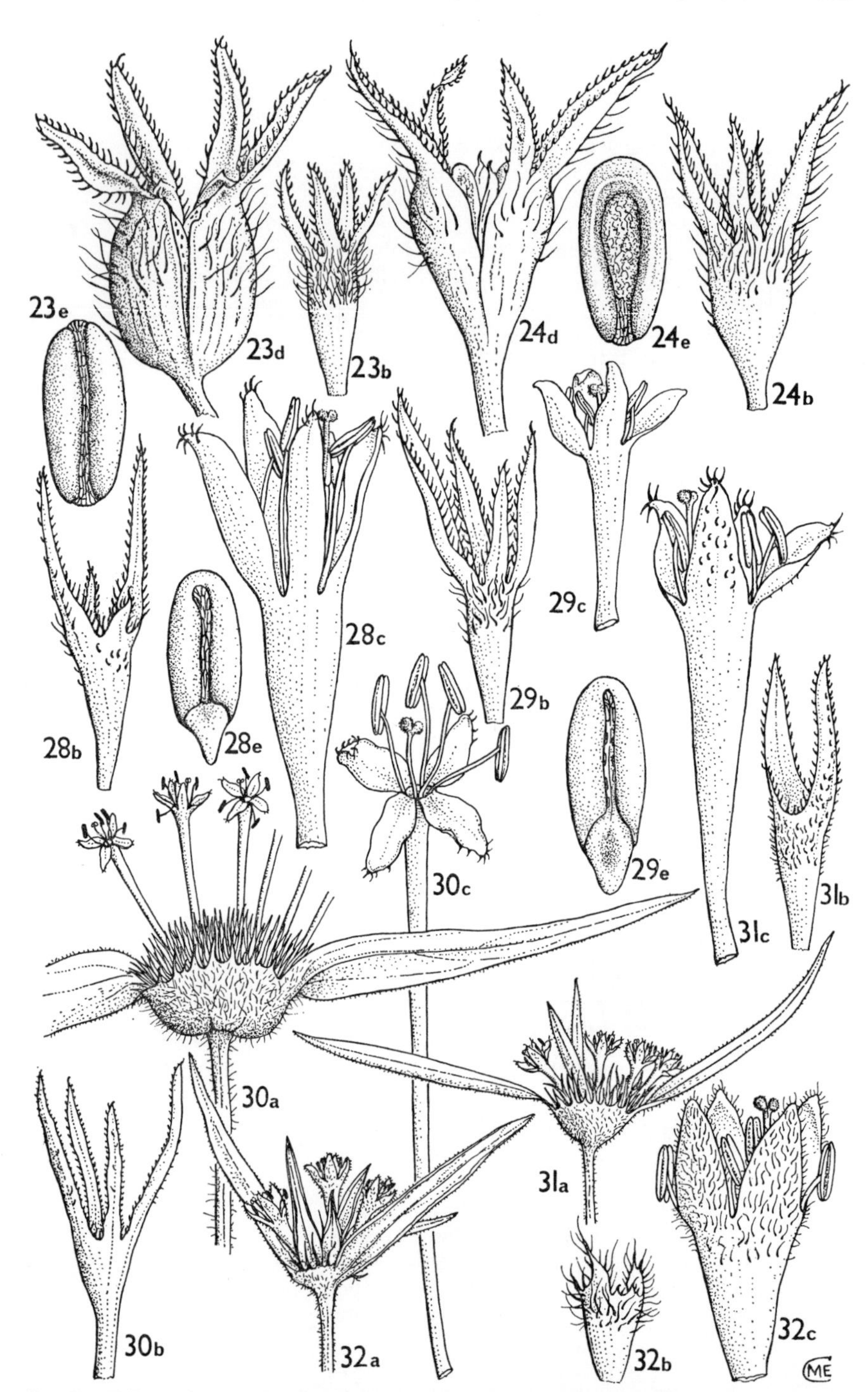

FIG. 51. *SPERMACOCE*—inflorescence (a), calyx (b), corolla (c), dehisced capsule (d), seed (e). Species numbered as in text. **23,** *S. senensis*, b, d × 6, e × 8; **24,** *S. ruelliae*, b, d × 6, e × 8; **28,** *S. congensis*, b, c × 4, e × 6; **29,** *S. milnei*, b, c, e × 6; **30,** *S. dibrachiata*, a × 1, b, c × 4; **31,** *S. phyteumoides* var. *phyteumoides*, a × 1, b, c × 6; **32,** *S. azurea*, a × 1, b, c × 4. 23b, from *Richards* 17959; 23d, e, from *Smith* 1117; 24, from *Angus* 5848; 29, from *Milne-Redhead & Taylor* 10082; 30, from *Richards* 8610; 31, from *Richards* 1152; 32, from *Richards* 8696. Drawn by Mrs M. E. Church.

decumbent; leaves elliptic-lanceolate, 2–7(–14) cm. long, 2·5–13 mm. wide . 31. *S. phyteumoides*
Corolla-lobes densely hairy outside all over; herb with strictly erect stems; leaves narrowly linear to linear-lanceolate, 2–5·8 cm. long, 1·5–6 mm. wide . . 32. *S. azurea*

13. Flowers in spicate or capitate heads 14
Flowers not in spikes or heads, the flowering nodes well separated 20

14. Flowering nodes condensed into heads supported by numerous associated leaves; corolla-tube 1–1·3 mm. long; calyx-lobes 4 15
Flowering nodes condensed into spikes; calyx-lobes 2 (rarely 3 or 2 large and 2 small in *S. subvulgata*) 17

15. Corolla-tube 5 mm. long; calyx-lobes 2·5–3 mm. long . . . . . . 11. *S. huillensis*
Corolla-tube 1–1·5 mm. long; calyx-lobes 1–2 mm. long 16

16. Leaf-blades without thick pale margins and midrib; heads with very numerous filiform bracteoles (fig. 49/10a) . . 10. *S. chaetocephala*
Leaf-blades (fig. 49/12d) with thick pale margins and midrib; heads without very numerous narrow bracteoles . . 12. *S. radiata*

17. Calyx-lobes 0·35–2 cm. long; valves of capsule almost round, opening right out; seeds with basal cream-coloured appendage (fig. 51/28e) . . . . . . 28. *S. congensis*
Calyx-lobes 3–7·5 mm. long; valves of capsule ± elliptic; seeds without an appendage 18

18. Tufts of radical leaves and cauline leaves 6–16 cm. long; corolla-tube 6–8 mm. long . . . . 8. *S. taylorii*
Tufts of radical leaves absent; cauline leaves 1–6 cm. long 19

19. Corolla-tube 9·5 mm. long, very narrowly cylindrical; leaf-blades 2–9 mm. wide . 5. *S. arvensis*
Corolla-tube 4 mm. long; leaf-blades 1–2(–4) mm. wide . . . . . . 7. *S. subvulgata*

20. Corolla-tube under 3 mm. long 21
Corolla-tube over 3 mm. long 29

21. Leaf-blades ± linear-lanceolate 22
Leaf-blades distinctly elliptic 25

22. Lower inflorescences unilateral, i.e. in only one axil at each node (fig. 50/17a); calyx-lobes 1–1·8 mm. long; capsule and whole plant ± glabrous . . . 17. *S. natalensis*
Lower inflorescences forming regular verticils in both axils at each node; calyx-lobes minute, up to 1·2 mm. long 23

23. Inflorescences 0·5–2 cm. wide, with very numerous reddish brown setiform bracteoles (fig. 49/10a); flower-clusters

always supported by several pairs of leaves . . . . 10. *S. chaetocephala*
Inflorescences mostly smaller and under 1 cm. wide, with far fewer flowers and less conspicuous bracteoles; flower-clusters supported by 1–several pairs of leaves 24

24. Plant of drier places, with leaf-blades up to 1·5–3 cm. long; calyx-lobes 0·6–1·2 cm. long; capsule glabrous or pubescent . 13. *S. pusilla*
Plant of marshy places, with leaf-blades up to 2·6–6 cm. long, glabrous; calyx-lobes minute, 0·25 mm. long; capsule glabrous 16. *S. quadrisulcata*

25. Inflorescences mostly with 10–20 or more flowers; calyx-lobes 0·6 mm. long; corolla-tube 1·2 mm. long; fruit distinctly capsular; erect or scrambling herb 0·3–1·2 m. tall . . . . . 14. *S. laevis*
Inflorescences with mostly less than 10 flowers 26

26. Weak herb, erect, decumbent or prostrate, 3–40 cm. tall, with fine fibrous roots; leaf-blades (fig. 50/19a) rounded to ± acute, mostly glabrous except for very short marginal hairs and sometimes a few on the midrib beneath; calyx-lobes 2(–4), 0·6–0·8 mm. long . . . . 19. *S. ocymoides*
Robuster herbs; leaf-blades tapering, acute; calyx-lobes 4 27

27. Prostrate or scrambling herb with stems 1–3·6 m. long; leaf-blades scabrid-pubescent above; calyx-lobes 1·5 mm. long; corolla-tube 1·8 mm. long; fruit distinctly dividing into 2 ± indehiscent cocci . . . . *Diodia sarmentosa* (p. 336)
More or less erect herbs; calyx-lobes ± 0·5 mm. long; corolla-tube 0·2–0·3 mm. long 28

28. Stems angular, with short stiff almost aculeate hairs on the angles, particularly beneath the nodes; leaves sparsely to distinctly hairy; capsule usually glabrous in Flora area but typically hairy . . 1. *S. confusa*
Stems less angular, glabrous; leaves glabrous except for margins; capsule glabrous . 2. *S. tenuior*

29. Prostrate or procumbent herbs, often with several to numerous stems from an annual root; calyx-lobes 4 30
More or less erect herbs 32

30. Leaf-blades (fig. 50/20a) smaller and more oblong, 0·6–1·5 cm. long, 1·5–6 mm. wide; corolla-tube (fig. 50/20c) narrowly cylindrical, 6 mm. long . . . . 20. *S. hispida*
Leaf-blades larger and more elliptic, 1·2–6·5 cm. long, 0·35–3 cm. wide 31

31. Leaves drying dark green; sepals 3–5(–9) mm. long; corolla-tube cylindrical or narrowly funnel-shaped, 6·5–10·5 mm. long . . . . 21. *S. princeae*
    Leaves drying a very characteristic pale yellow-green colour; sepals 1·2–2 mm. long; corolla-tube funnel-shaped, 5 mm. long (introduced) . . . . . 22. *S. latifolia*
32. Calyx-lobes 2 33
    Calyx-lobes (3–)4(–6) 38
33. Calyx-lobes elongated, 0·8–2 cm. long; seeds with a distinct basal appendage; corolla-tube 6–7 mm. long . . . . . 28. *S. congensis*
    Calyx-lobes 2–4 mm. long; seeds never with an appendage 34
34. Corolla-tube (fig. 49/3c) filiform, (0·5–)1(–1·2) cm. long, with short lobes 2–2·2 mm. long . . . . 3. *S. filituba*
    Corolla-tube more funnel-shaped and lobes 3·5–5 mm. long, or if cylindrical tube well under 0·5 cm. long 35
35. Corolla-tube 1–1·2 cm. long . . . . 4. *S. sp. A*
    Corolla-tube 3·5–5 mm. long 36
36. Strictly erect herb with stem leafless for basal 25–30 cm.; leaves up to 1–2 cm. wide, glabrous beneath but margins ciliate with short stiff hairs; flowers solitary or few in the axils . . . . 25. *S. sp. C*
    Stems not leafless for basal 25–30 cm.; leaves mostly narrower and hairy beneath; inflorescences of several to many flowers 37
37. Corolla-tube cylindrical, 3·5 mm. long . . 6. *S. sp. B*
    Corolla-tube 4–5 mm. long, narrowly funnel-shaped . . . . 7. *S. subvulgata*
38. Stems glabrous or, if hairy (in some extra-East African material) then inflorescences supported by small leaves under half the size of the stem leaves and style and stigma well exserted from the corolla, much over-topping the anthers 39
    Stems hairy; inflorescence leaves not markedly smaller; style and stigma not exserted much beyond the anthers 40
39. Corolla-tube distinctly funnel-shaped (fig. 49/9c); leaf-blades 1–4·7 cm. long . . 9. *S. latituba*
    Corolla-tube narrowly cylindrical at the base, more funnel-shaped above; leaf-blades 4–10 cm. long . . . . . . 15. *S. hepperana*
40. Calyx-lobes (fig. 51/29b) 7–9 mm. long; seeds with a distinct basal appendage (fig. 51/29e) . . . . 29. *S. milnei*
    Calyx-lobes small; seeds without a basal appendage 41
41. Corolla hairy outside except at the base; indumentum of more flattened hairs;

seeds (fig. 48/7, p. 338) characteristically lobed; plant of sand dunes or at least coastal . . . . . *Diodia aulacosperma* (p. 337)

Corolla glabrous outside or with a few hairs at the lobe-apices; general indumentum of fine hairs; seeds not lobed; usually not a dune plant 42

42. Leaves lanceolate, under 5 mm. wide; calyx-lobes 4, subequal 43

Leaves elliptic or elliptic-lanceolate, 0·25–2·2 cm. wide 44

43. More slender herb, with more numerous flowers in the inflorescences; corolla-tube slender, usually 0·5 mm. wide near the middle . . . . . . 11. *S. huillensis*

Coarser herb with fewer flowers in the inflorescences; corolla-tube more cylindrical, 1·5 mm. wide near the middle. . 7. *S. subvulgata* var. *quadrisepala*

44. Calyx-lobes (fig. 51/23b) more linear to narrowly oblong-lanceolate, mostly subequal; seeds (fig. 51/23e) with a narrow ventral groove filled with rhaphides . 23. *S. senensis*

Calyx-lobes (fig. 51/24b) narrowly triangular-lanceolate, 2 longer and broader alternating with 2 shorter and narrower 45

45. Leaves usually hairy beneath; stems branched or unbranched, not leafless at base; seeds (fig. 51/24e) with a wide ventral excavation usually filled with white cellular material for most of its length . . . . 24. *S. ruelliae*

Leaves glabrous beneath, the margins with short stiff spreading hairs; strictly erect unbranched stems with basal 25–30 cm. leafless; seeds not known . . . 25. *S. sp. C*

1. **S. confusa** *Gillis* in Phytologia 29: 185 (1974).* Type: Jamaica, Lower Clarendon Parish, Inverness, *W. Harris* 12749 (BM, holo. !, GH, NY, iso.)

Annual herb 30–90 cm. tall, with generally erect angled stems, the angles mostly rather rough, often with very short stiff papilla-like hairs, particularly just below the nodes. Leaf-blades linear-lanceolate to narrowly elliptic, 2·5–5 cm. long, 0·2–1·3 cm. wide, acute or acuminate at the apex, cuneate at the base, shortly to densely scabrid above and on the nerves beneath; petiole very short; stipule-sheath ± 1 mm. long, with several fimbriae 1–2 mm. long. Calyx-tube narrowly turbinate, 1·5 mm. long; lobes triangular, ± 0·5 mm. long. Corolla white, white and pink or pale mauve; tube ± 1·5 mm. long; lobes ovate, ± 0·5 mm. long. Capsule ellipsoid to subglobose, 2–2·5 mm. long, 1·5 mm. wide, finely wrinkled striate, glabrous to densely hairy, crowned with the short calyx-lobes. Seeds dark chestnut-brown, narrowly ellipsoid,

*The name was proposed earlier, but not validated by a latin diagnosis, as follows: Rendle in J.B. 74: 12, figs. D–F(1936) & in Fl. Jam. 7: 120 (1936); Adams, Fl. Pl. Jam.: 732 (1972).

1·6–1·8 mm. long, 0·8 mm. wide, scarcely pitted (*fide* Rendle), very distinctly reticulate-foveolate (in African material).

TANZANIA. Lushoto District: Sigi Railway Station, 7 Mar. 1917, *Peter* 56010! & Sigi to Mpandeni [Pandeni], 29 May 1917, *Peter* 56032! & Derema to Longusa, 12 May 1918, *Peter* 56127!; Zanzibar I., Mwera Swamp, 16 Nov. 1930, *Vaughan* 1682!; Pemba I., Chake Chake, 27 Aug. 1929, *Vaughan* 591!
DISTR. **T3**; **Z**; **P**; SE. U.S.A., Bahamas, W. Indies, Trindad and tropical America
HAB. Cultivated ground; 0–800 m.

SYN. [*S. tenuior* sensu Gaertn., Fruct. & Sem. 1: 122, t. 25/9 (1788) et auctt. mult. *non* L. (Rendle gives full list of refences where misapplied)]

NOTE. Bremekamp has annotated *Vaughan* 591 as *Diodia ocimifolia* (Roem. & Schultes) Bremek. (*Hemidiodia ocimifolia* (Roem. & Schultes) K. Schum.), but the corolla, glabrous stipule fimbriae, stiff stem hairs and ± glabrescent capsules with a different dehiscence are not correct for that species, although the general habit is extremely similar. Since it is a widespread weed it could easily occur.
No recent material of *S. confusa* has been seen from the Flora area although A. Peter, collected 9 specimens, most of which he annotated as *Borreria aspera* A. Peter, a name which has never been published. None of this material is typical of *S. confusa* as it occurs in the West Indies where the capsules are mostly densely hairy. The African material has almost or quite glabrous capsules and even the leaves are only sparsely hairy above, moreover the seeds are more strongly foveolate. Nevertheless similar specimens do occur in America and presumably the original introduction was atypical. I am, however, not altogether convinced that *S. confusa* and *S. tenuior* are as distinct as claimed by Rendle.

2. **S. tenuior** *L.*, Sp. Pl.: 102 (1753); Rendle in J.B. 72: 329, fig. A, t. 607/2, 608/3 (1934) & Fl. Jam. 7: 122 (1936); Adams, Fl. Pl. Jam.: 732 (1972). Type: Dillenius, Hortus Elthamensis: 370, t. 277 [Linnaeus gives 227 in error], fig. 359 (drawn from specimen grown at Eltham (OXF, typotype))

Straggling or suberect annual or perennial herb 0·7(–2) m. tall; stems glabrous, not markedly angular. Leaf-blades narrowly elliptic to lanceolate, 2–5(–8·5) cm. long, 0·4–1·35(–2·2) cm. wide, narrowly acute at the apex, narrowly cuneate at the base, glabrous save for slightly scabrid margins; petiole 1·5–2·5 mm. long; stipule-base 1–2·5 mm. long, glabrous or slightly pubescent with 7 setae 1·5–3 mm. long. Inflorescences rather few-flowered, congested in the axils. Calyx-tube narrowly obconic, 1·5 mm. long; 2 large calyx-lobes triangular, 0·5 mm. long, ± 3 small. Corolla white, papillate, very small; tube 0·2–0·5 mm. long; lobes triangular, 0·5–1·5 mm. long, 0·4 mm. wide, usually hairy at and above the throat. Anthers subsessile; style 0·5–2 mm. long. Capsule obovoid to ellipsoid, 2·5–3 mm. long, glabrous, crowned with ± persistent calyx-lobes. Seeds dark purple or chestnut-brown, ellipsoid, 2 mm. long.

TANZANIA. Morogoro District: 6.4 km. N. of Turiani, near Mtibwa sawmill, Lusunguru Forest Reserve, 31 Mar. 1953, *Drummond & Hemsley* 1933!
DISTR. **T6**; SE. U.S.A., West Indies, tropical S. America
HAB. Forest clearing; 500 m.

NOTE. I am far from convinced of the identity of this specimen; there are differences in style length, the indumentum inside the corolla-lobes, etc. but the specimen can be matched almost exactly with American material.

3. **S. filituba** (*K. Schum.*) *Verdc.* in K.B. 30: 302 (1975). Type: Kenya, Lamu, *F. Thomas* 213 (B, holo. †, BM, K, iso.!)

Branched annual herb 12–60 cm. tall, with a ± simple root; stem obscurely 4-ribbed, covered with ± spreading white hairs. Leaf-blades narrowly elliptic-lanceolate to narrowly ovate-lanceolate, 1·4–5·6 cm. long, 2·5–8·5 mm. wide, acute at the apex, often widened and rounded at the base, the petioles forming a sheath with the stipules, with scattered longer hairs on both surfaces and also with short stiff ones along the margins, sometimes saw-edged; stipules chestnut-brown; base 2–3 mm. long, hairy, bearing 8–10 filiform setae 6–9 mm. long. Flowers in axillary verticillate clusters scattered along the length of the stems and crowded towards the apex. Calyx-tube conic or ellipsoid, 1·8 mm. long, hairy; lobes 2, lanceolate, 2–3·5 mm. long, 0·6–0·8 mm. wide, with stiff hairs along the margins. Corolla white; tube very narrowly cylindrical, (0·5–)1–1·2 cm. long, glabrous; lobes elliptic-lanceolate to oblong-elliptic, 2–2·2 mm. long, 0·5–1 mm. wide, with spreading white hairs outside. Filaments exserted 1·5 mm. Style 1·2 cm. long, exserted about 2 mm.; stigma capitate, 0·4 mm. wide, hairy. Capsule ellipsoid, 2·2–3 mm. long, finely transversely wrinkled, hairy, each half bearing a persistent calyx-lobe. Seeds brown, ellipsoid-oblong, 1·8–2·3 mm. long, 1·1 mm. wide, 0·7–0·8 mm. thick, shiny, convex dorsally, with a narrow median groove packed with rhapides ventrally, the surface reticulation very obscure even at high power magnification. Fig. 49/3, p. 340.

KENYA. Kilifi District: Fundisa Is., *R. M. Graham* in *A.D.* 1629!; Tana River District: Kurawa, 24 Sept. 1961, *Polhill & Paulo* 550!; N. of Mombasa to Lamu and Witu, *Whyte*!

TANZANIA. Tanga District: Kange Estate, 2 Nov. 1951, *Faulkner* 804! & Moa, 4 Aug. 1953, *Drummond & Hemsley* 3632! & same locality, June 1893, *Holst* 2999!

DISTR. **K**1, 7; **T**3; Somali Republic (S.), see note

HAB. Grassland with scattered trees and shrubs particularly *Commiphora* and *Acacia*, sea-shore bushland on sand, also in plantations, coconut groves, etc.; 0–600 m.

SYN. *Borreria filituba* K. Schum. in E.J. 28: 110 (1899)

NOTE. *Borreria tenuiflora* Chiov. (types: Somali Republic (S.), Giumbo, *Paoli* 240 & Hididle to Hamagudu, *Paoli* 675 & Bur Meldac, *Paoli* 704 (FI, syn. !)) may be merely a variety of *S. filituba* with hairier leaves, but better flowering material is needed.

Several specimens from inland localities, e.g. *Greenway* 9772 (Tsavo National Park East, plot 1, 19 Jan. 1961), have the corolla-tube only 5–6 mm. long but are I think only minor variants. Field observations are needed to determine their relationship with *S. sp. A* and *S. sp. B*.

Gillett & Gachathi 20636 (Kenya, N. Frontier Province, Dadaab–Wajir road, just N. of Lagh Dera, 12 May 1974) is clearly this species but in some whorls the calyx is 3-lobed, eventually in fruit one coccus bearing 1 larger lobe and the other 2 smaller ones. In most whorls however the calyces are all normally 2-lobed.

4. **S. sp. A**

Probably perennial herb with ascending branches to 60 cm. tall; stems 4-angled, bifariously sparsely to fairly densely pilose. Leaves fairly thick; blades elliptic-lanceolate, 2–5 cm. long, 2·5–5 mm. wide, acute at the apex, narrowed at the base into the stipular cup, glabrous above, with scattered ± adpressed pubescence beneath; margins scabrid with short stiff hairs; petioles entirely joined to the stipule-sheath, which is 3 mm. long, pilose and produced into 5–7 fimbriae 6 mm. long. Flowers 4-merous, congested into few–10-flowered axillary clusters at most nodes 0·7–1·2 cm. wide; bracteoles few with fimbriae 6 mm. long. Calyx-tube obovoid, 1·2 mm. long, pilose; lobes 2, lanceolate, 3·5–6 mm. long, ciliate at the margins. Corolla-tube white tinged mauve and streaked purple in the throat, narrowly infundibuliform with a cylindrical tube, 1–1·2 cm. long; lobes white with mauve mid-veins outside (most marked at the tips), 5 mm. long, 2·5 mm. wide, with a few hairs outside at the apex. Filaments exserted 3–4 mm. Style exserted 3–5 mm.; stigma capitate, 0·5 mm. wide, with long papillae. Capsule transversely subglobose, 2 mm. long, 3 mm. wide, 1·5 mm. thick, ± hairy or pilose with long hairs towards the apex. Seeds purplish brown, $\frac{1}{2}$-ellipsoid-oblong, 1·5 mm. long, 1·1 mm. wide, 0·8 mm. thick, dorsally very convex with very sloping sides, ventrally with a narrow deep excavation, minutely rugulose.

KENYA. Teita District: Mackinnon Road–Buchuma, 9 Dec. 1961, *Polhill & Paulo* 930! & 40 km. E. of Voi on the Mombasa road, 2 Jan. 1962, *Greenway* 10431!
DISTR. **K7**; known only from the above gatherings
HAB. Deciduous bushland; 360–390 m.

NOTE. Further study of this is needed before it can be described.

5. **S. arvensis** (*Hiern*) *Good* in J.B. 65, Suppl. 2: 41 (1927). Types: Angola, Pungo Andongo, near Muta Lucala and Quisonde, *Welwitsch* 3236 & Huila, *Welwitsch* 3227 (both LISU, syn., BM, K, isosyn.!)

Annual herb (4·5–)8–30(–70) cm. tall, sometimes slender in habit but often coarse and appearing quite different; stems drying blackish purple, often with paler longitudinal lines, densely covered with spreading hairs, usually sparingly branched or unbranched. Leaf-blades lanceolate, 1·3–7 cm. long, 2–9 mm. wide, acute at the apex, narrowed at the base into the stipule-sheath, covered with long rather coarse hairs on both surfaces and also scabrid with shorter hairs near the margins; stipules with fimbriae crimson or red-brown in life but mostly blackish purple when dry; base 2 mm. long, with 5–7 fimbriae 3–8 mm. long, with long white hairs. Flowers in many-flowered verticillate clusters, mostly at the upper nodes which run together to form a dense coarse strobilate inflorescence 0·7–1·5 cm. wide, but usually with 1–2 clusters at the nodes beneath and well separated; bracteoles stipule-like, of 5–6 lanceolate fimbriae 5–6 mm. long, joined at the base, ciliate along the margins. Calyx-tube ellipsoid, 1·8 mm. long; lobes 2, lanceolate, 3 mm. long, 0·6 mm. wide at the base. Corolla white or with purplish tips to lobes outside or blue; tube very narrowly cylindrical, 9·5–12 mm. long; lobes narrowly elliptic or linear-oblong, 1·8–3 mm. long, 0·8–1·2 mm. wide, sparsely hairy outside. Filaments exserted ± 1–2 mm. Style exserted 1·2–2 mm.; stigma slightly bifid, 0·5 mm. wide. Capsule ellipsoid, 3–3·5 mm. long, 1·8–2 mm. wide, shortly hairy at the apex. Seeds chestnut-brown, semi-ellipsoid, 2·7 mm. long, 1·1–1·5 mm. wide, 0·9 mm. thick, shiny, flat and narrowly grooved ventrally, the groove opening out at each end, convex dorsally, rugulose. Fig. 49/5, p. 340.

TANZANIA. Tabora, 13 Apr. 1954, *Smith* 1137!; Dodoma District: 16 km. E. of Itigi Station, 10 Apr. 1964, *Greenway & Polhill* 11486! & Kazikazi, 12 June 1932, *B. D. Burtt* 3700!
DISTR. **T4, 5**; Malawi, Zambia, Rhodesia and Angola
HAB. Secondary deciduous woodland, bushland and grassland; 1200–1620 m.

SYN. *Tardavel arvensis* Hiern in Cat. Afr. Pl. Welw. 1: 504 (1898)
Borreria arvensis (Hiern) K. Schum. in Just's Bot. Jahresb. 26(I): 391 (1900)

NOTE. Included in the above is *Polhill & Paulo* 2258 (Singida District, Iramba Plateau, Kiomboi, 30 Apr. 1962) from overgrazed grassland with scattered small trees; it differs markedly in its condensed habit, being almost tufted, densely branched and leafy, showing none of the long internodes so typical of the cited specimens; nevertheless the inflorescence structure and corolla make me convinced it is the same species. There is also much variation in the coarseness of the stems and foliage among the cited specimens and a range of fresh material is needed to confirm my conclusions. *Tanner* 4308 (Tanzania, Musoma District, Lupa, Nyambone, 4 June 1959) has the appearance of *S. arvensis* but the corolla is small and glabrous. It could possibly be a form of *S. subvulgata*.

6. **S. sp. B**

Erect herb about 60 cm. tall; stems slightly angled, glabrous. Leaf-blades linear-lanceolate, 1·2(upper " floral bracts ")–6·5 cm. long, 0·8–4 mm. wide, tapering acute at the apex, narrowed to the base, glabrous; petiole combined with the stipular cup-like sheath, which is 3–4 mm. long, glabrescent or ± adpressed scabrid-pubescent, with 5 bristles 3–8·5 mm. long. Flowers in small axillary several-flowered clusters 0·5–1 cm. wide at most of the nodes (up to 12 upper nodes); stipule-like bracteoles 3–5 mm. long. Calyx-tube obconic or ellipsoid 1·5 mm. long, shortly pubescent; lobes 2, narrowly lanceolate, 4·5–5 mm. long, slightly ciliate towards the base. Corolla white; tube cylindrical, 3·5 mm. long; lobes linear-lanceolate, 3·5 mm. long, 0·9 mm. wide. Stamens exserted 2·5 mm. Style 8 mm. long; stigma ± globose. Capsule ellipsoid, 3 mm. long, 1·7 mm. wide, pubescent. Seeds dark brown, narrowly ellipsoid-oblong, 2·2 mm. long, 1 mm. wide, 0·7 mm. thick, very finely reticulate.

KENYA. Kilifi District: 40 km. inland from Kilifi, Bamba, 11 Feb. 1953, *Bally* 8546!
DISTR. **K7**
HAB. Clearings in light evergreen forest; 225 m.

NOTE. May prove to be an odd variant of *S. subvulgata* but more material is required. The capsule is slightly stipitate the stalks being about 1 mm. or so long but the description of *Borreria pedicellata* (see p. 374) indicates that the two are not closely related.

7. **S. subvulgata** (*K. Schum.*) *Garcia* in Mem. Junta Invest. Ultram. sér. 2, No. 6: 49 (1959). Types: Tanzania, Tabora District, Igonda [Gonda], *Böhm* 268 (B, syn. †) & Rhodesia, near Umtali, *Schlechter* 12184 (B, syn. †, BM, K, isosyn.!) & Angola, Malange, *Buchner* 3 & 21 & *Mechow* 579 (B, syn. †)

Annual or perennial erect herb 25–90 cm. tall, with ± woody base; stems glabrous except for pubescence just below the nodes or in vertical grooves on the internodes between the ribs, arranged bifariously. Leaf-blades linear, 3–6 cm. long, 1–2(–4) mm. wide, ± acute at the apex, narrowed to the base, ± scabrid on the margins and midrib beneath to distinctly shortly hairy; stipule base hairy, 3 mm. long, with 6–11 fimbriae 3–6·5 mm. long, often purplish. Flowers in clusters at the nodes, the apical 7 running together to form a sort of cone with the leaf-like bracts protruding and the stipules overlapping or, in some specimens, the clusters remaining quite separate. Calyx-tube ellipsoid, 1·5 mm. long; lobes 2 (rarely 3) or 2 large and 2 small or in one variety 4, linear, 3–4·5 mm. long. Corolla white; tube narrowly funnel-shaped, 4–5 mm. long; lobes lanceolate, 2·5–5 mm. long, mostly with a few hairs at the apex outside. Filaments exserted 1·5–2·5 mm. Style exserted ± 2–5 mm. Capsule oblong-ellipsoid, 3–4 mm. long, hairy or less often ± glabrous. Seeds dark brown, narrowly oblong, 3–3·3 mm. long, 1·2 mm. wide, 0·7 mm. thick, narrowed at one end, with a ventral groove which is opened out at basal end and bears a trace of a white aril; dorsal surface rugulose.

var. **subvulgata**

Calyx-lobes 2, or exceptionally sometimes 3 or 2 long and 2 short or ± vestigial. Fig. 49/7, p. 340.

UGANDA. Kigezi District: Ruzhumbura, Apr. 1939, *Purseglove* 672!; Mengo District: Entebbe, *Fyffe* 108!; Masaka District: Bukoto, N. of Lake Kayanja, 31 May 1972, *Lye* 7062!
KENYA. Machakos District: Saui, 18 May 1902, *Kassner* 749!; Masai District: Chyulu Hills, 16 June 1938, *Bally* 552 & 1043 in *C.M.* 8346!
TANZANIA. Mbulu District: 14.5 km. from Babati on the Singida road, 6 May 1962, *Polhill & Paulo* 2365!; Dodoma District: Manyoni, Kazikazi, 12 June 1932, *B. D. Burtt* 3659!; 4 km. N. of Njombe District boundary on road from Songea to Njombe, 6 July 1956, *Milne-Redhead & Taylor* 10758!
DISTR. **U**2, 4; **K**3, 4, 6; **T**1, 2, 4–8; Zaire, Burundi, Rwanda, Mozambique, Malawi, Zambia and Rhodesia
HAB. Grassland, roadsides, deciduous bushland, *Brachystegia* and other woodlands, also as a weed of cultivation; 840–1800 m.

SYN. *Borreria subvulgata* K. Schum. in E.J. 28: 111 (1899)
Tardavel kaessneri S. Moore in J.B. 43: 250 (1905). Type: Kenya, Machakos District, Saui, *Kassner* 749 (BM, holo. !) [227 & 228 mounted on same sheet are *Spermacoce filituba*]

NOTE. *S. subvulgata* is an exceptionally variable species. The typical rather coarse plant, with the upper whorls run together to form a spike-like inflorescence, is very easy to recognise, but slender forms with the whorls mostly separated are often difficult to place. These slender forms scarcely differ from *S. kirkii* (Hiern) Verdc. (*Borreria diodon* K. Schum.) known from a few specimens collected on the Mozambique coast at about 18°54′–23°S. Apart from the ecological differences the stipules are also different, the slender forms of *S. subvulgata* having much longer more slender fimbriae. There are also slight differences in the leaves and fruits but further work may show the two have to be considered variants of one species—in that case the species as a whole will have to be known by the much older name *S. kirkii* (type: Mozambique, mouth of W. R. Luabo, *Kirk* (K, holo. !)). *Borreria mohasiensis* K. Krause (type: Rwanda, by Lake Mohasi, *Mildbraed* 621 (B, holo.†)) may, judging by the description, be referable to *S. subvulgata*, but no authentic material seems to be extant.

Mwasumbi & Mhoro in *U.D.S.* 11075 (Tanzania, Uzaramo District, Mwandege, c. 30 km. S. of Dar es Salaam along Kilwa road) seems to be *S. subvulgata* but from a lower altitude than usual (80 m.) and with a short corolla-tube (3 mm.); there are 2(–3) calyx-lobes.

var. **quadrisepala** *Verdc.* in K.B. 30: 303 (1975). Type: Tanzania, 35 km. W. of Dar es Salaam, Kibaha, *Flock* 35 (K, holo. !, DAR, iso.)

Sepals 4, mostly equal.

TANZANIA. Dodoma District: Ndachi, May 1938, *Savile* 62 !; Morogoro District: NW. of Morogoro, Nguru ya Ndege Mt., 4 June 1972, *Pócs & A. Bjørnstad* 6704 E !; Newala, 9 Apr. 1959, *Hay* 54 !
DISTR. T1, 5, 6, 8; Mozambique
HAB. Sandy soil and wet slopes; 150–1080 m.

NOTE. It was at first thought this might be a narrow-leaved variant of *S. senensis*, and typical specimens are not closely similar to var. *subvulgata*, but many specimens of the latter exist, even those with the inflorescences congested into spikes, where there are 2–3 calyx-lobes or 2 large and 2 small. There is a possibility that var. *quadrisepala* is of hybrid origin. Coarse specimens of *S. huillensis* are also similar.

8. **S. taylorii** *Verdc.* in K.B. 30: 303 (1975). Type: Tanzania, 9·5 km. SW. of Songea, *Milne-Redhead & Taylor* 8821 (K, holo. !, EA, iso.)

Tufted erect perennial 0·7–1 m. tall from a thick tap root and underground stem; stems glabrous, with 4 reddish purple longitudinal lines or with minute hairs in lines just below the stipule-sheaths. Leaves apparently held erect, giving a strict look to the plant; blades linear-lanceolate or linear to linear-oblanceolate, 6–16 cm. long, 2·6–7 mm. wide, acute at the apex, very gradually attenuate at the base, glabrous save for the scabrid margins, densely minutely speckled; true petiole absent but tenuous part of blade up to 3 cm. long; stipule-sheath cylindrical, 1–1·2 cm. long, with 4 setae 4–6 mm. long. Flowers in clusters at the nodes, the apical 4 or 5 clusters running together to form a congested cone-like inflorescence, the lower nodes separated by internodes 0·5–9 cm. long; leafy bracts 2–7 cm. long; stipuliform bracteoles with fimbriae 8 mm. long, ciliate. Calyx-tube* narrowly obconic, 6 mm. long; lobes 2, linear-triangular, 7·5 mm. long, shortly ciliate. Corolla white; tube 6–8 mm. long, densely hairy at the throat; lobes triangular-lanceolate, 1–1·1 cm. long, 3–4 mm. wide. Filaments exserted 9 mm. Style exserted 1·2 mm., minutely roughened; stigma slightly clavate, 1 mm. wide. Capsule ellipsoid, straw-coloured, streaked with reddish brown, 5 mm. long, minutely transversely wrinkled, crowned with the persistent calyx-lobes. Seeds pale chestnut-brown, 3·8–4 mm. long, 1·8 mm. wide, 0·8 mm. thick with a marked

* Floral measurements are from spirit material.

ventral groove; a trace of an aril appears as a white patch between the groove and is just visible at the end of the seed from above. Fig. 49/8, p. 340.

TANZANIA. Songea District: 9.5 km. SW. of Songea, near R. Mtanda, 17 Feb. 1956, *Milne-Redhead & Taylor* 8821! & 8821A!
DISTR. T8; not known elsewhere
HAB. Boggy grassland; 990 m.

9. **S. latituba** (*K. Schum.*) *Verdc.* in KB. 30: 304 (1975). Type: Malawi, without locality, *Buchanan* 1138* (B, holo. †, K, iso.!)

Annual herb with ascending stems to over 50 cm. tall; stems 4-angled, glabrous to densely pilose. Leaf-blades linear-elliptic, 1–4·7 cm. long, 0·2–1·25 cm. wide, acute at the apex, narrowed at the base into the stipule-sheath, very shortly scabrid above and on the margin, or densely pilose above, glabrous to scabrid beneath; petioles ± obsolete; stipule-bases glabrous or scabrid-pubescent, 2 mm. long, with 4 fimbriae 1·5–3·5 mm. long. Flowers in well spaced clusters in the axils of the apical 3–5 nodes; leaf-like bracts 1–3·5 cm. long; stipuliform bracteoles with ciliate fimbriae 8 mm. long. Calyx-tube ellipsoid, glabrous or hairy above, 2 mm. long; lobes 4, triangular-lanceolate, 3 mm. long, 0·8 mm. wide, closely scabridulous-ciliate along the margins or sometimes quite hairy. Corolla white with faint mauve stripes; tube funnel-shaped, 8 mm. long, ± 7–8 mm. wide at the throat but strongly narrowed at the base to ± 1 mm.; lobes triangular, 3–4·5 mm. long, 3·2 mm. wide, glabrous to pubescent outside. Filaments exserted 2 mm. Style exserted 4–6 mm.; stigma capitate, 0·8 mm. wide. Inflorescences up to 1·8 cm. wide in fruit. Capsule oblong-ellipsoid, 4 mm. long and wide, finely transversely wrinkled, glabrous or hairy above, crowned with the persistent calyx-lobes. Seeds brown, oblong-ellipsoid, 3 mm. long, 1·5 mm. wide, 1 mm. thick, deeply grooved beneath. Fig. 49/9, p. 340.

TANZANIA. Songea District: shore of Lake Malawi [Nyasa] at Mbamba Bay, 5 Apr. 1956, *Milne-Redhead & Taylor* 9482!
DISTR. T8; Mozambique, Malawi
HAB. Sandy lake shore; 465 m.

SYN. *Borreria latituba* K. Schum. in E.J. 28: 110 (1899)

10. **S. chaetocephala** *DC.*, Prodr. 4: 554 (1830). Type: Senegal, Galam, Bakel, *Leprieur* (G-DC, holo., K, photo.!)

A rather coarse erect mostly sparsely branched annual herb 10–60 cm. tall, the stems often reddish or pinkish, 4-angled, hairy below the nodes. Leaf-blades linear-lanceolate, 2·5–6 cm. long, 3–8 mm. wide, acute at the apex, mostly broadened at the base, pubescent or ± scabrid above with white hairs and with longer white hairs beneath, particularly or sometimes only on the broadened base; stipule-bases 1–2 mm. long, hairy, with 7 setae 0·2–1·5 cm. long. Flowers (3–)4–merous, congested into usually quite large axillary heads 0·5–2 cm. wide, usually made up of normal axillary clusters together with those at the nodes of much reduced axillary shoots, the whole coalescing, each compound inflorescence thus supported by numerous leaves and also with leaves emerging from its centre; bracteoles very numerous, reddish brown, 3 mm. long; pedicels 0·5 mm. long. Calyx-tube ellipsoid, 1·4 mm. long; lobes 4, linear-lanceolate or slightly spathulate, 1–2 mm. long, scabrid on the margins. Corolla white; tube narrowly cylindrical with upper quarter funnel-shaped and ± 4 times as wide as basal part of tube, altogether 1·3 mm. long; lobes ovate-elliptic, 0·8 mm. long, 0·4 mm. wide. Filaments

* K. Schumann gives the number as 1198 but I am convinced this is a misreading of a poorly written 3.

exserted 0·6 mm. Style exserted 0·7 mm.; stigma with long papillae. Capsule oblong-ellipsoid, 2·5 mm. long, hairy at the apex. Seeds chestnut-brown, very shiny, narrowly oblong-ellipsoid, 1·8–2·2 mm. long, 0·8–0·9 mm. wide, 0·5 mm. thick, with a deep ventral groove but ± no sculpture. Fig. 49/10, p. 340.

UGANDA. W. Nile District: Terego, Nov. 1940, *Purseglove* 1078!; Karamoja District: 64.5 km. S. of Moroto, 13 Sept. 1956, *Hardy & Bally* 10817!; Teso District: Serere, Nov. 1931, *Chandler* 13!

KENYA. W. Suk District: 3 km. from Kacheliba, 9 Oct. 1964, *Leippert* 5071! & near Kongelai, Sept. 1966, *Tweedie* 3344! & 56 km. N. of Kitale, 27 Nov. 1962, *Bogdan* 5605!

TANZANIA. Shinyanga, *Koritschoner* 2167 (in part)! & near Shinyanga, *Bax* 190 (in part)! (in both cases mixed with *S. subvulgata*); Ufipa District: E. Rukwa, Apr. 1938, *MacInnes* 259!; Mbeya District: Great North Road, between Igawa and Chimala, 2 Mar. 1962, *Polhill & Paulo* 1984!

DISTR. **U**1, 3; **K**2, 3; **T**1, 4, ? 5, 7;* Senegal, Mali, N. Nigeria, Sudan and Ethiopia

HAB. Grassland, deciduous bushland and *Brachystegia* woodland; 1050–1440 m.

SYN. *S. kotschyana* Oliv. in Trans. Linn. Soc. 29: 88, t. 53 (1873); Hiern in F.T.A. 3: 239 (1877). Types: Sudan, Fasoglu, *Kotschy* 547 & Uganda, Madi, *Grant* (both K, syn.!)

S. compacta Hiern in F.T.A. 3: 239 (1877). Type: Sudan, Cordofan, *Kotschy* 240 (K, holo.!)

S. hebecarpa (A. Rich.) Oliv. var. *major* Hiern in F.T.A. 3: 237 (1877). Type: Ethiopia, near Gallabat, *Schweinfurth* 1479 (K, holo.!, BM, iso.!)

Borreria compacta (Hiern) K. Schum. in E. & P. Pf. IV.4: 144 (1891)

B. chaetocephala (DC.) Hepper in K.B. 14: 256 (1960) & in F.W.T.A., ed. 2, 2: 220 (1963)

NOTE. Hepper divides *S. chaetocephala* into two varieties. His *Borreria chaetocephala* DC. var *minor* Hepper (in K.B. 14: 256 (1960) & in F.W.T.A., ed. 2, 2: 220 (1963)) is based on *B. hebecarpa* A. Rich., Tent. Fl. Abyss. 1:347 (1848); Hiern in F.T.A. 3: 326 (1877); F.P.S. 2: 427 (1952); types: Ethiopia, Djeladjeranne, *Quartin-Dillon* (P, syn.) & *Schimper* 1712 (P, syn., K, isosyn.!). Others consider this a variant of *S. pusilla*.

11. **S. huillensis** (*Hiern*) *Good* in J.B. 65, Suppl. 2: 41 (1927); Verdc. in K.B. 30: 304 (1975). Types: Angola, Huila, near Lopollo, *Welwitsch* 3238 (LISU, syn., BM, isosyn.) & *Welwitsch* 3239 (LISU, syn., BM, isosyn.) & near Humpata, *Welwitsch* 3230 (LISU, syn., BM, K, isosyn.!)

Annual herb (10–)15–30(–70) cm. tall, mostly branched near the base or quite unbranched; stems purplish brown, usually densely covered with spreading white hairs or sometimes only slightly scabrid-pubescent or glabrescent. Leaf-blades lanceolate or lowermost elliptic, 1·3–2·6(–7) cm. long, 1·8–3·5(–5) mm. wide, acute at the apex, slightly narrowed at the base into the stipule-sheath, slightly shining and with scattered subscabrid hairs or glabrous above, hairy beneath at least on the midrib, margins slightly thickened, revolute; stipule-sheath 2·5–3 mm. long, hairy, with 7 usually reddish brown or purple setae 3–8 mm. long. Flowers in axillary verticillate congested clusters 0·8–1·3 cm. wide, the terminal cluster made up of the true terminal and 1–2 lower nodes condensed together subtended by 4(–8) leaves; below this compound terminal cluster and well separated is a further cluster subtended by 2–4 leaves and sometimes a third cluster, although sometimes only the terminal one is present; filiform bracteoles 3–6 mm. long. Calyx-tube ellipsoid, 1·5 mm. long, hairy; lobes 4, linear-lanceolate, 2·5–3 mm. long, often hairy, the margins scabrid-ciliate. Corolla white, often with bluish or purple marks at the throat in dried material; tube almost filiform beneath, narrowly funnel-shaped above, 5 mm. long, glabrous outside, with an internal ring of hairs at junction of the two parts; lobes elliptic-oblong,

* *Johnson* 4 from E. Coast of L. Nyasa may be from **T**7, **T**8 or Mozambique.

2 mm. long, 0·6 mm. wide, incurved at the apex and tipped with a few long hairs. Filaments exserted 2 mm. Style 7·5 mm. long; stigma small. Capsule ovoid, 3 mm. long, 2·2 mm. wide, spreading pubescent. Seeds dark chestnut-brown, ½-ellipsoid, slightly narrowed to the base, 2·5 mm. long, 1·3 mm. wide, 0·8 mm. thick, slightly rugulose, ventrally deeply grooved.

TANZANIA. Rungwe District: Ulambya, Mlale, 2 Apr. 1972, *Leedal* 1111!
DISTR. **T7**; Malawi, Zambia, Rhodesia and Angola
HAB. Unknown; 1560 m.

SYN. *Tardavel huillensis* Hiern in Cat. Afr. Pl. Welw. 1: 503 (1898)

NOTE. The specimen cited above is not typical and is the only one seen from the Flora area which is referable to this species.

12. **S. radiata** (*DC.*) *Hiern* in F.T.A. 3: 237 (1877). Types: Senegal, Galam, *Sieber* 8 (G, syn., K, isosyn. !) & Walo, *Perrottet* & *Leprieur* (G, syn.)

Branched or unbranched rigid annual herb 9–40(–100) cm. tall, with a ± simple root; stem densely covered with spreading white hairs. Leaf-blades very narrowly elliptic-lanceolate, 2–5·5 cm. long, 1·2–5·5 mm. wide, acute at the apex, narrowed at the base into the stipule-sheath, with scattered long hairs above and on the midnerve beneath or ± entirely glabrous save for the obscurely scabrid margins; midrib and margins distinctly thickened, white; stipules often white, base 4 mm. long, hairy, bearing 5–9 fimbriae 6·5 mm. long. Flowers in very dense terminal heads consisting of several condensed nodes, the leaves of which form radiating bracts, often without any further flowers at lower nodes but if so then these are smaller heads terminating undeveloped lateral branches; bracts of 2 sorts, one leaf-derived, ovate-lanceolate, 5–7 mm. long, 1·5–2 mm. wide, hyaline at the base, stiff and green at the apex, the others stipule-derived of narrow thin fimbriae 3 mm. long from a narrow base. Calyx-tube narrowly oblong, 2 mm. long, hairy above; lobes 4, filiform, 1·5 mm. long, hairy. Corolla greenish or white; tube funnel-shaped, 1 mm. long; lobes ovate-triangular, 0·6 mm. long, 0·4 mm. wide. Filaments exserted 0·5 mm. Style 1·3 mm. long; stigma 0·2 mm. wide. Capsule compressed-cylindrical, 3 mm. long, 1·2 mm. wide, with a median furrow, hairy in upper half. Seeds pale brown, oblong-ellipsoid, 1·8 mm. long. 0·7 mm. wide, 0·4 mm. thick, shiny. Fig. 49/12, p. 340.

UGANDA. W. Nile District: Koboko, Oct. 1938, *Hazel* 683!; Teso District: 24 km. from Soroti on Moroto road, 13 Oct. 1952, *Verdcourt* 828!
KENYA. Turkana District: Kacheliba, 9 Oct. 1924, *Leippert* 5071a!
DISTR. **U1**, 3; **K2**; Senegal to Cameroun and Sudan, also in Mozambique* and Angola
HAB. Roadsides, grassland with scattered trees; ± 1300 m.

SYN. *Borreria radiata* DC., Prodr. 4: 542 (1830); F.P.S. 2: 427 (1952); Hepper, F.W.T.A., ed. 2, 2: 219 (1963)

13. **S. pusilla** *Wall.* in Roxb., Fl. Indica, ed. Carey & Wall. 1: 379 (1820). Type: " Nepala Valley ", *Gardner* (ubi ?)

Erect or rarely prostrate annual usually branched herb (2–)7·5–60 cm. tall, with grooved ± glabrous or slightly papillate-puberulous often reddish stems. Leaf-blades linear-lanceolate to narrowly lanceolate, 1–5·3 cm. long, 2–5·5 mm. wide, acute at the apex, narrowed to the base, minutely scabrid above, glabrous beneath; true petiole absent; stipule-sheath 1·5–2 mm. long, glabrous or pubescent, 1·5–2 mm. long, bearing 7 setae 2–4 mm. Flowers in dense very compact spherical clusters at most nodes, 0·6–1(–1·5) cm. in

* A single specimen from Niassa, Cuamba, 13 May 1948, *Pedro & Pedrogão* 3358A is undoubtedly this species (EA !)

diameter; bracteoles filiform, numerous, 2 mm. long. Calyx-tube ovoid, 1 mm. long, pubescent; teeth 4, equal or slightly unequal, subulate, 0·6–1·2 mm. long, scabrid. Corolla white or pink; tube narrowly funnel-shaped, 1·3 mm. long; lobes 0·8–1·1 mm. long, 0·4 mm. wide, with some long flattened hairs at the apex. Filaments exserted ± 1 mm. Style exserted 0·5 mm. Capsule ellipsoid, 1·5 mm. long, glabrous or ± sparsely pubescent. Seeds chestnut-brown, shiny, oblong-ellipsoid or narrowly oblong, 1·3 mm. long, 0·55 mm. wide, 0·4 mm. thick, with a wide ventral groove. Fig. 50/13, p. 342.

UGANDA. Kigezi District: Kinkizi, Nyakinomi, Mar. 1951, *Purseglove* 3589 !; Mbale District: West Bugwe Local Forest Reserve, 0.8 km. NE. of Lugombe Hill, 28 May 1951, *G. H. Wood* 236 !; Masaka District: Sese Is., Kalangala, 28 June 1935, *A. S. Thomas* 1339 !

KENYA. Trans-Nzoia District: Kitale, 5 Oct. 1962, *Bogdan* 5558 !; Uasin Gishu District:? Kipkarren, *Brodhurst Hill* 136 !; Masai District: Chyulu North, 24 Apr. 1938, *Bally* 354 ! & 725 !

TANZANIA. Musoma District: Lupa, Nyambono, 4 June 1959, *Tanner* 4321 !; Buha District: Gombe Stream Reserve, Kasakela stream valley, 13 Apr. 1964, *Pirozynski* P. 683 !; Morogoro, May 1930, *Haarer* 1916 !

DISTR. **U**1–4; **K**3, 4–6; **T**1, 3, 4, 6–8; widespread in tropical Africa, Madagascar and tropical Asia to Japan and Philippines

HAB. Grassland, grassland with scattered trees, woodland, roadsides, stony areas, rocky places, also as a weed of cultivation; (150–)450–1920 m.

SYN. *Borreria pusilla* (Wall.) DC., Prodr. 4: 543 (1830); F.P.S. 2: 427 (1952); Hepper, F.W.T.A., ed. 2, 2: 220 (1963)
[*B. stricta* sensu F.W.T.A. 2: 135 (1931); U.K.W.F. 409, fig. (1974) et auct. mult. *non* (L.f.) K. Schum., *nec* G.F.W. Mey.*]
[*B. hebecarpa* sensu F.W.T.A. 2: 135 (1931), *non* A. Rich.]

NOTE. Drummond (annot. in herb.) treats *B. hebecarpa* as a synonym of this species but I think Hepper is probably correct in treating it as a variety of *B. chaetocephala* (p. 355).

14. **S. laevis** *Lam.*, Tab. Encycl. 1: 273 (1792), excl. ref. Sloane; Poiret in Lam., Encycl. Méth. Bot. 7: 313 (1806). Type: S. Domingo, *J. Martin* (P-LAM, holo.)

Erect or scrambling annual herb 0·3–1·2 m. tall; stems with slight ribs, glabrous or with lines of very short sparse hairs. Leaf-blades narrowly elliptic or elliptic-lanceolate to ovate, 0·8–5·5 cm. long, 0·3–2·5 cm. wide, acute at the apex, cuneate at the base into the stipule-sheath, the very narrow leaf-base resembling a short petiole ± 2 mm. long, entirely glabrous save for very short hairs at and near margins above giving a slightly scabrid feel; stipules drying a rather bright reddish brown; bases slightly pubescent, 2–3 mm. long, with 5–7 fine fimbriae 2·5 mm. long, or at inflorescence-bearing nodes ± 100·5 mm. long. Flowers in terminal and axillary many-flowered sessile clusters; stipule-derived bracteoles with fimbriae 3 mm. long. Calyx-tube narrowly obconic, 2–2·5 mm. long, 0·8 mm. wide, densely hairy above; limb obsolete; lobes triangular, 0·6 mm. long, ciliate, separated by a fringe.

* *Spermacoce stricta* L. f., Suppl. Pl.: 120 (1781), was based on *Crateogonum* (*amboinicum*) *minus s. verum* Rumph., Herb. Amboin. 6: 25 (1750) and possibly also a specimen. Bremekamp in de Wit, Rumphius Mem. Vol.: 404 (1959) equates this with *Oldenlandia verticillata L.* A specimen from Linnaeus' Uppsala Garden labelled *S. stricta* preserved at the Linnean Society (125·5) is not the plant either so it seems best to use the name *S. pusilla*. It must be noted that if *Borreria* is kept up the name *stricta* is not usable in any case. G. F. W. Meyer stated " forte *S. stricta* L. f." and must be assumed to have described a new plant—his specimens are in any case *B. verticillata* (L.) G. F. W. Mey.; K. Schumann's combination *B. stricta* (L. f.) K. Schum. is thus illegitimate being predated by *B. stricta* G. F. W. Mey.

Corolla white or lobes tipped with pink; tube 1·2 mm. long; lobes lanceolate, 1·3 mm. long, 0·6 mm. wide, slightly hairy at the apex outside. Stamens with filaments exserted 1–1·2 mm. Style exserted 1·3 mm.; stigma 0·5 mm. wide. Capsule ellipsoid or obovoid–fusiform, 2·5–4 mm. long, 1·8 mm. wide, 1 mm. thick, hairy above, the valves completely falling, each 2·2 mm. long, 1·8 mm. wide, bifid at the apex, crowned by the calyx-lobes. Seeds chestnut-brown, oblong-ellipsoid, 1·5–2 mm. long, 0·7–0·8 mm. wide, 0·5 mm. thick, divided into transverse areas by anastomosing transverse grooves, the actual areas reticulate. Fig. 50/14, p. 342.

KENYA. Kwale District: Shimba Hills, Kwale Boma area, 3 Apr. 1968, *Magogo & Glover* 661!

TANZANIA. Lushoto District: Amani old ornamental plots, 26 Dec. 1956, *Verdcourt* 1732!; Tanga, 14 Dec. 1949, *Verdcourt* 2 & 3!; Dar es Salaam, 13 Aug. 1967, *Harris* 769!; Zanzibar I., Kisimbani, 10 Mar. 1961, *Faulkner* 2775! & Kidichi, 9 July 1960, *Faulkner* 2633!

DISTR. **K**7; **T**3, 6; **Z**; Florida, Bermuda, W. Indies, Central and S. America to Bolivia and Galapagos Is., now widely spread as a weed and seen from Fiji, Samoa, Solomon Is., Papua, Java, Madagascar and Chagos Is.

HAB. Grassland, bushland, roadsides and waste land, often in old cultivations as a weed, now thoroughly naturalised in many places; 0–1130 m.

SYN. *Borreria laevis* (Lam.) Griseb., Fl. Brit. West Indies: 349 (1861); Bremek. in Pulle, Fl. Suriname 4: 289 (1934); Rendle, Fl. Jam. 7: 123, fig. 38 (1936); Gooding, Loveless and Proctor, Fl. Barbados: 400, fig. 25 (1963); Adams, Fl. Pl. Jam.: 733 (1972)

NOTE. Two MS names of A. Peter (who did not realise it was an introduced plant) will be found on many East African sheets namely *B. pubescens* and *B. aspera.*

15. **S. hepperana** *Verdc.* in K.B. 30: 305 (1975). Type: Sierra Leone, *Afzelius* (BM, lecto. !)

Herb 0·3–1·2 m. tall; stems decumbent at the base but ± strictly erect above and practically unbranched, drying purplish or brownish, glabrous. Leaf-blades lanceolate, 4–15 cm. long, 0·2–9(–2·1) cm. wide, narrowed to an acute apex and to the sheathing stipules; upper surface, particularly near the margins, with very short strongly scabrous emergences, glabrous beneath; stipules forming a sheath closely adhering to the stem at flowerless nodes, the sheath cylindrical, 8–9 mm. long with 3 unequal setae 3–6 mm. long, but at flowering nodes the stipules are stretched and funnel-shaped. Flowers in axillary verticillate few–many-flowered clusters, mainly at the top 2–3 nodes; filiform bracteoles 3·5 mm. long, joined at the base. Calyx-tube ovoid, 1·2 mm. long; lobes 4, linear-lanceolate, 1·5–2 mm. long, minutely scabrid along the margins. Corolla white, yellowish or tinged mauve; tube 4–5·5 mm. long, narrowly cylindrical but funnel-shaped above; lobes triangular, 2–3 mm. long, 1·6 mm. wide. Filaments exserted 2 mm. Style 6–7 mm. long; stigma capitate, 0·5 mm. wide, slightly bifid. Fruiting heads globose, 1·3–1·5 cm. wide. Capsule obovoid, 2–3(–5) mm. long, straw-coloured, densely reddish-brown spotted. Seeds dark purplish brown, 2·6–3·4 mm. long, 1·1–1·4 mm. wide, 0·8–1·1 mm. thick, laterally compressed, very shiny, very narrowly grooved ventrally and sometimes with a narrow white aril which scarcely projects.

TANZANIA. Songea District: SE. of Songea, by R. Likonde, 26 June 1956, *Milne-Redhead & Taylor* 10902!

DISTR. **T**8; Senegal to Ghana, ? Sudan

HAB. Dried up seasonal pool in riverine grassland; 750 m.

SYN. *Spermacoce compressa* Hiern in F.T.A. 3: 235 (1877), *non* G. Don (1834). Type: as for species

Borreria compressa Hutch. & Dalz., F.W.T.A. 2: 133, 135 (1931); Hepper in K.B. 14: 258 (1960) & in F.W.T.A., ed. 2, 2: 222 (1963)

NOTE. Hutchinson & Dalziel's name can be treated as a new name in *Borreria* if that genus is maintained; this is permissible under the rules of nomenclature. Berhaut has recently (Adansonia, sér. 2, 13 (4): 475 (1973)) split the old concept of *Borreria compressa* into two, mainly on fruiting characters. *Borreria bambusicola* Berhaut is easily distinguishable by the large membranous septum left after the valves dehisce and by the fact that the seeds have a small basal white conical aril. The seeds of the only specimen of the complex known from East Africa lacks the aril, but further comparisons are needed with West African material when more is available.

16. **S. quadrisulcata** (*Bremek.*) *Verdc.* in K.B. 30: 305 (1975). Type: Zaire, Sakadi–Haut Lomami, by R. Lubishi, *Mullenders* 1696 (BR, holo. !)

Annual ± glabrous herb 0·35–1·3 m. tall from a fibrous root; stems grooved, branched or ± simple. Leaf-blades linear to linear-lanceolate, 2·6–8 cm. long, 1·5–9(–11) mm. wide, acute at the apex, cuneate at the base, ± scabrid or glabrous above; stipule-sheath 3–5 mm. long, with (3–)5–7 setae 1–3 mm. long; petiole ± obsolete. Flowers in axillary clusters ± 6 mm. in diameter, 3–4-merous; bracteoles of fascicles of short fimbriae present. Calyx-tube ellipsoid, 1–1·5 mm. long, glabrous; lobes very small, triangular, 0·25–0·3 mm. long, with ± ciliolate margins. Corolla rose-violet or white with mauve lobes; tube funnel-shaped, 1·5–1·7 mm. long; lobes ovate-triangular, 1 mm. long. Stamens with filaments exserted 0·4–0·8 mm. Style exserted 0·8–1·2 mm.; stigma subcapitate. Capsule oblong-ellipsoid, 2–3 mm. long, 1 mm. wide, glabrous, crowned with the minute calyx-teeth. Seeds narrowly oblong, (1·8–)2–2·4 mm. long, (0·5–)0·75–0·8 mm. wide, 0·5–0·6 mm. thick, rugulose-pitted, grooved ventrally, the groove widened at the base and with a white granular aril. Fig. 50/16, p. 342.

UGANDA. Masaka District: Buddu County, Kalungu, 31 Dec. 1970, *Lye* 5840!
TANZANIA. Bukoba District: Karagwe, Sept. 1894, *Scott Elliot* 8183!; Buha District: near Mbirira [Birira], Lake Manyoni, 26 Feb. 1926, *Peter* 37800!; Songea District: about 2.5 km. SW. of Kitai, by R. Nakawali, 9 Mar. 1956, *Milne-Redhead & Taylor* 9072!
DISTR. **U**4; **T**1, 4, 8; Guinée to Nigeria, Central African Republic, Zaire, Burundi, Ethiopia, Malawi, Zambia, Angola and northern South West Africa
HAB. Disturbed wet places, dry (? seasonally) grassland; 900–1400 m.

SYN. [*S. compressa* sensu Hiern in F.T.A. 3: 235 (1877), quoad syntypum *Barter* 1231]
[*Tardavel stricta* sensu Hiern, Cat. Afr. Pl. Welw. 2: 503 (1898), quoad *Welwitsch* 3219 & 3220, *non* (L.f.) Hiern]
[*Borreria compressa* sensu Hutch. & Dalz., F.W.T.A. 2: 135 (1931), pro parte, *non* Hutch. & Dalz. sensu stricto]
B. quadrisulcata Bremek. in B.J.B.B. 22: 102 (1952)
B. paludosa Hepper in K.B. 14: 259 (1960) & in F.W.T.A., ed. 2, 2: 221 (1963). Type: Sierra Leone, Magbema, *Jordon* 873 (K, holo. !)

NOTE. During his preliminary examination of African *Spermacoce*, Drummond suggested that *Borreria tetraodon* K. Schum. in Schlechter, Westafr. Kautschuk-Exped.: 332 (1901), a nomen nudum based on Schlechter 12475 (Zaire, Dolo), was a synonym and the name has been used despite its invalidity. It is, however, a different species, the correct name for which is *S. bequaertii* (De Wild.) Verdc. (type: Zaire, Kinshasa [Leopoldville], *Bequaert* 7673 (BR, holo. !)); the corolla is larger with longer lobes, the calyx-limb is different and the habitat is wooded grassland not marshes.

17. **S. natalensis** *Hochst.* in Flora 27: 555 (1844); Sond. in Fl. Cap. 3: 24 (1865); Hiern in F.T.A. 3: 236 (1877), in obs. sub sp. 7. Type: South Africa, Natal Bay, *Krauss* 328 (TUB, holo., K, iso. !)

Perennial erect, prostrate or trailing herb 15–55 cm. long or tall; stems branched or ± unbranched, 4-ribbed, glabrous except at nodes. Leaf-blades ± linear, elliptic, or oblong-elliptic to narrowly elliptic-lanceolate, 0·8–4·5 cm. long, 1·5–8·5(–16) mm. wide, subacute at the apex, narrowed to the base, scabrid with very short hairs near the margins above but otherwise glabrous;

petiole obsolete; stipules with base ± 1 mm. long, pubescent, bearing 5 fimbriae 2–4·5 mm. long. Flowers in small clusters at most nodes, often appearing a little 1-sided due to the fact that they are present in 1 axil only; clusters to 8 mm. in diameter in fruiting stage; bracts stipule-like, with filiform fimbriae 2–2·5 mm. long. Calyx-tube obconic, 1·2–1·5 mm. long, glabrous; limb-tube ± 0·2 mm. long; lobes 4, linear-lanceolate to oblong-triangular, 1–1·8 mm. long, sometimes reddish brown. Corolla white; tube 1 mm. long; lobes triangular, 1·2–1·8 mm. long, 1·2 mm. wide, densely white hairy inside, obscurely papillate outside. Anthers borne at the sinuses. Style and stigma together very short, 0·3 mm. long. Capsule squarish with rounded sides in outline, 1·2–1·3 mm. long, 1·5–1·8 mm. wide, compressed, 1 mm. thick, glabrous, very finely transversely wrinkled, pitted inside the valves, crowned with the persistent lobes. Seeds blackish purple, 1·3 mm. long, 0·65 mm. wide, 0·4 mm. thick, deeply densely punctate. Fig. 50/17, p. 342.

UGANDA. Masaka District: Kabula County, 2–3 km. W. of Lyantonde, 16 May 1972, *Lye* 6886! & Lake Nabugabo, July 1937, *Chandler & Hancock* 1761! & Mawogola County, 17–18 km. SE. of Ntusi, 19 Oct. 1969, *Lye* 4493!

TANZANIA. Mbulu District: Issaras, Oct 1925, *Haarer* 43B!; Ufipa District: Sumbawanga, Chapota, 11 Mar. 1957, *Richards* 8631!; Mbeya Mt., 26 Jan. 1963, *Richards* 17540!; Njombe District: Ndumbi Forest, Feb. 1954, *Paulo* 256!; Songea District: Matengo Hills, Lupembe Hill, 29 Feb. 1956, *Milne-Redhead & Taylor* 8761!

DISTR. **U**4; **T**1–4, 6–8; Nigeria, Cameroun, Mozambique, Zambia, Malawi, Rhodesia and South Africa

HAB. Upland grassland (particularly where seasonally wet), rough grassland with scattered trees, roadsides, also swamps and old cultivations; 1080–2200 m.

SYN. [*S. stricta* sensu Hiern in F.T.A. 3: 236 (1877), pro parte, *non* L.f.]
Borreria natalensis (Hochst.) S. Moore in J.L.S. 40: 103 (1911); Hepper in F.W.T.A., ed. 2, 2: 220 (1963)
Diodia natalensis (Hochst.) Garcia in Mem. Junta Invest. Ultram., sér. 2, 6: 47 (1959)

18. **S. minutiflora** (*K. Schum.*) *Verdc.* in K.B. 30: 306 (1975). Type: Kenya, Nandi, *Scott Elliot* 6972 (B, holo. †, BM, iso.!)

Perennial herb with (5–)20–100 decumbent to suberect stems from a woody usually multi-headed rootstock; stems 4–20 cm. long or tall, strongly 4-ribbed, glabrous or with traces of minute emergences. Leaf-blades narrowly to broadly elliptic or oblong-elliptic, 0·6–2·1 cm. long, 1–8 mm. wide, rounded or obscurely acute at the apex, narrowed into the stipule-sheath or very short petiole at the base, entirely glabrous save for some minute hairs on the margins; petiole ± obsolete or apparently up to 1 mm. long; stipule-base 1–2 mm. long, bearing 3–5 fimbriae 0·5–2 mm. long. Flowers in small terminal and axillary clusters at the upper nodes, (3–)4–5 mm. in diameter. Calyx-tube obconic, 1 mm. long; lobes 4, sometimes with additional narrow ones (or abnormally ± 10), oblong, 0·8–0·9 mm. long. Corolla white; tube 0·5 mm. long; lobes oblong-triangular, 1 mm. long, 0·8 mm. wide. Anthers just protruding beyond the sinuses. Stigma minute, together with the style 0·4 mm. long. Capsule oblong in outline, 2 mm. long, 1·9 mm. wide, 1·2 mm. thick. Seed ½-ellipsoid, 1·8 mm. long, 1 mm. wide, 0·4–0·5 mm. thick.

UGANDA. W. Nile District: Logiri, *Eggeling* 1880!; Karamoja District: Lonyili [Longili] Mt., Apr. 1960, *J. Wilson* 896!; Mbale District: Bugisu, Bufumbo, July 1926, *Maitland* 1260!

KENYA. W. Suk District: Kapenguria, 15 May 1932, *Napier* 1973!; Trans-Nzoia District: Kitale, 5 Mar. 1953, *Bogdan* 3672!; Londiani District: Tinderet Forest, 26 June 1949, *Maas Geesteranus* 5190!

DISTR. **U**1, 3; **K**2, 3, 5, ?6; not known elsewhere

HAB. Grassland, grassland with scattered trees (*Combretum*, etc.), open woodland of *Acacia lahai*, *Syzygium* and *Erythrina*, also by roadsides and in rocky places, usually coming up after burning; 1440–2340 m.

SYN. *Borreria minutiflora* K. Schum. in E.J. 33: 373 (1903)
[*Oldenlandia hedyotoides* sensu Bullock in K.B. 1933: 80 (1933), *non* (Fisch. & Mey.) Boiss.]
Borreria sp. A sensu U.K.W.F.: 409, fig. (1974)

NOTE. Very closely allied to *S. natalensis* of which it may be no more than a habit form but typical material of each species is very different in habit and ecology.

19. **S. ocymoides** *Burm. f.*, Fl. Indica: 34 (1768). Type: Java, Sajor Babi, *Kleynhoff* (G, holo.)

A weak erect, decumbent or procumbent annual herb, usually ± well-branched, 3–40 cm. tall, with fine fibrous roots; stems with sparse to fairly dense crisped hairs on the ± wing-like prominent angles. Leaf-blades elliptic to elliptic-lanceolate, 0·4–3·6 cm. long, 0·25–1·6 cm. wide, rounded to ± acute at the apex, concavely narrowed into the petiole at the base, glabrous on both surfaces save for short marginal hairs or ± pubescent on the main nerve beneath; petiole 0–8 mm. long, with scattered hairs; stipules with base 2 mm. long, bearing ± 7 fimbriae 1–3 mm. long. Flowers in small few-flowered clusters at many of the nodes, attaining 3–6 mm. in diameter in the fruiting state; stipule-like bracteoles with fimbriae 1·8–2 mm. long. Calyx-tube transversely oblong, 0·5 mm. long; lobes 2(–4), 0·6–0·8 mm. long, ciliate. Corolla white; tube 0·3 mm. long; lobes triangular, 0·35 mm. long, 0·3–0·4 mm. wide, with a few short hairs inside. Anthers situated just above the sinuses of the corolla-lobes. Style ± 0·2 mm. long; stigma 0·2 mm. wide. Fruit oblong, 1 mm. long, 0·8 mm. wide, 0·5–0·7 mm. thick, compressed, finely transversely wrinkled and very shortly pubescent. Seeds chestnut-brown, oblong-ellipsoid, 0·7–0·8 mm. long, 0·4 mm. wide, 0·3–0·35 mm. thick, strongly reticulate with raised ribs, the foveae elongated in the direction of the short axis. Fig. 50/19, p. 342.

UGANDA. Toro District: Bwamba, Buyayu–Sempaya road, Oct. 1929, *Liebenberg* 926!; Masaka District: Sese Is., Bugalla I., 12 Oct. 1958, *Symes* 512!
TANZANIA. Buha District: Gombe Stream Reserve, Kakombe valley, 6 Feb. 1964, *Pirozynski* 326!; Rungwe District: Masoko [Massoka] Lake, 23 Mar. 1932, *St. Clair-Thompson* 1310!; Songea District: Lake Malawi, Liuli, 29 Apr. 1960, *Hay* 101!
DISTR. **U**2, 4; **T**4, 7, 8; Senegal to Cameroun, Fernando Po, Principe, St. Tomé, Annobon, Cabinda, Congo (Brazzaville), Zambia and Malawi; described from Malesia, now widely spread in Solomon Is., Fiji, Samoa, Philippines, India, Madagascar, Mauritius, Central America, West Indies and S. America to Argentina
HAB. Forest paths and leaf-litter, rocky paths in open grassland, also as an epiphyte on planted *Raphia* palms; 930–1700 m.

SYN. *Borreria ocymoides* (Burm. f.) DC., Prodr. 4: 544 (1830); Bremek. in Pulle, Fl. Suriname 4: 285 (1934); Rendle in Fl. Jam. 7: 125 (1936); Taylor in Exell, Cat. Vasc. Pl. S. Tomé: 219 (1944); Hepper, F.W.T.A., ed. 2, 2: 220 (1963)
B. ramisparsa DC., Prodr. 4: 544 (1830); Hutch. & Dalz., F.W.T.A. 2: 135 (1931). Type: Brazil, *Pohl* (G, holo.)
Spermacoce ramisparsa (DC.) Hiern in F.T.A. 3: 238 (1877)
Tardavel ocymoides (Burm. f.) Hiern, Cat. Afr. Pl. Welw. 2: 504 (1898)

20. **S. hispida** *L.*, Sp. Pl.: 102 (1753); Verdc. in K.B. 30: 307 (1975). Type: Ceylon, *Hermann* (BM-HERM, Vol. 1: 15, lecto.!)

Much-branched prostrate perennial herb forming mats up to 23–40 cm. wide; stems 10–20 cm. long, tinged reddish, covered with spreading white hairs, with short internodes and rather congested foliage. Leaf-blades elliptic or oblong-elliptic, 0·6–1·5 cm. long, 1·5–6 mm. wide, ± acute at the apex, cuneate at the base, scabrid with white hairs on upper surface and particularly along the margins, also with hairs on midrib and nerves beneath; stipule-base hairy, ± 2 mm. long, with 3–6 fimbriae 3–6 mm. long; free petioles ± 1 mm. long or obsolete. Flowers 1–several, sessile, in most leaf-axils. Calyx-tube oblong-ovoid, 2 mm. long, densely hairy; lobes 4, lanceolate, 1·5–2 mm. long, hairy along the margins. Corolla pale blue or lilac; tube narrowly cylindrical, 6 mm. long; lobes oblong, 2 mm. long, 1 mm. wide, hairy at the apex. Filaments exserted 1 mm. Style exserted 2·5–5 mm.; stigma shortly bifid, 0·8 mm. long. Capsule obovoid, 3 mm. long, 2·7 mm. wide, with spreading white hairs. Seeds chestnut-brown, oblong- ellipsoid, 2·5 mm. long, 1·2 mm. wide, 0·8 mm. thick, sometimes narrowed to one end, finely reticulate, ventrally grooved. Fig. 50/20, p. 342.

TANZANIA. Uzaramo District: Dar es Salaam, 11 July 1966, *Archbold* 831! & about 9.5 km. S. of Dar es Salaam, near Mbagala Mission, 1 Dec. 1955, *Milne-Redhead & Taylor* 7522! & Msasani, 24 Nov. 1968, *Batty* 302!

DISTR. **T6**; India, China, Malaysia, Indochina and Philippines

HAB. Secondary grassland on sandy soil, under coconut palms on sandy ground near beach; 0–80 m.

SYN. *S. articularis* L.f., Suppl. Pl.: 119 (1781). Type: specimen from "India orientalis" cultivated in Hort. Upsala (LINN, 125.6, holo.!)

Borreria hispida (L.) K. Schum. in E. & P. Pf. IV.4: 144 (1891), *non* Spruce ex K. Schum. (1888)

B. articularis (L.f.) F.N. Williams in Bull. Herb. Boiss., sér 2, 5: 956 (1905)

NOTE. This common Asian species has clearly been introduced quite recently into Tanzania by way of trade with some Asiatic port. There seem to be no records prior to 1955. The description above covers only African material; considerably more variation is displayed where the species is endemic.

21. **S. princeae** (*K. Schum.*) *Verdc.* in K.B. 30: 307 (1975). Type: Tanzania, Kilimanjaro, *Uhlig* 252 (EA, lecto.!)

Scrambling or decumbent herb with ascending branchlets, much branched, 0·3–0·6 m. long; stems often dark crimson, square, sparsely to densely hairy on the angles. Leaf-blades elliptic to ovate, 1·2–6·5 cm. long, 0·35–3 cm. wide, acute at the apex, cuneate at the base, glabrous to pubescent above, with sparse hairs to pubescent beneath and always scabrid on the margins; venation impressed, giving a bullate and plicate appearance to the blade; petiole 0–1 mm. long; stipule-base 4–6 mm. long, hairy, at least above, with 5–9 fimbriae 7–8·5 mm. long. Flowers in dense few–many-flowered axillary clusters at most nodes; fimbriae of stipuliform bracts 6 mm. long, ciliate. Calyx-tube fusiform or obconic, 1·2–3 mm. long; lobes foliaceous, lanceolate to linear-lanceolate, 3·5–5(–9) mm. long, 0·8–1 mm. wide, with pubescent margins. Corolla white; tube cylindrical or narrowly funnel-shaped, 6·5–10·5 mm. long, 0·4–2·5 mm. wide; lobes oblong or elliptic, 3–4 mm. long, 0·8–2·5 mm. wide. Filaments exserted 1–1·5 mm. Style exserted 2·5–4 mm.; stigma-lobes 0·5–1 mm. long. Capsule oblong-ellipsoid, 5–6 mm. long, 2 mm. wide, the valves opening widely, finely transversely wrinkled, glabrous or finely pubescent, crowned with the persistent calyx-lobes. Seeds dark purple-brown, fusiform, 4 mm. long, 1·6 mm. wide, 0·7–0·8 mm. thick, deeply grooved ventrally, narrowed at one end, shiny, finely rugulose and sometimes with a shallow dorsal groove as well.

var. **princeae**

Stems pubescent on the angles; leaf-blades glabrescent or with sparse hairs beneath and on margins; capsule ± glabrescent or sometimes pubescent. Fig. 52.

UGANDA. Kigezi District: near Bunagana Hill, 13 Dec. 1930, *B. D. Burtt* 2988!; Mbale District: Bugisu, Bulago, 28 Aug. 1932, *A. S. Thomas* 337! & Budadiri, Jan. 1932, *Chandler* 497!

KENYA. Trans-Nzoia District: Kitale, Grassland Research Station, 2 Oct. 1959, *Verdcourt* 2465!; Nyeri District: Iriaini, Githi Location, 11 Dec. 1963, *Kibui* 40!; Kiambu District: Uplands, 15 Oct. 1950, *Verdcourt* 356!

FIG. 52. *SPERMACOCE PRINCEAE* var. *PRINCEAE*—**1,** habit, × $\frac{2}{3}$; **2,** node showing stipule, × 4; **3,** flower, × 4; **4,** longitudinal section of ovary, × 8; **5,** fruiting shoot, × 2; **6,** capsule, × 6; **7,** seed, × 10. 1, 2, from *Kirrika* 237; 3, 4, from *Mathenge* 205; 5–7, from *Rose* 1145. Drawn by Mrs M. E. Church.

TANZANIA. Arusha District: Ngurdoto Crater National Park, Longil, East Lake, 24 Feb. 1966, *Greenway & Kanuri* 12390 !; Rungwe District: Kyimbila, 16 Oct. 1910, *Stolz* 349 !; Songea District: about 5 km. E. of Songea by R. Luhira, 15 Feb. 1956, *Milne-Redhead & Taylor* 8700 !

DISTR. **U2–4**; **K** ?1, 2–5; **T2–4**, 6–8; Cameroun, Zaire, Burundi, Ethiopia

HAB. Grassland, forest and woodland clearings, grassland with scattered trees, swampy places, cleared bamboo forest; 960–2600 m.

SYN. *Borreria princeae* K. Schum. in E.J. 34: 341 (1904); Hepper, F.W.T.A., ed. 2, 2: 222 (1963); Verdc. in K.B. 17: 500 (1964); U.K.W.F.: 407 (1974)
Diodia stipulosa S. Moore in J.L.S. 37: 310 (1906). Types: Kenya, Kikuyu, *Gregory* 92 (BM, syn. !) & Machakos District, Sani, *Kassner* 753 (BM, syn. !, K, isosyn.!) & Tanzania, Moshi District, Marangu, *Volkens* 413 (BM, syn. !, K, isosyn. !) & Cameroun, *Preuss* 682 (BM, syn. !)

var. **pubescens** (*Hepper*) *Verdc.* in K.B. 30: 307 (1975). Type: Fernando Po, Moka, *Boughey* 32 (K, holo. !)

Stems, both leaf surfaces and capsule densely hairy.

UGANDA. Ankole District: Kalinzu Forest Reserve, 31 July 1960, *Paulo* 623 !; Kigezi District: Echuya Forest Reserve, 8 Aug. 1960, *Paulo* 672 !; Masaka District: Sese Is., Bugala, 10 June 1932, *A. S. Thomas* 181 !

KENYA. Kericho District: Kericho, Sept. 1958, *Tweedie* 1716 ! & SW. Mau Forest Reserve, 10 Aug. 1949, *Maas Geesteranus* 5670 ! & Sotik, 15 June 1956, *Verdcourt* 947 !

TANZANIA. Bukoba District: Maruku, *Panayotis & G. K. R. Williams* 95 !; Ufipa District: Sumbawanga, Chapota, 7 Mar. 1957, *Richards* 8555 !; Iringa District: Ilongelo, *Carmichael* 405 !

DISTR. **U2**, 4; **K4** (± intermediate), 5; **T1**, 4, 7; Burundi, Cameroun, Fernando Po, Malawi, Zambia

HAB. Grassland, swampy areas, lake edges, roadsides, forest edges and clearings, also as a weed in ley plots, etc.; 1140–2300 m.

SYN. *Borreria princeae* K. Schum. var. *pubescens* Hepper in K.B. 14: 259 (1960) & F.W.T.A., ed. 2, 2: 222 (1963)

22. **S. latifolia** *Aubl.*, Hist. Pl. Guiane Fr. 1: 55, t. 19/1 (1775). Type: French Guiana, Cayenne, etc., *Aublet* (? P, holo.)

Straggling or prostrate annual herb 10–60 cm. long with bright pale yellow green stems and foliage; stems branched or unbranched, square with angles slightly to distinctly winged, glabrous or with sparse to fairly dense short hairs on the angles or stems hairy all over. Leaf-blades often red margined, elliptic, 1·2–5 cm. long, 0·8–2·9 cm. wide, acute at the apex, cuneate at the base, pubescent or ± scabrid above with tubercle-based hairs, pubescent beneath or almost glabrescent all over save for the scabrid margins; petiole 0·5–3 mm. long; stipule-sheath 1·5 mm. long, with 5–9 setae 1·5–3·5 mm. long. Flowers in axillary clusters ± 8 mm. wide. Calyx-tube pubescent, obconic, 2·5 mm. long; lobes 4, oblong to lanceolate, 1·2–2 mm. long. Corolla whitish, blue or pink; tube funnel-shaped, 5 mm. long; lobes ovate-triangular, 1·5 mm. long and wide. Filaments exserted ± 1 mm. Style exserted 2 mm.; stigma-lobes linear, 1 mm. long. Capsule ellipsoid or subglobose, 2·5–3 mm. long, wrinkled, hairy. Seeds yellow brown, ellipsoid, 2–2·8 mm. long, 1·3–1·7 mm. wide, 0·8–1 mm. thick, reticulate-rugulose, with a deep wide ventral excavation. Fig. 50/22, p. 342.

UGANDA. Mengo District: near Entebbe, Bukiberu, 9 May 1969, *Lye* 2824 ! & W. side of Kisubi, 9 Jan. 1969, *Lye* 1111 ! & near Entebbe, N. of Kisi, 31 Aug. 1969, *Lye* 3686 !

DISTR. **U4**; Sierra Leone, Liberia, Ivory Coast and Ghana; a native of S. America from Central America to Bolivia and West Indies but now a common casual in many parts of the world including India, Ceylon, Nepal, Malay Peninsula, Java and Australia

HAB. Grassland; 1140 m.

SYN. *Borreria latifolia* (Aubl.) K. Schum. in Mart., Fl. Bras. 6 (6): 61 (1888); Bremek. in Pulle, Fl. Suriname 4: 291 (1934); Hepper, F.W.T.A., ed. 2, 2: 219 (1963)

NOTE. The above description covers only African material. There is wider variation shown by the native S. American material.

23. **S. senensis** (*Klotzsch*) *Hiern* in F.T.A. 3: 236 (1877). Type: Mozambique, Tete, Rios de Sena, *Peters* (B, holo. †)

Annual unbranched or sparsely branched herb (3–)16–60 cm. tall, with 4-ribbed stems ± densely covered with spreading white hairs. Leaf-blades narrowly elliptic to elliptic-lanceolate, 2–7 cm. long, 0·35–2·2 cm. wide, acute at the apex, narrowly cuneate at the base, with ± long hairs on both surfaces and scabrid at the margins; true petiole absent; stipule-sheath 3 mm. long, densely hairy, with 7 setae 0·2–1·1 cm. long. Flowers in clusters at the nodes, up to 1·5–2 cm. in diameter in fruiting stage. Calyx-tube subcylindric, 2·5 mm. long, hairy above; lobes 4, ± equal, mostly lanceolate or narrowly oblong, 2–4 mm. long, 0·2–0·8 mm. wide, scabrid on the margins. Corolla white, sometimes with mauve streaks reaching into the throat, tube funnel-shaped, 3–8 mm. long; lobes triangular, 1·5–4·5 mm. long, 1–3 mm. wide, pubescent at the apex outside. Filaments exserted 1–3 mm. Style exserted 1–3·5 mm. Capsule pale, often streaked red-brown, ellipsoid, 3 mm. long, 2 mm. wide, densely hairy. Seeds pale chestnut-coloured to deep blackish red, oblong-ellipsoid, 2·6–2·8 mm. long, 1·2–1·7 mm. wide, 0·8–1 mm. thick, with a deep narrow ventral groove, covered with a fine shallow reticulation. Fig. 51/23, p. 344.

KENYA. Northern Frontier Province: Moyale, 10 July 1952, *Gillett* 13556!; Embu/Kitui Districts: slopes W. of Kindaruma Dam, Tana R., Seven Forks, 30 Apr. 1967, *Gillett & Faden* 18120!; Teita District: Tsavo National Park (East), Voi Lodge, 6 Jan. 1969, *Leuthold* 55!

TANZANIA. Mbulu District: Lake Manyara National Park, Bagoyo R., 20 Mar. 1964, *Greenway & Kanuri* 11393!; Kondoa District: Kolo, Chungai, 13 Jan. 1962, *Polhill & Paulo* 1162!

DISTR. **K**1, 4, 7; **T**1–8; Burundi, Mozambique, Malawi, Zambia, Rhodesia, Angola, Botswana, South West Africa and South Africa

HAB. Bare ground, grassland, scrub and woodland, often on sandy soil and on rocky hillsides; (150–)400–1600 m.

SYN. *Diodia senensis* Klotzsch in Peters, Reise Mossamb., Bot. 1: 289 (1861)

Mitracarpum dregeanum Harv. & Sond., Fl. Cap. 3: 25 (1864). Types: South Africa, near Durban [Port Natal], *Drège*, *Gerrard & McKen* (TCD, syn.)

Borreria senensis (Klotzsch) K. Schum. in Abh. Preuss. Akad. Wiss.: 23 (1894) & in P.O.A. C: 394 (1895)

Diodia benguellensis Hiern, Cat. Afr. Pl. Welw. 1: 502 (1898). Types: Angola, Bumbo, *Welwitsch* 3223 (LISU, syn., BM, K, isosyn.!) & Huila, Lopollo, *Welwitsch* 3224 (LISU, syn.)

Borreria stolzii K. Krause in E.J. 57: 52 (1920). Type: Tanzania, Rungwe District, Kiwira [Kibila] R., *Stolz* 1971 (B, holo.†, K, iso.!) [locality given as Kalalamuka on Kew sheet]

B. squarrosa Schinz in Viert. Nat. Ges. Zürich 68: 438 (1923). Type: South West Africa, Amboland, between Ondonga and Uukuambi, *Rautenen* 811 (Z, holo.)

B. rhodesica Suesseng. in Trans. Rhodes. Sci. Assoc. 43: 130 (1951). Type: Rhodesia, Marandellas, *Dehn* 98 (M, holo., K, iso.!) [Kew sheet bears locality Rusape]

[*B. ruelliae* sensu auctt. mult., *non* (DC.) H. Thoms]

[*B. scabra* sensu auctt. mult., *non* (Schumach. & Thonn.) K. Schum.]

NOTE. The differences between this and *S. ruelliae* are slight but in general throughout most of their ranges they are fairly easily separable by the seed and calyx differences; unfortunately in Uganda the calyx character often seems to be unreliable and it is difficult to believe the two can be distinct species. I am convinced, however, that they are best kept distinct. There has been much confusion with *S. sphaerostigma* when no fruit has been available but the more terminal nature of the inflorescences and larger flowers will usually separate that species; in fruit the two are utterly distinct. Hiern saw Peter's specimen but no authentic material is extant.

24. **S. ruelliae** *DC.*, Prodr. 4: 554 (1830); Hiern in F.T.A. 3: 238 (1877). Type: Senegal, Galam, Bakel, *Leprieur* (G, holo.)

Annual unbranched to branched herb 3–60 cm. tall; stems 4-ribbed, ± densely covered with spreading white hairs. Leaf-blades narrowly lanceolate to elliptic-lanceolate, 1·8–7(–9·5) cm. long, 0·25–1·9(–2·1) cm. wide, tapering acute at the apex, cuneate at the base, sparsely to densely ± scabridly hairy or rarely glabrous on both surfaces, the margins always very scabrid with short hairs, the nerves impressed above and surface somewhat bullate; true petiole absent; stipule-sheath 3·5–5 mm. long, densely hairy, with 5–9 setae 0·4–1·1 cm. long. Flowers in few–many-flowered clusters at the nodes, up to 1·3–1·8 cm. in diameter in fruiting stage. Calyx-tube obconic, 2·2 mm. long, hairy above; lobes 4, narrowly triangular-lanceolate, 2 shorter narrower ones alternating with 2 longer wider ones 2–4(–7) mm. long, 0·5–1·6 mm. wide, scabrid on the margins. Corolla white or pink; tube 3–5(–7) mm. long; lobes triangular or ovate-triangular, 1·8–2·5(–5) mm. long, 1·2–1·8 mm. wide, with a few hairs at the apex outside. Filaments exserted 1–2 mm. Style exserted 2–4 mm.; stigma capitate. Capsule ellipsoid, 3–4 mm. long, 2·1–2·5 mm. wide, hairy above or all over. Seeds chestnut-brown, oblong-ellipsoid, 2·4–2·8 mm. long, 1·3 mm. wide, 0·6 mm. thick, convex dorsally, with a wide ventral excavation which is usually filled with white cellular material for most of its length but basally narrowed and there filled with rhaphides, finely rugulose. Fig. 51/24, p. 344.

UGANDA. W. Nile District: W. of Mt. Wati [Eti], 25 July 1953, *Chancellor* 46!; Acholi/Bunyoro Districts: Kabalega [Murchison Falls] National Park, near Chobi Lodge, 5 Sept. 1967, *Angus* 5848! & Chobi, 8 Sept. 1967, *Angus* 5896!

HAB. Dry places, bare ground, grassland; 1200 m.

DISTR. **U**1, 2, 4 (N. Mengo); West Africa from Senegal to Angola, Chad, Central African Republic, Zaire, Sudan

SYN. *Diodia scabra* Schumach. & Thonn., Beskr. Guin. Pl.: 76 (1827), *non Spermacoce scabra* Willd. (1798), *nec* Ewart (1917). Type: Ghana [Guinea], *Thonning* (C, holo.)

Borreria scabra (Schumach. & Thonn.) K. Schum. in P.O.A. C: 385 (1895); Hepper, F.W.T.A., ed. 2, 2: 220 (1963)

B. ruelliae (DC.) H. Thoms in N.B.G.B. 5: 104 (1909); Hutch. & Dalz., F.W.T.A. 2: 135 (1931); F.P.S. 2: 428 (1952)

NOTE. *Buechner* 68, from Kabalega National Park, Paraa, Rhino Viewpoint, 18 June 1957, has 4 equal calyx-lobes 5 mm. long but is clearly the same species.

25. **S. sp. C**

Annual herb with strictly erect unbranched stem ± 70 cm. tall, the basal 25–30 cm. leafless, bifariously densely pubescent above, almost glabrous at base. Leaf-blades elliptic-lanceolate, ± 6 cm. long, 0·8–1·2 cm. wide, acute at the apex, narrowed at the base into the stipule-sheath, sparsely pubescent above, glabrous beneath, the margins ciliate with short stiff hairs; stipule-sheath 4–5 mm. long, with 2 lines of dense pubescence, bearing 5–7 ciliate pale yellow-brown setae 0·5–1·2 cm. long. Flowers solitary (or ? few) in the axils (only 3 seen in two specimens available); filiform bracteoles 7 mm. long. Calyx-tube deep inside the stipule-sheath, ± linear-obconic, 5 mm. long; lobes linear-lanceolate, 2 or 2 large and 2 small, 3–5·5 mm. long, ciliate. Corolla pale lilac; tube narrowly funnel-shaped, 5 mm. long; lobes ovate-oblong, 3·5 mm. long, 0·8–1·2 mm. wide, tipped with 2–3 long hairs. Filaments exserted 1–1·3 mm.; style exserted 2–5 mm., the stigma capitate, 0·4 mm. wide. Capsule not seen.

TANZANIA. Kigoma District: Uvinza, W. of Lugufu near km. 1176.5 on Central Rail way, 8 Feb. 1926, *Peter* 36555!

DISTR. **T**4; known only from the above gathering
HAB. Forest; 1060 m.

26. **S. sphaerostigma** (*A. Rich.*) *Vatke* in Linnaea 40: 196 (1876). Types: Ethiopia, Shire [Chiré], *Quartin Dillon* (P, syn.) & near Adua [Adowa], *Schimper* 100 (P, syn., BM, K, isosyn. !)

Annual erect herb 10–90 cm. tall, with unbranched to much-branched stems, sparsely to densely covered with spreading hairs. Leaf-blades lanceolate to elliptic-lanceolate, 1–10 cm. long, 0·4–3·5 cm. wide, acute at the apex, cuneate at the base, sparsely to distinctly hairy on both surfaces, sometimes very scabrid all over but nearly always so along the margin; free petiole 0–2 mm. long, joined to the stipule-sheath beneath; stipules hairy, base 1–4·5 mm. long, with 5–7 fimbriae 2–8 mm. long, ciliate. Flowers in axillary clusters up to 1·8 cm. in diameter in fruiting stage, at apical and lower nodes, those at the apex supported by usually 4 bract-like leaves; stipule-derived bracteoles filiform, 2–3 mm. long. Calyx-tube obconic, 2·5–3·5 mm. long, hairy at the apex; lobes 4, ovate to lanceolate, leafy, 2·8–5 mm. long, 1·3–1·8 mm. wide, scabrid-ciliate. Corolla white, blue, purple or white with blue streaks; buds mostly tipped with erect hairs; tube funnel-shaped, often quite broadly so, 5–8·5 mm. long, 0·8 mm. wide at the base, 5 mm. wide at the throat; lobes triangular, 3–4 mm. long, 1·5–2 mm. wide, usually with some long hairs outside at the apex, ± finely papillate inside. Filaments exserted 1·5–2·2 mm. Style exserted 2–4·5 mm., usually thickened upwards and ± roughened; stigma slightly bifid, 0·5 mm. wide. Capsule green, ovoid, 3·5–4 mm. long, 2–3 mm. wide, splitting from the base upwards but remaining in one piece, the split not crossing the apical rim bearing the persistent ± spreading calyx-lobes; valves shortly hairy in upper half, finely transversely rugulose; central septum a persistent white oblong left behind on the inflorescence. Seeds brown, narrowly oblong-ellipsoid, 3–3·2 mm. long, 1·3–1·5 mm. wide, 0·8 mm. thick, slightly narrowed to the base, shiny, finely obscurely reticulate, with a broad ventral groove and usually a small basal aril 0·5–1 mm. long. Fig. 53, p. 368.

UGANDA. Karamoja District: Lochoi (? Lochowai), 24 May 1940, *A. S. Thomas* 3540 !; Teso District: Lake Kyoga, 13 Oct. 1952, *Verdcourt* 833 !; Mengo District: Entebbe, Nov. 1924, *Maitland* 307 !
KENYA. Trans-Nzoia District: S. Cherangani Hills, 23 Aug. 1957, *Symes* 198 !; 16 km. NE. of Nakuru, 6 Sept. 1948, *Bogdan* 1970 !; S. Kavirondo District: Kuja R., July 1934, *Napier* 3424 in *C.M.* 6715 !
TANZANIA. Shinyanga, 12 Apr. 1932, *B. D. Burtt* 3752 !; Ufipa District: Ilemba Gap, road to Rukwa, 12 Mar. 1959, *Richards* 11181 !; Songea District: 3 km. NE. of Kigonsera, 12 Apr. 1956, *Milne-Redhead & Taylor* 9589 !
DISTR. **U**1–4; **K**1–3,5; **T**1, 2, 4–8; Nigeria, Cameroun, Zaire, Burundi, Sudan, Ethiopia, Mozambique, Malawi and Zambia
HAB. Coarse grassland, grassland with scattered trees, scrub, woodland (including *Brachystegia–Isoberlinia*) and particularly in disturbed areas, e.g. roadsides, cultivations, waste places, damp ditches, etc. on sandy and black loam soils; often a weed; (240–)780–2300 m.

SYN. *Hypodematium sphaerostigma* A. Rich. in Tent. Fl. Abyss. 1: 348 (1847); Hiern in F.T.A. 3: 240 (1877); Verdc. in K.B. 7: 360 (1952) & in B.J.B.B. 28: 278 (1958)
H. ampliatum A. Rich. in Tent. Fl. Abyss. 1: 349 (1847). Types: Ethiopia, Taccaze, near Djelajeranne, *Quartin Dillon* (P, syn., K, ? isosyn. !) & *Schimper* (P, syn.)
Spermacoce ampliata (A. Rich.) Oliv. in Trans. Linn. Soc., 29: 88, t. 54, excl. fig. 4–6 (1873)
Arbulocarpus sphaerostigma (A. Rich.) Tennant in K.B. 12: 386 (1958); Cufod., E.P.A.: 1023 (1965); U.K.W.F.: 409 (1974)

NOTE. The genus *Arbulocarpus* can be kept up on the technical character of the fruit splitting from the base upward insteads of from the apex downwards but the

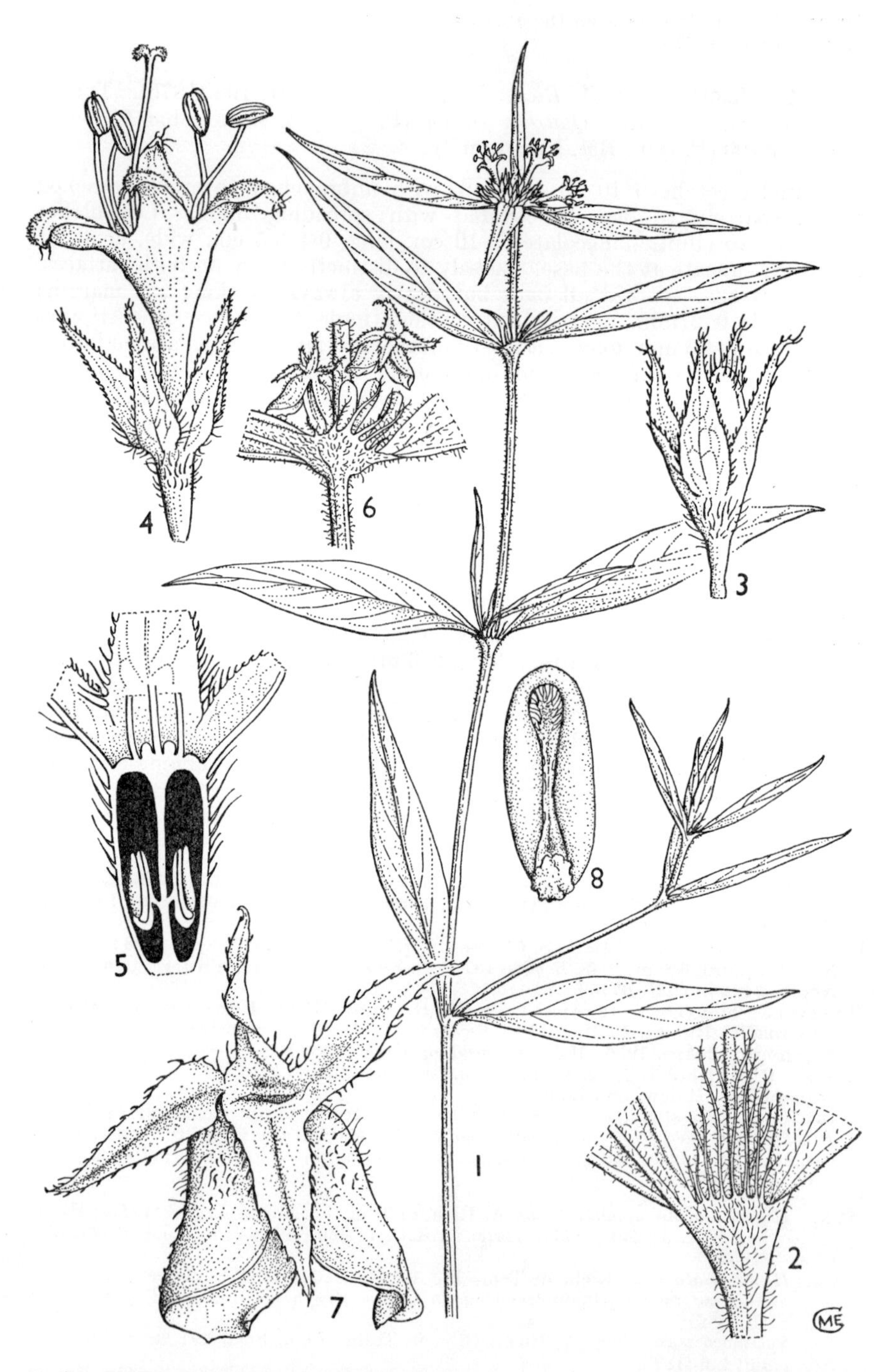

FIG. 53. *SPERMACOCE SPHAEROSTIGMA*—**1,** flowering branch, × ⅔; **2,** node showing stipule, × 4; **3,** flower bud × 6;, **4,** flower, × 4; **5,** longitudinal section of ovary, × 10; **6,** node of fruiting branch (note persistent septa), × 2; **7,** dehisced capsule, × 8; **8,** seed, × 8. 1–5, from *Chancellor* 142; 6–8, from *J. Wilson* 1029. Drawn by Mrs M. E. Church.

facies is exactly that of *Spermacoce*. My suggestion that *Arbulocarpus* should be merged with the American genus *Staelia* Cham. & Schlecht. (B.J.B.B. 28: 281 (1958)) is quite unfounded; the facies of the two is quite different; Hooker (G. P. 2: 145 (1873)) mentions *Hypodematium* A. Rich. (1847) (*non* Kunze (1833)) in a list after his description of *Spermacoce* and clearly did not consider it to merit generic rank.

27. S. sp. D

Probably perennial herb, with ± 10 probably prostrate or at least decumbent stems 30–90 cm. long, rather obscurely ribbed, glabrous. Leaf-blades elliptic-lanceolate, 1·8–2·3 cm. long, 3–8 mm. wide, acute at the apex, cuneate at the base into a short apparent petiole, glabrous; petiole ± 1–1·5 mm. long; stipule-bases 1–2·5 mm. long, with 1–3 fimbriae 1·5–4 mm. long; sometimes there are a few hairs on the ridged sides of the stipule-bases. Flowers 2–several in the axils of the leaves. Calyx-tube glabrous, oblong-obconic, 1·5 mm. long; lobes 4, lanceolate, 2–3 mm. long, 0·7–1 mm. wide. Buds glabrous. Corolla said to be white but not seen. Persistent style 1 mm. long; stigma 0·5 mm. wide. Capsule compressed subglobose, transversely wrinkled, crowned with the persistent calyx-lobes, eventually with the 2 valves completely opening so that they form an angle of ± 180°, each valve almost circular, 2·5–3 mm. long and wide, convex, appearing emarginate at base and apex. Immature seed oblong-ellipsoid, 2 mm. long, 1·3 mm. wide, 0·6 mm. thick, ventral edge winged, dorsal surface very distinctly reticulate foveolate.

TANZANIA. Rufiji District: Mafia I., 3 Apr. 1933, *Wallace* 800!
DISTR. T6; known only from the above gathering
HAB. Not known; near sea-level

NOTE. This had been compared with *S. ocimifolia* Roem. & Schultes. (*Hemidiodia ocimifolia* (Roem. & Schultes) K. Schum.) but the capsule is quite different in structure. Until more adequate material is available its identity will remain uncertain but it is probably an undescribed species.

28. S. congensis (*Bremek.*) *Verdc.* in K.B. 30: 308 (1975). Type: Zaire, Ruzizi Plain, *Germain* 6506 (BR, holo., EA, iso.)

Erect annual herb 17–60 cm. tall; stems ± simple or sparingly branched, 4-angled, bifariously shortly pubescent in the grooves. Leaf-blades linear-lanceolate to lanceolate, 1·3–8·5 cm. long, 0·22–1·25 cm. wide, acute at the apex, cuneate or ± rounded at the base, sessile, ± thick, with thickened whitish margins and midrib, glabrous save for the margins and main nerve beneath or scabrid to pilose on both surfaces, the margins nearly always ciliate with short stiff hairs; stipule-sheath united with petioles, 5–6 mm. long, ridged, hairy or shortly pubescent with 5–6 ciliolate setae 0·2–1·2 cm. long. Flowers sessile in axillary and pseudoterminal (1–)3–6-flowered clusters at many of the upper nodes, the apical 2–3 often run together to form a short spike-like inflorescence; bracteoles 7 mm. long. Calyx-tube glabrous, oblong, 1·5 mm. long; lobes 2, linear-lanceolate to lanceolate, similar to the leaves in texture and indumentum, 0·3–2 cm. long, 0·8–1·8 mm. wide, ciliate. Corolla white or blue; tube cylindrical to narrowly funnel-shaped, 6–7(–11) mm. long, 3·5 mm. wide at the mouth, with a ring of hairs inside 2·5–3 mm. above the base; lobes oblong-lanceolate, 3–6 mm. long, 1·2–2·2 mm. wide. Filaments 1–3·5 mm. long. Style exserted 1·5–6 mm., finely papillate; stigma capitate, ± 7 mm. wide. Capsule subglobose, 3·5–5 mm. long, 2·5 mm. wide, crowned with the calyx-lobes, opening right out, the valves joined only at the base, almost circular in outline, convex, 3·5 mm. long, 4 mm. wide, glabrous, transversely wrinkled; septum persistent, shiny. Seeds dark brown to black, oblong-ellipsoid, 3–3·6 mm. long, 1·5 mm. wide, 1·3 mm. thick, shining, very finely reticulate, with a ventral groove and a conical yellowish basal appendage (elaeosome) 1·1–1·5 mm. long. Fig. 51/28, p. 344.

TANZANIA. Kigoma District: Ujiji, Bikare to Kigoma, 18 Mar. 1926, *Peter* 38901B!; Ufipa District: road to Lake Rukwa, Ilemba Gap, 12 Feb. 1959, *Richards* 11152!; Iringa District: Mt Mferu, 22 Mar. 1932, *Lynes* I.h. 263d!
DISTR. **T**4, 7; Zaire, Zambia
HAB. Ditches, grassland and woodland; 1050–1920 m.

SYN. *Dichrospermum congense* Bremek. in B.J.B.B. 22: 75 (1952)

NOTE. The possession of a seed appendage does not seem sufficient reason for erecting a new genus; traces of such appendages are present in other species of *Spermacoce* and the facies is exactly that of *Spermacoce*. Whether or not the appendage has any connection with dispersal by ants must await observations by a field observer.

29. **S. milnei** *Verdc.* in K.B. 30: 308 (1975). Type: Tanzania, Mbeya District, 1·5 km. up the Tukuyu road, about 17·5 km. SW. of Mbeya, *Milne-Redhead & Taylor* 10082 (K, holo.!)

Strictly erect annual herb 20–35 cm. tall; stems branched, 4-angled and grooved, bifariously shortly pubescent with stiff downwardly directed hairs within the grooves. Leaf-blades lanceolate, 2–7 cm. long, 2·5–6·5 mm. wide, acute at the apex, rounded at the base, sessile, with ± thickened margins and midnerve, ± pilose above, glabrous beneath save for some short stiff hairs on the midnerve, the margins ciliate with short stiff hairs particularly towards the base; stipule-sheath 3·5–4 mm. long, hairy, with 5 ciliolate setae 3–6 mm. long. Flowers sessile in axillary several-flowered clusters at all but the lowest nodes; bracteoles similar to stipules. Calyx-tube narrowly oblong-obconic, hairy above, 3 mm. long; lobes 3(–? 4), unequal, lanceolate, (2·5–)5–8 mm. long, (0·5–)1–1·6 mm. wide, similar to leaves in texture, etc., margins ciliate. Corolla white; tube tubular, 3 mm. long; lobes triangular, 1·1 mm. long, 0·8 mm. wide, glabrous. Filaments 0·25 mm. long. Style exserted 0·5 mm.; stigma globose, 0·8 mm. diameter. Capsule oblong-ellipsoid, 6 mm. long, 3 mm. wide, crowned with the persistent calyx-lobes, hairy above. Seeds shiny, dark brown, oblong-ellipsoid, 4 mm. long, 1·9 mm. wide, 1 mm. thick, flat and narrowly grooved ventrally, convex and very slightly keeled dorsally, with a cream appendage 1·5–1·8 mm. long. Fig. 51/29, p. 344.

TANZANIA. Mbeya District: 1·5 km. up the Tukuyu road, about 17·5 km. SW. of Mbeya, 12 May 1956, *Milne-Redhead & Taylor* 10082!
DISTR. **T**7; not known elsewhere
HAB. Overgrazed grassland; 1530 m.

30. **S. dibrachiata** *Oliv.* in Trans. Linn. Soc. 29: 87, t. 52 (1873); Hiern in F.T.A. 3: 239 (1877). Type: Tanzania, Bukoba District, Karagwe, *Grant* 439 (K, holo.!)

Annual or perhaps biennial very variable herb 6–75 cm. tall, with simple or sparsely branched stems covered with dense to sparse spreading often asperous hairs and shorter hairs within the longitudinal grooves between the longitudinal ribs; the basal internodes are often reduced giving the appearance of a rosette of leaves. Leaf-blades lanceolate, 4·5–12 cm. long, 0·35–1·5(–2·8) cm. wide, acute at the apex, narrowed at the base into the stipule-sheath, with long hairs on both surfaces or only on the midrib or glabrous except for the ± scabrid margins; stipules hairy, base 7–9 mm. long, 4·5 mm. wide at normal nodes, 0·8–2 cm. wide at the apical inflorescence-supporting node; fimbriae ± 5, 3–8 mm. long at lower nodes, 8–11, 2 mm. long at apical node. Flowers in very dense mostly many-flowered apical heads, closely subtended by 2–4 leafy bracts, the heads varying greatly in size, 1–2·5 cm. wide; bracteoles linear, 5–7 mm. long, joined at the base, colourless, glabrous beneath, greenish and ciliate above; rarely an inflorescence at node below the apical one. Calyx-tube obconic or ellipsoid, 1·7–5 mm. long, 2·8 mm. wide, glabrous; free limb 0·7–1·5 mm. long; lobes 4, linear-lanceolate, 3–5·5 mm. long, 0·8–1·8 mm. wide at base but strongly tapering, margins and sinuses scabridulous; some small intermediate lobes may also be present. Corolla blue or violet-blue or with white tube and blue to violet limb with a white eye; tube filiform-cylindrical, 1–2·1 cm. long; lobes oblong-elliptic, 2·5–6 mm. long, 1·5–3 mm. wide, with a few hairs at the apex outside. Filaments exserted 2–4 mm. Style exserted 2–4·5 mm.; stigma capitate, 1 mm. wide, slightly bifid, the lobes ± 0·6 mm. long. Capsule straw-coloured with numerous red longitudinal streak-like marks, 3·5 mm. long, 3 mm. wide, 1·3 mm. thick, glabrous. Seeds dark brown, ellipsoid, 2·6 mm. long, 1·3 mm. wide, 0·95 mm. thick, finely reticulate and rugulose, with a ventral narrow groove which widens at base and apex. Fig. 51/30, p. 344.

Tanzania. Bukoba District: Kabirizi, Oct. 1931, *Haarer* 2235!; Mbulu District: Mbulumbulu, 16 July 1943, *Greenway* 6806!; Mbeya District: N. Usafwa Forest Reserve, May 1951, *Eggeling* 6118!

Distr. **T** 1–8; Zaire, Rwanda, Burundi, Mozambique, Malawi, Zambia, Rhodesia and Angola

Hab. Grassland, grassland with scattered shrubs, grassy clearings in *Brachystegia*, *Combretum* woodland, also by marshes and in cultivations; 840–2250 m.

Syn. *Borreria dibrachiata* (Oliv.) K. Schum. in E. & P. Pf. IV.4: 144 (1891)
Pentas involucrata Bak. in K.B. 1895: 66 (1895). Type: Zambia, Lake Tanganyika, Fwambo, *Carson* 40 of 1894 coll. (K, holo.!)

Note. A sheet from "Basutoland" in BM must I think be wrongly localized.

31. **S. phyteumoides** *Verdc.* in K.B. 30: 308, fig. 9 (1975). Type: Zambia, Mbala [Abercorn], *Bullock* 2636 (K, holo.!, EA, iso.)

Suberect or ± prostrate annual or perennial herb 15–40(–60) cm. tall; stems dark, covered with spreading white hairs. Leaf-blades elliptic-lanceolate, 2–7(–14) cm. long, 2·5–10(–13) mm. wide, acute at the apex, narrowed at the base into the stipule-sheath, glabrous on both surfaces save for hairs on the midrib beneath (but even these may be absent) or densely scabrid-hairy on both surfaces; margins always scabrid with short stiff hairs; stipule-bases very hairy, 2–3 mm. long, with 3–5 purple fimbriae 2·5–4·5 mm. long; stipules supporting the leaves which subtend the inflorescences with 7 fimbriae of varying thickness ± 5–6 mm. long and colleters 1·5 mm. long. Flowers in terminal heads very similar to those in species 30; bracteoles linear, 3–4 mm. long, ciliolate, hyaline. Calyx-tube oblong to obconic, 1–3 mm. long, without a free limb; lobes 2, linear to lanceolate, 3·5–5 mm. long, 0·6–1·3 mm. wide, with hairy margins. Corolla white or usually deep blue; tube 6–7 mm. long; lobes ovate-oblong, 2–4·5 mm. long, 1·2–2 mm. wide, with short to long hairs at the apex outside. Filaments exserted 2·5 mm. Style exserted 4 mm.; stigma 1·5 mm. wide. Capsule straw-coloured, streaked with brown or purplish, oblong-ellipsoid or obovoid, 2·5–3 mm. long, 1·5–2 mm. wide, 1 mm. thick, glabrous beneath, ± hairy at apex, crowned with the remains of the calyx-lobes. Seeds chestnut-brown, narrowly oblong-ellipsoidal, 2·5 mm. long, 1 mm. wide, 0·6 mm. thick, dorsally convex, ventrally sulcate, rugulose.

var. **phyteumoides**

Corolla white or very pale, rarely dark blue; mostly annual. Fig. 51/31, p. 344.

Tanzania. Mpanda District: Katabi Game Reserve, Starike, 19 Mar. 1973, *Ludanga* in *M.R.C.* 1479!; Ufipa Plateau, 27 Feb. 1957, *Whellan* 1205!; Songea District: Songea to Manda, Likuyu R., 31 Aug. 1930, *Migeod* 854!

Distr. **T**4, 8; Zaire (Katanga), Zambia

Hab. Wooded grassland; 1800 m.

SYN. [*Borreria phyteuma* sensu R. E. Fries, Wiss. Ergebn. Schwed. Rhod.-Kongo-Exped. 1911–12, Bot. Erg.: 17 (1921), *non* (Hiern) K. Schum.]
[*B. dibrachiata* sensu Hutch., Botanist in S. Afr.: 504 (1946), *non* (Oliv.) K. Schum.]

NOTE. Although bearing a close general resemblance to *S. phyteuma* Hiern the calyx is quite different in the two species, that of *S. phyteuma* having 4 equal lobes. Moreover the fruit of *S. phyteuma* shows it belongs to section *Arbulocarpus*.

var. **caerulea** *Verdc.* in K.B. 30: 311 (1975). Type: Tanzania, Ufipa, Sumbawanga, path to Mbisi Forest, *Richards* 8677 (K, holo.!)

Corolla deep azure blue; probably perennial.

TANZANIA. Ufipa District: Sumbawanga, path to Mbisi Forest, 13 Mar. 1957, *Richards* 8677! & Sumbawanga, Malonje Plateau, 4 Mar. 1957, *Richards* 8431! & Malonje Plateau, Nsanga Mt., 13 Mar. 1959, *Richards* 12136!
DISTR. **T4**; Zambia
HAB. Coarse grassland, edges of cultivations, rock crevices in rocky outcrops; 2100–2250 m.

NOTE. The character of flower colour is not constant, *Migeod* 854 having dark blue flowers, but the species is common in Zambia and the specimens I have separated as var. *caerulea* have a distinctly different facies. A specimen from Mbeya District, Mbosi, Feb. 1935, *Horsbrugh-Porter*, has the corolla-tube 1.1 cm. long and may perhaps be a third variant or an extreme form of var. *phyteumoides*.

32. **S. azurea** *Verdc.* in K.B. 30: 312, fig. 10 (1975). Type: Tanzania, Ufipa District, Sumbawanga, Mbisi Forest, *Richards* 8696 (K, holo.!)

Erect herb with several strict stems from a woody rootstock; stems 15–30 cm. tall, pubescent or ± densely covered with spreading white hairs which are thickened at the base. Leaf-blades linear to linear-lanceolate, 2–5·8 cm. long, 1·5–6 mm. wide, acute at the apex, slightly narrowed to the base, ± scabrid with forwardly directed hairs on both surfaces; free petiole absent; stipule-sheath 4–5 mm. long, hairy, with 3 fimbriae 3–6·5 mm. long or 5 fimbriae in the case of the flower-bearing leaves. Flowers borne in sessile clusters at the apex and sometimes at the node beneath; leafy bracts expanded at the base into the cup-like stipule-sheath; stipule-like bracts ± 8 mm. long, the base ± 4 mm. long, with a broad triangular central lobe and 2 lanceolate lobes on either side or reduced to a collection of fimbriae 4–5 mm. long. Calyx-tube obovoid, 3 mm. long, hairy at the apex; lobes 2, triangular at base, drawn out into filiform tips, 3 mm. long, hairy, or 0, reduced to a hairy limb ± 1 mm. high. Corolla deep blue; tube funnel-shaped, 5·5 mm. long, 4·5 mm. wide, hairy or glabrescent outside, strongly narrowed at the extreme base with a ring of hairs inside at the base; lobes triangular, 5 mm. long, 3 mm. wide, densely hairy outside. Filaments exserted 2 mm. Style exserted ± 4 mm. with some sparse hairs near the apex; stigma biglobose, 1·2 mm. wide. Capsule not seen. Fig. 51/32, p. 344, and 54.

TANZANIA. Ufipa District: Sumbawanga–Mbisi road, 25 Nov. 1958, *Napper* 1116! & top of Mbisi Forest, 13 Mar. 1957, *Richards* 8696! & track to Mbisi Forest, 16 Mar. 1959, *Richards* 12162!
DISTR. **T4**; not known elsewhere
HAB. Coarse upland grassland, grassland with scattered *Protea*, etc., often in rough or rocky places; 1650–2400 m.

33. **S. sp. E**

Unbranched annual herb 20 cm. tall; stem slender, leafless below, glabrous. Leaf-blades narrowly elliptic, 3–4 cm. long, 5–8 mm. wide, acute at the apex, cuneate at the base, glabrous; petiole 2 mm. long; stipule-base 2·5 mm. long, with ± 7 filiform fimbriae 3–7 mm. long. Flowers not seen, possibly solitary or very few in the axils. Capsules solitary in the axils (? always, only one

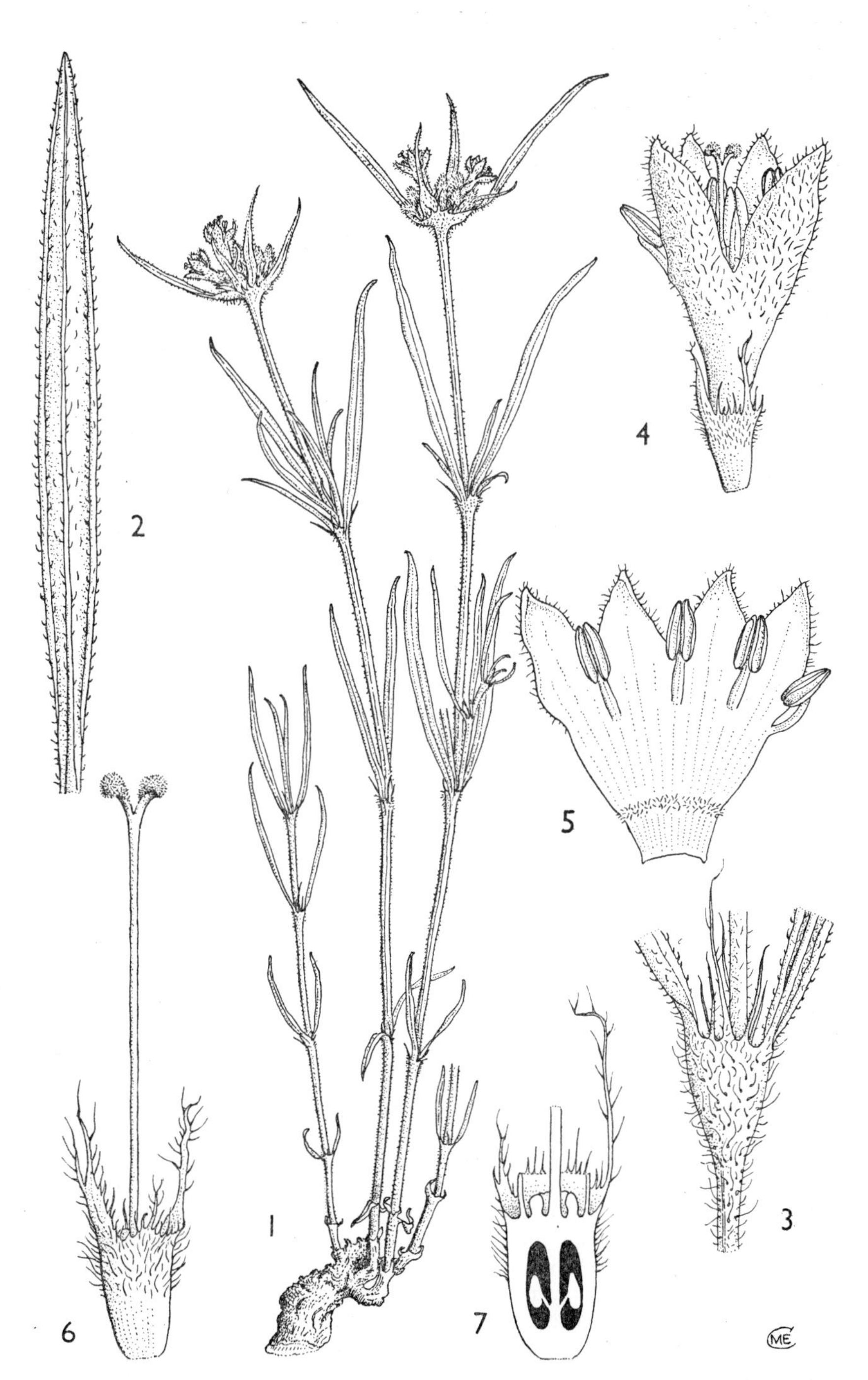

FIG. 54. *SPERMACOCE AZUREA*—**1,** habit, × $\frac{2}{3}$; **2,** lower surface of leaf, × 2; **3,** node showing stipule, × 4; **4,** flower, × 4; **5,** corolla, opened out, × 4; **6,** flower, with corolla removed, × 8; **7,** longitudinal section of ovary, × 10. All from *Richards* 12162. Drawn by Mrs M. E. Church.

seen), ellipsoid, 4 mm. long, 3 mm. wide, pubescent, crowned with the 4 lanceolate calyx-lobes, 2 mm. long.

TANZANIA. Mpwapwa District: Kongwa Ranch, 18 Feb. 1966, *Leippert* 6296!
DISTR. **T5**; not known elsewhere
HAB. Thicket; 1300 m.

NOTE. I have been unable to satisfactorily identify this with any species but better flowering material may enable it to be placed. It is not included in the key.

Species of Uncertain Position

Borreria pedicellata *K. Schum.* in E.J. 28: 111 (1899). Types: Kenya, Lamu District, Wange, *Tiede* 9 & 11 (B, syn. †)

Annual erect herb with strictly branched stems 35–40 cm. long, the epidermis greenish or brownish; stems obtusely 4-angled, alternately bisulcate, the margins of the sulci ciliolate, the rest glabrous. Leaves sessile, lanceolate or linear-lanceolate, 3–7 cm. long, 6–10 mm. wide at the middle, acute at the apex, attenuate at the base, margin white, spinulose-serrulate, scabrid, with 3 nerves on either side of the midrib; stipules sheathing, setose-lacerate, minutely pilosulose, 6–7 mm. long of which more than half is taken up by the longest of the 5 bristles. Flowers 4-merous, 3–4 in the axils of the leaves, distinctly pedicellate, the pedicels 2–3 mm. long. Calyx-tube 1 mm. long, glabrous; lobes 2, ovate-lanceolate, 2 mm. long. Corolla white; tube very slender, 9 mm. long; lobes reflexed, 2 mm. long. Anthers 1 mm. long, exserted 2 mm. from the tube. Style exserted; stigma capitate, bilobed. Capsule yellow-grey, 3 mm. long, hairy above. Seeds brown, shining, 2·5 mm. long.

KENYA. Lamu District: Wange, *Tiede* 9 & 11
DISTR. **K7**; not known elsewhere
HAB. Sandy cliffs (Sandklippen)

NOTE. This must be related to *Spermacoce filituba* (K. Schum.) Verdc. and might perhaps be an abnormal variant. The description above is translated more or less directly from the original.

Borreria stenophylla *K. Krause* in E.J. 43: 158 (1909). Type: Tanzania, Mwanza District, W. Ukerewe, *Uhlig* V.10 (B, holo. †)

Probably perennial herb, erect, branched from the base, 45 cm. tall; stems simple or sparsely branched, subangular, grey-green and shortly pilose above, glabrous and brownish beneath, ± compressed and bisulcate when young. Leaves sessile, narrowly linear or linear-lanceolate, 3–5 cm. long, 1–2 mm. wide, acute at the apex, narrowed towards the base, sparsely scabrid on both surfaces; stipules 8 mm. long, joined into a sheath, setaceous dilacerate. Flowers few, subsessile in the axils; bracteoles filiform. Calyx-tube elongate-turbinate, 1 mm. long; lobes (number not mentioned) subulate, 3 mm. long, very acute, ciliolate. Corolla white; tube cylindrical, 3–3·5 mm. long, dilated towards the apex (yet in diagnosis Krause states "slightly shorter than the calyx-lobes"); lobes ovate-oblong, 1·5 mm. long, subacute. Anthers 0·6 mm. long; filaments very short, inserted just below the throat. Style tenuous, 5 mm. long, shortly bifid at the apex.

TANZANIA. Mwanza District: W. Ukerewe, Apr. 1904, *Uhlig* V.10
DISTR. **T1**
HAB. "High grass steppe"

NOTE. No material of this is extant. From the description it appears that it could be *Spermacoce subvulgata* but the description of the stipules does not agree at all.

39. MITRACARPUS

Zucc. in Schultes & Schultes, Mant. Pl. 3: 210 (1827), in obs.; Anderson in Taxon 20: 643 (1971)*

Erect or prostrate annual or perennial herbs with 4-angled stems. Stipules connate with the petioles to form a fimbriated sheath. Leaves opposite; blades linear-lanceolate to ovate or broadly elliptic. Flowers not heterostylous or only slightly so, in dense spherical sessile terminal or axillary heads. Calyx-tube obconic, obovoid or subglobose; teeth 4–5, 2 often longer, sometimes with minute supplementary ones between. Corolla salver-shaped or funnel-shaped, the tube often with an internal ring of hairs; lobes 4; throat glabrous or hairy. Stamens inserted in the throat, anthers included or exserted. Disc fleshy. Style short or long, divided into 2 short linear branches. Ovary 2(–3)-locular; ovules solitary in each locule, attached by the middle to the septum. Fruit a thin circumscissile capsule, the upper part splitting off together with the calyx-lobes, the septum persistent. Seeds oblong or globose, the ventral face divided into 4 distinct areas; endosperm fleshy.

A genus of about 30–40 species mostly confined to tropical America, but 1 species now common throughout tropical Africa (see note at end of species) and also seen from Burma, Selangor, New Guinea, and Marianas Is.

M. villosus (*Sw.*) *DC.*,** Prodr. 4: 572 (1830); Rendle in Fl. Jam. 7: 127, fig. 39 (1936); Gooding, Loveless & Proctor, Fl. Barbados: 405 (1965); Adams, Fl. Pl. Jam.: 733 (1972); Verdc. in K.B. 30: 317–322 (1975). Type: Jamaica, *Swartz* (S, holo.!, BM, iso.!)

Erect or spreading annual herb (5–)9–40 cm. tall, with unbranched or sparsely to much-branched stems; branchlets pubescent with short curled ± adpressed hairs and often with spreading ones as well, the older with epidermis eventually peeling; sometimes quite woody at the base. Leaf-blades elliptic, 1–6 cm. long, 0·3–2·3 cm. wide, subacute at the apex, cuneate at the base, glabrescent to scabrid pubescent above, glabrescent or glabrous beneath save for hairs on the main nerves; margins often scabrid; petiole ± 1 mm. long, often densely pubescent and with ciliate margins; stipule-sheath 1–3 mm. long, divided into 6–9(–15) often colleter-tipped fimbriae, 1–5 mm. long, ciliate. Inflorescences numerous, present in most axils, subglobose, (0·5–)0·8–1·8 cm. in diameter; flowers sessile or almost so; bracteoles filamentous, white, 1–2 mm. long. Calyx-tube 1–1·4 mm. long; limb-tube 0·15–0·4 mm. long; lobes 4, 2 oblong-lanceolate, green with hyaline margins, rather thick, 1·3–2·3(–3) mm. long, and 2 hyaline, triangular-lanceolate, 0·55–1·5 mm. long, narrower than the others, all with usually ciliate margins and often hairy below. Corolla white, glabrous or slightly hairy outside; tube 1·4–1·9 mm. long; lobes ovate, 0·6–1 mm. long, 0·3–0·9 mm. wide. Flowers showing very slight heterostyly, the anthers varying in their degree of exsertion; style 1·1–1·6 mm. long; stigma 0·3–0·5 mm. long. Capsule straw-coloured, ± 1 mm. long and wide. Seeds pale yellow-brown, compressed ellipsoid-rectangular, of very characteristic appearance (see fig. 55/10), dorsally resembling a rectangle with a square portion removed from each

* The gender of this was commented on long ago by K. Schumann in Fl. Brasiliensis and N. Sandwith (pencil adnot. in Index Kewensis); in the original description the generic name occurs in the accusative as object of communicavit but is masculine in the index to the same work.

** Combination often given as (Sw.) Cham. & Schlecht. in Linnaea 3: 363 (1828), but not made properly there.

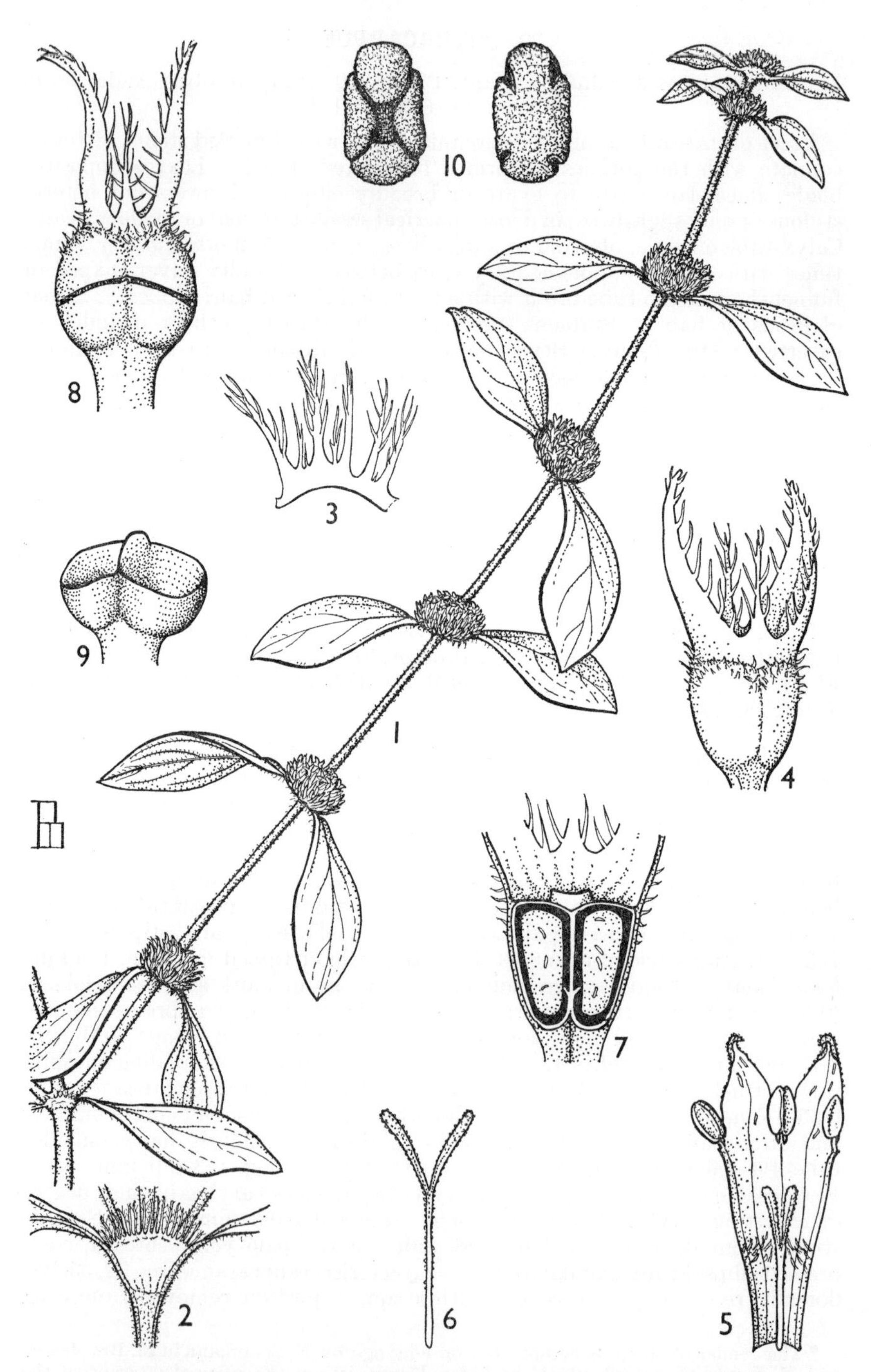

FIG. 55. *MITRACARPUS VILLOSUS*—**1**, habit, × ⅔; **2**, node showing stipule, × 3; **3**, bracteole, × 6; **4**, calyx, × 14; **5**, longitudinal section of short-styled flower, × 14; **6**, style from long-styled flower, × 14; **7**, longitudinal section of ovary, × 20; **8**, capsule, × 14; **9**, lower part of dehisced capsule, × 14; **10**, seed, two views, × 20. 1, 2, 3–5, 7–9, from *Chancellor* 24; 6, 10, from *Haarer* 2084. Drawn by Diane Bridson.

corner, ventrally separated into 4 distinct areas by 4 impressed lines radiating from the hilum, 0·8 mm. long, 0·5 mm. wide, rugulose and reticulate. Fig. 55.

UGANDA. W. Nile District: Maracha, 23 July 1953, *Chancellor* 24!; Teso District: Serere, Dec. 1931, *Chandler* 261!; Mengo District: near Entebbe, Bukiberu, 9 May 1969, *Lye* 2823!

KENYA. N. Kavirondo District: near Uganda border, Bungoma, Nov. 1961, *Tweedie* 2256!; Central Kavirondo District: about 6·4 km. N. of Kisumu, Kisumu–Mutet road, Escarpment, 27 Sept. 1953, *Drummond & Hemsley* 4486!

TANZANIA. Mwanza, 30 May 1931, *B. D. Burtt* 2482!; Tanga District: E. Usambara foothills, Mt. Mlinga, 17 Aug. 1950, *Verdcourt* 316!; Uzaramo District: Vikindu Forest Reserve, Aug. 1953, *Semsei* 1343!

DISTR. **U1–4; K5, 7; T1**, 3–6, 8; widespread in tropical Africa from Mauritania to Angola, Zaire & Central African Repuplic, Sudan, Malawi, Zambia, Seychelles and Cape Verde Is., also in India, Burma, Selangor, New Guinea and Marianas Is., West Indies and tropical S. America

HAB. Weed in cultivations, pathsides, fallowland, also in secondary scrub and thicket, open dry rocky areas, etc.; 0–1500 m.

SYN. *Spermacoce hirta* L., Sp. Pl., ed. 2: 148 (1762); Sw., Obs. Bot.: 45 (1791). Type: Jamaica, *P. Browne* (LINN, holo.!)

S. villosa Sw., Prodr. Veg. Ind. Occ.: 29 (1788)

Mitracarpus scaber Zucc. in Schultes & Schultes, Mant. Syst. Veg. 3: 210, 399 (1827); Hiern in F.T.A. 3: 243 (1877); Hepper, F.W.T.A., ed. 2, 2: 222 (1963). Type: ? country, Forte Louis, ? collector (not found)

Staurospermum verticillatum Schumach. & Thonn., Beskr. Guin. Pl.: 73 (1827); Type: Ghana [Guinea], *Thonning* (C, holo.)

Mitracarpus senegalensis DC., Prodr. 4: 572 (1839); Oliv. in Trans. Linn. Soc. 29: 89 (1873), *nom superfl.*, based on *Staurospermum verticillatum*

M. verticillatus (Schumach. & Thonn.) Vatke in Linnaea 40: 196 (1876); K. Schum. in P.O.A. C: 394 (1895); F.W.T.A. 2: 136 (1931); Verdc. in K.B. 7: 360 (1952); Sebastine & Ramamurthy in Bull. Bot. Surv. India 9: 291 (1968); U.K.W.F.: 409 (1974)

[*M. hirtus* sensu K. Schum. in Martius, Fl. Bras. 6(6): 84 (1888) & in E. & P. Pf. IV.4: 142, fig. 46/U (1891); Standley in Field Mus. Publ. Bot. 7: 157 (1930), 331 (1931) & 473 (1931), *non* (Sw.) DC.]

NOTE. This has always been considered a native African species, but I have been unable to separate it from *M. villosus*; it makes good sense that this sole African example of a genus widespread in tropical and subtropical America should have been introduced long ago (it has been known in Africa since the early part of the nineteenth century), probably from the West Indies. I am following Mr Dandy's advice on the nomenclature of this species. Those using the name *M. villosus* usually claim that there is a bar to using the Linnean epithet *hirtus* since *M. hirtus* (Sw.) DC. is different, but the reference to Swartz is to a paper where Swartz divides *Spermacoce hirta* into two. He refers directly back to Reichard, Syst. Pl. 291. 3 where the references are to Syst. Veg., ed. 13 (Murray): 124 (1774) and to Browne Jam.: 141, i.e. directly back to Linnaeus's original description. Under his treatment of *M. villosus* (Sw.) DC., DeCandolle does say "and almost certainly the same as *Spermacoce hirta* L." which must be taken as an exclusion of this from consideration as the true basionym of *M. hirtus* (Sw.) DC. A full account of this problem is given in my paper in the K.B.

40. RICHARDIA

L., Sp. Pl.: 330 (1753) & Gen. Pl., ed. 5: 153 (1754)

Richardsonia Kunth in Mém. Mus. Paris 4: 43 (1818)

Annual or perennial erect or prostrate hairy herbs. Leaves opposite, sessile or shortly petiolate, mostly ovate, elliptic or oblong, the lateral nerves evident; stipule-sheath connate with the petioles, bearing several fimbriae. Flowers hermaphrodite or said sometimes to be polygamo-dioecious, not heterostylous, small, in dense terminal heads enclosed in an involucre formed of (2–)4 leaves which are several-nerved from the base. Calyx-tube turbinate or subglobose, the limb deeply lobed; lobes 4–8, lanceolate, ovate or subulate, persistent. Corolla shortly funnel-shaped; lobes 3–5, ovate to lanceolate; throat glabrous but tube with a narrow area of hairs inside near the base.

Stamens with anthers exserted. Style filiform; stigmas 3–4, linear or spathulate, exserted. Ovary 3–4-locular; ovules solitary, affixed to the middle of the septum. Capsule 3–4-coccous, crowned by the persistent calyx, eventually splitting into separate cocci, sometimes leaving a very small persistent axis. Cocci mostly obovoid, smooth or more often muricate or papillose. Seeds oblong-ellipsoid or obovoid, dorsally convex, ventrally with 2 grooves; endosperm corneous.

A small genus of 15 species in Central and S. America but several now widely spread throughout the tropics and subtropics.

Cocci with inner face having smooth depressed area (i.e. actual septa of ovary) almost as broad as the face; leaf-blades with upper surface hairy . . 1. *R. brasiliensis*

Cocci with inner face having smooth depressed area very narrow; leaf-blades with upper surfaces practically glabrous save near the margins . . 2. *R. scabra*

1. **R. brasiliensis** *Gomes**, Mem. Ipecac.: 31, t. 2 (1801); Rendle in Fl. Jam. 7: 129, fig. 40 (1936); Hepper, F.W.T.A., ed. 2, 2: 216 (1963); Lewis & Oliver in Brittonia 26: 276 (1974); U.K.W.F.: 407, fig. (1974). Type: Brazil, Rio de Janeiro, *Gomes* (LISU, holo.)

Perennial (or ? annual) prostrate herb, often forming a mat, from a central taproot; stems 7–40 cm. long, densely covered with spreading hairs. Leaf-blades elliptic, 1–6·5 cm. long, 0·4–2·7 cm. wide, acute or subacute at the apex, very narrowly attenuated to the base, the apparent petiole up to 1·5 cm. long, mostly with short hairs all over the upper surface and on the margins and nerves beneath; basal narrowed part with longer hairs; stipular sheath 1–3·5 mm. long, with 3–5 fimbriae, 1–4 mm. long, usually with long hairs. Inflorescences 0·7–1·2 cm. in diameter; bracts ovate-elliptic, rounded at the base, the long ones 1·5–3·5 cm. long, 0·65–2 cm. wide, the short ones 1–1·7 cm. long, 4–9·5 mm. wide, or sometimes lacking, with a similar indumentum to that on the leaves, save that there are much longer hairs towards the base; basal part of bracts often subhyaline. Calyx-tube 1·2–1·7 mm. long; lobes 5–6, ovate-triangular, 1–1·5 mm. long, 0·3–1 mm. wide, the margins conspicuously ciliate, basal united part of the limb 0·5–1 mm. long. Corolla white, 2·7–3·2 mm. long; lobes 4–6, 1–1·4 mm. long, 0·5–0·8 mm. wide. Style 3–4 mm. long, the branches 0·2–0·5 mm. long; stigmas spathulate, 0·2–0·3 mm. long. Cocci brown, oblong-obovoid, 2–2·6 mm. long, 1·4–1·6 mm. wide, inner face with smooth depressed area (i.e. actual septum of ovary) almost as broad as the face, dorsal face covered with short flat hairs which are longer in the middle or a mixture of papilla-like hairs and longer hairs. Seeds brown, compressed oblong-obovoid, 2·5 mm. long, 1·8 mm. wide, with ventral face broadly grooved, with 2 short basal projections 0·1 mm. long. Fig. 56/9.

UGANDA. Mengo District: Mabira, 10 Sept. 1961, *Rose* 224! & Lugazi, 5 May 1971, *G.B.A. Ltd* A.1!

KENYA. Trans-Nzoia District: Kitale, Aug. 1964, *Tweedie* 2875!; Nairobi District: Thika Road House, 29 Apr. 1951, *Verdcourt* 492! & Nairobi, Riverside, 15 Apr. 1945, *Bally* 4402!

TANZANIA. Lushoto District: 6.4 km. NE. of Lushoto, Mkuzi, 10 Apr. 1953, *Drummond & Hemsley* 2070! & Korogwe, Magunga Estate, 12 May 1952, *Faulkner* 954! & Lushoto, Jaegertal, 11 Jan. 1966, *Archbold* 597!; Zanzibar I., Mazizini [Massazine], 2 Aug. 1959, *Faulkner* 2320!

DISTR. **U4**; **K3**, 4; **T3**; 6; **Z**; a South American species now widely naturalized, specimens having been seen from Mauritius, Hong Kong, S. India, Ceylon, Java, Australia,

* Usually misspelt Gomez.

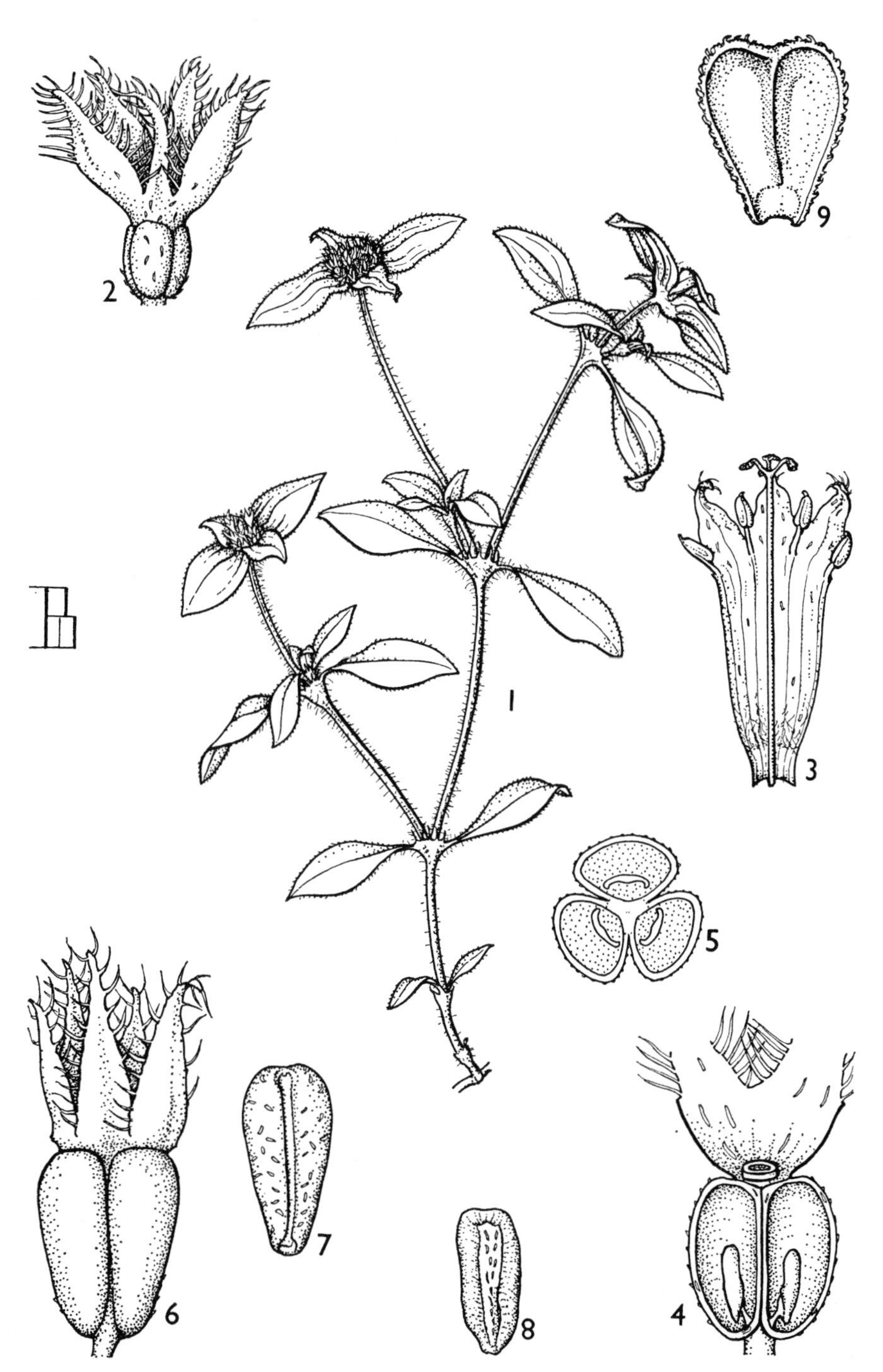

FIG. 56. *RICHARDIA SCABRA*—**1,** habit, × $\frac{2}{3}$; **2,** calyx, × 8; **3,** longitudinal section of flower, × 8; **4,** longitudinal section of ovary, × 14; **5,** transverse section of ovary, × 14; **6,** capsule, × 10; **7,** coccus, view of inner face, × 10; **8,** seed, ventral view, × 10. *R. BRASILIENSIS*—**9,** coccus, view of inner face, × 10. 1–8, from *Drummond & Hemsley* 2888; 9, from *Drummond & Hemsley* 2070. Drawn by Diane Bridson.

Hawaii, Jamaica, U.S.A. and Mexico; in Africa it is known from Ghana, Nigeria, Mozambique, Malawi, Zambia, Rhodesia and South Africa (Natal and Transvaal)
HAB. Gardens, roadsides, cleared areas in forest, grassland, etc.; 20–2000 m.

SYN. *Richardsonia brasiliensis* (Gomes) Hayne, Arzn. Gew. 8, t. 21 (1822); K. Schum. in Martius, Fl. Bras. 6 (6): 94 (1888)
[*Richardia scabra* sensu Hiern, F.T.A. 3: 242 (1877) et auctt. mult. *non* L.]

2. **R. scabra** *L.*, Sp. Pl.: 330 (1753); Lewis & Oliver in Brittonia 26: 282 (1974). Type: Mexico, Vera Cruz, *Houstoun* (LINN. 451.1, holo. fide Lewis & Oliver, BM, iso.!)

Very similar to previous species, with erect or ascending stems 5–55 cm. long. Leaf-blades (0·9–)1·3–5·5 cm. long, 0·4–2·3 cm. wide, the apparent petiole 0·2–1·2 cm. long, with short rather stiff conical hairs on the margins and marginal areas above and also on nerves beneath, but most of the upper surface is ± glabrous; stipule-sheath 2–4 mm. long, with 3–7 fimbriae 1–4 mm. long. Inflorescences 0·65–1·6 cm. in diameter; long bracts 1–2·7 cm. long, 0·6–1·5 cm. wide; short bracts 0·5–1·5 cm. long, 3–9 mm. wide. Calyx-lobes usually 6, triangular to oblong-lanceolate, 1·8–2·3 mm. long, 1 mm. wide, basal united part of the limb ± 1 mm. long. Corolla white or pale pink or mauve or only tips of lobes tinged pink; tube 5·8–6·3 mm. long; lobes ovate-triangular, 2·5 mm. long, 1·5 mm. wide, sometimes with a few hairs at the apex. Style 7·5 mm. long, the 3 short branches 0·7 mm. long; stigmas spathulate, 0·7 mm. long. Cocci grey-brown, 2·5–2·8 mm. long, 1·5 mm. wide; inner face with smooth depressed area very narrow, other parts particularly the dorsal area densely verrucose or less often practically or entirely smooth. Seeds purplish brown, almost oblong, ± 2·5 mm. long and 1·3 mm. wide; hilum very narrow. Fig. 56/1–8.

TANZANIA. Lushoto District: 6·4 km. NE. of Lushoto, Mkuzi, 11 June 1953, *Drummond & Hemsley* 2888! & Amani, 2 Jan. 1950, *Verdcourt* 14!; Bagamoyo District: Kidomole, Apr. 1964, *Semsei* 3787!
DISTR. T2–4, 6; Rhodesia, South Africa, S. and Central America, Jamaica, Cuba and U.S.A.
HAB. Cultivations, newly cleared ground, roadsides on red soil in forest clearings, sandy grassland and bushland; 0–1600 m.

SYN. *Richardsonia scabra* (L.) St.-Hil., Pl. Usuelles Brasiliens, t. 8 (1824) (as to name only, the description and figure refer to *R. brasiliensis* which St. Hilaire included in synonymy of *R. scabra*)

NOTE. Miss Archbold, who collected both species at Lushoto, says that *R. scabra* has larger flowers, darker green leaves, more reddish stems and that the sepals are more erect after the corolla falls whereas in *R. brasiliensis* the fruiting sepals are reflexed, adpressed to the coccus and star-like in plan view.

Tribe 11. **RUBIEAE**

Leaf-blades ovate to lanceolate, the petiole very well developed; corolla mostly 5-merous . . . 41. **Rubia**
Leaf-blades mostly narrow, linear or lanceolate or, if wider and ± elliptic then petiole very short; corolla mostly 4-merous . . . . . . 42. **Galium**

41. **RUBIA**

L., Sp. Pl.: 109 (1753) & Gen. Pl., ed. 5: 47 (1754)

Herbs, the stems sometimes woody towards the base, often stiffly hairy or with tiny prickles. Leaves in whorls of 4–6(–8), some of which may be stipular in origin; blades linear, lanceolate, obovate or less often cordate, often prickly. Flowers small, in axillary and/or terminal cymes, hermaphrodite. Calyx-tube ovoid or globose, the limb obsolete. Corolla rotate or subcampanulate, (4–)5(–6)-merous. Stamens inserted in the corolla-tube, the anthers exserted. Disc small, swollen. Ovary 1–2-locular; ovules solitary in each locule, erect, affixed to the septum, amphitropous; style 2-fid or styles 2, short; stigmas capitate. Fruit fleshy, of 2 parts or rarely 1-locular. Seeds suberect, with plane or concave ventral face, horny endosperm and membranous testa, adhering to the pericarp.

A genus estimated at 60 species but they are extremely difficult and critical and this is doubtless an overestimate; they occur in Europe, N. Africa, Asia (particularly the Far East) and Africa but are rare or absent elsewhere and apparently not in America although some species there have previously been referred to the genus. *Rubia* scarcely differs from *Galium*, neither the fleshy/dry fruits, nor 4/5-mery distinctions being constant, but it is scarcely feasible to merge such well-known European genera whilst even the feeblest distinction remains; they are likely to be kept separate for convenience. Only one variable species occurs in the Flora area.

R. cordifolia *L.*, Syst. Nat., ed. 12, 3 (app.): 229 (1768) & in Mantissa Altera: 197 (1771); DC., Prodr. 4: 588 (1830); Sond. in Harv. & Sond., Fl. Cap. 3: 35 (1864); Hiern in F.T.A. 3: 244 (1877); Hook., Fl. Brit. Ind. 3: 202 (1881); K. Schum. in P.O.A. C: 395 (1895); W.F.K.: 61, fig. 51 (1948); Pojark. in Fl. U.R.S.S. 23: 387 (1958); F.P.U., ed. 2: 166, fig. 107 (1972); U.K.W.F.: 412, fig. (1974). Type: Majorca, *L. Gérard* (ubi ?)*

Scrambling, climbing or creeping plant, less often tussock-forming, with branched stems 0·3–6 m. long; stems brittle, with quite strong curved prickles on the 4 ribs and either pubescent all over or at least with hairs beneath the nodes; roots usually quite woody, containing a red dye. Leaves in whorls of 4(–8) or sometimes paired; blades lanceolate to broadly ovate, 0·7–8·5 cm. long, 0·2–4·2 cm. wide, acuminate at the apex, rounded to cordate at the base, scabrid pubescent above, glabrescent to often densely covered with long white hairs beneath at least when young, margins often with curved prickles, 5–7-nerved from the base, the nerves sometimes with scattered curved prickles beneath; petioles 0·8–1·5 cm. long, with recurved prickles and often pubescent as well. Inflorescences usually numerous, scattered along the stem, fairly lax to rather dense, 0·5–2·5 cm. long; peduncles 1–2·5 cm. long; pedicels 0·2–6 mm. long; bracts elliptic, 1·2–1·5 mm. long, 0·3–0·4 mm. wide. Flowers glabrous. Calyx-tube (0·25–)0·5–0·8 mm. long and 0·8–1·4 mm. wide. Corolla yellowish, green, greenish cream or greenish yellow, often pink- or purple-tipped in bud, 4–6 mm. wide; tube 0·2–0·8 mm. long; lobes usually triangular, 1·5–3 mm. long, 0·6–1·3 mm. wide, apiculate, margins minutely papillate. Fruit glabrous, brownish black; lobes globose, 2·5–5 mm. in diameter; pyrenes globose, 3 mm. in diameter. Seeds globose, 1·2–2·9 mm. in diameter.

subsp. **conotricha** (*Gandoger*) *Verdc.* in K.B. 30: 323 (1975). Type: South Africa, E. Griqualand, near R. Umzimhlava [Umzimklowa], *Schlechter* 6550 (LY, holo., K, iso. !)

Leaves in whorls of 4; blades often hairy beneath. Fruits with cocci not exceeding 3 mm. in diameter in fresh material. Fig. 57, p. 382.

* I have been unable to trace this specimen or to prove that Louis Gérard ever visited Majorca although it is known that he corresponded with Linnaeus and sent plants from Provence. The species is not recorded for the Balearic Is. in Knocke's flora of the area nor is there any material of the species from these islands in any herbarium consulted by Mrs. Lorna Ferguson who has made a detailed study of their flora. It may be an error or the plant may have become extinct at an early date. There is no specimen of *Rubia cordifolia* in Gérard's herbarium preserved in the library at Draguignan.

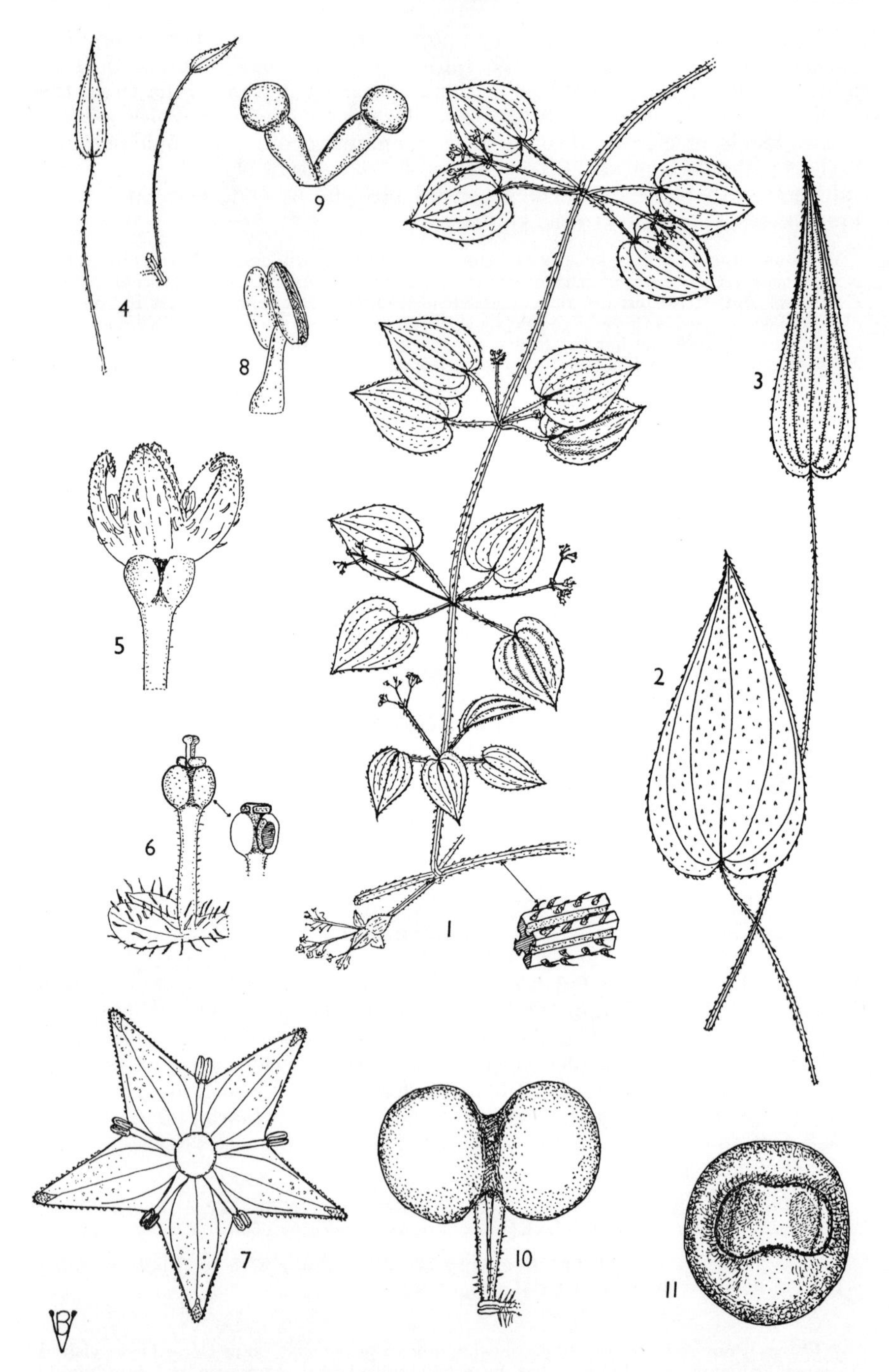

FIG. 57. *RUBIA CORDIFOLIA* subsp. *CONOTRICHA*—**1,** flowering branch, × ½, with enlargement of stem, × 4; **2, 3,** large leaves, × ½; **4,** small leaves, × ½; **5,** young flower, lateral view, × 7; **6,** young flower, with corolla removed, × 5; **7,** small flower, flattened out, × 10; **8,** stamen, lateral view, × 20; **9,** styles and stigmas, × 20; **10,** fruit, × 3½; **11,** seed, × 10. 1, from *Greenway* 3055; 2, from *Drummond & Hemsley* 3166; 3, from *Brodhurst Hill* 358; 4, from *Heriz-Smith & Paulo* 852 and *Newbould* 3400; 5–9, from *Verdcourt* 592; 10, 11, from *Milne-Redhead & Taylor* 9013 B. Drawn by B. Verdcourt.

UGANDA. Toro District: Ruwenzori, Nyinabitaba, Aug. 1937, *Eggeling* 1378!; Mbale District: S. Boundary of W. Bugwe Forest Reserve, 30 Apr. 1951, *G. H. Wood* 202!; Mengo District: Kampala, Kawanda, Feb. 1936, *Hazel* 370!
KENYA. Trans-Nzoia District: Kitale, 3 Oct. 1952, *Verdcourt* 731!; Kiambu District: Theta road, 5 July 1952, *Kirrika* 203!; Masai District: Narok, W. side of Nasampolai [Enesambulai] valley, 25 July 1970, *Greenway & Kanuri* 14532!
TANZANIA. Moshi District: Kilimanjaro, Lyamungu, 20 Aug. 1932, *Greenway* 3055!; Lushoto District: W. Usambara Mts. Mtai–Mlalo road, near Kidologwai, 19 May 1953, *Drummond & Hemsley* 2643!; Njombe District: Igawa–Njombe road, 27 Feb. 1963, *Richards* 17662!
DISTR. **U**1–4; **K**1–6; **T**1–8; Zaire, Sudan, Somali Republic (N.), Mozambique, Malawi, Zambia, Rhodesia, Angola and South Africa (mostly Natal and Transvaal)
HAB. Mostly at forest edges, in clearings and thickets or less often in denser forest (e.g. *Neoboutonia*), open grassland, and bushland, also in scrubland and rocky gullies; 1140–2640 m.

SYN. *R. conotricha* Gandoger in Bull. Soc. Bot. Fr. 65: 35 (1918)
R. longipetiolata Bullock in K.B. 1932: 497 (1932); Brenan in Mem. N.Y. Bot. Gard. 8: 455 (1954).* Type: Kenya, Elgon, *Lugard* 204 (K, holo.!, EA, iso.!)

NOTE. In tropical Africa *Rubia cordifolia* is variable enough but when its entire range is considered the variation is unbelievable. Attempts to split it into more reasonable taxa have failed and it would doubtless need a lifetime of experimental taxonomy to unravel the systematics of this species and its allies. The specific description has been drawn up deliberately to exclude some of the extreme Asiatic variants involving such characters as dark red flowers, under surfaces of the leaf-blades red, leaves in whorls of 6–8, leaf-blades of large size up to 11 × 8 cm., glabrous leaves and very large fruits 5.5 × 3.5 mm. in dry state.

The intense variation in East Africa in leaf-shape, length of petiole and indumentum seems to be related to features of the habitat, climatic conditions and the age of the plant; it is not of a kind related to geography. The plant would lend itself to studies of a similar nature to those carried out by Burkill on *Tamus*. Some specimens, e.g. Rungwe District, Poroto Mts., 13 Mar. 1932, *St. Clair-Thompson* 711, have been annotated by Deb as *R. cordifolia* L. var. *cordifolia* forma *strigosa* Deb & Malick (in Journ. Bombay Nat. Hist. Soc. 63: 782 (1967) based on Bhutan, Chumbi, Taesieu Doorm, *King* 482 (CAL, holo., K, iso.!)), but it is facile to suppose that the species can be divided into forms by technical characters. Almost certainly the African and Indian specimens referred to this form are not truly that closely genetically related. *R. discolor* Turcz. (type: Ethiopia, Scholoda, *Schimper* 24. (? CW, holo., BM, K, iso.!)) has the leaf-blades densely covered with white hairs beneath and also distinctly campanulate corolla-tubes, but many plants of *R. cordifolia* subsp. *conotricha* have exactly the same facies and there is a slight variation in the shape of the corolla; it has sometimes been regarded as a variety (var. *discolor* (Turcz.) K. Schum. in P.O.A. C: 395 (1895)). An additional problem concerns *R. petiolaris* DC., a S. African species, some variants of which closely resemble *R. cordifolia*.

DISTR. (of species as a whole). Greece, N. Africa, Siberia, Manchuria, China, Japan, Afghanistan, Baluchistan, India, Vietnam, Sumatra, Borneo, E., central and S. Africa

42. GALIUM**

L., Sp. Pl.: 105 (1753) & Gen. Pl., ed. 5: 46 (1754)

Annual or perennial herbs, sometimes slightly shrubby at the base, with erect, decumbent or ± climbing 4-angled stems, glabrous to hispid and sometimes with small prickles. Leaves in whorls of 3–10–many; blades setaceous to ovate, sometimes with prickly margins; stipules foliar and counted with the leaves. Flowers hermaphrodite or rarely unisexual, usually very small,

* The name *R. orientalis* written on many sheets appears never to have been published.

** This account is partly based on the work done on the Kew Herbarium material by J. P. M. Brenan who also borrowed some of the types involved—a (!) indicates a borrowed type seen by him but not by the author. The key is based on an unpublished key devised by Mr. Brenan which has been found invaluable during two decades of routine naming of this genus.

arranged in a basically thyrsoid inflorescence, but mostly in exinvolucrate ebracteate axillary or terminal cymes or rarely solitary; pedicel articulated with the calyx. Calyx-tube ovoid or globose; limb obsolete, but usually with a ring of sessile colleters. Corolla usually white, yellow or reddish, rotate; lobes (3–)4(–5). Stamens (3–)4; filaments short, inserted in the very short tube. Ovary 2-locular; ovules solitary in each locule, affixed to the septum, amphitropous; styles 2, short, sometimes connate below; stigmas capitate. Disc annular. Fruit 2-lobed, each lobe with 1 seed, leathery and ± dry or somewhat to quite distinctly fleshy, smooth or tuberculate, glabrous, hairy or with hooked bristles. Seeds with membranous testa adhering to the pericarp, dorsally convex, ventrally excavated; albumen horny.

A large almost cosmopolitan genus, usually (and probably exaggeratedly) estimated at 300–400 species, mostly in temperate regions but also in subtropical and tropical areas at higher altitudes. The species are difficult and often rather poorly defined, particularly at the edges of their ranges.

Margin of leaf-blades with ± reflexed prickles or denticles (often difficult to see if leaf-margin is recurved):
- Fruits densely clothed with long brownish hooked hairs 1 mm. long or rather more; leaf-blades 2–11 mm. wide; inflorescence short, 1–3-flowered; corolla 3–5 mm. wide, green to purplish, brownish or yellowish, the lobes not long-aristate . . . . 2. *G. chloroionanthum*
- Fruits and ovaries glabrous or with mostly sparser shorter hairs less than 0·5 mm. long:
  - Fruits with short hairs:
    - Many of the fruits borne on unbranched apparent peduncles arising from nodes of the main stem or main branches, the peduncles up to 3·7 cm. long, bent just below the fruit; leaf-blades narrow, 1–3(–6) mm. wide; corolla ± 1–2 mm. wide, whitish to greenish, the lobes not aristate . . . . . 4. *G. spurium* subsp. *africanum*
    - Fruits and flowers all in several–many-flowered terminal panicles or lateral leafy cymes arising from the main stem; corolla 2–5 mm. wide, purplish brown or bright yellow:
      - Flowers purplish brown, in several-flowered lateral cymes; leaf-blades (0·5–)3–8 mm. wide; stems often prickly; corolla ± 3–4 mm. wide, the lobes aristate; fruits black, clothed with hooked hairs . . 3. *G. aparinoides*
      - Flowers mostly bright yellow, in many-flowered terminal panicles; leaf-blades narrower, 0·5–2(–4) mm. wide; stems usually hairy or sometimes only slightly so; corolla ± 2–3 mm. wide, the lobes acuminate; fruits with sparse hairs . . 16. *G. scabrellum*

Fruits and ovaries glabrous, the former usually going black and fleshy when ripe:*
- Many of the fruits borne on unbranched apparent peduncles 1–3·5 cm. long, arising from the nodes on main stems or branches . . . . . 4. *G. spurium* ? var. *spurium*
- Fruits and flowers in 1–several-flowered terminal or lateral inflorescences but not on long unbranched peduncles:
  - Plants small, alpine, with leaves under 1 cm. long:
    - Leaf-blades oblanceolate to spathulate, usually very distinctly widest near the apex, 2·5–8 × 1–2·5 mm., glabrous above or (in var. *satimmae*) with numerous forwardly directed curved hairs, the margins smooth or with distinctly retrorse prickles; usually at 3510–4350 m. in East Africa . 8. *G. glaciale*
    - Leaf-blades linear-oblanceolate to obovate-lanceolate, (3–)5·5–9 × 1–1·4 mm., either ± glabrous or the upper surface with short forwardly directed hairs and margins with forwardly directed stiff hairs but midrib with retrorse prickles; usually at 2700–3200 m. in East Africa . . 9. *G. hochstetteri*
  - Plants larger, of lower altitudes, with larger leaves:
    - Leaf-blades (3–)4–11 mm. wide; petioles 1–3 mm. long; flowers white to yellowish white or pale green; inflorescences mostly 1(–3)-flowered . . . 7. *G. brenanii*
    - Leaf-blades narrower, 0·5–3(–4) mm. wide and without the other characters combined:
      - Nodes usually with a dense zone of short brownish hairs within the whorls of leaves; corolla large, ± 3–5 mm. wide, green to yellow; flowers in short 1–3-flowered lateral cymes; stems often minutely puberulous; leaf-blades narrow, 0·3–1·5(–3·5) mm. wide . . 6. *G. ruwenzoriense*
      - Nodes without such a zone or with only reduced traces; corolla small, ± 2–3 mm. wide, bright yellow, yellowish white or

* A W. African race of *G. simense* has hairy fruits.

white; flowers usually in several-flowered lateral cymes or terminal panicles or sometimes the inflorescence 1-flowered; leaf-blades 1–3(–4) mm. wide:

Leaf-margins and often stem also with conspicuous coarse prickles making the whole plant harshly adhesive but otherwise glabrous or glabrescent; fruit going black and fleshy; flowers yellow to white or greenish . . 5. *G. simense*

Leaf-margins with small closely set inconspicuous denticles; stem not or only slightly prickly, the plant thus much less adhesive; fruit dry, not going black and fleshy:

Flowers bright yellow; leaf-blades with the apical filiform acumen straight or but slightly twisted . 16. *G. scabrellum*

Flowers white; leaf-blades with short acute apical acumen which is often twisted or decurved . 13. *G. scioanum*

Margin of leaf-blades without reflexed prickles or denticles or with practically none (some species have been included in the key twice):

Leaves in 4's, elliptic to ovate, mostly 0·4–1 cm. wide; fruits densely clothed with hooked hairs; plant mostly hairy in Flora area . . . . 1. *G. thunbergianum*

Leaves in 6's or 8's, or if in 4's then neither elliptic nor ovate, usually setaceous with usually only the lower blades up to 2 mm. (rarely more) wide:

Leaf-blades with a distinct filiform acumen at the apex (sometimes knocked off on older leaves); flowers usually yellow, less often yellowish white:

Leaf-blades, at least those on the main stem, mostly conspicuously elongate and narrow, usually 1·5–3·5 cm. long, parallel-sided, 0·5–2(–3) mm. wide, only those on the weaker side branches as short as 1 cm.; internodes often elongate, ± 1·5–8·5 cm. long . . 14. *G. bussei*

Leaf-blades short, usually ± 1–1·2 cm. long, rarely up to 1·4 cm.:

Leaf-margins glabrous or hairy but without reflexed saw-like denticles; plants decumbent with erect shoots to 35 cm. or forming clumps; inter-

nodes often mostly shorter than the leaves . . . . . . 15. *G. ossirwaense*
Leaf-margins under a lens usually showing closely set reflexed saw-like denticles; plants trailing and scrambling to 3·6 m., the internodes mostly long . . . . . 16. *G. scabrellum*
Leaf-blades acute to obtuse at the apex but without a filiform acumen, although sometimes with a hair at the tip; flowers usually white:
Many or all of the leaves on the main stem exceeding 1·5 cm. in length, ± parallel-sided, often ascending, not reflexed; corolla ± campanulate; fruiting pedicels sometimes bent . 12. *G. stenophyllum*
Most leaves short, usually up to 1–1·2(–1·4) cm. long (rarely a few to 2·5 cm.) or if longer then ± elliptic:
Leaf-blades blunt at apex save sometimes for a hair:
Leaves ± 1·5–4 mm. long; flowers 1–2 on short lateral shoots, the inflorescence not exceeding the leaves; alpine plant at 3000–3300 m. . . . . . . 10. *G. kenyanum*
Leaves ± 7–8 mm. long; flowers in terminal ± dense several-flowered inflorescences exceeding the leaves; upland plant at 2000–2700 m. . . . . . . 11. *G. tanganyikense*
Leaf-blades acute or ± acuminate at the apex:
Leaf-blades acute at the apex, which is often deflexed or twisted, 0·4–1·7(–2·5) cm. × 1–2·8(or –7 in one variant) mm., linear to elliptic . . . . . . 13. *G. scioanum*
Leaf-blades with a short triangular acute acumen at the tip, 2·5–8 × 1–2·5 mm., mostly oblanceolate or spathulate, widest near the apex; alpine plant at 2800–4500 m. . . . . . . 8. *G. glaciale*

1. **G. thunbergianum** *Eckl. & Zeyh.*, Enum. Pl. Afr. Austr.: 369 (1837); A.V.P.: 178, 329 (1957); Hepper, F.W.T.A., ed. 2, 2: 223 (1963); E.P.A.: 1026 (1965); U.K.W.F.: 411, fig. (1974). Type: South Africa, Cape Province, Katriviersberg, *Ecklon & Zeyher* 2321 (S, iso.)

Straggling, climbing or procumbent plant, 7–75 cm. tall; stems and foliage either glabrous or pilose with quite long spreading hairs but not prickly. Leaves and stipules in whorls of 4; blades elliptic to ovate, (0·3–)0·7–1·8 cm. long, (0·2–)0·3–1 cm. wide, acute or rounded and shortly mucronate at the apex, cuneate at the base, the margins not prickly. Inflorescences very condensed or lax, 0·3–5·5 cm. long; pedicels 1–7 mm. long. Calyx-tube

0·3–0·6 mm. in diameter, densely covered with spreading bristly hairs. Corolla white, yellow or greenish white to yellowish white; lobes 0·7–1·5 mm. long, 0·5–0·9 mm. wide; tube 0·09–0·3 mm. long. Fruits 1·2–1·5 mm. long, 1·3–2 mm. wide, the cocci densely covered with spreading hooked bristly hairs ± 1 mm. long.

var. **thunbergianum**

Stems and foliage glabrous.

KENYA. Kericho District: SW. Mau Forest, Sambret, Oct. 1961, *Kerfoot* 2947!
DISTR. **K**5; Ethiopia, South Africa
HAB. ?Evergreen forest; 2160 m.

SYN. [*G. rotundifolium* sensu Sond. in Fl. Cap. 3: 39 (1865), *non* L.]
G. natalense Rouy in Rouy & Foucaud, Fl. Fr. 8: 9 (1903), adnot. Type: South Africa, Natal, Drakensberg Mts., Van Reenen's Pass, *Medley Wood* 5562 (LY (Herb. Rouy), holo. (!), K, photo. !)

var. **hirsutum** (*Sond.*) *Verdc.* in K.B. 30: 326 (1975). Type: South Africa, *Masson* in *Herb. Thunberg* (UPS, holo., ? TCD, iso.)

Stems and foliage sparsely to densely covered with spreading bristly white hairs.

UGANDA. Kigezi District: Mt. Mgahinga, June 1951, *Purseglove* 3698! & Mt. Muhavura, 11 Jan. 1953, *Norman* 194!; Mbale District: Bugisu, Bulambuli, 3 Sept. 1932, *A. S. Thomas* 495!
KENYA. NE. Elgon, Aug. 1951, *Tweedie* 929!; Naivasha District: Aberdare Mts., Nyeri Track, 28 Oct. 1934, *G. Taylor* 1380!; W. Mt. Kenya, 5 Jan. 1922, *Fries* 748!
TANZANIA. Arusha District: path to Meru Crater, 16 Feb. 1969, *Richards* 24080!; Rungwe District: Poroto Mts., Kikondo Camp, 20 Jan. 1961, *Richards* 13988!; Njombe District: Kipengere Mts., Mtorwi Peak, 13 Jan. 1957, *Richards* 7719!
DISTR. **U**2, 3; **K**3, 4; **T**2, 7; Cameroun Mt., Fernando Po, Zaire, Ethiopia, Sudan, eastern Rhodesia and South Africa
HAB. Upland grassland, open places in upland rain-forest and dry evergreen forest, also lower ericaceous belt, sometimes in stream gullies; 1950–3360(–3450) m.

SYN. [*G. rotundifolium* sensu Thunb., Fl. Cap. 1: 551 (1813); Hiern in F.T.A. 3: 245 (1877), quoad *Schimper* 370, 604 & 1219, *non* L.]
G. rotundifolium L. var. *hirsutum* Sond. in Fl. Cap. 3: 39 (1865)
G. dasycarpum Schweinf., Beitr. Fl. Aeth.: 135 (1867); K. Krause in N.B.G.B. 10: 610 (1929); F.P.N.A. 2: 379 (1947). Type: Ethiopia, Simien, Debra-Eski, *Schimper* (B, holo.†)
G. biafrae Hiern in F.T.A. 3: 245 (1877); F.P.N.A. 2: 378 (1947). Types: Fernando Po, *Mann* 605 & Cameroun Mt., *Mann* 1284 (both K, syn. !)

NOTE. Most tropical African specimens of this species are hairy and most South African glabrous but the character is sufficiently scattered geographically to retain the taxa at varietal rank.

2. **G. chloroionanthum** *K. Schum.* in E.J. 30: 417 (1901); Brenan in Mem. N.Y. Bot. Gard. 8: 456 (1954); U.K.W.F.: 411 (1974). Type: Tanzania, Rungwe Mt., *Goetze* 1162 (B, holo. †, BM, K, iso. !)

Climbing or procumbent perennial herb 0·15–1 m. long, with coarsely prickly stems and sometimes coarse hairs at the nodes. Leaves and stipules in whorls of 6; blades narrowly elliptic to elliptic-oblanceolate, 0·5–3·7 cm. long, 0·2–0·8(–1·1) cm. wide, sharply acuminate at the apex, cuneate at the base, margins and midnerve above and beneath with coarse prickles, glabrous on upper surface with rather scattered bristly hairs. Inflorescences short, 1–3-flowered; peduncles 0·8–2·6 cm. long; true pedicels short, 0–1 mm. long, but sometimes appearing up to 8 mm. long. Calyx-tube greenish, purplish or brownish, 0·4–0·8 mm. in diameter, hairy. Corolla yellowish, 3–5 mm. wide; lobes 1·2–2·1 mm. long, 0·9–1·2 mm. wide, subacute to apiculate; joined part of the tube 0·3–0·5 mm. long. Fruit, covered with brownish hooked hairs ± 1·2 mm. long; cocci 3 mm. in diameter.

UGANDA. Toro District: Ruwenzori, Bujuku valley, Aug. 1932, *Eggeling* 1274! & Ruwenzori, Aug. 1938, *Purseglove* 292!; Kigezi District: Virunga Mts., pass between Mgahinga and Sabinio, 21 Nov. 1934, *G. Taylor* 1876!
KENYA. Nakuru District: Mau Forest, Endabarra, 19 Jan. 1946, *Bally* 4891!; Kericho District: 10.4 km. E. of Kericho, along the S. edge of Sambret Tea Estate, 6 Dec. 1967, *Perdue & Kibuwa* 9283!; Masai District: Narok area, Nasampolai [Enesambulai] valley crest, 2 June 1969, *Greenway & Kanuri* 13636!
TANZANIA. Ufipa District: Mbisi Forest, 20 July 1962, *Richards* 16832!; Rungwe, 19 Sept. 1912, *Stolz* 1563!; Songea District: Matengo Hills, Liwiri Kiteza, 24 May 1956, *Milne-Redhead & Taylor* 10511!
DISTR. **U**2; **K**3, ?4, 5, 6; **T**4, 6–8; Rwanda, Burundi, Zaire, Sudan, Malawi, Rhodesia
HAB. Upland evergreen forest margins, bracken scrub, also in rocky places; 2000–3300 m.

SYN. [*G. hamatum* sensu Robyns, F.P.N.A. 2: 380 (1947), *non* A. Rich.]

3. **G. aparinoides** *Forssk.*, Fl. Aegypt.-Arab.: CV, No. 87, & 30 (1775); DC., Prodr. 4: 608 (1830); Schweinf., Arab.-Pflanzennamen: 114 (1912); Blatter, Fl. Arabica, in Rec. Bot. Survey India 8: 224 (1921); Schwartz, Fl. Trop. Arab., in Mitt. Inst. Bot. Hamburg 10: 264 (1939); E.P.A.: 1024 (1965); Verdc. in K.B. 28: 60 (1973); U.K.W.F.: 411 (1974). Type: Yemen, Hadie Mts., Mokhaja, *Forsskål* (C, lecto.!)

Scrambling perennial herb up to 1·8–2·1 m. long; stems usually densely covered with small prickles but otherwise mostly glabrous, less often ± hairy; roots said to contain a red dye. Leaves and stipules in whorls of (4–)6(–7); blades narrowly elliptic to oblanceolate, 0·4–3 cm. long, (0·5–)3–8 mm. wide, with an apical acumen 0·5–1·5 mm. long, cuneate at the base, sparsely hairy on both surfaces, particularly when young and with close recurved prickles on margins and midnerve beneath. Cymes lateral, several-flowered, 2·3–4·5 cm. long; pedicels 0·1–1·8 mm. long. Calyx-tube subglobose, (0·3–)0·5–0·9 mm. long, 0·7–1·5 mm. wide, appearing glabrous. Corolla brownish red to dark purple-brown or greenish, flushed brownish or purple, less often reported to be green, yellowish or white, 3–4(–6) mm. wide; tube 0·5–0·7 mm. long; lobes narrowly triangular, 1·4–1·7 mm. long, 1–1·2 mm. wide, with a long acumen 0·5–1 mm. long, making the total length usually ± 2·5 mm. Fruits with cocci 1·5–3 mm. in diameter, black, sparsely clothed with short white hooked hairs ± 0·5 mm. long.

UGANDA. Karamoja District: Moroto Mt., Feb. 1936, *Eggeling* 2879!; Mbale District: Elgon, near Bulambuli, 12 Nov. 1933, *Tothill* 2374! & from Bulambuli Camp to near Madangi Camp, *Liebenberg* 1630!
KENYA. Elgeyo District: Cherangani Hills, below Kaisungur, 1 Oct. 1959, *Verdcourt, G. Taylor et al.* 2434!; Nakuru District: Mau Forest Reserve, camp 9, 27 Aug. 1949, *Maas Geesteranus* 5913!; N. Nyeri District: Mt. Kenya, Sirimon Track, 22 Sept. 1963, *Verdcourt, Cooley & Howard* 3766G!
TANZANIA. Arusha District: Mt. Meru, 30 July 1970, *Hedberg* 4696!; N. slopes of Kilimanjaro, above Rongai, Kimengalia stream, 20 Feb. 1933, *C. G. Rogers* 449!; W. Usambara Mts., Mkusi, 19 Aug. 1950, *Verdcourt* 319!
DISTR. **U**1, 3; **K**3–6; **T**2, 3; Arabia, Ethiopia, ? Malawi
HAB. Upland evergreen forest, particularly edges and clearings amongst moist undergrowth, *Hypericum* thickets, by rocky streams also grassy slopes and amongst lava blocks; 1680–3700 m.

SYN. *G. hamatum* A. Rich., Tent. Fl. Abyss. 1: 345 (1847); K.Krause in N.B.G.B. 10: 611(1929). Types: Ethiopia, Wodjerat, *Petit* & Simien, Mt. Selki, *Schimper* 675 (both P, syn., K, isosyn.!)
G. aparine L. var. *hamatum* (A. Rich.) Hook. f. in J.L.S. 6: 11 (1861) & 7: 197 (1864); Hiern in F.T.A. 3: 246 (1877), excl. specim. cit.
[*G. mollugo* sensu K. Krause in N.B.G.B. 10: 611(1929), *non* L.]

NOTE. It is possible that there is an alpine race of this species on Kilimanjaro worthy of recognition, with small leaves, reduced inflorescences and ± glabrous ovaries and fruits, e.g. Shira Plateau, Ngare Nairobi R., 3420 m., 6 Nov. 1968, *Bigger* 2274! & Shira Plateau, 3383 m., 11 Feb. 1969, *Richards* 23994!

4. **G. spurium** *L.*, Sp. Pl.: 106 (1753); Pobed. in Fl. U.R.S.S. 23: 306 (1958); Clapham, Tutin & Warburg, Fl. Brit. Is., ed. 2: 785 (1962); U.K.W.F.: 411 (1974). Type: plant grown in Upsala Botanic Garden (no specimen in LINN)

Weak scrambling herb 0·3–1·8 m. long, with prickly but otherwise glabrous stems. Leaves and stipules in whorls of 6–8; blades linear-lanceolate or linear-oblanceolate, 1–4(–4·7) cm. long, 1–3(–6) mm. wide, with a filiform acumen at the apex, cuneate at the base; margins and midvein with coarse recurved prickles, making the whole plant harshly adhesive. Cymes 1–9-flowered, but sometimes with many of the fruits borne on apparent unbranched peduncles arising from nodes of the main stem or main branches; peduncles and pedicels 0–3·2(–3·7) cm. long. Calyx-tube 0·5–0·65 mm. long, 0·8–1·1 mm. wide, hairy or glabrous. Corolla greenish white, white, green or yellowish, 1–2 mm. wide; tube 0·3–0·4 mm. long; lobes triangular, 0·7–1·1 mm. long, 0·5–0·6 mm. wide. Cocci of fruit 1–3 mm. in diameter, glabrous or hairy.

var. **spurium**

Fruit glabrous.

KENYA. Nakuru District: Mau Narok, *G. R. Cunningham van Someren*!
DISTR. **K**3; widely distributed in Europe, W. Asia and N. Africa and as an introduced weed elsewhere
HAB. Weed in arable fields; ± 2400 m.

NOTE. I have accepted J. P. M. Brenan's determination of the above specimen, but agree with Ehrendorfer, who has also seen it, that it is by no means certain—the material is poor.

subsp. **africanum** *Verdc.* in K.B. 30: 324 (1975). Type: Kenya, Kiambu District, Muguga, *Milne-Redhead & Taylor* 7147 (K, holo.!)

Many fruits borne on what appear to be unbranched peduncles; fruit covered with white hooked hairs which do not have tuberculate bases.

UGANDA. Toro District: Kichwamba, 12 Mar. 1932, *Hazel* 197!; Kigezi District: Kinkizi road, about 1 km. from Kisizi road junction, 7 June 1952, *Norman* 133! & Kachwekano Farm, Mar. 1950, *Purseglove* 3332!
KENYA. Uasin Gishu District: 24 km. SE. of Eldoret, 31 Aug. 1948, *Bogdan* 1931!; Kiambu District: Muguga, 6 June 1952, *Verdcourt* 654!; Londiani, *R. M. Graham* LD.877 in *F.D.* 2834!
TANZANIA. Mbulu District: Mt. Hanang, above Katesh, 2 May 1962, *Polhill & Paulo* 2294!; Lushoto, Jaegertal, 16 Jan. 1966, *Archbold* 648!; Uluguru Mts., Chenzema, 2 Jan. 1934, *Michelmore* 868!
DISTR. **U**2, 3; **K**2–6; **T**2, 3, 6, 7; Zaire, Rwanda, Burundi, Sudan, Ethiopia, Socotra, Somali Republic, Malawi, Rhodesia and South Africa
HAB. Mostly in cultivations but also bushland, grassy slopes, swamps, roadsides and forest edges; 1250–2400(–2700) m.

SYN. [*G. spurium* L. var. *echinospermon* sensu auctt., *non* (Wallr.) Desportes]
[*G. aparine* sensu auctt. e.g. K. Krause in N.B.G.B. 10: 610(1929); F.P.N.A. 2: 379 (1947), *non* L.)]
[*G. aparinoides* sensu Cufod. in Phyton 1: 150(1949) & E.P.A.: 1024 (1965), *non* Forssk.]

NOTE. Ehrendorfer has added pencilled annotations throughout the African covers indicating that he thinks all the material is referable to *G. aparine* but I am following Brenan who looked into the matter very carefully in referring the plant to *G. spurium*, although I do not agree with him in calling it var. *echinospermon*. In the field it certainly does not seem to be identical with *G. aparine* as known in Europe. Cytological investigation is required.*

*A recent count carried out by Dr Krendl on material from Jimma in Ethiopia showed the chromosome number to be 2n = 44. Since many counts have shown that *G. spurium* is consistently 2n = 20 and that the basic number for *G. aparine* is 11 rather than 10 this subspecies will have to be transferred to *G. aparine*.

5. **G. simense** *Fresen.* in Mus. Senckenb. 2: 165 (1837); Hepper, F.W.T.A., ed. 2, 2: 223 (1963); E.P.A.: 1026 (1965), pro parte; F.P.U., ed. 2: 166 (1972); U.K.W.F.: 411, fig. (1974). Type: Ethiopia, Temben to Simien, *Rueppell* (FR, holo. (!), K, photo.!)

Straggling or climbing herb 0·5–2·4 m. long; stems weak, mostly with small prickles but otherwise glabrous (except for a hairy ring at the nodes), or with sparse curved hairs when young. Leaves and stipules in whorls of 4–8; blades linear-oblanceolate, 0·5–3·2(–4) cm. long, 1–2·5(–4) mm. wide, with a distinct acumen at the apex, often deflexed, cuneate at the base, glabrous and often somewhat shining but margins and midnerve with very recurved coarse prickles making the whole plant harshly adhesive. Cymes ± condensed, 0·3–0·5 cm. long, and some flowers solitary; pedicels 0–2 mm. long or, in case of solitary flowers, up to 0·8–1·4 cm. long in fruit. Calyx-tube biglobose, 0·35–0·6 mm. long, 0·4–0·85 mm. wide, glabrous, with clusters of large cells around the rim. Corolla white, green or yellowish green, ± 4 mm. wide; tube 0·3–0·4 mm. long; lobes narrowly triangular, 1·3–1·7 mm. long, 0·7–0·8 mm. wide, acute and with an inflexed acumen. Fruits black or purplish black and fleshy; cocci 2·5–3·5 mm. in diameter, glabrous (see note).

UGANDA. E. Ruwenzori, Bwamba Pass, July 1940, *Eggeling* 4023!; Kigezi District: Kanaba Gap, June 1946, *Purseglove* 2079! & Mt. Mgahinga, June 1951, *Purseglove* 3699! & Bufumbira, Nyamagana footpath to Lake Mutanda, 24 Feb. 1953, *Norman* 209!

KENYA. NE. Elgon, Aug. 1962, *Tweedie* 2412!; Naivasha, 27 Oct. 1930, *Napier* 462! & Lake Naivasha, near Lake Hotel, 7 July 1963, *Verdcourt* 3679!

TANZANIA. Mpanda District: Kungwe Mt., Kahoko, 23 July 1959, *Newbould & Harley* 4599!

DISTR. **U**2; **K**3, 4; **T**4; Cameroun, Fernando Po, Zaire, Burundi, Sudan, Ethiopia

HAB. Undergrowth of forest edges, bushland, woodland, etc., also in rough pasture, swamps with tall sedge borders; 1500–2700 m.

SYN. [*G. aparine* L. var. *hamatum* sensu Hook. f. in J.L.S. 6: 11(1861) & 7: 197 (1864), *non* (A. Rich.) Hook f. sensu stricto]

[*G. aparine* L. var. *spurium* sensu Hiern in F.T.A. 3: 245(1877), *non* (L.) Coss. & Germ.]

G. abyssinicum Chiov. in Ann. Bot. Roma 9: 68 (1911). Types: Ethiopia, Amhara, Semien, Debarek, *Chiovenda* 911, 2929 & without locality, *Chiovenda* in *Herb. Gavioli* 25667 (FI, syn. (!), K, photo.!)

[*G. spurium* sensu F.W.T.A. 2: 137 (1931); F.P.N.A. 2: 380(1947), pro parte, *non* L.]

NOTE. In West Africa true *G. simense* is restricted to the area of Cameroun Mt.; elsewhere (in Bamenda, etc.) the fruit is covered with hooked hairs and thus much resembles *G, aparine* L. except that the flowers are usually greenish (with the exception of a specimen from Fernando Po with white flowers). These may represent a western variety of *G. simense*, but until further study has been made I have not extended the description of the plant to cover these variants.

6. **G. ruwenzoriense** (*Cortesi*) *Chiov.* in Ann. Bot. Roma 9: 69(1911); Ehrend. in A.V.P.: 179, 329, fig. 10 (1957); U.K.W.F.: 411(1974). Types: Uganda, Ruwenzori, Valley of the Lakes & Bujuku valley, Bujongolo, *Abruzzi Exped.* (TO, syn.)

Scrambling or climbing herb 0·3–1·9(–5) m. long; stems usually minutely puberulous or distinctly pubescent or hairy, less often ± glabrous. Leaves and stipules in whorls of 6–8, the nodes nearly always with a dense zone of brownish hairs 1–2(–3) mm. long within the whorl; blades linear to narrowly elliptic, 0·2–1·9 cm. long, 0·3–1·5(–3·5) mm. wide, acuminate, the margins with very coarse curved prickles rendering the whole plant harshly adhesive. Flowers in short 1–3-flowered lateral cymes; peduncle 2–4 mm. long;

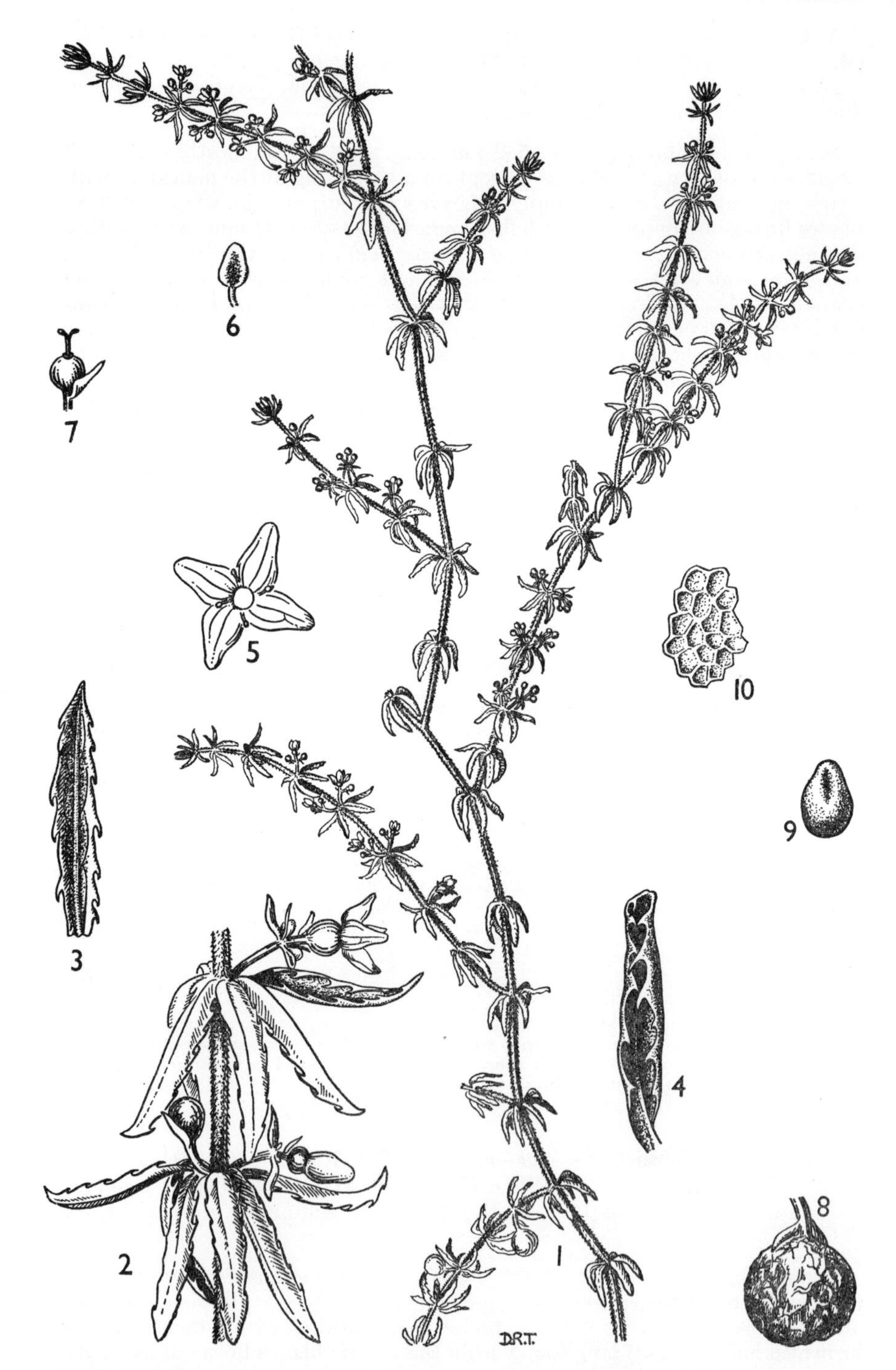

FIG. 58. *GALIUM RUWENZORIENSE*—**1,** habit, × 1; **2,** part of fertile branch, × 6; **3, 4,** leaves, × 6; **5,** corolla and stamens, × 6; **6,** stamen, × 12; **7,** pistil, × 6; **8,** fruit, × 4; **9,** seed, × 4; **10,** detail of testa, much enlarged. All from *C. G. Rogers* 448. Drawn by Miss D. R. Thompson.

pedicels 0–1·5(–2) mm. long. Calyx-tube glabrous, 0·5–0·7 mm. in diameter. Corolla green to yellow, sometimes purplish tinged in bud; tube 0·25–0·8 mm. long; lobes ovate, 1·4–2·2 mm. long, 0·8–1·2 mm. wide, acute. Fruit purplish black, black or blue, becoming distinctly fleshy, 3·5–7 mm. in diameter, the cocci not very clearly divided, glabrous. Fig. 58.

UGANDA. Ruwenzori, Aug. 1938, *Purseglove* 265! & above Lake Bujuku, Aug. 1933, *Eggeling* 1331!; Elgon, Bulambuli, 11 Nov. 1933, *Tothill* 2258!

KENYA. NE. Elgon, Feb. 1963, *Tweedie* 2543!; Aberdares National Park, N. Kinangop–Nyeri road, Cave Waterfall, 30 July 1960, *Polhill* 236!; Mt. Kenya, Naromoru Track, 12 Dec. 1957, *Verdcourt* 2007!

TANZANIA. Mbulu District: Loolmalassin Mt., 17 Sept. 1932, *B. D. Burtt* 4199 in part!; Arusha District: Mt. Meru, eastern slopes above Olkokola Estate, 1 Nov. 1948, *Hedberg* 2429!; Kilimanjaro, N. slopes above Rongai, Kimengelia stream, 29 Feb. 1933, *C. G. Rogers* 448!

DISTR. **U**2, 3; **K**3, 4, 6; **T**2; E. Zaire, Rwanda, ? Ethiopia

HAB. In open forest or bushland, extending from the upper part of the montane forest (in clearings, at edges, etc.) into the alpine moorland, sometimes also in rocky places; (2350–)2700–4050(–4200) m.

SYN. *Rubia ruwenzoriensis* Cortesi in Ann. Bot. Roma 6: 152(1907) & in Il Ruwenzori, Parte Scientifica 1: 448, t. 22 (1909); F.P.N.A. 2: 382 (1947)

Galium serratohamatum S. Moore in J.L.S. 38: 256(1908); K. Krause in N.B.G.B. 10: 612(1929). Type: Uganda, Ruwenzori, E. slope, *Wollaston* (BM, holo.!)

G. mildbraedii K. Krause in Z.A.E. 2: 342(1911). Type: Zaire, Ruwenzori, Butahu [Butagu] valley, *Mildbraed* 2553 (B, holo.†, BR, iso. fragment!)

[*G. spurium* sensu K. Krause in N.B.G.B. 10: 611(1929); F.P.N.A. 2: 380(1947), pro parte, *non* L.]

G. afroalpinum Bullock in K.B. 1932: 498(1932). Type: Kenya, Mt. Elgon, *E. J. & C. Lugard* 365 (K, holo.!, EA, iso.!)

NOTE. Chiovenda erected a section *Sarcogalium* for this and a related Ethiopian species by virtue of its fleshy fruits.

A curious cushion plant with leaf-blades about 3 mm. long and 1 mm. wide from a "Senecio-Heath bog" in small sheltered valley at 3150 m. at Sasa Camp, Mbale District, Bugisu, Uganda, 16 Apr. 1950, *Forbes* 273, may be a stunted form of the above but the plant is sterile and no decision can be made as to its identity. *G. simense* and *G. ruwenzoriense* are easy to distinguish in the case of typical material by means of the nodal hairs, the inflorescence structure, the corolla size and altitude of the habitat but in certain areas, notably mountain areas in E. Zaire and W. Uganda, many specimens are difficult to place. Some seemingly identical with *G. ruwenzoriense* lack the hairs within the nodes and others, having the ± open inflorescence of *G. simense*, possess a few nodal hairs; such intermediates could well be hybrids and the problem needs study in the field. The record from Ethiopia is based on such a difficult intermediate specimen which might be a form of *G. simense*.

7. **G. brenanii** *Ehrend. & Verdc.* in K.B. 28: 485, fig. 1/7–14 (1974). Type: Tanzania, Uluguru Mts., Salaza Forest, *Drummond & Hemsley* 1625 (K, holo.!)

Scrambling brittle herb to 3–4 m. or more; stems coarsely prickly, but otherwise glabrous except near the nodes, the older ones often becoming corky and thick, ± winged, yellowish, up to 6 mm. wide. Leaves and stipules in whorls of 4–6, distinctly petiolate, sometimes with a ring of stiffish reddish brown hairs within the whorl; petioles up to 3(–5) mm. long, often pubescent; blades elliptic, 0·9–3·6 cm. long, (0·3–)0·4–1·2 cm. wide, acute and apiculate at the apex, cuneate at the base, margins and midnerve very coarsely prickly so that plant is very harshly adhesive but otherwise glabrous except for a few hairs on midrib above. Cymes mostly 1(–3)-flowered; peduncles 0·3–1·3 cm. long; pedicels 0–9 mm. long. Calyx-tube 0·3–0·55 mm. in diameter, glabrous. Corolla green, white, greenish white or yellowish white; tube (0·5–)0·6–1·1 mm. long; lobes ovate, 1–1·5 mm. long, 0·7–1 mm. wide, very shortly acuminate. Fruits blackish purple; cocci 3–5·5 mm. in diameter, glabrous. Fig. 59/7–14, p. 396.

KENYA: Teita Hills, Vuria Peak, 17 Apr. 1960, *Verdcourt & Polhill* 2732! & 8 May 1972, *R. B. & A. J. Faden* 72/248!
TANZANIA. W. Kilimanjaro, Shira Mts., Feb. 1928, *Haarer* 1120!; Lushoto–Shume road, Magamba Forest, 1 Mar. 1953, *Drummond & Hemsley* 1361!; Uluguru Mts., 10 Oct. 1932, *Schlieben* 2777!
DISTR. **K**7; **T**2, 3, 6; not known elsewhere
HAB. Evergreen forest; 1100–3000 m.

SYN. [*G. glabrum* sensu K. Schum. in E.J. 28: 500(1900), *non* Thunb.]

NOTE. Collected as early as 1888 on Kilimanjaro by the Rev. W. E. Taylor. *Peter* 55644 (Tanzania, Pare District, S. Pare Mts., Tona) is perhaps an odd state of this species.

8. **G. glaciale** *K. Krause* in E.J. 43: 159 (1909); A.V.P.: 178, 329 (1957); Verdc. in K.B. 28: 492 (1974); U.K.W.F.: 411 (1974). Types: Tanzania, Kilimanjaro, Kibo, *Jaeger* 150 (B, holo. †); Kilimanjaro, saddle between Kibo and Mawenzi, *Hedberg* 1345 (UPS, neo.!, BR, EA!, K!, LD, S, isoneo.)

Dwarf ± erect or procumbent herb with prostrate or ascending stems 2·5–15(–20) cm. long, glabrous, pubescent or slightly prickly from a creeping lower stem. Leaves green to purplish, in whorls of 3–6; blades oblanceolate or spathulate, widest near the apex, 2·5–8 mm. long, 1–2·5 mm. wide, with a short triangular acute acumen at the tip, gradually attenuated to the base, glabrous or with a few sparse hairs, the margins only very slightly recurved, without or with only a few recurved prickles. Inflorescences 1–2(–4)-flowered, on short lateral shoots 3–4 mm. long, bearing 1–2 leaf-like bracts; flowers shortly pedicellate or sessile, but fruiting pedicels up to 2 mm. long. Calyx-tube ovoid, 0·6–0·9 mm. long, 0·6–0·8 mm. wide. Corolla greenish cream or pink; tube 0·15–0·3 mm. long; lobes ovate, 0·6–1·2 mm. long, 0·5–0·6(–0·8) mm. wide, acute. Fruit black, 1·5–2·3 mm. long, 2·5–3 mm. wide, glabrous, the epicarp papery in the dry state; cocci black, ovoid, 1·7 mm. long, 1·1 mm. wide, wrinkled or in case of fruit reduced to 1 subglobose coccus, 3 mm. long, 2·5 mm. wide.

var. **glaciale**, Verdc. in K.B. 28: 492, fig. 1/B, C (1974)

Leaf blades glabrous above.

KENYA. Elgon, W. slope of Koitobos, 11 May 1948, *Hedberg* 892!; Mt. Kenya, head of Teleki valley, Sept. 1948, *J. G. Williams* in *Bally* 6432! & upper Hinde valley, 2 Feb. 1965, *Allt* 47! & 50!
TANZANIA. Kilimanjaro, saddle between Kibo and Mawenzi, 23 June 1948, *Hedberg* 1345! & Mweka route at 3510 m., 14 Sept. 1966, *Bie* 212! & Shira Plateau, E. of Johnsell Point, 6 Nov. 1968, *Bigger* 2269!
DISTR. **K**3, 4; **T**2; not known elsewhere
HAB. Upland moor in grass tussocks, moss cushions, *Carex* bogs, etc. in alpine regions up to the limit of phanerogamic vegetation; 3510–4350 m.

var. **satimmae** *Verdc.* in K.B. 28: 493, fig.1/D(1974). Type: Kenya, Aberdare Mts., Satima, *Fries* 2680 (UPS, holo.!)

Leaf-blades ± densely covered with short curved hairs.

KENYA. N. Aberdare Mts., Satima, 27 Mar. 1936, *Meinertzhagen*! & 19 Mar. 1922, *Fries* 2680!
DISTR. **K**3/4; not known elsewhere
HAB. Boggy alpine slopes; 3600 m.

9. **G. hochstetteri** *Pichi-Serm.* in Webbia 7: 339 (1950); A.V.P.: 329 (1957); E.P.A.: 1025 (1965); Verdc. in K.B. 28: 491, fig. 1/A (1974). Type: Ethiopia, Semien, Bachit, *Schimper* 548 (P, holo., BM, K, iso.!)

Scrambling herb forming loose tufts, but individual stems weak and mostly elongate, up to 35 cm. long, without hairs but usually with weak downwardly directed prickles. Leaves and stipules in whorls of 6; leaf-blades linear-oblanceolate to obovate-oblanceolate, (3–)5·5–9 mm. long, 1–1·4 mm. wide, acuminate at the apex and often ending in a long point, the margins with short forwardly directed hairs or weak prickles and the upper surface ± glabrous to covered with short forwardly directed hairs, with some backwardly directed prickles on the midrib beneath, or rarely entirely glabrous. Inflorescences 1–3-flowered; peduncle or reduced peduncle-like slender side branches with reduced leaves up to 2 mm. long; pedicels 2·5–4 mm. long. Calyx-tube half-ellipsoid, 0·5–0·7 mm. long, glabrous. Corolla white; tube 0·4 mm. long; lobes triangular to ovate, 0·5–1 mm. long, 0·5–0·8 mm. wide. Fruits with rounded or ellipsoid cocci 1·5–3 mm. long and wide, reticulate when dry.

KENYA. N. Aberdares, 21 June 1969, *Kokwaro* 1947!; Mt. Kenya, W. side, 22 Jan. 1922, *Fries* 1187! & above Naromoru, 11 Dec. 1957, *Verdcourt* 1985!
TANZANIA. Kilimanjaro, N. side, Mar. 1894, *Volkens* 2019!
DISTR. **K**4; **T**2; Ethiopia
HAB. Clearings in bamboo and upland rain-forest; 2700–3200 m.

SYN. *G. simense* A. Rich., Tent. Fl. Abyss. 1: 344(1848); Hiern in F.T.A. 3: 246 (1877); K. Krause in N.B.G.B. 10: 612(1929), pro parte, *non* Fresen. (1837), *nom. illegit.* Type: as for species
[*G. vaillantii* sensu K. Krause in N.B.G.B. 10: 611(1929), quoad *Fries* 1187, *non* Lam. & DC.]
[*G. simense* sensu K. Krause in N.B.G.B. 10: 612 (1929), pro parte, *non* Fresen., *nec* A. Rich.]
G. simense A. Rich. var. *hypsophilum* R.E. Fr. in K. Svenska Vetensk.-Akad. Handl., ser. 3, 25: 79 (1948). Type: Mt. Kenya, W. slope, *Fries* 1276b (UPS, holo.!, BR, K!, S, iso.)
G. simense A. Rich. var. *keniense* R.E. Fr. in K. Svenska Vetensk.-Akad. Handl., ser. 3, 25: 79(1948). Type: Mt. Kenya, *Fries* 1368 (UPS, holo.!)

10. **G. kenyanum** *Verdc.* in K.B. 28: 493, fig. 1/E, F (1974); U.K.W.F.: 412 (1974). Type: Kenya, Mt. Aberdares National Park, near Fort Jerusalem, *Verdcourt* 3998 (K, holo.!, EA, iso.!)

Procumbent mat-forming herb (2–)7–60 cm. long; branches mostly short, 5–10 cm. long, glabrous or with a few scattered spreading hairs and with a ring of hairs at the nodes; plant smells of coumarin. Leaves and stipules in whorls of 6; blades linear-lanceolate, 1·5–4 mm. long, 0·3–0·8 mm. wide, obtuse or subacute at the apex, narrowed to the base, broadest about the middle, glabrous or with a few hairs on the midnerve, the margins plane or slightly recurved. Flowers 1–2 at the end of short lateral shoots; pedicels 0·5–2 mm. long in fruit. Calyx-tube ± biglobose, 0·5 mm. long, 0·65–0·85 mm. wide, glabrous. Corolla white or tinged pink, broadly campanulate-rotate; tube 0·5–0·7 mm. long; lobes ovate, 0·5–0·9 mm. long, 0·6–0·8 mm. wide, ± acute. Styles joined for most of their length. Fruits grey, 2·3 mm. long, 3·5–4 mm. wide, each coccus ellipsoid, 2·3 mm. long, 1·9 mm. wide (or about half these dimensions when dry), glabrous. Seeds slightly smaller than the cocci, closely reticulate.

KENYA. Elgeyo District: Cherangani Hills, Kamelogon, tributary of R. Embobut, 12 Dec. 1970, *Mabberley* 518! & near Kamelogon, Jan. 1971, *Tweedie* 3864!; Naivasha District: Aberdares National Park, Fort Jerusalem, 30 July 1960, *Polhill* 239!
TANZANIA. Kilimanjaro, SE side, 7 Mar. 1934, *Schlieben* 4955! (atypical)
DISTR. **K**2, 3, ? 4; ? **T**2; not known elsewhere
HAB. Upper margin of montane forest in *Hagenia* zone and in moorland; 2800–3240 (–3550) m.

SYN. [*G. simense* sensu K. Krause in N.B.G.B. 10: 612 (1929), quoad *Fries* 2377, *non* A. Rich., *nec* Fresen.]

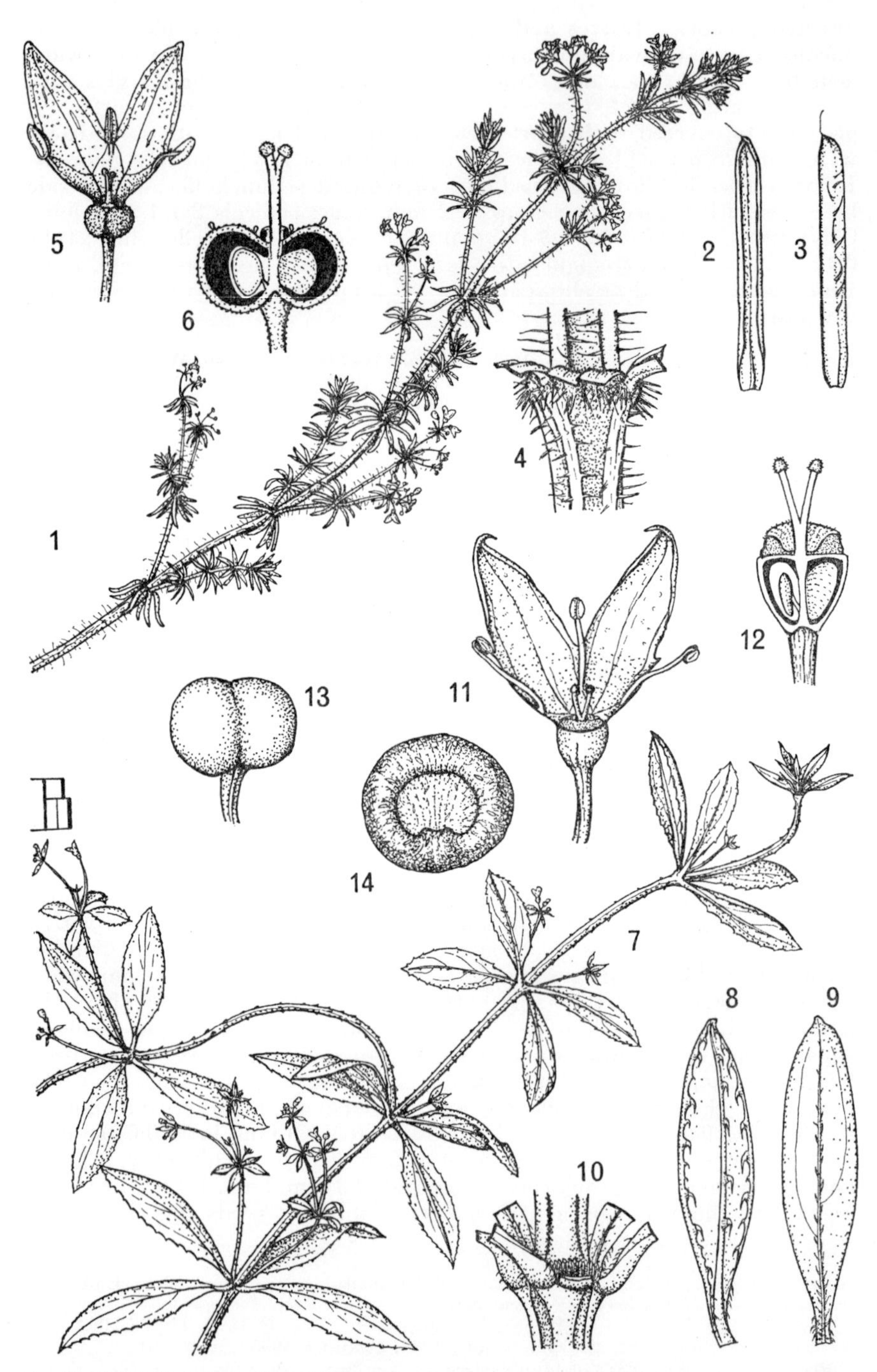

FIG. 59. *GALIUM TANGANYIKENSE*—**1,** flowering branch, × $\frac{2}{3}$; **2, 3,** leaf/stipule, lower and upper surfaces, × 6; **4,** node with leaves and stipules removed, × 10; **5,** flower, with half corolla removed, × 10; **6,** longitudinal section of ovary, × 20. *G. BRENANII*—**7,** flowering branch, × $\frac{2}{3}$; **8, 9,** leaf/stipule, lower and upper surfaces, × 2; **10,** node with leaves and stipules removed, × 10; **11,** flower, with part of corolla removed, × 10; **12,** longitudinal section through ovary, × 20; **13,** fruit, × 3; **14,** pyrene, × 6. 1–6, from *Richards* 7601; 7, 11, 12, from *Drummond & Hemsley* 1625; 8, 9, 13, 14, from *Haarer* 1120; 10, from *Thulin & Mhoro* 1022. Drawn by Diane Bridson.

11. **G. tanganyikense** *Ehrend. & Verdc.* in K.B. 28: 488, fig. 1/1–6 (1974). Type: Tanzania, Rungwe District, Mwakaleli, *Stolz* 2282 (K, holo.!, BM, iso.!)

Presumably a scrambling plant with stems 20–40 cm. long, either glabrous or sparsely to densely covered with spreading white hairs but not prickly; rootstock creeping. Leaves and stipules in whorls of 6; blades linear to narrowly linear-oblong, 2·3–8 mm. long, 0·5–1·5 mm. wide, ± obtuse at the apex and without an acumen but sometimes terminating in a hair, glabrous or with spreading white hairs, the margins slightly and regularly recurved. Cymes dense, terminal or axillary, several-flowered; peduncles 2·5–5 mm. long; pedicels 0–2·5 mm. long. Calyx-tube 0·4–0·5 mm. in diameter, glabrous. Corolla white to pale pink; tube slightly campanulate, 0·8–1·1 mm. long; lobes ovate, 1·2–1·5 mm. long, 0·9–1 mm. wide, subacute or acute but not acuminate at the apex. Unripe fruits with cocci 1·5 mm. in diameter, glabrous. Fig. 59/1–6.

TANZANIA. Njombe/Mbeya Districts: Kitulo [Elton] Plateau, 24 Nov. 1966, *Robertson* 217! & 11 Nov. 1931, *R. M. Davies* E. 18! & 29 Nov. 1963, *Richards* 18432!

DISTR. T7; not known elsewhere

HAB. Upland grassland, particularly among rocks and in short grass, also in boggy ground and coarse streamside grassland; (2000–)2400–2820 m.

NOTE. *Richards* 6622 (Njombe District, Ndumbi R., 19 Oct. 1956), *Richards* 18512 (Kitulo [Elton] Plateau, 1 Dec. 1963) and the *Richards* sheet cited above are either glabrous specimens or mixtures of glabrous and hairy specimens. The variation is too continuous to permit the recognition of varieties.

12. **G. stenophyllum** *Bak.* in K.B. 1895: 68 (1895); Brenan in Mem. N.Y. Bot. Gard. 8: 456 (1954). Types: Zambia, Fwambo, *Carson* 40, 41 & 80 of 1893 collection and s.n. of 1889 collection (K, syn.!)*

Procumbent straggling or subscandent herb 13–45 cm. long; stems rather numerous from a slightly woody base, glabrous to densely covered with rather rough spreading hairs but never with short forwardly directed hairs. Leaves and stipules in whorls of 6; blades linear to linear-elliptic, 0·8–3·5 cm. long, 0·5–1·5(–5) mm. wide, acute at the apex but lacking a distinct filiform acumen, glabrous to quite densely hairy, the margins usually in-rolled, not prickly. Inflorescences 0·5–2 cm. long; pedicels 1–4 mm. long, reflexed after flowering. Calyx-tube 0·4–0·5 mm. long, 0·6–0·9 mm. wide. Corolla white or greenish white, slightly sweetly scented; tube campanulate, 0·6–0·9 mm. long; lobes 1·1–1·7 mm. long, 0·6–1·1 mm. wide. Fruit with cocci 2–3 mm. in diameter, glabrous.

TANZANIA. Ufipa District: Malonje, 8 Feb. 1933, *Michelmore* 754! & below Chito Mt., 12 Dec. 1934, *Michelmore* 1053!; Mbeya District: Mbozi, 19 Nov. 1932, *R. M. Davies* 709!

DISTR. T4, 7; Malawi, Zambia

HAB. Damp or seasonally flooded grassland, also in *Brachystegia* woodland; 1560–2250 m.

NOTE. Brenan separates this from *G. bussei* on a number of characters—flaccid appearance, leaf apices obtuse or acute but lacking a long-attenuate acumen, widely spaced partial inflorescences, short thicker pedicels which become deflexed and arcuate after flowering and large cocci; moreover the flowers are predominantly white. Other characters are the campanulate corolla and lack of short forwardly directed hairs on the leaves (as distinct from longer hairs). Despite an occasional difficulty in identification the two species are not actually related, *G. stenophyllum* belonging to the *G. palustre* L. group. A specimen from just over the border (Zambia, Mbala District, Kasesha, 15 May 1969, *Sanane* 708) is said to have small purple flowers—perhaps just the buds are tinged purple.

* The syntype *Buchanan* 770 was excluded by Brenan as belonging to *G. bussei*.

13. **G. scioanum** *Chiov.* in Ann. Bot. Roma 9: 322 (1911); Brenan in K.B. 5: 372 (1951); U.K.W.F.: 412, fig. (1974). Type: Ethiopia, Addis Ababa, *Negri* 1410 (FI, holo.(!), K, photo. !)

Creeping, scrambling or climbing herb up to 0·9 m., the stems strongly ribbed, glabrous or quite densely covered with spreading hairs particularly on the ribs, when sparsely hairy the hairs often breaking to leave scabrid bases resembling small prickles. Leaves and stipules in whorls of 6, often strongly reflexed; blades linear-lanceolate to narrowly elliptic, 0·4–1·7(–2·5) cm. long, 1–2·8(–10) mm. wide, with a short acute apical acumen which is often twisted or decurved, cuneate at the base, glabrous to hairy, the margins revolute or flat, ciliate, often with small closely set inconspicuous denticles. Inflorescences many-flowered terminal and axillary panicles; peduncles 0·6–1·9 cm. long; pedicels 1–5 mm. long, reflexed in fruit. Calyx-tube 0·3–0·35(–0·6) mm. in diameter. Corolla white; tube (0·2–)0·5–0·7(–1) mm. long; lobes narrowly to broadly ovate-triangular, (0·8–)1·2–2·1 mm. long, (0·4–)0·6–1·2 mm. wide, subacute. Fruit reddish purple; cocci 1·5–2 mm. in diameter, glabrous.

KEY TO INFRASPECIFIC VARIANTS

Leaf-blades up to ± 3 mm. wide, the margin usually distinctly inrolled:
- Stems and foliage distinctly hairy . . . . var. **scioanum**
- Stems and foliage ± glabrescent except for small prickles on stem-ribs and leaf-margins . . var. **glabrum**

Leaf-blades wider, up to ± 7(–10) mm. wide, the margins mostly not distinctly inrolled . . var. **latum**

var. **scioanum**

Plant with hairy stems and foliage. Leaf-blades mostly under 3 mm. wide.

KENYA. Trans-Nzoia District: Koitobos R., Sandum's Bridge, July 1964, *Tweedie* 2864 !; Elgeyo/Trans-Nzoia District: Cherangani Hills, Kapolet, Aug. 1963, *Tweedie* 2699 !; N. Nyeri District: Liki R., 11 Feb. 1922, *Fries* 1475 !; Kericho District: SW. Mau Forest, Sambret, Oct. 1961, *Kerfoot* 2917 !
TANZANIA. Njombe District: Msima Stock Farm, *Emson* 279 ! & upper Ruhudji R., Lupembe area, May 1931, *Schlieben* 896 ! & Njombe, 1 Jan. 1932, *Lynes* V.54 !
DISTR. **K**2–5; **T**2, 7; Zaire, Ethiopia
HAB. Swamp edges, vleis, wet savanna, riversides, etc.; 1800–2490 m.

SYN. *G. homblei* De Wild. in F.R. 13: 140 (1914). Type: Zaire, Katanga, Lualaba area, Kapanda valley, *Homblé* 981 (BR, holo.(!))

NOTE. There are intermediates between the typical variety and var. *glabrum*, e.g. *Albrechtsen* in *C.M.* 6717 (Kenya, Kinangop, Sept. 1934).

var. **glabrum** *Brenan* in K.B. 5: 372(1951). Type: Zambia, Mwinilunga District, R. Lunga, just below R. Mudjanyama junction, *Milne-Redhead* 3401 (K, holo. !, BM, iso. !)

Plant glabrous except usually for minute prickles on the stem-ribs and marginal leaf-denticles, or stems quite glabrous except for hairy nodes. Leaf-blades mostly under 3 mm. wide

KENYA. Trans-Nzoia District: Endebess, Keben Springs, *Irwin* 28 ! & near Kitale, July 1954, *Tweedie* 1180A !; Uasin Gishu District: Eldoret area, 11 June 1951, *G. R. Williams* 231 !
TANZANIA. Mbeya Mt., 13 May 1956, *Milne-Redhead & Taylor* 10323 ! & 10323A ! & 3 Dec. 1961, *Kerfoot* 3205 ! (intermediate with var. *latum*); Iringa District: Iheme, 28 Jan. 1934, *Michelmore* 934 !
DISTR. **K**3; **T**7; Ethiopia, Zambia and Angola
HAB. Swampy grassy places, damp ground near rivers, streams and springs; 1800–2100(–2700) m.

var. **latum** (*De Wild.*) *Verdc.* in K.B. 30: 326(1975). Type: Zaire, Katanga, Kapanda valley, *Homblé* 981 bis (BR, holo.(!))

Stems and foliage hairy or glabrescent except for the nodes. Leaf-blades 0·8–2·5 cm. long, 2·5–7(–10) mm. wide.

TANZANIA. Ufipa District: Ufipa Plateau, edge of Rukwa escarpment, Mtumba, 20 Dec. 1934, *Michelmore* 1080 ! & Mbisi, 16 Jan. 1950, *Bullock* 2244 !; Mbeya District: Chunya Escarpment, 20 Jan. 1957, *Richards* 7947 !
DISTR. **T4**, 7; Zaire, Ethiopia (intermediate with var. *scioanum*)
HAB. Swamps, marshy streamsides, wet upland grassland, also in rocky grassland and grassland with scattered *Protea*, etc.; 1650–2350 m.

SYN. *G. latum* De Wild. in F.R. 13: 140(1914)

NOTE. *Richards* 13828, a sterile glabrescent specimen from Mbeya District, Mbosi area, Judyland Farm, with broad flat leaves having rather more coarsely denticulate margins may belong to some other species.

14. **G. bussei** *K. Schum. & K. Krause* in E.J. 39: 571 (1907); Brenan in Mem. N.Y. Bot. Gard. 8: 456 (1954). Type: Tanzania, Songea District, Ngaka [Mgaka] valley, *Busse* 941 (B, holo. †, EA, iso. !, K, photo. !)

Tufted erect or somewhat scrambling herb; stems several, 20–70 cm. tall from a slightly woody rootstock, glabrous to hairy. Leaves and stipules in whorls of 5–10; blades linear to linear-oblanceolate, 0·6–4 cm. long, 0·5–2(–3) mm. wide, with a distinct filiform acumen at the apex, glabrous to hairy but margins without prickles; surface always with some forwardly directed "emergences". Inflorescences lax to dense, few–many-flowered, 1–2 cm. long; pedicels 1–4(–9) mm. long. Calyx-tube subglobose, 0·5–0·8 mm. long, 0·5–1·1 mm. wide, glabrous. Corolla typically yellow, sometimes cream, white or greenish cream, 2·5–4 mm. wide; tube and joined part of lobes 0·4–0·5 mm. long; lobes (1–)1·4–2 mm. long, (0·5–)0·8–1·2 mm. wide, acute or acuminate. Fruits globose, ± 2 mm. wide, often reduced to 1 coccus.

KEY TO INFRASPECIFIC VARIANTS

Inflorescence dense and many-flowered:
  Stems and leaves ± pubescent . . . . var. **bussei**
  Stems and leaves glabrous . . . . . var. **glabrum**
Inflorescence often lax and diffuse, sometimes few-flowered:
  Stems and leaves ± pubescent . . . . var. **strictius**
  Stems and leaves glabrous . . . . . var. **glabrostrictius**

var. **bussei**; Brenan in Mem. N.Y. Bot. Gard. 8: 456(1954)

Stems and leaves with ± conspicuous spreading white hairs. Inflorescence dense and many-flowered.

TANZANIA. Ufipa District: Sumbawanga, 15 Jan. 1950, *Bullock* 2254 !; Kondoa District: scarp above Kandaga, 20 Jan. 1928, *B. D. Burtt* 1206 !; Songea District: Lupembe Hill, 29 Feb. 1956, *Milne-Redhead & Taylor* 8760A !
DISTR. **T2**, 4–8; Malawi, Rhodesia
HAB. Grassland, grassland with scattered trees, grassy places in *Brachystegia* woodland, also rocky hill slopes and old cultivations; 1200–2250 m.

NOTE. Has frequently been misidentified with *G. stenophyllum* Bak. Sometimes found together with var. *glabrum*.

var. **glabrum** *Brenan* in Mem. N.Y. Bot. Gard. 8: 456(1954). Type: Tanzania, Mbulu District, Ufiome Mt., *B. D. Burtt* 2728 (K, holo. !, EA, iso. !)

Stems and leaves glabrous. Inflorescences dense and many-flowered.

TANZANIA. Mbulu District: Kitingi, 1 Mar. 1965, *Hukui* 21!; W. Usambara Mts., escarpment near Gologolo–Mkumbara footpath, 4 June 1953, *Drummond & Hemsley* 2871!; Iringa District: Dabaga Highlands, Kilolo, 9 Feb. 1962, *Polhill & Paulo* 1397!
DISTR. T2–4, 7, 8; Mozambique, Malawi, Zambia and Rhodesia
HAB. Grassland with scattered trees, grassy slopes with many herbs, thickets, cultivated ground; 1500–2550 m.

NOTE. In T2 this merges with *G. ossirwaense* K. Krause var. *glabrum* to such an extent that specimens are difficult to name. It is possible all the material should be referred to *G. bussei*.

var. **strictius** *Brenan* in Mem. N.Y. Bot. Gard. 8: 457(1954). Type: Malawi, Zomba, *Purves* 59 (K, holo.!)

Stems and usually leaves pubescent. Inflorescences lax and often few-flowered.

TANZANIA. Mbeya District: Mbosi Circle, Boma Riva Estate, 15 Jan. 1961, *Richards* 13932 (in part)!; Iringa District: 11 km. Sao Hill to Igawa, 12 Dec. 1962, *Richards* 17005!; Songea District: about 32 km. E. of Songea by R. Mkurira, 26 Dec. 1955, *Milne-Redhead & Taylor* 7754!
DISTR. T7, 8; Malawi and Zambia
HAB. Open grassland and wooded grassland; 930–1920 m.

var. **glabrostrictius** *Brenan* in Mem. N.Y. Bot. Gard. 8: 457(1954). Type: Zambia, Mwinilunga District, just N. of Matonchi Farm, *Milne-Redhead* 4283 (K, holo.!, BM, iso.!)

Stems and leaves glabrous save for the nodes. Inflorescence lax and often few-flowered.

TANZANIA. Mbeya District: Poroto Mts., 18 Mar. 1932, *St. Clair-Thompson* 730!; Iringa District: 11 km. Sao Hill to Igawa, 12 Dec. 1962, *Richards* 17005A!; Songea District: about 32 km. E. of Songea, by R. Mkurira, 26 Dec. 1955, *Milne-Redhead & Taylor* 7754A!
DISTR. T2, 4, 7, 8; Zaire, Malawi and Zambia
HAB. Grassland, grassland with scattered trees, dambos, forest edges, *Brachystegia* woodland; 930–1920 m.

NOTE. Extremes of the above varieties are distinctive but many specimens are not easy to place. It might be better to recognize only varieties based on the type of inflorescence since the indumentum varies in both of these types.

15 **G. ossirwaense** *K. Krause* in E.J. 43: 159 (1909); A.V.P.: 177, 328 (1957); U.K.W.F.: 411 (1974). Type: Tanzania, Masai District, W. slope of Ossirwa, *Jaeger* 506 (B, holo. †)

Straggling herb with several to many stems 7·5–35 cm. tall from a slender root, sometimes forming clumps; stems decumbent at the base but with erect flowering shoots, covered with usually dense short spreading pubescence or less often glabrous or with coarser hairs; internodes mostly short, giving a compact appearance to the plant; whole plant said to smell of coumarin. Leaves and stipules in whorls of (7–)8(–9); blades linear, 0·3–1·5 cm. long, 0·5–1·5(–2·5) mm. wide, attenuate at the apex into a distinct ± straight filiform acumen, usually ± shiny above, glabrous or with a few stiff hairs near the inrolled margins but without prickles. Inflorescences dense, the numerous cymes packed along the length of the flowering shoots, sweet-smelling, scarcely if at all exceeding the leaves, 3–9 mm. long; pedicels 1–4 mm. long. Calyx-tube 0·4–0·85 mm. in diameter, glabrous. Corolla dull pale yellow to bright yellow; tube 0·3–0·6 mm. long; lobes ovate-triangular, 1·3–2 mm. long, 0·75–1·2 mm. wide, acute. Fruit glabrous.

var. **ossirwaense**

Stems mostly densely covered with short spreading pubescence or sometimes more coarsely hairy.

UGANDA. Elgon, Mudangi [Madangi], 5 Sept. 1932, *A. S. Thomas* 592! & W. slope above Butadiri, along track via Mudangi through the caldera, 5 Dec. 1967, *Hedberg* 4476! & Elgon, without locality, Apr. 1930, *Liebenberg* 1619!
KENYA. Naivasha District: Kinangop, Apr. 1938, *Chandler* 2314!; Mt. Kenya, N. slopes, 114 km. from Nyeri on Nanyuki–Meru road, 4 Feb. 1933, *C. G. Rogers* 401!; Masai District: Narok area, Nasampolai valley, 14 May 1971, *Greenway & Kanuri* 14860!
TANZANIA. Mbulu District: Elanairobi Volcano, 20 Sept. 1932, *B. D. Burtt* 4178! & Oldeani Mt., 12 Nov. 1932, *Geilinger* 3677!; Njombe District: Kitulo [Elton] Plateau, 5 Jan. 1957, *Richards* 7473A!
DISTR. **U3**; **K3–6**; **T2, 7**; Mozambique, Malawi
HAB. Essentially grassland, at montane forest edges and in forest clearings up to alpine moorland, sometimes a weed; (1900–)2100–3450(–3600) m.

SYN. [*G. verum* sensu K. Krause in N.B.G.B. 10: 611(1929), *non* L.]
G. mollicomum Bullock in K.B. 1932: 498(1932). Type: Kenya, Mt. Elgon, *Lugard* 400a (K, holo.!)
G. mollicomum Bullock var. *friesiorum* Bullock in K.B. 1932: 498(1932). Type: Mt. Kenya, *Fries* 479 (K, holo.!, S, UPS, iso.)

NOTE. The information on the label of *Chandler* 2314, cited above, indicating that the plant attains 5–6 feet and emerges from other vegetation must be an error.
G. ossirwaense is a well characterized species in Kenya but shows more variation in Tanzania. Apart from the glabrous variety mentioned below some specimens with hairy stems have a more open inflorescence, e.g. *Milne-Redhead & Taylor* 10068! (Mbeya District, about 1.5 km. up the Tukuyu road, 17·5 km. SW. of Mbeya, 12 May 1956). This had been identified as *Galium scabrellum* but although the leaf margins have a few rough hairs they are not close denticles as in that species. *Hindorf* 6814! is stated to have been collected at Amboseli, on sandy ground at 1200 m., but such a habitat and altitude need confirmation. *Bagshawe* 394! (Uganda, Ankole District, Irunga, 15 Nov. 1903) has leaf-blades up to 2 cm. long and approaches *G. bussei.*

var. **glabrum** *Verdc.* in K.B. 30: 474 (1975). Type: Tanzania, W. Porotos–Rungwe, *Greenway* 3553 (K, holo.!, EA, iso.!)

Stems glabrescent or glabrous except at the nodes.

TANZANIA. Mbulu District: Oldeani Mt., 19 Nov. 1957, *Tanner* 3803!; Rungwe District: Bundali [Undalis], 25 Feb. 1933, *R. M. Davies* 856! & W. Porotos–Rungwe, 13 Aug, 1933, *Greenway* 3553!
DISTR. **T2**, 7; not known elsewhere
HAB. Grassland; 1530–2400 m.

16. **G. scabrellum** *K. Schum.* in E.J. 28: 113 (1899); Brenan in Mem. N.Y. Bot. Gard. 8: 457 (1954). Type: Malawi, Nyika Plateau, *Whyte* 269 (B, holo. †, K, iso.!)*

Prostrate, trailing, scrambling or ± climbing plant 0·3–3·6 m. long, the stems very strongly ribbed, pubescent to densely hairy particularly on the ribs and sometimes minutely prickly. Leaves and stipules in whorls of (6–)8(–10); blades very narrowly elliptic, 0·4–1·7 cm. long (mostly under 1·2 cm.), 0·5–2(–4) mm. wide, produced into a distinct filiform acumen at the apex, cuneate at the base, glabrescent or with rather long hairs all over both surfaces or only on the midnerve above and beneath, sometimes also scabrid above, the margins usually with well-developed closely set reflexed saw-like denticles sometimes obscured by the inrolling of the margins. Inflorescences many-flowered terminal and axillary panicles; peduncles 1·5–2 mm. long; pedicels 1·5–3 mm. long. Calyx-tube 0·3–0·5 mm. in diameter, glabrous or hairy. Flowers dull, pale or bright yellow or cream-coloured; tube 0·2–0·4(–0·5) mm. long; lobes ovate-triangular, 1–1·5 mm. long, 0·45–0·8 mm. wide, acuminate. Fruit reddish, each coccus ± 1 mm. in diameter, glabrous or sparsely hairy.

* The collector is not "*Carsson*" as given by Schumann; see Brenan reference cited above.

UGANDA. Kigezi District: Virunga Mts., Muhavura, NW. face, *Eggeling* 1026! & Bufumbira, path to col between Mts. Muhavura and Mgahinga, 9 Sept. 1952, *Norman* 170! & Mabungo, 20 Dec. 1933, *A. S. Thomas* 1096!
TANZANIA. Ufipa District: Sumbawanga, Malonje Plateau, 4 Mar. 1957, *Richards* 8432!; Mbeya Mt., just N. of the peak, 13 May 1956, *Milne-Redhead & Taylor* 10227!; Rungwe Mt., S. slopes, 23 Mar. 1932, *St. Clair-Thompson* 1007!
DISTR. **U2**; **T1**, 4, 7; Zaire, Rwanda, ? Burundi, Malawi, NE. Zambia, ? Rhodesia
HAB. Short and coarse grassland in rocky and shrubby places, forest edges and clearings (including bamboo); 1500–3450 m.

SYN. *G. bequaertii* De Wild. in Rev. Zool. Afr. 9, Suppl. Bot.: 12(1921) & in Pl. Bequaert. 2: 307(1923); F.P.N..A.: 381(1947). Type: Zaire, Tshilirunge, *Bequaert* 6047 (BR, holo. (!))

NOTE. *Maitland* 846! (Masaka–Mbarara road, Oct. 1925) seems to be a variant of this species with practically no denticles on the leaf-margins. It resembles *G. scioanum* and *G. ossirwaense* in certain characters. Similar forms occur in **T7**, many of which are difficult to name and seem to be intermediate with *G. ossirwaense*.

17. G. sp. A

Scrambling or ? prostrate herb at least 60 cm. long; the stems prominently ribbed, glabrous. Leaves in whorls of 6, with a few colleter-like hairs within the whorl; blades curiously inflated, oblong, 4·5 mm. long, 1·3 mm. wide, acuminate, glabrous, the upper surface deep green, thick, shining, the margins strongly recurved, the lower surface white, thin, separated from the upper surface by a large air cavity. Inflorescences terminal, ± 4-flowered; pedicels 1–2·5 mm. long. Calyx-tube biglobose, 0·6 mm. long, 0·9 mm. wide, glabrous, tuberculate. Corolla white; tube 1 mm. long; lobes triangular, 1·7 mm. long, 1 mm. wide, slightly papillate inside. Fruit not seen.

TANZANIA. Njombe District: W. Kipengere Range, 14 Jan. 1957, *Richards* 7768!
DISTR. **T7**
HAB. Coarse streamside grassland; 2100 m.

NOTE. The curious leaves seem pathological but every leaf is identical and the flowers are in no way abnormal. The appearance is due to the separation of the abaxial epidermis from the rest of the leaf but the cause of this is not known.

UNIDENTIFIABLE RECORD

The identity of *Stuhlmann* 755, recorded by K. Schumann in E.J. 28: 500 (1900) as *Galium mollugo* L., is now impossible to discover.

INDEX TO RUBIACEAE (Part 1)

GEOGRAPHICAL DIVISIONS OF THE FLORA

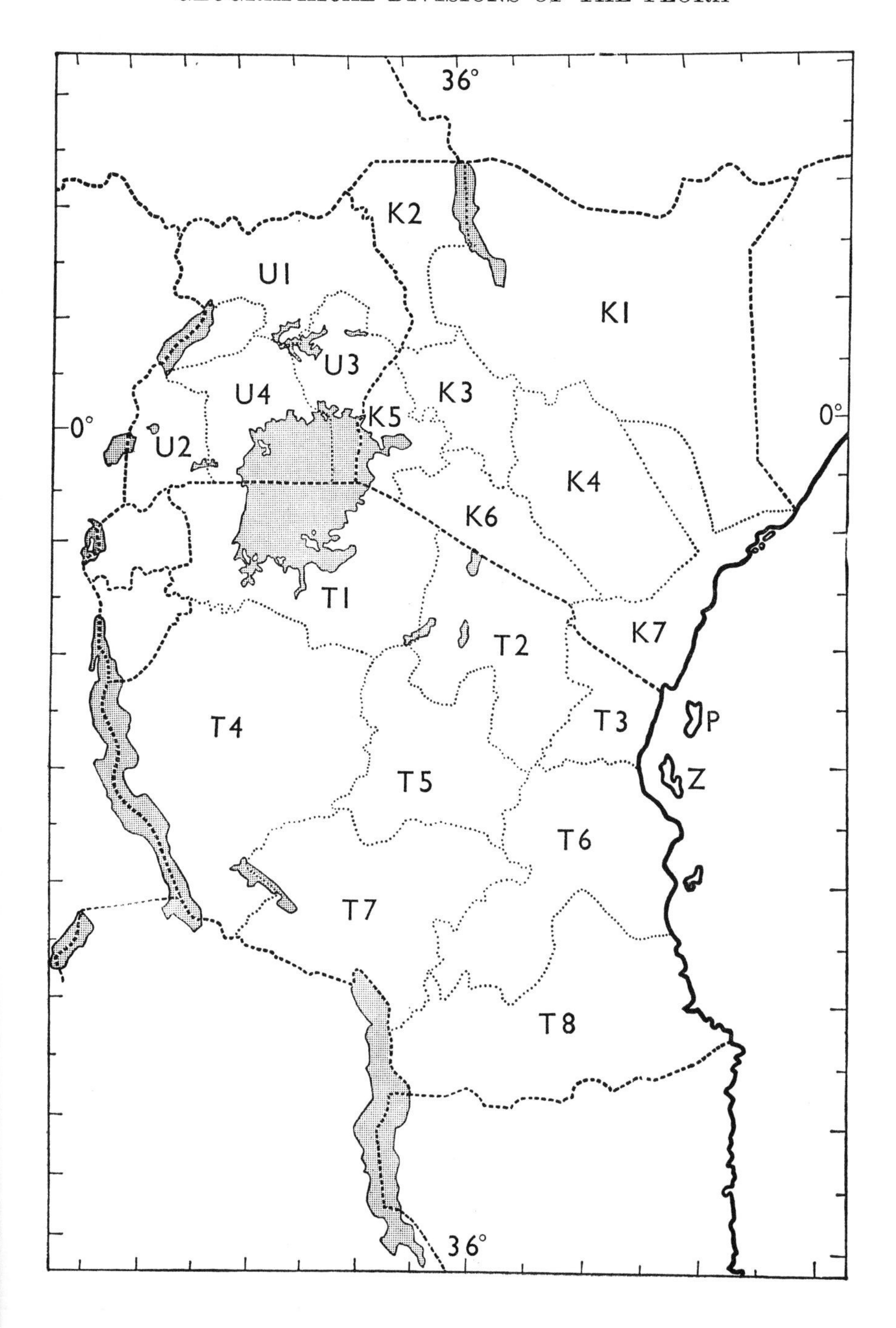